Plant Acclimation to Environmental Stress

Narendra Tuteja • Sarvajeet Singh Gill
Editors

Plant Acclimation to Environmental Stress

Springer

Editors
Narendra Tuteja
Plant Molecular Biology Group
International Centre for Genetic
Engineering & Biotechnology (ICGEB)
Aruna Asaf Ali Marg
New Delhi, India

Sarvajeet Singh Gill
Stress Physiology & Molecular Biology Lab
Centre for Biotechnology
MD University
Rohtak, India

ISBN 978-1-4614-5000-9 ISBN 978-1-4614-5001-6 (eBook)
DOI 10.1007/978-1-4614-5001-6
Springer New York Heidelberg Dordrecht London

Library of Congress Control Number: 2012951621

Printed on acid-free paper

Springer is part of Springer Science+Business Media (www.springer.com)

Professor E.A. Siddiq
(July 15, 1937 —)

Professor E.A. Siddiq was born in Ilayangudi, Tamil Nadu, India and received his B.Sc. from Madras University and his M.Sc. and Ph.D in Cytogenetics from Indian Agricultural Research Institute (IARI), New Delhi in 1958, 1964, and 1964, respectively. His main field of research is genetics, plant breeding, biotechnology, and his contributions in this field are significantly important. Professor Siddiq's research in the past three and half decades contributed significantly to the development of high yielding dwarf basmati and non-basmati varieties and hybrid rice, thereby boosting rice production in India. His scientific work is featured in the most prestigious international journals. He is a member of various national and international societies, including the Board of Trustees, International Rice Research Institute (IRRI), the Philippines, and has published more than 200 research papers. Professor Siddiq is Adjunct Professor, University of Hyderabad, 2008 to date; Adjunct Faculty, Centre for DNA Fingerprinting and Diagnostics, Hyderabad, 2008 to date; Hon. Professor—Biotechnology, Acharya N.G. Ranga Agricultural University, Hyderabad, 2002 to date. Professor Siddiq received many Awards and Honors from various scientific bodies such as Hari Om Ashram Trust National Award, 1975; VASVIK Foundation National Award, 1981; Amrik Singh Cheema Award, 1987–1988; Om Prakash Bhasin Award, 1994; Rafi Ahmed Kidwai Prize, 1995; Norman Borlaug Award, 1995; INSA Silver Jubilee Medal, 1997; ISCA G.P. Chatterjee Memorial Lecture Award, 2001; Agriculture Leadership Award for Agricultural Research and Development, 2008; Padma shree, 2011; INSA Sundarlal Hora Medal, 2011; and NASI Senior Scientist Fellow, 2012.

This book is dedicated to Professor E.A. Siddiq for nurturing plant genetics, plant breeding, and plant biotechnology.

Foreword

Agriculture is the base of food for all times. Thus, agricultural security is fundamental to the global food security. In agriculturally important countries, the sector is pivotal for generating economic growth and opportunity for overall livelihood security.

Towards the year 2050, the world population is projected to stabilize at around 9.2 billion. In order to adequately feed this population, the global agriculture must double its food production, and farm productivity would need to increase by 1.8 % each year—indeed a tall order. On the other hand, the natural resources—the agricultural production base, especially land, water, and biodiversity, are fast shrinking and degrading. For instance by 2025, 30 % of crop production will be at risk due to the declining water availability. Thus, in order to meet the ever-intensifying demand for food and primary production, more and more is to be produced from less and less of the finite natural and nonrenewable resource.

The bulk of the population increase will materialize in developing countries. Most of these are agriculture-dependent, and several of them are food deficit. Moreover, these countries have high concentration of smallholder resource-poor farmers and their agriculture is predominantly rainfed, which is inherently low-yielding and vulnerable to weather fluctuations.

The challenges of attaining sustainably accelerated and inclusive growth and comprehensive food security have been exacerbated by the global climate change and unusual fluctuations. The global warming due to rising concentration of greenhouse gases (GHGs) causing higher temperature, disturbed rainfall pattern causing frequent drought and flood, sea level rise, etc., is already adversely impacting productivity and stability of production, resulting in increased vulnerability, especially of resource-poor farmers. World Bank projects that the climate change will depress crop yields by 20 % or more by the year 2050. Recognizing that agriculture is both a victim of and a contributor to GHGs and other environmental pollutions, a two-pronged approach to reduce the emission and to develop adaptive measures to increase agricultural resilience will be needed.

Research and technology development (supported by policy and institutions) will need to be geared to meet the veritable challenges. We may recall, genetic alchemy of rice, wheat, and other major crops, which triggered the Green Revolution, is a shining example of science- and technology-led agricultural transformation immensely contributing to global food security and agrarian prosperity. With the greater emphasis on congruence of high productivity and sustainability in face of the intensifying volatilities due to climate change, biotic and abiotic stresses, and market instabilities, let alone the challenges of adequately feeding the swelling population from shrinking and degrading natural resources, new and modern sciences, and cutting-edge technologies, especially molecular breeding and genetic engineering for crop improvement and development of designer crops, will increasingly be called upon to provide the desired solutions.

This volume, *Plant Acclimatization to Environmental Stress*, ably compiled and edited by Drs. Narendra Tuteja and Sarvajeet Singh Gill, is a rich source of information on molecular, biochemical, microbial, and physiological bases of tolerance to abiotic stresses (drought, cold, salt, various toxicities, etc.) and on the development of tolerant varieties for adverse environmental conditions. It is hoped that researchers, scientists, and students, especially of crop biology, breeding, ecology, and production agronomy will greatly benefit from this volume. Most importantly, judicious use of this information should be used to develop crop varieties and management practices conducive to enhanced and resilient production leading to improved food, nutritional, ecological, and economic security.

I must congratulate the Editors and Springer for preparing this invaluable scientific resource.

New Delhi, India R.B. Singh

Preface

Abiotic stress factors mainly salinity, drought, flooding, and low and high temperature are the main elements which drastically limit the agricultural crop productivity globally. It has been estimated that salinity and drought are expected to cause serious salinization of more than 50 % of all available productive, arable lands by the year 2050. Extreme environmental events in the era of global climatic change further aggravate the problem and remarkably restrict the plant growth and development. Potential yield of economically important crops is drastically coming down every year just because of abiotic stresses. The mechanisms underlying endurance and adaptation to environmental stress factors have long been the focus of intense research. Plants overcome environmental stresses by the development of tolerance, resistance, or avoidance mechanisms. Plant acclimation to environmental stress is the process to adjust to a gradual change in its environment which allows the plants to maintain performance across a range of adverse environmental conditions.

In this book "Plant Acclimation to Environmental Stress," we present a collection of 17 chapters written by 50 experts in the field of crop improvement, genetic engineering, and abiotic stress tolerance. Plant Acclimation to Environmental Stress presents the latest ideas and trends on induced acclimation of plants to environmental stresses under changing environment. Various chapters included in this book provide a state-of-the-art account of the information is a resourceful guide suited for scholars and researchers working in the field of crop improvement, genetic engineering, and abiotic stress tolerance.

Chapter 1 deals with the use of priming agents towards plant acclimation to environmental stress. In this chapter, an up-to-date overview of the literature is presented in terms of some of the main priming agents commonly employed towards induced acclimation of plants to environmental challenges. Chapter 2 uncovers the sensing, signaling, and defending mechanisms in crop plants facing cold stress in the changing environment where authors discuss the status of effects of cold stress on plant metabolism, perception, and transduction of cold stress, genes expressed, defense mechanisms, and target genes for genetic engineering. Chapter 3 deals with drought and salinity tolerant biofuel crops for the Thar Desert. Chapter 4 covers strategies for the salt tolerance in bacteria and archeae and its implications in developing crops for adverse conditions. Chapter 5 deals with the adverse effects of abiotic stresses on medicinal and aromatic plants and their alleviation by calcium where authors emphasized that exogenously applied Ca can alleviate salt, heat, drought, high temperature, and cold stresses by regulation of antioxidant activities and discussed that in several plant cell-elicitor systems, the activation of defense responses depends on the presence of extracellular Ca. Thus, the growth, yield, and quality of the medicinal and aromatic plants could be improved under abiotic stress by supplying the plants with sufficient calcium nutrient. Chapter 6 discusses the role of DREB-like proteins in improving stress tolerance of transgenic crops. Chapter 7 focuses on Homeobox genes as potential candidates for crop improvement under abiotic stress. This chapter highlights the importance of homeobox genes in abiotic stress responses and their potential for engineering stress tolerance for crop improvement. Chapter 8 deals with APETALA2 gene family and its potential for crop improvement under adverse conditions. This chapter sheds light on transgenic expression of a single AP2 TF that has led to improved tolerance to multiple stresses like salinity, drought, and heat stress and pathogen infection, therefore emphasized that engineering of AP2 TFs seems to be a valuable tool towards achieving enhanced crop productivity under adverse conditions. Chapter 9 discusses the potential of osmoprotectants for crop improvement under adverse conditions. This chapter will encompass the potential role of osmoprotectants in plant stress adaptation and the possibilities for crop improvement. Chapter 10 deals with epigenetic modifications in plants under adverse conditions where authors discussed that epigenetic marks modify the properties of chromatin and change gene transcriptional states on the scale from the entire genome to a single specific gene. These marks allow for greater genome plasticity which results in better adaptation of plants to changing environmental conditions. Chapter 11 sheds light on the physiological role of nitric oxide in plants grown under adverse environmental conditions where authors reviewed recent progress in NO research in a broader context of abiotic stress tolerance and discussed its diverse roles in physiological and biochemical processes in plants and the protective mechanisms it exhibits towards abiotic stress tolerance. Chapter 12 deals with weeds, as a source of genetic material for crop improvement under adverse conditions. In this chapter an effort has been made to point out the useful traits of the weeds which can be transferred into crop plants for improvement along with the few successful case studies. Chapter 13 talks about sustainable agriculture practices for food and nutritional security and authors discussed the issues related to sustainability of existing agriculture; lessons learnt from green revolution; and possibility of new technologies so as to have sustainable ever green revolution. Chapter 14 deals with approaches for abiotic stress tolerance in crop plants for sustainable agriculture by the use of arbuscular mycorrhiza. Chapter 15 deals with the potential use of biofertilizers as a sustainable eco-friendly agricultural approach to crop improvement. Chapter 16 deals with plant–pathogen interactions and crop improvement under adverse conditions. Chapter 17 uncovers whether G-proteins may be the key elements for overcoming environmental stresses and increasing crop yield in plants? In this chapter the authors discuss the stress in general followed by the role of GPCR and G-proteins in biological processes including those that are related to environmental stresses.

The editors and contributing authors hope that this book will include a practical update on our knowledge for plant acclimation to environmental stress and lead to new discussions and efforts to the use of various tools for the improvement of crops for abiotic stress tolerance.

We are highly thankful to Dr. Ritu Gill, Centre for Biotechnology, MD University, Rohtak for her valuable help in formatting and incorporating editorial changes in the manuscripts. We would like to thank Professor R.B. Singh, President, National Academy of Agricultural Sciences, New Delhi for writing the foreword and Springer Science+Business Media, New York, particularly Daniel Dominguez, Developmental Editor/Project Manager; Andy Kwan, Assistant Editor, and Eric Stannard, Editor, Botany for their professional support and efforts in the layout. This book is dedicated to Professor E.A. Siddiq for nurturing plant genetics, plant breeding, and plant biotechnology.

New Delhi, India Narendra Tuteja
Rohtak, India Sarvajeet Singh Gill

Editors

Narendra Tuteja

Narendra Tuteja was born in 1955. Currently, Dr. Tuteja is working as Group Leader and Senior Scientist in Plant Molecular Biology Group, International Centre for Genetic Engineering and Biotechnology (ICGEB), New Delhi. Dr. Tuteja obtained his M.Sc., Ph.D., and D.Sc. in Biochemistry from the Lucknow University in 1977, 1982, and 2008, respectively. He is fellow of the Academies of Sciences: FNASc. (2003), FNA (2007), FASc. (2009), FNESA (2009), and FTWAS (2011).

Dr. Tuteja has made major contributions in the field of plant DNA replication and abiotic stress signal transduction, especially in isolating novel DNA/RNA helicases and several components of calcium and G-proteins signaling pathways. Initially he made pioneer contributions in isolation and characterization of large number of helicases from human cells while he was at ICGEB Trieste and published several papers in high impact journals including *EMBO J.* and *Nucleic Acids Res.* From India he has cloned the first plant helicase (*Plant J.* 2000) and presented the first direct evidence for a novel role of a pea DNA helicase (*Proc Natl Acad Sci U S A*. 2005) in salinity stress tolerance, pea heterotrimeric G-proteins (*Plant J*. 2007) in salinity, and heat stress tolerance. Dr. Tuteja has reported the first direct evidence in plant that PLC functions as an effector for Gα subunit of G-proteins. All the above work has received extensive coverage in many journals, including Nature Biotechnology, and bulletins all over the world. His group has also discovered novel substrate (pea CBL) for pea CIPK (*FEBS J*. 2006). He has already developed the salinity-tolerant tobacco and rice plants without affecting yield. Recently, few new high salinity stress-tolerant genes (e.g., Lectin receptor-like kinase, Chlorophyll a/b-binding protein, and Ribosomal L30E) have been isolated from *Pisum sativum* and have been shown to confer high salinity stress tolerance in bacteria and plant (*Glycoconjugate J*. 2010; *Plant Signal. Behav*. 2010). Recently, very high salinity stress-tolerant genes from fungus *Piriformospora indica* have been isolated and their functional validation in fungus and plants is in progress. Overall, Dr. Tuteja's research uncovers three new pathways to plant abiotic stress tolerance. His results are an important success and indicate the potential for improving crop production at suboptimal conditions.

Sarvajeet Singh Gill

Sarvajeet Singh Gill was born on January 21, 1979. Dr. Gill obtained his B.Sc. (1998) from Y.D. College, Kanpur University and M.Sc. (2001, Gold Medalist), M.Phil. (2003), and Ph.D (2009) from Aligarh Muslim University. Presently, Dr. Gill is working as Assistant Professor in Centre for Biotechnology, MD University, Rohtak, Haryana.

Dr. Gill's main area of research includes Genetic Engineering, Stress Physiology, and Molecular Biology (Development of abiotic stress-tolerant crop plants, the physiological, biochemical, and molecular characterization of agronomically important plants under abiotic stress factors, involvement of mineral nutrients, and other biotechnological approaches in the amelioration of abiotic stress effects in crop plants, use of a combination of genetic, biochemical, genomic, and proteomic approaches to understand the responses of various components of antioxidant machinery to abiotic stress and stress signaling, and stress tolerance in crop plants. Dr. Gill has several research papers, review articles, and book chapters to his credit in the journals of national and international repute and in edited books. He has edited four books namely *Sulfur Assimilation and Abiotic Stress in Plants*; *Eutrophication: Causes, Consequences and Control*; *Plant Responses to Abiotic Stress, Omics and Abiotic Stress Tolerance*; and *Improving Crop Resistance to Abiotic Stress*, published by Springer-Verlag (Germany), IK International, New Delhi, Bentham Science Publishers and Wiley-VCH, Verlag GmbH & Co. Weinheim, Germany, respectively. Dr. Gill is a regular reviewer of National and International journals and grants. He was awarded Junior Scientist of the year award by National Environmental Science Academy New Delhi in 2008. With Dr. Tuteja, Dr. Gill is working on heterotrimeric G-proteins Minichromosome maintenance (MCM) proteins and plant DNA helicases to uncover the abiotic stress tolerance mechanism in rice. The transgenic plants overexpressing heterotrimeric G-proteins and plant DNA helicases may be important for improving crop production at suboptimal conditions.

Contents

Contributors

Venkateswara R. Andugula National Institute of Plant Genome Research, New Delhi, India

Deepak Bhardwaj International Centre for Genetic Engineering and Biotechnology (ICGEB), New Delhi, India

Annapurna Bhattacharjee National Institute of Plant Genome Research (NIPGR), New Delhi, India

Alex Boyko Institute of Plant Biology, University of Zürich, Zürich, Switzerland

Debasis Chattopadhyay National Institute of Plant Genome Research, New Delhi, India

Gulshan K. Chhabra Synthetic Biology and Biofuel, International Center for Genetic Engineering & Biotechnology, Aruna Asaf Marg, New Delhi, India

Vibha Dhawan The Energy and Resources Institute (TERI), New Delhi, India

Heikham Evelin Department of Botany, University of Delhi, Delhi, India

Panagiota Filippou Department of Agricultural Sciences, Biotechnology and Food Science, Cyprus University of Technology, Limassol, Cyprus

Vasileios Fotopoulos Department of Agricultural Sciences, Biotechnology and Food Science, Cyprus University of Technology, Limassol, Cyprus

Masayuki Fujita Laboratory of Plant Stress Responses, Department of Applied Biological Science, Kagawa University, Kagawa, Japan

Sarvajeet Singh Gill Stress Physiology and Molecular Biology Lab, Centre for Biotechnology, MD University, Rohtak, India

Bhoopander Giri Department of Botany, Swami Shraddhanand College, University of Delhi, Delhi, India

Mirza Hasanuzzaman Laboratory of Plant Stress Responses, Department of Applied Biological Science, Kagawa University, Kagawa, Japan

Department of Agronomy, Sher-e-Bangla Agricultural University, Dhaka, Bangladesh

Deepti Jain National Institute of Plant Genome Research, New Delhi, India

Mukesh Jain National Institute of Plant Genome Research (NIPGR), New Delhi, India

Rachana Jain Department of Bioscience & Biotechnology, Banasthali University, Banasthali, Rajasthan, India

Rupam Kapoor Department of Botany, University of Delhi, Delhi, India

Harmeet Kaur National Institute of Plant Genome Research, New Delhi, India

Nat Kav Department Agricultural Food and Nutritional Science, University of Alberta, Edmonton, AB, Canada

M. Masroor A. Khan Plant Physiology Section, Department of Botany, Aligarh Muslim University, Aligarh, India

M. Nasir Khan Department of Biology, College of Science, University of Tabuk, Tabuk, Saudi Arabia

Igor Kovalchuk Department of Biological Sciences, University of Lethbridge, Lethbridge, AB, Canada

Sowmya Krishnaswamy Department Agricultural Food and Nutritional Science, University of Alberta, Edmonton, AB, Canada

Southern Crop Protection and Food Research Centre, Agriculture and Agri-Food Canada, London, ON, Canada

Bhumesh Kumar Directorate of Weed Science Research, Jabalpur, MP, India

Kamal Kumar Plant Immunity Laboratory, National institute of Plant Genome Research, New Delhi, India

Shashi Kumar Synthetic Biology and Biofuel, International Center for Genetic Engineering & Biotechnology, Aruna Asaf Marg, New Delhi, India

Suman Lakhanpaul Department of Botany, University of Delhi, New Delhi, India

Manoj Majee National Institute of Plant Genome Research, New Delhi, India

Karan Malhotra Synthetic Biology and Biofuel, International Center for Genetic Engineering & Biotechnology, Aruna Asaf Marg, New Delhi, India

Piyush Mathur Department of Botany, University of Delhi, Delhi, India

Moinuddin Plant Physiology Section, Department of Botany, Aligarh Muslim University, Aligarh, India

Athanassios Molassiotis Faculty of Agriculture, Aristotle University of Thessaloniki, Thessaloniki, Greece

M. Naeem Plant Physiology Section, Department of Botany, Aligarh Muslim University, Aligarh, India

Harsh Nayyar Department of Botany, Punjab University, Chandigarh, India

Bhanu P. Petla National Institute of Plant Genome Research, New Delhi, India

Megha K. Purohit Department of Biosciences, Saurashtra University, Rajkot, India

Muhammad H. Rahman Department Agricultural Food and Nutritional Science, University of Alberta, Edmonton, AB, Canada

A.R.G. Ranganatha Directorate of Weed Science Research, Jabalpur, MP, India

Meenal Rathore Directorate of Weed Science Research, Jabalpur, MP, India

Vikram Raval Department of Biosciences, Saurashtra University, Rajkot, India

Ranjan Kumar Sahoo International Centre for Genetic Engineering and Biotechnology (ICGEB), New Delhi, India

Saurabh C. Saxena National Institute of Plant Genome Research, New Delhi, India

Vinay Sharma Department of Bioscience & Biotechnology, Banasthali University, Banasthali, Rajasthan, India

Satya P. Singh Department of Biosciences, Saurashtra University, Rajkot, India

Georgia Tanou Faculty of Agriculture, Aristotle University of Thessaloniki, Thessaloniki, Greece

Prince Thakur Department of Botany, Punjab University, Chandigarh, India

Narendra Tuteja Plant Molecular Biology, International Centre for Genetic Engineering and Biotechnology (ICGEB), New Delhi, India

Pooja Verma National Institute of Plant Genome Research, New Delhi, India

Praveen Kumar Verma Plant Immunity Laboratory, National institute of Plant Genome Research, New Delhi, India

Shiv Verma Department Agricultural Food and Nutritional Science, University of Alberta, Edmonton, AB, Canada

Chapter 1
Plant Acclimation to Environmental Stress Using Priming Agents

Panagiota Filippou, Georgia Tanou, Athanassios Molassiotis, and Vasileios Fotopoulos

1 Introduction

Several reports have provided increasing evidence that plants can be "conditioned" for more rapid or intense induction of defense responses leading to enhanced resistance to biotic and abiotic stresses (Beckers and Conrath 2007). An analogy therefore exists with the concept of vaccination in animals, where the administration of antigenic material results in the stimulation of adaptive immunity to a disease and the ultimate prevention or amelioration of the effects of infection by pathogens. The physiological state in which plants are able to activate defense responses faster, better, or both, is called the primed state of the plant. Priming may be initiated in response to an environmental cue that reliably indicates an increased probability of encountering a specific stress factor, but a primed state may also persist as a residual effect following an initial exposure to the stress. The primed state can also be induced upon treatment with an acclimation-inducing agent, such as natural or synthetic compounds, as well as by colonization of plant tissues with beneficial microorganisms such as bacteria and arbuscular-mycorrhizal (AM) fungi. Under conditions of stress pressure, primed plants exhibit a higher fitness than non-primed plants or defense-expressing plants. Although priming has been known to occur in plants for several decades, most progress in the understanding of this phenomenon has been made over the past few years. The present chapter represents an up-to-date overview of the literature in terms of some of the main priming agents commonly employed toward induced acclimation of plants to environmental challenges. These include nitric oxide (NO),

P. Filippou • V. Fotopoulos (✉)
Department of Agricultural Sciences, Biotechnology and Food Science,
Cyprus University of Technology, Limassol PC 3036, Cyprus
e-mail: vassilis.fotopoulos@cut.ac.cy

G. Tanou • A. Molassiotis
Faculty of Agriculture, Aristotle University of Thessaloniki,
Thessaloniki 54 124, Greece

N. Tuteja and S. Singh Gill (eds.), *Plant Acclimation to Environmental Stress*,
DOI 10.1007/978-1-4614-5001-6_1, © Springer Science+Business Media New York 2013

hydrogen peroxide (H_2O_2), hydrogen sulfide (H_2S), polyamines and beneficial microorganisms. However, it should be pointed out that several more priming agents exist and are successfully employed toward induced acclimation of plants to environmental stress, including the quaternary amine glycine betaine and β-aminobutyric acid. Some of the key research carried out with the use of the specific priming agents under examination are summarized in Table 1.1.

2 Nitric Oxide

Nitric oxide (NO) is a redox-reactive, small, diffusible, ubiquitous, bioactive gaseous molecule that participates in a multitude of physiological and developmental processes in plants, including the response to environmental stimuli. For instance, heat, salt, and hyperosmotic stress induce NO production in tobacco (*Nicotiana tabacum*) cell suspensions (Gould et al. 2003). Also, NO metabolism is modulated during different abiotic stress conditions (high light intensity, low and high temperature, continuous light, continuous dark, and mechanical wounding) in pea plants (Corpas et al. 2008). It is well established that the endogenous NO can have two opposite physiological roles: a high cellular production of NO can provoke extensive cellular damage whereas NO at low levels participates in various signaling pathways (del Rio et al. 2004; Lamattina et al. 2003). However, many questions remain concerning exactly how NO is produced and scavenged, and how this signal is perceived and propagated in defined biological responses in stressed plant cells (Gupta et al. 2011).

Even though the details remain to be resolved, an increasing number of articles have been published during the last decade concerning the effects of exogenous NO on alleviating abiotic stress in plants (reviewed in Baudouin 2011; Besson-Bard et al. 2008). It has also been increasingly evident that prior exposure to NO renders plants more resistant to future environmental stress, thereby suggesting that NO acts as a priming agent (reviewed in Molassiotis et al. 2010). In pioneering reports, Uchida et al. (2002) showed that pre-exposure of rice seedlings to sodium nitroprusside (SNP; a NO donor) resulted in protection against salt and heat stress, preventing the impairment of photosystem II, activating the enzymatic antioxidant machinery, and increasing the transcriptional levels of genes encoding sucrose-phosphate synthase, D-pyrroline-5-carboxylate synthase, and small heat shock protein 26. The priming function of pretreatments with NO against salinity stress was also confirmed in other plant species, including maize (Zhang et al. 2004), *Arabidopsis* (Wang et al. 2009; Zhao et al. 2007), cucumber (Fan et al. 2007), citrus (Tanou et al. 2009a, b), and also in germinating seeds in saline environment (Kopyra and Gwóźdź 2003; Li et al. 2005). In experiments performed with callus cell cultures under salt conditions, it was found that NO could regulate plasma membrane H^+-ATPase activity, thus increasing K^+/Na^+ ratio leading to salt acclimation (Wang et al. 2009; Zhao et al. 2004). The hypothesis that NO-mediated regulation of Na^+ homeostasis and K^+ acquisition via ATPase is an important salt acclimation mechanism in plants was also supported in *Arabidopsis* plants (Zhao et al. 2007). Other NO-driven cellular

Table 1.1 Selected priming agents (compounds or beneficial organisms) inducing tolerance to abiotic stress factors in the greenhouse and field

Priming agent	Abiotic Stress	Plant	Reference(s)
Nitric oxide	Salt	Rice	Uchida et al. (2002)
		Maize	Zhang et al. (2004)
		Arabidopsis	Zhao et al. (2007), Wang et al. (2009))
		Cucumber	Fan et al. (2007)
		Citrus	Tanou et al. (2009a, b, 2010)
	Heat	Rice	Uchida et al. (2002)
		Reed	Song et al. (2008)
		Arabidopsis	Lee et al. (2008)
	Cold	Cucumber	Cui et al. (2011)
		Loquat	Wu et al. (2009)
		Arabidopsis	Zhao et al. (2009); Cantrel et al. (2011)
	Drought	Wheat	Garcia-Mata and Lamattina (2001)
		Rice	Farooq et al. (2009)
	UV-B radiation	Maize	An et al. (2005); Wang et al. (2006)
		Bean	Shi et al. (2005)
		Arabidopsis	Zhang et al. (2009d)
	Heavy metals	Rice	Singh et al. (2009); Xiong et al. (2009)
		Tomato	Wang et al. (2010b)
		Tobacco	Ma et al. (2010)
		Yellow lupin	Kopyra and Gwóźdź (2003)
		Arabidopsis	Graziano and Lamattina (2005)
Hydrogen peroxide	Cold	Maize	Prasad et al. (1994)
		Bean	Yu et al. (2003)
		Sweet potato	Lin and Block (2010)
		Mustard	Kumar et al. (2010)
	Salt	Rice	Uchida et al. (2002)
		Wheat	Wahid et al. (2007); Li et al. (2011)
		Citrus	Tanou et al. (2009a, b, 2010)
		Maize	Neto et al. (2005)
		Oat	Xu et al. (2008)
		Barley	Fedina et al. (2009)
		Pigeonpea	Chawla et al. (2010)
	Heat	Rice	Uchida et al. (2002)
		Bentgrass	Larkindale and Huang (2004)
		Cucumber	Gao et al. (2010)
	Heavy metals	Pigeonpea	Chawla et al. (2010)
		Wheat	Xu et al. (2011)
		Rice	Chao and Kao (2010)
Hydrogen sulfide	Salt	Strawberry	Christou, Fotopoulos et al. (unpublished data)
		Wheat	Zhang et al. (2010a)
		Sweet potato	Zhang et al. (2009c)
	Drought	Wheat	Shan et al. (2011)
		Arabidopsis	Garcia-Mata and Lamattina (2010)
		Broad bean	Garcia-Mata and Lamattina (2010)
		Soybean	Zhang et al. (2010b)
	Heavy metals	Cucumber	Wang et al. (2010a)
		Wheat	Zhang et al. (2008a, 2010a, c)
	Cold	Wheat	Stuiver et al. (1992)

(continued)

Table 1.1 (continued)

Priming agent	Abiotic Stress	Plant	Reference(s)
Polyamines	Salt	Oat	Besford et al. (1993)
		Rice	Maiale et al. (2004); Ndayiragije and Lutts (2006a, b, 2007); Quinet et al. (2010)
		Mustard	Verma and Mishra (2005)
		Spinach	Öztürk and Demir (2003)
		Arabidopsis	Kusano et al. (2007)
	Drought	Rice	Yang et al. (2007)
		Arabidopsis	Kusano et al. (2007)
Beneficial microorganisms	Heavy metals	*Arabidopsis*	Farinati et al. (2011)
	Cold	Blue mustard	Ding et al. (2011)
		Grapevine	Ait-Barka et al. (2006)
	Salt	Maize	Harman (2006); Abdelkader and Esawy (2011)
		Poplar	Luo et al. (2009)
		Tomato	Latef and He (2011)
		Olive	Porras-Soriano et al. (2009)
	Drought	Rice	Ruiz-Sánchez et al. (2010)
		Soybean	Porcel and Ruiz-Lozano (2004)
		Citrus	Fan and Liu (2011)
		Southern Beech	Alvarez et al. (2009)

responses toward salt stress acclimation involve the increase in chlorophyll content, the decrease in electrolyte leakage along with changes in polyamine metabolism (Zhang et al. 2004), the increase in the activities of endopeptidase and carboxypeptidase (Zheng et al. 2010), and the induction of ATP synthesis and the respiratory electron transport in mitochondria (Yamasaki et al. 2001; Zottini et al. 2002). In addition, Tanou et al. (2009b) provided evidence that NO exhibits a strong antioxidant role during the establishment acclimation of citrus plants to salinity. More interestingly, a proteomic study on citrus plants grown under salinity stress by Tanou et al. (2009a) provided a wide list of proteins whose accumulation levels are regulated by salt stress, whereas it was further shown that exogenous supply of NO via root pretreatment with SNP reversed a large part of the NaCl-responsive proteins. These SNP/NaCl-responsive proteins are mainly involved in photosynthesis, defense mechanism, and energy/glycolysis pathways. These data indicate that NO pre-exposure can specifically modify protein expression signatures, and that a NO-specific priming function is needed for a proper salt acclimation response.

In addition to NO-mediated priming phenomena against salinity stress, several studies demonstrated that NO is also involved in drought acclimation in many plant species, including wheat (Garcia-Mata and Lamattina 2001), reed cell suspension cultures (Zhao et al. 2008), and rice (Farooq et al. 2009). These results were followed by findings indicating the involvement of NO in the maintenance of tissue water potential through stomatal closure (Garcia-Mata et al. 2003), alleviation of oxidative damage via protein synthesis acceleration, photosynthesis rate enhancement, and stimulation of antioxidant enzymes activities (Tan et al. 2008). Apart from an osmotic stress alleviation inducing molecule, there is also evidence that NO is also an osmotic stress induced molecule in a biphasic manner through an early production phase followed by a lateral one (Kolbert et al. 2007). Notably, NO was previously shown to be involved in the ABA-induced stomatal closure (Bright et al. 2006) via activating mitogen-activated protein kinase (MAPK) (Zhang et al. 2007). In this sense, transgenic tobacco plants over expressing *SgNCED1* gene encoding the 9-*cis*-epoxy-carotenoid dioxygenase, which accounts for increased ABA biosynthesis, resulted in a NO-associated drought and salt stress acclimation (Zhang et al. 2009a).

Within the context of temperature stress, NO is also known to be involved in the plant response to high- and low-temperature stresses. There is evidence that NO exhibits priming phenomena under heat stress conditions (Uchida et al. 2002), but in an ABA-independent manner (Song et al. 2008). Ion leakage prevention, growth and cell viability retention, decreased H_2O_2 and MDA contents, and increase in antioxidant enzyme activities have been reported to be responses to heat acclimation via NO pretreatment (Song et al. 2008). In a study conducted with transgenic *Arabidopsis* plants impaired in NO synthesis, this was directly connected with lack of thermotolerance (Xuan et al. 2010) since the transgenic plant cannot accumulate a specific heat shock protein (Hsp18.2). However, the NO-induced priming against heat stress seems to be more a complicated scenario. For example, Lee et al. (Lee et al. 2008) showed that *S*-nitrosoglutathione reductase (GSNOR), the enzyme which metabolizes the NO adduct *S*-nitrosoglutathione, is necessary for the acclimation of *Arabidopsis* plants to high temperature. These authors also found that *Arabidopsis* mutants lacking HOT5 (encoding GSNOR) were thermosensitive but NO donors failed to rescue thermotolerance (Lee et al. 2008). Clearly, these observations need to be developed further to establish the specific roles of Hsp and especially of GSNOR in cold acclimation and, most importantly, their interplay with NO during this process. On the other hand, NO priming action against cold stress seems to be mediated by brassinosteroids (BRs). Indeed, pretreatment of cucumber plants with NO donors leads to cold acclimation and to the induction of antioxidant enzymes (Cui et al. 2011). Pharmacological studies with Arabidopsis plants using nitric reductase (NR) inhibitor, NO scavenger, and NO donor showed that NR-dependent NO production was linked with freezing acclimation via increasing the expression levels of *P5CS1* and *ProDH* genes and enhanced accumulation of proline (Zhao et al. 2009). Another report supports that NO-induced cold acclimation is associated with scavenging ability of NO against ROS (Wu et al. 2009), whereas a more recent study on *Arabidopsis* revealed that genetic impairment of NO accumulation upon chilling inhibited the expression of specific cold-responsive genes, phosphatidic acid synthesis, and sphingolipid phosphorylation (Cantrel et al. 2011).

Another severe abiotic damaging factor on plant metabolism is UV-B radiation (250–320 nm), resulting in disturbances in plant growth and development (Rozema et al. 1997). Early studies revealed that plants undergoing UV-B exposure produce NO leading also in the expression of UV-B radiation-specific defense genes

(Mackerness et al. 2001). The protective signaling effects of NO against UV-B treatment were previously established in maize seedling receiving NO hydroponically (via SNP application) (An et al. 2005), in bean leaves sprayed with SNP (Shi et al. 2005) or in SNP-treated *Cyanobacterium* cell suspension cultures (Xue et al. 2007). According to Shi et al. (2005), NO could partially alleviate the decrease in chlorophyll content and F_v/F_m ratio caused by UV-B exposure, probably by mediating the activities of antioxidant enzymes. Xue et al. (2007) suggested also the participation of reduced glutathione (GSH) in the NO-mediated acclimation against UV-B exposure. In addition, Wang et al. (2006) proposed that NO acts in the same direction or synergistically with ROS to induce ethylene biosynthesis in the leaves of maize seedlings under UV-B radiation; however, the connection of this crosstalk with acclimation against UV-B irradiation has not been established yet. Meanwhile, Zhang et al. (2009d), using AtNOS1 mutant Arabidopsis plants impaired in regular NO biosynthesis and containing lower amount of NO, showed that these plants were prone to UV-B damage, whereas NO supplementation could alleviate the oxidative damaged by increasing flavonoid and anthocyanin contents.

Recently, the alleviating effects of exogenous NO on heavy metal toxicity in plants are becoming apparent (reviewed in Hasanuzzaman et al. 2010; Xiong et al. 2010), including arsenic toxicity in roots of *Oryza sativa* (Singh et al. 2009), copper toxicity in tomato plants (Wang et al. 2010b), cadmium toxicity in tobacco (Ma et al. 2010), and zinc toxicity in *Solanum nigrum* (Xu et al. 2010). In addition, there is evidence that NO is involved in the acclimation of various cell systems against multiple heavy metals stress (Kopyra and Gwóźdź 2003). By contrast to these observations, Arasimowicz-Jelonek et al. (2011) found that NO contributes to Cd toxicity by promoting Cd uptake and participates in metal-induced reduction of root growth. In several studies conducted with various plant species exposed to single (Singh et al. 2009; Wang et al. 2010b) or combined multiple heavy metal stressors (Kopyra and Gwóźdź 2003), it was evidenced that exogenous application of NO resulted in the modulation of the antioxidant mechanism or in the scavenging of ROS. In addition, Xiong et al. (2009) demonstrated that NO-induced acclimation to Cd toxicity in rice is attributed to the decreased distribution of Cd in the soluble fraction of leaves and roots and in the increased distribution of Cd in the cell walls of roots. Furthermore, a NO-driven reduced transpiration rate with concomitant decreased heavy metal translocation from roots to shoots has also been proposed, but without further cross-verification of the effect of exogenously applied SNP to stomatal movement (Xiong et al. 2009). Another interesting mechanism through which NO could act as a signaling factor under heavy metal toxicity situations is its action as a regulator of stress-related genes. For example, the regulation of iron homeostasis by ferritin gene expression in Arabidopsis leaves has been proven to be attributed to exogenous NO application (Graziano and Lamattina 2005) whereas NO production in algae under Cu toxicity has been shown to be implicated in up-regulation of the expression of *P5CS* which encodes D1-pyrroline-5-carboxylate synthetase responsible for proline biosynthesis (Zhang et al. 2008b).

3 Hydrogen Peroxide

It is now well established that virtually all biotic and abiotic stresses induce or involve oxidative stress to some degree, and the ability of plants to control oxidant levels is highly correlated with stress acclimation (Gill and Tuteja 2010a). Hydrogen peroxide (H_2O_2) is usually the end step conversion of ROS, which, despite its reactive perception is the most stable molecule among them (Hung et al. 2005). In addition to being toxic and causing cell death at high concentrations (Dat et al. 2000), H_2O_2 has been regarded as a central player in growth and developmental processes of plants (Hung et al. 2005). Currently, H_2O_2 is considered to be a messenger molecule to various abiotic and biotic stress conditions when applied at low concentrations and a number of acclimation mechanisms based on the physiological and biochemical changes inquired have been proposed (Fukao and Bailey-Serres 2004; Mittler et al. 2004). These mechanisms normally include genes encoding antioxidants, cell rescue/defense proteins, and signaling proteins such as kinase, phosphatase, and transcription factors (Hung et al. 2005).

One of the first studies regarding chilling stress acclimation reported that chilling imposed oxidative stress in maize seedlings could be prevented by H_2O_2 pretreatment by increasing *cat3* transcripts and the enzymatic activities of catalase3 and guaiacol peroxidase (Prasad et al. 1994). In this report, the dual role of H_2O_2 at low temperatures was proposed, since H_2O_2 early accumulation triggered the production of antioxidant enzymes whereas H_2O_2 accumulated to damaging levels in the tissues at non-pretreated and therefore non-acclimated seedlings (Prasad et al. 1994). A H_2O_2-induced chilling acclimation mechanism was also reported in mung bean plants via the induction of GSH (Yu et al. 2003). In studies with sweet potato (*Ipomoea batatas*) and sweet peppers (*Capsicum annuum*), it was evidenced that exogenously applied H_2O_2 may lead to chilling acclimation; however, the practical benefits of exogenous H_2O_2 application could not be clearly observed under all experimental conditions tested (Lin and Block 2010). In addition, it was observed that H_2O_2 pretreatment had beneficial effects in sweet potato against chilling injury when the pretreatment was applied under long photoperiod whereas this was not the case when it was applied under short photoperiod. These observations revealed that the H_2O_2-mediated priming phenomena may also be regulated by other intra- or extracellular factors. Brassinosteroids (BRs), such as 24-epibrassinolide, have been proposed to be involved in the H_2O_2-induced acclimation against chilling stress in *B. juncea* L. seeds and seedlings (Kumar et al. 2010). In this study it was supported that 24-epibrassinolide helped in alleviating the toxic effect of H_2O_2 through modulation of the antioxidant enzymes such as catalase (CAT), ascorbate peroxidase (APX), and superoxide dismutase (SOD) (Kumar et al. 2010). Comparable results to those were shown in a study with tomato plants, where BRs applied exogenously could alleviate drought-induced oxidative stress via increase in the activity of antioxidant enzymes and antioxidant compounds such as ascorbate, carotenoids, and proline (Behnamnia et al. 2009). The regulation of the enzymatic antioxidant machinery under heat stress conditions by H_2O_2 pretreatment and a resulting reduction of the oxidative damage have also been observed (Larkindale and Huang 2004; Uchida et al. 2002). In conditions of heat stress, exogenous H_2O_2 also increased antioxidant enzyme activities in cucumber leaves, decreased lipid peroxidation, and thus protected the ultrastructure of chloroplasts (Gao et al. 2010). Apart from the antioxidant machinery, *Hsp20* genes, the small Hsps that act as chaperones and are involved in cellular protection under several environmental stress conditions (Liberek et al. 2008; Rhee et al. 2009), were up-regulated in response to H_2O_2 (Rhee et al. 2011). In fact, H_2O_2 is considered to be an essential prerequisite component in the heat stress signaling pathway for the effective expression of heat shock genes (Volkov et al. 2006).

It is also interesting to note that H_2O_2 priming activity does not seem to be stress-specific since commonly H_2O_2-modulated cellular responses were recorded in rice seedlings subjected to heat or to salt stress (Li et al. 2011; Uchida et al. 2002). The increased acclimation of H_2O_2-treated plants to salt stress has been attributed mostly to increased antioxidative enzyme activities (Fedina et al. 2009; Neto et al. 2005; Xu et al. 2008) and to other stress-related genes and proteins (Wahid et al. 2007), including the expression of transcripts for genes encoding sucrose-phosphate synthase, Δ-pyrroline-5-carboxylate synthase, and small Hsp 26 (Uchida et al. 2002). The priming effects of H_2O_2 pretreatment against salinity were well documented by Tanou et al. (2009b), who showed that pre-exposure to H_2O_2 elicits long-lasting, systemic antioxidant activity in citrus plants under both physiological and NaCl-stress conditions. In addition, the authors showed that H_2O_2 treatment at low concentrations in the absence or presence of salinity stress could modulate specific protein targets involved in photosynthesis, defense, and energy metabolism (Tanou et al. 2009a, 2010). Furthermore, a proteome-wide analysis revealed that the specific priming function of H_2O_2 in citrus plants against salt stress involves the depression of protein carbonylation and the stimulation of protein *S*-nitrosylation (Tanou et al. 2009a).

The time of intracellular H_2O_2 production following external H_2O_2 application is tightly associated with acclimation responses and plays a major role in H_2O_2 signaling. The determination of H_2O_2 endogenous levels in NaCl-stressed plant tissues after H_2O_2 pretreatment showed an early H_2O_2 peak (Uchida et al. 2002; Xu et al. 2008). Application of diphenylene iodonium (DPI) during H_2O_2 pretreatment in naked oat seedlings counteracted the beneficial effects of H_2O_2 toward salinity. The authors emphasized, in addition, that when DPI was applied at the immediate end of the H_2O_2 pretreatment, it did not alter the H_2O_2 protective role, indicating that the H_2O_2 formed early on during salt stress might play an important role in regulating plant acclimation to saline environments (Xu et al. 2008). In contrast to the aforementioned positive physiological effects of exogenously applied H_2O_2 on the growth and development of salt-stressed plants, even when salinity conditions occur simultaneously with other abiotic stresses like boron (B) toxicity (Chawla et al. 2010), H_2O_2 could not reduce the detrimental effects on the mitotic activity and the chromosomal aberrations of barley seeds exposed to NaCl (Cesur and Tabur 2011).

Heavy metal toxicity, which represents abiotic stress conditions with hazardous health effects to animals and plants, has been reported to act among others through the generation of reactive oxygen species, especially H_2O_2 (Maksymiec 2007). In a study conducted with wheat and rice seedlings, it was shown that the priming effects of H_2O_2 are extended also to Al and Cd stress alleviation (Chao and Kao 2010; Xu et al. 2011). In addition, H_2O_2 pretreatment could induce Al acclimation by enhancing the antioxidant defense capacity, which prevented the Al-caused ROS accumulation (Xu et al. 2011). Furthermore, the central role of the nonenzymatic antioxidant ascorbate in H_2O_2-signaling was reported by Chao and Kao (2010), who showed that H_2O_2-induced protection against subsequent Cd stress of rice seedlings is mediated through ascorbate production. In the same study, where heat priming effects on Cd stress were also examined, an early H_2O_2 endogenous production prior to ascorbate accumulation was observed (Chao and Kao 2010), thus confirming previous reports that demonstrated an active interplay between ascorbate and H_2O_2 signaling (Fotopoulos et al. 2006, 2008). Apart from ascorbate, there are many experimental data indicating that there is a connection between H_2O_2 and other signaling pathways during priming phenomena, especially as evidenced by the active interactions between H_2O_2 and NO (for a review see Molassiotis and Fotopoulos 2011). The connection between H_2O_2 and NO signaling networks during the establishment of priming-based acclimation against salinity has been extensively examined in citrus plants (see Tanou et al. 2009a, 2010).

4 Hydrogen Sulfide

Hydrogen sulfide (H_2S) is a colorless gas with a strong odor of rotten eggs. The toxicity of H_2S at high concentration has been substantiated for almost 300 years (Lloyd 2006). Hydrogen sulfide is often thought to be phytotoxic, being harmful to the growth and development of plants. It was found to inhibit oxygen release from young seedlings of six rice cultivars (Joshi et al. 1975), but it was also noted that, although in some cultivars nutrient uptake was reduced, in other cultivars it was increased. The impact of atmospheric H_2S on plants is paradoxical. On the one hand, it may be utilized as a sulfur nutrient source, and on the other hand, it may negatively affect plant growth and functioning above a certain threshold level (De Kok et al. 2002). The predominant natural sources of H_2S in terrestrial ecosystems are the biological decay of organic sulfur and the activity of dissimilatory sulfate-reducing bacteria (Bates et al. 1992; Beauchamp et al. 1984). H_2S is thought to be released from cysteine via a reversible *O*-acetylserine(thiol)lyase reaction in plants (Sekiya et al. 1982a; Wirtz et al. 2004). It was reported that higher plants could emit H_2S when exposed to excess sulfur and cysteine (Rennenberg 1983; Sekiya et al. 1982a). H_2S is endogenously synthesized in both animals and plants by enzymes with L-Cys desulfydrase activity in the conversion of L-Cys to H_2S, pyruvate, and ammonia. The fact that H_2S is also produced by cut branches, detached leaves, leaf discs, or tissue cultures, thus acting as evidence that green cells of higher plants can release H_2S into the atmosphere (Rennenberg 1983, 1984, 1990; Sekiya et al. 1982a, b; Wilson et al. 1978; Winner et al. 1981).

An early report by Thompson and Kats (1978), who treated a variety of plants with continuous fumigation of H_2S (3,000 ppb), resulted in the appearance of lesions on leaves, defoliation, and reduced growth of the plants supporting the role of H_2S as a phytotoxin. However, significantly lower levels of fumigation (100 ppb) caused a significant increase in the growth of *Medicago*, lettuce, and sugar beets.

In plants, it has been documented that H_2S can promote root organogenesis (Zhang et al. 2009b) and seed germination (Zhang et al. 2008a). It is conceivable that H_2S might serve as a signaling molecule to other parts of the plant, or to plants in the vicinity in a similar manner to NO and CO (Zhang et al. 2008a, 2009b, c). More recently, many studies have revealed that H_2S can act as a signaling molecule at lower concentrations and participate in several other key physiological processes (Hosoki et al. 1997; Li et al. 2006; Wang 2002; Yang et al. 2008).

Although at present there is no direct evidence that H_2S acts as an endogenous regulator or a signal molecule in plants, the induction of L-cysteine desulfhydrase upon pathogen attack (Bloem et al. 2004), emission of H_2S from plants exposed to SO_2 injury (Hällgren and Fredriksson 1982; Sekiya et al. 1982a), abiotic stress acclimation in plants supplied with exogenous H_2S donor (Stuiver et al. 1992; Zhang et al. 2008a, 2009c, 2010a, b, c, d), and its involvement in guard cell signaling (Garcia-Mata and Lamattina 2010) all suggest that this is indeed the case. At low H_2S concentration, it can promote the embryonic root length of *Pisum sativum* (Li et al. 2010). Rausch and Wachter (2005) reviewed sulfur metabolism, a versatile platform for launching defense operations and revisited the hypothesis of "sulfur-induced resistance", which may play an important role in the defense potential of plants.

NaHS is a commonly used H_2S donor in biological systems (Hosoki et al. 1997). NaHS dissociates to Na^+ and HS^- in solution and HS^- associates with H^+ to produce H_2S. Solutions of Na_2S, Na_2SO_4, Na_2SO_3, $NaHSO_4$, and $NaHSO_3$ were sometimes used and found ineffective (Zhang et al. 2010c). However, new compounds are now being developed which release H_2S in a more gradual manner (Li et al. 2008, 2009; Whiteman et al. 2010). The effects of novel compounds on plant tissues that mimic well the effects seen with NaHS, and could be used more extensively to study the effects of H_2S on plant function were also studied recently (Lisjak et al. 2010).

Some of the more recent reports on H_2S biology in plants have shown that H_2S counteracts the oxidative burst generated by H_2O_2 production upon different stresses by reducing H_2O_2 concentrations and increasing the activity of antioxidant enzymes (Zhang et al. 2008a, 2009c, 2010c).

More recently, it was demonstrated that H_2S is involved in the antioxidant response during wheat seeds germination against copper stress (Zhang et al. 2008a), chromium stress (Zhang et al. 2010a), drought stress (Zhang et al. 2010b) and in sweet potato seedlings growth under osmotic stress conditions (Zhang et al. 2009c). Moreover, the protective role of H_2S in seed germination and seedling growth was also studied in wheat seeds subjected to aluminum (Al^{3+}) stress (Zhang et al. 2010c). NaHS pretreatment significantly increased the activities of amylases and esterases and sustained much lower levels of MDA and H_2O_2 in germinating seeds under Al^{3+} stress, indicating that H_2S could increase antioxidant capability in wheat seeds leading to the alleviation of Al^{3+} stress. Similarly, boron toxicity was also shown to be alleviated by H_2S (Wang et al. 2010a).

It has also been proven that exogenous H_2S induces stomatal closure and participates in ABA dependent signaling, possibly through the regulation of ABC transporters in guard cells (Garcia-Mata and Lamattina 2010). As NO is involved in the signaling pathways which cause stomatal closure, it is tempting to speculate the interaction between NO and H_2S. Such an induction of stomatal closure potentially assists in the protection of the plant against low water supply by limiting water loss via reduced transpiration.

The effects of elevated atmospheric H_2S levels (0.25, 0.5, and 0.75 μL/L) have been investigated in a short-term exposure experiment (3–48 h) on the model plant *Arabidopsis thaliana* in comparison to untreated control plants. The most pronounced effects of H_2S fumigation could be observed on the metabolite levels: the contents of the thiols, cysteine and GSH, were increased up to 20- and 4-fold, respectively. In general, H_2S exposure of plants results in a slight overload of the plant sulfur supply, which is illustrated by an increased size and change in composition of the thiol pool in the shoots (De Kok et al. 2002). In *Arabidopsis* shoots there was a significant increase in cysteine and GSH levels upon H_2S fumigation. The amounts of cysteine in the H_2S-exposed plants could be directly correlated with increasing H_2S concentrations and with the duration of the treatment, as after 48 h the cysteine levels only slightly decreased. The same observations were true for GSH (Riemenschneider et al. 2005).

In addition, Shan et al. (2011) suggested that exogenously applied H_2S regulates the ascorbate and GSH metabolism by increasing the activities of APX, GR, DHAR, c-ECS and the contents of AsA, GSH, total ascorbate, and total GSH, which, in turn, enhances the antioxidant ability and protects wheat seedlings against oxidative stress induced by water stress.

As previously mentioned, H_2S promotes seed germination and root formation, and acts as an antioxidant signal counteracting heavy metal and other stresses in plants (Zhang et al. 2008a, 2009b, c, 2010a, b, c, d). The molecular mechanisms by which this signaling molecule acts are also being investigated. Quite recently, Zhang et al. (2009c) showed that the H_2S donor NaHS would alleviate the osmotic-induced decrease in chlorophyll concentration in sweet potato. Spraying NaHS increased the activity of the antioxidant enzymes such as SOD, CAT, and APX, while decreasing the concentration of hydrogen peroxide and lipoxygenase, suggesting the protective role of H_2S against oxidative stress. Supporting this hypothesis are the findings that fumigation of spinach increased GSH levels (De Kok et al. 1985), and it was estimated that approximately 40% of the H_2S was converted to GSH in the leaves. On cessation of fumigation GSH levels once again fell, with the levels being comparable to control levels after 48 h in the absence of H_2S application.

H_2S was also shown to increase drought acclimation in soybean seedlings by acting as an antioxidant signal molecule regulating the plant's response. Spraying soybean seedlings with exogenous H_2S donor NaHS prolonged the life and enlarged higher biomass of both leaf and root compared with non-sprayed controls under continuous drought stress. The drought-induced decrease in chlorophyll could be alleviated by spraying with a H_2S donor. It was also shown that spraying with NaHS dramatically retained higher activities of SOD, CAT, and suppressed activity of lipoxygenases, and delayed excessive accumulation of malondialdehyde, hydrogen peroxide, and superoxide anion (O_2^-) compared with the control (Zhang et al. 2010b).

In addition, recent results indicating the protective role of H_2S as a priming agent in strawberry (*Fragaria* x *ananassa* Camarosa) pretreated with 100 μM NaHS for 48 h demonstrated increased resistance to high salinity (100 mM NaCl) and hyperosmotic stress (10% PEG-6000 w/v). Meanwhile, pretreatment with NaHS decreased the malondialdehyde content, H_2O_2 and NO content compared with control and water stress without NaHS. Results suggested that exogenous hydrogen sulfide alleviated oxidative damage by regulating the ascorbate and GSH metabolism in strawberry under salinity and hyperosmotic stresses (Christou, Fotopoulos et al., unpublished data). All the above findings suggest that the study of H_2S as a priming agent in plants is just in its beginning, with several experimental data supporting the possible role of H_2S as a new antioxidant signal. However, its molecular mechanisms of antioxidant adaptation are still poorly understood and the signaling pathways involved need to be further investigated.

5 Polyamines

Polyamines (PAs), mainly putrescine (PUT), spermidine (SPD), and spermine (SPM), are polycationic compounds of low molecular weight that are present in all living organisms. They have been proposed as a new category of plant growth regulators that are purported to be involved in various physiological processes, such as embryogenesis, cell division, morphogenesis, and development (Bais and Ravishankar 2002; Liu et al. 2006).

The simplest polyamine, PUT, is derived either directly from ornithine by ornithine decarboxylase (ODC) or from arginine through several steps catalyzed by arginine decarboxylase (ADC), agmatine iminohydrolase, and *N*-carbamoylputrescine amidohydrolase. In contrast to animals and fungi, in which ODC is the first and rate-limiting enzyme in the synthesis of polyamines, plants typically use ADC. The *Arabidopsis thaliana* genome lacks a gene encoding ODC (Hanfrey et al. 2001). PUT is converted to SPD and SPM by successive activities of SPD synthase and SPM synthase with the use of decarboxylated *S*-adenosyl methionine (dcSAM) as an aminopropyl donor. The dcSAM is produced by *S*-adenosylmethionine decarboxylase (SAMDC) from SAM. Polyamines are further metabolized by oxidation and conjugation with other molecules (Bagni and Tassoni 2001; Cona et al. 2006; Moschou et al. 2008).

Polyamines in plants are preferentially detected in actively growing tissues as well as under stress conditions and have been implicated in the control of cell division, embryogenesis, root formation, fruit development, and ripening (Kumar et al. 1997). In the past decade, however, molecular and genetic studies with mutants and transgenic plants having no or altered activity of enzymes involved in the biosynthesis of polyamines have contributed much to a better understanding of the biological functions of polyamines in plants.

Plant polyamines frequently accumulate in response to abiotic and biotic stresses (Bouchereau et al. 1999; Urano et al. 2004; Walters 2003a, b). There is an extensive literature describing the correlation of changes in polyamine levels and physiological perturbations and on the protective effect of polyamines on environmental stresses (Alcázar et al. 2006; Groppa and Benavides 2008; Liu et al. 2007, and references therein). Classical approaches, using exogenous polyamine application and/or inhibitors of enzymes involved in polyamine biosynthesis, pointed to a possible role of these compounds in plant adaptation/defense to several environmental stresses (Alcázar et al. 2006; Bouchereau et al. 1999; Groppa and Benavides 2008).

Several lines of evidence have shown that the stimulatory effect of exogenous polyamines may be related to their multifaceted nature, which includes working as an antioxidant, a free radical scavenger, and a membrane stabilizer (Velikova et al. 2000). Polyamines act as antioxidants, and they counteract oxidative damage in plants, which, as a consequence, reduce free radicals and alleviate lipid peroxidation (Kramer and Wang 1989; Singh et al. 2002).

Verma and Mishra (2005) reported that exogenous PUT affected the activities of several antioxidant enzymes, such as SOD, CAT, POD, APX, and GR, when added to *Brassica juncea* seedlings treated with NaCl, which occurred concomitantly with a reduction of H_2O_2 and lipid peroxidation, implying that the positive effects of exogenous polyamines may be related to its antioxidant properties. In another study, Öztürk and Demir (2003) demonstrated that exogenous polyamines increased the activities of POD and CAT, along with the accruement of proline, an important osmoprotectant involved in the plant's response to abiotic stress.

More recent studies using either transgenic overexpression or loss-of-function mutants support the protective role of polyamines in plant response to abiotic stress (Alcázar et al. 2006; Gill and Tuteja 2010b; Kusano et al. 2008). Indeed, heterologous overexpression of *ODC*, *ADC*, *SAMDC*, and *SPDS* from different animal and plant sources in rice, tobacco, and tomato has shown acclimation traits against a broad spectrum of stress conditions. Enhanced acclimation always correlated with elevated levels of PUT and/or SPD and SPM (Liu et al. 2007). The results obtained from loss-of-function mutations in polyamine biosynthetic genes further support the protective role of polyamines in plant response to abiotic stress. EMS mutants of *Arabidopsis thaliana spe1-1* and *spe2-1* (which map to *ADC2*) displaying reduced ADC activity are deficient in polyamine accumulation after acclimation to high NaCl concentrations and exhibit more sensitivity to salt stress (Kasinathan and Wingler 2004). Moreover, *acl5/spms* Arabidopsis double mutants that do not produce SPM are hypersensitive to salt and drought stresses, and the phenotype is mitigated by application of exogenous SPM (Kusano et al. 2007).

Nitric oxide, polyamines, diamine oxidases, and polyamine oxidases play important roles in wide spectrum of physiological processes such as germination, root development, flowering, and senescence and in defense responses against abiotic and biotic stress conditions. This functional overlapping suggests interaction of NO and PA in signaling cascades (Wimalasekera et al. 2011). PA is related to NO through arginine, a common precursor in their biosynthetic pathways, in a similar way to that in animals (Palavan-Unsal and Arisan 2009; Yamasaki and Cohen 2006).

Previous reports present evidence that PA induces the production of NO (Arasimowicz-Jelonek et al. 2009; Groppa et al. 2008; Tun et al. 2006). Conversely, recent work by Filippou and Fotopoulos also indicates the reverse effect: NO application results in the induction of PAs (unpublished data).

In *A. thaliana*, SPD and SPM stimulate NO production whereas PUT has little effect (Tun et al. 2006). The promotion by SPD and SPM of the 14-3-3-dependent inhibition of phospho-NR (Athwal and Huber 2002), which down-regulates nitrate assimilation and NO production from nitrite, suggests the involvement of other sources for SPD and SPM-induced NO production (Yamasaki and Cohen 2006). In *Araucaria angustifolia*, SPD and SPM inhibited NO biosynthesis in both embryonic and suspensor cells, while PUT induced NO biosynthesis in embryonic cells (Silveira et al. 2006). Treatment with PUT significantly inhibits the softening of banana fruit with concomitant increases in endogenously formed NO as well as PUT, where the mechanism involved is as yet to be established (Manjunatha et al. 2010).

In addition, PUT modulates ABA biosynthesis in response to abiotic stress (Alcázar et al. 2010). It is therefore likely that polyamines participate in ABA-mediated stress responses involved in stomatal closure. In this regard, evidence points to an interplay between polyamines with ROS generation and NO signaling in ABA-mediated stress responses (Yamasaki and Cohen 2006). The generation of ROS is tightly linked to polyamine catabolic processes, since amino oxidases generate H_2O_2, which is a ROS associated with plant defense and abiotic stress responses (Cona et al. 2006). Both H_2O_2 and NO are involved in the regulation of stomatal movements in response to ABA, in such a way that NO generation depends on H_2O_2 production (Neill et al. 2008).

Stress responses involve the generation of secondary messengers such as Ca^{2+}. The increase in cytosolic Ca^{2+} modulates the stress signaling pathways controlling stress acclimation. In guard cells, the increase in cytosolic Ca^{2+} may activate different ion channels and induce stomatal closure (Blatt et al. 1990; Gilroy et al. 1990). Changes of free Ca^{2+} in the cytoplasm of guard cells are involved in stomatal movement that may explain drought acclimation induced by SPM (Maiale et al. 2004) indicating a possible link between polyamines, Ca^{2+} homeostasis and stress responses.

The application of exogenous PAs is one of the possible strategies to study the implication of those molecules in stress response, but some of the studies suggest that their impact may vary depending on the considered genotype. Lefèvre et al. (2001) showed that the roots of the salt-resistant rice cv Pokkali contain high amounts of PUT compared with the salt-sensitive cv IKP and it may thus be hypothesized that an exogenous application of PUT could help the salt-sensitive genotype to cope with high external doses of salt. Ndayiragije and Lutts (2006a), however, demonstrated that although PUT is efficiently absorbed and translocated to the shoots and had a positive impact on monovalent cation discrimination in this cultivar, the increase in PUT did not allow the plant to overcome the deleterious effect of salt stress and even reinforced the negative impact of NaCl in terms of both shoot and root growth.

In a recent study, Yang et al. (2007) demonstrated that drought acclimation of some rice cultivars was directly associated with their ability to increase bound PA fractions in the flag leaf, but no data are available concerning such an involvement in response to salinity. Nevertheless, in rice, application of PAs leads to an increase in ethylene production (Chen et al. 1991; Lutts et al. 1996), thus reinforcing the hypothesis of a specific metabolic behavior in rice. The impact of exogenously applied PAs on the endogenous PA pathway and the putative influence of salinity on this impact remained unknown since previous data demonstrated that long-term application of exogenous PUT reduced Na^+ and Cl^- accumulation in salt-treated rice calli (Ndayiragije and Lutts 2006b) and improved grain yield of a salt-sensitive cultivar exposed to NaCl (Ndayiragije and Lutts 2007).

Quinet et al. (2010) demonstrated that exogenous PUT reduces Na^+ accumulation in root of a salt-sensitive rice cultivar already after a few days of salt exposure. Moreover, the impact of exogenous PUT on salt-treated rice depends on the cultivar in relation to the influence of exogenous PUT on endogenous PA metabolism. It was suggested that salt resistance was associated with an ability to increase PUT synthesis as a consequence of higher ADC and ODC activities, and to maintain a high proportion of conjugated PAs within stressed tissues. PUT had no feedback effect on ADC and ODC activities and could induce a transcriptional activation of genes coding for amine oxidase in the shoot of salt-treated plants.

In plants, much data obtained through the exogenous supply of PAs or from loss-of-function mutants in PA metabolism genes show that different PAs may delay programmed cell death (PCD). Examples are offered by excised leaves and protoplasts (Besford et al. 1993; Galston and Kaur-Sawhney 1990) or aged barley leaf disks (Legocka and Zajcher 1999), as well as the different types of cell death of flowers (Della Mea et al. 2007; Serafini-Fracassini et al. 2002). The addition of PAs to osmotically stressed oat leaves prevented degradation of plastid proteins, such as Dl, D2, cytochrome *f*, and the large subunit of Rubisco, all typical phenomena associated with PCD (Besford et al. 1993).

Overall, as these "mysterious" molecules are versatile players, the exploitation of the information revealed using plant models and the transfer of knowledge to a wide range of crop species for breeding purposes is a current challenge for the improvement of plant acclimation by modulation of polyamine content. Genetic manipulation of polyamine metabolism has already given some valuable information regarding their roles in stress response. Moreover, as discussed above, overexpression or deletion of polyamine biosynthetic genes and pretreatment with polyamines could be exploited with biotechnological/biochemical purposes to obtain information regarding their roles in stress response with a detailed knowledge of signaling hierarchies and the impact of metabolic changes involved in this response.

6 Microorganisms

Several recent reports have demonstrated the potential for plant priming by colonization of plant tissues with beneficial microorganisms. Plants are naturally associated with microorganisms in various ways. Endophytic bacteria colonize inner host tissues, sometimes in high numbers, without damaging the host or eliciting strong defense responses (Reinhold-Hurek and Hurek 2011). Ample evidence exists demonstrating that many endophytic bacteria have beneficial effects on plants. Growth promotion of plants may be achieved by bacterial production of plant growth regulators such as auxins, cytokinins, and gibberellins, while nitrogen or other nutrients may be provided by biological nitrogen fixation or mobilized as is the case for phosphorus (Compant et al. 2010). Furthermore, plants have established a mutualistic association between their roots and soil-borne fungi known as arbuscular mycorrhiza (AM). The AM symbiosis is beneficial to both the host plants and the AM fungus (AMF). The host plants can provide the AMF with part of their photosynthetically fixed carbohydrates that are essential for the completion of the life cycle of the latter. In turn, the AMF brings about an array of favorable influences on the host plants, such as absorption of more water and access to poorly available nutrients due to the fine exploration of the rhizosphere by the hyphae (Navarro et al. 2009).

The alleviating effect of the symbiosis between symbiotic bacteria and plants toward abiotic stress factors has been shown in a number of reports. Work carried out by Farinati et al. (2011), who studied the interaction between selected bacterial strains and *Arabidopsis halleri*, suggested that cocultivation of certain bacterial strains with plants determined a lower Cd accumulation in the shoots, thus providing protection from soils contaminated with heavy metals. It is known that certain bacteria can solubilize metals and adsorb them to their biomass and/or precipitate them with a consequent decrease in metal bioavailability (Gadd 2000). Protection can also be achieved via direct modulation of the plant's antioxidant machinery. Inoculation of *Chorispora bungeana* plantlets with the endophyte *Clavibacter* sp. strain Enf12 stimulated their growth and resulted in the improvement of their acclimation to chilling stress as evidenced by increases in activities of antioxidant enzymes such as CAT, APX, and SOD (Ding et al. 2011). Inoculation also significantly attenuated the chilling-induced electrolyte leakage, lipid peroxidation, and ROS accumulation. Similar findings were reported by Ait-Barka et al. (2006), who observed increased chilling acclimation in grapevines inoculated with the rhizobacterium *Burkholderia phytofirmans* strain PsJN.

Priming of plants can also be achieved with the use of known bacterial biological control agents. A recent study by Abdelkader and Esawy (2011), who inoculated maize plants with *Geobacillus caldoxylosilyticus* IRD, resulted in the plants being protected from severe salt stress. In addition to the induction observed in the enzymatic antioxidant machinery of the plant, the authors proposed that *Geobacillus* sp. must have utilized NaCl to successfully carry out key cellular activities necessary for its growth thus playing a role in ion exclusion important for the plant's acclimation to increased salt levels. Similarly, Harman (2006) also concluded that root inoculation of maize plants with *Trichoderma harzianum* strain T-22 resulted in enhanced concentration of antioxidant enzymes (like peroxidases, chitinases, etc.). These antioxidant enzymes act as scavengers of ROS and thus cause membrane stability, while playing a major role in protecting the cell from subsequent oxidative damage.

In addition, a growing body of studies has demonstrated that AMF inoculation confers acclimation to either biotic or abiotic stress. So far, AM-induced acclimation has been shown to be involved in the enhanced tolerance to drought, high salinity, chemical pollution, and oxidative stress, among others, in numerous plant species

(Alvarez et al. 2009; Bressano et al. 2010; Debiane et al. 2009; Latef and He 2011; Porras-Soriano et al. 2009). The mechanisms underlying the protective roles of AM are ascribed to alleviation of oxidative stress (Bressano et al. 2010; Debiane et al. 2009; Latef and He 2011), stimulation of water uptake and/or nutrient absorption (Alvarez et al. 2009; Porras-Soriano et al. 2009), and change of transcript levels of genes involved in signaling pathway or stress response (López-Ráez et al. 2010; Luo et al. 2009).

To date, the most extensive attempts have focused on the elucidation of the mechanisms pointing to the effect of AMF inoculation on water and nutrient uptake and the enhanced acclimation to drought (Smith and Read 2008). Ruiz-Sánchez et al. (2010) came to the conclusion that AM symbiosis in rice enhances the photosynthetic efficiency and the antioxidative response of rice plants subjected to drought stress. Similarly, Porcel and Ruiz-Lozano (2004) demonstrated that AM inoculation greatly influences leaf water potential, while it results in solute accumulation and oxidative stress alleviation in soybean plants subjected to drought stress. Subsequent molecular analyses on the same model system revealed the involvement of PIP aquaporin gene expression in the regulation of inoculated plants' response to drought stress acclimation (Porcel et al. 2006). Recent findings by Fan and Liu (2011) provide supporting evidence on the priming effect of AM fungi toward protection from drought stress, as the authors observed increased acclimation of *Poncirus trifoliata* seedlings to drought stress, correlating with significant induction in the expression levels of antioxidant genes and proteins such as SOD and POD.

Furthermore, modern "omics" approaches have allowed the global examination of the plant's transcriptome and metabolome, thus allowing us to decipher the molecular mechanisms involved in the improvement of stress acclimation in host plants primed with microorganisms. Transcriptome analyses on a whole genome poplar microarray revealed activation of genes related to abiotic and biotic stress responses as well as of genes involved in auxin-related pathways. Comparative transcriptome analysis in salt-stressed poplar plants indicated AM-related genes whose transcript abundances were independent of salt stress and a set of salt stress-related genes that were common to AM non-salt-stressed and non-AM salt-stressed plants. Salt-exposed AM roots showed stronger accumulation of myoinositol, abscisic acid, and salicylic acid and higher K^+ to Na^+ ratio than stressed non-AM roots. These findings lead to the conclusion that AMs activated stress-related genes and signaling pathways, apparently leading to priming of pathways conferring acclimation to abiotic stress (Luo et al. 2009).

7 Conclusions

In a constantly changing environment, the plant has to be able to adapt by quickly altering their physiology and metabolism in response to prior experience. Priming is an important mechanism of various induced acclimation phenomena in plants. Over the past few years, priming has emerged as a promising strategy in modern crop production management because it protects plants against both pathogens and

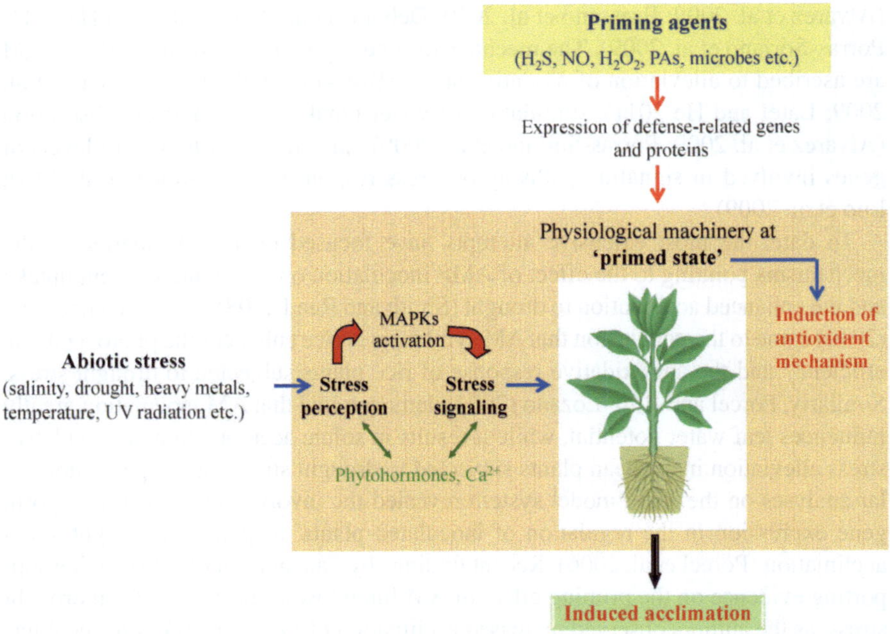

Fig. 1.1 Model of action of priming agents during the acclimation of plants to abiotic stress conditions. Induced acclimation theoretically involves the induction of defense-related genes/proteins and antioxidants ultimately leading to a specific cellular status, the so-called "primed state." Environmental stimuli are effectively perceived and sensed in primed plants through a complex crosstalk that involves various signaling compounds, such as MAPKs, phytohormones, and Ca^{2+}

abiotic stresses. A graphical overview of the key processes involved during priming for protection against abiotic stress factors is shown in Fig. 1.1. Better information on plant stress and associated signaling would facilitate the development of priming treatments for crops to enhance yields under conditions of stress. On the basis of the up-to-date findings outlined in this chapter, it is safe to conclude that priming plants toward an induced acclimation in response to environmental stress is one of the most promising areas of research for several years to come.

Acknowledgments V.F. would like to acknowledge financial support from C.U.T. Internal Grant EX032 and Grants-in-Aid from COST Action FA0605.

References

Abdelkader AF, Esawy MA (2011) Case study of a biological control: geobacillus caldoxylosilyticus (IRD) contributes to alleviate salt stress in maize (Zea mays L.) plants. Acta Physiol Plant 33(6):2289–99. doi:10.1007/s11738-011-0769-x

Ait-Barka E, Nowak J, Clement C (2006) Enhancement of chilling resistance of inoculated grapevine plantlets with a plant growth-promoting rhizobacterium, *Burkholderia phytofirmans* strain PsJN. Appl Environ Microbiol 72:7246–7252

Alcázar R, Altabella T, Marco F, Bortolotti C, Reymond M, Koncz C, Carrasco P, Tiburcio AF (2010) Polyamines: molecules with regulatory functions in plant abiotic stress tolerance. Planta 231:1237–1249
Alcázar R, Marco F, Cuevas JC, Patron M, Ferrando A, Carrasco P, Tiburcio AF, Altabella T (2006) Involvement of polyamines in plant response to abiotic stress. Biotechnol Lett 28:1867–1876
Alvarez M, Huygens D, Olivares E, Saavedra I, Alberdi M, Valenzuela E (2009) Ectomycorrhizal fungi enhance nitrogen and phosphorus nutrition of *Nothofagus dombeyi* under drought conditions by regulating assimilative enzyme activities. Physiol Plant 136:426–436
An L, Liu Y, Zhang M, Chen T, Wang X (2005) Effects of nitric oxide on growth of maize seedling leaves in the presence or absence of ultraviolet-B radiation. J Plant Physiol 162:317–326
Arasimowicz-Jelonek M, Floryszak-Wieczorek J, Gwóźdź EA (2011) The message of nitric oxide in cadmium challenged plants. Plant Sci 181(5):612–20. doi:10.1016/j.plantsci.2011.03.019
Arasimowicz-Jelonek M, Floryszak-Wieczorek J, Kubis J (2009) Interaction between polyamine and nitric oxide signaling in adaptive responses to drought in cucumber. J Plant Growth Regul 28:177–186
Athwal GS, Huber SC (2002) Divalent cations and polyamines bind to loop 8 of 14-3-3 proteins, modulating their interaction with phosphorylated nitrate reductase. Plant J 29:119–129
Bagni N, Tassoni A (2001) Biosynthesis, oxidation and conjugation of aliphatic polyamines in higher plants. Amino Acids 20:301–317
Bais HP, Ravishankar GA (2002) Role of polyamines in the ontogeny of plants and their biotechnological applications. Plant Cell Tissue Org Cult 69:1–34
Bates TS, Lamb BK, Guenther A, Dignon J, Stoiber RE (1992) Sulfur emissions to the atmosphere from natural sources. J Atmos Chem 14:315–337
Baudouin E (2011) The language of nitric oxide signaling. Plant Biol (Stuttg) 13:233–242
Beauchamp RO, Bus JS, Popp JA, Boreiko CJ, Andjelkovich DA (1984) A critical review of the literature on hydrogen sulfide toxicity. Crit Rev Toxicol 13:25–48
Beckers GJM, Conrath U (2007) Priming for stress resistance: from the lab to the field. Curr Opin Plant Biol 10:1–7
Behnamnia M, Kalantari KM, Rezanejad F (2009) Exogenous application of brassinosteroid alleviates drought-induced oxidative stress in *Lycopersicon esculentum* L. Gen Appl Plant Physiol 35:22–34
Besford RT, Richardson CM, Campos JL, Tiburcio AF (1993) Effect of polyamines on stabilization of molecular complexes in thylakoid membranes of osmotically stressed oat leaves. Planta 189:201–206
Besson-Bard A, Pugin A, Wendehenne D (2008) New insights into nitric oxide signaling in plants. Annu Rev Plant Biol 59:21–39
Blatt MR, Thiel G, Trentham DR (1990) Reversible inactivation of K+channels of Vicia stomatal guard-cells following the photolysis of caged Inositol 1, 4, 5-trisphosphate. Nature 346:766–769
Bloem E, Riemenschneider A, Volker J, Papenbrock J, Schmidt A, Salac I et al (2004) Sulphur supply and infection with *Pyrenopeziza brassica* influence L-cysteine desulfhydrase activity in *Brassica napus* L. J Exp Bot 55:2305–2312
Bouchereau A, Aziz A, Larher F, Martin-Tanguy J (1999) Polyamines and environmental challenges: recent development. Plant Sci 140:103–125
Bressano M, Curetti M, Giacheroa L, Gil SV, Cabello M, March G et al (2010) Mycorrhizal fungi symbiosis as a strategy against oxidative stress in soybean plants. J Plant Physiol 167:1622–1626
Bright J, Desikan R, Tancock JT, Weir IS, Neill SJ (2006) ABA-induced NO generation and stomatal closure in *Arabidopsis* are dependent on H_2O_2 synthesis. Plant J 45:113–122
Cantrel C, Vazquez T, Puyaubert J, Rezé N, Lesch M, Kaiser WM et al (2011) Nitric oxide participates in cold-responsive phosphosphingolipid formation and gene expression in *Arabidopsis thaliana*. New Phytol 189:415–427
Cesur A, Tabur S (2011) Chromotoxic effects of exogenous hydrogen peroxide (H_2O_2) in barley seeds exposed to salt stress. Acta Physiol Plant 33:705–709

Chao Y, Kao CH (2010) Heat shock-induced ascorbic acid accumulation in leaves increases cadmium tolerance of rice (*Oryza sativa* L.) seedlings. Plant Soil 336:39–48

Chawla S, Goyal SC, Angrish R, Rani C, Arora V, Datta KS et al (2010) Acclimatory response to hydrogen peroxide and glutathione under salt-boron stress through their impact on mineral nutrition and antioxidant defense system in pigeonpea (*Cajanus cajan* L.). Physiol Mol Biol Plants 16:295–304

Chen SL, Chen CT, Kao CH (1991) Polyamines promote the biosynthesis of ethylene in detached rice leaves. Plant Cell Physiol 100:238–245

Compant S, Clement C, Sessitsch A (2010) Plant growth-promoting bacteria in the rhizo- and endosphere of plants: their role, colonization, mechanisms involved and prospects for utilization. Soil Biol Biochem 42:669–678

Cona A, Rea G, Angelini R, Federico R, Tavladoraki P (2006) Functions of amine oxidases in plant development and defence. Trends Plant Sci 11:80–88

Corpas FJ, Chaki M, Fernandez-Ocan A, Valderrama R, Palma JM, Carreras A, Begara-Morales JC, Airaki M, del Rio LA, Barroso JB (2008) Metabolism of reactive nitrogen species in pea plants under abiotic stress conditions. Plant Cell Physiol 49:1711–1722

Cui J, Zhou Y, Ding J, Xia X, Shi K, Chen S et al (2011) Role of nitric oxide in hydrogen peroxide-dependent induction of abiotic stress tolerance by brassinosteroids in cucumber. Plant Cell Environ 34:347–358

Dat J, Vandenabeele S, Vranova E, Montagu M, Inze D, Van Breusegem F (2000) Dual action of the active oxygen species during plant stress responses. Cell Mol Life Sci 57:779–795

De Kok LJ, Bosma W, Maas FM, Kuiper PJC (1985) The effect of short-term H2S fumigation on water-soluble sulphydryl and glutathione levels in spinach. Plant Cell Environ 8:189–194

De Kok LJ, Stuiver CEE, Westerman S, Stulen I (2002) Elevated levels of hydrogen sulfide in the plants environment: nutrient or toxin. In: Omasa K, Saji H, Youssefian S, Kondo N (eds) Air pollution and plant biotechnology. Springer, Tokyo, pp 201–219

Debiane D, Garçon G, Verdin A, Fontaine J, Durand R, Shirali P et al (2009) Mycorrhization alleviates benzo[a]pyrene-induced oxidative stress in an in vitro chicory root model. Phytochemistry 70:1421–1427

del Rio LA, Corpas FJ, Barroso JB (2004) Nitric oxide and nitric oxide synthase activity in plants. Phytochemistry 65:783–792

Della Mea M, De Filippis F, Genovesi V, Serafini-Fracassini D, Del Duca S (2007) The acropetal wave of developmental cell death (DCD) of *Nicotiana tabacum* corolla is preceded by activation of transglutaminase in different cell compartments. Plant Physiol 144:1–13

Ding S, Huang CL, Sheng HM, Song CL, Li YB, An LZ (2011) Effect of inoculation with the endophyte *Clavibacter* sp. strain Enf12 on chilling tolerance in *Chorispora bungeana*. Physiol Plant 141:141–151

Fan GJ, Liu JH (2011) Colonization with arbuscular mycorrhizal fungus affects growth, drought tolerance and expression of stress-responsive genes in *Poncirus trifoliata*. Acta Physiol Plant 33:1533–1542

Fan H, Guo S, Jiao Y, Zhang R, Li J (2007) Effects of exogenous nitric oxide on growth, active oxygen species metabolism, and photosynthetic characteristics in cucumber seedlings under NaCl stress. Front Agric China 1:308–314

Farinati S, DalCorso G, Panigati M, Furini A (2011) Interaction between selected bacterial strains and *Arabidopsis halleri* modulates shoot proteome and cadmium and zinc accumulation. J Exp Bot 62(10):3433–47. doi:10.1093/jxb/err015

Farooq M, Basra SMA, Wahid A, Rehman H (2009) Exogenously applied nitric oxide enhances the drought tolerance in fine grain aromatic rice (*Oryza sativa* L.). J Agron Crop Sci 195:254–261

Fedina IS, Nedeva D, Çiçek N (2009) Pre-treatment with H_2O_2 induces salt tolerance in barley seedlings. Biol Plant 53:321–324

Fotopoulos V, De Tullio MC, Barnes J, Kanellis AK (2008) Altered stomatal dynamics in ascorbate oxidase over-expressing tobacco plants suggest a role for dehydroascorbate signalling. J Exp Bot 59:729–737

Fotopoulos V, Sanmartin M, Kanellis AK (2006) Effect of ascorbate oxidase over-expression on ascorbate recycling gene expression in response to agents imposing oxidative stress. J Exp Bot 57:3933–3943

Fukao T, Bailey-Serres J (2004) Plant responses to hypoxia-is survival a balancing act? Trends Plant Sci 9:449–456

Gadd GM (2000) Bioremedial potential of microbial mechanisms of metal mobilization and immobilization. Curr Opin Biotechnol 11:271–279

Galston AW, Kaur-Sawhney R (1990) Polyamines in plant physiology. Plant Physiol 94:406–410

Gao Y, Guo Y, Lin S, Fang Y, Bai J (2010) Hydrogen peroxide pretreatment alters the activity of antioxidant enzymes and protects chloroplast ultrastructure in heat-stressed cucumber leaves. Sci Hortic 126:20–26

Garcia-Mata C, Gay R, Sokolovski S, Hills A, Lamattina L, Blatt MR (2003) Nitric oxide regulated K^+ and Cl^- channels in guard cells through a subset of abscisic acid-evoked signaling pathways. Proc Natl Acad Sci U S A 100:11116–11121

Garcia-Mata C, Lamattina L (2001) Nitric oxide induces stomatal closure and enhances the adaptive plant responses against drought stress. Plant Physiol 126:1196–1204

Garcia-Mata C, Lamattina L (2010) Hydrogen sulphide, a novel gasotransmitter involved in guard cell signalling. New Phytol 188:977–984

Gill SS, Tuteja N (2010a) Reactive oxygen species and antioxidant machinery in abiotic stress tolerance in crop plants. Plant Physiol Biochem 48:909–930

Gill SS, Tuteja N (2010b) Polyamines and abiotic stress tolerance in plants. Plant Signal Behav 5:26–33

Gilroy S, Read ND, Trewavas AJ (1990) Elevation of cytoplasmic calcium by caged calcium or caged inositol trisphosphate initiates stomatal closure. Nature 346:769–771

Gould KS, Lamotte O, Klinguer A, Pugin A, Wendehenne D (2003) Nitric oxide production in tobacco leaf cells: a generalized stress response? Plant Cell Environ 26:1851–1862

Graziano M, Lamattina L (2005) Nitric oxide and iron in plants: An emerging and converging story. Trends Plant Sci 10:4–8

Groppa MD, Benavides MP (2008) Polyamines and abiotic stress: recent advances. Amino Acids 34:35–45

Groppa MD, Rosales EP, Iannone MF, Benavides MP (2008) Nitric oxide, polyamines and Cd-induced phytotoxicity in wheat roots. Phytochemistry 69:2609–2615

Gupta KJ, Fernie AR, Kaiser WM, van Dongen JT (2011) On the origins of nitric oxide. Trends Plant Sci 16:160–168

Hällgren JE, Fredriksson SA (1982) Emission of hydrogen sulphide from sulfate dioxide-fumigated pine trees. Plant Physiol 70:456–459

Hanfrey C, Sommer S, Mayer MJ, Burtin D, Michael AJ (2001) Arabidopsis polyamine biosynthesis: absence of ornithine decarboxylase and the mechanism of arginine decarboxylase activity. Plant J 27:551–560

Harman GE (2006) Overview of mechanisms and uses of Trichoderma spp. Phytopathology 96:190–194

Hasanuzzaman M, Hossain MA, Fujita M (2010) Physiological and biochemical mechanisms of nitric oxide induced abiotic stress tolerance in plants. Am J Plant Physiol 5:295–324

Hosoki R, Matsuki N, Kimura H (1997) The possible role of hydrogen sulfide as an endogenous smooth muscle relaxant in synergy with nitric oxide. Biochem Biophys Res Commun 237:527–531

Hung S, Yu C, Lin CH (2005) Hydrogen peroxide functions as a stress signal in plants. Bot Bull Acad Sinica 46:1–10

Joshi MM, Ibrahium IKA, Hollis JP (1975) Hydrogen sulphide: effects on the physiology of rice plants and relation to straighthead disease. Phytopathology 65:1170–1175

Kasinathan V, Wingler A (2004) Effect of reduced arginine decarboxylase activity on salt tolerance and on polyamine formation during salt stress in *Arabidopsis thaliana*. Physiol Plant 121:101–107

Kolbert Z, Bartha B, Erdei L (2007) Osmotic stress- and indole-3-butyric acid-induced NO generation are partially distinct processes in root growth and development in *Pisum sativum*. Physiol Plant 133:406–416

Kopyra M, Gwóźdź EA (2003) Nitric oxide stimulates seed germination and counteracts the inhibitory effect of heavy metals and salinity on root growth of *Lupinus luteus*. Plant Physiol Biochem 41:1011–1017

Kramer GF, Wang CY (1989) Correlation of reduced chilling injury with increased spermine and spermidine levels in zucchini squash. Physiol Plant 76:479–484

Kumar A, Altabella T, Taylor MA, Tiburcio AF (1997) Recent advances in polyamine research. Trends Plant Sci 2:124–130

Kumar M, Sirhindi G, Bhardwaj R, Kumar S, Jain G (2010) Effect of exogenous H_2O_2 on antioxidant enzymes of *Brassica juncea* L. seedlings in relation to 24-epibrassinolide under chilling stress. Ind J Biochem Biol 47:378–382

Kusano T, Berberich T, Tateda C, Takahashi Y (2008) Polyamines: essential factors for growth and survival. Planta 228:367–381

Kusano T, Yamaguchi K, Berberich T, Takahashi Y (2007) The polyamine spermine rescues *Arabidopsis* from salinity and drought stresses. Plant Signal Behav 2:250–251

Lamattina L, Garcia-Mata C, Graziano M, Pagnussat G (2003) Nitric oxide: the versatility of an extensive signal molecule. Annu Rev Plant Biol 54:109–136

Larkindale J, Huang B (2004) Thermotolerance and antioxidant systems in *Agrostis stolonifera*: Involvement of salicylic acid, abscisic acid, calcium, hydrogen peroxide, and ethylene. J Plant Physiol 161:405–413

Latef AA, He CX (2011) Effect of arbuscular mycorrhizal fungi on growth, mineral nutrition, antioxidant enzymes activity and fruit yield of tomato grown under salinity stress. Sci Hortic 127:228–233

Lee U, Wie C, Fernandez BO, Feelisch M, Vierling E (2008) Modulation of nitrosative stress by S-nitrosoglutathione reductase is critical for thermotolerance and plant growth in *Arabidopsis*. Plant Cell 20:786–802

Lefèvre I, Gratia E, Lutts S (2001) Discrimination between the ionic and osmotic components of salt stress in relation to free polyamine level in rice (*Oryza sativa*). Plant Sci 161:943–952

Legocka J, Zajcher I (1999) Role of spermidine in stabilization of the apoprotein of the lightharvesting chlorophyll a/b-protein complex of photosystem II during leaf senescence process. Acta Physiol Plant 21:127–132

Li D, Xiao Z, Liu L, Wang J, Song G, Bi Y (2010) Effects of exogenous hydrogen sulfide (H_2S) on the root tip and root border cells of *Pisum sativum*. Chin Bull Bot 45:354–362 [Chinese with English abstract]

Li J, Qiu Z, Zhang X, Wang L (2011) Exogenous hydrogen peroxide can enhance tolerance of wheat seedlings to salt stress. Acta Physiol Plant 33:835–842

Li L, Bhatia M, Moore PK (2006) Hydrogen sulphide-a novel mediator of inflammation? Curr Opin Pharmacol 6:125–129

Li L, Whiteman M, Salto-Tellez M, Tan CH, Moore PK (2009) GYY4137, a novel hydrogen sulfide-releasing molecule, protects against endotoxic shock in the rat. Free Radic Biol Med 47:103–113

Li L, Whiteman M, Tan CH, Guan YY, Neo KL, Cheng Y et al (2008) Characterization of a novel, water-soluble hydrogen sulfide releasing molecule (GYY4137): new insights into the biology of hydrogen sulphide. Circulation 117:2351–2360

Li W, Liu X, Khan MA, Yamaguchi S (2005) The effect of plant growth regulators, nitric oxide, nitrate, nitrite and light on the germination of dimorphic seeds of *Suaeda salsa* under saline conditions. J Plant Res 118:207–214

Liberek K, Lewandowska A, Ziętkiewicz S (2008) Chaperones in control of protein disaggregation. EMBO J 27:328–335

Lin WC, Block GS (2010) Can H_2O_2 application reduce chilling injury of horticultural crops? Acta Hort 875:33–36

Lisjak M, Srivastava N, Teklic T, Civale L, Lewandowski K, Wilson I et al (2010) A novel hydrogen sulfide donor causes stomatal opening and reduces nitric oxide accumulation. Plant Physiol Biochem 48:931–935
Liu JH, Honda C, Moriguchi T (2006) Involvement of polyamine in floral and fruit development. JARQ 40:51–58
Liu JH, Kitashiba H, Wang J, Ban Y, Moriguchi T (2007) Polyamines and their ability to provide environmental stress tolerance to plants. Plant Biotechnol 24:117–126
Lloyd D (2006) Hydrogen sulfide: clandestine microbial messenger? Trends Microbiol 14:456–462
López-Ráez JA, Verhage A, Fernandez I, Garcıa JM, Azcon-Aguilar C, Flors V, Pozo MJ (2010) Hormonal and transcriptional profiles highlight common and differential host responses to arbuscular mycorrhizal fungi and the regulation of the oxylipin pathway. J Exp Bot 61:2589–2601
Luo ZB, Janz D, Jiang X, Gobel C, Wildhagen H, Tan Y et al (2009) Upgrading root physiology for stress tolerance by ectomycorrhizas: insights from metabolite and transcriptional profiling into reprogramming for stress anticipation. Plant Physiol 151:1902–1917
Lutts S, Kinet JM, Bouharmont J (1996) Ethylene production by leaves of rice (*Oryza sativa* L.) in relation to salinity tolerance and exogenous putrescine application. Plant Sci 116:15–25
Ma W, Xu W, Xu H, Chen Y, He Z, Ma M (2010) Nitric oxide modulates cadmium influx during cadmium-induced programmed cell death in tobacco BY-2 cells. Planta 232:325–335
Mackerness SAH, John CF, Jordan B, Thomas B (2001) Early signaling components in ultraviolet-B responses: Distinct roles for different reactive oxygen species and nitric oxide. FEBS Lett 489:237–242
Maiale S, Sanchez DH, Guirado A, Vidal A, Ruiz OA (2004) Spermine accumulation under salt stress. J Plant Physiol 161:35–42
Maksymiec W (2007) Signaling responses in plants to heavy metal stress. Acta Physiol Plant 29:177–187
Manjunatha G, Lokesh V, Neelwarne B (2010) Nitric oxide in fruit ripening: trends and opportunities. Biotechnol Adv 28:489–499
Mittler R, Vanderauwera S, Gollery M, Van Breusegem F (2004) Reactive oxygen gene network of plants. Trends Plant Sci 9:490–498
Molassiotis A, Fotopoulos V (2011) Oxidative and nitrosative signaling in plants: two branches in the same tree? Plant Signal Behav 6:210–214
Molassiotis A, Tanou G, Diamantidis G (2010) NO says more than 'YES' to salt tolerance: Salt priming and systemic nitric oxide signaling in plants. Plant Signal Behav 5:1–4
Moschou PN, Paschalidis KA, Roubelakis-Angelakis KA (2008) Plant polyamine catabolism: the state of the art. Plant Signal Behav 3:1061–1066
Navarro A, Sanchez-Blanco MJ, Mortec A, Banon S (2009) The influence of mycorrhizal inoculation and paclobutrazol on water and nutritional status of *Arbutus unedo* L. Environ Exp Bot 66:362–371
Ndayiragije A, Lutts S (2006a) Do exogenous polyamines have an impact on the response of a salt-sensitive rice cultivar to NaCl? J Plant Physiol 163:506–516
Ndayiragije A, Lutts S (2006b) Exogenous putrescine reduces sodium and chloride accumulation in NaCl-treated calli of the salt-sensitive rice cultivar I Kong Pao. Plant Growth Regul 48:51–63
Ndayiragije A, Lutts S (2007) Long term exogenous putrescine application improves grain yield of a salt-sensitive rice cultivar exposed to NaCl. Plant Soil 291:225–238
Neill S, Barros R, Bright J, Desikan R, Hancock J, Harrison J et al (2008) Nitric oxide, stomatal closure, and abiotic stress. J Exp Bot 59:165–176
Neto A, Prisco JT, Filho JE, Medeiros J, Filho EG (2005) Hydrogen peroxide pre-treatment induced salt-stress acclimation in maize plants. J Plant Physiol 162:1114–1122
Öztürk L, Demir Y (2003) Effects of putrescine and ethephon on some oxidative stress enzyme activities and proline content in salt stressed spinach leaves. Plant Growth Regul 40:89–95

Palavan-Unsal N, Arisan D (2009) Nitric oxide signalling in plants. Bot Rev 75:203–229
Porcel R, Aroca R, Azcon R, Ruiz-Lozano JM (2006) PIP aquaporin gene expression in arbuscular mycorrhizal *Glycine max* and *Lactuca sativa* plants in relation to drought stress tolerance. Plant Mol Biol 60:389–404
Porcel R, Ruiz-Lozano JM (2004) Arbuscular mycorrhizal influence on leaf water potential, solute accumulation, and oxidative stress in soybean plants subjected to drought stress. J Exp Bot 55:1743–1750
Porras-Soriano A, Soriano-Martin ML, Porras-Piedra A, Azcon R (2009) Arbuscular mycorrhizal fungi increased growth, nutrient uptake and tolerance to salinity in olive trees under nursery conditions. J Plant Physiol 166:1350–1359
Prasad TK, Anderson MD, Martin BA, Stewart CR (1994) Evidence for chilling-induced oxidative stress in maize seedlings and a regulatory role for hydrogen peroxide. Plant Cell 6:65–74
Quinet M, Ndayiragije A, Lefèvre I, Lambillotte B, Dupont-Gillain CC, Lutts S (2010) Putrescine differently influences the effect of salt stress on polyamine metabolism and ethylene synthesis in rice cultivars differing in salt resistance. J Exp Bot 61:2719–2733
Rausch T, Wachter A (2005) Sulfur metabolism: a versatile platform for launching defence operations. Trends Plant Sci 10:503–509
Reinhold-Hurek B, Hurek T (2011) Living inside plants: bacterial endophytes. Curr Opin Plant Biol 14:1–9
Rennenberg H (1983) Role of O-acetylserine in hydrogen sulphide emission from pumpkin leaves in response to sulfate. Plant Physiol 73:560–565
Rennenberg H (1984) The fate excess of sulfur in higher plants. Annu Rev Plant Physiol 35:121–153
Rennenberg H, Huber B, Schroder P, Stahl K, Haunold W, Georgii HW, Slovik S, Pfanz H (1990) Emission of volatile sulfur compounds from spruce trees. Plant Physiol 92:560–564
Rhee J, Kim R, Choi H, Lee J, Lee Y, Lee J (2011) Molecular and biochemical modulation of heat shock protein 20 (Hsp20) gene by temperature stress and hydrogen peroxide (H_2O_2) in the monogonont rotifer, *Brachionus* sp. Comp Biochem Physiol C-Toxicol Pharmacol 154:19–27
Rhee JS, Raisuddin S, Lee KW, Seo JS, Ki JS, Kim IC et al (2009) Heat shock protein (Hsp) gene responses of the intertidal copepod *Tigriopus japonicus* to environmental toxicants. Comp Biochem Physiol C-Toxicol Pharmacol 149:104–112
Riemenschneider A, Nikiforova V, Hoefgen R, De Kok LJ, Papenbrock J (2005) Impact of elevated H_2S on metabolite levels, activity of enzymes and expression of genes involved in cysteine metabolism. Plant Physiol Biochem 43:473–483
Rozema J, van de Staaij JWM, Björn LO, Caldwell MM (1997) UV-B as an environmental factor in plant life: stress and regulation. Trends Ecol Evol 12:22–28
Ruiz-Sánchez M, Aroca R, Munoz Y, Polon R, Ruiz-Lozano JM (2010) Arbuscular mycorrhizal symbiosis enhances the photosynthetic efficiency and the antioxidative response of rice plants subjected to drought stress. J Plant Physiol 167:862–869
Sekiya J, Schmidt A, Wilson LG, Filner P (1982a) Emission of hydrogen sulfide by leaf tissue in response to L-cysteine. Plant Physiol 70:430–436
Sekiya J, Wilson LG, Filner P (1982b) Resistance to injury by sulphur dioxide: correlation with its reduction to, and emission of hydrogen sulfide in Cucurbitaceae. Plant Physiol 70:437–441
Serafini-Fracassini D, Del Duca S, Monti F, Poli F, Sacchetti G, Bregoli AM et al (2002) Transglutaminase activity during senescence and programmed cell death in the corolla of tobacco (*Nicotiana tabacum*) flowers. Cell Death Differ 9:309–321
Shan CJ, Zhang SL, Li DF, Zhao YZ, Tian XL, Zhao XL et al (2011) Effects of exogenous hydrogen sulfide on the ascorbate and glutathione metabolism in wheat seedlings leaves under water stress. Acta Physiol Plant 33(6):2533–2540. doi:10.1007/s11738-011-0746-4
Shi S, Wang G, Wang Y, Zhang L, Zhang L (2005) Protective effect of nitric oxide against oxidative stress under ultraviolet-B radiation. Nitric Oxide. Biol Chem 13:1–9
Silveira V, Santa-Catarina C, Tun NN, Scherer FE, Handro W, Guerra MP, Floh IS (2006) Polyamine effects on the endogenous polyamine contents, nitric oxide release, growth and differentiation of embryogenic suspension cultures of *Araukaria angustifolia* (Bert.) O. Ktze. Plant Sci 171:91–98

Singh DB, Verma S, Mishra SN (2002) Putrescine effect on nitrate reductase activity, organic nitrogen, protein, and growth in heavy metal and salinity stressed mustard seedlings. Biol Plant 45:605–608

Singh HP, Kaur S, Batish DR, Sharma VP, Sharma N, Kohli RK (2009) Nitric oxide alleviates arsenic toxicity by reducing oxidative damage in the roots of *Oryza sativa* (rice). Nitric Oxide. Biol Chem 20:289–297

Smith SE, Read DJ (2008) Mycorrhizal symbiosis, 3rd edn. Academic, Elsevier, London

Song L, Ding W, Shen J, Zhang Z, Bi Y, Zhang L (2008) Nitric oxide mediates abscisic acid induced thermotolerance in the calluses from two ecotypes of reed under heat stress. Plant Sci 175:826–832

Stuiver CEE, De Kok LJ, Kuiper PJC (1992) Freezing tolerance and biochemical changes in wheat shoots as affected by H_2S fumigation. Plant Physiol Biochem 30:47–55

Tan J, Zhao H, Hong J, Han Y, Li H, Zhao W (2008) Effects of exogenous nitric oxide on photosynthesis, antioxidant capacity and proline accumulation in wheat seedlings subjected to osmotic stress. World J Agric Sci 4:307–313

Tanou G, Job C, Belghazi M, Molassiotis A, Diamantidis G, Job D (2010) Proteomic signatures uncover hydrogen peroxide and nitric oxide cross-talk signaling network in citrus plants. J Proteome Res 9:5994–6006

Tanou G, Job C, Rajjou L, Arc E, Belghazi M, Diamantidis G et al (2009a) Proteomics reveals the overlapping roles of hydrogen peroxide and nitric oxide in the acclimation of citrus plants to salinity. Plant J 60:795–804

Tanou G, Molassiotis A, Diamantidis G (2009b) Hydrogen peroxide- and nitric oxide-induced systemic antioxidant prime-like activity under NaCl-stress and stress-free conditions in citrus plants. J Plant Physiol 166:1904–1913

Thompson CR, Kats G (1978) Effects of continuous hydrogen sulphide fumigation on crop and forest plants. Environ Sci Technol 12:550–553

Tun NN, Santa-Catarina C, Begum T, Silveira V, Handro W, Floh EIS, Scherer GFE (2006) Polyamines induce rapid biosynthesis of nitric oxide (NO) in *Arabidopsis thaliana* seedlings. Plant Cell Physiol 47:346–354

Uchida A, Jagendorf AT, Hibino T, Takabe T, Takabe T (2002) Effects of hydrogen peroxide and nitric oxide on both salt and heat stress tolerance in rice. Plant Sci 163:515–523

Urano K, Yoshiba Y, Nanjo T, Ito T, Yamaguchi-Shinozaki K, Shinozaki K (2004) Arabidopsis stress-inducible gene for arginine decarboxylase *AtADC2* is required for accumulation of putrescine in salt tolerance. Biochem Biophys Res Commun 313:369–375

Velikova V, Yordanov I, Edreva A (2000) Oxidative stress and some antioxidant systems in acid rain-treated bean plants: Protective role of exogenous polyamines. Plant Sci 151:59–66

Verma S, Mishra SN (2005) Putrescine alleviation of growth in salt stressed *Brassica juncea* by inducing antioxidative defense system. J Plant Physiol 162:669–677

Volkov RA, Panchuk II, Mullineaux PM, Schöffl F (2006) Heat stress-induced H_2O_2 is required for effective expression of heat shock genes in Arabidopsis. Plant Mol Biol 61:733–746

Wahid A, Perveen M, Gelani S, Basra SMA (2007) Pretreatment of seed with H_2O_2 improves salt tolerance of wheat seedlings by alleviation of oxidative damage and expression of stress proteins. J Plant Physiol 164:283–294

Walters DR (2003a) Polyamines and plant disease. Phytochemistry 64:97–107

Walters DR (2003b) Resistance to plant pathogens: possible roles for free polyamines and polyamine catabolism. New Phytol 159:109–115

Wang BL, Shi L, Li YX, Zhang WH (2010a) Boron toxicity is alleviated by hydrogen sulfide in cucumber (*Cucumis sativus* L.) seedlings. Planta 231:1301–1309

Wang H, Liang X, Wan Q, Wang X, Bi Y (2009) Ethylene and nitric oxide are involved in maintaining ion homeostasis in *Arabidopsis callus* under salt stress. Planta 230:293–307

Wang L, Yang L, Yang F, Li X, Song Y, Wang X et al (2010b) Involvements of H2O2 and metallothionein in NO-mediated tomato tolerance to copper toxicity. J Plant Physiol 167:1298–1306

Wang R (2002) Two's company, three's a crowd: can H_2S be the third endogenous gaseous transmitter? FASEB J 16:1792–1798

Wang Y, Feng H, Qu Y, Cheng J, Zhao Z, Zhang M et al (2006) The relationship between reactive oxygen species and nitric oxide in ultraviolet-B-induced ethylene production in leaves of maize seedlings. Environ Exp Bot 57:51–61

Whiteman M, Ling L, Rose P, Tan HC, Parkinson DB, Moore PK (2010) The effect of hydrogen sulfide donors on lipopolysaccharide-induced formation of inflammatory mediators in macrophages. Antiox, Redox Signal 12:1147–1154

Wilson LG, Bressan RA, Filner P (1978) Light-dependent emission of hydrogen sulfide from plants. Plant Physiol 61:184–189

Wimalasekera R, Tebartz F, Scherer GFE (2011) Polyamines, polyamine oxidases and nitric oxide in development, abiotic and biotic stresses. Plant Sci 181(5):593–603. doi:10.1016/j.plantsci.2011.04.002

Winner WE, Smith CL, Koch GW, Mooney HA, Bewley JD, Krouse HR (1981) Rates of emission of H_2S from plants and patterns of stable sulfur isotope fractionation. Nature 289:672–673

Wirtz M, Droux M, Hell R (2004) O-Acetylserine (thiol) lyase: an enigmatic enzyme of plant cysteine biosynthesis revisited in Arabidopsis thaliana. J Exp Bot 55:1785–1798

Wu J, Chen J, Liang J, Yang W, Wu J, Chen L et al (2009) Effects of exogenous NO on ascorbate-glutathione cycle in loquat leaves under low temperature stress. Chin J Appl Ecol 20:1395–1400

Xiong J, An L, Lu H, Zhu C (2009) Exogenous nitric oxide enhances cadmium tolerance of rice by increasing pectin and hemicellulose contents in root cell wall. Planta 230:755–765

Xiong J, Fu G, Tao L, Zhu C (2010) Roles of nitric oxide in alleviating heavy metal toxicity in plants. Arch Biochem Biophys 497:13–20

Xu FJ, Jin CW, Liu WJ, Zhang YS, Lin XY (2011) Pretreatment with H_2O_2 alleviates aluminum-induced oxidative stress in wheat seedlings. J Integr Plant Biol 53:44–53

Xu J, Yin H, Li Y, Liu X (2010) Nitric oxide is associated with long-term zinc tolerance in *Solanum nigrum*. Plant Physiol 154:1319–1334

Xu Q, Xu X, Zhao Y, Jiao K, Herbert SJ, Hao L (2008) Salicylic acid, hydrogen peroxide and calcium-induced saline tolerance associated with endogenous hydrogen peroxide homeostasis in naked oat seedlings. Plant Growth Regul 54:249–259

Xuan Y, Zhou S, Wang L, Cheng Y, Zhao L (2010) Nitric oxide functions as a signal and acts upstream of AtCaM3 in thermotolerance in Arabidopsis seedlings. Plant Physiol 153:1895–1906

Xue L, Li S, Sheng H, Feng H, Xu S, An L (2007) Nitric oxide alleviates oxidative damage induced by enhanced ultraviolet-B radiation in cyanobacterium. Curr Microbiol 55:294–301

Yamasaki H, Cohen MF (2006) NO signal at the crossroads: polyamine-induced nitric oxide synthesis in plants? Trends Plant Sci 11:522–524

Yamasaki H, Shimoji H, Ohshiro Y, Sakihama Y (2001) Inhibitory effects of nitric oxide on oxidative phosphorylation in plant mitochondria. Nitric Oxide 5:261–270

Yang G, Wu L, Jiang B, Yang W, Qi J, Cao K et al (2008) H_2S as a physiologic vasorelaxant: hypertension in mice with deletion of cystathionine *c*-lyase. Science 322:587–590

Yang J, Zhang J, Liu K, Wang Z, Liu L (2007) Involvement of polyamines in the drought resistance of rice. J Exp Bot 58:1545–1555

Yu C, Murphy TM, Lin C (2003) Hydrogen peroxide-induced chilling tolerance in mung beans mediated through ABA-independent glutathione accumulation. Funct Plant Biol 30:955–963

Zhang A, Jiang M, Zhang J, Ding H, Xu S, Hu X, Tan M (2007) Nitric oxide induced by hydrogen peroxide mediates abscisic acid-induced activation of the mitogen-activated protein kinase cascade involved in antioxidant defense in maize leaves. New Phytol 175:36–50

Zhang H, Hu LY, Hu KD, He YD, Wang SH, Luo JP (2008a) Hydrogen sulfide promotes wheat seed germination and alleviates the oxidative damage against copper stress. J Integr Plant Biol 50:1518–1529

Zhang H, Hu LY, Li P, Hu KD, Jiang CX, Luo JP (2010a) Hydrogen sulfide alleviated chromium toxicity in wheat. Biol Plant 54:743–747

Zhang H, Jiao H, Jiang CX, Wang SH, Wei ZJ, Luo JP, Jones RL (2010b) Hydrogen sulfide protects soybean seedlings against drought-induced oxidative stress. Acta Physiol Plant 32:849–857

Zhang H, Tan ZQ, Hu LY, Wang SH, Luo JP, Jones RL (2010c) Hydrogen sulfide alleviates aluminum toxicity in germinating wheat seedlings. J Integr Plant Biol 52:556–567
Zhang H, Tang J, Liu XP, Wang Y, Yu W, Peng WY et al (2009a) Hydrogen sulfide promotes root organogenesis in *Ipomoea batatas*, *Salix matsudana* and *Glycine max*. J Integr Plant Biol 51:1084–1092
Zhang H, Wang MJ, Hu LY, Wang SH, Hu KD, Bao LJ, Luo JP (2010d) Hydrogen sulfide promotes wheat seed germination under osmotic stress. Russ J Plant Physiol 57:532–539
Zhang H, Ye YK, Wang SH, Luo JP, Tang J, Ma DF (2009b) Hydrogen sulfide counteracts chlorophyll loss in sweetpotato seedling leaves and alleviates oxidative damage against osmotic stress. Plant Growth Regul 58:243–250
Zhang L, Zhou S, Xuan Y, Sun M, Zhao L (2009c) Protective effect of nitric oxide against oxidative damage in Arabidopsis leaves under ultraviolet-B irradiation. J Plant Biol 52:135–140
Zhang LP, Mehta SK, Liu ZP, Yang ZM (2008b) Copper-induced proline synthesis is associated with nitric oxide generation in *Chlamydomonas reinhardtii*. Plant Cell Physiol 49:411–419
Zhang Y, Tan J, Guo Z, Lu S, He S, Shu W et al (2009d) Increased abscisic acid levels in transgenic tobacco over-expressing 9 cis-epoxycarotenoid dioxygenase influence H_2O_2 and NO production and antioxidant defences. Plant Cell Environ 32:509–519
Zhang YY, Liu J, Liu YL (2004) Nitric oxide alleviates the growth inhibition of maize seedlings under salt stress. J Plant Physiol Mol Biol 30:455–459
Zhao L, He J, Wang X, Zhang L (2008) Nitric oxide protects against polyethylene glycol-induced oxidative damage in two ecotypes of reed suspension cultures. J Plant Physiol 165:182–191
Zhao L, Zhang F, Guo J, Yang Y, Li B, Zhang L (2004) Nitric oxide functions as a signal in salt resistance in the calluses from two ecotypes of reed. Plant Physiol 134:849–857
Zhao MG, Chen L, Zhang LL, Zhang WH (2009) Nitric reductase dependent nitric oxide production is involved in cold acclimation and freezing tolerance in Arabidopsis. Plant Physiol 151:755–767
Zhao MG, Tian QY, Zhang WH (2007) Nitric oxide synthase dependent nitric oxide production is associated with salt tolerance in *Arabidopsis*. Plant Physiol 144:206–217
Zheng C, Jiang D, Dai T, Jing Q, Cao W (2010) Effects of nitroprusside, a nitric oxide donor, on carbon and nitrogen metabolism and the activity of the antioxidation system in wheat seedlings under salt stress. Acta Ecol Sinica 30:1174–1183
Zottini M, Formentin E, Scattolin M, Carimi F, Schiavo FL, Terzi M (2002) Nitric oxide affects plant mitochondrial functionality in vivo. FEBS Lett 515:75–78

Chapter 2
Facing the Cold Stress by Plants in the Changing Environment: Sensing, Signaling, and Defending Mechanisms

Prince Thakur and Harsh Nayyar

1 Introduction

Growth constraints and stress result in significant crop losses and therefore the mechanisms underlying endurance and adaptation to these changes have long been the focus of intense research (Bray 2004). Kültz (2005) elaborated two types of responses to a particular kind of stress (1) stress specific adaptive responses and (2) general responses that confer basic protection. Temperature is one of the important factors, which determine the distribution of plants geographically in an optimum environment where they can survive and complete their life cycle. Chilling stress (<20 °C) is a direct result of low temperature effects on cellular macromolecules, which leads to slowing of metabolism, solidification of cell membranes, and loss of membrane functions (Jewell et al. 2010). Chilling has been known to severely inhibit plant reproductive development in many crop plant species such as rice displaying sterility when exposed to chilling temperatures during anthesis (Jiang et al. 2002). The sudden changes in the plant's environment also lead to the slower growth and low yield because of the shunting of the resources from reproductive processes to metabolic process to achieve tolerance (Smith and Stitt 2007). Chilling stress effects include reduced leaf expansion and growth (Sowinski et al. 2005; Rymen et al. 2007), wilting (Bagnall et al. 1983), chlorosis (Yoshida et al. 1996), and may lead to necrosis and impaired development of reproductive components, restricted seed, and pod development in sensitive plants species (Kaur et al. 2008; Ohnishi et al. 2010; Kumar et al. 2011), which ultimately reduces the yield of grain crops (Suzuki et al. 2008).

The plants are of sessile nature and so they have developed some specific mechanisms to deal with temperature changes in their environment. Hällgren and Öquist

P. Thakur (✉) • H. Nayyar
Department of Botany, Punjab University, Chandigarh 160 014, India
e-mail: harshnayyar@hotmail.com

N. Tuteja and S. Singh Gill (eds.), *Plant Acclimation to Environmental Stress*,
DOI 10.1007/978-1-4614-5001-6_2, © Springer Science+Business Media New York 2013

(1990) have divided the plants into chilling sensitive, chilling resistant, and freezing tolerant types; however the term chilling resistant may be treated as a misnomer because it implies that the these plants are able to regulate their temperature. However, the plants are poikilotherms so the term may be modified to a more elaborative 'chilling tolerant'. In the first category, 'chilling sensitive', the plants show metabolic dysfunctions at the exposure of temperatures slightly below the optimum. The chilling tolerant plants are those ones, which survive the lower range of temperatures but nonfreezing, than optimum. The freezing tolerant plants also survive the freezing conditions and are most hardy of the above classes of plants.

Plants experience a wide range of temperature fluctuations in natural environments. Thus, they have evolved mechanisms to minimize cellular damage at temperature extremes. Growth at low temperatures (cold acclimation) enables plants to initiate signaling cascades and metabolic alterations, which enhance tolerance to freezing temperatures (Chinnusamy et al. 2003).

Temperature change in the micro or macro-environment is a very critical factor, which determines the growth, development, and physiology of the plant. Some of the alterations are visible to us as cold stress symptoms but the main role players are always behind the curtain, which take part in various biochemical and molecular processes in response to cold temperature exposure. These processes together can be termed as 'low temperature induced signal transduction (LTST). These processes are decidedly beneficial to the plant because these are the strategy measures to cope with the stress conditions. Besides the plants also get a very useful character out of these processes i.e. stress memory or cold stress acclimation. The LTST leads to the expression of certain genes of interest in the nucleus, which through central dogma results in the synthesis of some specific proteins. These proteins either structural or enzymes work for the survival of plant during stress conditions and the plant acquires stress tolerance. All these behind the curtain processes are described in detail in the coming heads of this review.

Chilling has been known to cause disruption of DNA strands, reductions in enzymatic activity, rigidification of membranes, destabilization of protein complexes, stabilization of RNA secondary structure, accumulation of reactive oxygen intermediates (ROIs), impairment of photosynthesis, and leakage across membranes ((Nayyar et al. 2005a, b, c, d) also (Nayyar and Chander 2004). Different methods have been used to quantify the cold tolerance in plants like electrolyte leakage (Patterson et al. 1976; (Nayyar et al. 2005a, b, c, d), LT_{50}, percent survival, and chlorophyll fluorescence imaging (Ehlert and Hincha 2008). It has been reported in different scientific writings that cold tolerance in plants comes from two ways (1) it is inherent and (2) after cold acclimation. For a better understanding of the cold tolerance in plants through cold acclimation, a detailed knowledge of biochemical and molecular methods involved in low temperature sensing and signal transduction is required, which is the earliest and most important stage in cold acclimation and development of cold tolerance. The main aim of this review is to discuss the mechanisms of cold sensing mechanisms in plants, the signaling processes and their components, which commence thereafter and the resulting tolerance mechanisms. Baena-Gonzalez (2010) has reviewed the various mechanisms that subsequently become engaged upon exposure of plants to stress to modulate gene expression in response to energy signals (Fig. 2.1).

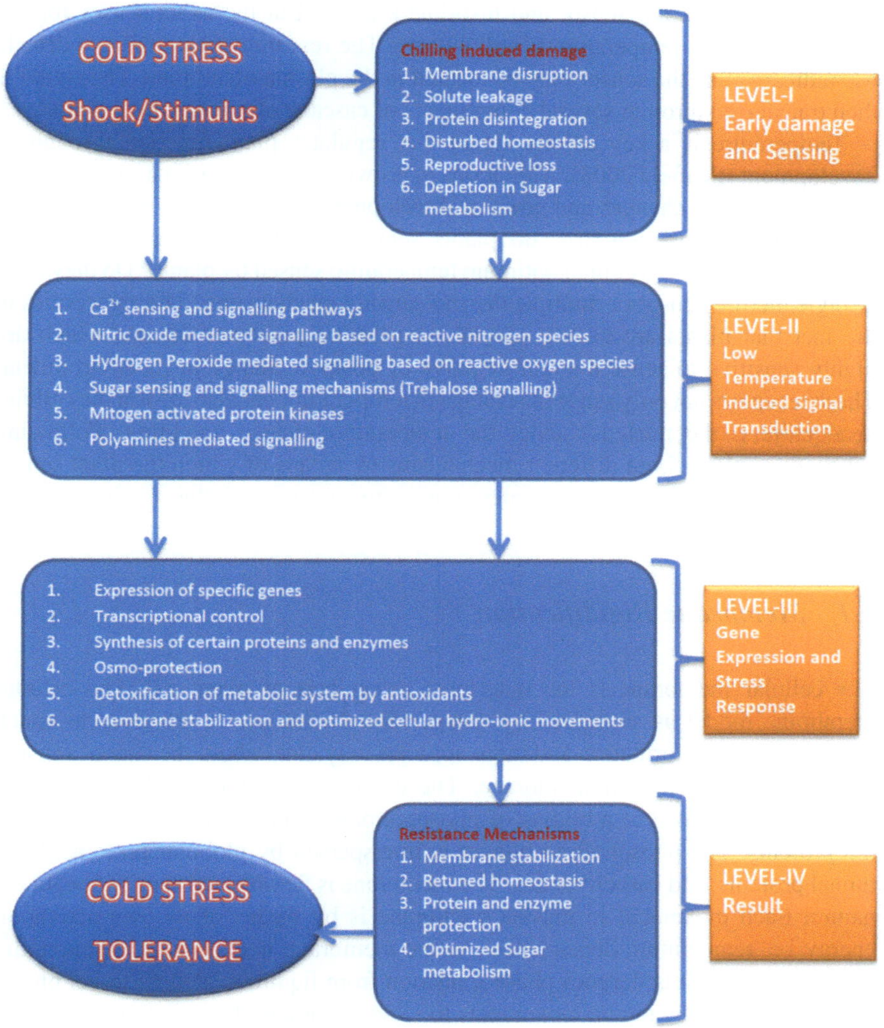

Fig. 2.1 A schematic four tier mechanism of cold stress response and acquired tolerance in plants. Cold tolerance is also known as cold hardening or cold acclimation and it is described as the development or increase in tolerance to cold temperatures over time by means of adaptive and resistive cellular mechanisms, which are activated in response to stressful cold temperature conditions

2 Low Temperature Sensing

Any type of environmental stimulus is sensed by the receptor/osmosensor molecules, which perceive the signal and transmit it to the suitable signal transduction pathways. In plants, the identified receptor/osmosensor molecules include ROP10 (a small G protein from ROP family) (Zheng et al. 2002), ATHK1 (a homologue of yeast SLN1) (Urao et al. 1999), NtC7 (a membrane protein) (Tamura et al. 2003) and Cre1 (a cytokinin sensor and histidine kinase). But the exact sensor for the perception of low temperature is still elusive. The research during the last decade has indicated that the sensor may be located in the plasma membrane. The signal is then transferred through several components of cascade of transduction pathways.

Temperature is a key abiotic signal that regulates plant function throughout development (Penfield 2008). Alterations in growth temperature act as a stimulus to initiate metabolic changes and promote developmental switches. In the first sight a simple question arises—How do plants sense change in temperature or more specifically how is lower than optimum temperature sensed by plants? Do they have a single thermo-sensor or multiple thermo-sensing mechanisms? The answer lies in the fact that plants are having special temperature preceptor organs, which are highly sensitive to sense a slight negative or positive change in its environment. These receptors not only sense the change in temperature but also inform the cellular headquarters (the nucleus) about the temperature-change condition. Following subheads explain about different mechanisms by means of which the plant senses temperature and subsequently frames strategy to cope up with the conditions.

2.1 Membrane Rigidification

The cellular membrane is the outermost living part of plant cell. The cellular membrane model as suggested by (Singer and Nicholson 1972) gives us much narrowed down information and more appropriately clues about the way of sensing temperature through cell membrane. The cellular membrane is fluid-mosaic in nature and is formed of a bilayer of phospholipids, which is sandwiched between the proteins. The phospholipid bilayer is interspersed by globular proteins, large tunnel proteins, and carbohydrates. The membrane is flexible and semipermeable in nature. Each movement in plasma membrane is by means of its own activation energy i.e. temperature dependence. As the membrane is exposed to temperature below optimum, it undergoes phase transition from liquid crystalline to gel phase. This causes the membrane movements to slow down and the membrane becomes more static than dynamic or rigid (Vigh et al. 2007). Therefore, it may be implicated that plasma membrane is a highly organized system, which plays an important role as communication interface between the cell and extracellular environment. Generally, chilling stress results in loss of membrane integrity and solute leakage. During the last few years, these observations have been documented as the same responses can be mimicked by plants in response to certain agents like DMSO at the ambient temperature. It has also been observed that the membrane fluidizing chemicals like benzyl alcohol, inhibit the responses of plants at considerably low temperatures also (Orvar et al. 2000; Sangwan et al. 2001, 2002; Vaultier et al. 2006). Therefore, it may be suggested here that the primary reception or perception is at the membrane level (Örvar et al. 2001). Injuries due to low temperature are mostly due to decrease in membrane fluidity; this is called rigidification (Hayashi and Maeda 2006). Alterations in the membrane fluidity have been demonstrated to initiate temperature-signaling pathways in a variety of organisms, tempting speculation that similar mechanisms may operate in plants (reviewed in Samach and Wigge 2005). The effects of low temperature on plasma membrane have been demonstrated by many authors in different experiments and in different organisms e.g. in fish (Cossins et al. 1978; Pehowich et al. 1988), in bacteria (Sinensky 1974), and in blue green algae (reviewed in Mikami and Murata 2003; Los et al. 2010). This results in considerable reduction in growth rate and increase in electrolyte leakage (Nayyar et al. 2005a, b, c, d, 2007) and leaf chlorosis (Murata 1989). Wada et al. (1990) have studied the role played by membrane rigidification in cold stress by cloning the desaturase gene *des*Aftoma from chilling tolerant cyanobacterium *Synechocystis* PCC6803, and then transferring it into the chilling sensitive cyanobaeterium *Anacyslisnidulans*. The activity of this gene caused membrane lipid desaturation in the sensitive species subsequently causing an increase in low temperature tolerance. Therefore, it may be anticipated that saturation of membrane lipids is expected to rigidify the membranes. It has also been postulated in this context that the variations in the membrane phospholipids leads to the generation of a signal phosphatidic acid (PtdOH) within the first one minute of cold exposure as was observed by (Ruelland et al. 2002) in *Arabidopsis thaliana* culture. This phosphatidic acid (PtdOH) formation is one of the earliest response of plants to cold stress and it acts as a signaling molecule in response to cold stress mediating the NO signaling cascade (Fig. 2.3) (Testerink and Munnik 2005). This leads to the conclusion that membrane rigidification activates the downstream low temperature induced signaling pathways (Suzuki et al. 2000a, b).

2.2 Configurational Changes in Proteins

The changes in the membrane fluidity also cause confirmatory changes in the membrane proteins, which starts the signaling cascade. The temperature downshift causes unfolding of proteins (Pastore et al. 2007). Xue (2003) has observed that DNA-binding activity of *CBF2* (*CBF* proteins are transcription factors) in barley (*Hordeum vulgare*) is also temperature dependent and CBF/CRT regulon is a major genetic regulon in cold response by plants (Nakashima et al. 2009; Ruelland et al. 2002). Bae et al. (2003) found 54 nuclear proteins in *Arabidopsis thaliana* and Cui et al. (2005) spotted 60 proteins, which are up- or down-regulated by cold temperature exposure in rice.

2.3 Changes in Cytoskeleton

The low temperature has also been known to affect the multimeric polypeptides. It was reported a long time ago that a drop in temperature causes depolymerization of microtubules and actin microfilaments (Ilker et al. 1979). Pokorna et al. (2004)

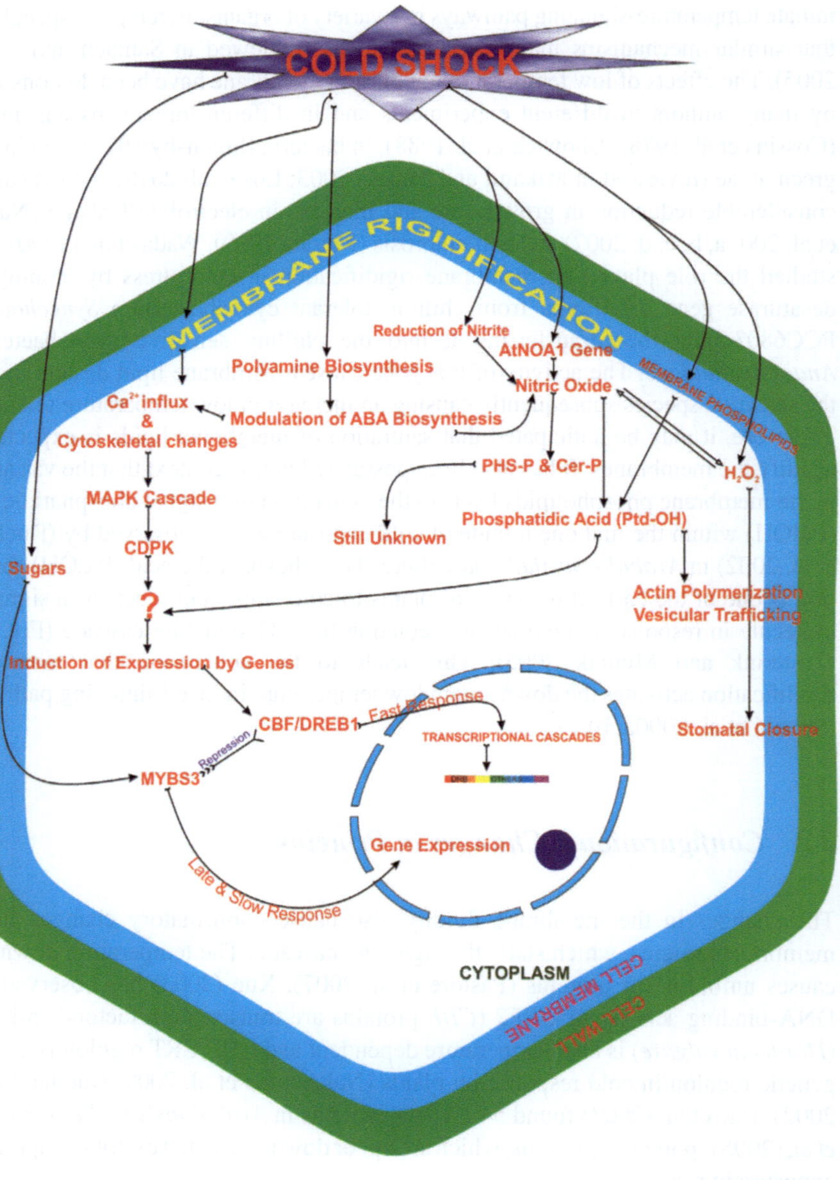

Fig. 2.3 The schematic representation showing interplay of different signaling pathways in response to cold stress in plants. Cod shock in plants starts signaling cascades in plants which also cross talk among themselves. The figure above shows the major pathways like Ca^{2+} signaling which is started as the membrane phase transition occurs. Polyamines are important osmoprotectants which act as signaling compounds and stimulate the ABA synthesis in response to cold exposure. Increased ABA biosynthesis increases the Ca^{2+} influx and biosynthesis of nitric oxide (NO) in turn. NO starts a signaling cascade involving the production of phosphatidic acid (PtdOH) which interacts with Ca^{2+} signaling pathway at some unknown point (?). H_2O_2 also interplays with NO in the web of signaling induced by cold stress exposure. Major converging point of different signaling pathways is CBF/DREB1 which is suppressed by MYBS3 induced in sugar signaling. The gene expression is ultimate and thus cold acclimation is achieved

observed that microtubules disassemble after an exposure of 0 °C for only 20 min. In *Medicago sativa* cells calcium influx and *cas30* expression at 4 °C were also prevented by jasplakinolide (an actin stabilizer) but induced at 25 °C by cytocalacin D a microfilament destabilizer (Orvar 2000). Therefore, this implies that the cytoskeleton assembly is necessary for defending the cold response (Sangwan et al. 2001; Al-Fageeh and Smales 2006)

2.4 Sugar Sensing in Cold Signaling

Sugars play a central regulatory role in many vital processes of photosynthetic plants besides serving the energetic function and are considered as important signals which regulate plant metabolism and development. Plants have the capacity to sense the presence as well as levels of sugars through various pathways that directly or indirectly recognize trehalose, fructose, glucose, or sucrose (Rolland et al. 2006). The basic mechanism behind the sugar sensing phenomena is still not clearly understood. The research in this area has led to the recognition of a hexokinase from *Arabidopsis thaliana* (AtHXK1) which is supposed to be a core component in plant sugar sensing and signaling pathways and plays vital functions as the glucose sensor that integrates the nutrient and hormone signals to govern the gene expression and plant growth in response to environmental aberrations such as cold (Moore et al. 2003; Cho et al. 2006). Cho et al. (2006) have elucidated that AtHXK1 functions to mediate the sugar repression like the photosynthetic *CAB* genes. Independent of the signaling function of HXK1, the metabolism of glucose through it induces the expression of defense-related genes (Xiao et al. 2000). We have recently reviewed the sugar sensing with respect to stressful conditions in grain crops (Thakur et al. 2010).

2.5 Reactive Species' (ROS and RNS) Role in Sensing

Reactive oxygen species (ROS) are toxic oxygen free radicals, which are produced in the plants out of phytoreactions and cellular oxidation byproducts under normal conditions (Finkel and Holbrook 2000). One of the earliest responses of plant cells under various abiotic and biotic stresses is the generation of the oxidative burst, during which large quantities of ROS like superoxide, hydrogen peroxide, hydroxyl radicals, peroxy radicals, alkoxy radicals, singlet oxygen, etc. are generated (Bhattacharjee 2005). They are having the potential to cause cellular damage when they accumulate to certain toxic levels. However, these ROS are also having an important role as their accumulation activated defense-signaling pathways thus mitigating cellular damage. It has been estimated that both resistance responses to stresses and normal physiological metabolism can lead to ROS production (Van Breusegem et al. 2001). These beneficial reactive species include nitric oxide and hydrogen peroxide and both of these are involved in stress response in plants.

2.6 Low Temperature Induced Signal Transduction

According to Kultz (2005) two types of stress responses exist, specific and general; the specific ones are against some unique stressful condition like lowered oxygen tension characteristic of hypoxic stress in flooded roots (Magneschi and Perata 2009) and general responses include signals and signaling components which are shared by multiple pathways (Bowler and Fluhr 2000; Kultz, 2005). Duie to this reason, the acclimation to one type of stress in plants may also confer tolerance to other types of stresses also. In plants, the homeostasis is constantly under threat by environmental variables. Hence, for the adaptation and survival, the plants have evolved sensitive and complex mechanisms, which modify their growth and metabolic patterns since to achieve the target of acclimation it must be immediate to reestablish homeostasis, repair damaged cellular components, and reprogram the altered metabolic system (Wang et al. 2003).

The earlier signaling events start with slight perturbation in optimal environment. In order for a plant to respond to low temperature stress conditions, the plant must have the ability to sense the slightest temperature change in the environment so that it may prepare for the larger change in temperature conditions that may follow and which may cause irreversible damage. There are two components of this LTST (1) a mechanism of sensing the low temperature i.e. sensing mechanism and (2) a series of events that transmit the information from sensor to the nucleus, where specific genes need to be activated (Zeller et al. 2009) (Fig. 2.2). The key to understanding plant cold response lies in the identification of new components involved in those processes and the elucidation of the signaling pathways.

3 Signal Transduction Mechanism

As stated earlier, the membranes are primary receptors for the low temperature signal and the proteins embedded in the plasma membrane transmit these signals to cellular machinery through signaling cascade. This ultimately results in gene expression (Murata and Loss, 1997; Loss and Murata, 2000) so this implies that physical state of plasma membrane lipids also regulates the activities of membrane proteins (Sukharev et al. 1999), receptor-associated protein kinases (Wood 1999; Hohmann, 2003) and sensor proteins (Tokishita and Mizuno 1994; Sugiura et al. 1999). Monroy et al. (1998) have elucidated the LTST in six steps, (1) sensing of low temperature (2) transduction of signal into biochemical processes via secondary messengers such as Ca^{2+}, (3) activation/deactivation of kinases and phosphatases (4) transfer of signal to the nucleus (5) activation of specific genes in response to signal {more accurately cold acclimation specific genes *cas* genes and (6) development of cold/freezing tolerance. Mostly these events are studied in isolation for simplicity but a complex set of biochemical and molecular reactions is activated in response to the input signal, which in turn activates many signaling pathways and these pathways

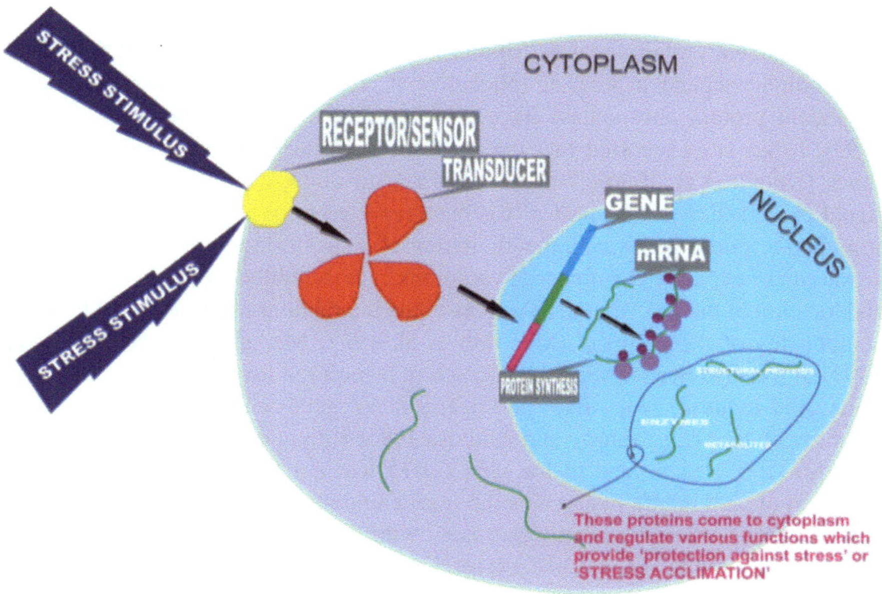

Fig. 2.2 A generalized and simplified scheme of stress-induced signal transduction consequently providing stress tolerance or stress acclimation

cross talk with each other i.e. these are interrelated to each other at various components and stages (Jenkins 1998; Trewavas and Malhó 1998).

Two methods are generally adopted for studying LTST. In the first method, the early signaling events, which take place upon exposure to cold, are studied and the cold inducible genes are investigated as end-point markers to understand the overall progress. In the second method, the mutants involved in the low temperature induced signaling pathway are identified and the role of each component and its sequence in the cascade is established.

3.1 Role of Ca^{2+} in Low Temperature Induced Signal Transduction

Ca^{2+} has been reported to regulate several important cellular functions. It also acts as secondary messenger in the signal transduction system. Whenever Ca^{2+} homeostasis is disturbed inside the cytoplasm, it leads to interference with cellular response, even when Ca^{2+} does not play a direct role in the mediation of cellular processes. The inward flow of Ca^{2+} into the cytosol has been reported to play a crucial role in signal transduction, where it acts as secondary messenger. In plant cells, Ca^{2+} is largely stored in the apoplast, where its concentration is at least 10^{-5} to 10^{-4} M (Cleland et al. 1990; Evans, et al. 1991). Regulation of many protein kinase activities occurs through binding of Ca^{2+} to key regulatory proteins like Ca^{2+}-dependent protein kinases (CDPKs) (Roberts and Harmon 1992; Cheng et al. 2002), Ca^{2+}/CaM-dependent protein kinases (Shimazaki et al. 1992; Pandey et al. 2002), Ca^{2+}/phospholipid-dependent kinases (Nickel et al. 1991), and a homologue of Ca^{2+}-dependent protein phosphatase have also been identified in plants (Kudla et al. 1999). It has been reported that the cytosolic Ca^{2+} increases in response to cold shock (Knight et al. 1996). This increase in cytosolic Ca^{2+} amplifies the stimulus signal perceived by the plant. Ca^{2+}has been reported to be involved in a variety of stimulus–response pathways., the elicitation of a specific response from a general signal can be explained by means of variations in the amplitude, duration, frequency, and location of the Ca^{2+} signal, as well as in the interactions of this signal with other components of the pathway (McAinsh et al. 1997; McAinsh and Hetherington, 1998). It has also been observed that calcium is required for the total expression of some cold-induced genes like *COR6* and *KIN1* genes of *Arabidopsis thaliana* (Monroy et al. 1993; Monroy and Dhindsa 1995; Knight et al. 1996). Monroy and Dhindsa (1995) elaborated that a gene *Cas15* was not fully expresses due to the chelation of Ca^{2+} in alfalfa and thus the plant could not acclimate to the cold conditions. Later on when the plant was treated with A23187 (A Ca^{2+} ionophore which increases the influx of Ca^{2+}), the expression of *Cas15* was achieved even at very high stressful temperature. This shows that Ca^{2+} is very important in temperature induced gene expression. Some workers have suggested that exposure to one kind of or a specific amount of stress leads to a specific Ca^{2+} influx and signal kinetics, but subsequent exposure to some different amount of stress causes a different Ca^{2+} signal kinetics, than observed previously. This establishes the hypothesis of "stress memory" which modulates plant stress responses. The strength of stimulus response is determined by the extent of Ca^{2+} influx which may be responsible for the specificity of the response. Another factor which confers specificity to the response is the destination of Ca^{2+} influx. Ca^{2+} sensors are also important as they couple the extracellular signaling to intercellular responses and comprise calmodulin- and CaM-related proteins (Snedden and Fromm 2001),calcineurin B-like proteins (Kudla et al. 1999) and CDPKs (Harmon et al. 2000). It has been noted by many authors that cytosolic calcium is involved in signaling pathways induced by various kinds of stresses like heat, cold, drought, and salinity etc. (Trewavas and Malhó 1998). Hence, it may be concluded that the cytosolic calcium acts as a convergence point and plays a central role in the integration of different signal transduction pathways.

3.2 Role of Nitric Oxide

Nitric oxide (NO) has emerged as a key signaling molecule in animals as well as plants during the last decade and its role has been implicated in number of physiological and developmental processes as well as response to abiotic stresses including heat and cold stress (Qiao and Fan 2008; Corpas et al. 2001). In recent years, NO has been shown to be involved in seed germination and reduction of seed dormancy (Bethke et al. 2006a, b, 2007; Libourel et al. 2006), photo-morphogenesis, leaf expansion, root growth, regulation of plant maturation and senescence (Mishina et al. 2007), suppression of floral transition (He et al. 2004), phytoalexin production (Noritake et al. 1996; Beligni and Lamattina 2000) and as an intermediate downstream of ABA signaling (Bright et al. 2006; Garcia-Mata and Lamattina 2007). NO is a free radical reactive gas with many physiological functions (Besson-Bard et al. 2008a, b; Neill et al. 2008a, b). It has been recognized as an important biological mediator in animals because of its role in certain important functions like neurotransmission, inflammatory responses, and relaxation of smooth muscles (Schmidt and Walter 1994). But its role in plant metabolic system was very much unknown till recently. Kopyra and Gwó (2004) have reported that NO alleviates the deleterious effects of ROS and establishes a stress resistance response. Corpas et al. (2008) have observed the involvement of NO as reactive nitrogen species in case of pea. They found that pea plants in response to stressful conditions activated the metabolism of reactive nitrogen species and that low and high temperature, continuous and high light intensity induced the overproduction of these reactive nitrogen species thereby consequently causing nitrosative stress which is although a cytotoxic effect of NO. The recent investigations on its relative role in plant regulatory and signaling mechanisms have spanned the part of the fissure, and the picture that came out of these investigations shows that it has got many important functions to play in plant system like ubiquitous signal involved in diverse physiological processes that include germination, root growth, stomatal closing, and adaptive response to biotic and abiotic stresses (reviewed in Stuehr et al. 2004; Besson-Bard et al. 2008a, b). But its generation mechanism in plant system is still controversial (Corpas et al. 2004, 2006; Crawford 2006; Zemojtel et al. 2006; Neill et al. 2008a, b). Zhao et al. (2009a, b) has also demonstrated that nitric oxide production in plants is involved in acquiring cold acclimation or cold tolerance. Guillas et al. (2011) have evidenced that NO is produced immediately as a plant response to cold stress and it participates in the regulation of cold-responsive gene expression. They also showed the presence of a novel downstream elements which were identified as phosphosphingolipid metabolic species i.e. phytosphingosine-phosphate (PHS-P) and a ceramide phosphate (Cer-P). Cantrel et al. (2011) have also stated that PHS-P and a Cer-P are transiently synthesized upon chilling. They also stressed that these two phosphosphingolipid species are negatively regulated by NO. NO mediates signaling in response to various abiotic stresses by involving ABA, calcium, and hydrogen peroxide which are also suggested to function in cold response too. The involvement of NO in imparting cold tolerance has been indicated by its exogenous application in certain cold-sensitive plant species such as maize and tomato (Neill et al. 2003). Its mechanism in providing protection against cold has been attributed to its antioxidative feature and suppression of peroxidative metabolism caused by stress (Neill et al. 2002). The role of NO in cold signaling has recently received attention. The cellular metabolism can be affected by NO through S-nitrosylation of protein thiols to form S-nitrosothiols and moreover it can lead to activation or inhibition of protein functions. In one of the studies, a brief cold stress (1–6 h) to *Brassica juncea* seedlings generated many S-nitrosothiols resulting in proteins modifications involving those in antioxidant metabolism (Abat and Deswal 2009). NO levels are reported to be restricted by non-symbiotic haemoglobins (nHb) that can scavenge NO and keeps its levels below toxic range to act as signaling molecule in cold response (Dordas et al. 2003a, b; Gupta et al. 2011a, b). Additionally, NO may mediate lipid-based signaling in cold response. The sphingolipids produced during cold stress are transiently phosphorylated while NO may prevent this step to facilitate lipid-based cold signaling. It has been reported that NO transduces signals through cGMP as its downstream mediator and also it may interact with other signaling molecules such as H_2O_2, Ca^{2+}, and salicylic acid directly or indirectly (Neil et al. 2003; Lamotte et al. 2004; Wendehenne et al. 2006). Recently, it has been found that NO downstream cascade involves the cytoskeletal proteins as these proteins are involved in many processes regulated by NO in plants (Yemets et al. 2011). Besides the roles of NO as signaling molecule it has also been reported to be a regulator in gene expression (Kopyra and Gwozdz 2004).

3.3 *Role of Polyamines*

As we know that the plants which are having inherent characteristic of low temperature tolerance, in response to low temperature exposure, start regulatory and molecular mechanisms that are triggered to optimize the metabolic parameters which make sure the survival of the plant under suboptimal temperatures (Stitt and Hurry 2002). Polyamines are the low molecular weight organic polycations having two or more amino ($^-NH_2$) groups. The role of polyamines has been implicated in growth and developmental processes in higher plants especially in response to stressful conditions like senescence and biotic or abiotic stresses. As such they have been reported to encourage DNA replication, transcription, and translation. It has been observed in different plants species hat during exposure to stressful conditions polyamines' biosynthesis is enhanced. Polyamines have also been known to be involved in the plant defense system against environmental changes (reviewed in Alcázar et al. 2006; Groppa and Benavides 2008; Liu and Moriguchi 2007; Hussain et al. 2011). Because polyamines act as the scavengers of ROIs so these confer the protection from the oxidative stress. In low temperature stress the role of polyamines has been studied in detail by many workers in different plant species (Nayyar and Chander 2004). It has been shown that putrescine accumulates in plants under cold stress regimes (Kaplan et al. 2004; Cook et al. 2004; Cuevas et al. 2008) and it is very important for their survival as the *Arabidopsis* mutants with defective putrescine synthesis were having reduced cold tolerance than wild ones. They also demonstrated that alterations in the levels of ABA caused depletion in the putrescine levels which was drastic to plant survival in cold stress.

Plant polyamines have also been known to function as secondary messengers and modulate various anatomical, biochemical, physiological, and molecular processes in intracellular as well as extracellular areas under stress (Kuznetsov et al. 2006; Alcázar et al. 2006, 2010; Cavusoglu et al. 2008). The polyamine-related metabolic enzymes are also associated with cell wall including apoplast, where lignification, suberization, and wall stiffening occur (Kujnetsov et al. 2006). Aronova et al. (2005) have elucidated that polyamines in apoplast are also related to the generation of H_2O_2 in the apoplast, where it is required for the formation of suberin, lignin, and oxyproline proteins. The role of polyamines has also been reported to be very important and it has been demonstrated that it functions during environmental insult as a part of antioxidative system and protects the membranes from oxidative damage (Kim et al. 2002; Verma and Mishra 2005; Kuznetsov et al. 2006; Shevyakova et al. 2006). In recent years, it has also been proved that polyamines are also having an important role in the regulation of structure and function of photosynthetic apparatus under low temperature stress conditions (Urao et al. 1999).

As already stated that polyamine biosynthesis is increased during exposure to stressful conditions including cold stress so it may be implicated that the initiation of polyamine biosynthesis requires a stress signal in the form of stimulus (Imai et al. 2004). The higher endogenous levels of polyamines may be positively correlated with the increased amount of antioxidants so it can be suggested that polyamines accumulation is able to optimize the metabolic rate and subsequently ensuring the growth and survival of the plant under stress i.e. it increases the stress tolerance of plants (Alcázar et al. 2006). But the exact mechanisms through which the polyamines act as a defense line against stress are still ill defined (Nayyar and Chander 2004; Aronova et al. 2005). Nayyar and Chander (2004) have found beneficial the effects of exogenous application of polyamines on chickpea (*Cicer arietinum* L.) in low temperature stress conditions. Kovacs et al. (2010) observed in case of wheat the effects of cold stress, osmotic stress, and abscisic acid (ABA) on polyamine accumulation and it was found that the levels of putrescine and spermidine levels were higher during the exposure to above stated stress conditions whereas the ABA treatment increased the levels of cadverine. Cuevas et al. (2008) have established that in response to low temperature stress putrescine is synthesized in plants and it modulates ABA biosynthesis at transcriptional levels and demonstrated that polyamines function as regulators of phytohormone biosynthesis. Despite all these findings which put forth a protective role for polyamines, it has still not been established how polyamines modulate biosynthesis of phytohormones like ABA and more studies are needed in this context.

3.4 Role of Trehalose

Trehalose is a nonreducing disaccharide in which the two glucose units are linked in an α, α-1, 1-glycosidic linkage. Although there are three possible anomers of trehalose, that is, α,β-1,1-, β,β-1,1-, and α,α1,1-, only the α,α-trehalose has been isolated from and biosynthesized in living organisms. It is synthesized in two steps (1) trehalose-6-phosphate synthase synthesizes trehalose-6-phosphate (T6P) and (2) T6P is converted to trehalose by trehalose6-phosphate phosphatase. It has also been reported to serve as a signaling molecule to direct or control certain metabolic pathways or even to affect growth in plants and yeast. Besides this, trehalose has also been reported to serve as protective guard for proteins and cellular membranes from inactivation and denaturation by various kinds of environmental constraints including cold stress as it accumulates in traces in plants in response to stress (Ramon and Rolland 2007; Paul et al. 2008; Paul 2008). Still the accurate role of trehalose in plants is still unclear and needs further research (Fernandez et al. 2010). Its role has also been implicated in the mediation of sugar metabolism in plants as it controls the rate of starch synthesis by means of redox modification of ADP-glucose pyrophosphorylase (Kolbe et al. 2005; Lunn et al. 2006). Trehalose may also stabilize cell membranes whose fluidity decreases during temperature downshift. And thus exogenous application of trehalose has also been observed to confer stress tolerance against cold temperatures (Su et al. 2010). As we have already reviewed that trehalose accumulates rapidly in response to cold shock in plants. This is followed by the transient induction of *TPP* activity (Pramanik and Imai 2005). *TPP* overexpression boosts the trehalose accumulation and confers cold tolerance (Jang et al. 2003; Ge et al. 2008).

3.5 Mitogen-Activated Protein Kinases

These are the proteins which catalyze reverse phosphorylations, which is very necessary for relaying signals. The MAPKKKs (MAPK kinase kinase) function by means of cascades which involves the sequential phosphorylation of a kinase by its upstream kinase (Xiong and Shitani 2006). MAPK pathways are activated by various abiotic stresses (Ligterink and Hirt 2001) and they also introduce the characteristic of specificity into the system. In *Arabidopsis thaliana,* three kinds of MAPKKKs have been found (1) CTR1 (2) ANP1-3 and (3) AtMEKK. Out of these, three AtMEKK are expressed in response to different abiotic stresses including cold (Knight and Knight 2001).

3.6 Transcription Factors

The process of acquiring tolerance to chilling (freezing or nonfreezing) temperatures can be achieved by exposure of plants to positive low temperatures. This is called cold acclimation. However, it has been experimentally proved that the cold acclimation can also be achieved by exposure to drought or application of ABA (Thomashow 1999). This is because many genes that are induced by cold temperatures are also expressed by application of ABA or exposure to drought stress. Moreover, these genes encode for proteins, which provide tolerance against both drought induced dehydration as well as cold stress. One of these common cold and drought-regulated genes is *RD29A* in *Arabidopsis thaliana.* This gene has been found to contain DRE or CRT (drought-responsive or C-repeat element) in their promoters (Kasuga et al. 2004). It has been noticed in *Arabidopsis thaliana* that two groups of transcription factors are present (1) *DREB1* (also called CBF) and (2) *DREB2.* These transcription factors induce the expression of specific genes for cold stress and other drought or salinity stress respectively. This also makes it clear that DRE transcription factor is a point at which different stress (cold drought or salt) induced pathways converge (Fig. 2.3). So it can be said that DRE has the capability to integrate the information from two or more stress stimuli and it plays an important role in cross talk of stress signaling pathways.

Transcript profiling experiments revealed that multiple regulatory pathways are activated during cold acclimation, and that one such important pathway involves the c-repeat binding factor (CBF) regulon (Thomashow 1999, 2001). The c-repeat/dehydration-responsive-element binding factor genes (*CBF1-3*) are transcriptional activators involved in governing the plant's responses to low temperatures (Schwager et al. 2011). These include *CBF1*, *CBF2,* and *CBF*3 (Gilmour et al. 2004). Several studies have reported that ectopic overexpression of some CBFs resulted in both activation of target genes and enhanced freezing, salt, or dehydration tolerance of transgenic plants (Jaglo-Ottosen et al. 1998; Liu et al. 1998; Kasuga et al. 1999; Haake et al. 2002). CBF pathway is a central component of cold response, but CBF-independent pathways might also be necessary for the cold stress response (Zhu et al. 2004). Hsieh et al. (2002) suggested that overexpression of *CBF1* increased chilling tolerance in tomato by enhancing *CATALASE1* gene expression and enzyme activity, and oxidative stress tolerance (Hsieh et al. 2002). Direct evidence exists for the activities of some cold-regulated transcription factors (TFs) not participating in the CBF cold-response pathway (Fowler and Thomashow 2002), which suggests that TFs play a crucial role in controlling downstream gene expression as well as the regulation of cross talk between different signaling pathways (reviewed in Knight and Knight 2001). Over-expression of AtCBF1/3 enhanced tolerance against cold, drought, and salt stress in *Brassica* species (Jaglo et al. 2001), wheat (Pellegrineschi et al. 2004) and rice (Oh et al. 2005).

Another transcription factor termed as inducer of CBF expression 1 (ICE1) acts as a key regulator of cold-induced gene expression and is present upstream of CBF. ICE 1 is an *MYC*-type basic helix-loop-helix (bHLH) transcription factor that binds to MYC-cis element in the CBF 3 promoter and may be able to activate the expression of CBF3 upon cold stress. The constitutive expression of *ICE1* enhanced the expression of *CBFs* and *COR* genes leading to increased cold tolerance (Chinnusamy et al. 2003). On the other hand, *ice1* mutant showed impaired chilling tolerance as well as cold acclimation. Moreover in such mutants, a large number of cold-induced genes were either not induced or their induction was 50 % than that of wild-type plants. These findings indicated that ICE1 acts as a key regulator of several cold-responsive CBF-dependent and independent regulons.

3.7 Role of Abscisic Acid

ABA is a phytohormone critical for plant growth and development and plays an important role in integrating various stress signals and controlling downstream stress responses. ABA has been reported to act as an endogenous messenger and regulates the water status of the plant (Swamy and Smith 1999). As various stresses have been known to induce ABA synthesis, it is now considered as a plant stress hormone (Swamy and Smith 1999; Mahajan and Tuteja 2005). Because the phytohormones mainly function as the regulators of adaptive response, the main function of ABA is to maintain and to optimize the plant water status (Swamy et al. 1999) by means of acting as endogenous messenger. In an experiment involving ABA deficient mutants of *Arabidopsis thaliana,* it was found that these mutants wilt and die readily under stress as compared to their wild counterparts (Shinozaki and Yamaguchi-Shinozaki 2000). Very recently (Nguyen et al. 2009) established by their experiments on maize (*Zea mays*) low temperature response that the genes induced by low temperature stress (*ZmCOI6.1, ZmACA1, ZmDREB2A, and ZmERF3*) are also induced by ABA application so it may be implicated that ABA synthesis regulates the induction and expression of specific cold-responsive genes in plants. Low temperatures have been reported to exert their effect on gene expression in ABA-independent pathways (Thomashow 1999; Shinozaki and Yamaguchi-Shinozaki 2000). Genetic analytical studies have shown that there is no clear line of demarcation between ABA-dependent and ABA-independent pathways and the components involved may often cross talk or even converge in the signaling pathway. Ca^{2+} has been found to mediate this cross talk (Fig. 2.3).

The expression of CBF1, CBF2, and CBF3 genes is induced by ABA but to a lower extent than that caused by cold acclimation (Knight et al. 2004). ABA has been also reported to induce the expression of *ICE1* (Chinnusamy et al. 2003). In this way, ICE 1 can also govern the ABA-mediated expression of *CBF* genes. Since cold-induced expression of *CBFs* is transient, ABA may activate ICE1-CBF-dependent and independent pathways that may be required to maintain the expression of *COR* genes during prolonged cold conditions. It has been reported that both ABA-independent and dependent pathways regulate cold-responsive genes (Xiong et al. 1999). ABA-dependent gene expression is regulated by transcription factors that belong to bZIP (ABRE-binding factors or AREB's), MYC and MYB families. *ABRE*-binding factor 1 (*ABF1*) was cloned from *Arabidopsis* (Choi et al. 2000) while its target genes are not known. However, ABAE elements can regulate the *COR* gene expression by involving a C2H2-type zinc finger protein which activates a bZIP transcription factor. *COR* gene expression can also be mediated by ABA by involving a cold inducible bZIP transcription factor in case of soybean (Kim et al. 2001). Cold stress has also been known to affect the auxin transport system in plants and inhibit basipetal auxin transport by blocking the intracellular trafficking of auxin efflux carrier PIN2 (Shibasaki et al. 2009).

3.8 Role of H_2O_2

The univalent reduction of $O_2^{\cdot-}$ produces H_2O_2. Hydrogen peroxide is considered a versatile actor of plant metabolic system. Because it plays dual roles as per its concentration, at low concentration it acts as mediator of signaling pathways leading to stress acclimation and at higher concentration it orchestrates the cellular damage and death. Low temperature stress has also been shown to induce H_2O_2 accumulation in cells (O'kane et al. 1996). H_2O_2 was disregarded as a cellular toxic metabolite for many years is a ROS, because its accumulation causes oxidative stress and it can lead to damage as well as death of plant. Plants were able to achieve a high degree of control over H_2O_2 accumulation in due course of evolution (Dröge 2002). But it has come to be known now that H_2O_2 is a key signaling molecule in plants under stressful conditions and modulates the expression of various genes (Neill et al. 2002). During the last decade H_2O_2 has been given due attention as a kind of reaction oxygen species which acts as secondary messenger in stress signaling pathways as it is having a long life and high permeability across membranes (Neill et al. 2002; Huang et al. 2002; Yang and Poovaiah 2002). Dat et al. (2000) demonstrated that H_2O_2 plays an important role in plants during biotic and abiotic stress conditions whereas Laloi et al. (2004) have observed that hydrogen peroxide is produced in plants in response to various biotic as well as abiotic stresses. Many physiological as well as biochemical processes in plants including systemic acquired resistance (SAR) and hypersensitive resistance (HR) (Melillo et al. 2006) senescence (Hung et al. 2006), programmed cell death (Houot et al. 2001), stomatal movements (Pei et al. 2000; Zhang et al. 2001; Bright et al. 2006), gravitropism property of roots (Joo et al. 2001), development of lateral secondary and tertiary roots (Su et al. 2006), cell wall formation (Potikha et al. 1999), and pollen–pistil interactions (Mcinnis et al. 2006a, b). Now it has been experimentally proved that proteins functioning in metabolism, energy movement, protein translocation and transport, cellular organization and defense and transcription are encoded by transcripts induced by H_2O_2 (Desikan et al. 2001). Studies have provided evidence that H_2O_2 itself is a key signal molecule, which mediates a series of responses (Desikan et al. 2003) and activates many other important signal molecules such as Ca^{2+}, salicylic acid, ABA, jasmonic acid, ethylene and nitric oxide of plants (Liu et al. 2004; Desikan et al. 2004; Wendehenne et al. 2004). H_2O_2 has also been reported to work in coordination with NO, ABA, jasmonic acid, and ethylene in response to cold stress. Especially, in cold response, ROS such as H_2O_2 may alter calcium expression (signatures) and activate mitogen protein kinases (MAPK) and redox-responsive transcription factors. The expression of *COR* (cold-responsive) genes is reported to be regulated by ROS (Lee et al. 2002). Under cold stress, ROS activate the AtMEKK1/ANP (MAPKKK)-AtMKK2(MAPKK)-AtMPK4/6 (MAPK) MAPK cascade that is imperative for cold acclimation in plants (Teige et al. 2004)

3.9 Role of Cytoskeleton

The eukaryotic cytoskeleton consists of tubulin dimers which form microtubules (MTs), actin monomers which form actin microfilaments (AFs) and vimentin and related proteins that constitute intermediate filaments. MTs and

AFs are both implicated in signaling, and are discussed in the following sections.

3.10 Microtubules

Microtubules are thought to transmit signals from the receptor to the nucleus, since they span the cell from the nucleus to the plasma membrane (Gundersen and Cook 1999). The minus ends of microtubules associate with the microtubule organizing center (MTOC, or centrosome in most animal cells) near the nucleus, and the plus ends terminate near the plasma membrane. This gives microtubules a defined polarity and enables directional transport via the motor molecules kinesin and dynein. As the microtubules provide a surface area ten times larger than the nuclear envelope, there is ample space for protein–protein interactions on their surface. So it should not be surprising that microtubules have been associated with various signaling pathways (Volkmann and Baluska 1999). Microtubules act as a scaffold, bringing components of the signaling pathways together. Another example of microtubules' involvement in signaling is the interaction between microtubules and ERK1/2, both in vitro (Mandelkow et al. 1992) and in vivo (Reszka et al. 1995; Morishima-Kawashima and Kosik 1996; Reszka et al. 1997), where microtubule association could retain some activated MAPKs in the cytoplasm. Another MAPK, ERK5, possesses C terminal sequences that suggest that it may also be targeted to the cytoskeleton (Zhou et al. 1995). In the G-protein signaling pathway, tubulins have been identified as secondary substrates for G-protein-coupled receptor kinases (Haga et al. 1998; Pitcher et al. 1998). However, no single mechanism for the modulation of G-protein signaling has been identified, since breakdown of microtubules and G-protein subunit-microtubule interactions leads to a multiplicity of events.

Spatial orientation of microtubules is generated by their interaction with proteins such as those found in MTOCs (Marc 1997; Vaughn and Harper 1998). However, since centrosomes are not found in higher plants (Vaughn and Harper 1998), the origin, identity, and precise locations of MTOCs is not known. Microtubules have been shown to play a role in growth orientation in plants (Williamson 1991; Joshi 1998). Mathur and Chua (2000) using transgenic plants expressing a fusion of green fluorescent protein and microtubule-associated protein 4 have shown that MT stabilization leads to growth reorientation in *Arabidopsis* trichomes. The role of MTs in Ca^{2+} channel opening was examined by Thion et al. (1996). When cold-shocked *Nicotiana plumbagnifolia* protoplasts were treated with oryzalin and cytochalasin D, destabilizers of MTs and actin microfilaments, respectively, a synergistic increase of Ca^{2+} influx was observed (Mazars et al. 1997). Thus, both MTs and actin microfilaments are speculated to be involved in Ca^{2+} influx in cold signaling.

3.11 Actin Microfilaments

Actin occurs in plant cells in two forms: globular actin (G-actin), which comprises actin monomers, and filamentous actin (F-actin), which consists of assemblies of G-actin and other proteins. Plant actin gene families are more abundant and diverse than those found in other organisms. It has been demonstrated that *Arabidopsis* has ten genes which code for actin (McDowell et al. 1996) whereas *Vicia faba* has five isoforms of actin gene (Janben et al. 1996).

The cell signaling processes are thought to be mediated by the balance between F and G-actin, alterations in the relative amounts of actin binding proteins and their binding abilities, and formation of actin-associated myosin filaments. Recent studies have shown that dynamic interconversions of F- and G-actin play a major role in the regulation of ion channels in the plasma membrane, controlling osmoregulation (Schwiebert et al. 1994; Tilly et al. 1996), as well as cell polarity (Drubin and Nelson 1996), cell growth and proliferation, secretion and cell wall interactions (Grabski et al. 1998).

Plasma membrane-associated actin is involved in the phosphoinositide signaling pathway (Tan and Boss 1992). Actin also plays a role in intracellular movement, including the endocellular localization of ER and Golgi elements, which are fully under F-actin control (Lichtscheidl et al. 1990).

4 Cold Stress Defense/Tolerance Mechanisms in Plants

The outcome of the signal perception, transduction and transcriptional up or down regulation of genes is the production of some metabolites which have plant protection, repair, and stabilizing functions. All these result into acquired tolerance against one or more abiotic stresses. Cold acclimation also known as cold hardening is one such responses that refers to increase in tolerance over time to cold temperatures and results from changes in gene expression and physiology (Xin and Browse 2000; Kalberer et al. 2006).

4.1 Cold Stress Proteins

Proteomic studies have revealed differential expression of proteins in some plant species exposed to cold stress. In pea mitochondria, 33 proteins showed either up-or down regulation under different stress conditions, 20 of which appeared to respond to low temperature of 4 °C for 36 h (Taylor et al. 2005). In rice anthers, a cold treatment for 4 days at 12 °C induced differential expression of 70 proteins out of which 47 were up-regulated, 12 were new, and 11 were down-regulated with a positive identification for 18 of them (Imin et al. 2004). In leaves of poplar seedlings subjected to 4 °C for 2 weeks, 26 proteins were identified that were COR proteins of which 21 were overexpressed and 5 were repressed (Renaut et al. 2004). Broadly, the types of proteins expressed in response to cold stress are antifreeze proteins (AFP's), dehydrins and late embryogenesis abundant (LEA) proteins, heat shock proteins (HSPs), chaperonins, pathogenesis-related (PR) proteins and those related to transduction, transcription and signaling pathways.

AFP's lower the freezing temperatures in cold acclimated leaves and after the leaves have been frozen and also prevent the growth of ice crystals by binding with them (Griffith et al. 2005). Thus, these proteins protect the cells from the mechanical injury by preventing the size of the individual ice crystals to increase as well as to inhibit the growth of ice crystals into the intercellular spaces. Though these proteins are primarily extracellular in location and activity but intracellular dehydrin in case of peach was found to have AFP proteins like activity (Wisniewski et al. 1999). AFP's are suggested to be homologous of PR proteins such as B 1, 3 glucanases, chitinases or thaumatin like proteins (Griffith and Yaish 2004).

LEA proteins though originally shown to be accumulated in plant embryos during the later stages of embryogenesis (Dure 1993) but now have been found to be expressed in response to osmotic stress, cold and ABA (Wise and Tunnacliffe 2004). These proteins have subclasses and many roles have been suggested for them such as chaperones, DNA-binding and repair, being a structural component of cytoskeleton. Dehydrins, a subgroup of LEA proteins are stable to heat, rich in glycine are expressed in response to abiotic stresses causing dehydration have role in stabilization of membranes and protection of other proteins from denaturation due to water loss induced by the stresses (Allagulova et al. 2003). Dehydrins have been reported to be accumulated due to cold stress in case of herbs and woody plants (Wisniewski et al. 2004). In poplar, the expression of a single 100 kDa LEA protein was documented (Renaut et al. 2004).

HSPs is another category of stress proteins which though originally discovered for their expression in response to heat stress are now reported to be generated due to drought, salt, and cold stresses (Sabehat et al. 1998). HSP's are also referred to as stress-related molecular chaperones. Especially, families of HSP90, HSP70 and small HSP's have been shown to accumulate due to cold stress (Lopez-Matas et al. 2004). HSP's have a role in translation, translocation into organelles, refolding of denatured proteins, prevention of aggregation of denatured proteins and protection of membranes (Tsvetkova et al. 2002).

Pathogenisis-related (PR) proteins, which are expressed due to pathogenic attack, are also produced in response to mechanical injury, xenobiotic compounds, and environmental stresses. There are 14 groups of PR proteins identified that represent B 1,3 glucanases, chitinases, thumatin-like proteins, and lipid transfer proteins (Liu et al. 2003). They are speculated to have a role in signal transduction pathway in reaction to abiotic stresses including cold stress (Hoffmann-Sommergruber 2000).

Besides these proteins, the expression and activity of several enzymes pertaining to various metabolic pathways are either up-regulated or down-regulated depending upon the severity and duration of the cold stress (Hurry et al. 1995). Among several enzymes, those related to photosynthesis (rubisco subuits, rubisco activase, polypeptides of the PSII O_2 evolving complex), carbohydrate metabolism (sucrose phosphate synthase, invertase, sucrose synthase, and enolase), and detoxification enzymes (antioxidants), those of proline metabolism (proline dehydrogenase), and lignin metabolism (caffeic acid 3-*O*-methyltransferase).

4.2 Metabolic Modifications

Exposure of plants to stress conditions results in the alteration of their metabolic activities. This happens by means of two ways (1) adjustment/restoring of the low temperature induced alterations in metabolic parameters like the structure and functional catalytic properties of enzymes by regulatory mechanisms as soon as it occurs (Schwender et al. 2004; Fernie et al. 2004) and (2) modification of the metabolic parameters according to the stress conditions (adaptive mechanisms) such as the production of some metabolites, osmolytes, and phytohormones have been reported to increase during stress conditions (Nayyar 2003a, b; Nayyar et al. 2005a, b, c, d, 2007; Farooq et al. 2008; Kaur et al. 2011). These include sugars, amino acids, organic acids, polyamines and lipids (Nayyar and Chander 2004; Nayyar et al. 2005a, b, c, d; Farooq et al. 2009; Kaur et al. 2011), which eventually assist in cellular protection from cold-induced damage by various mechanisms.

4.3 Antioxidant Systems

Much of the injury to plants caused by chilling stress is associated with oxidative damage at cellular level (Bowler et al. 1992). Inherent metabolic homeostasis of plants is disturbed due to adverse environmental factors, which results in the production of ROS (Suzuki and Mittler 2006).

Protective mechanisms against stressful low temperature conditions can be divided into two separate categories, those involved in removing reactive oxygen intermediates and those involved in reducing production of reactive oxygen intermediates. Generally, the defense system against reactive oxygen intermediates in plant cells is a net result of suppression mechanisms, scavenging, and repair systems. Higher plants have active oxygen scavenging systems consisting of several antioxidant enzymes, such as superoxide dismutase (SOD), ascorbate peroxidase (APX), catalase (CAT), glutathione reductase (GR), and some non-enzymatic antioxidants, such as ascorbic acid, α-tocopherols, phenolic compounds, and reduced glutathione (Bowler et al. 1992). In recent years, it has become apparent that plants actively produce ROIs as signaling molecules to control processes such as programmed cell death, abiotic stress responses, pathogen defense, and systemic signaling. Higher plants contain numerous enzymatic and non-enzymatic reactive oxygen intermediate scavengers and antioxidants, both water- and lipid-soluble, localized in different cellular compartments. Non-enzymatic antioxidants include

(1) pigments, (2) reduced glutathione (GSH), (3) ascorbate (AsA), (4) vitamin E, and many others.

α-Tocopherol is one of the most acknowledged antioxidant (Polle and Rennenberg 1994). α-Tocopherol is the most abundant tocopherol of the four forms found in plants (α-, β-, γ-, and δ-tocopherol). Its main location is within the chloroplast. Ascorbate, and enzymes that metabolize AsA-related compounds, are involved in the control of several plant growth processes (Cordoba and Gonzalez-Reyes 1994). The most abundant thiol in higher plants is glutathione (Foyer and Halliwell 1976; Foyer 1997; Mullineaux and Creissen 1997). The general picture is that the levels of glutathione in its reduced form (GSH) increase several fold during the chilling conditions in evergreens (Wingsle and Hällgren, 1993; Wildi and Lütz 1996). Many factors, including low temperature and other environmental stresses have been shown to change the ratio or redox status of glutathione (GSH/(GSSG+GSH)) (Karpinski et al. 1997).

4.4 *Enzymatic Antioxidants*

In plant cells the enzymatic scavenging system for reactive oxygen intermediates consists of such enzymes as, SOD, CATs, APX, monodehydroascorbatereductase (MDAR), dehydroascorbatereductase (DHAR), glutathione peroxidase (GPX), and glutathione reductase (GR) (Inzé and Montagu 1995; Foyer et al. 1997; Mullineaux and Creissen 1997). Following are some examples where the expression of these antioxidants has been genetically engineered to achieve cold tolerance (Table 2.1).

The enzyme SOD can be taken as an example of the complexity in studying the role of the enzymatic defense system. Different SOD isoforms in plants are differentially expressed and also localized in different compartments within and outside the cell (Schinkel et al. 1998). SOD mRNA levels have been observed to increase during recovery from naturally established winter stress, a combination of high light and low temperature stress (Karpinski et al. 1993, 1994). SOD isoforms are differentially expressed during recovery from winter stress. A comparison of chloroplastic and cytosolic CuZn-SOD mRNA levels showed a 4-fold higher transcript level for the chloroplastic form until mid-May (Karpinski et al. 1993). This higher transcript level was also associated with a higher chloroplastic CuZn-SOD activity. Transcript levels were reduced for both chloroplastic and cytosolic CuZn-SODs and reached similar low levels after the repair process of the photosynthetic apparatus was completed and photosynthetic capacity had fully recovered from winter stress (Karpinski et al. 1993, 1994). In alfalfa plants, Camp et al. (1994) demonstrated that Fe-SOD and Mn-SOD have different protective properties in response to chilling treatment.

In Arabidopsis, a network of at least 152 genes has been reported to be involved in managing the level of H_2O_2 (Davletova et al. 2005). The key enzymes involved in H_2O_2 scavenging are CAT and APX, which catalyze the following reactions:

Table 2.1 Genetic engineering approaches to achieve cold tolerance by overexpression of antioxidant enzymes

Enzyme	Reaction catalyzed	Transgenic plant against cold stress	Reference
Superoxide dismutase (SOD)	$O_2^- + O_2^- + 2H^+ \leftrightarrow 2H_2O + O_2$	Cu/Zn-SOD from *Pisum sativum* to *Nicotiana tobaccum*	Gupta et al. (1993)
		Fe-SOD from *Arabidopsis thaliana* to *Madicago sativa*	McKersie et al. (2000)
		Mn-SOD in *Gossypium hirsutum*	Payton et al. (2001)
Catalase (CAT)	$2H_2O_2 \leftrightarrow 2H_2O + O_2$	Rice (*Oryza sativa*)	Matsumura et al. (2002)
Glutathione reductase (GR)	NADPH+GSSG$\leftrightarrow$NADP +2GSH	Tobacco (*Nicotiana tabacum*)	Le Martret et al. (2011)
		From *Arabidopsis thaliana* to *Gossypium hirsutum*	Payton et al. (2001); Kornyeyev et al. (2003b)
Dehydroasco-rbate reductase (DHAE)		From human to Tobacco (*Nicotiana tabacum*)	Kwon et al. (2003)
		Tobacco(*Nicotiana tabacum*)	Le Martret et al. (2011)
Ascorbate peroxidase (APX)	$AA + H_2O_2 \leftrightarrow DHA + 2H_2O$	From *Pisum sativum* to *Gossypium hirsutum*	Kornyeyev et al. (2001, 2003a, b)
		From *Spinacea oleracea* to *Nicotiana tobaccum*	Yabuta et al. (2002)
		From *Pisums sativum* to *Lycopersicon esculentum*	Wang et al. (2005)
		Tomato (*Lycopersicum esculentum*) St*APX*gene in Tobacco(*Nicotiana tabacum*)	Sun et al. (2010)

$$[APX] \rightarrow 2AsA + H_2O_2 \rightarrow 2monodehydroascorbate(mdAsA) + 2H_2O$$

$$[CAT] \rightarrow H_2O_2 \rightarrow 2H_2O + O_2$$

CAT acts as a major scavenger of H_2O_2 generated during mitochondrial electron transport, β-oxidation of the fatty acids, and most importantly in photorespiratory oxidation (Scandalios et al. 1997). GPX has generated much attention as an important enzyme in the scavenging of H_2O_2 or the products of lipid peroxidation. The role and function of the chloroplastic GPX during cold hardening and low temperature-induced oxidative stress in trees is under investigation. Expression of genes encoding different isoforms of the same ROI scavenging enzyme are regulated differently in response to low temperature-induced oxidative stress (Karpinski et al. 1993). CATs have also received much attention in respect of plants response to chilling and are thought to play a major role in inducing chilling tolerance (Prasad 1996).

4.5 Other Involvements (Compatible Solutes, Phytohormones, and Others)

In response to almost all the stresses, the increase has been observed in the levels of compatible solutes, which implies that they are having significant role in stress defense/tolerance. The compatible solutes are organic compounds belonging to a chemically diverse small group and are highly soluble. These are also known as osmolytes. These molecules are considered perfectly compatible to cellular functioning, since these do not interfere with cellular metabolism, even at higher concentrations (reviewed in Sung et al. 2003). Proline is one of the most studied and extensively reported cryo- and osmoprotectant, and has been found to accumulate in response to almost all the kinds of abiotic stress conditions like drought, salinity, high temperature, chilling, UV radiation, and heavy metals (Rhodes and Hanson 1993; Nayyar and Walia 2003). In case of *Arabidopsis,* it has been seen that proline levels are increased and accumulated to considerable level during the stress conditions. In our case of review, the integrity of plasma membrane is vital for the low temperature tolerance and it has been suggested that proline may interact with the enzymes to protect the membrane-proteins' structure and activity (Hamilton and Heckathorn 2001). Proline accumulation has been experimentally observed in cold-shocked greenbean plants along with ornithine-δ-aminotransferase and proline dehydrogenase enzymes (Ruiz et al. 2002).

Glycine betaine is another osmolyte coming from the group of betaines (the quarternary ammonium compounds in which the nitrogen atom is fully methylated). In higher plants glycine betaine is synthesized from choline using two enzymes (1) choline monoxygenase and (2) betaine aldehyde dehydrogenase (Rathinasabapathi et al. 1997). Glycinebetaine has been seen to be synthesized at increased levels and accumulated in many plant species in response to various stresses and thereby providing tolerance to the stress (Hincha et al. 2006), although its presence in plants is not universal as it is not reported to be accumulated in *Arabidopsis,* rice, and tobacco. Transgenics having over-accumulation of glycine betaine have been reported to have tolerance against different stresses including chilling (Sakamoto and Murata 2002). Its exogenous application has also been known to confer stress tolerance and increase growth and survival (Chen et al. 2011). Trehalose is also a compatible solute but its roles are still not much unblemished in plant exposed to cold shock. Some authors have reviewed that it may be considered as a double-faced molecule with both negative as well as positive effects (Fernandez et al. 2010)

Compatible solutes also include sugar alcohols that are acyclic polyols containing three or more hydroxyl groups, which include erythritol, D-arabitol, ribitol, xylitol, sorbitol, D-mannitol, galactinol, and rhamnitol (Ahmad et al. 1979).

5 Modification in Gene Expression Pattern and Synthesis of Stress Responsive Genes

At low temperature conditions the plants reorganize their patterns of gene expression and try to maintain homeostasis for obtaining cold stress tolerance (Cook et al. 2004). A number of genes have been identified and reported which express during low temperature stress conditions (Mantri et al. 2007 reviewed in Yadav 2010). The recent DNA microarray technique has made it possible to analyze large scale gene expression and in last few years numerous stress-induced genes have been identified in different crops not only in chilling stress but in other abiotic stresses also (Bray 2004; Maruyama et al. 2004; Seki et al. 2004; Vogel et al. 2005; Mantri et al. 2007). Some drawbacks of microarray like analysis of arbitrarily selected gene segments have been overcome by another technique i.e. serial analysis of gene expression. This technique allows the identification of novel genes under various physiological states of plants. These two methods have helped us to reveal that under stress conditions, some new genes are expressed and in some cases the expression patterns of some genes are altered. Either now they produce the protein products, which directly take part in processes against stress, or they regulate the expression of other genes. Based on these documentations the product proteins can also be classified into two types. First are those which are involved directly in the processes against the stress e.g. LEA proteins, antifreezins, osmotins, chaperones, mRNA binding proteins, enzymatic proteins for osmolytes (proline, trehalose, transporter proteins for proline, sugars, and lipids), detoxification processes and fatty acid metabolism, proteinase inhibiter proteins, and water channel proteins(Kreps et al. 2002; Seki et al. 2002). The functioning of these genes has been proved as in case of some transgenic plants in which these genes are over expressed are considerably stress tolerant (Cushman and Bohnert 2000). The second type of proteins are those which themselves do not take part directly in the stress tolerance mechanisms but further regulate the other signal transduction pathways. The examples of these types of proteins are some transcription factors

(Seki et al. 2003). These transcription factors are involved in the regulation of expression of other stress responsive genes. Some other examples of stress responsive proteins are kinases, phosphatases, calmodulin binding proteins, and 14-3-3 proteins. It has been elaborated by some authors that transgenics having these genes overexpressed in them are tolerant to stress conditions (Zhang et al. 2004; Tester and Bacic 2005; Vinocur and Altman 2005).

The expression of these stress responsive genes (genes expressed in response to cold stress are called cold-responsive genes) is vital for the tolerance and acclimation to the low temperature conditions. The vitality of these genes has been proved by the help of producing transgenics with overexpression of these genes (Ma et al. 2009; Sanghera et al. 2011).

6 Genetic Engineering Against Cold Stress

On exposure of plants to low temperature, a series of genes are induced, the products of which may either directly protect against stress or further control the expression of other target genes (Yamaguchi-Shinozaki and Shinozaki 2006). Transgenic plants have also been and are also being prepared against cold stress i.e. to achieve cold tolerance. These plants have one or more alien genes from stranger or their wild relatives, which over-express and regulate the functioning of metabolic process in a positive manner against stressful temperatures. The analysis of transgenic plants overexpressing one or other genes provides us an understanding of basic mechanism of functioning of stress genes during cold stress exposure (Tayal et al. 2005). (Table 2.2)

7 Conclusion

The study of plant temperature interactions is of great relevance with respect to the global climate change. Even after two decades of molecular and biochemical plant metabolomics research we are not yet able to clearly identify the plant thermo-sensors. However, we have considerably grown in the field of knowledge of various cross talking signaling pathways and responses of plants in respect of changes in their micro- as well as macro environments. The deeper analysis of these responses will bring new insights about the thermo-sensing mechanisms in these sessile poikilotherms. The observable phenological changes are very informative about the small periodic responses of plants to the temperature changes. These phenological alterations must be studied as a link to the temperature changes to facilitate the molecular, physiological, and biochemical studies related to cold tolerance. These will help to reveal about the probable stimulating inputs of the temperature. The genetic studies besides this will help to produce the computer models to understand the problem in digitized way, which will take the comprehensive approaches in hypothesis making

Table 2.2 Shows a list of transgenic plants produced for cold tolerance

Transgenic crop	Gene engineered	Effect of the gene engineering	Reference
Nicotiana tobaccum	wheat *TaSOD1.1* and *TaSOD1.2* genes	Increased SOD activities and decreased MDA content, lessened degree of over-oxidation of the cellular membrane system, the enhancement of physiological functions	Hai Na et al. (2009)
N. tobaccum	*OsSPX1*	Better seedling survival and reduced cellular electrolyte leakage, decreased total leaf Pi content and accumulation of free proline and sucrose	Zhao et al. (2009a, b)
Arabidopsis thaliana	*Coda*	Accumulation of glycinebetaine	Hayashi et al. (1997)
N. tobaccum	(*AtP5Cs* and *VacP5Cs* for Δ1-pyroline-5-carboxylate synthetase production) from *Arabidopsis* or *Vigna SacB for* levansucrase from *Bacillus subtilis* or the *codA* gene coding for choline oxidase from *Arthrobacter globiformis*	Accumulation of osmoprotectants like proline, fructan or glycine betaine	Konstantinova et al. 2002
Oryza sativa	*TERF2*	Increased accumulation of osmotic substances and chlorophyll, reduced ROS & MDA content and decreased electrolyte leakage	Tian et al. (2011)
Triticum aestivum	DRE-binding transcription factor gene, *GhDREB* from *Gossypium hirsutum*	Improved tolerance to drought, high salt, and freezing stresses through accumulating higher levels of soluble sugar and chlorophyll in leaves after stress treatments	Gao et al. (2009)
O. sativa	*OsRAN2*	maintained cell division, decreased proportion of cells with intranuclear tubulin and formation of a normal nuclear envelope under the cold condition	Chen et al. (2011)

(continued)

Table 2.2 (continued)

Transgenic crop	Gene engineered	Effect of the gene engineering	Reference
O. sativa	*MYBS3*, Single DNA-binding repeat MYB transcription factor	Repressed the well-known DREB1/CBF-dependent cold signaling pathway in rice, and the repression appears to act at the transcriptional level	Su et al. (2010)
O. sativa	*OsMYB3R-2* transcription factor	higher transcript levels of several G2/M phase-specific genes, including *OsCycB1;1*, *OsCycB2;1*, *OsCycB2;2*, and *OsCDC20.1* increased cell mitotic index, level of cellular free proline was increased	Ma et al. (2009)
O. sativa	*OsMYB3R-2*	Stress tolerance	Dai et al. (2007)
E. Coli	*OtsA/OtsB*	Ability to synthesize trehalose	Kandror et al. (2002)
O. sativa	*OsTPP1*	Trehalose synthesis	Ge et al. (2008)
N. tobaccum	*FAD7*	ω-3-fatty acid desaturase gene; survival in chilling conditions	Khodakovskaya et al. (2006)

and testing to new horizons. Our group is working to explore these issues based on study of phenology, physiology, biochemistry, and molecular biology of different crops under thermal stresses.

References

Abat JK, Deswal R (2009) Differential modulation of S-nitrosoproteome of *Brassica juncea* by low temperature: change in S-nitrosylation of Rubisco is responsible for the inactivation of its carboxylase activity. Proteomics 9:4368–4380

Ahmad I, Larher F, Stewart GR (1979) Sorbitol, a compatible osmotic solute in Plantago maritima. New Phytol 82:671–678

Alcázar R, Marco F, Cuevas JC, Patron M, Ferrando A et al (2006) Involvement of polyamines in plant response to abiotic stress. Biotechnol Lett 28:1867–1876

Alcázar R, Altabella T, Marco F, Bortolotti C, Reymond M et al (2010) Polyamines: molecules with regulatory functions in plant abiotic stress tolerance. Planta 231:1237–1249

Al-Fageeh MB, Smales CM (2006) Control and regulation of the cellular responses to cold shock: the responses in yeast and mammalian systems. Biochem J 397:247–259

Allagulova CHR, Gimalov FR, Shakirova FM, Vakhitov VA (2003) The plant dehydrins: structure and putative functions. Biochemistry (Mosc) 68:945–951

Aronova EE, Shevyakova NI, Stetsenko LA, Kuznetsov VV (2005) Cadaverine-induced induction of superoxide dismutase gene expression in *Mesembryanthemum crystallinum* L. Dokl Biol Sci 403:257–259

Bae MS, Cho EJ, Choi EY, Park OK (2003) Analysis of the Arabidopsis nuclear proteome and its response to cold stress. Plant J 36:652–663

Baena-Gonzalez E (2010) Energy signaling in the regulation of gene expression during stress. Mol Plant 3:300–313

Bagnall D, Wolfe JOE, King RW (1983) Chill-induced wilting and hydraulic recovery in mung bean plants. Plant Cell Environ 6:457–464

Beligni MV, Lamattina L (2000) Nitric oxide stimulates seed germination and de-etiolation, and inhibits hypocotyl elongation, three light-inducible responses in plants. Planta 210:215–221

Besson-Bard A, Pugin A, Wendehenne D (2008a) New insights into nitric oxide signaling in plants. Annu Rev Plant Biol 59:21–39

Besson-Bard A, Courtois C, Gauthier A, Dahan J, Dobrowolska G et al (2008b) Nitric oxide in plants: production and cross-talk with Ca^{2+} signaling. Mol Plant 1:218–228

Bethke PC, Libourel IG, Jones RL (2006a) Nitric oxide reduces seed dormancy in Arabidopsis. J Exp Bot 57:517–526

Bethke PC, Libourel IG, Reinohl V, Jones RL (2006b) Sodium nitroprusside, cyanide, nitrite, and nitrate break Arabidopsis seed dormancy in a nitric oxide-dependent manner. Planta 223:805–812

Bethke PC, Libourel IG, Aoyama N, Chung YY, Still DW et al (2007) The Arabidopsis aleurone layer responds to nitric oxide, gibberellin, and abscisic acid and is sufficient and necessary for seed dormancy. Plant Physiol 143:1173–1188

Bhattacharjee S (2005) Reactive oxygen species and oxidative burst: roles in stress, senescence and signal transduction in plant. Curr Sci 89:1113–1121

Björn Lárus O, Veena S, Franz O, Dhindsa RS. Early steps in cold sensing by plant cells: the role of actin cytoskeleton and membrane fluidity. *Plant J*, <START>September 2000, vol. 23, no. 6, p. 785-794.

Bowler C, Fluhr R (2000) The role of calcium and activated oxygens as signals for controlling cross-tolerance. Trends Plant Sci 5:241–246

Bowler C, Montagu MV, Inze D (1992) Superoxide dismutase and stress tolerance. Annu Rev Plant Physiol Plant Mol Biol 43:83–116

Bray EA (2004) Genes commonly regulated by water-deficit stress in Arabidopsis thaliana. J Exp Bot 55:2331–2341

Bright J, Desikan R, Hancock JT, Weir IS, Neill SJ (2006) Aba-induced No generation and stomatal closure in Arabidopsis are dependent on H_2O_2 synthesis. Plant J 45:113–122

Camp WV, Willekens H, Bowler C, Montagu MV, Inze D et al (1994) Elevated levels of superoxide dismutase protect transgenic plants against ozone damage. Nat Biotechnol 12:165–168

Cantrel C, Vazquez T, Puyaubert J, Rezé N, Lesch M et al (2011) Nitric oxide participates in cold-responsive phosphosphingolipid formation and gene expression in Arabidopsis thaliana. New Phytol 189:415–427

Cavusoglu K, Kilic S, Kabar K (2008) Effects of some plant growth regulators on leaf anatomy of radish seedlings grown under saline conditions. J Appl Biol Sci 2:47–50

Chen N, Xu Y, Wang X, Du C, Du J et al (2011) OSRAN2, essential for mitosis, enhances cold tolerance in rice by promoting export of intranuclear tubulin and maintaining cell division under cold stress. Plant Cell Environ 34:52–64

Cheng S-H, Willmann MR, Chen H-C, Sheen J (2002) Calcium signaling through protein kinases. The Arabidopsis calcium-dependent protein kinase gene family. Plant Physiol 129:469–485

Chinnusamy V, Ohta M, Kanrar S, Lee BH, Hong X et al (2003) Ice1: a regulator of cold-induced transcriptome and freezing tolerance in Arabidopsis. Genes Dev 17:1043–1054

Cho YH, Yoo SD, Sheen J (2006) Regulatory functions of nuclear hexokinase1 complex in glucose signaling. Cell 127:579–589

Choi H, Hong J, Ha J, Kang J, Kim SY (2000) Abfs, a family of Aba-responsive element binding factors. J Biol Chem 275:1723–1730

Cleland RE, Virk SS, Taylor D, Bjiirkman T (1990) Calcium, cell walls and growth. In: Leonard RT, Hepler PK (eds) Calcium in plant growth and development. American Society of Plant Physiologists, Rockville, MD, pp 9–16

Cook D, Fowler S, Fiehn O, Thomashow MF (2004) A prominent role for the Cbf cold response pathway in configuring the low-temperature metabolome of Arabidopsis. Proc Natl Acad Sci U S A 101:15243–15248

Cordoba F, Gonzalez-Reyes JA (1994) Ascorbate and plant cell growth. J Bioenerg Biomembr 26:399–405

Corpas FJ, Barroso JB, Del Río LA (2001) Peroxisomes as a source of reactive oxygen species and nitric oxide signal molecules in plant cells. Trends Plant Sci 6:145–150

Corpas FJ, Barroso JB, Carreras A, Quiros M, Leon AM et al (2004) Cellular and subcellular localization of endogenous nitric oxide in young and senescent pea plants. Plant Physiol 136:2722–2733

Corpas FJ, Barroso JB, Carreras A, Valderrama R, Palma JM et al (2006) Constitutive arginine-dependent nitric oxide synthase activity in different organs of pea seedlings during plant development. Planta 224:246–254

Corpas FJ, Chaki M, Fernandez-Ocana A, Valderrama R, Palma JM et al (2008) Metabolism of reactive nitrogen species in pea plants under abiotic stress conditions. Plant Cell Physiol 49:1711–1722

Cossins AR, Christiansen J, Prosser CL (1978) Adaptation of biological membranes to temperature. The lack of homeoviscous adaptation in the sarcoplasmic reticulum. Biochim Biophys Acta 511:442–452

Crawford NM (2006) Mechanisms for nitric oxide synthesis in plants. J Exp Bot 57:471–478

Cuevas JC, Lopez-Cobollo R, Alcazar R, Zarza X, Koncz C et al (2008) Putrescine is involved in Arabidopsis freezing tolerance and cold acclimation by regulating abscisic acid levels in response to low temperature. Plant Physiol 148:1094–1105

Cui S, Huang F, Wang J, Ma X, Cheng Y et al (2005) A proteomic analysis of cold stress responses in rice seedlings. Proteomics 5:3162–3172

Cushman JC, Bohnert HJ (2000) Genomic approaches to plant stress tolerance. Curr Opin Plant Biol 3:117–124

Dai X, Xu Y, Ma Q, Xu W, Wang T et al (2007) Overexpression of an R1R2R3 Myb gene, OSMYB3R–2, increases tolerance to freezing, drought, and salt stress in transgenic Arabidopsis. Plant Physiol 143:1739–1751

Dat J, Vandenabeele S, Vranova E, Van Montagu M, Inze D et al (2000) Dual action of the active oxygen species during plant stress responses. Cell Mol Life Sci 57:779–795
Davletova S, Rizhsky L, Liang H, Shengqiang Z, Oliver DJ et al (2005) Cytosolic ascorbate peroxidase 1 is a central component of the reactive oxygen gene network of Arabidopsis. Plant Cell 17:268–281
Desikan R, A-H-Mackerness S, Hancock JT, Neill SJ (2001) Regulation of the Arabidopsis transcriptome by oxidative stress. Plant Physiol 127:159–172
Desikan R, Hancock JT, Neill SJ (2003) Oxidative stress signalling. In: Hirt H, Shinozaki K (eds) Topics in current genetics. Springer, London, pp 121–150
Desikan R, Cheung MK, Bright J, Henson D, Hancock JT et al (2004) Aba, hydrogen peroxide and nitric oxide signalling in stomatal guard cells. J Exp Bot 55:205–212
Dordas C, Hasinoff BB, Igamberdiev AU, Manac'h N, Rivoal J et al (2003a) Expression of a stress-induced hemoglobin affects No levels produced by alfalfa root cultures under hypoxic stress. Plant J 35:763–770
Dordas C, Rivoal J, Hill RD (2003b) Plant haemoglobins, nitric oxide and hypoxic stress. Ann Bot 91 Spec No:173–178
Dröge W (2002) Free radicals in the physiological control of cell function. Physiol Rev 82:47–95
Drubin DG, Nelson WJ (1996) Origins of cell polarity. Cell 84:335–344
Dure L III (1993) A repeating 11-mer amino acid motif and plant desiccation. Plant J 3:363–369
Ehlert B, Hincha DK (2008) Chlorophyll fluorescence imaging accurately quantifies freezing damage and cold acclimation responses in Arabidopsis leaves. Plant Methods 4:12
Evans DP, Corbin JR, Tomasovic SP (1991) Effects of calcium buffering on the synthesis of the 26-kDa heat-shock protein family. Radiat Res 127:261–268
Farooq M, Aziz T, Hussain M, Rehman H, Jabran K et al (2008) Glycinebetaine improves chilling tolerance in hybrid maize. J Agron Crop Sci 194:152–160
Farooq M, Aziz T, Wahid A, Lee D, Siddique KHM (2009) Chilling tolerance in maize: agronomic and physiological approaches. Crop Pasture Sci 60:501–516
Fernandez O, Bethencourt L, Quero A, Sangwan RS, Clement C (2010) Trehalose and plant stress responses: friend or foe? Trends Plant Sci 15:409–417
Fernie AR, Carrari F, Sweetlove LJ (2004) Respiratory metabolism: glycolysis, the TCA cycle and mitochondrial electron transport. Curr Opin Plant Biol 7:254–261
Finkel T, Holbrook NJ (2000) Oxidants, oxidative stress and the biology of ageing. Nature 408:239–247
Fowler S, Thomashow MF (2002) Arabidopsis transcriptome profiling indicates that multiple regulatory pathways are activated during cold acclimation in addition to the Cbf cold response pathway. Plant Cell 14:1675–1690
Foyer CH (1997) Oxygen metabolism and electron transport in photosynthesis. In: Scandalios JG (ed) Oxidative stress and the molecular biology of antioxidant defences. Cold Spring Harbor Laboratory Press, New York, pp 587–622
Foyer CH, Halliwell B (1976) The presence of glutathione and glutathione reductase in chloroplasts: a proposed role in ascorbic acid metabolism. Planta 133:21–25
Foyer CH, Lopez-Delgado H, Dat JF, Scott IM (1997) Hydrogen peroxide- and glutathione-associated mechanisms of acclimatory stress tolerance and signalling. Physiol Plant 100:241–254
Gao SQ, Chen M, Xia LQ, Xiu HJ, Xu ZS et al (2009) A cotton (Gossypium hirsutum) Dre-binding transcription factor gene, GHDREB, confers enhanced tolerance to drought, high salt, and freezing stresses in transgenic wheat. Plant Cell Rep 28:301–311
Garcia-Mata C, Lamattina L (2007) Abscisic acid (Aba) inhibits light-induced stomatal opening through calcium- and nitric oxide-mediated signaling pathways. Nitric Oxide 17:143–151
Ge L-F, Chao D-Y, Shi M, Zhu M-Z, Gao J-P et al (2008) Overexpression of the trehalose-6-phosphate phosphatase gene *OSTPP1* confers stress tolerance in rice and results in the activation of stress responsive genes. Planta 228:191–201
Gilmour SJ, Fowler SG, Thomashow MF (2004) Arabidopsis transcriptional activators Cbf1, Cbf2, and Cbf3 have matching functional activities. Plant Mol Biol 54:767–781

Grabski S, Arnoys E, Busch B, Schindler M (1998) Regulation of actin tension in plant cells by kinases and phosphatases. Plant Physiol 116:279–290

Griffith M, Yaish MWF (2004) Antifreeze proteins in overwintering plants: a tale of two activities. Trends Plant Sci 9:399–405

Griffith M, Lumb C, Wiseman SB, Wisniewski M, Johnson RW et al (2005) Antifreeze proteins modify the freezing process in planta. Plant Physiol 138:330–340

Groppa MD, Benavides MP (2008) Polyamines and abiotic stress: recent advances. Amino Acids 34:35–45

Guillas I, Zachowski A, Baudouin E (2011) A matter of fat: interaction between nitric oxide and sphingolipid signaling in plant cold response. Plant Signal Behav 6:140–142

Gundersen GG, Cook TA (1999) Microtubules and signal transduction. Curr Opin Cell Biol 11:81–94

Gupta AS, Webb RP, Holaday AS, Allen RD (1993) Overexpression of superoxide dismutase protects plants from oxidative stress (induction of ascorbate peroxidase in superoxide dismutase-overexpressing plants). Plant Physiol 103:1067–1073

Gupta KJ, Fernie AR, Kaiser WM, Van Dongen JT (2011a) On the origins of nitric oxide. Trends Plant Sci 16:160–168

Gupta KJ, Hincha DK, Mur LA (2011b) No way to treat a cold. New Phytol 189:360–363

Haake V, Cook D, Riechmann JL, Pineda O, Thomashow MF et al (2002) Transcription factor Cbf4 is a regulator of drought adaptation in Arabidopsis. Plant Physiol 130:639–648

Haga K, Ogawa H, Haga T, Murofushi H (1998) Gtp-binding-protein-coupled receptor kinase 2 (Grk2) binds and phosphorylates tubulin. Eur J Biochem 255:363–368

Hai-NA Z, Jun-Tao G, Wen-Jing L, Cun-Dong L, Kai X (2009) Improvement of low-temperature stress tolerant capacities in transgenic tobacco plants from overexpression of wheat *TASOD1.1.* and *TASOD1.2* genes. Sci Agric Sin 42:10–16

Hamilton EW III, Heckathorn SA (2001) Mitochondrial adaptations to NaCl. Complex I is protected by anti-oxidants and small heat shock proteins, whereas complex Ii is protected by proline and betaine. Plant Physiol 126:1266–1274

Harmon AC, Gribskov M, Harper JF (2000) Cdpks—a kinase for every Ca^{2+} signal? Trends Plant Sci 5:154–159

Hayashi M, Maeda T (2006) Activation of the Hog pathway upon cold stress in Saccharomyces cerevisiae. J Biochem 139:797–803

Hayashi H, Alia L, Mustardy P, Deshnium MI et al (1997) Transformation of Arabidopsis thaliana with the codA gene for choline oxidase; accumulation of glycinebetaine and enhanced tolerance to salt and cold stress. Plant J 12:133–142

He Y, Tang RH, Hao Y, Stevens RD, Cook CW et al (2004) Nitric oxide represses the Arabidopsis floral transition. Science 305:1968–1971

Hincha DK, Popova AV, Cacela C (2006) Effects of sugars on the stability and structure of lipid membranes during drying (Chapter 6). In: Liu AL (ed) Advances in planar lipid bilayers and liposomes. Academic, New York, pp 189–217

Hoffmann-Sommergruber K (2000) Plant allergens and pathogenesis-related proteins. What do they have in common? Int Arch Allergy Immunol 122:155–166

Houot V, Etienne P, Petitot AS, Barbier S, Blein JP et al (2001) Hydrogen peroxide induces programmed cell death features in cultured tobacco BY-2 cells, in a dose-dependent manner. J Exp Bot 52:1721–1730

Hsieh TH, Lee JT, Yang PT, Chiu LH, Charng YY et al (2002) Heterology expression of the Arabidopsis C-repeat/dehydration response element binding factor 1 gene confers elevated tolerance to chilling and oxidative stresses in transgenic tomato. Plant Physiol 129:1086–1094

Huang C, Li J, Ke Q, Leonard SS, Jiang BH et al (2002) Ultraviolet-induced phosphorylation of p70(S6K) at Thr(389) and Thr(421)/Ser(424) involves hydrogen peroxide and mammalian target of rapamycin but not Akt and atypical protein kinase C. Cancer Res 62:5689–5697

Hung KT, Hsu YT, Kao CH (2006) Hydrogen peroxide is involved in methyl jasmonate-induced senescence of rice leaves. Physiol Plant 127:293–303

Hussain SS, Ali M, Ahmad M, Siddique KH (2011) Polyamines: natural and engineered abiotic and biotic stress tolerance in plants. Biotechnol Adv 29:300–311

Ilker R, Breidenbach RW, Lyons JM (1979) Sequence of ultra-structural changes in tomato cotyledons during short periods of chilling. In: Lyons JM, Graham D, Raison JK (eds) Low temperature stress in crop plants. Academic, New York, pp 97–114

Imai A, Matsuyama T, Hanzawa Y, Akiyama T, Tamaoki M et al (2004) Spermidine synthase genes are essential for survival of Arabidopsis. Plant Physiol 135:1565–1573

Imin N, Kerim T, Rolfe BG, Weinman JJ (2004) Effect of early cold stress on the maturation of rice anthers. Proteomics 4:1873–1882

Inzé D, Montagu MV (1995) Oxidative stress in plants. Curr Opin Biotechnol 6:153–158

Jaglo-Ottosen KR, Gilmour SJ, Zarka DG, Schabenberger O, Thomashow MF (1998) Arabidopsis Cbf1 overexpression induces Cor genes and enhances freezing tolerance. Science 280:104–106

Janben M, Hunte C, Schulz M, Schnabl H (1996) Tissue specification and intracellular distribution of actin isoforms *Vicia faba* L. Protoplasma 191:158–163

Jang IC, Oh SJ, Seo JS, Choi WB, Song SI et al (2003) Expression of a bifunctional fusion of the Escherichia coli genes for trehalose-6-phosphate synthase and trehalose-6-phosphate phosphatase in transgenic rice plants increases trehalose accumulation and abiotic stress tolerance without stunting growth. Plant Physiol 131:516–524

Jenkins GI (1998) Signal transduction networks and the integration of responses of environmental stimuli. In: Callow JA (ed) Advances in botanical research. Academic, New York, pp 53–73

Joo JH, Bae YS, Lee JS (2001) Role of auxin-induced reactive oxygen species in root gravitropism. Plant Physiol 126:1055–1060

Joshi HC (1998) Microtubule dynamics in living cells. Curr Opin Cell Biol 10:35–44

Kalberer SR, Wisniewski M, Arora R (2006) Deacclimation and reacclimation of cold-hardy plants: current understanding and emerging concepts. Plant Sci 171:3–16

Kandror O, Deleon A, Goldberg AL (2002) Trehalose synthesis is induced upon exposure of Escherichia coli to cold and is essential for viability at low temperatures. Proc Natl Acad Sci U S A 99:9727–9732

Kaplan F, Kopka J, Haskell DW, Zhao W, Schiller KC et al (2004) Exploring the temperature-stress metabolome of Arabidopsis. Plant Physiol 136:4159–4168

Karpinski S, Wingsle G, Karpinska B, Hallgren JE (1993) Molecular responses to photooxidative stress in Pinus sylvestris (L.) (II. differential expression of CuZn-superoxide dismutases and glutathione reductase. Plant Physiol 103:1385–1391

Kasuga M, Liu Q, Miura S, Yamaguchi-Shinozaki K, Shinozaki K (1999) Improving plant drought, salt, and freezing tolerance by gene transfer of a single stress-inducible transcription factor. Nat Biotechnol 17:287–291

Kasuga M, Miura S, Shinozaki K, Yamaguchi-Shinozaki K (2004) A combination of the Arabidopsis Dreb1a gene and stress-inducible rd29A promoter improved drought- and low-temperature stress tolerance in tobacco by gene transfer. Plant Cell Physiol 45:346–350

Kaur G, Kumar S, Nayyar H, Upadhyaya HD (2008) Cold stress injury during the pod-filling phase in chickpea (*Cicer arietinum* L.): effects on quantitative and qualitative components of seeds. J Agron Crop Sci 194:457–464

Kaur G, Kumar S, Thakur P, Malik JA, Bhandhari K et al (2011) Involvement of proline in response of chickpea (Cicer arietinum L.) to chilling stress at reproductive stage. Sci Hortic 128:174–181

Khodakovskaya M, Mcavoy R, Peters J, Wu H, Li Y (2006) Enhanced cold tolerance in transgenic tobacco expressing a chloroplast ω-3 fatty acid desaturase gene under the control of a cold-inducible promoter. Planta 223:1090–1100

Kim JC, Lee SH, Cheong YH, Yoo CM, Lee SI et al (2001) A novel cold-inducible zinc finger protein from soybean, Scof-1, enhances cold tolerance in transgenic plants. Plant J 25:247–259

Kim TE, Kim SK, Han TJ, Lee JS, Chang SC (2002) Aba and polyamines act independently in primary leaves of cold-stressed tomato (Lycopersicon esculentum). Physiol Plant 115:370–376

Knight H, Knight MR (2001) Abiotic stress signalling pathways: specificity and cross-talk. Trends Plant Sci 6:262–267

Knight H, Trewavas AJ, Knight MR (1996) Cold calcium signaling in Arabidopsis involves two cellular pools and a change in calcium signature after acclimation. Plant Cell 8:489–503

Knight H, Zarka DG, Okamoto H, Thomashow MF, Knight MR (2004) Abscisic acid induces Cbf gene transcription and subsequent induction of cold-regulated genes via the Crt promoter element. Plant Physiol 135:1710–1717
Kolbe A, Tiessen A, Schluepmann H, Paul M, Ulrich S et al (2005) Trehalose 6-phosphate regulates starch synthesis via posttranslational redox activation of Adp-glucose pyrophosphorylase. Proc Natl Acad Sci U S A 102:11118–11123
Konstantinova T, Parvanova D, Atanassov A, Djilianoiv D (2002) Freezing tolerant tobacco, transformed to accumulate osmoprotectants. Plant Sci 163:157–164
Kopyra ME, Gwó D (2004) The role of nitric oxide in plant growth regulation and responses to abiotic stresses. Acta Physiol Planta 26:459–473
Kornyeyev D, Logan BA, Payton P, Allen RD, Holaday AS (2001) Enhanced photochemical light utilization and decreased chilling-induced photoinhibition of photosystem Ii in cotton overexpressing genes encoding chloroplast-targeted antioxidant enzymes. Physiol Plant 113:323–331
Kornyeyev D, Logan BA, Allen RD, Holaday AS (2003a) Effect of chloroplastic overproduction of ascorbate peroxidase on photosynthesis and photoprotection in cotton leaves subjected to low temperature photoinhibition. Plant Sci 165:1033–1041
Kornyeyev D, Holaday S, Logan B (2003b) Predicting the extent of photosystem Ii photoinactivation using chlorophyll a fluorescence parameters measured during illumination. Plant Cell Physiol 44:1064–1070
Kovacs Z, Simon-Sarkadi L, Szucs A, Kocsy G (2010) Differential effects of cold, osmotic stress and abscisic acid on polyamine accumulation in wheat. Amino Acids 38:623–631
Kreps JA, Wu Y, Chang HS, Zhu T, Wang X et al (2002) Transcriptome changes for Arabidopsis in response to salt, osmotic, and cold stress. Plant Physiol 130:2129–2141
Kudla J, Xu Q, Harter K, Gruissem W, Luan S (1999) Genes for calcineurin B-like proteins in Arabidopsis are differentially regulated by stress signals. Proc Natl Acad Sci U S A 96:4718–4723
Kumar S, Malik J, Thakur P, Kaistha S, Sharma K et al (2011) Growth and metabolic responses of contrasting chickpea (*Cicer arietinum* L.) genotypes to chilling stress at reproductive phase. Acta Physiol Plant 33:779–787
Kuznetsov V, Radyukina N, Shevyakova N (2006) Polyamines and stress: biological role, metabolism, and regulation. Russian J Plant Physiol 53:583–604
Kwon SY, Choi SM, Ahn YO, Lee HS, Lee HB et al (2003) Enhanced stress-tolerance of transgenic tobacco plants expressing a human dehydroascorbate reductase gene. J Plant Physiol 160:347–353
Laloi C, Apel K, Danon A (2004) Reactive oxygen signalling: the latest news. Curr Opin Plant Biol 7:323–328
Lamotte O, Gould K, Lecourieux D, Sequeira-Legrand A, Lebrun-Garcia A et al (2004) Analysis of nitric oxide signaling functions in tobacco cells challenged by the elicitor cryptogein. Plant Physiol 135:516–529
Le Martret B, Poage M, Shiel K, Nugent GD, Dix PJ (2011) Tobacco chloroplast transformants expressing genes encoding dehydroascorbate reductase, glutathione reductase, and glutathione-S-transferase, exhibit altered anti-oxidant metabolism and improved abiotic stress tolerance. Plant Biotechnol J 9:661–673
Lee BH, Lee H, Xiong L, Zhu JK (2002) A mitochondrial complex I defect impairs cold-regulated nuclear gene expression. Plant Cell 14:1235–1251
Libourel IG, Bethke PC, De Michele R, Jones RL (2006) Nitric oxide gas stimulates germination of dormant Arabidopsis seeds: use of a flow-through apparatus for delivery of nitric oxide. Planta 223:813–820
Lichtscheidl IK, Url WG (1990) Organization and dynamics of cortical endoplasmic reticulum in inner epidermal cells of onion bulb scales. Protoplasma 157:203–215
Ligterink W, Hirt H (2001) Mitogen-activated protein (Map) kinase pathways in plants: versatile signaling tools. Int Rev Cytol 201:209–275
Liu JH, Moriguchi T (2007) Changes in free polyamine titers and expression of polyamine biosynthetic genes during growth of peach in vitro callus. Plant Cell Rep 26:125–131

Liu J-J, Ekramoddoullah AKM, Yu X (2003) Differential expression of multiple Pr10 proteins in western white pine following wounding, fungal infection and cold-hardening. Physiol Plant 119:544–553
Liu Y, Schiff M, Dinesh-Kumar SP (2004) Involvement of Mek1 Mapkk, Ntf6 Mapk, Wrky/Myb transcription factors, Coi1 and Ctr1 in N-mediated resistance to tobacco mosaic virus. Plant J 38:800–809
Lopez-Matas MA, Nunez P, Soto A, Allona I, Casado R et al (2004) Protein cryoprotective activity of a cytosolic small heat shock protein that accumulates constitutively in chestnut stems and is up-regulated by low and high temperatures. Plant Physiol 134:1708–1717
Lunn JE, Feil R, Hendriks JH, Gibon Y, Morcuende R et al (2006) Sugar-induced increases in trehalose 6-phosphate are correlated with redox activation of Adpglucose pyrophosphorylase and higher rates of starch synthesis in Arabidopsis thaliana. Biochem J 397:139–148
Ma Q, Dai X, Xu Y, Guo J, Liu Y et al (2009) Enhanced tolerance to chilling stress in OSMYB3R–2 transgenic rice is mediated by alteration in cell cycle and ectopic expression of stress genes. Plant Physiol 150:244–256
Magneschi L, Perata P (2009) Rice germination and seedling growth in the absence of oxygen. Ann Bot 103:181–196
Mahajan S, Tuteja N (2005) Cold, salinity and drought stresses: an overview. Arch Biochem Biophys 444:139–158
Mandelkow EM, Drewes G, Biernat J, Gustke N, Van Lint J et al (1992) Glycogen synthase kinase-3 and the Alzheimer-like state of microtubule-associated protein tau. FEBS Lett 314:315–321
Mantri NL, Ford R, Coram TE, Pang EC (2007) Transcriptional profiling of chickpea genes differentially regulated in response to high-salinity, cold and drought. BMC Genomics 8:303
Marc J (1997) Microtubule-organizing centres in plants. Trends Plant Sci 2:223–230
Maruyama K, Sakuma Y, Kasuga M, Ito Y, Seki M et al (2004) Identification of cold-inducible downstream genes of the Arabidopsis Dreb1a/Cbf3 transcriptional factor using two microarray systems. Plant J 38:982–993
Mathur J, Chua NH (2000) Microtubule stabilization leads to growth reorientation in Arabidopsis trichomes. Plant Cell 12:465–477
Matsumura T, Tabayashi N, Kamagata Y, Souma C, Saruyama H (2002) Wheat catalase expressed in transgenic rice can improve tolerance against low temperature stress. Physiol Plant 116:317–327
Mazars C, Thion L, Thuleau P, Graziana A, Knight MR et al (1997) Organization of cytoskeleton controls the changes in cytosolic calcium of cold-shocked Nicotiana plumbaginifolia protoplasts. Cell Calcium 22:413–420
McDowell JM, An YQ, Huang S, McKinney EC, Meagher RB (1996) The arabidopsis Act7 actin gene is expressed in rapidly developing tissues and responds to several external stimuli. Plant Physiol 111:699–711
Mcinnis SM, Emery DC, Porter R, Desikan R, Hancock JT et al (2006a) The role of stigma peroxidases in flowering plants: insights from further characterization of a stigma-specific peroxidase (Ssp) from Senecio squalidus (Asteraceae). J Exp Bot 57:1835–1846
Mcinnis SM, Desikan R, Hancock JT, Hiscock SJ (2006b) Production of reactive oxygen species and reactive nitrogen species by angiosperm stigmas and pollen: potential signalling crosstalk? New Phytol 172:221–228
McKersie BD, Murnaghan J, Jones KS, Bowley SR (2000) Iron-superoxide dismutase expression in transgenic alfalfa increases winter survival without a detectable increase in photosynthetic oxidative stress tolerance. Plant Physiol 122:1427–1438
Melillo MT, Leonetti P, Bongiovanni M, Castagnone-Sereno P, Bleve-Zacheo T (2006) Modulation of reactive oxygen species activities and H_2O_2 accumulation during compatible and incompatible tomato–root-knot nematode interactions. New Phytol 170:501–512
Mikami K, Murata N (2003) Membrane fluidity and the perception of environmental signals in cyanobacteria and plants. Prog Lipid Res 42:527–543

Mishina TE, Lamb C, Zeier J (2007) Expression of a nitric oxide degrading enzyme induces a senescence programme in Arabidopsis. Plant Cell Environ 30:39–52

Monroy AF, Dhindsa RS (1995) Low-temperature signal transduction: induction of cold acclimation-specific genes of alfalfa by calcium at 25 degrees C. Plant Cell 7:321–331

Monroy AF, Sarhan F, Dhindsa RS (1993) Cold-induced changes in freezing tolerance, protein phosphorylation, and gene expression (evidence for a role of calcium). Plant Physiol 102:1227–1235

Monroy AF, Sangwan V, Dhindsa RS (1998) Low temperature signal transduction during cold acclimation: protein phosphatase 2A as an early target for cold-inactivation. Plant J 13:653–660

Moore B, Zhou L, Rolland F, Hall Q, Cheng W-H et al (2003) Role of the Arabidopsis glucose sensor Hxk1 in nutrient, light, and hormonal signaling. Science 300:332–336

Morishima-Kawashima M, Kosik KS (1996) The pool of map kinase associated with microtubules is small but constitutively active. Mol Biol Cell 7:893–905

Mullineaux PM, Creissen GP (1997) Glutathione reductase: regulation and role in oxidative stress. In: Scandalios J (ed) Oxidative stress and the molecular biology of antioxidant defenses (Monograph 34). Cold Spring Harbor Laboratory Press, New York, pp 667–714

Murata N (1989) Low-temperature effects on cyanobacterial membranes. J Bioenerg Biomembr 21:61–75

Nakashima K, Ito Y, Yamaguchi-Shinozaki K (2009) Transcriptional regulatory networks in response to abiotic stresses in Arabidopsis and grasses. Plant Physiol 149:88–95

Nayyar H (2003a) Accumulation of osmolytes and osmotic adjustment in water-stressed wheat (Triticum aestivum) and maize (Zea mays) as affected by calcium and its antagonists. Environ Exp Bot 50:253–264

Nayyar H (2003b) Calcium as environmental sensor in plants. Curr Sci 84:893–902

Nayyar H, Chander S (2004) Protective effects of polyamines against oxidative stress induced by water and cold stress in chickpea. J Agron Crop Sci 190:355–365

Nayyar H, Walia DP (2003) Water stress induced proline accumulation in contrasting wheat genotypes as affected by calcium and abscisic acid. Biologia Planta 46:275–279

Nayyar H, Chander K, Kumar S, Bains T (2005a) Glycine betaine mitigates cold stress damage in Chickpea. Agron Sustain Dev 25:381–388

Nayyar H, Bains T, Kumar S (2005b) Low temperature induced floral abortion in chickpea: relationship to abscisic acid and cryoprotectants in reproductive organs. Environ Exp Bot 53:39–47

Nayyar H, Bains TS, Kumar S (2005c) Chilling stressed chickpea seedlings: effect of cold acclimation, calcium and abscisic acid on cryoprotective solutes and oxidative damage. Env Exp Bot 54:275–285

Nayyar H, Bains TS, Kumar S, Kaur G (2005d) Chilling effects during seed filling on accumulation of seed reserves and yield of chickpea. J Sci Food Agric 85:1925–1930

Nayyar H, Kaur G, Kumar S, Upadhyaya HD (2007) Low temperature effects during seed filling on chickpea genotypes (Cicer arietinum L.): probing mechanisms affecting seed reserves and yield. J Agron Crop Sci 193:336–344

Neill SJ, Desikan R, Clarke A, Hurst RD, Hancock JT (2002) Hydrogen peroxide and nitric oxide as signalling molecules in plants. J Exp Bot 53:1237–1247

Neill S, Bright J, Desikan R, Hancock J, Harrison J et al (2008a) Nitric oxide evolution and perception. J Exp Bot 59:25–35

Neill S, Barros R, Bright J, Desikan R, Hancock J et al (2008b) Nitric oxide, stomatal closure, and abiotic stress. J Exp Bot 59:165–176

Nguyen HT, Leipner J, Stamp P, Guerra-Peraza O (2009) Low temperature stress in maize (Zea mays L.) induces genes involved in photosynthesis and signal transduction as studied by suppression subtractive hybridization. Plant Physiol Biochem 47:116–122

Nickel R, Schutte M, Hecker D, Scherer GFE (1991) The phospholipid platelet-activating factor stimulates proton extrusion in cultured soybean cells and protein phosphorylation and Atpase activity in plasma membranes. J Plant Physiol 139:205–211

O'kane D, Gill V, Boyd P, Burdon R (1996) Chilling, oxidative stress and antioxidant responses in *Arabidopsis thaliana* callus. Planta 198:371–377
Ohnishi S, Miyoshi T, Shirai S (2010) Low temperature stress at different flower developmental stages affects pollen development, pollination, and pod set in soybean. Environ Exp Bot 69:56–62
Orvar BL, Sangwan V, Omann F, Dhindsa RS (2000) Early steps in cold sensing by plant cells: the role of actin cytoskeleton and membrane fluidity. Plant J 23:785–794
Pandey S, Tiwari SB, Tyagi W, Reddy MK, Upadhyaya KC et al (2002) A Ca^{2+}/CaM-dependent kinase from pea is stress regulated and in vitro phosphorylates a protein that binds to AtCaM5 promoter. Eur J Biochem 269:3193–3204
Pastore A, Martin SR, Politou A, Kondapalli KC, Stemmler T et al (2007) Unbiased cold denaturation: low- and high-temperature unfolding of yeast frataxin under physiological conditions. J Am Chem Soc 129:5374–5375
Patterson BD, Murata T, Graham D (1976) Electrolyte leakage induced by chilling in Passiflora species tolerant to different climates. Aust J Plant Physiol 3:435–442
Paul MJ (2008) Trehalose 6-phosphate: a signal of sucrose status. Biochem J 412:e1–e2
Paul MJ, Primavesi LF, Jhurreea D, Zhang Y (2008) Trehalose metabolism and signaling. Annu Rev Plant Biol 59:417–441
Payton P, Webb R, Kornyeyev D, Allen R, Holaday AS (2001) Protecting cotton photosynthesis during moderate chilling at high light intensity by increasing chloroplastic antioxidant enzyme activity. J Exp Bot 52:2345–2354
Pehowich DJ, Macdonald PM, Mcelhaney RN, Cossins AR, Wang LC (1988) Calorimetric and spectroscopic studies of lipid thermotropic phase behavior in liver inner mitochondrial membranes from a mammalian hibernator. Biochemistry 27:4632–4638
Pei ZM, Murata Y, Benning G, Thomine S, Klusener B et al (2000) Calcium channels activated by hydrogen peroxide mediate abscisic acid signalling in guard cells. Nature 406:731–734
Penfield S (2008) Temperature perception and signal transduction in plants. New Phytol 179:615–628
Pitcher JA, Hall RA, Daaka Y, Zhang J, Ferguson SS et al (1998) The G protein-coupled receptor kinase 2 is a microtubule-associated protein kinase that phosphorylates tubulin. J Biol Chem 273:12316–12324
Pokorna J, Schwarzerova K, Zelenkova S, Petrasek J, Janotova I et al (2004) Sites of actin filament initiation and reorganization in cold-treated tobacco cells. Plant Cell Environ 27:641–653
Potikha TS, Collins CC, Johnson DI, Delmer DP, Levine A (1999) The involvement of hydrogen peroxide in the differentiation of secondary walls in cotton fibers. Plant Physiol 119:849–858
Pramanik MH, Imai R (2005) Functional identification of a trehalose 6-phosphate phosphatase gene that is involved in transient induction of trehalose biosynthesis during chilling stress in rice. Plant Mol Biol 58:751–762
Prasad TK (1996) Mechanisms of chilling-induced oxidative stress injury and tolerance in developing maize seedlings: changes in antioxidant system, oxidation of proteins and lipids, and protease activities. Plant J 10:1017–1026
Qiao W, Fan L-M (2008) Nitric oxide signaling in plant responses to abiotic stresses. J Integr Plant Biol 50:1238–1246
Ramon M, Rolland F (2007) Plant development: introducing trehalose metabolism. Trends Plant Sci 12:185–188
Rathinasabapathi B, Burnet M, Russell BL, Gage DA, Liao PC et al (1997) Choline monooxygenase, an unusual iron-sulfur enzyme catalyzing the first step of glycine betaine synthesis in plants: prosthetic group characterization and cDNA cloning. Proc Natl Acad Sci U S A 94:3454–3458
Renaut J, Lutts S, Hoffmann L, Hausman JF (2004) Responses of poplar to chilling temperatures: proteomic and physiological aspects. Plant Biol (Stuttg) 6:81–90
Reszka AA, Seger R, Diltz CD, Krebs EG, Fischer EH (1995) Association of mitogen-activated protein kinase with the microtubule cytoskeleton. Proc Natl Acad Sci U S A 92:8881–8885
Reszka AA, Bulinski JC, Krebs EG, Fischer EH (1997) Mitogen-activated protein kinase/extracellular signal-regulated kinase 2 regulates cytoskeletal organization and chemotaxis via catalytic and microtubule-specific interactions. Mol Biol Cell 8:1219–1232

Rhodes D, Hanson AD (1993) Quaternary ammonium and tertiary sulphonium compounds in higher plants. Ann Rev Plant Physiol Plant Mol Biol 44:357–384

Roberts DM, Harmon AC (1992) Calcium-modulated proteins: targets of intracellular calcium signals in higher plants. Ann Rev Plant Physiol Plant Mol Biol 43:375–414

Rolland F, Baena-Gonzales E, Sheen J (2006) Sugar sensing and signalling in plants: conserved and novel mechanisms. Ann Rev Plant Biol 57:675–709

Ruelland E, Cantrel C, Gawer M, Kader JC, Zachowski A (2002) Activation of phospholipases C and D is an early response to a cold exposure in Arabidopsis suspension cells. Plant Physiol 130:999–1007

Ruiz JM, Sanchez E, Garcia PC, Lopez-Lefebre LR, Rivero RM et al (2002) Proline metabolism and NAD kinase activity in greenbean plants subjected to cold-shock. Phytochemistry 59:473–478

Rymen B, Fiorani F, Kartal F, Vandepoele K, Inze D et al (2007) Cold nights impair leaf growth and cell cycle progression in maize through transcriptional changes of cell cycle genes. Plant Physiol 143:1429–1438

Sabehat A, Lurie S, Weiss D (1998) Expression of small heat-shock proteins at low temperatures. A possible role in protecting against chilling injuries. Plant Physiol 117:651–658

Sakamoto A, Murata N (2002) The role of glycine betaine in the protection of plants from stress: clues from transgenic plants. Plant Cell Environ 25:163–171

Samach A, Wigge PA (2005) Ambient temperature perception in plants. Curr Opin Plant Biol 8:483–486

Sanghera GS, Wani SH, Wasim B, Singh N (2011) Engineering Cold Stress Tolerance in Crop Plants. Current Genomics 12:30–43

Sangwan V, Foulds I, Singh J, Dhindsa RS (2001) Cold-activation of *Brassica napus* BN115 promoter is mediated by structural changes in membranes and cytoskeleton, and requires Ca^{2+} influx. Plant J 27:1–12

Sangwan V, Orvar BL, Beyerly J, Hirt H, Dhindsa RS (2002) Opposite changes in membrane fluidity mimic cold and heat stress activation of distinct plant MAP kinase pathways. Plant J 31:629–638

Schmidt HH, Walter U (1994) NO at work. Cell 78:919–925

Schwender J, Ohlrogge J, Shachar-Hill Y (2004) Understanding flux in plant metabolic networks. Curr Opin Plant Biol 7:309–317

Schwiebert EM, Mills JW, Stanton BA (1994) Actin-based cytoskeleton regulates a chloride channel and cell volume in a renal cortical collecting duct cell line. J Biol Chem 269:7081–7089

Seki M, Narusaka M, Ishida J, Nanjo T, Fujita M et al (2002) Monitoring the expression profiles of 7000 Arabidopsis genes under drought, cold and high-salinity stresses using a full-length cDNA microarray. Plant J 31:279–292

Seki M, Kamei A, Yamaguchi-Shinozaki K, Shinozaki K (2003) Molecular responses to drought, salinity and frost: common and different paths for plant protection. Curr Opin Biotechnol 14:194–199

Seki M, Satou M, Sakurai T, Akiyama K, Iida K et al (2004) RIKEN Arabidopsis full-length (RAFL) cDNA and its applications for expression profiling under abiotic stress conditions. J Exp Bot 55:213–223

Shevyakova N, Shorina M, Rakitin V, Kuznetsov V (2006) Stress-dependent accumulation of spermidine and spermine in the halophyte *Mesembryanthemum crystallinum* under salinity conditions. Russian J Plant Physiol 53:739–745

Shibasaki K, Uemura M, Tsurumi S, Rahman A (2009) Auxin response in Arabidopsis under cold stress: underlying molecular mechanisms. Plant Cell 21:3823–3838

Shimazaki K, Kinoshita T, Nishimura M (1992) Involvement of calmodulin and calmodulin-dependent myosin light chain kinase in blue light-dependent H + pumping by guard cell protoplasts from Vicia faba L. Plant Physiol 99:1416–1421

Shinozaki K, Yamaguchi-Shinozaki K (2000) Molecular responses to dehydration and low temperature: differences and cross-talk between two stress signaling pathways. Curr Opin Plant Biol 3:217–223

Sinensky M (1974) Homeoviscous adaptation–a homeostatic process that regulates the viscosity of membrane lipids in Escherichia coli. Proc Natl Acad Sci U S A 71:522–525
Singer S, Nicholson G (1972) The fluid mosaic model of the structure of cell membranes. Science 175:720–731
Smith AM, Stitt M (2007) Coordination of carbon supply and plant growth. Plant Cell Environ 30:1126–1149
Snedden WA, Fromm H (2001) Calmodulin as a versatile calcium signal transducer in plants. New Phytol 151:35–66
Sowinski P, Rudzinska-Langwald A, Adamczyk J, Kubica I, Fronk J (2005) Recovery of maize seedling growth, development and photosynthetic efficiency after initial growth at low temperature. J Plant Physiol 162:67–80
Stitt M, Hurry V (2002) A plant for all seasons: alterations in photosynthetic carbon metabolism during cold acclimation in Arabidopsis. Curr Opin Plant Biol 5:199–206
Stuehr DJ, Santolini J, Wang ZQ, Wei CC, Adak S (2004) Update on mechanism and catalytic regulation in the NO synthases. J Biol Chem 279:36167–36170
Su G-X, Zhang W-H, Liu Y-L (2006) Involvement of hydrogen peroxide generated by polyamine oxidative degradation in the development of lateral roots in soybean. J Integr Plant Biol 48:426–432
Su C-F, Wang Y-C, Hsieh T-H, Lu C-A, Tseng T-H et al (2010) A novel MYBS3-dependent pathway confers cold tolerance in rice. Plant Physiol 153:145–158
Sugiura R, Toda T, Dhut S, Shuntoh H, Kuno T (1999) The MAPK kinase Pek1 acts as a phosphorylation-dependent molecular switch. Nature 399:479–483
Sukharev SI, Sigurdson WJ, Kung C, Sachs F (1999) Energetic and spatial parameters for gating of the bacterial large conductance mechanosensitive channel, MscL. J Gen Physiol 113:525–540
Sun WH, Duan M, Shu DF, Yang S, Meng QW (2010) Over-expression of StAPX in tobacco improves seed germination and increases early seedling tolerance to salinity and osmotic stresses. Plant Cell Rep 29:917–926
Sung DY, Kaplan F, Lee KJ, Guy CL (2003) Acquired tolerance to temperature extremes. Trends Plant Sci 8:179–187
Suzuki N, Mittler R (2006) Reactive oxygen species and temperature stresses: a delicate balance between signaling and destruction. Physiol Plant 126:45–51
Suzuki I, Los DA, Murata N (2000a) Perception and transduction of low-temperature signals to induce desaturation of fatty acids. Biochem Soc Trans 28:628–630
Suzuki I, Los DA, Kanesaki Y, Mikami K, Murata N (2000b) The pathway for perception and transduction of low-temperature signals in Synechocystis. EMBO J 19:1327–1334
Suzuki K, Nagasuga K, Okada M (2008) The chilling injury induced by high root temperature in the leaves of rice seedlings. Plant Cell Physiol 49:433–442
Swamy PM, Smith BN (1999) Role of abscisic acid in plant stress tolerance. Curr Sci 76:1220–1227
Tan Z, Boss WF (1992) Association of phosphatidylinositol kinase, phosphatidylinositol monophosphate kinase, and diacylglycerol kinase with the cytoskeleton and F-Actin fractions of carrot (Daucus carota L.) cells grown in suspension culture: response to cell wall-degrading enzymes. Plant Physiol 100:2116–2120
Tayal D, Srivastava PS, Bansal KC (2005) Transgenic crops for abiotic stress tolerance. In: Srivastava PS, Narula A, Srivastava S (eds) Plant biotechnology and molecular markers. Springer, Netherlands, pp 346–365
Taylor TM, Davidson PM, Bruce BD, Weiss J (2005) Ultrasonic spectroscopy and differential scanning calorimetry of liposomal-encapsulated nisin. J Agric Food Chem 53:8722–8728
Teige M, Scheikl E, Eulgem T, Doczi R, Ichimura K et al (2004) The MKK2 pathway mediates cold and salt stress signaling in Arabidopsis. Mol Cell 15:141–152
Tester M, Bacic A (2005) Abiotic stress tolerance in grasses. From model plants to crop plants. Plant Physiol 137:791–793
Testerink C, Munnik T (2005) Phosphatidic acid: a multifunctional stress signaling lipid in plants. Trends Plant Sci 10:368–375

Thakur P, Kumar S, Malik JA, Berger JD, Nayyar H (2010) Cold stress effects on reproductive development in grain crops: an overview. Environ Exp Bot 67:429–443
Thion L, Mazars C, Thuleau P, Graziana A, Rossignol M et al (1996) Activation of plasma membrane voltage-dependent calcium-permeable channels by disruption of microtubules in carrot cells. FEBS Lett 393:13–18
Thomashow MF (1999) Plant cold acclimation: freezing tolerance genes and regulatory mechanisms. Annu Rev Plant Physiol Plant Mol Biol 50:571–599
Thomashow MF (2001) So what's new in the field of plant cold acclimation? Lots! Plant Physiol 125:89–93
Tian Y, Zhang H, Pan X, Chen X, Zhang Z et al (2011) Overexpression of ethylene response factor TERF2 confers cold tolerance in rice seedlings. Transgenic Res 20(4):857–866
Tilly BC, Edixhoven MJ, Tertoolen LG, Morii N, Saitoh Y et al (1996) Activation of the osmosensitive chloride conductance involves P21rho and is accompanied by a transient reorganization of the F-actin cytoskeleton. Mol Biol Cell 7:1419–1427
Tokishita S, Mizuno T (1994) Transmembrane signal transduction by the Escherichia coli osmotic sensor, EnvZ: intermolecular complementation of transmembrane signalling. Mol Microbiol 13:435–444
Trewavas AJ, Malhó R (1998) Ca^{2+} signalling in plant cells: the big network! Curr Opin Plant Biol 1:428–433
Tsvetkova NM, Horvath I, Torok Z, Wolkers WF, Balogi Z et al (2002) Small heat-shock proteins regulate membrane lipid polymorphism. Proc Natl Acad Sci U S A 99:13504–13509
Urao T, Yakubov B, Satoh R, Yamaguchi-Shinozaki K, Seki M et al (1999) A transmembrane hybrid-type histidine kinase in Arabidopsis functions as an osmosensor. Plant Cell 11:1743–1754
Van Breusegem F, Vranová E, Dat JF, Inzé D (2001) The role of active oxygen species in plant signal transduction. Plant Sci 161:405–414
Vaughn KC, Harper JD (1998) Microtubule-organizing centers and nucleating sites in land plants. Int Rev Cytol 181:75–149
Vaultier MN, Cantrel C, Vergnolle C, Justin AM, Demandre C et al (2006) Desaturase mutants reveal that membrane rigidification acts as a cold perception mechanism upstream of the diacylglycerol kinase pathway in Arabidopsis cells. FEBS Lett 580:4218–4223
Verma S, Mishra SN (2005) Putrescine alleviation of growth in salt stressed Brassica juncea by inducing antioxidative defense system. J Plant Physiol 162:669–677
Vigh L, Torok Z, Balogh G, Glatz A, Piotto S et al (2007) Membrane-regulated stress response: a theoretical and practical approach. Adv Exp Med Biol 594:114–131
Vinocur B, Altman A (2005) Recent advances in engineering plant tolerance to abiotic stress: achievements and limitations. Curr Opin Biotechnol 16:123–132
Vogel JT, Zarka DG, Van Buskirk HA, Fowler SG, Thomashow MF (2005) Roles of the CBF2 and ZAT12 transcription factors in configuring the low temperature transcriptome of Arabidopsis. Plant J 41:195–211
Volkmann D, Baluska F (1999) Actin cytoskeleton in plants: from transport networks to signaling networks. Microsc Res Tech 47:135–154
Wada H, Gombos Z, Murata N (1990) Enhancement of chilling tolerance of a cyanobacterium by genetic manipulation of fatty acid desaturation. Nature 347:200–203
Wang YJ, Zhang ZG, He XJ, Zhou HL, Wen YX et al (2003) A rice transcription factor OsbHLH1 is involved in cold stress response. Theor Appl Genet 107:1402–1409
Wang FZ, Wang QB, Kwon SY, Kwak SS, Su WA (2005) Enhanced drought tolerance of transgenic rice plants expressing a pea manganese superoxide dismutase. J Plant Physiol 162:465–472
Wendehenne D, Durner J, Klessig DF (2004) Nitric oxide: a new player in plant signalling and defence responses. Curr Opin Plant Biol 7:449–455
Wendehenne D, Courtois C, Besson A, Gravot A, Buchwalter A et al (2006) NO-based signaling in plants. In: Lamattina L, Polacco JC (eds) Nitric oxide in plant growth, development and stress physiology. Springer, Berlin, pp 35–51
Wildi B, Lütz C (1996) Antioxidant composition of selected high alpine plant species from different altitudes. Plant Cell Environ 19:138–146

Williamson RE (1991) Orientation of cortical microtubules in interphase plant cells. Int Rev Cytol 129:135–208
Wise MJ, Tunnacliffe A (2004) POPP the question: what do LEA proteins do? Trends Plant Sci 9:13–17
Wisniewski M, Webb R, Balsamo R, Close TJ, Yu X-M et al (1999) Purification, immunolocalization, cryoprotective, and antifreeze activity of PCA60: a dehydrin from peach (Prunus persica). Physiol Plant 105:600–608
Wisniewski M, Bassett C, Arora R (2004) Distribution and partial characterization of seasonally expressed proteins in different aged shoots and roots of 'Loring' peach (Prunus persica). Tree Physiol 24:339–345
Wood JM (1999) Osmosensing by bacteria: signals and membrane-based sensors. Microbiol Mol Biol Rev 63:230–262
Xiao W, Sheen J, Jang JC (2000) The role of hexokinase in plant sugar signal transduction and growth and development. Plant Mol Biol 44:451–461
Xin Z, Browse J (2000) Cold comfort farm: the acclimation of plants to freezing temperatures. Plant Cell Environ 23:893–902
Xiong L, Ishitani M (2006) Stress signal transduction: components, pathways, and network integration. In: Rai AK, Takabe T (eds) Abiotic stress tolerance in plants. Springer, Dordrecht
Xiong L, Ishitani M, Zhu JK (1999) Interaction of osmotic stress, temperature, and abscisic acid in the regulation of gene expression in Arabidopsis. Plant Physiol 119:205–212
Xue GP (2003) The DNA-binding activity of an AP2 transcriptional activator HvCBF2 involved in regulation of low-temperature responsive genes in barley is modulated by temperature. Plant J 33:373–383
Yadav SK (2010) Cold stress tolerance mechanisms in plants. A review. Agron Sustain Dev 30:515–527
Yamaguchi-Shinozaki K, Shinozaki K (2006) Transcriptional regulatory networks in cellular responses and tolerance to dehydration and cold stresses. Annu Rev Plant Biol 57:781–803
Yang T, Poovaiah BW (2002) Hydrogen peroxide homeostasis: activation of plant catalase by calcium/calmodulin. Proc Natl Acad Sci U S A 99:4097–4102
Yemets AI, Krasylenko YA, Lytvyn DI, Sheremet YA, Blume YB (2011) Nitric oxide signalling via cytoskeleton in plants. Plant Sci 181(5):545–554
Yoshida R, Kanno A, Kameya T (1996) Cool temperature-induced chlorosis in rice plants. II. Effects of cool temperature on the expression of plastid-encoded genes during shoot growth in darkness). Plant Physiol 112:585–590
Zeller G, Henz SR, Widmer CK, Sachsenberg T, Ratsch G et al (2009) Stress-induced changes in the Arabidopsis thaliana transcriptome analyzed using whole-genome tiling arrays. Plant J 58:1068–1082
Zemojtel T, Frohlich A, Palmieri MC, Kolanczyk M, Mikula I et al (2006) Plant nitric oxide synthase: a never-ending story? Trends Plant Sci 11:524–525; author reply 526–528
Zhang X, Miao YC, An GY, Zhou Y, Shangguan ZP et al (2001) K+ channels inhibited by hydrogen peroxide mediate abscisic acid signaling in Vicia guard cells. Cell Res 11:195–202
Zhao L, Liu F, Xu W, Di C, Zhou S et al (2009a) Increased expression of OsSPX1 enhances cold/subfreezing tolerance in tobacco and Arabidopsis thaliana. Plant Biotechnol J 7:550–561
Zhao MG, Chen L, Zhang LL, Zhang WH (2009b) Nitric reductase-dependent nitric oxide production is involved in cold acclimation and freezing tolerance in Arabidopsis. Plant Physiol 151:755–767
Zheng L, Shan J, Krishnamoorthi R, Wang X (2002) Activation of plant phospholipase Dβ by phosphatidylinositol 4,5-bisphosphate: characterization of binding site and mode of action. Biochemistry 41:4546–4553
Zhou G, Bao ZQ, Dixon JE (1995) Components of a new human protein kinase signal transduction pathway. J Biol Chem 270:12665–12669
Zhu J, Shi H, Lee BH, Damsz B, Cheng S et al (2004) An Arabidopsis homeodomain transcription factor gene, HOS9, mediates cold tolerance through a CBF-independent pathway. Proc Natl Acad Sci U S A 101:9873–9878

Chapter 3
Drought and Salinity Tolerant Biofuel Crops for the Thar Desert

Karan Malhotra, Gulshan K. Chhabra, Rachana Jain, Vinay Sharma, and Shashi Kumar

1 Introduction

The depleting fossil fuel reserves and concomitant upward trend in the oil prices has made energy security an important global policy issue for more than four decades. The heavy dependence on nonrenewable sources of energy like oil, natural gas, and coal, which fulfill almost 80% of world's supply of primary energy needs (IEA 2007), has posed serious environmental concerns and has threatened energy security. One of the major problems for developing nations like India is to strike a balance between their growing energy demands and economic growth without affecting the environment. Alternative sources of biofuels can play an important role to meet India's future energy needs and reducing the dependence on oil imports. The biofuel policy encourages use of renewable resources as alternate fuel to supplement transport fuel and proposed an indicative target of 20% blending of biofuel (biodiesel and bioethanol) by 2017 (Pohit et al. 2011). In India, bioethanol and biodiesel is primarily produced from fermentation of sugar molasses and from seeds of *Jatropha* and *Pongamia*. An estimate suggests that around 60% of the ethanol produced in the world comes from corn and rest from sugarcane (ADB report 2011; Shinoj et al. 2011).

Growing bioenergy crops on marginal lands will be crucial in future for food security as well as reducing the liability of fuel import, and it offers an attractive way for retaining the arable land for food crops and offers a new source of income for poor farmers. The Thar Desert also known as "The Great Indian Desert" is world's ninth largest subtropical desert in the north-western part of India, is spread over an area of 0.2 million km^2, and is the one of the most heavily populated desert

K. Malhotra • G.K. Chhabra • S. Kumar (✉)
Synthetic Biology and Biofuel, International Center for Genetic Engineering & Biotechnology, Aruna Asaf Marg, New Delhi 110067, India
e-mail: skrhode@icgeb.res.in

R. Jain • V. Sharma
Department of Bioscience and Biotechnology, Banasthali University, Banasthali, Tonk, Rajasthan 304022, India

N. Tuteja and S. Singh Gill (eds.), *Plant Acclimation to Environmental Stress*,
DOI 10.1007/978-1-4614-5001-6_3,

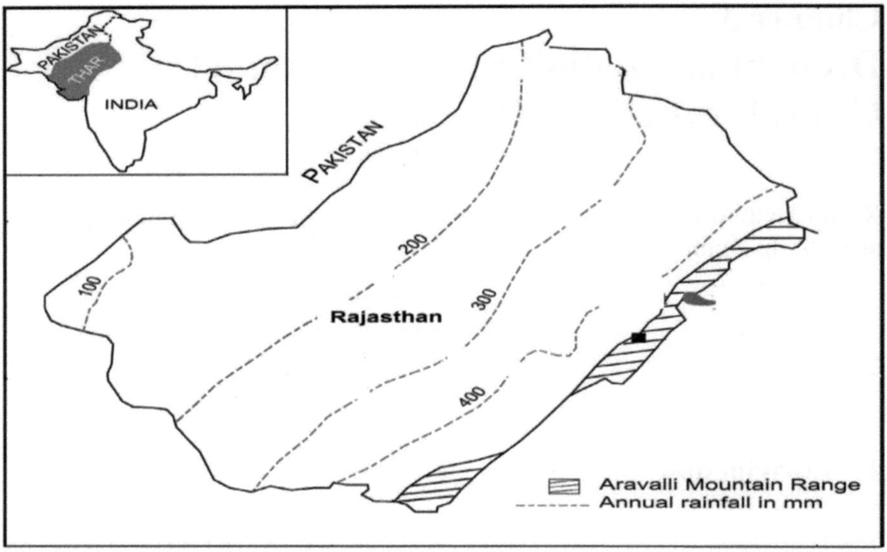

Fig. 3.1 Topo-geographical map showing location of Thar Desert in India

in the world (Fig. 3.1). About 720,000 ha desert area is saline and is used for production of table salt through open pits (subsoil) or wells (underground). Due to the high salt conditions, plants in this region have adapted to withstand high salt concentrations. It could be potentially used for the cultivation of bioenergy crops like *Jatropha*, Jojoba, Guayule, Sweet Sorghum, and Pearl millet, as these plants are well suited to grow in arid deserts and have special growth characteristics such as low water consumption under high salinity and require less nitrogen for their growth. Further, salt tolerance in these bioenergy crops can be enhanced using the approach of genetic engineering technology; however mechanism of salt tolerance differs in different species (Munns and Tester 2008; Arora et al. 2010). The prospects of these bioenergy crops are discussed as following.

1.1 *Jatropha*

It is a multipurpose, drought resistant perennial plant that is native to Mexico and Central America and now cultivated panatropically (GRIN 2000; Jules and Paull 2008). *Jatropha* is non-feed, poisonous crop, especially used as a hedge crop in countryside. Its tap and lateral roots help in preventing the soil erosion (Achten et al. 2007) and its farming on wasteland reduces the surface run-off by a rapid increase in evaporation and transpiration (Heuvelmans et al. 2005). *Jatropha* is gaining interest as a renewable source of energy as it can easily survive in areas of low rainfall (200 mm per year), marginal productivity lands, and even grow on alkaline soils. It has about 27–40% inedible oil in seed, mainly composed of oleic

Fig. 3.2 Biofuels crops suitable for growing in the Thar Desert. (**a**) Jatropha fruits, (**b**) Sweet sorghum, (**c**) Perl millet, (**d**) Jojoba seeds and fruits, (**e**) Guayule

acid, linoleic acid, and palmitic acid (Pramanik and Tripathi 2005; Achten et al. 2007, 2008), which can be easily converted to biodiesel. The collection of seeds is easy as the plant is not too tall. The *Jatropha curcas* has been already under cultivation in some areas of Rajasthan (Fig. 3.2a) from past 7 years with the help of Indian government. Indian government has identified 400,000 km^2 (98 million acres) of land where *Jatropha* can be grown that will provide much needed employment to the rural poor of India. Interestingly, both public and private sector companies such as Indian oil corporation (IOC), IFFCO, ONGC, Emami, and TERI have approached Chhattisgarh and Andhra Pradesh governments for contract farming of *Jatropha*. Mission New Energy operates in over 15,000 villages across five states in India, cultivating over 194,000 acres of *jatropha* (Nobrega and Sinha 2008; Mariano 2011). The initiative taken for the large-scale *Jatropha* biodiesel production on barren lands could be one of the most practical options for increasing India's share in biofuels and transportation sector.

1.2 Sweet Sorghum (Sorghum bicolor (L.) Moench)

There is a variety of sorghum with high content of sugar and has a good potential as a feedstock that can be used as a fuel. Sweet sorghum can thrive under drier and warmer conditions, grown primarily for forage, silage, and syrup production. C_4 biochemical mechanism of this plant makes it more efficient to survive in drought conditions (Billa et al. 1997; Reddy and Reddy 2003). Sweet Sorghum is widely cultivated in United States on marginal lands for the purpose of ethanol production. The life cycle of the plant is short and it attains a height of 3–4 m within a period of

4 months (Dajue 2009). The stalk of the plant is rich in fermentable sugar and consists of glucose, fructose, and sucrose (Sipos et al. 2009). Ethanol production by yeast fermentation of sorghum juice has also been reported (Laopaiboon et al. 2007, 2009). It is also being considered as a substrate for hydrogen production (Antonopoulou et al. 2008).

1.3 Guayule (*Parthenium argentatum* Gray)

It is a perennial shrub belonging to the family Asteraceae and is native to Chihuahuan desert in the south-western United States and northern Mexico (Foster and Coffelt 2005; Jasso Cant et al. 1996). Commercially, *P. argentatum* is an excellent source of natural rubber and its latex finds wide use as medical products like surgical balloons. Other uses include production of termite resistant wood and resin based products (Nakayama et al. 2001). Recent energy crisis has gathered attention on evaluating its potential as an energy crop. The biomass of the entire plant has higher energy values (21.77 MJ/kg) compared to other plant biomass like corn and switch grass. Moreover, resins produced by guayule have energy values (37.90 MJ/kg) comparable to most other oilseed crops (Nakayama et al. 2003). In that way, guayule offers an economically viable biofuel option which does not compete with the food production and could be easily produced in desert area.

1.4 Pearl Millet (*Pennnisetum glaucum*)

This crop is well suited to grow in arid regions of India and grows well on sandy soils with low pH (Kumar 1989). It has higher protein and oil content than maize and sorghum (Burton et al. 1992). The crop can supplement maize feedstock for fuel production; however amount of ethanol produced is less but higher protein content makes it economically more feasible than corn as fermentation rates in pearl millet are 30% higher than corn (Wu et al. 2006).

1.5 Jojoba (*Simmondsia chinensis*)

It is native to Sonoran desert (Southwest USA and Northern Mexico) and is now widely used as a commercial crop. Seeds contain 50% of a light yellow, odorless wax commonly referred to as Jojoba oil (Toress et al. 2006). The oil consists of straight chain esters of monounsaturated long-chain fatty acids and long-chain primary fatty alcohols, in particular two ester molecules containing 40 and 42 carbon atoms which make up to 80% of the oil (Tobares et al. 2004). Minor amounts of fatty acids and alcohols, phytosterols, tocopherols, and trace amounts of free triacylglycerol have also been reported. The plants' drought and salt tolerant nature makes it a viable alternative in arid/semi-arid regions of the world (Mills et al. 1997) (Fig. 3.2d). It is increasingly being used as a constituent of biofuel. Selim et al. (2008) tested the efficacy of Jojoba as a fuel and the performance was compared by connecting an array of sensors to diesel engine which was run separately on regular diesel fuel and another on jojoba methyl ester made by adding a dash of methanol and a catalyst to raw jojoba oil. The fuel matched diesel for torque and power for engine speeds; they tested between 1,000 and 2,000 rpm. Jojoba offers as a more reliable source of biofuel because its oil contains less carbon than current fuel which means less green house gas emission. Moreover, the oil is completely devoid of sulphur which results in long-lasting engines. The oil has a higher flash point which makes it convenient to be stored and transportation (Selim et al. 2008).

2 Genetic Improvement for Salt and Drought Tolerance

High salinity and drought conditions are the major abiotic stresses that prevail in Thar Desert. Soil salinity, one of the most severe abiotic stresses, hampers agricultural productivity on nearly 20% of irrigated land worldwide (Rhoades et al. 1990). A saline soil typically has high sodium ion concentrations which are toxic to the plant and disrupts the ionic and osmotic equilibrium of the plant (Fig. 3.3). The response of plants to these stresses involves variations in the activity of numerous stress responsive genes which function in coordination to reestablish the homeostatic conditions. These harsh conditions permit the growth of only selected crops that are adapted to these stresses.

There are two approaches being used to improve salt tolerance in plants: one is through direct selection of plants capable of growing in stressful environments and another through development of transgenic plants by introduction of novel genes having role in improving salt tolerance or modifying the expression levels of existing

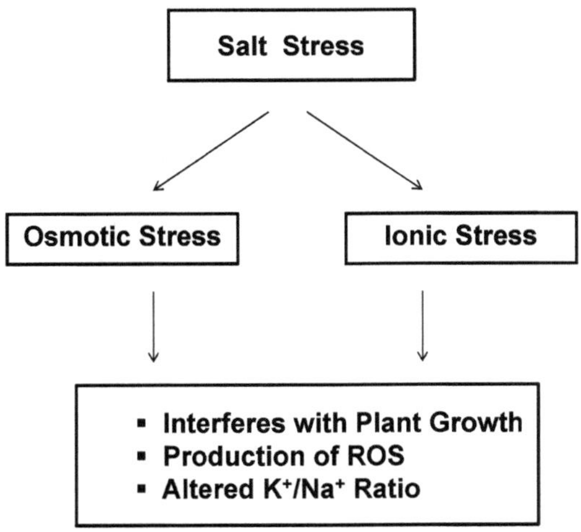

Fig. 3.3 Impact of salt stress: Saline soil negatively impacts crop growth by disrupting the ionic and osmotic equilibrium. Altered K^+/Na^+ ratio and production of reactive oxygen species (ROS) retard the growth of plant

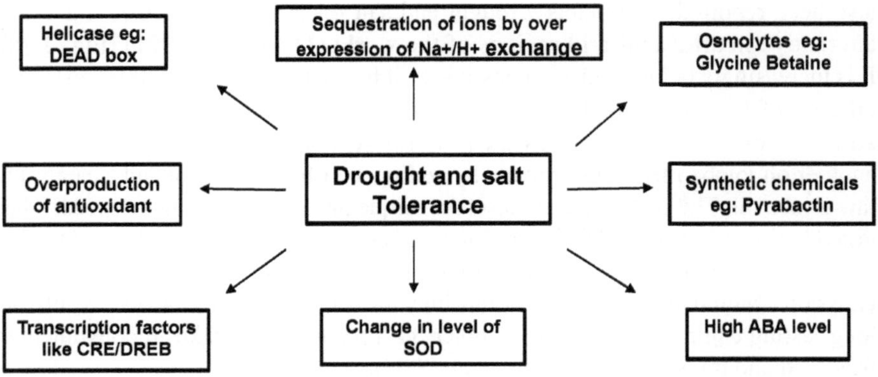

Fig. 3.4 Various components involved in drought and salt tolerance. Halophytes and glycophytes exhibit a variety of adaptations like vacuolar sequestration of ions, production of osmolytes, superoxide dismutase (SOD), and antioxidants. High level of abscisic acid (ABA) increases tolerance to drought. Engineering plants with transcription factors like CRE/DREB improves drought tolerance. Pyrabactin, a synthetic chemical that mimics ABA, helps in drought tolerance. The role of RNA helicases (DEAD-box) in abiotic stress is also beginning to emerge

genes to improve salt tolerance (Yamaguchi and Blumwald 2005). Ongoing research reveals that salt tolerance in halophytes and glycophytes involves a multitude of adaptations ranging from production of compatible solutes (osmolytes), sequestration of ions, changes in the level of superoxide dismutase (SOD), peroxidases, transcription factors, and preferential uptake of desirable ions or removal of the toxic ones. The role of RNA helicases in salt and drought stress is also emerging (Fig. 3.4).

As a result of salinity, there is a decrease in water flow towards plant roots due to reduction in the permeability of soil. Consequently, the permeability of plasmalemma drops resulting in fall of water potential. Under such dehydrating conditions, glycine betaine (GB), a quaternary ammonium compound encoded by betaine aldehyde dehydrogenase (BADH) gene, was found to accumulate in certain halophytes (Yancey 1994). This novel gene was also reported in *Jatropha curcas* (JcBD1) and expression levels were higher in leaves undergoing environmental stress. GB has dual functions acting not only as an osmoprotectant but also in maintaining protein and membrane conformations (Papageorgiou and Murata 1995; Hamilton and Heckathorn 2001). Overexpressing this enzyme in transformed *E. coli* has conferred resistance to salt (Zhang et al. 2008). Therefore, engineering glycine betaine in biofuel crops may offer a possible way to improve their efficiency to grow in harsh conditions of Thar Desert.

Saline conditions may also subject plants to oxidative stress leading to formation of reactive oxygen species (ROS) which have inhibitory effects on cell metabolism (Bowler et al. 1992). As a defence mechanism, the level of SOD increases to eliminate ROS. The activity of SOD increases about 1.5-fold in response to salt stress in transgenic as compared to control, implicating its role in abiotic stress (Tanaka et al. 1999).

In the absence of any salt regulating mechanisms, the internal salt concentrations can reach three times than that of external medium (Munns et al. 1983). The concentration of sodium ions above a certain threshold level is toxic to both halophytes and

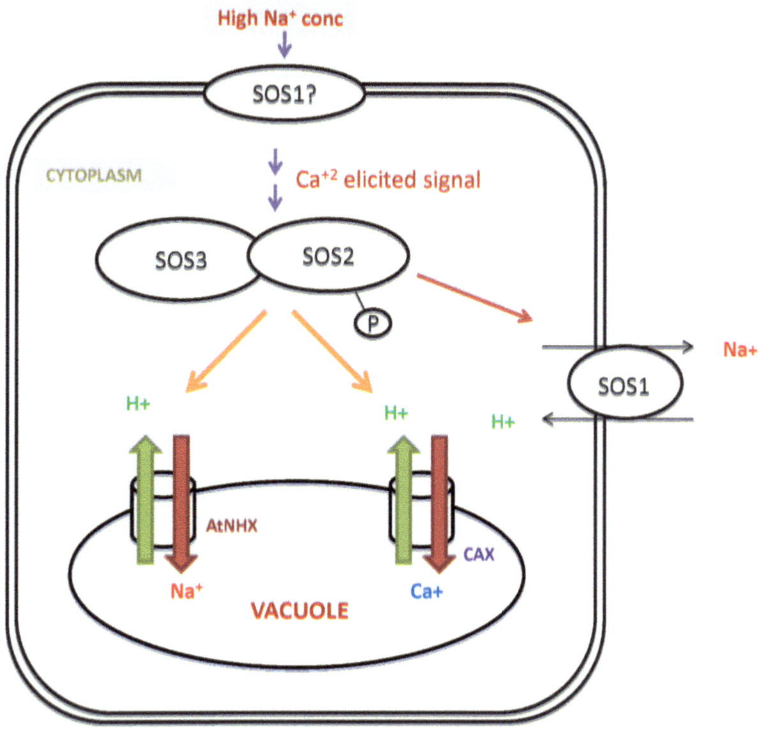

Fig. 3.5 Overview of SOS pathway in *Arabidopsis*. High Na_+ concentration sensed by plasma membrane transporter (probably SOS1) elicits a Ca^{+2} signal that activates SOS3–SOS2 complex, which stimulates Na^+/H^+ activity of SOS1 at plasma membrane. Further, SOS2 activates Na^+/H^+ and Ca^{+2}/H^+ activity on vacuolar membrane

glycophytes. Moreover, the enzymes present in the cytosol of both halophytes and glycophytes are sensitive to high salt concentrations suggesting that maintaining a high K^+/Na^+ cytosolic concentration is very essential for survival in saline soils. Halophytes achieve this ratio by increasing the efflux of sodium through ion channels or by compartmentalizing sodium ions in the vacuole. The genes involved in these processes have strong roles in improving salt tolerance in plants and appear as an attractive option for improving the growth of biofuel crops in Thar Desert.

The mechanism of salt overly sensitive (SOS) pathway (Fig. 3.5) and the genes involved in genetic improvement of plants for salt and drought tolerance is better studied in the mutants of *Arabidopsis*. In this pathway, SOS3 interacts with SOS2 and the complex of these acts on SOS1 which is a plasma membrane Na^+/H^+ antiporter that removes sodium out of the cell. SOS2 also activates Na^+/H^+ exchangers (AtNHX) on the vacuolar membrane. Under salt stress conditions, SOS1 mRNA accumulates in the cell. Overexpression of SOS1 gene results in crops that are more resistant to salt stress (Table 3.1).

Agriculture requires irrigation water in bulk amounts. The Thar Desert witnesses very less rainfall; therefore it is essential to develop crop varieties with increased

Table 3.1 Engineering salt tolerance in plants

Gene/protein	Gene/protein source	Transgenic species	Function	References
ABA3	*Arabidopsis*	*Nicotiana tabacum* cv. *Xanthi-nc*	Drought tolerance	Yue et al. (2011)
JcDREB	*Jatropha curcas*	*Arabidopsis*	Salt and freezing tolerance	Tang et al. (2011)
codA	*Arthrobacter globiformis*	*Lycopersicon esculentum*	Salt and water stress	Goel et al. (2011)
OsHAK5	Rice	Tobacco BY-2 cells	Salt tolerance	Horie et al. (2011)
nhaA	*Escherichia coli*	Rice	Salt and drought tolerance	Wu et al. (2005)
OsDREB1A	Rice	*Arabidopsis*	Drought, salt, and freeze tolerance	Dubouzet et al. (2003)
AtNHX1 (vacuolar Na^+/H^+ antiporter)	*Arabidopsis*	*B. napus*	Salt tolerance	Zhang et al. (2001)
AtSOS1 (PM Na^+/H^+ antiporter)	*Arabidopsis*	*Arabidopsis*	Salt tolerance	Shi et al. (2000)
AtNHX1 (vacuolar Na^+/H^+ antiporter)	*Arabidopsis*	*Arabidopsis*	Salt tolerance	Apse et al. (1999)
CBF1, DREB1	*Arabidopsis*	*Arabidopsis*	Drought and salt tolerance	Jaglo-Ottosen et al. (1998)

drought tolerance by using conventional breeding methods or by employing genetic engineering strategy. Conventional breeding, although a slow process, involves identifying variability in genes in relation to drought and introducing this tolerance into lines with suitable agronomic characteristics.

2.1 Abscisic Acid Dependent Approaches to Drought Tolerance

Development of drought tolerant varieties by genetic engineering involves identifying genes whose expression profiles change during drought. Research reveals that more than 100 genes are either induced or repressed in a plant during drought conditions (Sahi et al. 2006). The levels of abscisic acid (ABA) are enhanced in response to water stress resulting in closure of guard cells of stomata. This phenomenon is reversible as the level of ABA drops as soon as the water becomes available. Thus, ABA is an important target for improving drought tolerance in crops.

Most of the drought tolerant genes have been isolated from *Arabidopsis*. One example is *ERA1* which encodes for β (beta)-subunit of farnesyl transferase and has been known to play an important role in suppressing ABA signal (Pei et al. 1998). Mutants of *Arabidopsis* deficient in ERA1 showed increased tolerance to drought but the yields were severely compromised (Wang et al. 2005).

2.2 ABA Independent Approach

There is an ABA independent approach to drought tolerant signal pathway where transcription factors are involved in regulating the stress responsive genes. A family of transcription factors, called the C-repeat/dehydration responsive elements binding factor (CBF), converts environment signals into gene responses and recognizes cold and dehydration responsive elements (CRT/DRE). These elements have conserved 5-bp sequences that are found in the promoters of genes induced by cold and drought stress (Jaglo et al. 2001). Increase in drought tolerance of transgenic *Arabidopsis* was observed on overexpressing native form of DREB1 and constitutively active form of DREB2 (Table 3.1). These genes were found to be ubiquitous in many other crop plants indicating its conserved nature and make it an appropriate target for crop improvement through genetic engineering (Shinozaki and Yamaguchi-Shinozaki 2007).

2.3 Chemical Approach

Spraying ABA directly onto crops is being used to enhance their protection in times of stress. But ABA suffers from certain drawbacks of being a light sensitive chemical apart from high cost. A recent report suggests that a synthetic chemical Pyrabactin is able to enhance the drought tolerance in plants. As compared to ABA, it is far more stable and easy to produce and offers a highly effective chemical strategy for improving plants' ability to survive in drought areas. It mimics the effect of ABA by activating some of the ABA receptors in plants and also turns on ABA signaling pathway (Peterson et al. 2010; Rock et al. 2010).

Desert plants may also hold the key to biofuel crop farming in arid/semi-arid regions. *Kalanchoe fedtschenkoi*, a Madagascan plant, is known to survive in harsh conditions by capturing most of the carbon dioxide at night when the air is cooler and humid. This "nocturnal-CO_2-capturing ability" makes the plants ten times more water efficient compared to other crops (Borland et al. 2009). The novel genes present in this plant could be engineered in other biofuel crops too for improving their growth capability under low water conditions of the Thar Desert.

2.4 Role of RNA Helicases

Latest research reveals the role of RNA helicases in processes related to abiotic stresses. Though RNA helicases have been portrayed as molecular motors having roles in cellular process concerning RNA metabolism, their involvement in salinity and drought stress is also beginning to emerge (Owttrim 2006). Reports of two DEAD-box related helicases viz. pea DNA helicases 45 (PDH45) and PDH47 being induced in response to salinity stress suggest their roles in general stress response mechanism. Induction of PDH45 transcript in pea seedlings in response to salt

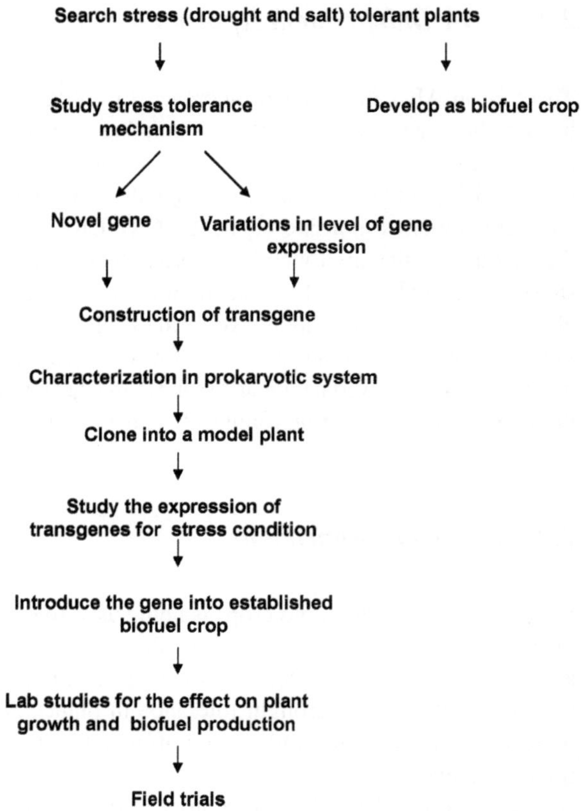

Fig. 3.6 Strategy for developing salt and drought tolerant crops

stress, dehydration, and low temperature suggests its role in abiotic stress. Moreover, constitutive expression of PDH45 in tobacco was found to make the plant resistant against salt stress (Sanan-Mishra et al. 2005; Tuteja 2007).

Another approach could involve studying gene expression and other change at cellular level in abiotic stress tolerant plant. By this method, we can identify genes that play critical role in stress tolerance and these genes can further be used to develop stress tolerant variety or in enhancement of stress tolerance in existing biofuel plants (Fig. 3.6).

3 Conclusions and Future Prospects of Desert Biofuel Crops

The paradigm for energy supply is shifting. To meet the growing demand for energy worldwide, we must identify regional biofuel solutions that are not only sustainable, but can actually regenerate the ecosystems where they are produced. We cannot afford to use arable lands for biofuel production on the compensation of food crops. The continuous increase in the world's population poses great pressure on increasing the production of food crops. Moreover, the transport sector is emerging as the largest consumer of liquid fuel worldwide. India is not self-sufficient to meet this growing demand and subsequently has to rely on oil import which is adversely affecting the country's foreign exchange.

The Thar Desert is one of the world's most densely populated desert and holds good prospects for growing biofuel crops like *Jatropha*, Jojoba, Sweet Sorghum, Pearl millet, etc. The region experiences frequent droughts, so majority of the commonly cultivated crops fail to grow. Growing biofuel crops that can withstand harsh conditions of drought and salinity of the Thar Desert will not only restore the ecosystem but will also bring back money into rural communities. However, growing crops that are better adapted to the abiotic stresses requires a deeper assessment. Most of the experiments for salt tolerance in transgenic were conducted in laboratory with limited number of plants and under controlled conditions. On the contrary, salinity in the field is variable. The plants are not exposed to alkaline soil pH, high diurnal temperatures, etc. that are typical to a saline soil. Apart from that, the performance of plant under salt stress needs to be evaluated with respect to yield.

Despite the slow progress in developing plants that are tolerant to salt and drought, there are several reasons for optimism. Development of several functional tools like molecular maps, express sequence tags (ESTs), availability of complete genome sequence of *Arabidopsis*, and understanding the mechanism of transgenes expression in chloroplasts have led to significant understanding of salt tolerance in plants. These findings can be directly applied for engineering the future biofuel crops relevant to Thar Desert.

References

Achten WMJ, Mathijs E, Verchot L et al (2007) Perspective: *Jatropha* biodiesel fueling sustainability? Biofuels Bioprod Bioref 1:283–291

Achten WMJ, Verchot L, Franken YJ et al (2008) Jatropha bio-diesel production and use. (a literature review). Biomass Bioenergy 32(12):1063–1084

ADB report (2011) http://re.indiaenvironmentportal.org.in/files/cross%20sectoral%20implications%20of%20biofuel.pdf

Antonopoulou G, Gavala HN, Skiadas IV et al (2008) Biofuels generation from sweet sorghum: fermentative hydrogen production and anaerobic digestion of the remaining biomass. Biores Technol 99:110–119

Apse MP, AharonGS SWA, Blumwald E (1999) Salt tolerance conferred by overexpression of a vacuolar Na^+/H^+antiport in Arabidopsis. Science 285:1256–1258

Arora J, Goyal S, Ramawat KG (2010) Biodiversity, biology and conservation of medicinal plants of the Thar Desert. In: Ramawat KG (ed) Desert plants biology and biotechnology. Springer, Heidelberg, pp 3–7

Billa E, Koullas DP, Monties B et al (1997) Structure and composition of sweet sorghum stalk components. Ind Crops Prod 6:297–302

Borland AM, Griffiths H, Hartwell J et al (2009) Exploiting the potential of plants with crassulacean acid metabolism for bioenergy production on marginal lands. J Exp Bot 60:2879–2896

Bowler C, Van Montagu M, Inze D (1992) Superoxide dismutase and stress tolerance. Annu Rev Plant Physiol 43:83–116

Burton GN, Wallace A, Rachie KO (1992) Chemical composition and nutritive value of pearl millet *Pennisetumtyphoides* (Burn) Staph and E C Hubbard grain. Crop Sci 12:187–188

Dajue L (2009) Sweet sorghum. In: El Bassam N (ed) Handbook of bioenergy crops, a complete reference to species, development and applications. Earthscan, London, pp 324–330

Dubouzet JG, Sakuma Y, Ito Y et al (2003) OsDREB genes in rice, Oryza sativa L., encode transcription activators that function in drought-, high-salt- and cold-responsive gene expression. Plant J 33:751–763

Foster MA, Coffelt TA (2005) Guayule agronomics: establishment, irrigated production, and weed control. Ind Crops Prod 22:27–40

Goel D, Singh AK, Yadav V et al (2011) Transformation of tomato with a bacterial codA gene enhances tolerance to salt and water stresses. J Plant Physiol 168:1286–1294

GRIN (2000) *Jatrophacurcas* information from Germplasm Resources Information Network. http://www.ars-grin.gov/cgi-bin/npgs/html/taxon.pl?20692. Accessed September 20, 2011

Hamilton EW, Heckathorn SA (2001) Mitochondrial adaptations to NaCl: Complex I is protected by anti-oxidants and small heat shock proteins, whereas complex II is protected by proline and betaine. Plant Physiol 126:1266–1274

Heuvelmans G, Muys B, Feyen J (2005) Extending the life cycle methodology to cover impacts of land use systems on the water balance. Int J Life Cycle Assoc 10:113–119

Horie T, Sugawara M, Okada T et al (2011) Rice sodium-insensitive potassium transporter, OsHAK5, confers increased salt tolerance in tobacco BY2 cells. J Biosci Bioeng 111:346–356

International Energy Agency (IEA), 2007. Key world energy statistics. Paris Cedex 15, France. http://www.iea.org/textbase/nppdf/free/2007/key_stats_2007.pdf. Accessed October 10, 2011

Jaglo KR, Kleff S, Amundsen KL, Zhang X et al (2001) Components of the Arabidopsis C-repeat/dehydration-responsive element binding factor cold-response pathway are conserved in Brassica napus and other plant species. Plant Physiol 127(3):910–917

Jaglo-Ottosen KR, Gilmour SJ, Zarka DG, Schabenberger O, Thomashow MF (1998) Arabidopsis CBF1 overexpression induces COR genes and enhances freezing tolerance. Science 280:104–106

Jasso Cantú D, Angulo Sánchez JL, Rodriguez García R (1996) Identification of guayule regions in northern Mexico, based on rubber yield and coproducts quality. In: Janick J (ed) Progress in new crops. Alexandria, ASHS, pp 336–339

Jules J, Paull RE (2008) The encyclopedia of fruits and nuts. CABI, Cambridge, pp 371–372

Kumar KA (1989) Pearl millet: current status and future potential. Outlook Agric 18:46–53

Laopaiboon L, Thanonkeo P, Jaisil P et al (2007) Ethanol production from sweet sorghum juice in batch and fed-batch fermentations by *S. cerevisiae*. World J Microbiol Biotechnol 23:1497–1501

Laopaiboon L, Nuanpeng S, Srinophakun P et al (2009) Ethanol production from sweet sorghum juice using very high gravity technology: effects of carbon and nitrogen supplementations. Bioresource Technol 100:4176–4182

Mariano KD. http://www.ecoseed.org/latest-news/article/102-latest-news/10652-indian-farmers-find-jobs-from-jatropha-production-for-biofuel. Accessed October 30, 2011.

Mills D, Wenkart S, Benzioni A (1997) Micropropagation of Simmondsia chinensis (Jojoba). In Biotechnology in Agriculture and Forestry. Ed., YPS, Bajaj. Springer-Verlag, Berlin. 40: 370–393

Munns R, Tester M (2008) Mechanisms of salinity tolerance. Annu Rev Plant Biol 59:651–681

Munns R, Greenway H, Kirst GO (1983) Halotolerant eukaryotes. In: Lange OL, Nobel PS, Osmond CB, Ziegler H (eds) Encyclopaedia of plant physiology, new series, vol 12C. Springer, New York, pp 60–131

Nakayama FS, Vinyard SH, Chow P et al (2001) Guayule as a wood preservative. Ind Crops Prod 14:105–111

Nakayama FS, Holt GA, Coffelt TA et al (2003) Pellet fuel from guayule plant material. Assoc. Advancement Industrial Crops. Annual Meeting, Portland, p. 62

Nobrega W, Sinha A (ed) (2008) Riding the Indian tiger: understanding India—the world's fastest growing market. Wiley, Hoboken, pp. 19–20

Owttrim GW (2006) RNA helicase and abiotic stress. Nucleic Acids Res 34:3220–3230

Papageorgiou GC, Murata N (1995) The unusually strong stabilizing effects of glycinebetaine on the structure and function in the oxygen-evolving photosystem II complex. Photosynthetica 44:243–252

Pei ZM, Ghaaaemian M, Kwak CM, McCourt P, Schroeder JI (1998) Role of farnesyltransferase in ABA regulation of guard cell anion channels and plant water loss. Science 282(5387):287–90

Peterson FC, Burgie ES, Park SY et al (2010) Structural basis for selective activation of ABA receptors. Nat Struct Mol Biol 17:1109–1113

Pohit S, Ashra S, Biswas PK (2011) Incentive structure of India's biofuel programs: status, shortcomings and implications. Biofuels 2:389–403

Pramanik T, Tripathi S (2005) Biodiesel—clean fuel of the future. Hydrocarbon Processing 84(2):49–54

Reddy BVS, Reddy SP (2003) Sweet sorghum: characteristics and potential. Int Sorghum Millets Newslett 44:26–28

Rhoades JD, Loveday J (1990) Salinity in irrigated agriculture. In: Steward BA, Nielsen DR (eds) American Society of Civil Engineers, irrigation of agricultural crops (Monograph 30). Madison, American Society of Agronomists, pp 1089–1142

Rock CD, Sakata Y, Quatrano RS (2010) Stress signaling I: the role of abscisic acid (ABA). In: Pareek A, Sopory SK, Bohnert HJ, Govindjee (eds). Abiotic stress adaptation in plants: Physiological, Molecular and Genomic Foundation. Springer, Dordrecht, pp. 33–73

Sahi C, Singh A, Blumwald E et al (2006) Beyond osmolytes and transporters: novel plant salt-stress tolerance-related genes from transcriptional profiling data. Physiol Plant 127(1):1–9

Sanan-Mishra N, Pham XH, Sopory SK, Tuteja N (2005) Pea DNA helicase 45 overexpression in tobacco confers high salinity tolerance without affecting yield. Proc Natl Acad Sci USA 102:509–514

Selim MYE, Radwan MS, Saleh HE (2008) Improving the performance of dual fuel engines running on natural gas/LPG by using pilot fuel derived from jojoba seeds. Renew Energy 33:1173–1185

Shi H, Ishitani M, Kim C et al (2000) The Arabidopsis thaliana salt tolerance gene SOS1 encodes a putative Na^+/H^+ antiporter. Proc Natl Acad Sci USA 97:6896–6901

Shinoj P, Raju SS, Joshi PK (2011) India's biofuels production programme: need for prioritizing the alternative options. Indian J Agric Res 81:391–397

Shinozaki K, Yamaguchi-Shinozaki K (2007) Gene networks involved in drought stress response and tolerance. J Exp Bot 58:221–227

Sipos B, Reczey J, Somorai Z et al (2009) Sweet sorghum as feedstock for ethanol production: enzymatic hydrolysis of steam-pretreated bagasse. ApplBiochemBiotechnol 153:151–162

Tang M, Liu X, Deng H et al (2011) Over-expression of JcDREB, a putative AP2/EREBP domain-containing transcription factor gene in woody biodiesel plant *Jatropha curcas*, enhances salt and freezing tolerance in transgenic Arabidopsis thaliana. Plant Sci 181:623–631

Tobares L, Frati M, Guzmán CA, Maestri DM (2004) Agronomic and chemical traits as descriptors for discrimination and selection of Jojoba (*Simmondsia chinensis*) clones. Ind Crops Prod 19:107–111

Tanaka Y, Hibino T, Hayashi Y, et al (1999) Salt tolerance of transgenic rice overexpressing yeast mitochondrial Mn-SOD in chloroplasts. Plant Sci 148:131–138

Torres MA, Jones JDG, Dang JL (2006) Reactive oxygen species signaling in response to pathogens. Plant Physiol 141:373–378

Tuteja N (2007) Mechanisms of high salinity tolerance in plants. Methods Enzymol 428:419–438

Wang Y, Ying J, Kuzma M et al (2005) Molecular tailoring of farnesylation for plant drought tolerance and yield protection. Plant J 43:413–424

Wu X, Wang D, Bean SD et al (2006) Ethanol production from pearl millet using *Saccharomyces cerevisiae*. Cereal Chem 83:127–131

Yamaguchi T, Blumwald E (2005) Developing salt-tolerant crop plants: Challenges and opportunities. Trends Plant Sci 10:615–620

Yancey PH (1994) Compatible and counteracting solutes. In: Strange K (ed) Cellular and molecular physiology of cell volume regulation. CRC, Boca Raton, pp 82–109
Yue Y, Zhang M, Zhang J, Duan L et al (2011) Arabidopsis LOS5/ABA3 overexpression in transgenic tobacco (*Nicotiana tabacum* cv. *Xanthi-nc*) results in enhanced drought tolerance. Plant Sci 181:405–411
Zhang HX, Hodson JN, Williams JP, Blumwald E (2001) Engineering salt tolerant Brassica plants: characterization of yield and seed oil quality in transgenic plants with increased vacuolar sodium accumulation. Proc Natl Acad Sci USA 98:12832–12836
Zhang FL, Niu B, Wang YC, et al (2008) A novel betaine aldehyde dehydrogenase gene from Jatropha curcas, encoding an enzyme implicated in adaptation to environmental stress. Plant Sci 174:510–518

Chapter 4
Strategies for the Salt Tolerance in Bacteria and Archeae and its Implications in Developing Crops for Adverse Conditions

Satya P. Singh, Vikram Raval, and Megha K. Purohit

1 Introduction

Life is based on chemistry which must be allowed to function for life to continue. Extremophiles adopt two distinct approaches within extreme environments; they might adapt to function in the physical and chemical limits of their environment or maintain mesophilic conditions intracellular, withstanding the external pressures. Among the extremophiles, halophiles are an interesting class of organisms adapted to moderate and hyper saline environments.

Halophiles are represented by all three domains of life. Among the bacteria; the phyla Cyanobacteria, Proteobacteria, Firmicutes, Actinobacteria, Spirochaetes, and Bacteroidetes are commonly reported. In Archaea, the most salt-requiring microorganisms belong to Halobacteria. Halobacterium and most of its relatives require over 100–150 g/L salt for growth and structural stability. Halophilic microorganisms use two strategies to balance their cytoplasm osmotically with their medium. The first involves accumulation of molar concentrations of KCl. This strategy requires adaptation of the intracellular enzymatic machinery, as proteins should maintain their proper conformation and activity at near-saturating salt concentrations (Oren 2008).

The minimum salt concentration required for growth, the salinity optimum, and the upper salt limit tolerated—within the microbial world highlighted towards a continuum of properties, which makes it nearly impossible to define by sharp boundaries. Moreover, the minimum, optimum, and maximum salt concentrations often depend on the medium composition and growth temperatures. The most widely used definitions were formulated 30 years ago (Kushner 1978) which distinguished halophiles into following categories: extreme halophiles (growing best in

S.P. Singh (✉) • V. Raval • M.K. Purohit
Department of Biosciences, Saurashtra University,
Rajkot 360 005, India
e-mail: satyapsingh@yahoo.com

N. Tuteja and S. Singh Gill (eds.), *Plant Acclimation to Environmental Stress*,
DOI 10.1007/978-1-4614-5001-6_4, © Springer Science+Business Media New York 2013

media containing 2.5–5.2 M salt), borderline extreme halophiles (growing best is media containing 1.5–4.0 M salt), moderate halophiles (growing best in media containing 0.5–2.5 M salt), and halotolerant microorganisms that do not require salt for growth but grow well up to reasonably high salt concentrations. The extreme halotolerant groups have growth range above 2.5 M salt. These definitions, though with loose boundaries, are valuable in the classification of microorganisms according to their relationship with salt (Oren 2002, 2006; Ventosa et al. 1998). Majority of halophilic organisms accumulate organic compounds, e.g. betaine, ectoines, and glycerol. These solutes could be quite useful for commercial applications.

Only few reports on the occurrence of plasmids and their physiological and ecological significance from haloalkaliphiles are available. Attempts are being made towards developing vectors and expression systems of halophilic origin to express their genes and investigate the regulation of gene expression.

Towards this end, exploration of diversity, phylogeny and biochemical and genetic characteristics of extracellular enzymes from haloalkaliphilic bacteria and actinomycetes dwelling in relatively moderate saline habitats have generated interesting clues on the functioning of enzymes under multitude of extremities (Dodia et al. 2008a, b; Gupta et al. 2005; Joshi et al. 2008; Nowlan et al. 2006; Patel et al. 2005, 2006; Purohit and Singh 2009; Ram et al. 2010; Siddhapura et al. 2010; Thumar and Singh 2007, 2009).

2 Strategies for Salt Adaptation

The halophilic proteins are highly acidic in nature and majority would denature in low salt concentration. The other strategies include exclusion of salt from the cytoplasm and to synthesize and accumulate organic "compatible" solutes that do not interfere with enzymatic activity. The organisms using the "organic-solutes-in strategy" often adapt to a surprisingly broad salt concentration range (Cho 2005). Most halophilic bacteria and the halophilic methanogenic archaea a number of such organic solutes like glycine betaine, ectoine and other amino acid derivatives, sugars and sugar alcohols. The *high-salt-in strategy* is not limited to the Halobacteriaceae. The Halanaerobiales (Firmicutes) also accumulate salt rather than organic solutes. A third, phylogenetically unrelated group of organisms accumulates KCl: the extremely halophilic *Salinibacter* (Bacteroidetes), recently isolated from saltern crystallizer brines. Analysis of its genome revealed its resemblance with Halobacteriaceae, which probably is a result of horizontal gene transfer. The example of *Salinibacter* indicates towards further discovery of unusual halophiles.

Bacteria and Archaea have developed two basic mechanisms to cope with osmotic stress. The "salt-in-cytoplasm mechanism" involves adjusting the salt concentration in the cytoplasm. The "organic-osmolyte mechanism" involves accumulation of uncharged and highly water-soluble organic compounds to maintain an osmotic equilibrium with the surrounding medium. The osmo-adaptation of prokaryotes through the organic-osmolyte strategy introduces a model of the fine-tuning of osmo regulatory osmolyte synthesis (Kunte 2006).

Mechanisms of adaptation of halophilic microorganisms at high salt points out towards Trüper's four "postulates," as presented in Alicante symposium (Truper et al. 1991). It deals with the presence, distribution and biosynthesis of organic osmotic solutes. A basic property of all halophilic microorganisms relate to the fact that their cytoplasm has to be atleast iso-osmotic with their surrounding medium. Biological membranes are permeable to water, and active energy-dependent inward transport of water to compensate for water loss by osmotic processes is energetically not feasible. Moreover, cells that keep a turgor need even to maintain their intracellular osmotic pressure higher than that of their environment (Brown 1976, 1990). There are two fundamentally different strategies used by halophilic microorganisms to balance their cytoplasm osmotically within their medium. The first involves accumulation of molar concentrations of potassium and chloride. This strategy *high-salt-in strategy* requires extensive adaptation of the intracellular enzymatic machinery to the presence of salt, as the proteins should maintain their proper conformation and activity at near-saturating salt concentrations (Lanyi 1974). The second strategy is to exclude salt from the cytoplasm and to synthesize and/or accumulate organic "compatible" solutes that do not interfere with enzymatic activity. Few adaptations of the cell's proteome are needed, and organisms using the *organic-solutes-in strategy* often adapt to a surprisingly broad salt concentration range. Most halophilic bacteria and some halophilic methanogenic archaea use such organic solutes. A variety of such solutes are known, including glycine betaine, ectoine and other amino acid derivatives, sugars and sugar alcohols. Far more widespread in nature is the second strategy of halo-adaptation based on the biosynthesis and/or accumulation of organic osmotic solutes. Cells that use this strategy exclude salt from their cytoplasm as much as possible. The high concentrations of organic "compatible" solutes do not greatly interfere with normal enzymatic activity. Such organisms can often adapt to a surprisingly broad salt concentration range (Ventosa et al. 1998). The list of organic compounds known to act as osmotic solutes in halophilic microorganisms—prokaryotic as well as _eukaryotic—is extensive. Most compatible solutes are based on amino acids and amino acid derivatives, sugars, or sugar alcohols. Many are either uncharged or zwitter ionic (Galinski 1986; Roberts 2005, 2006).

Although the "high-salt-in strategy" is energetically less costly to the cell than the biosynthesis of large amounts of organic osmotic solutes (Oren 1999), this strategy is not widely used among the different phylogenetic and physiological groups of halophiles. It is best known from the extremely halophilic Archaea of the family Halobacteriaceae, and species such as *Halobacterium salinarum* and *Haloarcula marismortui*. These organisms have emerged as popular model organisms to examine the implications of the accumulation of high intracellular KCl concentrations.

Our understanding of the biology of the Halobacteriaceae has greatly increased in recent years due to the elucidation and analysis of the genome sequences of *Halobacterium* NRC-1 (Kennedy et al. 2001; Ng et al. 2000), *Haloarcula marismortui* (Baliga et al. 2004), *Natronomonas pharaonis* (Falb et al. 2005), and *Halquadratum walsbyi* (Bolhuis et al. 2006). The strategy of salt adaptation is not limited to the aerobic halophilic archaea. The anaerobic fermentative Halanaerobiales

(Bacteria, Firmicutes) also use potassium chloride (KCl) rather than organic solutes to osmotically balance their cytoplasm and are also have adapted their intracellular machinery to tolerate the salt (Oren 1986, 2006).

3 Mechanisms for Stress Tolerance

The organisms living in extreme conditions possess special adaptation strategies that make them interesting not only for fundamental research but also towards exploration of their applications (Horikoshi 2008). These organisms may hold secret, for the origin of life and unfold many basic questions about the stability of the macromolecules, under extreme conditions. Therefore, their studies would provide important clues for adaptation under salinity. To cope with high and often changing salinity of their environment, the aerobic halophilic bacteria, similar to all other microorganisms, need to balance their cytoplasm with the osmotic pressure exerted by the external medium (Oren 2008, 2010). Osmotic balance can be achieved by the accumulation of salts, organic molecules, or similar mechanism. Alternatively, the cell is able to control water movement in and out and maintain a hypo osmotic state of their intracellular space.

The extremely halophilic archaea and bacteria, adopt various strategies; molar concentrations of chloride is pumped into the cells by co-transport with sodium ions and/or using the light-driven primary chloride pump halorhodopsin (Shazia 2004). Distribution of charged amino acids could also serve as one of the major approaches. Certain organisms show a specific requirement of chloride for growth, endospore germination, motility and flagellar synthesis, and glycine betaine transport (Muller and Oren 2003).

3.1 Chloride Pumps

A high requirement for chloride was demonstrated in two groups of bacteria; anaerobic *Halanaerobiales* and the aerobic extremely halophilic *Salinibacter rubber*, that accumulate inorganic salts intracellularly rather than using organic osmotic solutes. Thus, it is clear that chloride has specific functions in halo-adaptation in different groups of halophilic microorganisms (Muller and Oren 2003).

3.2 Osmoregulation

Osmoregulation is a fundamental phenomenon developed by bacteria, fungi, plants, and animals to overcome osmotic stress. The most widely distributed strategy of response to hyperosmotic stress is the accumulation of compatible solutes, which protect the cells and allow growth. Adaptation of bacteria to high solute concentrations involves intracellular accumulation of organic compounds called *osmolytes*.

Osmolytes are referred as compatible solutes because they can be accumulated to high intracellular concentrations without adversely affecting cellular processes. The solutes can be either taken up from the environment or synthesized *de novo*, and they act by counterbalancing external osmotic strength and thus preventing water loss from the cell and plasmolysis. Since cytoplasmic membrane is highly permeable to water, the imbalances imposed between turgor pressure and the osmolality gradient across the bacterial cell wall are of shorter duration. Bacteria respond to osmotic upshifts in three overlapping phases: dehydration (loss of some cell water); adjustment of cytoplasmic solvent composition and rehydration and cellular remodeling. Responses to osmotic downshifts also proceed in three phases: water uptake (phase I), extrusion of water and co-solvents (phase II), and cytoplasmic co-solvent re-accumulation and cellular remodeling (phase III) (Munns 2005).

3.3 Compatible Solutes

The accumulation of organic solutes is a prerequisite for osmotic adjustment of the organism. Archaea synthesize unusual solutes, such as β-amino acids, Nε-acetyl-β-lysine, mannosylglycerate, and di-*myo*-inositol phosphate. Among them, uptake of solutes such as glycine betaine is preferred over *de novo* synthesis. Most interestingly, some solutes are not only produced in response to salt but also to temperature stress (Muller et al. 2005).

3.4 Glycine Betaine

The ability of the organism to survive in both high salt concentrations and low temperatures is attributed mainly to the accumulation of the compatible solute glycine betaine, one of the most effective compatible solutes widely used by bacteria. This solute is N-trimethyl derivative of glycine and can be accumulated intracellularly at high concentration through synthesis, uptake, or both. *Bacillus subtilis* has been shown to possess three transport systems for glycine betaine: the secondary uptake system opuD and two binding-protein-dependent transport systems, opuA and opuC (proU). The secondary transport system betP is involved in glycine betaine accumulation in *Corynebacterium glutamicum* (Sleator et al. 1999). Further, characterization and disruption of betL, a gene which plays an important role in *glycine betaine* uptake in *L. monocytogenes* has been studied. Studies on some of the candidate genes from microbes for salinity tolerance highlight their functions. *L. monocytogenes* can survive a variety of environmental stresses. Growth at 10 % NaCl concentrations and temperature, as low as 20 °C has been reported. The ability of the organism to survive both high salt and low temperatures is attributed mainly to the accumulation of the compatible solute glycine betaine.

The genetic basis of glycine betaine uptake in other gram-positive bacteria has been studied extensively (Boscari et al. 2002).

3.5 *Distribution of Amino Acids*

The cell wall of halophilic archaea, *Halobacterium* has a high proportion of the acidic amino acids; aspartate and glutamate as sodium salts. Interestingly, this sodium binding is essential to maintain the cell wall and dilution of the medium leads to repulsion between the free carboxylate groups leading to cell wall disintegration and cell lysis (Bullock 2000). *Halococcus*, on the other hand, incorporates regular uronic acid residues, bearing charged sulfate groups (Madigan and Marrs 1997).

4 Molecular Aspects of Salinity

Marine microbes are known to play an essential role in the global cycling of nitrogen, carbon, oxygen, phosphorous, iron, sulfur, and trace elements (Nada et al. 2011). Salinity tolerance stems from the genetic regulation that limits the rate of salt uptake from the soil or water and the transport of salt adjust the ionic and osmotic balance of cells in roots and shoots and regulate leaf development (Munns and Tester 2008). However, only limited progress has been made with respect to the gene expression. Most of the sequenced culturable microorganisms from the deep-sea are *Alteromonadales* from the *Gammaproteobacteria.* Unique properties of sequenced deep-sea microbes indicated towards a high ratio of rRNA operon copies per genome size, and the fact that their intergenic regions are larger than average (Lauro and Bartlett 2008). These features characteristically relate to bacteria with an opportunistic lifestyle and gene regulation to respond to the rapidly changing environmental conditions.

Studies on the molecular basis of osmo-adaptation and its regulation in archaea are quite in infancy. However, genomics and functional genome analyses in conjunction with biochemistry shed light on the processes conferring to osmo-adaptation in archaea. The molecular characterization and disruption of betS, a gene which plays an important role in high-affinity Na-coupled glycine betaine and proline betaine transport in *S. meliloti* (Boscari et al. 2002), has been described. Furthermore, it has been shown that BetS is constitutively expressed, while its activity depends on posttranslational activation by high osmolarity. The emergency system transporting betaines for immediate osmotic protection appears to be quite significant. Many microorganisms possess two or more glycine betaine transport systems. *Salmonella typhimurium,* for example, operates two genetically distinct pathways; a constitutive low affinity system (ProP) and an osmotically induced high-affinity system (ProU), while *B. subtilis* has three glycine betaine transport systems, OpuD, OpuA, and OpuC (Kappes et al. 1996)

5 Glycine Betain: Synthesis and Cloning for Salt-Induced Stress Tolerance in Plants

Glycine betaine, a compatible solute, has ability to restore and maintain osmotic balance of living cells. It is synthesized and accumulated in response to abiotic stress. Betaine also acts as a methyl group donor and has a number of important applications including its use as a feed additive. The biosynthetic pathways of betaine are universal and well characterized. A number of enzymes catalyzing the two-step oxidation of choline to betaine have been isolated and characterized. Novel betaine biosynthetic pathway in two phylogenically distant extreme halophiles, *Actinopolyspora halophila* and *Ectothiorhodospira halochloris* have been described (Galinski and Truper 1994). There are three-step series of methylation reactions from glycine to betaine, catalyzed by two methyltransferases—glycine sarcosine methyltransferase and sarcosine dimethylglycine methyltransferase, with partially overlapping substrate specificity. The methyltransferases from the two organisms show high sequence homology. *E. halochloris* methyltransferase genes were successfully expressed in *E. coli*, leading to betaine accumulation and improved salt tolerance (Nyyssola et al. 2000).

The betaine biosynthetic pathway of *E. halochloris* expressed in *E. coli* led to the accumulation of betaine under moderate osmolarity. In addition, the cells were capable to grow with a higher cell density. The results clearly indicated that the methyltransferase pathway can be used to improve the osmotic tolerance of heterologous organisms. Drought and soil salinity are among the most important factors limiting crop productivity. Although certain plants synthesize betaine, several commercially important crops such as potato, rice, tomato and tobacco, do not accumulate betaine. The introduction of the choline oxidation pathway has been shown to increase salt and freezing tolerance of many plants (Holmberg 1996; Lilius et al. 1996). Thus, it will be interesting to compare the efficiency of the methyltransferase pathway with the choline oxidation pathway for improving stress tolerance in plants.

Genetic introduction of the betaine pathway into non-halotolerant plants would result in the accumulation of betaine and enhancement of their tolerance to salt stress (Rathinasabapathi et al. 1994). Transformation of tobacco with cDNA for betaine aldehyde dehydrogenase of spinach was carried out (Holmstrom et al. 1994). However, these transgenic plants required betaine aldehyde in the medium for the production of betaine (Lilius et al. 1996) reported transformation of tobacco with the gene for choline dehydrogenase from *E. coli*, which synthesizes betaine aldehyde from choline. Although the transformed plants demonstrated enhanced tolerance against salt stress, the accumulation of betaine in the transgenic plant was not indicated. Thus, further work would be required to substantiate the accumulation of organic solutes linked to salt tolerance.

Osmotic stress, being one of the most important environmental factors limiting plant productivity, is mainly caused by drought and salinity. Although irrigation increases crop yields, its application is not without problems. Water is a limited resource, which is also needed for other purposes. Furthermore, the accumulation of salts in soil due to irrigation has been a problem for agriculture for thousands of years

(Boyer, 1982). Consequently, there has been considerable interest in the genetic engineering of stress-tolerant plants. Single genes coding for the synthesis of compatible solutes, such as proline, polyols, trehalose, and betaine have therefore been introduced into many plants (Nuccio et al. 1999). Salt, drought (Gorham 1996; Rhodes and Hanson 1993), and cold stress (Kishitani et al. 1994; Naidu et al. 1991) have enhanced betaine accumulation in plants capable of its synthesis. In addition, there exists substantial experimental evidence indicating that betaine protects plant macromolecules against various stress factors. These findings have led to the assumption that the engineering of betaine synthesis into crop plants unable to synthesize it could be used to improve their stress tolerance (McCue and Hanson 1990; Park et al. 2004; Rudulier et al. 1984; Sakamoto and Murata 2000). Spinach choline mono-oxygenase and *E. coli* choline dehydrogenase have been introduced into tobacco (Lilius et al. 1996; Nuccio et al. 1998). Similarly, bacterial choline oxidases have been introduced into tobacco *Brassica napus* (Huang et al. 2000), *Arabidopsis* (Alia et al. 1998; Hayashi et al. 1997; Huang et al. 2000), and rice (Sakamoto et al. 1998). However, the levels of betaine in the transgenic plants have been significantly lower than their native sources. Although in some cases, improved stress tolerance has been reported. The supply of choline for betaine synthesis is limited, because it is converted almost exclusively to phosphatidyl choline. The choline synthesis itself is constrained at the first methylation step of phosphoethanolamine. As suggested by Nuccio et al. (1998, 1999). It is assumed that the main source of choline in non-producers is from the turnover of phosphatidylcholine. However, the free choline thus formed is rapidly and virtually irreversibly converted into phosphocholine, which acts as a reserve for the synthesis of phosphatidylcholine (Nuccio et al. 1998, 1999).

The codA gene for choline oxidase, the enzyme that converts choline into glycinebetaine, was earlier cloned from a soil bacterium, *Arthrobacter globiformis* into cyanobacterial species, *Synechococcus PCC7942* (Deshnium et al. 1995) but not into higher plants due to lack of sufficient information about the plant genes and promoter sequences. Transformation of *Arabidopsis thaliana* with the cloned codA gene under the control of the 35S promoter of cauliflower mosaic virus (CMV) enabled the plant to accumulate glycinebetaine with the enhanced tolerance against salt and cold stress. At 300 mM NaCl, considerable proportions of seeds of transformed plants germinated well, whereas seeds of wild-type plants failed to germinate. At 100 mM NaCI, transformed plants grew well whereas wild-type plants did not. The transformed plants tolerated 200 mM NaCI, which was lethal to wild-type plants. After incubating plants with 400 mM NaCI for 2 days, the photosystem II activity of wild-type plants disappeared almost completely whereas that of transformed plants remained at more than 50 % of the original level. When exposed to a low temperature in the light, leaves of wild-type plants exhibited symptoms of chlorosis, whereas those of transformed plants did not. These observations demonstrated that the genetic modification of *Arabidopsis thaliana* that allowed it to accumulate glycine-betaine enhanced its ability to tolerate salt and cold stress (Hayashi et al. 1997)

Brassica compestris L. spp. *Chinensis,* a vegetable crop widely cultivated in South China, does not synthesize betaine in vivo, and is sensitive to salt, drought, and high temperature stresses. Through the *Agrobacterium tumefaciens*-mediated transformation, the *coda* gene was transferred into the genome of *Brassica compestris* L. spp. *chinensis* var *Aikangqing* (Wang et al. 2010). The transgenic plants were evaluated and reported for their tolerance to temperatures and high salinity stresses by the assessment of their photosynthetic performance at the growth stage (Wang et al. 2010).

6 Ectoine and Hydroxyectoine: Synthesis and Cloning for Salt Tolerance in Plants

Ectoine, a cyclic tetrahydropyrimidine (1, 4, 5, 6-tetrahydro-2-methyl-4-pyrimidinecarboxylic acid) can be considered as marker for halophilic bacteria. It is synthesized by a wide range of bacteria, both halotolerant and halophilic. This solute was first detected in the halophilic, phototrophic *Halorhodospora halochloris* (Galinski et al. 1985). The intracellular ectoine concentration increased with increasing extracellular NaCl. Screens of a number of microorganisms have shown that ectoine is the major osmolyte in aerobic chemoheterotrophic bacteria (Galinski 1995). It is also the major solute in bacterial strains isolated from alkaline, hypersaline Mono Lake (Ciulla et al. 1997). More recently, it has been observed in the moderately halophilic methylotrophic bacteria; *Methylarcula marina*, *M. terricola*, and *Methylophaga* sp. (Doronina et al. 2000, 2003). A variant of this solute, hydroxyectoine, has been detected in halotolerant *Sporosarcina pasteurii* grown in high osmolarity medium (Kahulmann and Bremer 2002).

1, 4, 5, 6-Tetrahydro-2-methyl-4-pyrimidinecarboxylic acid (ectoine) functions as a compatible osmolyte in the moderate halophile *Halomonas elongata* OUT30018. Ectoine is biosynthesized by three successive enzyme reactions from aspartic-semialdehyde. The genes encoding the enzymes involved in the biosynthesis, ectA, ectB, and ectC, encoding L-2,4-diaminobutyric acid acetyltransferase, L-2,4-diaminobutyric acid transaminase, and L-ectoine synthase, respectively, have been previously cloned (Nakayama et al. 2000). To investigate the function of ectoine as a compatible solute in plant cells, the three genes were individually placed under the control of the CMV 35S promoter and introduced together into cultured tobacco (*Nicotiana tabacum L.*) cv Bright Yellow 2 (BY2) cells. The transgenic BY2 cells accumulated a small quantity of ectoine with the increased tolerance to hyperosmotic shock. Further, the transgenic BY2 cells exhibited normal growth under hyperosmotic conditions, in which the growth of wild was delayed. The results indicated that transgenic expressing ectoine resisted hyperosmotic conditions even at low level of the solute (Nakayama et al. 2000).

7 Nε-Acetyl-β-lysine and β-Glutamine (Nε-Acetyl-Beta-Lysine and Beta-Glutamine)

Methanogens have different strategy than many other organisms. They accumulate several β-amino acids (beta amino acids) to maintain osmotic balance. β-amino acids (beta amino acids) are not incorporated into proteins or other macromolecules.

At higher external NaCl concentrations, two zwiterionic β-amino acids (beta amino acids) get accumulated. Nε-acetyl-β-lysine (Nε-acetyl-beta-lysine and beta-glutamine) detected in a wide range of mesophilic and a few thermophilic methanogens (Robertson et al. 1992a, 1992b; Sowers et al. 1990; Sowers and Gunsalus 1995). β-Glutamine (beta-glutamine) has been detected in *Methanohalophilus* species where it is synthesized and accumulated along with Nε-acetyl-β-lysine and betaine (Lai et al. 1991). Methanogens tend to accumulate β-glutamate (beta-glutamate) and α-glutamate (alpha-glutamate) for osmotic balance.

Over the past few years, genes have been identified, cloned, and proteins isolated. The cloning and expression of genes confirmed pathways initially proposed on the basis of ^{13}C isotopic labeling of the solutes. The pathway originally proposed for biosynthesis of Nε-acteyl-β-lysine has two key enzymes: isomerization of α-lysine (alpha-lysine) to β-lysine (beta-lysine) catalyzed by a lysine aminomutase and acetylation of the ε-amino group (Roberts et al. 1992; Robertson et al. 1992b). Genes coding for these two enzymes were identified in *Methanosarcina mazei Gö;* ablA codes for the aminomutase while ablB codes for the β-lysine acetyltransferase (Pfluger et al. 2003). Expression of the two genes, organized in an operon, is salt dependent in *M. mazei*. Several other methanogens, including *Methanococcus maripaludis,* have homologous genes. Deletion of the *abl* operon in *M. maripaludis* hampered the growth in high salt medium. It will be interesting to characterize the methanogen lysine amino-mutase to compare it with the catabolic enzyme from bacteria having the same chemistry.

8 Conclusion

Environmental stresses such as drought, high salinity, and low temperature are major factors limiting plant growth and productivity by disturbing the intracellular water balance. Salinity is one of the major environmental factors that limit the worldwide productivity and distribution of cereal crops. Thus, the development of genetically engineered plants with enhanced tolerance to salt would be an important strategy. Among the major responses of plants towards acclimatization against unfavorable environments, such as water deficiency and high salinity is the accumulation of low molecular weight organic compounds, collectively known as compatible solutes. Most plants synthesize and accumulate osmolytes in response to these abiotic stresses. The osmolytes, or the compatible solutes, are neutral under physiological pH and have high solubility in water, besides being nontoxic to the organisms. Polyols (e.g. glycerol, sorbitol, and mannitol), nonreducing sugars (e.g. Sucrose and trehalose) and amino acids (e.g. Glutamate, Proline, and Betaine) are some of the known organic compatible solutes. Transgenic plants harboring genes for the biosynthesis of mannitol, ononitol, trehalose, proline, betaine, or fructan exhibited significant improvement in stress tolerance.

Majority of halophilic organisms accumulate organic compounds e.g. betaine, ectoines, and glycerol. These solutes could be quite useful for commercial applications. However, very few reports on plasmids from haloalkaliphiles are available but their physiological and ecological significance is yet to be established. Attempts are also being made towards developing vectors and expression systems of halophilic origin to express their genes and investigate the regulation of gene expression. Towards this end, exploration of diversity, phylogeny and biochemical and genetic characteristics of extracellular enzymes from haloalkaliphilic bacteria and actinomycetes dwelling in relatively moderate saline habitats have generated interesting clues on the functioning of enzymes under multitude of extremities The on-going studies on the cloning of multi enzyme systems into plants and regulation of gene expression relating to the synthesis of biocompatible solutes are likely to open new area for improved and large-scale production of these molecules in plants of economic interest helping them to fight against salt-induced stresses.

Archaea and Eubacteria produce several unique compatible solutes. The genes responsible for the synthesis of these solutes, if cloned, may prove to be of immense significance for developing draught and salinity stress tolerance in plants. However, it's important to know the detailed genetics under water stress. Although there are no substantial differences among the compatible solutes accumulated in bacteria and archaea, the regulation of osmo-adaptation is quite distinct due to unique transcription machineries among them. The increasing number of whole genome sequences, subsequent analysis by functional genomics, proteomics, and biochemical studies will pave way to better understanding of the regulatory network of osmo-adaptation and stress responses.

References

Alia HH, Chen THH, Murata N (1998) Transformation with a gene for choline oxidase enhances the cold tolerance of *Arabidopsis* during germination and early growth. Plant Cell Environ 21:232–239

Baliga NS, Bonneau R, Facciotti MT (2004) Genome sequence of *Haloarcula marismortui*: a halophilic archaeon from the Dead Sea. Genome Res 14:2221–2234

Bolhuis H, Palm P, Wende A (2006) The genome of the square archaeon *Haloquadratum walsbyi*: life at the limits of water activity. BMC Genomics 7:169

Boscari A, Mandon K, Dupont L (2002) *BetS* is a major glycine betaine/proline betaine transporter required for early osmotic adjustment in *Sinorhizobium meliloti*. J Bacteriol 184:2654–2663

Boyer JS (1982) Plant productivity and environment. Science 218:443–448

Brown AD (1976) Microbial water stress. Bacteriol Rev 40:803–846

Brown AD (1990) Microbial Water Stress Physiology. Principles and Perspectives. John Wiley & Sons, Chichester England

Bullock C (2000) The Archaea: a biochemical perspective. Biochem Mol Biol Edu 28:186–190

Cho BC (2005) Heterotrophic flagellates in hypersaline waters. In: Gunde-Cimerman N, Oren A, Plemenitas A (eds) Adaptation to life at high salt concentrations in Archaea, Bacteria, and Eukarya, vol 9, Cellular origin, life in extreme habitats and astrobiology. Springer, Netherlands, pp 541–549

Ciulla RA, Diaz MR, Taylor BF (1997) Organic osmolytes in aerobic bacteria from Mono Lake, an alkaline, moderately hypersaline environment. Appl Environ Microbiol 63:220–226

Deshnium P, Los DA, Hayashi H (1995) Transformation of *Synechococcus* with a gene for choline oxidase enhances tolerance to salt stress. Plant Mol Biol 29:897–907

Dodia MS, Rawal CM, Bhimani HG, Joshi RH, Khare SK, Singh SP (2008a) Purification and stability characteristics of an alkaline serine protease from a newly isolated *Haloalkaliphilic bacterium* sp. AH-6. J Ind Microbiol Biotechnol 35(2):121–132

Dodia MS, Bhimani HG, Rawal CM, Joshi RH, Singh SP (2008b) Salt dependent resistance against chemical denaturation of alkaline protease from a newly isolated *Haloalkaliphilic Bacillus sp.* Bioresour Technol 99:6223–6227

Doronina NV, Trotsenko YA, Tourova TP (2000) *Methylarcula marina gen. nov.* sp. *nov. and methylarcula terricola sp., nov.*: novel aerobic moderately halophilic facultatively methylotrophic bacteria from coastal saline environments. Int J Syst Evol Microbiol 50:1849–1859

Doronina NV, Darmaeva TD, Trotsenko YA (2003) *Methylophagaalcalica sp. nov.* A novel alkaliphilic and moderately halophilic obligately methylotrophic bacterium from an East Mangolian Saline soda lake. Int J Syst Evol Microbiol 53:223–229

Falb M, Pfeiffer F, Palm P (2005) Living with two extremes: conclusions from the genome sequence of *Natronomonas pharaonis*. Genome Res 15:1336–1343

Galinski EA (1986) Osmoadaptation in bacteria. Adv Microb Physiol 37:273–328

Galinski EA (1995) Osmoadaptation in bacteria. Adv Microb Physiol 37:272–328

Galinski EA, Truper HG (1994) Microbial behaviour in salt-stressed ecosystems. FEMS Microbiol Rev 15:95–108

Galinski EA, Pfeiffer HP, Truper HG (1985) 1, 4, 5, 6-Tetrahydro-2- methyl-4-pyrimidinecarboxylic acid. A novel cyclic amino acid from halophilic phototrophic bacteria of the genus *Ectothiorhodospira*. Eur J Biochem 149:135–139

Gorham J (1996) Mechanisms of salt tolerance of halophytes. In: Choukr-Allah R, Malcolm CV, Hamdy A (eds) Halophytes and biosaline agriculture. Dekker, New York, pp 31–53

Gupta A, Roy I, Patel RK, Singh SP, Khare SK, Gupta MN (2005) One-step purification and characterization of an alkaline protease from Haloalkaliphilic *Bacillus sp.* J Chromatogr A 1075:103–108

Hayashi H, Alia, Mustardy L (1997) Transformation of *Arabidopsis thaliana* with the *codA* gene for choline oxidase; accumulation of glycine-betaine and enhanced tolerance to salt and cold stress. Plant J 12(1):133–142

Holmberg N (1996) Metabolic engineering: approaches toward improved stress tolerance in microorganisms and plants. Ph.D. thesis, Lund University, Lund

Holmstrom KO, Welin B, Mandal A, Kristiansdottir I, Teeri TH, Lamark T, Strum AR, Palva ET (1994) Production of the *Escherichia coil* betaine-aldehyde dehydrogenase, an enzyme required for the synthesis of the osmoprotectant glycine betaine, in transgenic plants. Plant J 6:749–758

Horikoshi K (2008) Past, present and future of extremophiles. Extremophiles 12:1–2

Huang J, Hirji R, Adam L (2000) Genetic engineering of glycine-betaine production toward enhancing stress tolerance in plants: metabolic limitations. Plant Physiol 122:747–756

Joshi RH, Dodia MS, Singh SP (2008) Optimization of culture parameters for production of commercially valuable alkaline protease from a Haloalkaliphilic bacterium isolated from sea water. Biotechnol Bioprocess Eng 13:552–559

Kahulmann AU, Bremer E (2002) Osmotically regulated synthesis of compatible solute ectoine in *Bacillus pasteruii* and related *bacillus spp*. Appl Environ Microbiol 68:772–783

Kappes RM, Kempf B, Bremer E (1996) Three transport systems for the osmoprotectant glycine betaine operate in *Bacillus subtilis*: characterization of *OpuD*. J Bacteriol 178:5071–5079

Kennedy SP, Ng WV, Salzberg SL (2001) Understanding the adaptation of *Halobacterium species NRC-1* to its extreme environment through computational analysis of its genome sequence. Genome Res 11:1641–1650

Kishitani S, Watanabe K, Yasuda S (1994) Accumulation of glycine-betaine during cold-acclimation and freezing tolerance in leaves of winter and spring barley plants. Plant Cell Environ 17:89–95

Kunte HJ (2006) Osmoregulation in bacteria: compatible solute accumulation and osmosensing. Environ Chem 3:94–99
Kushner DJ (1978) Life in high salt and solute concentrations. In: Kushner DJ (ed) Microbial life in extreme environments. Academic, London, pp 317–368
Lai MC, Sowers KR, Robertson DE, Roberts MF, Gunsalus RP (1991) Distribution of compatible solutes in the halophilic methanogenic archaebacteria. J Bacteriol 173:5352–5358
Lanyi JK (1974) Salt dependent properties of proteins from extremely halophilic bacteria. Bacteriol Rev 38:272–290
Lauro FM, Bartlett DH (2008) Prokaryotic lifestyles in deep sea habitats. Extremophiles 12:15–25
Lilius G, Holmberg N, Bulow L (1996) Enhanced NaCl stress tolerance in transgenic tobacco expressing bacterial choline dehydrogenase. Biotechnology 14:177–180
Madigan MT, Marrs BL (1997) Extremophiles. Sci Am 276(4):66–71
McCue KF, Hanson AD (1990) Drought and salt tolerance: towards understanding and application. Trends Biotechnol 8:358–362
Muller V, Oren A (2003) Metabolism of chloride in halophilic prokaryotes. Extremophiles 7(4):261–266
Muller V, Spanheimer R, Santos H (2005) Stress response by solute accumulation in archaea. Curr Opin Microbiol 8(6):729–736
Munns R (2005) Genes and salt tolerance: bringing them together. New Phytol 167:645–663
Munns R, Tester M (2008) Mechanisms of Salinity Tolerance. Annu Rev Plant Biol 59:651–681
Nada AMK, Refaat MH, Abdel-Sabour MS (2011) Molecular studies on *EctC* gene (Ectoine) in some halophilic bacterial isolates. Researcher 3(2):34–42
Naidu BP, Paleg LG, Aspinall D (1991) Amino-acid and glycine betaine accumulation in cold-stressed wheat seedlings. Phytochemistry 30:407–409
Nakayama H, Yoshida K, Ono H (2000) Ectoine, the compatible solute of *Halomonas elongata*, confers hyperosmotic tolerance in cultured tobacco cells. Plant Physiol 122:1239–1247
Ng WV, Kennedy SP, Mahairas GG, Berquist B, Pan M, Shukla HD, Lasky SR, Baliga NS, Thorsson V, Sbrogna J et al (2000) Genome sequence of *Halobacterium species NRC-1*. Proc Natl Acad Sci 97(22):12176–12181
Nowlan B, Dodia MS, Singh SP, Patel BKC (2006) *Bacillus okhensis* sp. nov., a halotolerant and alkalitolerant bacterium from an Indian saltpan. Int J Syst Evol Microbiol 56(5):1073–1077
Nuccio M, Russell B, Nolte K, Rathinasabapathi B, Gage D, Hanson A (1998) The endogenous choline supply limits glycine betaine synthesis in transgenic tobacco expressing choline monooxygenase. Plant J 16(4):487–496
Nuccio ML, Russell BL, Nolte KD (1999) Metabolic engineering of plants for osmotic stress resistance. Curr Opin Plant Biol 2(2):128–134
Nyyssola A, Kerovuo J, Kaukinen P (2000) Extreme halophiles synthesize betaine from glycine by methylation. J Biochem 275(29):22196–22201
Oren A (1986) Intracellular salt concentrations of the anaerobic halophilic eubacteria *Haloanaerobium praevalens* and *Halobacteroides halobius*. Can J Microbiol 32:4–9
Oren A (1999) Bioenergetic aspects of halophilism. Microbiol Mol Biol Rev 63:334–348
Oren A (2002) Halophilic microorganisms and their environments. Kluwer Scientific, Dordrecht
Oren A (2006) Life at high salt concentrations. In: Dworkin M, Falkow S, Rosenberg E, Schleifer KH, Stackebrandt E (eds) The prokaryotes: a handbook on the biology of bacteria: ecophysiology and biochemistry, vol 2. Springer, New York, pp 263–282
Oren A (2008) Microbial life at high salt concentrations: phylogenetic and metabolic diversity. Saline Syst 4:2
Oren A (2010) Industrial and environmental applications of halophilic microorganisms. Environ Technol 31(8):825–834
Park EJ, Jeknic Z, Sakamoto A (2004) Genetic engineering of glycinebetaine synthesis in tomato protects seeds, plants, and flowers from chilling damage. Plant J 40:474–487

Patel RK, Dodia MS, Singh SP (2005) Extracellular alkaline protease from a newly isolated *Haloalkaliphilic Bacillus sp.*: production and optimization. Process Biochem 40(11): 3569–3575

Patel RK, Dodia MS, Singh SP (2006) Purification and characterization of alkaline protease from a newly isolated *Haloalkaliphilic Bacillus sp.* Process Biochem 41(9):2002–2009

Pfluger K, Baumann S, Gottschalk G (2003) Lysine-2, 3-aminomutase and β-lysine acetyltransferase genes of methanogenic archaea are salt induced and are essential for the biosynthesis of Nε-acetyl-β-lysine and growth in high salinity. Appl Environ Microbiol 69:6047–6055

Purohit MK, Singh SP (2009) Assessment of various methods for extraction of Metagenomic DNA from saline habitats of Coastal Gujarat (India) to explore Molecular Diversity. Lett Appl Microbiol 49:338–344

Ram K, Singh SP, Kapoor S, Khare SK (2010) A novel organic solvent tolerant alkaline protease from a newly isolated *Geomicrobium* sp. EMB2 (MTTC 10310): production optimization by response surface methodology. N Biotechnol 28(2):136–145

Rathinasabapathi B, McCue KR, Gage DA (1994) Metabolic engineering of glycine betaine synthesis: plant betaine aldehyde dehydrogenases lacking typical transit peptides are targeted to tobacco chloroplasts where they confer betaine aldehyde resistance. Planta 193:155–162

Rhodes D, Hanson AD (1993) Quaternary ammonium and tertiary sulfonium compounds in higher plants. Annu Rev Plant Physiol Plant Mol Biol 44:357–384

Roberts MF (2005) Organic compatible solutes of halotolerant and halophilic microorganisms. Saline Systems 1:5

Roberts MF (2006) Characterization of organic compatible solutes of halotolerant and halophilic microorganisms. In: Rainey FA, Oren A (eds) Methods in microbiology, extremophiles, vol 35. Academic, Amsterdam, pp 615–647

Roberts MF, Lai MC, Gunsalus RP (1992) Biosynthetic pathway of the osmolytes Nε-acetyl-β-lysine, β-glutamine, and betaine in *Methanohalophilus strain FDF1* suggested by nuclear magnetic resonance analyses. J Bacteriol 174:6688–6693

Robertson DE, Lai MC, Gunsalus RP, Roberts MF (1992a) Composition, variation, and dynamics of major compatible solutes in *Methanohalophilus strain FDF1*. Appl Environ Microbiol 58:2438–2443

Robertson DE, Noll D, Roberts MF (1992b) Free amino acid dynamics in marine methanogens. β-Amino acids as compatible solutes. J Biol Chem 267:14893–14901

Rudulier DL, Strom AR, Dandekar AM (1984) Molecular biology of osmoregulation. Science 224(4653):1064–1068

Sakamoto A, Murata N (2000) Genetic engineering of glycinebetaine synthesis in plants: current status and implications for enhancement for stress tolerance. J Exp Biol 51(342):81–88

Sakamoto A, Alia, Murata N (1998) Metabolic engineering of rice leading to biosynthesis of glycine-betaine and tolerance to salt and cold. Plant Mol Biol 38:1011–1019

Shazia A (2004). Evaluation of the role of salt tolerant bacteria in development resistance of plants to salt stress conditions. Ph.D. thesis, University of Punjab, Lahore.

Siddhapura P, Vanparia S, Purohit M, Singh SP (2010) Comparative studies on the extraction of metagenomic DNA from the saline habitats of Coastal Gujarat and Sambhar Lake, Rajasthan (India) in prospect of molecular diversity and search for novel biocatalysts. Int J Biol Macromol 47:375–379

Sleator RD, Gahan CGM, Abee T (1999) Identification and disruption of *BetL*, a secondary glycine betaine transport system linked to the salt tolerance of *Listeria monocytogenes LO28*. Appl Env Microbiol 65:2078–2083

Sowers KR, Gunsalus RP (1995) Halotolerance in *Methanosarcina spp.*: role of Nε-acetyl-β-lysine, β-glutamate, glycine betaine and K + as compatible solutes for osmotic adaptation. Appl Environ Microbiol 61:4382–4388

Sowers KR, Robertson DE, Noll D, Gunsalus RP, Roberts MF (1990) Nε-acetyl-β-lysine: an osmolyte synthesized by methanogenic bacteria. Proc Natl Acad Sci 87:9083–9087

Thumar JT, Singh SP (2007) Two - step purification of a highly thermostable alkaline protease from salt-tolerant alkaliphilic *Streptomyces clavuligerus* strain Mit-1. J Chromatogr B 854:198–203

Thumar JT, Singh SP (2009) Organic solvent tolerance of an alkaline protease from salt-tolerant alkaliphilic *Streptomyces clavuligerus* strain Mit-1. J Ind Microbiol Biotechnol 36:211–218

Truper HG, Severin J, Wolhfarth A, Muller E, Galinksi EA (1991) Halophily, taxonomy, phylogeny and nomenclature. In: Rodriguez-Valera F (ed) General and applied aspects of halophilism. Plenum, New York, pp 3–7

Ventosa A, Nieto JJ, Oren A (1998) Biology of aerobic moderately halophilic bacteria. Microbiol Mol Biol Rev 62:504–544

Wang Q, Xu W, Xue Q (2010) Transgenic *Brassica chinensis* plants expressing a bacterial *codA* gene exhibit enhanced tolerance to extreme temperature and high salinity. J Zhejiang Univ Sci B 11:851–861

Chapter 5
Adverse Effects of Abiotic Stresses on Medicinal and Aromatic Plants and Their Alleviation by Calcium

M. Naeem, M. Nasir Khan, M. Masroor A. Khan, and Moinuddin

1 Introduction

The role of mineral nutrition is of paramount importance for the cultivation of plants. In fact, yield of most crop plants increases linearly within limits with the amount of mineral nutrients that they absorb (Franz 1983; Loomis and Conner 1992). Using balanced mineral nutrition, the plant's maximum genetic potential can be realized successfully (Wallace and Wallace 2003; Khan and Mohammad 2006). The scientific cultivation may improve the yield and quality of plants including those of medicinal value. Application of optimal amounts of fertilizer may help to meet the increasing demands of medicinal plants to a great extent. This would augment the yield and quality of the medicinal herbs ensuring their steady supply in the market.

India is called the botanical garden of the world and the treasure house of plant biodiversity (Ahmedullah and Nayar 1999). It also displays a rich repository of medicinal plants and, hence, is the largest producer of medicinal herbs. Around 70% of India's medicinal plants are found in tropical areas, mostly in forests spread across the Western and Eastern Ghats, the Vindhyas, the Chota Nagpur plateau, the Aravalis, and the Himalyas. Rest of the Indian medicinal plants grow in temperate and alpine areas and on higher altitudes (Purohit and Vyas 2004; Seth and Sharma 2004). According to WHO, the international market of herbal products is US$62 billion, which is expected to reach US$5 trillion by the year 2050. However, India has presently less than 0.5% share in the global export market of medicinal plants in spite of being one of the largest producers of medicinal plants (Purohit and Vyas 2004; Bhattacharjee 2004).

M. Naeem (✉) • M.M.A. Khan • Moinuddin
Plant Physiology Section, Department of Botany, Aligarh Muslim University, Aligarh, 202 002, India
e-mail: naeem_phd@yahoo.co.in

M.N. Khan
Department of Biology, College of Science, University of Tabuk, Tabuk 71491, Kingdom of Saudi Arabia

N. Tuteja and S. Singh Gill (eds.), *Plant Acclimation to Environmental Stress*,
DOI 10.1007/978-1-4614-5001-6_5,

Because of no side effects, the interest in medicinal products of plant has increased considerably all over the world (Groenewegen et al. 1992; Lipp 1996). Due to a high demand of medicinal plants and herbal remedies in the world, the cultivation of these plants has significantly increased in the recent years. Since most medicinal plants are consumed raw, proper management of crop production is needed to achieve high quality plants (Hyden 2006). Among the factors responsible for achieving higher yield, adequate nutrient supply is considered one of the most effective tools (Munsi, 1992; Yazdani et al., 2004). Therefore, optimum concentrations of mineral nutrients should be set up to achieve the desired yield of MAPs.

Of the essential mineral nutrients, nitrogen, phosphorus, potassium, and calcium are considered to be of prime importance as they are required by plants in large quantities. These nutrients, by virtue of their function in the generation of energy, producing the building molecules, participating in the repair of protoplasm and regulation of metabolic process, and maintaining the physical organization and function of livings cells, play several important roles in enhancing crop production and maintaining the soil fertility. Balanced fertilizer application can play a vital role in sustaining high yield of medicinal plants and maintaining the fertility status of soils on long-term basis. Of the macro nutrient elements, calcium plays structural as well as physiological roles in plants. It is important in maintaining the stability of the cell walls and cell membranes and is also required in the maintenance of cell integrity with respect to membrane-bound proteins. Plants need calcium to continue normal function of their membranes, growth of the meristematic tissues and the youngest leaves, and to send signals in response to internal and external cues (Dordas, 2009). In fact, calcium (Ca) has attracted much interest in plant physiology and molecular biology because of its function as a second messenger in the signal conduction between environmental factors and plant responses in terms of growth, differentiation, and development (Price et al. 1994; Sanders et al. 1999; Rudd and Franklin-Tong 2001; Naeem et al. 2009a, b). Furthermore, calcium is highly required by medicinal legumes during nitrogen fixation processes.

Calcium is known to protect the integrity of cell membranes, reduce membrane permeability, and prevent ion leakage caused by environmental stresses. It has emerged as a key secondary messenger and signal transducer of various environmental stress stimuli. The effect of calcium and the changes in its signature depend on the particular environmental stress, the rate of stress development, previous exposure to stress conditions, and the tissue type (Bin 1995). The effects of Ca on shoot elongation, senescence process, and photosynthetic activity are dependent on its cytosolic concentration, which is governed by the activity of Ca channels in the plasma membrane (Knight 2000). When taken up into the plant system, Ca is involved in the regulation of plant responses to various abiotic stresses by contributing either directly or indirectly in plant defense mechanisms. Exogenously applied Ca alleviates the stresses caused by salt, heat, and drought by regulation of antioxidant activities such as GR, superoxide dismutase (SOD), AP, and CAT, decreasing the membrane lipid peroxidation (LPO), and helping the plant cells to survive during stress conditions (Jiang and Huang 2001a, b; Nayyar and Kaushal 2002; Fu and Huang 2003; Khan et al. 2010).

The literature regarding importance of calcium nutrient in crop improvement (medicinal legumes and other MAPs) under normal and abiotic stress conditions has been reviewed in this article.

2 Functions of Calcium in Plants

A wide variety of diverse biochemical functions are initiated by the changes in cellular Ca. Calcium is particularly important for structural stability and functional integrity of biological membranes and tissues. An adequate supply of Ca maintains membrane integrity and contributes to ion selectivity of membranes (Marschner 2002; Grattan and Grieve 1999; Mengel and Kirkby 2001; White and Broadley 2003). Calcium binds as pectate in middle lamella to strengthen the cell walls and plant tissues. The concentration of Ca in the cytosol is extremely low, ranging from 0.1 to 1.2 μM as free Ca. However, this low calcium concentrations is much essential for the functioning of certain key enzymes, including ATPases at the plasma membrane of roots and NADPH oxidases bound to plasma membrane (Marschner 2002). Calcium also regulates various responses of plants to biotic stresses caused by microbial or pathogen attacks (Yang and Poovaiah 2002; Grant and Loake 2000; Sagi and Fluhr 2001). During pathogen attacks in plants, a membrane-bound enzyme, resembling the neutrophil NADPH oxidase, was identified to contribute to the pathogen-induced oxidative burst of O_2^- in plants that indirectly leads to the generation of other reactive oxygen species (ROS). NADPH oxidases are the major source of ROS produced during the oxidative stress in plants (Lamb and Dixon 1997; Sagi and Fluhr 2001). Coordination between Ca and ROS production during pathogen attack is mediated mostly through NADPH oxidases. The activity of NADPH oxidase and the production of both O_2^- and H_2O_2 is enhanced in the presence of Ca either directly through the Ca binding domains (EF hands) located on the NADPH subunit (gp91phox) or indirectly by activating NAD kinase (Sagi and Fluhr 2001).

There are several climatic factors that either influence the availability of Ca in soil or restrict the transport of Ca inside the plant cells. The most common factor that influences the availability of Ca in soils is soil acidity (Goenaga and Smith 2002). Acid soils usually contain very high concentration of toxic Al, which limits availability as well as uptake of Ca and enhances precipitation of Ca in soils, causing occurrence of Ca deficiency in plants (Marschner, 1995; Mengel and Kirkby 2001; Goenaga and Smith 2002). Interestingly, under certain conditions, plants may suffer from Ca deficiency stress despite huge amounts of Ca in soils and high Ca concentration in plants. In such cases, the absorbance of Ca by roots or the translocation of Ca to the sink organs is prevented either by high humidity or due to the high salt concentration in the growth medium (Choi et al. 1997). For the production of fruits and vegetables with high quality, an adequate Ca supply is particularly important. If the xylem sap is low in Ca or the rate of transpiration of the fruits is poor, as occurs under humid conditions, inadequate levels of Ca might move into the fruits resulting in formation of Ca deficiency symptoms. In tomato, Ca deficiency disease is known as "blossom-end rot" and is characterized by a cellular breakdown of tissue at the distal end of the fruit (Saure 2001; Schmitz-Eiberger et al. 2002). In apple, Ca deficiency problem is called "bitter pit" that is characterized by the occurrence of small brown necrotic pitted spots (Mengel and Kirkby 2001). In vegetables, Ca deficiency causes the "tipburn" disorder in leaves (Saure 1998).

3 Importance of Calcium in Medicinal Legumes

Majority of the medicinal plants belong to angiospermic families, of which legume family (Fabaceae) is the third largest one, distributed in approximately 650 genera and 20,000 species. A handsome number of medicinal legumes are potential sources of glycosides (e.g. aloe-emodin, chrysophenol, emodin, and rhein, etc.), antibiotics, flavonoids, alkaloids, and phytochemicals (Fig. 5.1), which are used in drug manufacturing by various pharmaceutical industries (Tyler et al. 1976; Morris 1996, 1997, 1999, 2003). Legumes produce primary and secondary metabolites and the phytochemicals such as nutraceuticals, pharmaceuticals, pesticides, and industrial products. The natural products of leguminous plant have been and will continue to be important sources of forage, gums, insecticides, phytochemicals, and other industrial, medicinal, and agricultural raw materials. Many legumes have been used as folk medicines too (Duke 1992).

Generally legumes require high amount of phosphorus and calcium for their growth, nodule formation, and N_2-fixation. Hence, intensive research efforts should

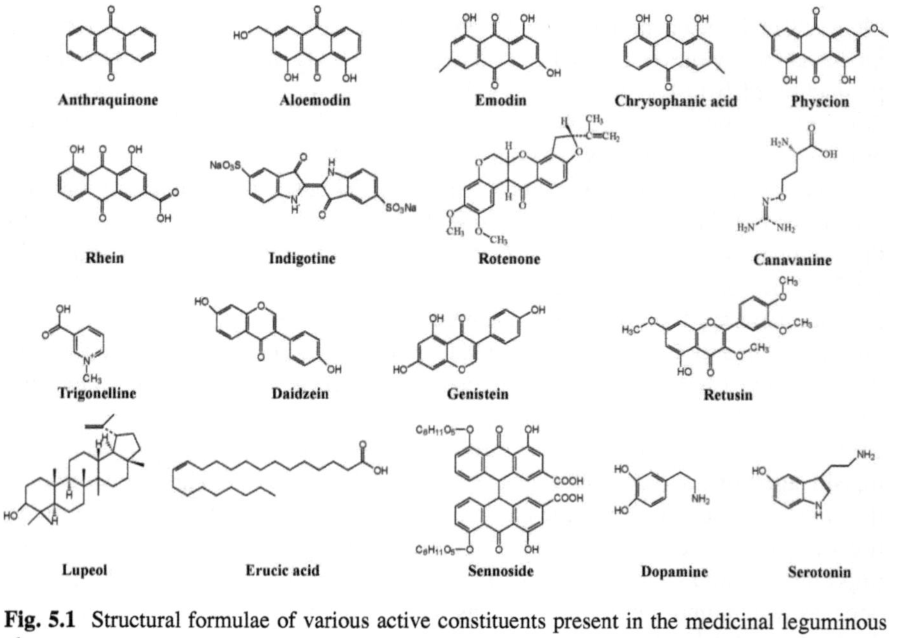

Fig. 5.1 Structural formulae of various active constituents present in the medicinal leguminous plants

be expanded for maximizing the yield of potentially useful leguminous medicinal plants particularly through optimal supply of the required fertilizers (phosphorus and calcium). For the cultivation of these medicinal legumes, the role of mineral nutrition is of paramount importance. Using optimal quantity of fertilizer application, the yield and quality of these medicinally important leguminous plants could be improved in order to meet their increasing demands.

3.1 *Cassia tora* L. (*Cassia obtusifolia* L.)

Naeem and Khan (2006) studied the influence of calcium application (0 (control), 40, 80, 120, and 160 mg Ca per kg soil) on growth, yield, and quality attributes of *Cassia tora* L. The results showed that calcium significantly enhanced most of the attributes studied. Calcium, applied at 120 mg Ca per kg soil, gave the highest values of these attributes (plant fresh and dry weights, number of leaves, leaf area, 100-seed weight, yield of the pods, and seed quality as well as leaf-nitrogen, -phosphorus, -potassium, and -calcium contents). Chlorophyll content and nitrate reductase activity were also increased over their respective controls. This treatment enhanced the seed yield, seed-yield merit, and seed-protein content by 31.6, 33.1, and 10.3%, respectively, over the control. The soil-applied Ca also stimulated the production of anthraquinone and sennoside content, increasing the therapeutic property of the plant.

3.2 Cluster Bean (*Cyamopsis tetragonoloba* L.) Taub

Garg et al. (1997) studied the effect of sodium chloride (NaCl) on cluster bean (*Cyamopsis tetragonoloba* L.). They observed that increasing NaCl concentrations (0, 50, 100, and 150 mM) decreased the growth and seed yield of the crop progressively, which was associated with decreased concentration of potassium and calcium and increased concentration of sodium in the shoots. Exogenous application of calcium (0, 2.5 and 5.0 mM) significantly ameliorated the adverse effects of NaCl due to enhanced uptake of Ca and K and reduced uptake of sodium. Calcium also alleviated the negative effects of NaCl on enzyme activities of nitrogen metabolism and contents of soluble protein and free amino acids.

3.3 Coffee Senna (*Senna occidentalis* L.)

Naeem et al. (2010) explored that plant biological yield was comparatively low in calcium-deficient soil of Aligarh, western Uttar Pradesh, India. Ca deficiency poses a serious yield and quality limitation for several crops, including medicinal herbs, in this region of India. Calcium application through soil enhanced crop productivity, photosynthetic efficiency, enzymatic activities, and contents of nutraceuticals in coffee senna, a medicinal legume. Plants were grown in pots containing soil supplied with five levels of calcium, viz. 0, 40, 80, 120 and 160 mg Ca per kg soil applied as calcium chloride ($CaCl_2$). Calcium application (120 mg Ca per kg soil) effectively increased various growth, physiological, biochemical, yield, and quality attributes studied at three growth stages (120, 270, and 300 DAS). Further, it increased the seed-yield and seed-protein content by 27.6 and 10.6%, respectively, compared to the control.

3.4 French Bean (*Phaseolus vulgaris L.*)

Ssali (1981) studied the effect of various levels of $CaCO_3$, inoculation and lime-pelleting on nodulation, dry matter yield, and nitrogen content of bean plant (*Phaseolus vulgaris*) in green house conditions, using five acid soils. The range of soil pH was 3.89–5.1, with exchangeable aluminium ranging from 0.0 to 4.0 mg per 100 g, exchangeable manganese from 0.35 to 2.32 mg per 100 g, and organic carbon from 6.69 to 5.60 mg per 100 g. Application of $CaCO_3$ increased the soil pH and exchangeable calcium in all the soils, but exchangeable aluminium and manganese decreased with increasing application of $CaCO_3$. Seed-inoculation with *Rhizobium* increased the nodule weight, dry matter yield, and the nitrogen content, particularly at low pH. However, when the seeds were not inoculated, the soil pH increased the nodule weight and dry matter yield. At low organic matter content and with substantial amounts of aluminium and/or manganese, liming increased the nodule weight and dry matter yield and decreased exchangeable aluminium and/or manganese. Low lime rates had little effects on exchangeable aluminium and calcium as well as on dry matter yield; however, higher lime rates decreased the exchangeable aluminium and dry matter yield, increasing the exchangeable calcium.

3.5 Hyacinth Bean (*Lablab purpureus L.*)

Khan et al. (2005) studied the effect of Ca application on vegetative growth, physiological attributes, seed yield, and quality of hyacinth bean in response to application of calcium at various rates (0, 0.2, 0.4, 0.6, and 0.8 g per pot). Ca application at 0.6 g per pot improved the vegetative growth as well as most of the physiological parameters and yield and quality attributes. It also enhanced the seed-yield and seed-protein content by 30.3 and 11.6%, respectively, over the control. Ca application also increased the pH of the soil that presumably enhanced the root-nodulation capability. The addition of Ca to soil also significantly enhanced the nodulation in the roots and activity of nitrate reductase in the leaves.

In another pot experiment, Naeem et al. (2009a, b) studied the effect of soil-applied Ca on nitrogen fixation, photosynthesis, enzymatic activities, contents of nutraceuticals, and yield and quality attributes of hyacinth bean. Calcium was applied to the soil as $CaCl_2$ at five levels, viz. 0, 40, 80, 120, and 160 mg Ca kg. The performance of the crop was assessed at 60, 90, 120, and 150 days after sowing (DAS). Ca application proved to be significantly effective on most of the parameters studied. Of the five Ca levels, 120 mg/kg soil showed the best results, improving most of the attributes studied significantly at all the sampling dates. It increased the seed yield, seed-protein content, and seed tyrosinase activity by 30.3, 16.6, and 20.3%, respectively, compared to the control. This study depicted that an optimum Ca dose should be included in the fertilizer recommendations for hyacinth and other beans in this region.

3.6 Lucern (Medicago sativa L.)

Pijnenborg and Lei (1990) investigated the effect of EGTA, a specific calcium chelator used at Wageningen (The Netherlands), on root-nodulation of lucerne (*Medicago sativa* L.). Calcium, given at 0.2 mM in the form of $CaCl_2$, improved the number of nodules in the root. In the absence of calcium, the number of nodules decreased by 70%. A similar decrease in nodule number was observed with the application of 0.2 mM calcium along with 0.2 mM EGTA. However, nodulation was restored by the addition of a higher dose of calcium, viz. 0.8–0.2 mM Ca. They concluded that depletion of soil calcium had different modes of action in the symbiotic process during the initiation and formation of the nodules.

Pijnenborg and Lei (1990) studied the effect of lime-pelleting on nodulation of lucerne (*Medicago sativa* L.) in an acid soil. Their comparative study was carried out in the field, in pots, and in rhizotrons at Wageningen (The Netherlands). The seeds were either inoculated with *Rhizobium meliloti* (R) only or the seed-inoculation was followed by pelleting with lime (RP). They observed that in the field conditions, inoculation and lime-pelleting (PR) was superior to mere inoculation (R) with regard to seedling establishment and nitrogen yield. The number of seedlings, carrying crown nodules, increased from 18% (R) to 56% (PR) at 26 DAS. PR treatment also increased crown nodulation in pots and rhizotrons. In pots, the increase in nodulation ranged from 32% (R) to 60% (PR) and in rhizotrons from 5% (R) to 90% (PR).

Grewal and Williams (2003) conducted a field experiment in Australia to investigate the cultivar variation in lucerne (*Medicago sativa* L.) with respect to soil acidity and lime application. They tested ten cultivars of lucerne (Hunter River, Hunter field, Seeptre, Aurora, Genesis, Aquarius, Venus, PL 90, PL 55 and breeding line Y 8804) using two levels of lime (0 and 2 t per ha). Lime application significantly increased the root growth, nodulation, leaf retention, leaf to stem ratio, herbage yield, crude protein content, and element composition of lucerne shoots. They reported that liming also improved the Ca concentrations of shoots, while there was a little effect of liming on P and Zn concentrations in the shoots. The effect of lime application was greater on ‘Y 8804’ and ‘Aurora’ cultivars; however, their yield increased by 32 and 31%, respectively. The yield increase was 11–22% in the case of other cultivars.

3.7 Mung Bean (*Vigna radiata* L.)

Naeem et al. (2005) performed a pot experiment to study the effect of four levels of calcium (0, 15, 30, and 45 kg Ca per ha) on fresh and dry weights per plant, number and dry weight of nodules per plant, nodule-nitrogen content, and total chlorophyll and carotenoids contents of mungbean (*Vigna radiata* L. Wilczek). Among the four levels of calcium, 45 kg Ca per ha proved the best for most parameters studied. It significantly enhanced the fresh and dry weights, number and dry weight of nodules per plant, nodule-nitrogen content, total chlorophyll content, and total carotenoids content over the control.

3.8 Senna (*Cassia angustifolia* Vahl.)

Arshi et al. (2006) conducted pot experiments to study vegetative growth, photosynthetic capacity, sennoside concentration, and yield attributes of senna (*Cassia angustifolia* Vahl.) under the individual as well as combined influence of NaCl and $CaCl_2$ at pre-flowering (45 DAS), flowering (75 DAS), and post-flowering (90 DAS) stages. As a result of NaCl treatments, significant reductions were observed in pod biomass, leaf area, stomatal conductance, photosynthetic rate, sennoside concentration, and seed yield. On the contrary, individual $CaCl_2$ treatments showed the favorable effect. Under the combined treatments, although the values of the parameters were reduced, the extent of reduction was much less than the one caused by NaCl treatments applied without $CaCl_2$. The combined treatments thus mitigated the adverse effects caused by NaCl.

3.9 Senna Sophera (*Cassia sophera* L.)

Naeem et al. (2009a, b) studied the effect of basal application of calcium on photosynthetic efficiency, enzymatic activities, nitrogen assimilation, and yield and quality attributes of *Cassia sophera* L. The plants were grown in pots containing soil supplied with five levels of calcium, viz. 0, 40, 80, 120, and 160 mg Ca kg soil (Ca_0, Ca_1, Ca_2, Ca_3, and Ca_4, respectively) applied as calcium chloride ($CaCl_2$). Calcium application at Ca_3 level proved significantly effective on the growth, physiological, biochemical, yield, and quality attributes studied at 120, 150, 180, and 210 days after sowing.

3.10 Soybean (*Glycine max* L.)

Bell et al. (1989) investigated the effect of calcium on five tropical food legumes, viz. peanut (*Arachis hypogea* (L.)), pigeonpea (*Cajanus cajan* (L). Millsp.), guar (*Cyamopsis tetragonoloba* (L.) Taub.), soybean (*Glycine max* (L.) Merr.), and cowpea (*Vigna unguiculata* (L.) Walp.) They applied six levels of calcium (2, 12, 50, 100, 500, and 2,500 μM) in the flowing solution culture at pH 5.5±0.1, using adequate inorganic nitrogen and controlled nutrient concentrations at appropriate rates. Increase of calcium concentration in the solution from 2 to 12 μM generally increased the rate of absorption of nitrogen, phosphorus, and potassium. Further increases in calcium concentration from 12 to 2,500 μM generally had no effect on potassium absorption rate, but it further increased the absorption rates of nitrogen (20–130%) and phosphorus (90–500%). Increases in the rates of nutrient absorption and their transport were associated with the Ca-applied alleviation of severe calcium deficiency. Further increases in the concentrations of solution Ca from 12 to 2,500 mM generally had no effect on K absorption rate. Progressive increase in the Ca concentration increased the absorption rates of N (20–130%) and P (90–500%), while it decreased the absorption rates of Mn and Zn. It also decreased rates of transport of Fe and Mg to plant tops.

Keiser and Mullen (1993) studied the response of soybean to calcium and relative humidity. Plants were grown in solution culture in an environment-controlled growth chamber. They applied Ca levels at 0, 0.6, 1.2, and 2.5 mM to the root medium from the beginning of seed growth stage (R_5) to the beginning of seed maturity stage (R_7). Seed-Ca concentration increased with increased Ca supply to the plant. Relative humidity had no effect on either seed-Ca concentration or germination. A decrease in the percentage of normal stand of seedlings from 96.7 to 41.8% coincided with a decrease in seed-Ca content from 2.37 to 0.87 mg per g. Reduced Ca supply to the plant reduced the seed-Ca concentration in addition to altering other seed nutrients. Poor seed germination was ascribed to the reduced Ca concentration in the substrate.

3.11 Cowpea (*Vigna unguiculata* L. Walp.)

Murillo-Amador et al. (2006) investigated the effects of foliar application of $Ca(NO_3)_2$ on cowpea plants grown under NaCl stress. The salt-stressed plants had low values of root and shoot dry matter and poor concentration of sodium and chloride both in the root and shoot. The concentrations of Ca_2^+, Mg_2^+, and K^+ also decreased significantly at high NaCl treatments. No significant differences were noted among the CA treatments regarding stomatal conductance, transpiration rate, net photosynthesis, and intercellular CO_2. However, chlorophyll fluorescence parameters of the NaCl-stressed plants had higher values following foliar $Ca(NO_3)_2$ sprays. It was concluded that supplemental calcium did not ameliorate the inhibitory effects of NaCl stress in cowpea plants.

Murillo-Amador et al. (2006) studied the effect of calcium silicate on cowpea and kidney bean under salt stress, using nutrient solution as substrate. Salinity stress reduced the values of all the growth attributes; however, calcium silicate supplementation partly overcame this growth reduction. Net photosynthesis, chlorophyll content, stomatal conductance, and transpiration rate were comparatively higher in control plants as a result of inclusion of calcium silicate in the nutrient solution. Shoot and root concentration of calcium was slightly higher in plants treated with calcium silicate. In the absence of calcium silicate, shoot and root concentration of potassium was reduced in salt-treated plants. Sodium and chloride concentration of shoot and root was slightly higher in the presence of NaCl, but it decreased slightly in plants treated with calcium silicate. It was suggested that in hydroponically grown cowpea and kidney bean plants, the inclusion of calcium silicate in the nutrient solution might enhance the nutrient uptake under NaCl stress, thus improving the plant growth and physiology.

4 Influence of Abiotic Stresses on Medicinal and Aromatic Plants

5 Salinity Stress

Salinity has highly adverse impacts on growth and productivity of agricultural plants. Soil salinity, resulting from natural processes or from crop irrigation with saline water, occurs in many arid and semi-arid regions of the world (Laüchli and Epstein, 1990). According to Tanji (1990), 20% of the cultivated lands are salt-affected worldwide. High concentrations of salts inhibit plant growth and productivity adversely. Amelioration of salinity stress is the main issue in salt-affected regions to ensure agricultural sustainability. An excess of soluble salts in the soil leads to osmotic stress, specific ion toxicity, and ionic imbalances (Munns 1993); Chartzoulakis et al. 2002; Zhu 2003; Arshi et al. 2002; Arshi et al 2004) leading to plant death or yield losses both in conventional crop species and medicinal plants (Rout and Shaw 2001). Salinity stress generates oxidative stress in plant tissues (Abel et al. 2003), which is manifested by ROS such as singlet oxygen, superoxide anion, hydrogen peroxide, and hydroxyl radicals (Gosset et al. 1994). Plants have evolved various protective mechanisms to eliminate or reduce ROS (Zhu 2003; Mittler et al. 2004). Free radical reactions, in participation with oxidative radicals, have been shown to be involved in many biological reactions, causing damage to lipids, proteins, membranes, and nucleic acids, thus giving rise to a variety of metabolic disorders (Cavalcanti et al. 2006). Plant stress responses involve the synthesis of several secondary metabolites of phenylpropanoid pathway to defend the plant metabolism. Phenolic compounds are intermediates in the phenylpropanoid pathway and play important roles in flavonoid production and lignin biosynthesis (Zheng et al. 2001). In fact, it has been determined that the antioxidant effect of plant products is mainly due to phenolic compounds, such as flavonoids, phenolic acids, tannins, and phenolic diterpenes (Lee et al., 2004). Phenolic compounds play an important role in adsorbing and neutralizing free radicals, quenching singlet oxygen, or decomposing peroxide radical (Ksouri et al. 2007).

In fundamentally agrarian nations, soil salinity has a remarkable negative impact on economy (Kumar and Abrol 1994; Kumar and Gill 1995; Sangwan et al. 1982). The National Academy of Sciences of the United States includes desalinization of soils and water as one of the leading programs contributing to the worldwide catastrophe. Utilization and subsequent improvement of salt-affected wastelands, which occupy more than 10 million hectares of land, is of significant importance in India. It requires suitable reclamation and management strategies. Preliminary appraisal indicated that some medicinal and aromatic crops might be suitable in salt-affected soils. Aromatic grasses like palmarosa (*Cymbopogon martinii*) and lemongrass (*Cymbopogon flexuosus*) are reported to withstand salinity to a great extent as compared to traditional agricultural crops. Since there is already much pressure on good quality fertile lands for the production of major agricultural crops, it will be beneficial to determine if grasses like palmarosa and lemongrass could be grown in salt-affected soils including those affected by sodicity.

Increasing frequency of dry periods in many regions of the world and the problems associated with salinity in irrigated areas frequently result in the consecutive incidence of drought and salinity on cultivated land. Currently, approximately 50% of the irrigated land in the world, which has at least twice the productivity of rainfed land and may produce one third of the world's food, is affected by salinization (Ghassemi et al. 1995; Hillel 2000). Water deficit and salt stress are probably the major physiological causes for growth reduction in plants as both the stresses lower the soil water potential. Drought and salinity may, however, differentially affect the mineral nutrient relations of plants. In general, drought reduces nutrient uptake by the roots as well as their transport from roots to shoots because of restricted transpiration rates, impaired active transport, and poor membrane permeability (Alam 1999). Contrarily, soils contain extreme ratios of Na/Ca, Na/K, Ca/Mg, and Cl/NO_3 under saline conditions, which cause reductions in plant growth because of specific ion toxicities (e.g. Na^+ and Cl^-) and ionic imbalance acting on biophysical and/or metabolic components of plant growth (Grattan and Grieve 1999). To date, it is not clear whether water deficit or ionic effects are major limiting factors for plant growth under salt stress. Munns (1993) proposed a two-phase model for the inhibition of growth by salinity. The first phase of growth reduction in this model results from the osmotic effect of soil salinity that may cause water deficit, and under extended periods (the second phase), salt begins to accumulate in older leaves and salt injury becomes apparent. To differentiate the effect of osmotic stress from that of ionic one during the first phase, a comparison between NaCl and polyethyleneglycol (PEG) treatments was carried out in iso-osmotic conditions, growing plants of barley (Kawasaki and Moritsugu 1983; Storey and Wyn Jones

1978), tomato (Perez-Alfocea et al. 1993), and maize (Sümer et al. 2004) hydroponically.

Literature is rich in the data concerning salt tolerance of conventional crops, but similar reports on volatile oil plants are scanty (Kumar and Abrol 1994). While comparing the performance of two *Cymbopogon* species (*C. winterianus* and *C. flexuosus*) under increasing salinity and sodic conditions, Kumar and Gill (1995) recorded that *C. flexuosus* suffered less from reduced shoot and root yield. Increasing stress of salinity and sodicity caused a reduction both in shoot and root yield of citronella, lemongrass, and vetiver (Kumar and Gill 1995). Ansari et al. (1998) studied the performance of three *Cymbopogon* grasses, viz. *C. winterianus, C. flexuosus,* and *C. martini*, at different levels of NaCl stress. Salinity resulted in suppression of plant growth and significant decline in content and yield of essential oil in the *Cymbopogon* species studied. The decline in oil yield was the greatest in *C. winterianus* and the least in *C. flexuosus*. Compositionally, proportions of citral and geraniol increased under salt stress in the oil of lemongrass and palmarosa, respectively. Patnaik and Debata (1997) reported that a salt tolerant line of palmarosa, developed by an in vitro procedure, has a regeneration potential, even under high salt concentrations (up to 200 mM).

Similar to majority of the cultivated crop plants, growth and yield of MAPs can be affected by environmental constraints such as salinity and drought. Changes in essential oil (EO) yield and its composition have been reported to be influenced by environmental conditions (Gil et al. 2002). Some studies showed a decrease in EO yield under salinity coupled with an alteration in EO composition (Dow et al. 1981). In EO plants, an enhancement in the content of secondary metabolites was found by Hendawy and Khalid (2005) and Abou El-Fadl et al. (1990) under salt stress. An increase in EO of coriander (*Coriandrum sativum*) at 25 and 50 mM NaCl, but a decrease under high salinity was signaled by Neffati and Marzouk (2008). Salt affects growth, mineral nutrition, and yield and composition of EO in marjoram (*Origanum majorana* L.) as reported by Dorman et al. (2000). The adverse effect of salt stress on the growth and chemical composition of EO of peppermint (*Mentha × piperita* L.), pennyroyal (*Mentha pulegium* L.), and apple mint (*Mentha suaveolens Ehrh.*) has been investigated by Aziza et al. (2008). EO yields per plant were reduced under salt stress in all the three EO species, as compared with untreated controls. Similar effect of salt stress on EO yield and composition was observed by El-Danasoury et al. (2010) and Khadhri et al. (2011) regarding lemongrass and spearmint.

Growth and essential oil content of peppermint (*Menhta piperita* var. officinalis) and lemon verbena (*Lipia citriodora* var. verbena) were evaluated in response to salinity and nutrient solution concentrations, measured as electrical conductivity (EC 0.7, 1.4, 2.8, 5.6, and 5.6 Na dS/m) (Tabatabaie and Nazari 2007). The rate of photosynthesis (P_N) was higher in treatments with EC levels of 1.4 and 2.8 as compared to the other treatments. In both peppermint and lemon verbena, the concentrations of N, P, and K increased as the EC of the solution increased from 0.7 to 5.6 dS/m, but in the 5.6 Na treatment their concentrations fell. The total content of essential oil was reduced by increasing the EC of the solution. In peppermint, the essential oil content in the 1.4 dS/m treatment was 55.0 and 40.5% higher than those of both 5.6 and 5.6 Na treatments, respectively. The increased P_N and leaf area in moderate EC level led to improved plant growth. Jaleel and Azooz (2009) carried out the experiment on *alba* and *rosea* varieties of *Catharanthus roseus* (L.) G. Don. using different concentrations of sodium chloride (25, 50, 100, and 200 mM NaCl) in order to determine the changes occurring in antioxidant potentials and to evaluate the possibility of increasing the anti-hypertension alkaloid, ajmalicine, present in the roots. High salinity caused an enhancement in total ascorbic acid (AA) contents. GSH (Glutathione) content decreased insignificantly only at 25 mM NaCl concentration, while it increased at higher salt concentration. NaCl treatment significantly increased the content of root alkaloid ajmalicine in both the varieties (*alba* and *rosea*). Furthermore, he reported the effect of salinity on *Catharanthus roseus* regarding enzymatic (SOD, peroxidase (POX), catalase (CAT)) antioxidant potentials. The SOD activity increased at level of 50 mM NaCl, but was reduced at higher treatment levels. There were no significant changes of POX activity at 25 mM NaCl level and significant increases of this activity at next higher levels of NaCl. Proline (PRO) accumulation in salt-stressed cells may occur because of decreased oxidation of PRO to glutamic acid as a result of PRO oxidase activity (Jaleel et al. 2007a). Such a reduction in PRO oxidase activity and a simultaneous increase in PRO level also occurred following low temperature and drought (Jaleel et al. 2007b).

6 Drought Stress

Drought is a major environmental stress that affects plant morphology, physiology, and biochemistry, causing a significant reduction in agricultural production (Hsiao 1973; Tyree and Karamanos 1981). Drought stress limits the production of 25% of the world's land (Delfine et al. 2005). Although the effects of drought on many plants have been widely investigated, less is known about the biosynthesis and accumulation of oil in aromatic plants under water deficit conditions (Sangwan et al. 1993; Farahani et al. 2009).

The effect of drought stress on growth and development of MAPs has earlier been studied. The results indicate that water deficit during the vegetative period (before flowering stage) might result in shorter plants and smaller leaf area of mint (Abbaszadeh et al. 2008), yarrow (Sharifi et al. 2005), and chicory (Taheri et al. 2008), reduced water use due to the reduction in plant size of calendula (Rahmani et al. 2008), and decreased vegetative dry matter of balm (Aliabadi et al. 2009). A number of studies indicate drastic reduction in grain yield as a result of water deficit during the reproductive period of coriander (Aliabadi et al. 2008a, b), Mexican marigold (Mohamed et al. 2002), and grapevine (Scalabrelli et al. 2007). Consequently, drought stress reduced vegetative growth period and plant moved to flowering stage. Thus, quantity characteristics of MAPs decreased under drought conditions sorely. There was found a significant reduction in dry matter and relative growth rate of *Thymus vulgaris* grown under drought stress (Letchamo et al. 1995).

Similarly, growth and oil content of fennel plants were adversely affected by irregular irrigation schedules (Patel et al. 2000; Mohamed et al. 2002; Manivannan et al. 2004). Sangwan et al. (1994) exposed lemongrass species, *Cymbopogon nardus* and *C. pendulus,* to water stress. They observed that water deficit reduced plant height, leaf length, leaf area, fresh and dry weight, and leaf moisture content. The oil content was significantly affected depending on water stress treatment. *Origanum majorana* had a higher oil content and leaf dry weight with increasing soil moisture deficit (Rhizopoulou and Diamantoglou 1991).

Limited water supply is a major environmental constraint to the productivity of crop plants. Moisture deficiency not only limits plant growth and survival, but also induces various physiological and metabolic responses like stomatal closure, decline in growth rate, accumulation of solutes (osmolytes) and antioxidants, following the expression of stress-specific genes (Hughes et al. 1989). The production of secondary metabolite is believed to be stimulated by stressful environment. However, there is too little experimental data to support this notion. Sangwan et al. (1993b, 1994) conducted short- as well as long-term experiments on aromatic grasses (*Cymbopogon* species), exploring the effect of moisture deprivation. The plant responses were studied in excised systems (leaf disks) as well as in *planta.* Short-term water stress substantially affected the EO biosynthesis, the response being different in various *Cymbopogon* species. Similar alterations in the terpene content of some other volatile oil plants have also been reported (Penka 1978). It has been suggested that greater oil gland density in mints and sweet basil, as a consequence of water-stress-induced reduction in leaf area, results in greater oil accumulation. However, experiments with *Cymbopogons* have demonstrated that water stress alters merely the oil biogenetic capacity. This happens without any change in the oil gland count, as observed in the experiments conducted with the excised systems subjected to short-term water-stress conditions. In another experiment, effects of water stress were studied, of long term, in two lemon grasses *C. nardus* var. *confertiflorus* and *C. pendulus*. Sangwan et al. (1993a, 1994) revealed that the amounts of EO produced under drought conditions was either maintained or enhanced, depending on the species and magnitude of the stress. However, the major oil constituents, viz. geraniol and citral, increased in both the *Cymbopogon* species. The activity of geraniol dehydrogenase was also modulated under moisture stress. Based on the physiological and oil data, Sangwan et al. (1993a, 1994) demonstrated that the relatively water-stress-resistant *Cymbopogon* species might restrict their growth under water-stress, accumulating elevated amounts of EO in their leaves. Charles et al. (1990) and Simon et al. (1992) suggested that a comparatively higher density of oil glands might result in an elevated amount of oil accumulation under water-stress owing to the reduction in leaf area. Compositional alterations in EO content, occurring as a consequence of drought stress, have been elucidated in mints (Chattopadhyay and Subramanyam 1993) and sweet basil (Simon et al. 1992). In lemongrasses, the stress-mediated changes in oil composition were more prominently reflected in the major oil constituent, i.e. citral and geraniol. Because of reduced vegetative growth of plants under water stress, closer planting of aromatic grasses has been suggested by Sangwan et al. (1994) in order to maintain the EO production in droughted areas. Farooqi et al. (1998) reported an increased oil biogenesis under water-stress in different genotypes of *C. winterianus*. According to Shabih et al. (1999), water-stress conditions promoted either an increase, decrease, or no change in the oil yield of *C. martinii* in a genotype-specific manner. Water-stress had a negative impact on green yield and EO yield as the intervals between stress applications increased in geranium (Putievsky et al. 1986). The citronellol to geraniol ratio increased as the interval between final irrigation and harvest was extended. Chalchat et al. (1994) observed that the amount of water received during vegetative growth hardly influenced the oil composition in *Artemisia annua*, while water stress strongly depressed oil yield and the plentiful irrigation raised it. Aromatic plants also differ in their performance under water stress. For example, pubescent leaf type strain of *Eucalyptus citriodora* is less drought-tolerant than the less valuable non-pubescent type. Besides, the performance of the same aromatic plant may differ under different drought conditions. For example, the oil produced after 3 wet months in drought-tolerant geranium is slightly milder than that produced after 3 dry months (Weiss 1997).

The content of EO and its composition are affected by different factors, including genetic makeup (Muzik et al. 1989) and cultivation conditions such as climate, habitat, harvesting time, water stress, use of fertilizer, etc. (Min et al. 2005; Stutte 2006). Plant *reactions* are affected by the amount of soil water directly or indirectly. Water-stress in plants influences many metabolic processes, and the extent of its effects depends on drought severity. The optimization of irrigation for the production of fresh herbs and essential oils is important, since water is a major component of the fresh produce and affects both weight and quality (Jones and Tardien 1998). Water deficit in plants may lead to physiological disorders, such as a reduction in the rate of photosynthesis and transpiration (Sarker et al. 2005), and in the case of aromatic plants it may cause changes in the yield and composition of EO. In fact, water deficit decreased the oil yield of rosemary (*Rosmarinus officinalis* L.) (Singh and Ramesh 2000) and anise (*Pimpinella anisum* L.) (Zehtab-Salmasi et al. 2001). By contrast, water stress caused an increase in oil production of thyme (Aziz et al. 2008) and citronella grass (*Cymbopogon winterianus* Jowitt.) expressed on the basis of plant fresh weight. Severity of water stress response varies with cultivar and plant density (Fatima et al. 2000, 2006). Plant nutrition might contribute to increased resistance in plants and their productivity when the crop is submitted to water stress (Cakmak 2005). Water stress is well known as one of the most significant factors affecting plant growth, photosynthesis, productivity, seed yield, and seed quality for most crops (Alderfasi and Alghamdi 2010). Jaleel et al. (2007a, b) suggested that the cultivation of medicinal plants such as *Catharanthus roseus* under water-deficit areas increases its PRO metabolism, osmoregulation, defense system, and the level of active principles. Jaleel et al. (2008a, b) explained the changes in the reactive oxygen metabolism of *Catharanthus roseus* (L.) G. Don. under drought stress in terms of H_2O_2 content, LPO and the free radical quenching systems (non-enzymatic (ascorbic acid, α-tocopherol, and reduced glutathione contents) and enzymatic (SOD, ascorbate peroxidase (APX), and CAT) antioxidants) and root alkaloid ajmalicine. They concluded that the water-deficit areas might be well used for the cultivation of medicinal plants like *C. roseus* and the production of economically important alkaloids could be enhanced accordingly. Manivannan et al. (2007a) tested two water treatments, viz., 100 and 60% of field capacity (FC), to understand the effects of water deficit on early growth, biomass allocation, contents of pigments and biochemical constituents, and PRO metabolism in five varieties of sunflower (*Helianthus annuus* L.). They found that there were significant differences among the five varieties regarding vegetative growth, dry matter accumulation, pigment contents, biochemical constituents, and PRO metabolism. The root length, shoot length, total leaf area, fresh and dry weight, chlorophyll *a*, *b*, total chlorophyll, and carotenoids were significantly reduced under water stress treatments. However, water stress increased the contents of PRO, free amino acid, and glycinebetaine. It increased the activity of γ-glutamyl kinase, but the activity of PRO oxidase was reduced as a consequence of water stress.

7 Essential Oil Content of Medicinal and Aromatic Plants under Drought Stress

The effect of water stress on essential oil was previously studied in excised leaves of palmarosa (*Cymbopogon martinii* var. motia) and citronella java (*C. winterianus*). Drought stress increased the percentage of EO, but the EO content/yield per plant was reduced under drought stress in palmarosa and citronella (Fatima et al. 2006) as well as in balm (Aliabadi et al. 2009). Khalid (2006) evaluated the influence of water stress on the EO content of *Ocimum basilicum* L. (sweet basil) and *Ocimum americanum* L. (American basil). In both of the *Ocimum* species, essential oil percentage and the main constituents of EO was increased under water stress. Parsley cultivars, viz. plain-leafed, curly-leafed, and turnip-rooted, were grown under conditions of 35–40 and 45–60% water deficit in order to evaluate the effect of water stress on the EO yield and its composition. Water stress increased the yield of EO (on a fresh weight basis) in plain-leafed and curly-leafed cultivars, but not in turnip-rooted cultivars of parsley. Water stress also caused changes in the relative contribution of certain aroma constituents of the essential oils (principally 1, 3, 8-*p*-menthatriene, myristicin, terpinolene+*p*-cymenene), but these changes varied between cultivars. The oil yield of roots was low, but the water deficit stress had relatively little effect on the root oil composition. Since the biomass of plants subjected to water deficit was reduced, it was concluded that increase in oil yield was possible by increasing the plant density per unit land area. Petropoulos et al. (2007) suggested that likely changes in oil composition must also be taken into account while applying water deficit stress for enhanced production of EO in parsley cultivars. Idrees et al. (2010) also reported the significant effect of water stress on growth parameters and biochemical constituents, PRO metabolism, and quality attributes including essential oil yield and citral content of two varieties (Neema and Krishna) of lemongrass (*Cymbopogon flexuosus* Steud. Wats.). The plants under water stress exhibited a significant increase in activities of nitrate reductase and carbonic anhydrase, and electrolyte leakage, PRO content, free amino acid, and in PEP carboxylase activity. Content and yield of essential oil also significantly decreased in plants that faced water stress. Thus, it was concluded that variety Neema is the more tolerant variety as compared to Krishna on the basis of content and oil yield and well adapted to drought stress conditions. Rahmani et al. (2008) showed that the highest oil yield was achieved under non-drought condition and the highest oil percentage was achieved under drought condition in calendula. Bettaieb et al. (2008) reported that drought stress significantly decreased the foliar fatty acid content and the double bond index (DBI) degree of *Salvia officinalis*. The latter was provoked mainly by a strong reduction of linolenic acid proportion and the disappearance of palmitoleic acid. Moreover, moderate water deficit increased the EO yield (expressed as g/100 g on the basis of dry weight) and the main constituents of EO (camphor, α-thujone, and 1, 8-cineole). According to Baher et al. (2002), the accumulation of oil increased significantly under severe water stress in Iranian *Satureja hortensis* L. at the flowering stage, when the mean leaf water potential (LWP) decreased from –0.5 to −1.6 MPa. This moisture deficit decreased the quantity of EO more than the moderate water deficit during the vegetative and flowering stages. The main oil constituents, carvacrol and -terpinene, were also affected by the water deficit. The amount of carvacrol was increased under moderate stress, while terpinene content decreased under moderate as well as severe water stress treatments. Singh-Sangwan et al. (2006) reported that the level of EO was maintained or enhanced under drought condition in lemongrass (*Cymbopogon nardus* and *Cymbopogon pendulus*), with the major oil constituents, geraniol and citral, being increased substantially. The activity of geraniol dehydrogenase was also modulated under moisture stress. Taheri et al. (2008) noticed that drought stress significantly increased the percentage of EO and the contents of its compounds (such as kaempferol) in chicory; the EO yield per plant was, however, significantly increased under no stress conditions. Similarly, Aliabadi et al. (2008a, b) reported the highest flowering shoot yield and the EO yield of flowering shoot in coriander under no stress conditions, while the highest oil percentage of flowering shoot was achieved under water stress. Thus, drought stress might be considered to reduce essential oil content of several MAPs coupled with an increase in the EO percentage.

Drought stress also encourages a dramatic increase in the PRO concentration of phloem sap in MAPs, suggesting that increased deposition of PRO at the root apex in water stressed plants could, in part, occur via phloem transport of PRO. In fact, the PRO transporter gene, $ProT_2$, is strongly induced by water and salt stress in *Arabidopsis thaliana* (Rentsch et al. 1996), indicating that the stress-hypersensitive mutants of MAPs, which exhibit disturbed PRO metabolism, can contribute significantly to the elucidation of signals to which PRO accumulation may respond. Petunia (*Petunia hybrida*) has been reported to accumulate free PRO under drought-stress conditions (Yamada et al. 2005). There were noted alterations in the metabolism of PRO in cassava and the extent of alteration varied between drought-susceptible and -tolerant cultivars during an imposed water stress (Sundaresan and Sudhakaran 2006). A study was conducted to determine the response of date palm (*Phoenix dactylifera* L., cvs. Barhee and Hillali) calli to water stress. Increasing PEG concentration was associated with the progressive reduction in the water content and increased content of endogenous free PRO in date palm (Al-Khayri and Al-Bahrany 2004).

Choline chloride (CC) pre-treatment accelerated accumulation of PRO in *Rehmannia glutinosa* seedlings during drought stress and retarded the drop in PRO concentration after dewatering (Zhao et al. 2007). Mazzafera and Teixeira (2006) subjected the seedlings of *Coffee arabica* genotypes (Catuai and BA10C1110-10), with different drought tolerance levels, to controlled water stress. They suggested that PRO accumulation was related to the injury to plants caused by the imposed water stress. Aliabadi et al. (2008a, b) investigated the effects of arbuscular mycorrhizal fungi, different levels of phosphorus, and drought stress on PRO accumulation rate in coriander (*Coriandrum sativum* L.). Their results indicated that drought stress had significant effect on PRO accumulation rate, with the highest PRO accumulation rate being achieved under stress conditions. Similar results were obtained by Baher et al. (2002) in the case of *S. hortensis* L.

8 Effect of Salt and Drought Stresses on Secondary Metabolites

Different growth conditions have significant impact on the biosynthesis and accumulation of secondary plant products. In this regard, there are various indications that stress reactions, especially those belonging to responses to salt and drought stress, might be responsible for the increase or decrease in the content of relevant secondary natural products. However, consequent research with respect to the scientific background is lacking or very rare.

In fact, the concentrations of various secondary plant products strongly depend on the growing conditions. Further, stress situations have a strong impact on the metabolic pathways responsible for the accumulation of the related natural products. In many studies, the corresponding results are not conclusive; nevertheless, decisive inferences regarding the effects of drought and salt stress on the accumulation of secondary plant products can be made. In the whole array of experiments, it could be shown that plants which are exposed to drought stress produce comparatively higher amounts of secondary metabolites. This counts for phenols and terpenes as well as for nitrogen containing substances, such as alkaloids, cyanogenic glucosides, or glucosinolates, etc. Undoubtedly, the application of drought stress enhances the concentration of secondary plant products; however, drought stress also reduces the growth of most plants. For example, the contents of various phenols and betulinic acid are drastically higher in *Hypericum brasiliense* plants grown under drought stress, with the vegetative growth being correspondingly decreased (de Abreu and Mazzafera 2005). Further, they reported that the total amount of some secondary plant products were significantly higher in plants grown under drought stress than in those cultivated under normal conditions. Although stressed plants were quite short in height, the product of biomass and the substance concentration yields in a 10% higher amount of phenolic compounds; however, the total content of betulinic acid was nearly the same in normal as well as stressed plants. Also, Nogueés et al. (1998) found a massive increase of phenolic compounds in stressed pea plants.

On the other hand, the overall yield of total flavonoids was nearly the same in *Pisum sativum* plants grown under drought-stressed or -unstressed conditions. Brachet and Cosson (1986) observed a strong increase in the concentration of tropane alkaloids in salt-stressed plants. Similarly, there were detected significantly higher concentrations of flavonoids in *Hordeum vulgare* plants under salt stress (Ali and Abbas 2003). Although massive amounts of $NADPH+H^+$ are reoxidized by photorespiration and the xanthophyll cycle under stress conditions, the corresponding strong reducing power ($NADPH+H^+$) seems to enhance the synthesis of highly reduced compounds, like isoprenoids, phenols, or alkaloids. Consequently, the synthesis and accumulation of highly reduced secondary plant products reveals a means within the metabolism to prevent too massive generation of oxygen radicals and the corresponding damage by photoinhibition (Selmar 2008).

The plants grown under Mediterranean or semi-arid climate conditions have enhanced amounts of secondary plant products and are much more pronounced in taste and aroma than those cultivated in moderate climate. In contrast, the data on plants grown under salt stress were not conclusive since the corresponding enhancement in secondary metabolites could not be obtained unequivocally (Selmar 2008). Hence, in contrast to the usage of drought stress, the application of salt stress seems to be unsuitable to increase the concentration of secondary metabolites in the MAPs. However, a concentration increase of active compounds of the spices and MAPs induced by moderate drought stress in general is associated with a reduction of biomass production. In the case of MAPs, which are used directly as pharmaceuticals, the quality and, thus, the concentration of active compounds is much more relevant than the total yield, whereas in all other cases, where the desired compounds are extracted, the overall yield needs to be very high. A successful and effective application of deliberate drought stress for quality improvement, e.g. by applying special watering regimes in combination with efficient soil draining by supplementation of sand, is an encouraging new tool for the production of spice and pharmaceutically relevant plants, but it requires solid and comprehensive database on the entire field concerned.

9 Heavy Metals Stress

Heavy metal pollution is an increasing problem in agricultural soils. Heavy metals are found naturally in the soils as rare elements and also might be added to the environment by traffic, refuse dumping, and metal working industries. Heavy metal stress poses several undesirable effects on plants leading to plant death and human health hazards.

Of the heavy metals (HMs), aluminium (Al) is the third most abundant metal in the earth's crust, comprising about 7% of its mass. The Al toxicity has been a major factor which limits growth and yield of crops in many acid soils throughout the world (Foy 1988). The Al toxicity is manifested in acid conditions, in which phytotoxic form of Al (Al^{3+}) predominates. Roots are the first and most affected part of plants to show Al toxicity symptoms. Al causes inhibition of root elongation, disruption of root cap process by decreasing the synthesis of non-cellulosic polysaccharides by dictyosome (Bennet et al. 1987a, b; Buchanan et al. 2000), decline in cell division (Clarkson 1965; Pan et al. 2004), suppressed nutrient uptake and translocation (Raynal et al. 1990; Foy 1988), restricted water uptake and transport (Sivaguru et al. 2003; Ohki 1986a, b; Arp and Strucel 1989; Rengel 1994), disruption of cellular calcium homeostasis (Huang et al. 1992), and suppressed photosynthesis (McCanny et al. 1995; Ridolfi and Garrec 2000). Aluminium toxicity is aggravated by soil acidification; therefore, strategies to maintain production on these soils include raising the soil pH and the use of plants that are tolerant of acid soils. A number of studies show that Al induces LPO and causes oxidative damage in various plant systems, acting as catalysts in ROS production (Cakmak and Horst 1991; Subrahmanyam 1998).

Panda et al. (2003) studied the effect of different Al concentrations (0, 0.001, 0.01, 0.1, 1.0 mM) on root and shoot growth and Al-induced oxidative stress in *Vigna radiata* L. They reported a concomitant decrease in root and shoot elongation with increasing Al concentration. The increasing Al concentrations resulted in progressive increase in total peroxide content, TBARS (thio barbituric acid reactive substances) content, and membrane injury index, which confirmed the production of ROS under Al stress. Plants have developed complex mechanisms to cope with excess of ROS under normal as well as stressful conditions. A uniform increase in SOD, POX, and GR activities paralleled with a gradual decrease in CAT activity was detected with the increase in Al concentrations. On the other hand, the changes in non-enzymic antioxidants like ascorbate and glutathione showed a gradual decrease with increasing aluminium concentrations (Panda et al. 2003), suggesting their inability to detoxify the ROS directly.

Ikegawa et al. (2000) exposed the tobacco plants to excess of Al and Fe. They found that Al-enhanced Fe (II)-mediated peroxidation of lipids, causing cell death. On the other hand, exposure of tobacco suspension cultures to Al alone for 24 h resulted in Al accumulation in cells with no significant cell death. Treatment with an antioxidant prevented the Al-enhanced LPO in pea root cells, reduced callose formation, but did not prevent Al-induced inhibition of root elongation (Yamamoto et al. 2001). The most observable symptoms of Al toxicity is inhibition of root elongation. There has been recorded suppressed elongation of adventitious roots of onion (*Allium cepa* L.) (Clarkson 1965), and those of primary roots of soybean (Horst et al. 1992; Lazof et al. 1994), corn (Blancafor et al. 1998; Sivaguru and Horst 1998), and wheat (Parker 1995; Ryan et al. 1992; Sasaki et al. 1997) within 1–3 h of Al exposure. Llugany et al. (1995) showed that shortest duration of Al exposure required to inhibit elongation rates in seminal roots of an Al-sensitive corn cultivar (BR 201 F) was just 30 min. A disruption in the activity of Golgi apparatus of peripheral root cap cells at 18 μM of Al was recorded by 68 Bennet et al. (1987a, b) in corn. Generally, Al toxicity has been shown to reduce root biomass to a greater degree than shoot biomass, contributing to decreased root/shoot ratio (Raynal et al. 1990).

It has often been demonstrated that a number of physiological and biochemical processes are affected within minute/hours of Al exposure. An Al-induced reduction in Ca and Mg concentrations of tree roots and needles was the main reason behind the forest decline in North America and Europe (Godbold et al. 1988). Ridolfi and Garrec (2000) reported that Al toxicity reduced Ca and Mg leaf concentrations in beech (*Fagus sylvatica* L.). McCanny et al. (1995) reported a decrease in photosynthetic rate of red spruce exposed to 250 μM of Al for 6–8 weeks. Similarly, exposure of beech seedlings to 0.37 mM of Al significantly decreased the net rate of CO_2 assimilation (Ridolfi and Garrec 2000). Al reduced membrane permeability to water as shown by plasmometric method using root disks of red oak (Chen et al. 1991). Sivaguru et al. (2003) reported the closure of stomata in *Arabidopsis* after exposure to 100 μM of Al at pH 4. 90. Ohki (1986a, b) reported a decrease in transpiration in wheat after 28 days of exposure to 148 μM of Al. On the other hand, they reported an increase in transpiration rate in sorghum after 28 days of Al treatment. Nichol et al. (1993) showed that presence of Al decreased the influx of cations (K^+, $NH4^+$, and Ca^{2+}) in the cells, whereas the influx of anions (NO_3^-, and HPO_4^{2-}) increased in the presence of Al. According to them, binding of Al^{3+} to the exterior of plasma membrane formed a positively charged layer that retarded the movement of cations to the membrane surface, while increasing the movement of anions to the membrane surface. Uptake of Ca by roots and its subsequent translocation to shoots was decreased at an exposure of wheat plants to 100 μM of Al (Johnson and Jackson 1964). Similar results were reported with 4-week-old Norway spruce seedlings, in which uptake of Ca was reduced by 77–92% at an exposure of 100–800 μM of Al (Godbold et al. 1988). One of the most prominent effects of Al toxicity on plants is inhibition of cell elongation. According to Rengel (1992), interaction of Al^{3+} with the cell wall components could be responsible for the observed Al toxicity symptoms in plants. Ca^{2+} also plays a key role in cross-linking the pectic materials in the cell wall (Carpita and Gibeaut 1993). Reid et al. (1995) and Taylor et al. (2000) reported a displacement of cell-wall-bound Ca^{2+} by Al in the internodal cells of *Chara*. Zhang et al. (1999) reported that Al rapidly induced the bursting of pollen tube tip in *Chamelaucium uncinatum* that was markedly reduced by increasing the Ca^{2+} activity of the incubation medium. Blamey (2001) demonstrated that Al binded more strongly to pectin than did the Ca^{2+}. They suggested that the displacement of pectin-bound Ca^{2+} would inevitably alter physical properties of the cell wall, such as extensibility, rigidity, and permeability, harming the cell extension as well as division (Horst 1995; Blamey 2001).

Most plant species are susceptible to cadmium toxicity. In agricultural soils, the intense use of phosphate fertilizers that contain cadmium (Cd) as a contaminant (Polle and Schuetzenduebel 2003) serves as a source of Cd pollution, beside the anthropogenic activities like traffic, refuse dumping, and metal industries. Being highly mobile and water soluble, Cd readily enters the roots through the cortical tissue and can reach the xylem via an apoplastic and/or symplastic pathway (Salt et al. 1995). Once loaded into the tracheary elements, Cd complexes spread throughout the entire plant following the water stream. The Cd is considered as non-nutrient element, since it has no known function in plant growth development, with the exception of the Cd-carbonic anhydrase of marine diatoms (Lane and Morel 2000). The most evident symptoms of Cd toxicity are leaf-roll, chlorosis, water uptake imbalance, and stomatal closure (Clemens 2006). Khalighi et al. (2007) reported chlorosis and necrosis in Cd-treated wheat leaves. At the cellular level, Cd damages the photosynthetic apparatus, particularly the light harvesting complex II, and causes a decrease in chlorophyll and carotenoid contents, leading to increased nonphotochemical quenching (Sanità di Toppi and Gabrielli 1999). By inhibiting enzymes involved in CO_2 fixation, Cd decreases carbon assimilation (Perfus-Barbeoch et al. 2002). Chen et al. (2011) studied the effects of different concentrations of soil-Cd (0–24 mg/kg) on growth and photosynthetic activities in leaves of pakchoi and mustard. They reported a concomitant decrease in the weight of root and shoot, chlorophyll content, photosynthetic rate, and stomatal conductance.

Cd causes imbalance in water uptake and nutrient metabolism by interfering with the uptake of Ca, Mg, K, and P (Benavides et al. 2005). Alcantara et al. (1994) showed that Cd-induced inhibition of the Fe (III) reductase in the roots leads to Fe deficiency in cucumber and sugarbeet. Cd alters the activity of different enzymes involved in nitrogen metabolism (Nussbaum et al. 1988; Boussama et al. 1999). Chaffei et al. (2004) reported a depression in the activity of nitrate and nitrite reductases in roots and leaves of Cd-treated tomato plants. While Sanità di Toppi and Gabrielli (1999) registered a decrease in nitrate transport from roots to shoots, leading to reduced nitrate assimilation by the plant, Ciećko et al. (2004) reported low content of potassium in oat grains and in the roots and above-ground parts of yellow lupine and radish grown in the soil contaminated by Cd. Jing et al. (2005) studied the effect of five Cd levels (0, 0.1, 1.0, 5.0, and 10.0 μmol/L) on growth and photosynthesis of two tomato cultivars showed that the addition of 0.1 μmol/L of Cd induced a slight increase in plant height in both the cultivars. However, at higher Cd levels (1 and 10 μmol/L), volume and length of root as well as plant height was significantly reduced. The addition of Cd in the growth medium also had significant deleterious effect on photosynthetic rate and intracellular CO_2 concentration.

As Cd ions do not alter their oxidation state because of not taking part in Fenton and Haber-Weiss reactions (Clemens 2006), Cd does not participate directly in cellular redox reactions. However, its exposure induces oxidative injuries, such as LPO, which leads to protein carbonylation and alteration in the membranes functionality, ultimately converging into general redox homeostasis impairment (Schützendübel et al. 2001). Cd triggers an over-accumulation of ROS through impairing the activities of antioxidative enzymes such as CAT and superoxide dismutase (Romero-Puertas et al. 2004). M-Kalantari and Oloumi (2005) treated *Brassica napus* seedlings with different Cd concentrations (0, 10, 50, and 100 μM). They reported that high concentration of Cd caused an increase in malondialdehyde (MDA) content in leaves, confirming the involvement of Cd in the induction of oxidative stress and LPO. Khalighi et al. (2007) registered an increase in MDA content along with the increase in the activities of detoxifying enzymes, guaiacol peroxidase, and APX, in wheat leaves grown under Cd stress.

Lead (Pb) is considered as one of the major pollutants both in aquatic and terrestrial ecosystems. Pb concentrations in uncontaminated soils are generally in the range of 20–50 mg/kg (Nriagu 1978). Soils that contain 400–800 mg/kg of Pb are regarded as significantly contaminated. Pb contamination of soil and plants mainly originates from coal-fired power plants and different industrial processes (Joshi and Mohanty 2004). Increasing levels of Pb in the soil environment exert a wide range of adverse effects on growth and metabolism of plants. The visual symptoms of Pd toxicity appear in the form of inhibited root growth, stunted growth of the plant, and chlorosis (Burton et al. 1984). Breckle (1991) reported the inhibition of root growth at a concentration of 10–2 and 10–6 M of Pb or at a soil-Pb content of above 10 mg/kg. Eun et al. (2000) suggested the inhibition of cell division in root tips as a result of inhibition of root growth under Pb toxicity. This proclamation was further confirmed by Wierzbicka (1994), who reported mitotic irregularities, chromosomal stickiness, irregularly shaped nuclei, and nuclei with decomposed nuclear material in the cells of Pb-treated root tips of onion (*Allium cepa*) that showed reduced growth. Yang et al. (2000) showed that Pb disturbed the alignment of microtubules in a concentration-dependent manner. Thus, it might be postulated that the damage to microtubules by Pb is an important component of Pb-induced injury in plants. Fargasova (2001) observed a very strong inhibitory effect of Pb on the content of chlorophyll a and b in mustard. The studies on photosynthesis may illustrate the direct and indirect effects of Pb. One of the direct effects includes the inhibition of chlorophyll synthesis that is often manifested by chlorosis (Akinci et al. 2010). Małkowski et al. (2002) observed the leakage of K ions from root cells of corn seedlings under Pb toxicity. Stefanov et al. (1995) reported the decrease in protein content and significant alterations in lipid compositions of cells due to Pb toxicity.

The Pb toxicity leads to inhibition of enzyme activities, disturbed mineral nutrition, water imbalance, change in hormonal status, and membrane permeability (Sharma and Dubey 2005). In most of the cases, Pb interacts with free –SH groups that are present in the active site of the enzyme. Prasad and Prasad (1987) reported Pb-induced inhibition of amino laevulinate dehydrogenase, the key enzyme of chlorophyll biosynthesis. Moustakas et al. (1994) reported the reduced activity of key enzyme of CO_2 fixation, i.e. Ribulose-1, 5-bisphosphate carboxylase oxygenase (Rubisco), in *Avena sativa.* On the other hand, Verma and Dubey (2003) noticed an increase in the activities of APX, SOD, and glutathione reductase and a decrease in CAT activity of *Oryza sativa* grown under Pb stress. Verma and Dubey (2003) tested the effect of 500 and 1,000 μM of $Pb(NO_3)_2$ on two rice (*Oryza sativa* L.) cultivars. Pb treatment significantly decreased the growth of seedlings and induced oxidative stress and LPO in rice plants. It was suggested that SOD, peroxidases, and GR (Glutathione reductase) might serve as important components of antioxidative defense mechanism against Pb-induced oxidative injury in rice. Kibria et al. (2010) studied the effects of six levels of Pb, viz. 0, 20, 40, 60, 80, 100 mg/kg, on growth and partitioning of ion concentration in *Spinacea oleracea*. They observed a gradual decrease in dry weight both of shoot and root with increasing Pb levels. Besides, K concentration both in shoots and roots and the concentration of Mg, Zn, and Fe in roots were decreased with Pb application. On the contrary, Pb application significantly increased N and Mn concentrations in shoots and Ca and Mn concentrations in roots. Whereas, Orhue and Inneh (2010) reported a decrease in N and P concentrations in Pb-treated *Celosia argentea.* Akinci et al. (2010) studied the effect of various Pb levels (0, 75, 150 and 300 mg/L) on *Solanum lycopersicum* L. seedlings. Increased values for Pb concentrations were recorded in root, shoot, and leaves. On the contrary, element uptake, root elongation, plant height, leaf fresh weight, dry biomass of root and shoot, and leaf area were negatively affected by increasing lead concentrations. Tissue water content, growth tolerance index, chlorophyll a and b, and total chlorophyll content were depressed by lead toxicity. Stefanov et al. (1995), in a long-term field experiment (41 years) involving regular application of mineral fertilizers to crops of sunflower or barley followed by oat-winter rye in rotation, revealed that the fertilizers increased the level of mobile forms of Pb in the soil and also its uptake by the crops.

10 Temperature Stress

Temperature stress affects several facets of cellular components and plant metabolism. Generally four kinds of temperature stress exist in nature: (1) sustained high temperature, (2) heat shock, (3) chilling at temperatures above 0 °C, and (4) freezing at temperatures below 0 °C. High and low temperature provoke the physiological and biochemical changes, including water deficit and oxidative stress leading to LPO and membrane damage, degradation of chlorophyll and protein, and reduced water status of plants (Xin and Browse 1998; Jiang and Huang 2001a, b; Suzuki and Mittler 2006).

An irreversible damage to plant growth and development due to rise in temperature beyond a threshold level for a period of time is termed as heat stress. In general, a transient elevation in temperature, usually 10–15 °C above ambient, is considered heat shock or heat stress. The highest photosynthetic rate is generally recorded at 30 °C. The optimum temperature for photosynthesis depends on the temperature in early growth of the plants (Hikosaka et al. 1999). Increase in the optimum temperature ranges was reported to inhibit the photosynthetic activity and plant growth (Berry and Björkman 1980). Before the detection of other symptoms, inhibition in photosynthesis is the most prominent symptom of high temperature stress (Camejo et al. 2005). Generally, increased temperature above the optimum limit may reduce photosynthetic rate (Wahid et al. 2007). Mohammad and Tarpley (2009) indicated that high temperatures had a negative effect on photosynthesis as well as on the activities of various enzymes involved in the process in rice (*Oryza sativa* L.). Decrease in photosynthesis could also result from structural and functional disruptions of chloroplasts and reduction in chlorophyll accumulation under high temperature stress (Xu et al. 1995; Dekov et al. 2000). Malgorzata et al. (2008) studied the effect of high temperature on various physiological parameters of tomato (*Lycopersicon esculentum* L.). They suggested that rate of photosynthesis and transpiration, stomatal conductance, and contents of chlorophyll a and b could be considered as reliable indicators of high temperature stress. Law and Crafts-Brandner (1999) suggested the decrease in the activity of Rubisco enzyme as a reason for the inhibition of photosynthesis under high temperature stress. Guchou et al. (2007) reported a slight increase in photosynthesis under moderate high temperature, but a significant decrease under high temperature. Haldimann and Feller (2005) in *Pisum sativum* L. and Langjun et al. (2006) in *Festuca arundinacea* reported that high temperature stress caused a reduction in Rubisco activity and net photosynthetic rate. In addition, it increased photoinhibition and caused modification in PSII functionality. They also reported an increase in LPO, decrease in cell membrane thermostability, and changes in the activities of APX and SOD under high temperature. Xu and Zhou (2006) studied the effect of high temperature on the performance of *Leymus chinensis* Trin. They observed that high temperature significantly decreased plant biomass, leaf green area, LWP, photosynthetic rate, maximal efficiency of PSII, actual efficiency of PSII, the activities of nitrate reductase (EC: 1.6.6.1), and glutamine synthetase (EC: 6.3.1.2), but markedly induced an increase in the ratio of leaf area to leaf weight (SLA), activity of endopeptidase (EC: 3.4.24.11), and content of MDA. Sucrose synthase (EC: 2.4.1.13) plays an important role in sucrose degradation and starch synthesis. Tian et al. (2006) studied the effect of high temperature on the sucrose content and the activity of sucrose cleaving enzyme (Sucrose synthase) in rice grain. They reported that high temperature caused a significant increase in grain sucrose content without any increase in fructose and glucose contents, suggesting that the high temperature treatment enhanced sucrose accumulation, while it diminished the sucrose degradation in rice grains. A decrease in the activities of sucrose synthase, vacuolar invertase, and cell wall bound invertase in the plants treated with high temperature indicated that decrease in sucrose degradation was related to the decrease in activities of sucrose synthase and invertase. Ali et al. (2005) reported that high temperature stress (at 25–40 °C) induced not only the activities of ROS scavenging enzymes, but also increased the activity of lipoxygenase and the content of MDA and cysteine as well as that of protein and non-protein-thiol (NP-SH) in the leaf and root segments of *Phalaenopsis*. However, the activities of dehydroascorbate reductase (EC: 1.8.5.1), glutathione peroxidase (EC: 1.11.1.9), and glutathione-S-transferase (EC: 2.5.1.18) in leaf and root, while those of glutathione reductase (EC: 1.6.4.2) in leaf and guaiacol peroxidase (EC: 1.11.1.7) in root, were induced significantly at 40 °C compared to that at 25 °C and in the normal greenhouse conditions, suggesting that these enzymes play protective roles at high temperature. In contrast, the activity of SOD and monodehydroascorbate reductase (EC: 1.6.5.4) in leaf and root and that of CAT and GR in root, and the contents of protein, cysteine, and NP-SH both in root and leaf in addition to the ratio of variable chlorophyll fluorescence to maximum chlorophyll fluorescence (Fv: Fm) were diminished significantly at 40 °C compared to that at 25 °C and in the normal greenhouse conditions in *Phalaenopsis*.

Low temperature (LT) is a major factor limiting the growth, productivity, and distribution of many important agricultural crops. Plants show damage to the cell membrane. The membrane is ruptured due to expanding ice crystals or due to lamination of the membrane following thawing. When plants are subjected to low temperature stress, ROS are produced frequently. The ROS are highly reactive in nature, and in the absence of any protective mechanism, they may disrupt normal metabolism through oxidative damage to lipids, protein, and nucleic acids (Allen 1995). Carbohydrate metabolism has been reported to have greater sensitivity to instantaneous low temperature than the other components of photosynthesis (Leegood and Edwards 1996). However, many plant species are able to show freezing tolerance (FT) in response to low or non-freezing temperature. This process, referred to as cold-acclimation, is associated with various physiological and biochemical changes like those in lipid composition, activities of ROS scavenging enzymes, accumulation of anthocyanin, and plant morphology owing to the alterations in the expression of a number of cold-responsive genes (Sarnighausen et al. 2004; Eris et al. 2007). Cao et al. (2011) showed that low temperature inhibited the growth of oil palm seedlings. Whereas, relative conductivity, injury index, MDA, and PRO content in the leaves were increased to different degrees with the extension of low temperature stress. SOD and peroxidase (POX) activities increased and then decreased gradually with the duration of treatment time.

Li et al. (2008) studied the effect of low temperature stress on various physiological and biochemical attributes in Croftonweed (*Eupatorium adenophorum*). Under low temperature, there were observed several physiological changes, including increases in MDA and total soluble protein contents, reductions in the contents of total soluble sugars and chlorophyll, changes in the Fv: Fm ratio, and fluctuations in SOD activity. However, different degrees of physiological responses were found. According to Islam et al. (2011), exposure of sweet potato (*Ipomoea batatas* L.) to chilling stress decreased transpiration rate and stomatal conductance but increased the electrolyte leakage and peroxidase activity. Naumburg et al. (2004) showed the inhibitory effect of night-time freezing air temperatures on the subsequent day-time photosynthetic rates, stomatal conductance, and the maximum quantum yield of PSII. Moreover, they found no significant relationship between leaf temperature and photosynthetic enhancement. Increase in membrane stability, which expresses enhanced abiotic stress tolerance, has been considered essential during cold-acclimation both under natural and synthetic environmental conditions (Premachandra et al. 1992; Uemura et al. 2006). For instance, Li et al. (2008) reported enhanced content of MDA in low temperature-sensitive populations, while there was found a low content of MDA in low temperature-tolerant populations of Croftonweed (*Eupatorium adenophorum*).

11 Role of Calcium in the Mitigation of Abiotic Stresses in Plants

Study of plant mineral nutrients relations is required to understand the effect of nutrient on plants under abiotic stresses. Steady supply of mineral nutrients via the roots is restricted under drought and salinized situations because of a negative effect of drought and salinity on nutrient availability. Efficacy of foliar fertilization is higher than that of fertilizer application through soil in these situations. In fact, foliar nutrient application under drought and salinity conditions may be able to exclude or include a water deficit or nutrient effect under short-term drought or salt stress.

Plant scientists have studied the effects of calcium on growth of plant and production of secondary metabolites in several medicinal plants (Lee and Yang 2005, Supanjani et al. 2005, Khan and Naeem 2006, Karaivazoglou et al. 2007, Dordas 2009). The effect of $CaCl_2$ on growth parameters, ionic relations, and PRO level was investigated by Arshi et al. (2005) in senna (*Cassia angustifolia*) plants under salinity stress. Combined treatments of $NaCl+CaCl_2$ applied at different growth stages reduced the biomass, but this reduction was less than that observed with NaCl treatments alone. They suggested that calcium could alleviate the NaCl-induced inhibition of plant growth via the maintenance of net K^+ to Na^+ selectivity and the enhancement of PRO accumulation in the leaves. In another study, Arshi et al. (2006) assessed the extent of growth, photosynthetic efficiency and nitrogen assimilation of chicory (*Cichorium intybus* L.) as affected by NaCl and $CaCl_2$ applied alone as well as in combination. Application of NaCl caused a significant reduction in total plant biomass, photosynthetic rate, stomatal conductance, total chlorophyll content, soluble protein content, NR activity, and nitrogen content, although the leaf-nitrate content was increased. However, the application of $CaCl_2$ mitigated the adverse effects caused by NaCl stress. Patra et al. (2002) studied the effect of three levels of gypsum (calcium sulfate), viz. no gypsum (control), GR25 (25% of the gypsum requirement), and GR50 (50% of the gypsum requirement), on lemongrass (*Cymbopogon flexuosus*). GR25 was optimum for the highest production of lemongrass and accumulation of major as well as trace elements in leaves. Goharrizi et al. (2011) investigated the effect of different salinity levels both on unstressed and salt-stressed walnut plants, using $CaCl_2$ as stress amelioration source. They found that the activities of peroxidase (POD) and CAT enzymes increased in parallel with the increase in salt levels in the leaves of tolerant genotypes of walnut treated with $CaCl_2$. However, in all the treatments, POX and CAT activities were reduced significantly from 6 to 10 days after beginning the salt stress. As per the study of Jaleel and Azooz (2009), the applied treatments comprising sodium chloride and calcium chloride (100 mM NaCl, 5 mM $CaCl_2$, and 100 mM $NaCl+5$ mM $CaCl_2$) altered the PRO metabolism of *Withania somnifera* plants. Combined application of NaCl and $CaCl_2$ resulted in decreased contents of chlorophyll a, chlorophyll b, total chlorophyll, and carotenoids, decreased activity of PRO oxidase, and increased activity of γ-glutamyl kinase activity. Application of $CaCl_2$ to NaCl-treated plants resulted in an increase in the content of chlorophyll a, chlorophyll b, and total chlorophyll along with an increase in PRO oxidase activity and a decrease in γ-glutamyl kinase activity, when compared with the plants treated with NaCl only. Gobinathan et al. (2009) showed that application of $CaCl_2$ increased the glycine betain (GB) contents in *Pennisetum typoidies*. In fact, $CaCl_2$ appeared to confer a greater osmoprotection to cells because of its additive role in GB accumulation in NaCl-treated plants of *Pennisetum typoidies*. Bolat et al. (2006) applied $CaSO_4$ at 2.5 and 5 mM to salt-treated plants of plum grown in sand culture with complete nutrient solution. They showed that $CaSO_4$ application ameliorated the negative effects of salinity on plant dry matter and chlorophyll content. In addition, salt-treatment impaired the membrane permeability by increasing electrolyte leakage; however, the addition of calcium sulfate partially maintained membrane permeability. The application of

$CaSO_4$ also lowered significantly the concentrations of Na both in leaves and roots. Abo-Kassem (2007) studied seed germination, seedling growth, and the activities of some of the enzymes of nucleic acid metabolism in five species of Chenopodiaceae (*Beta vulgaris* L., *Chenopodium quinoa* Willd., *Spinacea oleracea* L., *Allenrolfia occidentalis* (S. Watson) Kuntze, and *Atriplex hortensis* L.) grown with NaCl alone or in combination with 0.5 mM of $CaSO_4$. According to them, high salinity delayed radical emergence and decreased the germination percentage in all the species studied. However, application of $CaSO_4$ reduced inhibition of seed germination in *B. vulgaris*, *S. oleracea*, and *A. hortensis*, but increased it in *C. quinoa* and *A. occidentalis*. Salinity progressively activated deoxyribonuclease (DNase I) and $CaSO_4$ enhanced its activation in all the species, except *B. vulgaris*. Salinity increased endonuclease activity, except in *S. oleracea* and *A. occidentalis*. Addition of $CaSO_4$ increased endonuclease activity in *C. quinoa, S. oleracea,* and *A. occidentalis.* Salinity inhibited ribonuclease A (RNase A) activity, but increased it in *C. quinoa* and *S. oleracea*, whereas $CaSO_4$ alleviated such inhibition in *A. occidentalis and A. hortensis*. Salinity increased ribonuclease T (RNase T) activity in all the species, especially in *C. quinoa, S. oleracea,* and *A. occidentalis*.

Using nutrient solution, Tuna et al. (2007) investigated the effects of supplementary $CaSO_4$ on tomato (*Lycopersicon esculentum* Mill.) grown at high NaCl concentration (75 mM). The plants grown under salt stress gave lower values of dry matter, fruit weight, and relative water content (RWC) than those grown in standard nutrient solution without salt stress. Application of $CaSO_4$ to the nutrient solution containing high salt concentration significantly improved the attributes that were adversely affected by salt stress (e.g. plant growth, fruit yield, and membrane permeability). In addition, the supplemental $CaSO_4$ increased the concentrations of K^+, Ca^{2+}, and N and reduced the concentration of Na^+ in the leaves. According to Jaleel et al. (2007c), supplementing the growth medium with Ca alleviated the salt-inhibited plant growth in glycophytic plants. Further, Jaleel et al. (2008b) maintained that Ca sustained K^+ transport and K^+/Na^+ selectivity in Na^+ challenged plants. Manivannan et al. (2007b) assessed the ameliorating effect of $CaCl_2$ on NaCl-stressed plants of *Vigna radiata* (L.) Wilczek. The stressed plants showed reduction in growth as indicated by diminished root length, stem length, total leaf area, and dry weight. The contents of PRO and glycinebetaine and the activities of the SOD, APX, and CAT enzymes were increased by the treatment of NaCl or $CaCl_2$ applied alone. However, when $CaCl_2$ was combined with NaCl, it altered the overall plant metabolism, ameliorating the deleterious effects of NaCl stress and increasing the vegetative growth of plants.

Application of $CaCl_2$ increased the drought tolerance of *Catharanthus roseus* with favorable changes in oxidative stress, osmoregulation, and indole alkaloid accumulation (Jaleel et al. 2007a, b). The plants were grown under water-deficit environments with or without $CaCl_2$. Drought-stressed plants showed increased contents of LPO, H_2O_2, glycinebetaine (GB), and PRO coupled with decreased proline oxidase (PROX) activity and increased gamma-glutamyl kinase (gamma-GK) activity. Application of $CaCl(_2)$ lowered the PRO concentration in drought-stressed plants while increasing the activity of PROX and decreasing that of gamma-GK.

Application of CaCl($_2$) appeared to confer osmoprotection to cells owing to its additive role with regard to GB accumulation under drought-stress. The drought-stressed *C. roseus* plants treated with CaCl($_2$) showed an increase in total content of indole alkaloids in shoots and roots when compared to drought-stressed and well-watered plants. Jaleel et al. (2008a, b) studied the effect of $CaCl_2$ application on NaCl-stressed plants of *Dioscorea rotundata*. NaCl-stressed plants showed decreased contents of protein and total sugars together with increased contents of free amino acid and PRO. When NaCl treatment was combined with $CaCl_2$, the overall plant metabolism was altered, with the plants showing increased activity of antioxidant enzymes and partial amelioration of oxidative stress caused by salinity stress. Lin et al. (2008) pre-treated the sweet potato varieties, with four levels of $CaCl_2$ (0, 60, 120, and 180 kg/ha), subjecting the plants to non-flooding (control) and flooding conditions. The level of activity of the enzymes of antioxidative system in the leaves was related to the $CaCl_2$ pre-treatment level during flooding. The $CaCl_2$ pre-treatment at 60 and 120 kg/ha amplified the activities of APX, SOD, and glutathione reductase significantly in addition to enhancing the contents of total ascorbate, glutathione, and MDA under flooding stress. Thus, pre-treatment with $CaCl_2$ enhanced the flooding tolerance of the sweet potato varieties, mitigating the adverse effects of flooding stress.

Chowdhury and Choudhuri (1986) observed a significant decrease in the RWC and LWP in two species of jute (*Corchorus capsularis* L. and *C. olitorius* L.) subjected to water stress. Pre-treatment of seeds with 5 mM of $CaCl_2$ improved the water uptake capacity of plants without altering the stomatal movement. Further, they observed a greater decrease in the uptake of phosphate (^{32}P) in *C. olitorius* than in *C. capsularis* under water-deficit stress, with Ca^{2+} application counteracting the adverse effects of drought stress in this regard. Calcium may be involved in drought tolerance of plants because of regulating antioxidant metabolism. Application of Ca considerably increased the fresh weight and RWC of liquorice (*Glycyrrhiza glabra*) cells after 10-days water stress (Li et al. 2003). Compared to untreated cells (control), lesser amounts of MDA and H_2O_2 were accumulated in Ca-treated cells in addition to increased activities of CAT, SOD, and POD in Ca-treated cells during the water-stress period. Water stress induced oxidative stress in liquorice cells; however, application of external calcium (40 mmol/L) improved it significantly. They concluded that extracellular Ca improved the adaptation of liquorice cells to drought stress, mitigating the oxidative stress thereof.

A considerable number of studies have been conducted to overcome the detrimental effects of Al toxicity in crop plants. Nevertheless, beneficial effects of calcium (Ca^{2+}) have long been implicated to combat Al phytotoxicity (Kinraide et al. 1992; Rengel 1992; Zhang et al. 1999). Aluminium affects symplastic Ca^{2+} homeostasis by modifying Ca^{2+} fluxes across the plasma membrane. Supplementary Ca^{2+} has been reported to alleviate deleterious effects of soil-Al (Rengel 1992). Silva et al. (2005) reported ameliorative effect of applied Ca on Al-induced rhizotoxicity. Ikegawa et al. (2000) reported that Al sensitizes the membranes to the Fe(II)-mediated peroxidation of lipids and Al-enhanced peroxidation of lipids serves as a direct cause for the loss of integrity of plasma memebrane in Ca medium. Decreased concentrations of Ca in soybean tops and roots were also associated with Al toxicity (Foy et al. 1969). Amelioration of Al toxicity by Ca has been attributed to two basic mechanisms: (1) restoration of Al-displaced Ca ions to non-limiting levels and (2) increased root cell potential with a decrease in Al ions at the membrane surface (Rengel 1992; Horst 1995; Kinraide 1998). Nonetheless, Silva et al. (2001) reported a less negative electrical potential at the membrane surface and decreased Al^{3+} activity at the root cell surface with increasing levels of Ca in the growth solution. Al toxicity rapidly induced the bursting of pollen tube tip in *Chamelaucium uncinatum*; while the tip-bursting was markedly reduced by increasing the Ca^{2+} activity of the incubation medium (Zhang et al. 1999). An increase in Ca^{2+} concentration from 0.25 to 5 mM significantly reduced the Al-induced bursting of pollen tube (Rout et al. 2001). In the presence of Al^{3+}, H^{+}, or Na^{+}, supplementation of the growth medium with higher levels of Ca alleviated the growth inhibition due to Ca-mediated displacement of cell-surface toxic cations (Kinraide and Parker 1987; Yan et al. 1992; Yermiyahu et al. 1997). High Al/Ca ratio resulted in lowered photosynthetic capacity and increased respiration rate in *Pinus sylvestris* (Reich et al. 1994). Ma (2005, 2007) reported an excess accumulation of ROS under elevated Al^{3+} toxicity in acid soils. ROS scavenging mechanism in plants is regulated by non-enzymatic and enzymatic antioxidants. The antioxidants of enzymatic nature such as SOD, CAT, peroxidase (POX), and APX and those of non-enzymatic nature such as ascorbate, α-tocopherol, β-carotene, and flavonoids (Stahl and Sies 2003; Shao et al. 2008) are involved in ROS scavenging mechanism as per need. Calcium has been shown to stimulate the enzymatic antioxidant activity of SOD, CAT, and POX in Al-suffered barley plants (Guo et al. 2006) and in Japanese cedar needles (Takami et al. 2005). Exogenous application of Ca increased the activity of CAT, APX, and GR, reduced the LPO, and raised the RWC and chlorophyll content (Jiang and Huang 2001a, b) in grasses. Additionally, external application of $CaCl_2$ was associated with relatively higher activities of antioxidant enzymes and lower levels of LPO compared with Ca-deficient maize seedlings (Gong et al. 1997; Jiang and Huang 2001a, b). Akaya and Takenaka (2001) reported that Al toxicity effects on the photosynthesis of *Quercus glauca* could be ameliorated by Ca application.

The inhibitory effect of Ca on cadmium (Cd) uptake and toxicity has also been reported. Ca concentration in the cells greatly increased during Cd stress in plants (DalCorso et al. 2008), leading to the stimulation of calmodulin-like proteins that interacted with Ca ions. These calmodulin proteins, by modulating their conformation in response to Ca binding, are reported to regulate ion transport, gene regulation, metabolism, and stress tolerance (Yang and Poovaiah 2003). The Ca/calmodulin system was also involved in sensing other heavy metals. Supplementary Ca was able to restore the Cd-induced inhibition of root elongation (Suzuki 2005). Wan et al. (2011) suggested that presence of Ca^{2+} in the proximity of plasma membrane is proficient in alleviating Cd toxicity by reducing the cell-surface negativity and by competing for Cd^{2+} ion influx, leading to the least alteration in seedling growth and photosynthetic traits of *Brassica napus* L. under Cd-toxicity. A proposed mechanism regarding the Ca-mediated alleviation of mineral toxicity is the displacement of cell-surface toxic cations by Ca. As plasma membrane surfaces are usually negatively charged, high-level Ca^{2+} reduces cell-surface negativity and, thus, alleviates the harmful effects of cationic toxicants (Kinraide 1998; Wan et al. 2011). However,

Perfus-Barbeoch et al. (2002) proposed the mechanism of the uptake of Cd through calcium channels to mimic Ca because the uptake of Cd is inhibited by the Ca channel blockers, diltiazem, verapmil, nifedipine, and nitrendipine (Blazka and Shaikh 1991). According to Suzuki (2005), the presence of high concentrations of Ca around Ca channels decreased the Cd uptake by competition for metal ion influx. Several studies showed that applied Ca alleviated the Cd-induced oxidative stress by reducing membrane damage coupled with enhanced activities of antioxidative enzymes (Wang and Song 2009).

Calcium has also been shown to mitigate the adverse effects of Pb toxicity. In fact, Garland and Wilkins (1981) reported a linear increase in root growth coupled with a decrease in Pb uptake with increasing concentration of Ca in *Hordeum vulgare* L. and *Festuca ovena* L. In addition, Antosiewicz (2005) in maize, rye, tomato, and mustard and Wojas et al. (2007) in tobacco plants reported Ca-mediated regulation of internal lead detoxification.

Several evidences suggest that Ca is involved in the regulation of plants to high and low temperature stress and enable the plant cells to survive comparatively better by alleviating the adverse effects of temperature injuries. An increase in the cell-sap Ca^{2+} is recorded in the plants grown under heat stress; even so, the supplementary Ca treatment enhances the Ca^{2+} level of the heat-stressed plants (Gong et al. 1998). This change in Ca^{2+} level affects the binding activity of heat stress transcription factors (HSF) to heat shock elements (HSE) (Mosser et al. 1990). However, Liu et al. (2003) proposed that heat stress signals are perceived by an unidentified receptor, which upon activation increases the concentration of Ca^{2+} through opening of calcium channels. This elevated calcium directly activates calmodulin protine which, in turn, activates the DNA binding ability of heat shock factors, thus initiating heat-stress-related proteins. Saidi et al. (2009) have found a Ca^{2+} permeable channel in the plasma membrane that was activated by high temperature and could also modulate the intensity of heat shock response. Applied Ca also improved heat stress-induced reduction in RWC and chlorophyll content together with reducing the LPO by enhancing the activities of antioxidant enzymes (Jiang and Huang 2001a, b). Heat stress caused impairs leaf expansion, which was overcome by increasing root zone calcium levels (Kleinhenz and Palta 2002).

Cold stress mainly results in disruption of membrane integrity leading to severe cellular dehydration and osmotic imbalance. Cold-acclimation results in triggering of various genes, which result in restructuring of the cellular membranes by change in the lipid composition and generation of osmolytes, which prevent cellular dehydration, leading to cold stress tolerance. One of the earliest consequences of the temperature sensing mechanism is Ca^{2+} influx into the cytosol (Chinnusamy et al. 2006; Kaplan et al. 2006), which enables the plant to withstand cold stress in a better way. The increase in cytosolic Ca^{2+} $[(Ca^{2+})_{cyt}]$ is considered to be the prime agent in the events resulting in protein phosphorylation, altered gene activity, formation of new gene products, alterations in plant membrane properties, and modifications in secondary metabolism, leading to the adaptation and acclimation of plants to cold stress. It has been demonstrated that Ca is required for the full expression of some of the cold-induced genes, including *CRT/DRE* controlled *COR6* and *KIN1* genes of *Arabidopsis* (Knight et al. 1996). The role of $[(Ca^{2+})_{cyt}]$ has further been confirmed through experiments in which the administration of calcium chelators and calcium channel blockers has been shown to prevent cold-acclimation (Monroy et al. 1997). Ruiz et al. (2002) suggested that acclimation of plants to cold stress is preceded by increased $[(Ca^{2+})_{cyt}]$ and found a significant and strong relationship between Ca^{2+}-Calmodulin-dependent NAD kinase activity and PRO metabolism.

12 Conclusion

Plant scientists are exploring different ways to improve yield limits of existing demand of MAPs. The most feasible and successfully adopted technique is the basal or foliar application of mineral nutrients for exploiting the genetic potential of a crop under normal as well as adverse conditions. There is thus an urgency to give special emphasis and to clearly define the policies to regulate conservation, cultivation, quality control standards, processing, preservation, marketing, and export of medicinal plants. Out of these useful steps and policies regarding sustenance of aromatic and medicinal plants, the cultivation of medicinal plants on scientific lines appears to be extraordinarily effective to obtain authentic, standard, and fresh herbal materials. This would be a safeguard against unauthentic, spurious, denatured, fake, and soiled drugs.

In several studies, calcium has been shown to ameliorate the growth, yield, and quality attributes of various medicinal plants including medicinal legumes under normal and adverse conditions occurring due to biotic and abiotic stresses. In fact, it is the need of the hour to explore the possibilities to ameliorate and enhance the biosynthesis of secondary plant metabolite in MAPs through the application of calcium under abiotic stresses. Secondary metabolites production of the MAPs is strongly influenced by environmental factors such as climate, plant density, use of fertilizers, etc. However, the effect of different levels of chemical fertilizers on bioactive components and secondary metabolites in the MAPs has attracted less attention. Exogenous application of Ca regulates growth and development of plants through the mediation of genes that may determine their orientation, physiology, and productivity.

Abiotic stresses are major limiting factors of crop yields and cause losses of billions of dollars annually around the world. Understanding as to how the plants respond to adverse conditions and adapt to a changing environment at the molecular level might help the plants to cope better with the abiotic stresses. Researches conducted during the last two decades have established that different stresses cause signal-specific changes in cellular Ca level, which functions as a second messenger in modulating diverse physiological processes that are important for stress adaptation. Ca-regulated transcriptional genes play key roles in stress-signaling pathways that might be involved in plant tolerance to abiotic stresses by regulating antioxidant metabolism in plants.

References

Abbaszadeh B, Sharifi Ashoorabadi E, Ardakani MR, Aliabadi FH (2008) Effect of drought stress on quantitative and qualitative of mint. Abstracts Book of 5th International Crop Science Congress & Exhibition, Korea. p. 23

Abel AJ, Sutherland MW, Guest DI (2003) Production of reactive oxygen species during non-specific elicitation, non-host resistance and field resistance expression in cultures of tobacco cells. Func Plant Biol 30:91–99

Abou El-Fadl IA, Abd-Ella MK, Hussein EH (1990) Effect of irrigation by saline water on 255 the growth and some principal compounds of peppermint and spearmint in two types of soil. J Agric Res Tanta Univ 16:276–295

Ahmedullah M, Nayar MP (1999) Red data book of Indian plants, vol 4 (Peninsular India). Botanical Survey of India, Calcutta

Akaya M, Takenaka C (2001) Effects of aluminum stress on photosynthesis of *Quercus glauca* Thumb. Plant Soil 237:137–146

Akinci IE, Akinci S, Yilmaz K (2010) Response of tomato (*Solanum lycopersicum* L.) to lead toxicity: Growth, element uptake, chlorophyll and water content. African J Agric Res 5:416–423

Alam SM (1999) Nutrient uptake by plants under stress conditions. In: Pessarakli M (ed) Handbook of plant and crop stress. Marcel Dekker, New York, pp 285–314

Alcantara E, Romera FJ, Canñete M, De La Guardia MD (1994) Effects of heavy metals on both induction and function of root Fe(III) reductase in Fe-deficient cucumber (*Cucumis sativus* L.) plants. J Exp Bot 45:1893–1898

Alderfasi AA, Alghamdi SS (2010) Integrated water supply with nutrient requirements on growth, photosynthesis productivity, chemical status and seed yield of Faba Bean. American-Eurasian J Agron 3:08–17

Aliabadi FH, Lebaschi MH, Hamidi A (2008a) Effects of arbuscular mycorrhizal fungi, phosphorus and water stress on quantity and quality characteristics of coriander. J Adv Nat Appl Sci 2:55–59

Aliabadi FH, Lebaschi MH, Shiranirad AH, Valadabadi AR, Daneshian J (2008b) Effects of arbuscular mycorrhizal fungi, different levels of phosphorus and drought stress on water use efficiency, relative water content and proline accumulation rate of coriander (*Coriandrum sativum* L.). J Med Plant Res 2:125–131

Aliabadi FH, Valadabadi SAR, Daneshian J, Khalvati MA (2009) Evaluation changing of essential oil of balm (*Melissa officinalis* L.) under water deficit stress conditions. J Med Plant Res 3:329–333

Ali RM, Abbas HM (2003) Response of salt stressed barley seedlings to phenylurea. Plant Soil Environ 49:158–162

Ali MB, Hahn EJ, Paek KY (2005) Effects of temperature on oxidative stress defense systems, lipid peroxidation and lipoxygenase activity in Phalaenopsis. Plant Physiol Biochem 43: 213–223

Al-Khayri JM, Al-Bahrany AM (2004) Growth, water content, and proline accumulation in drought-stressed callus of date palm. Biol Planta 48:105–108

Allen R (1995) Dissection of oxidative stress tolerance using transgenic plants. Plant Physiol 107:1049–1054

Ansari SR, Frooqi AHA, Sharma S (1998) Interspecific variation in sodium and potassium ion accumulation and essential oil metabolism in three Cymbopogon species raised under sodium chloride stress. J Essential Oil Res 10:413–418

Antosiewicz DM (2005) Study of calcium-dependant lead-tolerance on plants differing in their level of Ca-deficiency tolerance. Environ Pollut 134:23–34

Arp PA, Strucel I (1989) Water uptake by black spruce seedlings from rooting media (solution, sand, peat) treated with inorganic and oxalated aluminum. Water Air Soil Pollut 44:57–70

Arshi A, Abdin MZ, Iqbal M (2004) Changes in biochemical status and growth performance of Senna (*Cassia angustifolia* Vahl.) grown under salt stress. Phytomorphology 54:109–124

Arshi A, Abdin MZ, Iqbal M (2006) Effect of $CaCl_2$ on growth performance, photosynthetic efficiency and nitrogen assimilation of *Cichorium intybus* L. grown under NaCl stress. Acta Physiol Plant 28:137–147

Arshi A, Abdin MZ, Iqbal M (2005) Ameliorative effect of $CaCl_2$ on growth, ionic relation and proline content of senna under salinity stress. J Plant Nutrition 28:101–125

Arshi A, Abdin MZ, Iqbal M (2002) Growth and metabolism of senna as affected by salt stress. Biol Plant 45:295–298

Aziz EE, Hendawi SF, Ezz El Din A, Omer EA (2008) Effect of soil type and irrigation intervals on plant growth, essential oil yield and constituents of Thymus vulgaris plant. American-Eurasian J Agric Environ Sci 4:443–450

Aziza EE, Al-Amier H, Craker LE (2008) Influence of salt stress on growth and essential oil production in peppermint, pennyroyal, and apple mint. J Herb Spice Med Plant 14:77–87

Baher Z, Mirza M, Ghorbanli M, Rezaii MB (2002) The influence of water stress on plant height, herbal and essential oil yield and composition in *Satureja hortensis* L. Flav Frag J 17:275–277

Bell RW, Edwards DG, Asher CJ (1989) External calcium requirements for growth and nodulation of six tropical food legumes grown in flowering solution culture. Aust J Agric Res 40:85–96

Benavides MP, Gallego SM, Tomaro ML (2005) Cadmium toxicity in plants. Braz J Plant Physiol 17:21–34

Bennet RJ, Breen CM, Fey MV (1987a) The effects of aluminium on root cap function and root development in *Zea mays* L. Environ Exp Bot 27:91–104

Bennet RJ, Breen CM, Fey MV (1987b) The effects of aluminium on root cap function and root development in *Zea mays* L. Environ Exptl 27:91–104

Berry JA, Björkman O (1980) Photosynthetic response and adaptation to temperature in higher plants. Annu Rev Plant Physiol 31:411–534

Bettaieb I, Zakhama N, Wannes WA, Kchouk ME, Marzouk B (2008) Water deficit effects on *Salvia officinalis* fatty acids and essential oils composition. Sci Hort 120:271–275

Bhattacharjee SK (2004) Handbook of medicinal plants. Pointer, Jaipur (4th revised and enlarged edition)

Bin Y (1995) A study of increasing salt resistance in rice using calcium. Acta Agron Sinica 21:685–690

Blamey FPC (2001) The role of the root cell wall in aluminum toxicity. In: Ae N, Arihara J, Okada K, Srinivasan A (eds) Plant nutrient acquisition: new perspectives. Springer, Tokyo, Japan, pp 201–226

Blancafor EB, Jones DL, Gilroy S (1998) Alterations in the cytoskeleton accompany aluminum-induced growth inhibition and morphological changes in primary roots of maize. Plant Physiol 118:159–172

Blazka ME, Shaikh ZA (1991) Differences in cadmium and mercury uptakes by hepatocytes: role of calcium channels. Toxicol Appl Pharmacol 110:355–363

Bolat I, Kaya C, Almaca A, Timucin S (2006) Calcium sulfate improves salinity tolerance in rootstocks of Plum. J Plant Nutr 29:553–564

Boussama N, Ouariti O, Suzuki A, Ghorbel MH (1999) Cd-stress on nitrogen assimilation. J Plant Physiol 155:310–317

Brachet J, Cosson L (1986) Changes in the total alkaloid content of Datura innoxia Mill. subjected to salt stress. J Experimental Bot 37:650–656

Breckle SW (1991) Growth under stress: heavy metals. In: Waisel Y, Eshel A, Kafkafi U (eds) Plant roots—the hidden half. Marcel Dekker, New York, pp 351–373

Buchanan BB, Gruiseem W, Jones RL (2000). Biochemistry and molecular biology of plants. American Society of Plant Physiologists, Rockville. pp. 2–50, 52–108, 110–158, 202–258, 930–987,1204–1249, 1250–1318

Burton KW, Morgan E, Roig A (1984) The influence of heavy metals on the growth of sitka-spruce in SouthWales forests. II. Greenhouse experiments. Plant Soil 78: 271–282
Cakmak I (2005) Role of mineral nutrition in tolerance to crop plants to environmental stress factors. Fertigation: optimizing the utilization of water and nutrients. Fertigation proceedings: Selected papers of the IPI-NATESE-CAU-CAAS. International Symposium on Fertigation, Beijing, 20–24 September 2005, pp. 35–48
Cakmak I, Horst WJ (1991) Effect of aluminium on lipid peroxidation, superoxide dismutase, catalase, and peroxidase activities in root tips of soybean (*Glycine max*). Physiol Plant 83:463–468
Camejo D, Rodriguez P, Angeles Morales M, Dell Amico JM, Torrecillas A, Alarcón JJ (2005) High temperature effects on photosynthetic activity of two tomato cultivars with different heat susceptibility. J Plant Physiol 162:281–289
Cao HX, Sun CX, Shao HB, Lei XT (2011) Effects of low temperature and drought on the physiological and growth changes in oil palm seedlings. African J Biotech 10:2630–2637
Carpita NC, Gibeaut DM (1993) Structural models of primary cell walls in flowering plants: consistency of molecular structure with the physical properties of the walls during growth. Plant J 3:1–30
Cavalcanti BC, Costa-Lotufo LV, Moraes MO, Burbano RR, Silveira ER, Cunha KM, Rao VS, Moura DJ, Rosa RM, Henriques JA, Pessoa C (2006) Genotoxicity evaluation of kaurenoic acid, a bioactive diterpenoid present in Copaiba oil. Food Chem Toxicol 44:388–392
Chaffei C, Pageau K, Suzuki A, Gouia H, Ghorbel HM, Mascalaux-Daubresse C (2004) Cadmium toxicity induced changes in nitrogen management in *Lycopersicon esculentum* leading to a metabolic safeguard through an amino acid storage strategy. Plant Cell Physiol 45:1681–1693
Chalchat JC, Garry RP, Lamy J (1994) Influence of harvest time on yield and composition of Artemisia annua oil produced in France. J Essential Oil Res 6:261–268
Charles DJ, Joly RJ, Simon JE (1990) Effect of osmotic stress on the essential oil content and comparison of peppermint. Phytochemistry 2:2837–2840
Chartzoulakis K, Loupassaki M, Bertaki M, Androulakis I (2002) Effects of NaCl salinity on growth, ion content and CO2 assimilation rate of six olive cultivars. Sci Hort 96:235–247
Chattopadhyay A, Subramanyam K (1993) Changes in oil yield of *C. wintrianus* suffering from iron chlorosis. J Ind Soc Soil 41:166–167
Chen J, Sucoff EI, Stadelmann EJ (1991) Aluminum and temperature alteration of cell membrane permeability of *Quercus rubra*. Plant Physiol 96:644–649
Choi JH, Chung GC, Suh SR (1997) Effect of night humidity on the vegetative growth and the mineral composition of tomato and strawberry plants. Sci Hort 70:293–299
Chen X, Wang J, Shi Y, Zhao MQ, Chi GY (2011) Effects of cadmium on growth and photosynthetic activities in pakchoi and mustard. Botanical Studies 52:41–46
Chinnusamy V, Zhu J, Zhu JK (2006) Gene regulation during cold acclimation in plants. Physiol Plant 126:52–61
Chowdhury SR, Choudhuri MA (1986) Effects of calcium ions on responses of two jute species under water-deficit stress. Physiol Plant 68:86–92
Ciećko Z, Kalembasa S, Wyszkowski M, Rolka E (2004) Effect of soil contamination by cadmium on potassium uptake by plants. Polish J Environ Studies 13:333–337
Clarkson DT (1965) The effect of aluminium and some other trivalent metal cations on cell division in the root apices of *Allium cepa*. Ann Bot 29:309–315
Clemens S (2006) Evolution and function of phytochelatin synthases. J Plant Physiol 163:319–332
DalCorso G, Farinati S, Maistri S, Furini A (2008) How plants cope with cadmium: staking all on metabolism and gene expression. J Integr Plant Biol 50:1268–1280
de Abreu IN, Mazzafera P (2005) Effect of water and temperature stress on the content of active constituents of *Hypericum brasiliense Choisy*. Plant Physiol Biochem 43:241–248
Dekov I, Tsonev T, Yordanov I (2000) Effects of water stress and high temperature stress on the structure and activity of photosynthetic apparatus of *Zea mays* and *Helianthus annuus*. Photosynthetica 38:361–366

Delfine S, Loreto F, Pinell P, Tognetti R, Alvino A (2005) Isoprenoids content and photosynthetic limitations in rosemary and spearmint plants under water stress. Agric Ecosystem Environ 106:243–252

Dordas C (2009) Foliar application of calcium and magnesium improves growth, yield, and essential oil yield of oregano (*Origanum vulgare* ssp. hirtum). Ind Crop Prod 29: 599–608

Dorman HJD, Figueiredo AC, Barraso G, Deans SG (2000) In vitro evaluation of antioxidant activity of essential oils and their components. Flavour Fragr J 15:12–16

Dow AI, Cline TA, Horning EV (1981) Salt tolerance studies on irrigated mint. Bulletin of Agriculture Research Center, Washington State University, Pullman, no. 906, p. 11

Duke JA (1992) Handbook of legumes of economic importance. Plenum, New York

El-Danasoury M, Al-Amier H, El-Din Helaly A, Aziz EE, Craker L (2010) Essential oil and enzyme activity in spearmint under salt stress. J Herb Spice Med Plant 16:136–145

Eris A, Gulen H, Barut E, Cansev A (2007) Annual patterns of total soluble sugars and proteins related to cold-hardiness in olive (*Olea europaea* L. 'Gemlik'). J Hort Sci Biotech 82:597–604

Abo-Kassem EEM (2007) Effects of salinity: Calcium interaction on growth and nucleic acid metabolism in five species of Chenopodiaceae. Turk J Bot 31:125–134

Eun SO, Youn HS, Lee Y (2000) Lead disturbs microtubule organization in the root meristem of *Zea mays*. Physiol Plant 110:357–365

Farahani HA, Valadabadi SA, Shiranirad JDAH, Khalvati MA (2009) Medicinal and aromatic plants farming under drought conditions. J Hort Forest 1:086–092

Fargasova A (2001) Phytotoxic effects of Cd, Zn, Pb, Cu and Fe on Sinapis alba L. seedlings and their accumulation in roots and shoots. Biol Plant 44:471–473

Farooqi AHA, Ansari SR, Kumar R, Sharma S, Fatima S (1998) Response of different genotypes of citronella Java (*C. winterianus*) to water stress. Plant Physiol Biochem 25:172–175

Fatima S, Farooqi AHA, Sangwan RS (2006) Water stress mediated modulation in essential oil, proline and polypeptide profile in palmarosa and citronella java. Physiol Mol Biol Plant 11:153–156

Fatima SF, Farooqi AHA, Srikant S (2000) Effect of drought stress and plant density on growth and essential oil metabolism in citronella java (*Cymbopogen winterianus* Jowitt). J Med Aromatic Plant Sci 22:563–567

Foy CD (1988) Plant adaptation to acid, aluminum-toxic soils. Commun Soil Sci Plant Anal 19:959–987

Foy CD, Fleming AL, Armiger WJ (1969) Aluminium tolerance of soybean varieties in relation to calcium nutrition. Agron J 61:505–511

Franz C (1983) Nutrient and water management for medicinal and aromatic plants. Acta Hort 132:203–215

Fu J, Huang B (2003) Effects of foliar application of nutrients on heat tolerance of creeping bentgrass. J Plant Nutr 26:81–96

Garg BK, Kathju S, Vyas SP, Lahiri AN (1997) Alleviation of sodium chloride induced inhibition of growth and nitrogen metabolism of clusterbean by calcium. Biol Plant 39:395–401

Garland CJ, Wilkins DA (1981) Effect of calcium on the uptake and toxicity of lead in *Hordeum vulgare* L. and *Festuca ovina* L. New Phytol 87:581–593

Ghassemi F, Jakeman AJ, Nix HA (1995) Salinisation of land and water resources. University of New South Wales Press, Canberra

Gil A, De La Fuente EB, Lenardis AE, Lopez Pereira M, Suarez SA, Bandoni A, Van Baren C, Di LeoLira P, Ghersa CM (2002) Coriander essential oil composition from two genotypes grown in different environmental conditions. J Agric Food Chem 50:2870–2877

Gobinathan P, Murali PV, Panneerselvam R (2009) Interactive effects of calcium chloride on salinity-induced proline metabolism in *Pennisetum typoidies*. Advance Biol Res 3:168–173

Godbold DL, Fritz E, Huttermann A (1988) Aluminum toxicity and forest decline. Proc Natl Acad Sci (USA) 85:3888–3892

Goharrizi AASB, Ziaee M, Mohammadpoor MR, Abdollahi J (2011) Influence of seedling treatment of calcium on some enzyme activity in walnut under salinity. J Bacteriol Res 3:10–15

Gong M, Chen SN, Song YQ, Li ZG (1997) Effect of calcium and calmodulin on intrinsic heat tolerance in relation to antioxidant systems in maize seedlings. Aus J Plant Physiol 24:371–379

Gong M, Li YJ, Chen SZ (1998) Abscisic acid induced thermo-tolerance in maize seedlings mediated by calcium and associated with antioxidant system. J Plant Physiol 153:488–496

Goenaga R, Smith JR (2002) Dry matter production and leaf elemental concentrations of common bean grown on an acid ultisol. J Plant Nutr 25:103–112

Gosset DR, Millhollon EP, Lucas MC (1994) Antioxidant response to NaCl stress in salt-tolerant and salt sensitive cultivars of cotton. Crop Sci 34:706–714

Grant JJ, Loake GJ (2000) Role of reactive oxygen intermediates and cognate redox signaling in disease resistance. Plant Physiol 124:21–29

Grattan SR, Grieve CM (1999) Salinity mineral nutrient relations in horticultural crops. Sci Hort 78:127–157

Grewal HS, Williams R (2003) Liming and cultivars affect root growth, nodulation, leaf to stem ratio, herbage yield, and elemental composition of alfalfa on an acid soil. J Plant Nutr 26:1683–1696

Groenewegen WA, Knight DW, Heptinstall S (1992) Progress in the medicinal chemistry of the herb feverfew. In: Ellis GP, Luscombe DK (eds) Progress in medicinal chemistry. Elsevier Science, Amsterdam, pp 217–238

Guchou S, Xiaoping Z, Xiaojing L, Ping Z (2007) Effects of moderate high-temperature stress on photosynthesis in three saplings of the constructive tree species of subtropical forest. Acta Ecol Sinica 27:1283–1290

Guo T, Chen Y, Zhang Y, Jin Y (2006) Alleviation of Al-toxicity in barley by addition of calcium. Agr Sci China 5:828–833

Haldimann P, Feller U (2005) Growth at moderately elevated temperature alters the physiological response of photosynthetic apparatus to heat stress in pea (*Pisum sativum*) leaves. Plant Cell Environ. 28: 302–317

Hendawy SF, Khalid KA (2005) Response of sage (*Salvia officinalis* L.) plants to zinc application under different salinity levels. J Appl Sci Res 1:147–155

Hikosaka K, Muradami A, Hirose T (1999) Balancing carboxylation and regeneration of ribulose-1,5-bisphosphate in leaf photosynthesis: temperature acclimation of an evergreen tree, *Quereus myrsinaefolia*. Plant Cell Environ 22:841–849

Hillel D (2000) Salinity management for sustainable irrigation. The World Bank, Washington

Horst WJ (1995) The role of the apoplast in aluminium toxicity and resistance of higher plants: a review. Zeitschrift für Pflanzenernährung und Bodenkunde 158:419–428

Horst WJ, Asher CJ, Cakmak I, Szulkiewicz P, Wissemeier AH (1992) Short-term responses of soybean roots to aluminium. J Plant Physiol 140:174–178

Hsiao TC (1973) Plant responses to water stress. Ann Rev Plant Physiol 24:519–570

Huang JW, Shaff JE, Grunes DL, Kochian LV (1992) Aluminium effects on calcium fluxes at the root apex of aluminium-tolerant and aluminium-sensitive wheat cultivars. Plant Physiol 98:230–237

Hughes SG, Bryant JA, Smirnoff N (1989) Molecular biological application to studies of stress tolerance. In: Hamlyn GJ, Flowers TJ, Jones MB (eds) Plants under stress. Cambridge University Press, New York, pp 131–135

Hyden A (2006) Aeroponic and hydroponic systems for medicinal herb, rhizome, and root crops. HortSci 41:536–538

Idrees M, Khan MMA, Aftab T, Naeem M, Hashmi N (2010) Salicylic acid-induced physiological and biochemical changes in lemongrass varieties under water stress. J Plant Interact 5:293–303

Ikegawa H, Yamamoto Y, Matsumoto H (2000) Responses to aluminum of suspension-cultured tobacco cells in a simple calcium solution. Soil Sci Plant Nutr 46:503–514

Islam S, Ezakor E, Garner JO (2011) Effect of chilling stress on the chlorophyll fluorescence, peroxidase activity and other physiological activities in Ipomoea batatas L. genotypes. Am J Plant Physiol 6(2):72–82. doi:10.3923/ajpp. 2011.72.82

Jaleel CA, Azooz MM (2009) Exogenous calcium alters pigment composition, α-glutamyl kinase and proline oxidase activities in salt-stressed *Withania somnifera*. Plant Omic J 2:85–90

Jaleel CA, Gopi R, Gomathinayagam M, Panneerselvam R (2008a) Effects of calcium chloride on metabolism of salt-stressed *Dioscorea rotundata*. Acta Biol Cracoviensia S Bot 50(1):63–67

Jaleel CA, Manivannan P, Kishorekumar A, Sankar B, Gopi R, Somasundaram R, Panneerselvam R (2007a) Alterations in osmoregulation, antioxidant enzymes and indole alkaloid levels in *Catharanthus roseus* exposed to water deficit. Colloid Surf B: Biointerf 59:150–157

Jaleel CA, Manivannan P, Kishorekumar A, Sankar B, Panneerselvam R (2007c) Calcium chloride effects on salinity induced oxidative stress, proline metabolism and indole alkaloid accumulation in *Catharanthus roseus*. Comptes Rendus Biologies.330 9: 674–683

Jaleel CA, Sankar B, Murali PV, Gomathinayagam M, Lakshmanan GMA, Panneerselvam R (2008b) Water deficit stress effects on reactive oxygen metabolism in *Catharanthus roseus*; impacts on ajmalicine accumulation. Colloid Surf B: Biointerf 62:105–111

Jaleel CA, Manivannan P, Sankar B, Kishorekumar A, Gopi R, Somasundaram R, Panneerselvam R (2007b) Water deficit stress mitigation by calcium chloride in Catharanthus roseus: effects on oxidative stress, proline metabolism and indole alkaloid accumulation. Colloids Surf B Biointerf 60:110–116

Jiang Y, Huang B (2001a) Effects of calcium on antioxidant activities and water relations associated with heat tolerance in two cool-season grasses. J Exp Bot 52:341–349

Jiang Y, Huang B (2001b) Physiological responses to heat stress alone or in combination with drought: A comparison between tall fescue and perennial ryegrass. Hort Sci 36:682–686

Jing D, Fei-bo WU, Guo-ping Z (2005) Effect of cadmium on growth and photosynthesis of tomato seedlings. J Zhejiang Univ 6:974–980

Johnson RE, Jackson WA (1964) Calcium uptake and transport by wheat seedlings as affected by aluminum. Soil Sci Soc Am Proc 28:381–386

Jones HG, Tardien F (1998) Modelling water relations of horticultural crops: a review. Sci Hort 74:21–46

Joshi MK, Mohanty P (2004) Chlorophyll *a* fluorescence as a probe of heavy metal ion toxicity in plants. In: Papageorgiou GC (ed) Chlorophyll *a* fluorescence: a signature of photosynthesis, advances in photosynthesis and respiration. Springer, Dordrecht, pp 637–661

Kaplan B, Davydov O, Knight H, Galon Y, Knight MR, Fluhr R, Fromm H (2006) Rapid transcriptome changes induced by cytosolic Ca^{2+} transients reveal ABRE-related sequences as Ca^{2+}-responsive cis elements in Arabidopsis. Plant Cell 18:2733–2748

Karaivazoglou NA, Tsotsolis NC, Tsadilas CD (2007) Influence of liming and form of nitrogen fertilizer on nutrient uptake, growth, yield, and quality of Virginia (flue-cured) tobacco. Field Crop Res 100:52–60

Kawasaki TT, Moritsugu M (1983) Effects of high concentrations of sodium chloride and polyethylene glycol on the growth and ion absorption in plants. I. Water culture experiments in a greenhouse. Plant Soil 75:75–85

Keiser JR, Mullen RE (1993) Calcium and relative humidity effects on soybean seed nutrition and seed quality. Crop Sci 33:1345–1349

Khadhri A, Neffati M, Smiti S, Nogueira JMM, Araujo MEM (2011) Influence of salt stress on essential oil yield and composition of lemon grass (*Cymbopogon schoenanthus* L. Spreng. ssp. Laniger (Hook) Maire et Weil). Natural Product Research: Former Natur Product Lett 25:108–117

Khalid KA (2006) Influence of water stress on growth, essential oil, and chemical composition of herbs (*Ocimum* sp.). Int Agrophys 20:289–296

Khalighi A, Abad J, Khara J (2007) Effect of cadmium toxicity on the level of lipid peroxidation and antioxidative enzyme activity in wheat plants colonized by arbuscular mycorrhizal fungi. Pakistan J Biol Sci 10:2413–2417

Khan MMA, Mohammad F (2006) Mineral nutrition of medicinal plants – a review. In: Trivedi PC (ed) Medicinal plants: an ethnobotanical approach. Agrobios, Jodhpur, pp 347–357

Khan MMA, Naeem M, Siddiqui MH (2005) Calcium fertilization ameliorates growth, yield and quality of hyacinth bean (*Lablab purpureus* L.). Proceedings of 1st International edible legume conference & 4th world cowpea congress, Durban, South Africa, April 17–21

Khan MN, Naeem M (2006) Supplementary calcium ameliorates growth and biochemical attributes of mung bean under NaCl stress. Biosci Biotech Res Asia 3:159–160
Khan MN, Siddiqui MH, Mohammad F, Naeem M, Khan MMA (2010) Calcium chloride and gibberellic acid protect linseed (*Linum usitatissimum* L.) from NaCl stress by inducing antioxidative defence system and osmoprotectant accumulation. Acta Physiol Plant 32:121–132
Kibria MG, Maniruzzaman M, Islam M, Osman KT (2010) Effects of soil-applied lead on growth and partitioning of ion concentration in *Spinacea oleracea* L. tissues. Soil Environ 29:1–6
Kinraide TB (1998) Three mechanisms for the calcium alleviation of mineral toxicities. Plant Physiol 118:513–520
Kinraide TB, Parker DR (1987) Cation amelioration of aluminum toxicity in wheat. Plant Physiol 83:546–551
Kinraide TB, Ryan PR, Kochian LV (1992) Interactive effects of Al^{3+}, H^+ and other cations on root elongation considered in terms of cell-surface electrical potential. Plant Physiol 99:1461–1468
Kleinhenz MD, Palta JP (2002) Root zone calcium modulates the response of potato plants to heat stress. Physiol Plant 115:111–118
Knight H (2000) Calcium signaling during abiotic stress in plants. Internat Rev Cytol 195:269–325
Knight H, Trewavas AJ, Knight MR (1996) Cold calcium signaling in *Arabidopsis* involves two cellular pools and a change in calcium signature after acclimation. Plant Cell 8:489–503
Ksouri R, Megdiche W, Debez A, Falleh H, Grignon C, Abdelly C (2007) Salinity effects on polyphenol content and antioxidant activities in leaves of the halophyte Cakile maritima. Plant Physiol Biochem 45:244–249
Kumar A, Abrol YP (1994) Effect of gypsum on five tropical grasses grown in normal and extensive sodic soil. Exp Agricul 19:169–177
Kumar AA, Gill KS (1995) Performance of aromatic grasses under saline and sodic stress condition. Salt tolerance of aromatic grasses. Indian Perfum 39:39–44
Lamb C, Dixon RA (1997) The oxidative burst in plant disease resistance. Annu Rev Plant Physiol Mol Biol 48:251–275
Langjun CUL, Jianlong LI, Yamin FAN, Sheng XU, Zhang Z (2006) High temperature effects on photosynthesis, PS II functionality and antioxidant activity of two *Festuca arundinacea* cultivars with different heat susceptibility. Bot Studies 47:61–69
Lane TW, Morel FM (2000) A biological function for cadmium in marine diatomeas. Proc Natl Acad Sci (USA) 97:4627–4631
Läuchli A, Epstein E (1990) Plant responses to saline and sodic conditions. In: KK Tanji (ed). Agricultural salinity assessment and management. ASCE manuals and reports on engineering practice No, 71. pp 113–137 ASCE New York
Law R, Crafts-Brandner SJ (1999) Inhibition and acclimation of photosynthesis to heat stress is closely correlated with activation of ribulose-1,5-bisphosphate carboxylase/oxygenase. Plant Physiol 120:173–181
Lazof DB, Goldsmith JG, Rufty TW, Linton RW (1994) Rapid uptake of aluminum into cells of intact soybean root tips: A microanalytical study using secondary ion mass spectrometry. Plant Physiol 106:1107–1114
Lee CK, Lu CK, Kuo YH, Chen JZ, Sun GZ (2004) New prenylated Xavones from the roots of *Ficus beecheyana*. J Chin Chem Soc 51: 437–442
Lee KD, Yang MS (2005) Changes in mineral and terpene concentration following calcium fertilization of *Chrysanthemum boreale* M Res J Agric Biol Sci 1: 222–226
Leegood RC, Edwards GE (1996) Carbon metabolism and photorespiration: temperature dependence in relation to other environmental factors. In: Baker NR, (ed) Photosynthesis and the environment. Dordrecht, Kluwer Academic Publishers 191:221–17
Letchamo W, Hi XU, Gosselin A (1995) Photosynthetic potential of *Thymus vulgaris* selection under 2 light regimes and 3 soil-water levels. Sci Hort 62:89–101
Li H, Qiang S, Qian Y (2008) Physiological response of different croftonweed (*Eupatorium adenophorum*) populations to low temperature. Weed Sci 56:196–202

Li M, Wang GX, Lin JS (2003) Application of external calcium in improving the PEG-induced water stress tolerance in liquorice cells. Bot Bull Acad Sin 44:275–284

Lin KH, Chiou YK, Hwang SY, Chen LFO, Lo HF (2008) Calcium chloride enhances the antioxidative system of sweet potato (*Ipomoea batatas*) under flooding stress. Ann Appl Biol 152:1–12

Lipp FJ (1996) The efficacy, history, and politics of medicinal plants. Alternative Therap Health Med 2:36–41

Liu HT, Li B, Shang ZL, Li XZ, Mu RL, Sun DY, Zhou RG (2003) Calmodulin is involved in heat shock signal transduction in wheat. Plant Physiol 132:1186–1195

Llugany M, Poschenrieder C, Barcelo J (1995) Monitoring of aluminium-induced inhibition of root elongation in four maize cultivars differing in tolerance to aluminium and proton toxicity. Physiol Plant 93:265–271

Loomis RS, Conner DJ (1992) Crop ecology: productivity and management in agricultural systems. Cambridge University Press, Cambridge

Ma B, Wan J, Shen Z (2007) H_2O_2 production and antioxidant responses in seeds and early seedlings of two different rice varieties exposed to aluminum. Plant Growth Regul 52:91–100

Ma JF (2005) Physiological mechanisms of Al resistance in higher plants. Soil Sci Plant Nutr 61:609–612

Malgorzata B, Nevena S, Zlatko Z, Daniela G (2008) Physiological response of some genotypes (*Lycopersicon esculentum* L.) to high temperature stress. J Cent Europ Agri 9:723–732

Małkowski E, Kita A, Galas W, Karcz W, Kuperberg JM (2002) Lead distribution in corn seedlings (*Zea mays* L.) and its effect on growth and the concentrations of potassium and calcium. Plant Growth Regulation 37:69–76

Mengel K, Kirkby EA (2001) Principles of plant nutrition, 5th edn. Kluver Academic, Netherlands

Manivannan P, Jaleel CA, Sankar B, Kishorekumar A, Somasundaram R, Mohamed M, Abdu M (2004) Growth and oil production of fennel (*Foeniculum vulgare* Mill): Effect of irrigation and organic fertilization. Biol Agric Hort 22:31–39

Manivannan P, Jaleel CA, Sankar B, Kishorekumar A, Somasundaram R, Lakshmanan GMA, Panneerselvam R (2007a). Growth, biochemical modifications and proline metabolism in *Helianthus annuus* L. as induced by drought stress. Colloid Surf B: Biointerf 59:141–149

Manivannan P, Jaleel CH, Sankar B, Somasundaram R, Murali PV, Sridharan R, Panneerselvam R (2007b) Salt stress mitigation by calcium chloride in *Vigna radiata* (L.) Wilczek. Acta Biol Cracoviensia S Bot 49(2):105–109

Marschner H (1995) Mineral Nutrition of Higher Plant. 2nd Edn., Academic Press Ltd., London, UK

Marschner H (2002) Mineral nutrition of higher plants. Academic, London

Mazzafera P, Teixeira JPF (2006) Proline in coffee subjected to drought stress. Turrialba. Universidade Estadual de Campinas, IB, Departamento de Fisiologia Vegetal, Campinas, Brazil

McCanny SJ, Hendershot WH, Lechowicz MJ, Shipley B (1995) The effects of aluminum on *Picea rubens*: factorial experiments using sand culture. Can J For Res 25:8–17

Min SY, Tawaha ARM, Lee KD (2005) Effects of ammonium concentration on the yield, mineral content and active terpene components of *Chrysanthemum coronarium* L. in a hydroponic system. Res J Agric Biol Sci 1:170–175

Mittler R, Vanderauwera S, Gallery M, van Breusegem F (2004) Reactive oxygen gene network of plants. Trends Plant Sci 9:490–498

M-Kalantari KH, Oloumi H (2005) Study the effects of $CdCl_2$ on lipid peroxidation and antioxidant compounds content in *Brassica napus*. Iranian J Sci Tech 29:201–208

Mohamed MAH, Harris PJC, Henderson J, Senatore F (2002) Effect of drought stress on the yield and composition of volatile oils of droughttolerant and non-drought-tolerant clones of *Tagetes minuta*. Planta Med 68:472–474

Mohammed AR, Tarpley L (2009) Effects of enhanced UV-B radiation on productive tillers, main-stem panicle and grain characteristics of southern US rice (*Oryza sativa* L.) cultivars. J Agron Crop Sci 195:292–300

Monroy AF, Labbe E, Dhindsa RS (1997) Low temperature perception in plants: effects of cold on protein phosphorylation in cell-free extracts. FEBS Lett 410:206–209

Morris JB (1996) Special legumes may be an unopened medicinal chest. Agric Res 4:13–15

Morris JB (1997) Special-purpose legume genetic resources conserved for agricultural, industrial and pharmaceutical use. Economic Bot 51:251–263

Morris JB (1999) Legume genetic resources with novel "value added" industrial and pharmaceutical use. In: Janick J (ed) Perspectives on new crops and new uses. ASHS, Alexandria, pp 196–200

Morris JB (2003) Bio-functional legumes with nutraceutical, pharmaceutical, and industrial uses. Economic Bot 57:254–261

Mosser DD, Kotzbauer PT, Sarge KD, Morimoto RI (1990) In vitro activation of heat shock transcription factor DNA-binding by calcium and biochemical conditions that affect protein conformation. Proc Natl Acad Sci (USA) 87:3748–3752

Moustakas M, Lanaras T, Symeonidis L, Karataglis S (1994) Growth and some photosynthetic characteristics of field grown *Avena sativa* under copper and lead stress. Photosynthetica 30:389–396

Munns R (1993) Physiological processes limiting plant growth in saline soils: some dogmas and hypotheses. Plant Cell Environ 16:15–24

Munsi PS (1992). Nitrogen and phosphorus nutrition response in Japan's mint cultivation. Acta Hort 306:436–443

Murillo-Amador B, Jones HG, Kaya C, Aguilar RL, García-Hernández JL, Troyo-Diéguez E, Ávila-Serrano NY, Rueda-Puente E (2006) Effects of foliar application of calcium nitrate on growth and physiological attributes of cowpea (*Vigna unguiculata* L. Walp.) grown under salt stress. Environ Exp Bot 58:188–196

Muzik RM, Pregitzer KS, Hanover JW (1989) Changes in terpene production following nitrogen fertilization of grand fir [*Abies grandis* (Dougl.)Lindl.] seedlings. Oecologia 80:485–489

Naeem M, Idrees M, Khan MMA (2009a) Calcium ameliorates photosynthetic capacity, nitrate reductase and carbonic anhydrase activities, nitrogen assimilation, yield and quality attributes of *Cassia sophera* L. Physiol Mol Biol Plant 15:237–247

Naeem M, Khan MMA (2006) Influence of calcium on crop yield and biochemical attributes, anthraquinone and sennoside contents of *Cassia tora* L. – a medicinal legume. J Herb Spice Med Plant 12:57–67

Naeem M, Khan MMA, Moinuddin, Siddiqui MH (2009b) Role of calcium in ameliorating photosynthetic capacity, nitrogen-fixation, enzyme activities, nutraceuticals and crop productivity of hyacinth bean (*Lablab purpureus* L.) under calcium deficient soil. Med Arom Plant Sci Biotechnol 3:64–73

Naeem M, Khan MMA, Moinuddin SM (2010) Calcium chloride stimulates crop yield, photosynthetic efficiency, enzyme activities and nutraceuticals of coffee senna (*Senna occidentalis* L.) under calcium deficient soil. Asian Australian J Plant Sci Biotechnol 4:30–37

Naeem M, Khan MN, Singh M (2005) Effect of calcium fertilization on growth, photosynthesis pigments and nodulation of mung bean (*Vigna radiata* L. Wilczek). Indian J App Pure Biol 20:253–254

Naumburg E, Loik ME, Smith SD (2004) Photosynthetic responses of *Larrea tridentate* to seasonal temperature extremes under elevated CO_2. New Phytol 162:323–330

Nayyar H, Kaushal SK (2002) Chilling induced oxidative stress in germinating wheat grains as affected by water stress and calcium. Biol Plant 45:601–604

Nichol BE, Oliveira LA, Glass ADM, Siddiqi MY (1993) The effects of aluminum on the influx of calcium, potassium, ammonium, nitrate, and phosphate in an aluminum-sensitive cultivar of barley (*Hordeum vulgare* L.). Plant Physiol 101:1263–1266

Neffati M, Marzouk B (2008) Changes in essential oil and fatty acid composition in coriander (*Coriandrum sativum* L) leaves under saline conditions. Ind Crops Prod 28:137–142. doi:10.1016/j.indcrop. 2008.02.005

Noguees S, Allen DJ, Morison JIL, Baker NR (1998) Ultraviolet-B radiation effects on water relations, leaf development, and photosynthesis in droughted pea plants. Plant Physiol 117: 173–181

Nriagu JO (1978) Lead in soils, sediments and major rock types. In: Nriagu JO (ed) The biogeochemistry of lead in the environment, vol 1. Elsevier, Amsterdam, pp 15–72
Nussbaum S, Schmutz D, Brunold C (1988) Regulation of assimilatory sulfate reduction by cadmium in *Zea mays* L. Plant Physiol 88:1407–1410
Ohki K (1986) Photosynthesis, chlorophyll, and transpiration responses in aluminum stressed wheat and sorghum. Crop Sci 26:572–575
Orhue ER, Inneh S (2010) Uptake of lead by *Celosia Argentea* in an ultisol. Res J Agric Biol Sci 6:103–107
Pan JW, Ye D, Wang LL, Hua J, Hua GF, Pan WH, Han N, Zhu MY (2004) Root border cell development is a temperature-insensitive and Al-sensitive process in barley. Plant Cell Physiol 45:751–760
Panda SK, Singha LB, Khan MH (2003) Does aluminium phytotoxicity induce oxidative stress in green gram (*Vigna radiata*)? Bulg J Plant Physiol 29:77–86
Parker DR (1995) Root growth analysis: An underutilized approach to understanding aluminium rhizotoxicity. Plant Soil 171:151–157
Patel BS, Patel KP, Patel ID, Patel M (2000) Response of fennel (*Foeniculum vulgare*) to irrigation, nitrogen and phosphorus. Indian J Agron 45:429–432
Patnaik J, Debata BK (1997) Regeneration of plantlets from NaCl tolerant callus lines of C. martinii (Roxb). Wats. Plant Sci 128:67–74
Patra DD, Prasad A, Anwar M, Singh D, Chand S, Ram B, Katiyar RS, Kumar S (2002) Performance of lemongrass cultivars intercropped with chamomile under sodic soils with different levels of gypsum application. Commun Soil Sci Plant Anal 33:1707–1721
Penka M (1978) Influence of irrigation on the contents of effective substances in officinal plant. Acta Hort 73:181–198
Perez-Alfocea F, Estan MT, Caro M, Guerrier G (1993) Osmotic adjustment in Lycopersicon esculentum and L. pennellii under NaCl and polyethylene glyco 6,000 iso-osmotic stresses. Physiol Plant 87:493–498
Perfus-Barbeoch L, Leonhardt N, Vavasseur A, Forestier C (2002) Heavy metal toxicity: cadmium permeates through calcium channels and disturbs the plant water status. Plant J 32:539–548
Petropoulos SA, Daferera D, Polissiou MG, Passam HC (2007) The effect of water deficit stress on the growth, yield and composition of essential oils of parsley. Sci Hort 115:393–397
Pijnenborg JWM, Lei TA (1990) Effect of lime-pelleting on the nodulation of lucerne (*Medicago sativa* L.) in an acid soil: A comparative study carried out in the field, in pots and in rhizotrons. Plant Soil 121:225–234
Polle A, Schuetzenduebel A (2003) Heavy metal signalling in plants: linking cellular and organismic responses. In: Hirt H, Shinozaki K (eds) Plant responses to abiotic stress. Springer, Berlin-Heidelberg, pp 187–215
Prasad DPH, Prasad ARK (1987) Effects of lead and mercury on chlorophyll synthesis in mungbean seedlings. Phytochemistry 26:881–884
Premachandra GS, Saneoka H, Fujita K, Ogata S (1992) Leaf water relations, osmotic adjustment, cell membrane stability, epicuticular wax load and growth as affected by increasing water deficits in Sorghum. J Exp Bot 43:1569–1576
Price AH, Taylor A, Ripley SJ, Griffiths A, Trewavas AJ, Knight MR (1994) Oxidative signals in tobacco increase cytosolic calcium. Plant Cell 6:1301–1310
Purohit SS, Vyas SP (2004) Medicinal plant cultivation—a scientific approach. Agrobios, Jodhpur
Putievsky E, Ravid U, Dudai N (1986) The influence of season and harvest frequency on essential oil and herbal yields from a pure clone of sage grown under cultivated conditions. J Nat Prod 49:326–329
Rahmani N, Aliabadi Farahani H, Valadabadi SAR (2008) Effects of nitrogen on oil yield and its component of Calendula (*Calendula officinalis* L.) in drought stress conditions. Abstracts Book of the world congress on medicinal and aromatic plants, South Africa, p. 364
Raynal DJ, Joslin JD, Thornton FC, Schaedle M, Henderson GS (1990) Sensitivity of tree seedlings to aluminum: III. Red spruce and loblolly pine. J Environ Qual 19:180–187

Reich PB, Oleksyn J, Tjoelker MG (1994) Relationship of aluminum and calcium to net CO_2 exchange among diverse scots pine provenances under pollution stress in Poland. Oecologia 97:82–92
Reid RJ, Tester MA, Smith FA (1995) Calcium/aluminium interactions in the cell wall and plasma membrane of *Chara*. Planta 195:362–368
Rengel Z (1992) Role of calcium in aluminium toxicity. New Phytol 121:499–513
Rengel Z (1994) Effects of Al, rare earth elements and other metals on net 45 Ca^{2+} uptake by *Amaranthus* protoplasts. Plant Physiol 98:632–638
Rentsch D, Hirner B, Schmeizer E, Frommer WB (1996) Salt stressinduced proline transporters and salt stress-repressed broad specificity amino acid permeases identified by suppression of a yeast amino acid permease-targeting mutant. Plant Cell 8:1437–1446
Rhizopoulou S, Diamantoglou S (1991) Water stress induced diurnal-variations in leaf water relations, stomatal conductance, soluble sugars, lipids and essential oil content of *Origanum maforanal*. J Horticul Sci 66:119–125
Ridolfi M, Garrec JP (2000) Consequences of an excess Al and a deficiency in Ca and Mg for stomatal functioning and net carbon assimilation of beech leaves. Ann For Sci 57:209–218
Romero-Puertas MC, Rodrúguez-Serrano M, Corpas FJ, del Río LA (2004) Cadmium-induced subcellular accumulation of O^{2-} and H_2O_2 in pea leaves. Plant Cell Environ 27:1122–1134
Rout GR, Samantaray S, Das P (2001) Aluminium toxicity in plants: a review. Agronomie 21:3–21
Rout NP, Shaw BP (2001) Salt tolerance in aquatic macrophytes: Ionic relation and interaction. Biol Plant 55:91–95
Rudd JJ, Franklin-Tong VE (2001) Unravelling response-specificity in Ca2+ signaling pathways in plant cells. New Phytol 151:7–33
Ruiz JM, Sánchez E, García PC, López-Lefebre LR, Rivero RM, Romero L (2002) Proline metabolism and NAD kinase activity in greenbean plants subjected to cold-shock. Phytochem 59:473–478
Ryan PR, Shaff JE, Kochian LV (1992) Aluminum toxicity in roots: Correlation between ionic currents, ion fluxes, and root elongation in aluminum-sensitive and aluminum-tolerant wheat cultivars. Plant Physiol 99:1193–1200
Sagi M, Fluhr R (2001) Superoxide production by plant homologues of the gp91phox NADPH oxidase. Modulation of the activity by calcium and by tobacco mosaic virus infection. Plant Physiol 126:1281–1290
Saidi Y, Finka A, Muriset M, Bromberg Z, Weiss YG, Maathuis FJM, Goloubinoff P (2009) The heat shock response in moss plants is regulated by specific calcium-permeable channels in the plasma membrane. Plant Cell 21:2829–2843
Salt DE, Prince RC, Pickering IJ, Raskin I (1995) Mechanisms of cadmium mobility and accumulation in Indian mustard. Plant Physiol 109:1427–1433
Sanders D, Brownlee C, Harper JF (1999) Communicating with calcium. Plant Cell 11:691–706
Sangwan NK, Dhindsa KS, Malik OP, Sharma GD, Paroda RS (1982) Quantitative changes in levels of essential oil in C. martinii var motia during different growth stages and on ageing the harvested crop in field and Laboratory. In: Proc Nat Sem Med Arom Plants, India
Sangwan NS, Abd Farooqi AH, Sangwan RS (1994) Effect of drought stress on growth and essential oil metabolism in lemongrasses. New Phytol 128:173–179
Sangwan NS, Farooqi AHA, Shabih F, Sangwan RS (2001). Regulation of essential oil production in plants. Plant Growth Regul 34:3–21
Sangwan RS, Abd Farooqi AH, Bansal RP, Sangwan NS (1993) Interspecific variation in physiological and metabolic responses of five species of Cymbopogon to water stress. J Plant Physiol 142:618–622
Sanità di Toppi L, Gabrielli R (1999) Response to cadmium in higher plants. Environ Exp Bot 41:105–130
Sarker BC, Hara M, Uemura M (2005) Proline synthesis, physiological responses and biomass, yield of eggplants during and after repetitive soil moisture stress. Sci Hort 103:387–402
Sarnighausen E, Karlson DT, Zeng Y, Goldsbrough PB, Raghothama KG, Ashworth EN (2004) Characterization of a novel YnSKn class of dehydrin-like cDNAs from cold acclimated

red-oiser dogwood (*Cornus sericea* L.) xylem. In: Arora R (ed) Adaptations and responses of woody plants to environmental stresses. The Haworth Press, New York, pp 17–36

Sasaki M, Yamamoto Y, Ma JF, Matsumoto H (1997) Early events induced by aluminum stress in elongating cells of wheat root. Soil Sci Plant Nutr 43:1009–1014

Saure MC (1998) Causes of the tipburn disorder in leaves of vegetables. Scientia Hort 76:131–147

Saure MC (2001) Blossum-end rot of tomato (Lycopersicon esculentum Mill.)-a calcium –or a stress-related disorder? Sci Hort 90:193–208

Scalabrelli G, Saracini E, Remorini D, Massai R, Tattini M (2007) Changes in leaf phenolic compounds in two grapevine varieties (*Vitis vinifera* L.) grown in different water conditions. Acta Hort 754:295–300

Schmitz-Eiberger M, Haefs R, Noga G (2002) Calcium deficiency- Influence on the antioxidaitne defense system in tomato plants. J Plant Physiol 159:733–742

Schützendübel A, Schwanz P, Teichmann T, Gross K, Langenfeld- Heyser R, Godbold DL, Polle A (2001) Cadmium-induced changes in antioxidative systems, hydrogen peroxide content, and differentiation in Scots pine roots. Plant Physiol 127:887–898

Selmar D (2008) Potential of salt and drought stress to increase pharmaceutical significant secondary compounds in plants. Agric Forestry Res 58:139–144

Seth SD, Sharma B (2004) Medicinal plants in India. Indian J Med Res 120:9–11

Shabih F, Farooqi AHA, Ansari SR, Sharma S (1999) Effect of water stress on growth and essential oil metabolism in Cymbopogon martinii cultivars. J Essential Oil Res 11:491–496

Shao HB, Chu LY, Shao MA, Jaleel CA, Mi HM (2008) Higher plant antioxidants and redox signaling under environmental stresses. CR Biol 331:433–441

Sharifi AE, Matin M, Lebaschi H, Abbaszadeh B, Naderi B (2005) Effects of water stress on quantity yield in *Achillea millefolium.* Abstracts Book of the first international conference on the theory and practices in biological water saving, p. 211

Sharma P, Dubey RS (2005) Lead toxicity in plants. Braz J Plant Physiol 17:35–52

Silva IR, Ferrufino A, Sanzonowicz C, Smyth TJ, Israel DW, Carter TE Jr (2005) Interactions between magnesium, calcium, and aluminum on soybean root elongation. R Bras Ci Solo 29:747–754

Silva IR, Smyth TJ, Israel DW, Raper CD, Rufty TW (2001) Magnesium is more efficient than calcium in alleviating aluminum rhizotoxicity in soybean and its ameliorative effect is not explained by the Gouy–Chapman–Stern model. Plant Cell Physiol 42:538–545

Simon JE, Bubenheim DR, Joly RJ, Charles DJ (1992) Water stress induced alterations in essential oil content and composition of sweet basil. J Essential Oil Res 4:71–75

Singh-Sangwan N, Farooqi AHA, Singh-Sangwan R (2006) Effect of drought stress on growth and essential oil metabolism in lemon grasses. New Phytol 128:173–179

Singh M, Ramesh S (2000) Effect of irrigation and nitrogen on herbage, oil yield and water- use efficiency in rosemary grown under semi- arid tropical conditions. J Med Aromatic Plant Sci 22:659–662

Sivaguru M, Ezaki B, He ZH, Tong H, Osawa H, Baluska F, Volkmann D, Matsumoto H (2003) Aluminum-induced gene expression and protein localization of a cell wall-associated receptor kinase in Arabidopsis. Plant Physiol 132:2256–2266

Sivaguru M, Horst WJ (1998) The distal part of the transition zone is the most aluminum-sensitive apical root zone of maize. Plant Physiol 116:155–163

Ssali H (1981) The effect of level of $CaCO_3$, inoculation and lime pellating on the nodulation and growth of beans of five acid soils. Plant Soil 62:53–63

Stahl W, Sies H (2003) Antioxidant activity of carotenoids. Mol Aspects Med 24:345–351

Stefanov K, Seizova K, Yanishlieva N, Marinova E, Popov S (1995) Accumulation of lead, zinc and cadmium in plant seeds growing in metalliferous habitats in Bulgaria. Food Chem 54:311–313

Storey R, Wyn Jones RG (1978) Salt stress and comparative physiology in the gramineae. I. Ion relations of two salt and water-stressed barley cultivars, California Mariout and Arimar. Aust J Plant Physiol 5:801–816

Stutte GW (2006) Process and product: recirculation hydroponics and bioactive compounds in a controlled environment. Hortsci 41:526–530

Subrahmanyam D (1998) Effect of aluminium on growth, lipid peroxidation, superoxide dismutase and peroxidase activities in rice and french bean seedlings. Indian J Plant Physiol 3:240–242

Sümer A, Zörb C, Yan F, Schubert S (2004) Evidence of sodium toxicity for the vegetative growth of maize during the first phase of salt stress. J App Bot 78:135–139

Sundaresan S, Sudhakaran PR (2006) Water stress-induced alterations in the proline metabolism of drought-susceptible and tolerant cassava (*Manihot esculenta*) cultivars. Physiol Planta 94:635–642

Supanjani, Tawaha ARM, Yang MS, Lee KD (2005) Calcium effects on yield, mineral uptake and terpene components of hydroponic *Chrysanthemum coronarium* L. Int J Bot 1:196–200

Suzuki N (2005) Alleviation by calcium of cadmium-induced root growth inhibition in *Arabidopsis* seedlings. Plant Biotechnol 22:19–25

Suzuki N, Mittler R (2006) Reactive oxygen species and temperature stresses: a delicate balance between signaling and destruction. Physiol Plant 126:45–51

Tabatabaie J, Nazari J (2007) Influence of nutrient concentrations and NaCl salinity on the growth, photosynthesis and essential oil content of peppermint and *Lemon Verbena*. Turk J Agric 31:245–253

Taheri AM, Daneshian J, Valadabadi SAR, Aliabadi FH (2008) Effects of water deficit and plant density on morphological characteristics of chicory (*Cichorium intybus* L.). Abstracts Book of 5th International Crop Science Congress & Exhibition. p. 26

Takami C, Takenaka C, Tezuka T (2005) Mitigation of aluminum toxicity by calcium and magnesium in Japanese cedar (*Cryptomeria japonica*). J For Res 10:9–14

Tanji KK (1990) Nature and extent of agricultural salinity. In: Tanji KK (ed) Agricultural salinity assessment and management. ASCE, New York, pp 1–13

Taylor GJ, McDonald-Stephens JL, Hunter DB, Bertsch PM, Elmore D, Rengel Z, Reid RJ (2000) Direct measurement of aluminum uptake and distribution in single cells of *Chara corallina*. Plant Physiol 123:987–996

Tian L, Qi-hua L, Ohsugi R, Yamagishi T, Sasaki H (2006) Effect of high temperature on sucrose content and sucrose cleaving enzyme activity in rice grain during the filling stage. Rice Sci 13:205–210

Tuna AL, Kaya C, Ashraf M, Altunlu H, Yokas I, Yagmur B (2007) The effects of calcium sulphate on growth, membrane stability and nutrient uptake of tomato plants grown under salt stress. Environ Exp Bot 59:173–178

Tyler VE, Brady LR, Robbers JE (1976) Glycosides. In: Pharmacognosy. Lea & Febiger, Philadelphia, pp. 76–103

Tyree MT, Karamanos AJ (1981) Water stress as an ecological factor. In: Grace J, Ford ED (eds) Plants and their atmospheric environment. Blackwell Scientific, Oxford, pp 237–261

Uemura M, Tominaga Y, Nakagawara C, Shigematsu S, Minami A, Kawamura Y (2006) Responses of the plasma membrane to low temperatures. Physiol Plant 126:81–89

Verma S, Dubey RS (2003) Lead toxicity induces lipid peroxidation and alters the activities of antioxidant enzymes in growing rice plants. Plant Sci 164:645–655

Wahid A, Gelani S, Ashraf M, Foolad MR (2007) Heat tolerance in plants: An overview. Environ Exp Bot 61:199–223

Wallace A, Wallace GA (2003) Closing the crop-yield gap through better soil and better management—the law of the maximum. Wallace Laboratories, El. Segundo, CA

Wan G, Najeeb U, Jilani G, Naeem MS, Zhou W (2011) Calcium invigorates the cadmium-stressed *Brassica napus* L. plants by strengthening their photosynthetic system. Environ Sci Pollut Res 18(9):1478–1486. doi:10.1007/s11356-011-0509-1

Wang CQ, Song H (2009) Calcium protects *Trifolium repens* L. seedlings against cadmium stress. Plant Cell Rep 28:1341–1349

Weiss EA (1997) Essential oil crops. CAB International, Oxon, pp 24–58

White PJ, Broadley MR (2003) Calcium in plants. Ann Bot 92:487–511

Wierzbicka M (1994) The resumption of metabolic activity in *Allium sepa* L. root tips during treatment with Lead salt. Environ Exp Bot 34:173–180

Wojas S, Ruszczyńska A, Bulska E, Wojciechowski M, Antosiewicz DM (2007) Ca^{2+}-dependent plant response to Pb^{2+} is regulated by *LCT1*. Environ Pollut 147:584–592

Xin Z, Browse J (1998) Eskimo1 mutants of Arabidopsis are constitutively freezing-tolerant. Proc Natl Acad Sci U S A 95:7799–7804

Xu Q, Paulsen AQ, Guikema JA, Paulsen GM (1995) Functional and ultrastructural injury to photosynthesis in wheat by high temperature during maturation. Environ Exp Bot 35:43–54

Xu ZZ, Zhou GS (2006) Combined effects of water stress and high temperature on photosynthesis, nitrogen metabolism and lipid peroxidation of a perennial grass *Leymus chinensis*. Planta 224:1080–1090

Yang YY, Jung JY, Song WY, Suh HS, Lee Y (2000) Identification of rice varieties with high tolerance or sensitivity to lead and characterization of the mechanism of tolerance. Plant Physiol 124:1019–1026

Yamada M, Morishita H, Urano K, Shiozaki N, Yamaguchi-Shinozaki K, Shinozaki K, Yoshiba Y (2005) Effects of free proline accumulation in petunias under drought stress. J Exp Bot 56:1975–1981

Yamamoto Y, Kobayashi Y, Matsumoto H (2001) Lipid peroxidation is an early symptom triggered by aluminum, but not the primary cause of elongation inhibition in pea roots. Plant Physiol 125:199–208

Yan F, Schubert S, Mengel K (1992) Effect of low root medium pH on net proton release, root respiration, and root growth of corn (*Zea mays* L.) and broad bean (*Vicia faba* L.). Plant Physiol 99:415–421

Yang T, Poovaiah BW (2002) Hydrogen peroxide homeostasis: activation of plant catalase by calcium/calmodulin. Plant Biol 99:4097–4102

Yang T, Poovaiah BW (2003) Calcium/calmodulin-mediated signal network in plants. Trends Plant Sci 8:505–512

Yazdani D, Shahnazi S, Seifi H (2004). Cultivation of medicinal plants medicinal plant Research Center press, Tehran, Iran p. 169

Yermiyahu U, Nir S, Ben-Hayyim G, Kafkafi U, Kinraide TB (1997) Root elongation in saline solution related to calcium binding to root cell plasma membranes. Plant Soil 191:67–76

Zehtab-Salmasi S, Javanshir A, Omidbaigi R, Aly-Ari H, Ghassemi-Golezani K (2001) Effects of water supply and sowing date on performance and essential oil production of anise (*Pimpinella anisum* L.). Acta Agron Hung 49:75–81

Zhang W-H, Rengel Z, Kuo J, Yan G (1999) Aluminium effects on pollen germination and tube growth of *Chamelaucium uncinatum*. A comparison with other Ca^{2+} antagonists. Ann Bot 84:559–564

Zhao H, Tan J, Qi C (2007) Photosynthesis of *Rehmannia glutinosa* subjected to drought stress is enhanced by choline chloride through alleviating lipid peroxidation and increasing proline accumulation. Plant Growth Regul 51:255–262

Zheng Z, Sheth U, Nadiga M, Pinkham JL, Shetty K (2001) A model for the role of the proline-linked pentose phosphate pathway in polymeric dye tolerance in oregano. Proc Biochem 36:941–946

Zhu JK (2003) Regulation of ion homeostasis under salt stress. Curr Opin Plant Biol 6:441–445

Chapter 6
Role of DREB-Like Proteins in Improving Stress Tolerance of Transgenic Crops

Deepti Jain and Debasis Chattopadhyay

1 Introduction

Plants are often exposed to several environmental stresses that adversely affect various stages of their growth and development. It has been estimated that potential yield of annual crops is lost up to 82 % due to abiotic stress every year. Drought and salinity are already spreading worldwide, and are expected to cause serious salinization of more than 50 % of all available productive, arable lands by the year 2050 (Ashraf 1994). In a world where population growth exceeds food supply, plant breeders and biologists need to fully implement the biotechnologies and agricultural practices to overcome these serious issues.

Plants overcome environmental stresses by development of tolerance, resistance or avoidance mechanisms. Tolerance allows an organism to withstand the assault. Resistance involves active countermeasures, while avoidance prevents exposure to the stress. Partly due to their sessile nature, plants have developed sophisticated metabolic responses, various strategies and pathways to tolerate or resist different forms of stress. Plant's tolerance or susceptibility to abiotic stresses is a complex phenomenon. Therefore, intense research has been focused on the mechanisms underlying abiotic stress tolerance and adaptation. Though plants have gradually evolved a remarkable ability to cope with such highly variable environmental onslaughts, the stresses nevertheless represent a primary cause of crop loss worldwide. Therefore, to meet the increasing demands for agricultural commodities it would be imperative to either enhance cultivable land in current use to expand agricultural lands or to create genetically redesigned crops to cope better with the environmental changes.

Drought, cold, and high-salinity stresses generate complex stimuli that have different yet related attributes and may deliver quite different information to the

D. Jain • D. Chattopadhyay (✉)
National Institute of Plant Genome Research, Aruna Asaf Ali Marg,
New Delhi 110 067, India
e-mail: deeptivjain@gmail.com; debasis@nipgr.ac.in

N. Tuteja and S. Singh Gill (eds.), *Plant Acclimation to Environmental Stress*,
DOI 10.1007/978-1-4614-5001-6_6,

plant cells (Xiong et al. 2002). A mild abiotic stress may induce an adaptive response in the plant, allowing it to grow with a greater tolerance to the same or different stresses (Knight et al. 1998; Lang et al. 1994; Mantyla et al. 1995; Siminovitch and Cloutier 1982). The mechanism through which plants perceive environmental signals and transmit them to cellular machinery to generate adaptive response is of fundamental importance to biology. It is the coordinated action of various genes in a pathway that bring about the requisite phenotype and lead to plant tolerance. Plant stress adaptation/tolerance not only involves physiological changes but also the changes at cellular and molecular levels. The ability of the plant to sustain itself under unfavorable environmental conditions determines by the manifestation of a single or a combination of these inherent changes (Farooq et al. 2009).

2 Abiotic Stress Response

Drought, salinity, extreme temperatures, and oxidative stress are often interconnected and may induce similar cellular damage. For example, plants suffer from dehydration under high salinity and drought, as well as low-temperature conditions, all of which cause hyperosmotic stress characterized by metabolic and osmotic imbalance in plants. This leads to turgor loss and closure of the stomata, followed by repression of cell growth and inadequate photosynthesis (Shinozaki and Yamaguchi-Shinozaki 2007). Oxidative stress, which frequently accompanies high temperature, salinity, or drought stress, may cause denaturation of functional and structural proteins and lipids (Smirnoff 1998). As a consequence, these diverse environmental stresses often activate similar cell signaling pathways (Knight et al. 1998; Shinozaki and Yamaguchi-Shinozaki 2000; Zhu 2001, 2002) and cellular responses, such as the production of stress proteins, up-regulation of anti-oxidants, and accumulation of compatible solutes. The switching 'on' of such cellular and molecular responses include perception of stress signal by membrane receptors, which then activate cytoplasmic Ca^{2+} and signaling pathways in cytoplasm and nucleus. This eventually leads to modification in the stress-responsive gene expression and physiological changes (Bressan et al. 1998; Xiong et al. 2002). Also, accumulation of abscisic acid (ABA) plays an important role in abiotic stress signaling and transduction pathways, mediating many responses (Wasilewska et al. 2008). The products of these genes ultimately lead to plant adaptation and/or tolerance and help the plant to survive and surpass the unfavorable conditions. The mechanism by which plants perceive environmental stress signals and transmit to the cellular machinery for activation of adaptive responses is of critical importance. This knowledge can be implied for the development of rational breeding and transgenic strategies leading to alleviate stress tolerance in crops.

In the past decades, a number of stress-inducible genes have been identified by transcriptome analyses using microarray technology in several plant species, like *Arabidopsis*, rice, etc. (Bohnert et al. 2001; Seki et al. 2001; Zhu 2001) by several research groups. Several genes that are induced by abiotic stresses have been classified into two major groups (Bray 1993; Shinozaki and Yamaguchi-Shinozaki 2000; Thomashow 1999). One group encodes for functional proteins such as membrane proteins (membrane channel, transporter proteins, etc.), key enzymes for osmolyte biosynthesis (proline, glycine betain, sugars, etc.), the detoxification enzymes (catalase, hydrolase, superoxide dismutase, etc.), and the proteins for the protection of macromolecules (LEA protein, chaperone, osmotin, etc.). Whilst, other group includes regulatory proteins such as transcription factors (bZIP, MYC, MYB, DREB, etc.), protein kinases (MAP kinase, receptor protein kinase, etc.), proteinases (phosphoesterases, phospholipase C, etc.) that regulate gene expression and signal transduction in stress responses and enzymes involved in phospholipids metabolism and ABA biosynthesis (Chen and Murata 2002; Shinozaki and Yamaguchi-Shinozaki 2007; Yamaguchi-Shinozaki and Shinozaki 2006). Plant engineering strategies for abiotic stress tolerance (Wang et al. 2003) rely on the expression of genes that are involved in signaling and regulatory pathways (Seki et al. 2003; Shinozaki et al. 2003), genes that encode proteins conferring stress tolerance (Wang et al. 2004) or enzymes present in pathways leading to the synthesis of functional and structural metabolites (Apse and Blumwald 2002; Park et al. 2004; Rontein et al. 2002). Many genes encoding enzymes related to functional metabolites are induced by stress. There is a practical limitation of overexpressing multiple genes in a plant in a tissue specific manner to improve stress tolerance. Therefore, early responsive genes that regulate a number of functionally related downstream genes could be attractive targets for engineering stress tolerance since they may regulate quantitative traits. Intuitively, genetic engineering would be a faster way to insert beneficial genes than through conventional or molecular breeding and thus seems to be a viable option to hasten the breeding of "improved" plants. Also, it would be the only option when genes of interest originate from cross-barrier species, distant relatives, or from non-plant sources.

3 Role of ABA in Stress Response

ABA is an important phytohormone that plays a pivotal role in various physiological processes during the plant life cycle, including seed dormancy, germination, and adaptive responses to various environmental stress conditions (Himmelbach et al. 2003; Schroeder et al. 2001; Shinozaki and Yamaguchi-Shinozaki 2000; Zhu 2002). Also, the application of ABA to plant mimics the effect of a stress condition. ABA level is increased in response to various stress signals, particularly when there are changes in the environment that result in cellular dehydration. Accumulation of ABA in leaves induces stomatal closure and inhibits opening (Wilkinson et al. 2001), thereby maintaining plant water potential under conditions of low soil moisture content or high evaporative demand and thus ABA is aptly called as a stress hormone. A rapid and sensitive increase in the ABA production is essential for instant cellular as well as long distance regulations. Also, a rapid ABA bleaching is needed when the stress is relieved.

Water deficit induces activation of genes that encode enzymes for both ABA biosynthesis and hydrolysis of ABA conjugates releasing the active hormone from an inactive pool (Iuchi et al. 2000; Lee et al. 2006; Schwartz et al. 2003; Xiong and Zhu 2003). The phenomenon of ABA biosynthesis in response to osmotic stress is well known (Finkelstein et al. 2002); however, the signaling networks orchestrated by ABA responses are highly complex. ABA is hydroxylated by cytochrome P450-monooxygenase to form unstable hydroxyl-ABA, which is subsequently converted to phaseic acid. This pathway is also regulated by environmental conditions (Kushiro et al. 2004). Also, ABA and hydroxyl-ABA are conjugated with glucose to form inactive ABA-glucose ester (ABA-GE) (Cutler and Krochko 1999) which can be converted into an active form by apoplastic and endoplasmic reticulum β (beta)-glucosidases. Thus, plants have to maintain a proper balance of active and inactive forms of ABA, which is critical for plants growth and development (Chinnusamy et al. 2004).

3.1 ABA Signaling Pathways

Main function of ABA seems to be the regulation of plant water balance and osmotic stress tolerance. Several ABA deficient mutants namely *aba1*, *aba2*, and *aba3* have been reported for *Arabidopsis* (Koornneef et al. 1998). ABA deficient mutants for tobacco, tomato, and maize have also been reported (Liotenberg et al. 1999). Without any stress treatment the growth of these mutants is comparable to wild type plants. Under drought stress, ABA deficient mutants readily wilt and die if stress persists. Under salt stress also ABA deficient mutants show poor growth (Xiong et al. 2001).

Several different sets of *cis*- and *trans*-acting factors are known to be involved in stress-responsive transcription. Some of them are controlled by ABA but others are not responsive to exogenous ABA treatment, indicating the involvement of both ABA-dependent and -independent signal transduction cascades for stress-responsive gene expression (Bray 1993; Shinozaki and Yamaguchi-Shinozaki 2000; Thomashow 1999; Xiong et al. 2002). Abiotic stress response involves at least four different regulons in plants. A regulon is a group of genes controlled by a certain type of transcription factor. These regulons are (a) AREB/ABF regulon, (b) MYC/MYB regulon, (c) NAC (NAM/ATAF and CUC) and ZF-HD regulon and (d) DREB/CBF regulon. The former two are ABA-dependent regulons and later two are ABA-independent (Fig. 6.1).

3.2 ABA-Dependent Signaling Cascade

The ABA-dependent pathway may follow either of the two routes, either requiring new protein synthesis or not (Ingram and Bartels 1996; Shinozaki and Yamaguchi-Shinozaki 1997). Route independent of new protein synthesis, includes

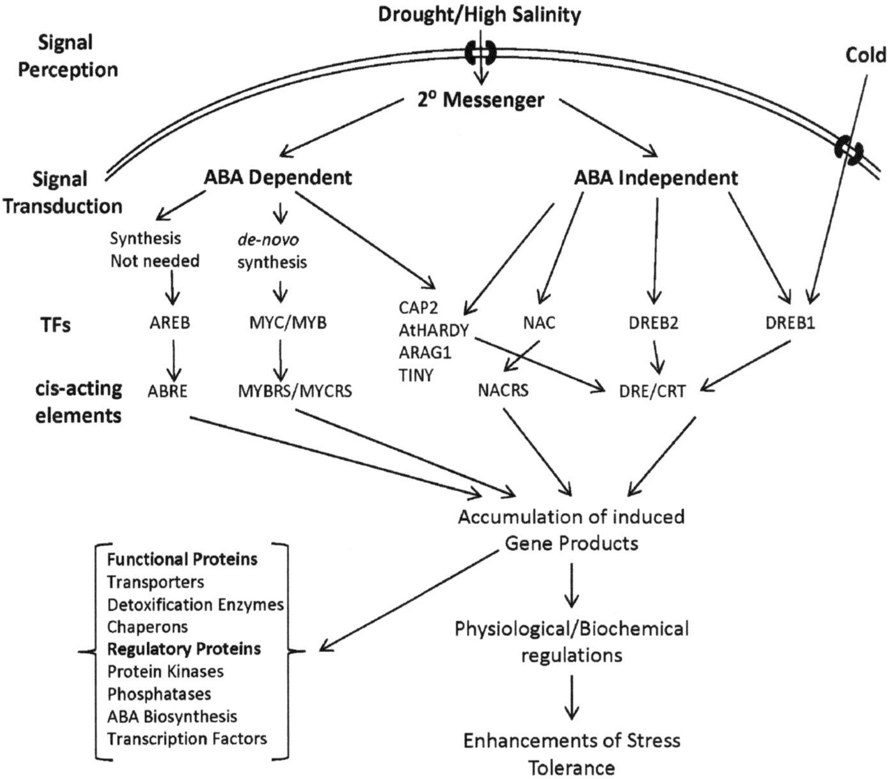

Fig. 6.1 A schematic representation of cellular signaling pathways. Stress signal is perceived by membrane receptors, activating secondary messenger molecules, signal is transduced by a set of transcription factors via two different signaling branches. The *cis-* and *trans-*elements involved in stress-responsive gene expression are also shown. Two different signal transduction pathways were followed by DREB proteins in response to cold and drought stresses. *ABRE* abscisic acid responsive binding element; *DRE*: drought responsive element; *MYBRS* MYB recognition site; *MYCRS* MYC recognition site; *NACRS* NAC recognition sequences

ABA-responsive genes with ABA-responsive element ABRE, (C/T)ACGTG(G/T)C, in their promoter domain (Himmelbach et al. 2003). ABRE is recognized by members of bZip-transcription factor family, AREB/ABF (ABA-responsive element-binding proteins/factors), and activate ABA-induced gene expression. ABA-dependent phosphorylation is required for the activation of AREB/ABF protein. Overexpression of ABF3/ABF4 enhances drought tolerance in *Arabidopsis* (Kang et al. 2002). Similarly ABF2/AREB1 overexpression improved stress tolerance to heat, drought, and oxidative stresses (Kim et al. 2004). Genes involved in the route where new protein synthesis is required for ABA-induced expression have no ABRE in their promoter region, else having *cis*-acting elements with binding affinity to MYC/MYB transcription factor family (Bray 2002; Shinozaki and Yamaguchi-Shinozaki 1997).

3.3 ABA-Independent Signaling Cascade

Existence of an ABA-independent pathway was unveiled when ERD1 (early responsive to dehydration1, the Clp protease regulatory subunit encoding gene) transcript was found to accumulate even before the accumulation of ABA during the dehydration and salinity stress responses in *Arabidopsis* (Nakashima et al. 1997). Promoter analysis of ERD1 gene shows the existence of DNA-binding domains for transcription factors (TFs) belonging to NAC and ZF-HD (zinc-finger homeodomain) family. *Arabidopsis* plants coexpressing these TFs activate the ERD1 gene expression (Tran et al. 2007). NAC family proteins bind specifically to the NAC recognition site (CATGTG) (Tran et al. 2004). OsNAC6 overexpressing transgenic plants are reportedly tolerant to salt and dehydration stress.

During dehydration stress, endogenous ABA began to accumulate 2 h after dehydration stress started and reached maximum in 10 h but rd29A transcript was found to accumulate within 20 min after dehydration and is followed by a secondary induction phase that begins after ~3 h (Yamaguchi-Shinozaki et al. 1992). This differential behavior suggests that the first rapid induction of *rd29A* is not mediated by endogenous ABA but by an ABA-independent pathway. Promoter analysis of rd29A revealed the presence of two *cis*-acting elements, one of which is ABRE, which is responsible for ABA-dependent late induction, and the other is a new *cis*-acting element (TACCGACAT). This new element was named as dehydration-responsive element (DRE)/C-repeat (CRT) and responsible for very early ABA-independent induction of rd29A. Many dehydration and low-temperature stress-inducible genes were found to have DRE in their promoter (Thomashow 1999; Yamaguchi-Shinozaki and Shinozaki 1994). A lot of efforts were made to identify the transcription factor that regulates DRE with the intuition that the DRE-binding protein would be a very early regulatory factor of stress response and, therefore, would be a potential candidate for genetic manipulation. The first DRE-binding protein identified was named CBF (CRT-binding factor) and was found to improve freezing tolerance in the overexpressing non-acclimated plants (Kasuga et al. 1999). Consequently several other TFs that bind and activate DRE/CRT were discovered (Jaglo-Ottosen et al. 1998). Those were commonly named as DRE-binding (DREB) proteins or CBF. One set of DREB genes were found to be induced at low temperature and named DREB1 and other group was found induced by dehydration and salinity and named DREB2 genes. All the DREB proteins have a conserved AP2/ERF (APETALA2/ethylene-responsive element-binding factor) DNA-binding domain (Riechmann and Meyerowitz 1998) of 50–60 amino acids in length.

AP2/ERF gene family is one of the most important multigene families of TFs. Recently, the AP2 domain was reportedly found in other proteins outside the plant kingdom too (Magnani et al. 2004). AP2/ERF family is classified into three groups based on the number of AP2/ERF domains and gene function. Class I members encode a protein containing two ERF/AP2 domains and includes APETALA2 (AP2), AINTEGUMENTA (ANT), and Glossy15 (Elliott et al. 1996; Jofuku et al. 1994; Klucher et al. 1996; Moose and Sisco 1996). Class II proteins have only one

ERF/AP2 domain and include EREBPs, TINY, DREB1/CBF, DREB2, Ptis, EBP, ERF, AtEBP, AtERFS, and ABI4 (Buttner and Singh 1997; Finkelstein et al. 1998; Fujimoto et al. 2000; Ohme-Takagi and Shinshi 1995; Solano et al. 1998; Vergani et al. 1997; Wilson et al. 1996; Zhou et al. 1997). The third class includes RAV1 and RAV2, with two different DNA-binding domains, ERF/AP2. Of all, DREB TFs are presumably the most promising candidates for genetic manipulation for attaining tolerance against drought, high salinity, low-temperature, and other abiotic stresses. This chapter emphasize on the role of DREB- and DREB-like proteins in the stress responses.

4 DRE-Binding Proteins in Stress Responses

In *Arabidopsis*, DREB/CBF-like proteins can be classified into 6 small subgroups (A-1 to A-6) based on similarities in the binding domain. Subgroups (A-1, A-2) include the DREB1/CBF and DREB2 gene families, respectively. The DREB1 family proteins are relatively short and of about 220 amino acids in size while the DREB2-family proteins are of about 330 amino acids in size. They show variation of amino acid sequences in the DNA-binding domains (Liu et al. 1998). Subgroup A-3 has only ABI4, A-4 includes TINY, A-5 contains RAP2.1, RAP2.9, and RAP2.10, and A-6 includes RAP2.4. DREB1/CBF and DREB2 are two independent DREB family members which function via two separate signaling pathways under stress (Liu et al. 1998). DREB1 proteins play major role in cold-induced gene expression and the DREB2 proteins are involved in high-salinity- and drought-induced gene expression. However, for exception, DREB1-related genes DREB1D/CBF4 expression is induced by osmotic stress (Haake et al. 2002) and DREB1F, DREB1E are induced by high-salinity stress (Magome et al. 2004). This suggests the cross-talk between the CBF/DREB1 and the DREB2 pathways. Overexpression of the DREB transcription factors activate a number of downstream genes leading to enhanced abiotic stress tolerance.

4.1 DREB1/CBF Regulates Cold-Inducible Gene Expression

In *Arabidopsis,* three genes encoding DREB1B/CBF1, DREB1A/CBF3, and DREB1C/CBF2 have been mapped on chromosome 4. Transgenic *Arabidopsis* plants overexpressing CBF/DREB1 genes showed improved survival rates under low temperatures (Jaglo-Ottosen et al. 1998; Kasuga et al. 1999). However, the DREB1A overexpression caused growth retardation of transgenic plants under normal growth conditions. Replacing the constitutive *Ca*MV 35S promoter with stress-inducible rd29A promoter minimizes the plant growth defect without compromising with the yield and also imparted tolerance to drought and salinity (Fowler and Thomashow 2002; Kasuga et al. 1999). Similarly, *Medicago* *trancatula* DREB1C gene when over expressed into China Rose under *Arabidopsis* rd29A promoter, show normal growth phenotype and enhance freezing tolerance in transgenic plants (Chen et al. 2010). Gene Chip and cDNA microarrays have identified more than 40 genes downstream of DREB1/CBF (Fowler and Thomashow 2002; Maruyama et al. 2004; Seki et al. 2002; Vogel et al. 2005). Categorically these genes belong to LEA proteins, osmoprotectant biosynthesis proteins, RNA-binding proteins, sugar transport proteins, desaturases, carbohydrate metabolism-related proteins, KIN (cold-inducible) proteins, protease inhibitors. These gene products are probably responsible for the stress tolerance of the transgenic plants. Transcription factors like C2H2 zinc-finger-type and AP2/ERF-type, act downstream to CBF/DREB suggesting existence of further regulation of gene expression downstream of the DRE/DREB regulon (Maruyama et al. 2004; Sakamoto et al. 2004). Whilst, CBF3/DREB1 regulates the expression of a number of downstream genes during stress response, it itself is controlled by the Inducer of CBF expression 1 (ICE1) protein, a MYC-type bHLH (basic helix-loop-helix) TF (Chinnusamy et al. 2003). The ICE1 protein is negatively regulated by the higher expression of Osmotically responsive genes 1 (HOS1) protein, a RING E3 ligase, which is responsible for ubiquitination and subsequent degradation of ICE1 protein (Dong et al. 2006). ICE1 ubiquitination can be blocked by SIZ1-dependent sumoylation (Miura et al. 2007). SIZ1 is a SUMO E3 ligase that mediates ICE1 sumoylation which activates and/or stabilizes ICE1 protein, thus facilitating its activity controlling the expression of the CBF3/DREB1A gene. However, ICE1 does not regulate the expression of CBF2/DREB1C. The CBF2 gene is activated by members of calmodulin binding transcription activators, CAMTA (Hua 2009). ICE2, an ICE1 homolog, probably regulates the CBF1 gene (Fursova et al. 2009). Also, CBF/DREB1 is negatively regulated by MYB15 TF. So, different pathways are involved in activation of different DREB1 proteins.

5 DREB2 Protein-Mediated Stress Response Under Osmotic Stress

DREB2 protein subfamily has two main members, DREB2A and DREB2B (Furihata et al. 2006). *Arabidopsis* genome encodes at least six DREB2 homologues other than DREB2A and DREB2B. DREB2A and DREB2B are induced strongly by drought and high salinity, but the others are not (Sakuma et al. 2006a, b). Unlike CBF/DREB1, transgenic plants overexpressing DREB2A did not show growth retardation. Also, overexpression of AtDREB2A and OsDREB2A in *Arabidopsis* was insufficient for stress-inducible gene expression (Dubouzet et al. 2003; Liu et al. 1998). This suggested that some posttranslational modifications might be needed by DREB2A for the activation of stress-inducible gene expression. The amino acid and domain analyses of AtDREB2A gene revealed the presence of a predicted nuclear localization signal in its N-terminal region, a transcriptional activation domain in the C-terminal region between amino acids 254–335, AP2/

ERF domain from amino acids 78–135 and a negative regulatory domain in the central region of the protein (136–165). Deletion of this negative regulatory domain resulted in significant increase of its activity. Sequence analysis shows the presence of a PEST sequence (RSDASEVTSTSSQSEVCTVETPGCV) in this region. The PEST sequence is rich in proline (P), glutamic acid (G), serine (S), and threonine (T). PEST sequence is often associated with proteins of short intracellular half-life, hence PEST sequences are hypothesized to act as a signal peptide for protein degradation (Rogers et al. 1986). The removal of negative regulatory domain containing the PEST sequence transforms DREB2A in a constitutively active form (DREB2-CA) (Sakuma et al. 2006a, b). The DREB2A protein containing the PEST sequence is degraded rapidly by the ubiquitin-proteasome system, whereas DREB2A-CA has a long lifetime in the nucleus. It was recently reported that over-expression of DREB2A-CA gene induces not only drought- and salt-responsive genes but also heat-shock (HS)-related genes (Sakuma et al. 2006a, b). Thus the DREB2A up-regulated genes can be classified into three groups: genes induced by HS, genes induced by drought stress, and genes induced by both HS and drought stress. HS stress induces HS proteins (HSPs). The expression of DREB2A itself was found to be induced by HS transiently and significantly. Microarray analysis of DREB2A-CA overexpressing *Arabidopsis* plants revealed the up-regulation of a number of drought, salt and heat responsive downstream genes even under non-stressed conditions (Sakuma et al. 2006a, b). Surprisingly, it shows the induction of a gene for heat shock factor (HSF), AtHsfA3, along with genes for LEA proteins, dehydrins, and COR15A, which function in acquisition of stress tolerance to drought and high-salinity genes. The heat induction of HSFA3 is directly regulated by the transcription factor DREB2A. Heat-inducible DREB2A can bind to the DRE element in the promoter of HSFA3 under heat shock to induce HSFA3 and further activate HSP expression (Schramm et al. 2008; Yoshida et al. 2008). Overexpression of DREB2C, another CRT/DRE-binding ERF member, enhances thermo-tolerance of transgenic *Arabidopsis* plants (Lim et al. 2007). Taken together, these results indicated that DREB2A plays a critical role in regulating drought- and heat-stress-responsive gene expression (Sakuma et al. 2006a, b).

Though both DREB1A and DREB2A recognize DRE, however, differences were observed between the DREB1A and DREB2A downstream genes. Moreover, some common downstream genes, such as COR15A, COR15B, KIN1, and KIN2, are recognized by both DREB1A and DREB2A, but their expression levels in the DREB2A-CA overexpressing plants were significantly lower than those in the DREB1A transgenic plants (Sakuma et al. 2006a, b). These differences in expression level of the downstream genes between the two DREB proteins explain the reason for the less freezing tolerance of the DREB2A-CA transgenic plants. Extensive promoter analysis of the DREB1A- and DREB2A-up-regulated genes demonstrated that the DREB2A protein could recognize both DRE/CRT variants, A/GCCGACNT and A/GCCGACNA/G/C, but prefers ACCGAC to GCCGAC. DREB1A protein has the highest affinity to the A/GCCGACNT sequence These different binding specificities between DREB1A and DREB2A may explain why these proteins control some different downstream genes.

Besides the major two subfamilies of DREB proteins, much less is known about a lot of small proteins (200aa and less) that bear DREB2-like AP2 domain. In this chapter we summarize these DREB2-like small subgroup proteins.

6 DREB2-Like Small Proteins

DREB proteins play important roles in plant morphology, development, and stress responses. A chickpea cDNA library of dehydration-induced transcripts was constructed and a novel DREB2-like TF, CAP2 was isolated (Shukla et al. 2006). In comparison to most of the well-known DREB2 family members, CAP2 was found to be relatively small (202 amino acids), though it qualifies for being an AP2 transcription factor. Since then, like CAP2, a number of small ORFs containing DREB2-like AP2 domain have been reported in GenBank from different plants. Like DREB2A of *Arabidopsis* and rice (*Oryza sativa*) expression of CAP2 is induced by dehydration and salt but not by cold. But, unlike AtDREB2A, CAP2 transcript was induced by ABA and auxin. Overexpression of CAP2 in transgenic tobacco caused increase in the leaf size and number of lateral roots, also promoted tolerance to salt, osmotic and heat stresses. This suggested that CAP2 is involved in both stress response and development. Another DREB2-like protein ZmDREB2A from *Zea mays* showed enhanced tolerance to drought, high salt and heat stress in the transgenic plants without growth penalty. A novel DREB-like gene, GmDREB2, was isolated from soybean. It has an open reading frame of 159 amino acids. GmDREB2 was classified into A-5 subgroup in DREB subfamily. GmDREB2 gene expression was induced by drought, high salt, and low temperature stresses and ABA treatment. Transgenic *Arabidopsis* plants overexpressing GmDREB2 has activated expression of downstream genes resulting in enhanced tolerance to drought and high-salt stresses. GmDREB2 overexpression did not cause growth retardation (Chen et al. 2007).

Similarly, A-4 subgroup member GhDBP3 (226 amino acids) is an abiotic stress and ABA-induced transcriptional activator (Huang and Liu 2006). A member of A-6 subgroup, ZmDBF1 (222 amino acids) of maize, is also induced by drought, NaCl, and ABA treatments in plant seedlings (Kizis and Pagès 2002). A gain-of-function mutant, *hardy* was identified in a phenotypic screen of an activation tagged mutant collection in *Arabidopsis* with robust roots and dark green leaves (Karaba et al. 2007). The HRD (HARDY) gene belongs to a class of AP2/ERF-like TFs. This gene has an ORF of 184 amino acids. HRD is probably involved in the maturation of inflorescence stage processes that needs tissue protection against desiccation. *Arabidopsis* plants overexpressing HRD gene confers drought and salt tolerance (Verslues et al. 2006).

In rice, a DREB2-like gene was isolated and named as ABA-responsive AP2-like gene (ARAG1). Its ORF is capable of encoding a 225 amino acid protein. Expression of ARAG1 was reportedly up-regulated by drought and ABA treatment as an early stress response as compared with the control. ARAG1-transgenic seeds were able to germinate in the presence of ABA application. It suggests that ARAG1 was involved with the tolerance-associated processes in the seedlings.

7 Conclusion

Taken together, a number of recent reports suggest that unlike DREB genes in A-1 and A-2 subgroups, which are involved only in ABA-independent pathway, these DREB2-like proteins are involved in both ABA-dependent and ABA-independent pathways. They may act as an overlap point and/or might take part separately in both ABA-independent and ABA-dependent pathways. Unlike the authentic DREB genes these DREB2-like genes do not show any growth defect when overexpressed. Some of them promoted growth and development in addition to abiotic stress tolerance when expressed in other plants. Since, little is known about the role of DREB2-like proteins during seed germination and seedling growth, and how they are regulated by ABA; a detailed and comprehensive functional study on these small AP2 proteins may give us an insight in plant developmental process and stress adaptation/tolerance mechanisms. They may also provide an attractive and complementary option for improving a plant's performance under stress conditions and emerge as an important future strategy for facilitating the crop yield in drought-prone environments for sustainable agricultural production.

References

Apse MP, Blumwald E (2002) Engineering salt tolerance in plants. Curr Opin Biotechnol 13(2):146–150

Ashraf M (1994) Breeding for salinity tolerance in plants. Crit Rev Plant Sci 13:17–42

Bohnert HJ, Ayoubi P, Borchert C et al (2001) A genomics approach towards salt stress tolerance. Plant Physiol Biochem 39(3–4):295–311

Bray EA (1993) Molecular responses to water deficit. Plant Physiol 103(4):1035–1040

Bray EA (2002) Abscisic acid regulation of gene expression during water-deficit stress in the era of the Arabidopsis genome. Plant Cell Environ 25(2):153–161

Bressan RA, Hasegawa PM, Pardo JM (1998) Plants use calcium to resolve salt stress. Trends Plant Sci 3(11):411–412

Buttner M, Singh KB (1997) Arabidopsis thaliana ethylene-responsive element binding protein (AtEBP), an ethylene-inducible, GCC box DNA-binding protein interacts with an ocs element binding protein. Proc Natl Acad Sci U S A 94(11):5961–5966

Chen TH, Murata N (2002) Enhancement of tolerance of abiotic stress by metabolic engineering of betaines and other compatible solutes. Curr Opin Plant Biol 5(3):250–257

Chen M, Wang QY, Cheng XG et al (2007) GmDREB2, a soybean DRE-binding transcription factor, conferred drought and high-salt tolerance in transgenic plants. Biochem Biophys Res Commun 353(2):299–305

Chen JR, Lü JJ, Liu R, Xiong XY, Wang TX, Chen SY, Guo LB, Wang HF (2010) DREB1C from *Medicago truncatula* enhances freezing tolerance in transgenic *M. truncatula* and China rose (*Rosa chinensis* Jacq.). Plant Growth Regulation 60(3):199–211

Chinnusamy V, Ohta M, Kanrar S et al (2003) ICE1: a regulator of cold-induced transcriptome and freezing tolerance in Arabidopsis. Genes Dev 17(8):1043–1054

Chinnusamy V, Schumaker K, Zhu JK (2004) Molecular genetic perspectives on cross-talk and specificity in abiotic stress signalling in plants. J Exp Bot 55(395):225–236

Cutler AJ, Krochko JE (1999) Formation and breakdown of ABA. Trends Plant Sci 4(12):472–478

Dong C-H, Agarwal M, Zhang Y, Xie Q, Zhu J-K (2006) The negative regulator of plant cold responses, HOS1, is a RING E3 ligase that mediates the ubiquitination and degradation of ICE1. Proc Natl Acad Sci 103(21):8281–8286

Dubouzet JG, Sakuma Y, Ito Y et al (2003) OsDREB genes in rice, Oryza sativa L., encode transcription activators that function in drought-, high-salt- and cold-responsive gene expression. Plant J 33(4):751–763

Elliott RC, Betzner AS, Huttner E et al (1996) AINTEGUMENTA, an APETALA2-like gene of Arabidopsis with pleiotropic roles in ovule development and floral organ growth. Plant Cell 8(2):155–168

Farooq M, Wahid A, Kobayashi N, Fujita D, Basra SMA (2009) Plant drought stress: effects, mechanisms and management. In: Lichtfouse E, Navarrete M, Debaeke P, Véronique S, Alberola C (eds) Sustainable agriculture. Springer, Netherlands, pp 153–188

Finkelstein RR, Wang ML, Lynch TJ, Rao S, Goodman HM (1998) The Arabidopsis abscisic acid response locus ABI4 encodes an APETALA 2 domain protein. Plant Cell 10(6):1043–1054

Finkelstein RR, Gampala SSL, Rock CD (2002) Abscisic acid signaling in seeds and seedlings. Plant Cell 14(Suppl 1):S15–S45

Fowler S, Thomashow MF (2002) Arabidopsis transcriptome profiling indicates that multiple regulatory pathways are activated during cold acclimation in addition to the CBF cold response pathway. Plant Cell 14(8):1675–1690

Fujimoto SY, Ohta M, Usui A, Shinshi H, Ohme-Takagi M (2000) Arabidopsis ethylene-responsive element binding factors act as transcriptional activators or repressors of GCC box-mediated gene expression. Plant Cell 12(3):393–404

Furihata T, Maruyama K, Fujita Y et al (2006) Abscisic acid-dependent multisite phosphorylation regulates the activity of a transcription activator AREB1. Proc Natl Acad Sci U S A 103(6):1988–1993

Fursova OV, Pogorelko GV, Tarasov VA (2009) Identification of ICE2, a gene involved in cold acclimation which determines freezing tolerance in Arabidopsis thaliana. Gene 429(1–2):98–103

Haake V, Cook D, Riechmann JL, Pineda O, Thomashow MF, Zhang JZ (2002) Transcription factor CBF4 is a regulator of drought adaptation in Arabidopsis. Plant Physiol 130(2):639–648

Himmelbach A, Yang Y, Grill E (2003) Relay and control of abscisic acid signaling. Curr Opin Plant Biol 6(5):470–479

Hua J (2009) From freezing to scorching, transcriptional responses to temperature variations in plants. Curr Opin Plant Biol 12(5):568–573

Huang B, Liu J-Y (2006) Cloning and functional analysis of the novel gene GhDBP3 encoding a DRE-binding transcription factor from Gossypium hirsutum. Biochim Biophys Acta 1759(6):263–269

Ingram J, Bartels D (1996) The molecular basis of dehydration tolerance in plants. Annu Rev Plant Physiol Plant Mol Biol 47:377–403

Iuchi S, Kobayashi M, Yamaguchi-Shinozaki K, Shinozaki K (2000) A stress-inducible gene for 9-cis-epoxycarotenoid dioxygenase involved in abscisic acid biosynthesis under water stress in drought-tolerant cowpea. Plant Physiol 123(2):553–562

Jaglo-Ottosen KR, Gilmour SJ, Zarka DG, Schabenberger O, Thomashow MF (1998) Arabidopsis CBF1 overexpression induces COR genes and enhances freezing tolerance. Science 280(5360):104–106

Jofuku KD, den Boer BG, Van Montagu M, Okamuro JK (1994) Control of Arabidopsis flower and seed development by the homeotic gene APETALA2. Plant Cell 6(9):1211–1225

Kang JY, Choi HI, Im MY, Kim SY (2002) Arabidopsis basic leucine zipper proteins that mediate stress-responsive abscisic acid signaling. Plant Cell 14(2):343–357

Karaba A, Dixit S, Greco R et al (2007) Improvement of water use efficiency in rice by expression of HARDY, an Arabidopsis drought and salt tolerance gene. Proc Natl Acad Sci U S A 104(39):15270–15275

Kasuga M, Liu Q, Miura S, Yamaguchi-Shinozaki K, Shinozaki K (1999) Improving plant drought, salt, and freezing tolerance by gene transfer of a single stress-inducible transcription factor. Nat Biotechnol 17(3):287–291

Kim S, Kang JY, Cho DI, Park JH, Kim SY (2004) ABF2, an ABRE-binding bZIP factor, is an essential component of glucose signaling and its overexpression affects multiple stress tolerance. Plant J 40(1):75–87

Kizis D, Pagès M (2002) Maize DRE-binding proteins DBF1 and DBF2 are involved in rab17 regulation through the drought-responsive element in an ABA-dependent pathway. Plant J 30(6):679–689

Klucher KM, Chow H, Reiser L, Fischer RL (1996) The AINTEGUMENTA gene of Arabidopsis required for ovule and female gametophyte development is related to the floral homeotic gene APETALA2. Plant Cell 8(2):137–153

Knight H, Brandt S, Knight MR (1998) A history of stress alters drought calcium signalling pathways in Arabidopsis. Plant J 16(6):681–687

Koornneef M, Alonso-Blanco C, Blankestijn-de Vries H, Hanhart CJ, Peeters AJ (1998) Genetic interactions among late-flowering mutants of Arabidopsis. Genetics 148(2):885–892

Kushiro T, Okamoto M, Nakabayashi K et al (2004) The Arabidopsis cytochrome P450 CYP707A encodes ABA 8'-hydroxylases: key enzymes in ABA catabolism. EMBO J 23(7):1647–1656

Lang V, Mantyla E, Welin B, Sundberg B, Palva ET (1994) Alterations in water status, endogenous abscisic acid content, and expression of rab18 gene during the development of freezing tolerance in Arabidopsis thaliana. Plant Physiol 104(4):1341–1349

Lee KH, Piao HL, Kim H-Y et al (2006) Activation of glucosidase via stress-induced polymerization rapidly increases active pools of abscisic acid. Cell 126(6):1109–1120

Lim CJ, Hwang JE, Chen H et al (2007) Over-expression of the Arabidopsis DRE/CRT-binding transcription factor DREB2C enhances thermotolerance. Biochem Biophys Res Commun 362(2):431–436

Liotenberg S, North H, Marion-Poll A (1999) Molecular biology and regulation of abscisic acid biosynthesis in plants. Plant Physiol Biochem 37(5):341–350

Liu Q, Kasuga M, Sakuma Y et al (1998) Two transcription factors, DREB1 and DREB2, with an EREBP/AP2 DNA binding domain separate two cellular signal transduction pathways in drought- and low-temperature-responsive gene expression, respectively, in Arabidopsis. Plant Cell 10(8):1391–1406

Magnani E, Sjolander K, Hake S (2004) From endonucleases to transcription factors: evolution of the AP2 DNA binding domain in plants. Plant Cell 16(9):2265–2277

Magome H, Yamaguchi S, Hanada A, Kamiya Y, Oda K (2004) Dwarf and delayed-flowering 1, a novel Arabidopsis mutant deficient in gibberellin biosynthesis because of overexpression of a putative AP2 transcription factor. Plant J 37(5):720–729

Mantyla E, Lang V, Palva ET (1995) Role of abscisic acid in drought-induced freezing tolerance, cold acclimation, and accumulation of LT178 and RAB18 proteins in Arabidopsis thaliana. Plant Physiol 107(1):141–148

Maruyama K, Sakuma Y, Kasuga M et al (2004) Identification of cold-inducible downstream genes of the Arabidopsis DREB1A/CBF3 transcriptional factor using two microarray systems. Plant J 38(6):982–993

Miura K, Jin JB, Lee J et al (2007) SIZ1-mediated sumoylation of ICE1 controls CBF3/DREB1A expression and freezing tolerance in Arabidopsis. Plant Cell 19(4):1403–1414

Moose SP, Sisco PH (1996) Glossy15, an APETALA2-like gene from maize that regulates leaf epidermal cell identity. Genes Dev 10(23):3018–3027

Nakashima K, Kiyosue T, Yamaguchi-Shinozaki K, Shinozaki K (1997) A nuclear gene, erd1, encoding a chloroplast-targeted Clp protease regulatory subunit homolog is not only induced by water stress but also developmentally up-regulated during senescence in Arabidopsis thaliana. Plant J 12(4):851–861

Ohme-Takagi M, Shinshi H (1995) Ethylene-inducible DNA binding proteins that interact with an ethylene-responsive element. Plant Cell 7(2):173–182

Park EJ, Jeknic Z, Sakamoto A et al (2004) Genetic engineering of glycinebetaine synthesis in tomato protects seeds, plants, and flowers from chilling damage. Plant J 40(4):474–487

Riechmann JL, Meyerowitz EM (1998) The AP2/EREBP family of plant transcription factors. Biol Chem 379(6):633–646

Rogers S, Wells R, Rechsteiner M (1986) Amino acid sequences common to rapidly degraded proteins: the PEST hypothesis. Science 234(4774):364–368

Rontein D, Basset G, Hanson AD (2002) Metabolic engineering of osmoprotectant accumulation in plants. Metab Eng 4(1):49–56

Sakamoto H, Maruyama K, Sakuma Y et al (2004) Arabidopsis Cys2/His2-type zinc-finger proteins function as transcription repressors under drought, cold, and high-salinity stress conditions. Plant Physiol 136(1):2734–2746

Sakuma Y, Maruyama K, Osakabe Y et al (2006a) Functional analysis of an Arabidopsis transcription factor, DREB2A, involved in drought-responsive gene expression. Plant Cell 18(5):1292–1309

Sakuma Y, Maruyama K, Qin F, Osakabe Y, Shinozaki K, Yamaguchi-Shinozaki K (2006b) Dual function of an Arabidopsis transcription factor DREB2A in water-stress-responsive and heat-stress-responsive gene expression. Proc Natl Acad Sci U S A 103(49):18822–18827

Schramm F, Larkindale J, Kiehlmann E et al (2008) A cascade of transcription factor DREB2A and heat stress transcription factor HsfA3 regulates the heat stress response of Arabidopsis. Plant J 53(2):264–274

Schroeder JI, Allen GJ, Hugouvieux V, Kwak JM, Waner D (2001) Guard cell signal transduction. Annu Rev Plant Physiol Plant Mol Biol 52:627–658

Schwartz SH, Qin X, Zeevaart JAD (2003) Elucidation of the indirect pathway of abscisic acid biosynthesis by mutants, genes, and enzymes. Plant Physiol 131(4):1591–1601

Seki M, Narusaka M, Abe H et al (2001) Monitoring the expression pattern of 1300 Arabidopsis genes under drought and cold stresses by using a full-length cDNA microarray. Plant Cell 13(1):61–72

Seki M, Narusaka M, Ishida J et al (2002) Monitoring the expression profiles of 7000 Arabidopsis genes under drought, cold and high-salinity stresses using a full-length cDNA microarray. Plant J 31(3):279–292

Seki M, Kamei A, Yamaguchi-Shinozaki K, Shinozaki K (2003) Molecular responses to drought, salinity and frost: common and different paths for plant protection. Curr Opin Biotechnol 14(2):194–199

Shinozaki K, Yamaguchi-Shinozaki K (1997) Gene expression and signal transduction in water-stress response. Plant Physiol 115(2):327–334

Shinozaki K, Yamaguchi-Shinozaki K (2000) Molecular responses to dehydration and low temperature: differences and cross-talk between two stress signaling pathways. Curr Opin Plant Biol 3(3):217–223

Shinozaki K, Yamaguchi-Shinozaki K (2007) Gene networks involved in drought stress response and tolerance. J Exp Bot 58(2):221–227

Shinozaki K, Yamaguchi-Shinozaki K, Seki M (2003) Regulatory network of gene expression in the drought and cold stress responses. Curr Opin Plant Biol 6(5):410–417

Shukla RK, Raha S, Tripathi V, Chattopadhyay D (2006) Expression of CAP2, an APETALA2-family transcription factor from chickpea, enhances growth and tolerance to dehydration and salt stress in transgenic tobacco. Plant Physiol 142(1):113–123

Siminovitch D, Cloutier Y (1982) Twenty-four-hour induction of freezing and drought tolerance in plumules of winter rye seedlings by desiccation stress at room temperature in the dark. Plant Physiol 69(1):250–255

Smirnoff N (1998) Plant resistance to environmental stress. Curr Opin Biotechnol 9(2):214–219

Solano R, Stepanova A, Chao Q, Ecker JR (1998) Nuclear events in ethylene signaling: a transcriptional cascade mediated by ethylene-insensitive3 and ethylene-response-factOR1. Genes Dev 12(23):3703–3714

Thomashow MF (1999) Plant cold acclimation: freezing tolerance genes and regulatory mechanisms. Annu Rev Plant Physiol Plant Mol Biol 50:571–599

Tran LS, Nakashima K, Sakuma Y et al (2004) Isolation and functional analysis of Arabidopsis stress-inducible NAC transcription factors that bind to a drought-responsive cis-element in the early responsive to dehydration stress 1 promoter. Plant Cell 16(9):2481–2498

Tran LS, Nakashima K, Sakuma Y et al (2007) Co-expression of the stress-inducible zinc finger homeodomain ZFHD1 and NAC transcription factors enhances expression of the ERD1 gene in Arabidopsis. Plant J 49(1):46–63
Vergani P, Morandini P, Soave C (1997) Complementation of a yeast delta pkc1 mutant by the Arabidopsis protein ANT. FEBS Lett 400(2):243–246
Verslues PE, Agarwal M, Katiyar-Agarwal S, Zhu J, Zhu JK (2006) Methods and concepts in quantifying resistance to drought, salt and freezing, abiotic stresses that affect plant water status. Plant J 45(4):523–539
Vogel JT, Zarka DG, Van Buskirk HA, Fowler SG, Thomashow MF (2005) Roles of the CBF2 and ZAT12 transcription factors in configuring the low temperature transcriptome of Arabidopsis. Plant J 41(2):195–211
Wang W, Vinocur B, Altman A (2003) Plant responses to drought, salinity and extreme temperatures: towards genetic engineering for stress tolerance. Planta 218(1):1–14
Wang W, Vinocur B, Shoseyov O, Altman A (2004) Role of plant heat-shock proteins and molecular chaperones in the abiotic stress response. Trends Plant Sci 9(5):244–252
Wasilewska A, Vlad F, Sirichandra C et al (2008) An update on abscisic acid signaling in plants and more …. Mol Plant 1(2):198–217
Wilkinson S, Clephan AL, Davies WJ (2001) Rapid low temperature-induced stomatal closure occurs in cold-tolerant Commelina communis leaves but not in cold-sensitive tobacco leaves, via a mechanism that involves apoplastic calcium but not abscisic acid. Plant Physiol 126(4):1566–1578
Wilson K, Long D, Swinburne J, Coupland G (1996) A Dissociation insertion causes a semidominant mutation that increases expression of TINY, an Arabidopsis gene related to APETALA2. Plant Cell 8(4):659–671
Xiong L, Zhu J-K (2003) Regulation of abscisic acid biosynthesis. Plant Physiol 133(1):29–36
Xiong L, Gong Z, Rock CD et al (2001) Modulation of abscisic acid signal transduction and biosynthesis by an Sm-like protein in Arabidopsis. Dev Cell 1(6):771–781
Xiong L, Schumaker KS, Zhu JK (2002) Cell signaling during cold, drought, and salt stress. Plant Cell 14(Suppl):S165–S183
Yamaguchi-Shinozaki K, Shinozaki K (1994) A novel cis-acting element in an Arabidopsis gene is involved in responsiveness to drought, low-temperature, or high-salt stress. The Plant Cell Online 6(2):251–264
Yamaguchi-Shinozaki K, Shinozaki K (2006) Transcriptional regulatory networks in cellular responses and tolerance to dehydration and cold stresses. Annu Rev Plant Biol 57:781–803
Yamaguchi-Shinozaki K, Koizumi M, Urao S, Shinozaki K (1992) Molecular cloning and characterization of 9 cDNAs for genes that are responsive to desiccation in Arabidopsis thaliana: sequence analysis of one cDNA clone that encodes a putative transmembrane channel protein. Plant Cell Physiol 33(3):217–224
Yoshida T, Sakuma Y, Todaka D et al (2008) Functional analysis of an Arabidopsis heat-shock transcription factor HsfA3 in the transcriptional cascade downstream of the DREB2A stress-regulatory system. Biochem Biophys Res Commun 368(3):515–521
Zhou J, Tang X, Martin GB (1997) The Pto kinase conferring resistance to tomato bacterial speck disease interacts with proteins that bind a cis-element of pathogenesis-related genes. EMBO J 16(11):3207–3218
Zhu JK (2001) Cell signaling under salt, water and cold stresses. Curr Opin Plant Biol 4(5):401–406
Zhu JK (2002) Salt and drought stress signal transduction in plants. Annu Rev Plant Biol 53:247–273

Chapter 7
Homeobox Genes as Potential Candidates for Crop Improvement Under Abiotic Stress

Annapurna Bhattacharjee and Mukesh Jain

1 Introduction

Various abiotic stresses such as drought, salinity, and extremes of temperature pose a major challenge for survival of plants and have a great impact on crop productivity. The world population is increasing at an alarming rate and on the contrary, availability of food resources is decreasing due to the abiotic stress factors. Thus, there is a great need to generate stress-tolerant crop plants with improved sustenance and better yield. Plants elicit several responses to combat the adverse effects of various abiotic stresses, including production and accumulation of osmolytes, maintenance of intracellular ion homeostasis, and scavenging of reactive oxygen species. Understanding plant responses to abiotic stresses at the molecular level provides an essential foundation for future breeding and genetic engineering programs. Abiotic stress response is a complex trait, which involves interplay of numerous regulatory molecules at the cellular level. Various approaches have provided a holistic view of the ongoing cellular activities in response to abiotic stresses.

Numerous genes which are induced in response to abiotic stresses have been identified and the products of these genes are supposed to enhance stress tolerance in plants. The role of several stress-inducible genes has been explored, which regulate gene expression via various signal transduction pathways. Among these genes, transcription factors represent master switches controlling several target genes and are considered most important for regulation of gene expression. Although several transcription factors have been implicated in abiotic stress responses, only a few master switches/regulons have been identified and characterized in detail so far. The identification of master switches, which control stress-inducible genes, seems to be the most challenging task. Serious endeavor needs to be made in this direction to

A. Bhattacharjee • M. Jain (✉)
National Institute of Plant Genome Research (NIPGR), Aruna Asaf Ali Marg,
New Delhi 110 067, India
e-mail: mjain@nipgr.res.in

N. Tuteja and S. Singh Gill (eds.), *Plant Acclimation to Environmental Stress*,
DOI 10.1007/978-1-4614-5001-6_7,

identify promising master regulator genes and eventually gain insights into the complex gene regulatory network, which leads to abiotic stress responses.

Several excellent comprehensive reviews are available on various aspects of complex signal transduction pathways controlling abiotic stress responses, emerging trends in genomics of abiotic stress responses and gene regulatory networks involved in abiotic stress responses and tolerance (Bartels and Sunkar 2005; Chinnusamy et al. 2004; Hirayama and Shinozaki 2010; Mahajan and Tuteja 2005; Nakashima et al. 2009; Urano et al. 2010; Vij and Tyagi 2007; Yamaguchi-Shinozaki and Shinozaki 2006). In this review, we provide a brief overview of the role of transcription factors in abiotic stress responses and majorly focus on the emerging role of homeobox transcription factors in abiotic stress responses and their potential as target genes for engineering stress tolerance.

2 Overview of Stress-Responsive Genes

Several studies in *Arabidopsis* and rice have revealed that quite a large proportion of the genome is involved in abiotic stress responses. The global expression profiling in various plant species has revealed that expression of thousands of genes is altered in response to various abiotic stress conditions (Hadiarto and Tran 2011; Hirayama and Shinozaki 2010; Nakashima et al. 2009; Urano et al. 2010; Vij and Tyagi 2007). These stress-responsive genes have been broadly categorized in two groups (Yamaguchi-Shinozaki and Shinozaki 2006). The first group is comprised of genes involved directly in abiotic stress tolerance, including those encoding for metabolic proteins like water channel proteins, enzymes required for synthesis of various osmoprotectants (like sugars, glycine-betaine, and proline), proteins that aid in protecting macromolecules and membranes (for example, LEA proteins, osmotin, chaperones), proteases for protein turnover and detoxification enzymes like glutathione-S-transferases. The second group is comprised of genes encoding for regulatory proteins like protein kinases, transcription factors, protein phosphatases, and other signaling molecules. Overall, the expression of genes involved in diverse cellular processes is altered in response to abiotic stresses. The comparative analysis has revealed that a larger fraction of stress-responsive genes are common among various plant species, indicating the conserved mechanism of abiotic stress response. Further, many of these genes have been analyzed for their exact function and ability to provide stress tolerance in transgenic plants (Vij and Tyagi 2007).

The emerging role of various plant hormones has also been explored in the context of abiotic stress responses. Abscisic acid (ABA) is found to be the key hormone produced in plants under stress conditions and is crucial for abiotic stress responses (Hirayama and Shinozaki 2007). The application of exogenous ABA often mimics the abiotic stress responses. The expression of several stress-responsive genes is also induced by ABA (Shinozaki et al. 2003; Zhu 2002). However, several studies report that many stress-responsive genes do not show response to ABA, suggesting the existence of two signal transduction pathways, namely ABA-dependent and ABA-independent pathways, in abiotic stress responses (Nakashima et al. 2009; Zhu 2002). Other phytohormones, such as salicylic acid, ethylene, and jasmonic acid have also been shown to play important roles in abiotic stress responses directly or via interplay with ABA (Fujita et al. 2006; Grant and Jones 2009; Pieterse et al. 2009). Recently, auxin has also been implicated in abiotic stress responses. Quite a large number of auxin-responsive genes have been shown to be differentially expressed under various abiotic stress conditions (Jain and Khurana 2009).

Although a comprehensive knowledge has accumulated about the stress-responsive genes, the biggest challenge is to decipher their functions and logically integrate the available knowledge to understand the mechanism underlying abiotic stress response and selection of most suitable target gene(s) for improving stress tolerance.

3 Overview of Gene Regulatory Network Involved in Abiotic Stress Response

The differential expression of a large number of genes indicates that the gene regulatory network operative during abiotic stress is very complex in plants. The expression of several transcription factor encoding genes is also induced in response to various abiotic stresses. The transcription factors (TFs) are considered as master regulators of the gene expression. An individual TF can govern the expression of numerous target genes by binding to specific *cis*-regulatory motifs present in their promoters either independently or in coordination with other proteins constituting gene regulatory network (Nakashima et al. 2009; Urano et al. 2010). Some of the stress-associated TFs are themselves regulated at the transcriptional level. This type of transcriptional regulatory system is called regulon and is required for fine-tuned gene expression in response to abiotic stresses.

TFs and their regulons involved in both ABA-dependent and ABA-independent pathways have been identified and characterized in plants (Nakashima et al. 2009). The regulons involving dehydration-responsive element (DRE) binding protein 1 (DREB1)/C-repeat binding factor (CBF) and DREB2 TFs regulate stress response via ABA-independent pathway. These TFs bind to the conserved DRE motif A/GCCGACNT sequence in the promoter region of their target genes. The regulons involving ABA-responsive element (ABRE) binding protein (AREB)/ABRE binding factor (ABF) TFs act via ABA-dependent pathway (Nakashima et al. 2009). AREB/ABF TFs harboring a bZIP type DNA-binding domain binds to the ABRE (PyACGTGG/TC) and plays a pivotal role in ABA-dependent gene activation (Choi et al. 2000). Other regulons comprised of NAC and MYB/MYC TFs are also supposed to regulate the abiotic stress response via ABA-independent pathway. NAC proteins recognize a MYC-like target sequence and activate downstream gene expression (Tran et al. 2006). The components of these regulons and their functions have been found to be conserved in dicots and monocots indicating the common regulatory mechanisms of gene expression among them in response to stress

(Nakashima et al. 2009). Recently, another regulon involving the transcriptional regulator, multiprotein bridging factor 1 (MBF1), has been identified in *Arabidopsis*, which regulates heat-response (Suzuki et al. 2011).

Although most of the TF regulons are functional and have overlapping roles in response to multiple stresses, some of them are specific to particular abiotic stress condition(s) only. DREB1/CBF regulon responds to cold stress and controls the expression of several downstream genes. The overexpression of DREB1/CBF of *Arabidopsis* and homologous genes from other plants in transgenics resulted in strong tolerance to abiotic stresses, including drought, high salinity, and freezing (Dubouzet et al. 2003; Kasuga et al. 1999; Qin et al. 2004). DREB2 regulon functions in both osmotic and heat stress responses in *Arabidopsis*, whereas in dehydration and high salinity in grasses (Nakashima et al. 2009). *Arabidopsis* AREB/ABFs are induced in response to ABA, dehydration, and high salinity. The overexpression of AREB1 resulted in ABA hypersensitivity and drought tolerance (Fujita et al. 2005). It has also been suggested that phosphorylation may be responsible for the activation of AREB/ABFs (Fujii et al. 2007; Furihata et al. 2006; Uno et al. 2000). The response of AREB/ARF regulon to dehydration and high salinity has been found to be conserved in rice also (Kagaya et al. 2002; Kobayashi et al. 2005). Further, the overexpression of NAC genes conferred tolerance to drought stress in transgenic *Arabidopsis* plants and up-regulated several stress-inducible genes (Fujita et al. 2004; Tran et al. 2004). It has also been demonstrated that NAC TFs act along with ZF-HD proteins to activate the expression of downstream target gene, *EARLY RESPONSE TO DEHYDRATION 1* (*ERD1*) (Tran et al. 2006). In rice, one of the NAC TFs has been found to be responsive to ABA, abiotic stresses and biotic stresses, and its overexpression imparted enhanced stress tolerance in transgenic plants (Nakashima et al. 2007).

Taken together, TFs play important roles in abiotic stress responses and are powerful targets for engineering stress tolerance in transgenic plants because the overexpression of a single TF may lead to induction of diverse stress-responsive genes. Further, the role of many other stress-responsive TFs and their regulons need to be identified for understanding the molecular mechanisms of abiotic stress responses.

4 Homeobox Genes

Homeobox genes represent a class of TFs containing a conserved 180 bp long DNA sequence, which encodes for a 60 amino acid long DNA-binding domain termed as homeodomain (HD). The HD consists of three alpha helices forming a helix-turn-helix, which binds to specific DNA sequence and regulates the expression of target genes. The first homeobox genes were identified in *Drosophila melanogaster* and thereafter in all the eukaryotes. Homeobox genes are known to be the key regulators of various aspects of development, including cell fate determination and body plan specification.

Homeobox genes are represented by a large multigene family in plants, which have been classified into several distinct classes based on the amino acid sequence of HD and presence of other conserved domains. Initially, homeobox genes were classified into seven classes, including KNOX, BEL, ZM-HOX, HAT1, HAT2, ATHB8, and GL2 (Bharathan et al. 1997). Later on, Chan et al. (1998) classified homeobox genes into five groups (HD-ZIP, GLABRA, KNOTTED, PHD, and BEL). However, based on genome-wide analysis, 107 homeobox genes were classified into ten distinct subfamilies, including HD-ZIP I, HD-ZIP II, HD-ZIP III, HD-ZIP IV, ZF-HD, PHD, BLH, KNOXI, KNOXII, and WOX, in rice (Jain et al. 2008). Among these, HD-ZIP represented the largest family comprising of at least 48 members. The expansion of homeobox gene family has been attributed due to the chromosomal segmental duplications in rice, which might be responsible for the diversification of their function (Jain et al. 2008). Most recently, a comprehensive classification of plant homeobox genes based on the characterization of new motifs has been accomplished from the analysis of ten complete genomes of flowering plants, mosses, *Selaginella*, unicellular green algae, and red algae (Mukherjee et al. 2009). A total of 14 classes were identified across various plant species, namely HD-ZIP I-IV, PLINC, WOX, KNOX, BEL, PHD, DDT, NDX, LD, SAWADEE, and PINTOX. The conservation of homeobox genes across lineages emphasized their functional significance. In a recent study, it has also been shown that uncharacterized conserved motifs outside the HD-ZIP domain of HD-ZIP I subfamily confers functional diversity to members of this group of homeobox genes (Arce et al. 2011). Further, a greater number of homeobox genes in flowering plants have been related to their higher developmental and organizational complexity (Mukherjee et al. 2009).

5 Role of Homeobox Genes in Plant Development

The role of homeobox genes in plant developmental patterns has been extensively explored. The homeobox genes belonging to different subfamilies exhibit distinct expression patterns indicating their specific regulatory roles in tissue/organ differentiation and development (Chan et al. 1998). The molecular genetic analyses of several mutants have revealed that the KNOX family homeobox genes (*SHOOTMERISTEMLESS*, *BREVIPEDICELLUS*, *KNAT2* and *KNAT6* in *Arabidopsis*) are the key determinants in the maintenance of shoot apical meristem (Hake et al. 2004). The role of KNOX genes as versatile regulators of plant development and diversity have been comprehensively reviewed recently (Hay and Tsiantis 2010). KNOX proteins interact with HD proteins of BELL family to regulate the target genes that control hormone homeostasis (Hake et al. 2004; Hay et al. 2004; Hay and Tsiantis 2010; Smith et al. 2002). BELL family HD proteins are involved in pattern formation, stem-cell fate determination and tuber formation either independently or in coordination with KNOX proteins (Byrne et al. 2003; Chen et al. 2003; Reiser et al. 1995). WUSCHEL and other WOX family homeobox genes are involved in embryonic patterning, stem-cell maintenance, and organ formation (van der Graaff et al. 2009). WUSHEL has also been reported to be involved in regulation of cell differentiation during anther development (Deyhle et al. 2007). Very recently, it has been shown that the overexpression of WOX1 leads to meristem developmental defects in *Arabidopsis* (Zhang et al. 2011). Further, ZF-HD family proteins play a critical role in floral development in *Arabidopsis* (Tan and Irish 2006). The members of HD-ZIP family have been implicated in several developmental processes (Ariel et al. 2007). The ectopic expression of members of HD-ZIP I class suggested their role in the regulation of cotyledon development, leaf cell fate determination and blue-light perception signaling (Aoyama et al. 1995; Henriksson et al. 2005; Wang et al. 2003). HD-ZIP II class proteins have a role in plant development associated with shade avoidance responses (Sessa et al. 2005). The class III HD-ZIP proteins have been well characterized as regulators of apical meristem formation, vascular development, and maintenance of adaxial or abaxial polarity of leaves and embryo (Prigge et al. 2005; Talbert et al. 1995). HD-ZIP IV proteins are supposed to play specific roles in the outer cell layer of the plant organs (Nakamura et al. 2006). The presence of a large number of HD-ZIP proteins in plants may be responsible for the fine regulation of the developmental program as a result of the evolutionary pressure (Ariel et al. 2007).

HD-ZIP III class HD proteins are known to be the targets of miRNAs in *Arabidopsis*. It has been demonstrated that their miRNA-mediated post-transcriptional regulation controls the establishment of adaxial–abaxial polarity (McConnell et al. 2001; Rhoades et al. 2002; Tang et al. 2003). Recently, small RNA signatures were found to be associated with a significantly large fraction of homeobox genes in rice (Jain and Khurana 2008). Further, the results suggested an unusually high degree of small RNA regulation of rice homeobox genes during panicle development (Jain and Khurana 2008). This study provided evidence for a highly complex small RNAs-mediated regulation of homeobox genes involved in various cellular processes.

6 Role of Homeobox Genes in Abiotic Stress Responses

6.1 Differential Expression Under Abiotic Stress Conditions

The differential expression of individual homeobox genes in response to abiotic stresses has been reported in many plant species (Deng et al. 2002; Frank et al. 1998; Gago et al. 2002; Lee and Chun 1998; Soderman et al. 1996, 1999; Tran et al. 2006). In rice and *Arabidopsis*, the expression levels of quite a large number of homeobox genes have been found to be altered in response to various abiotic stress conditions. *ATHB7* and *ATHB12* transcripts were present in all organs at a basal level but there was phenomenal increase in the transcript level once the plant was exposed to drought conditions (Hjellstrom et al. 2003; Olsson et al. 2004; Soderman et al. 1996). Microarray analysis revealed that among a total of 107 homeobox genes, 37 were differentially expressed under desiccation, salt and/or cold stress conditions in rice seedlings (Jain et al. 2008). The expression of some of these genes was regulated by a specific stress and others by multiple stresses. A few HD-ZIP genes showed differential expression in the flowering stage of drought-sensitive and drought-tolerant rice cultivars also (Agalou et al. 2008; Bhattacharjee and Jain, unpublished), which further confirmed their role in abiotic stress responses. The activity of HD TFs was found to be regulated by cellular redox status also (Tron et al. 2002). Many of the homeobox genes, whose expression is regulated by abiotic stresses, were found to be preferentially expressed during specific developmental stage(s) (Jain et al. 2008). This suggested the role of homeobox TFs as mediators of plant growth response to different abiotic stress conditions during various stages of development. A novel homeobox gene, *GhHB1* (HD-ZIP I class member) from cotton has also been identified, which was specifically expressed in roots (Ni et al. 2008). The expression of this gene was found to be up-regulated in the presence of exogenous salt and ABA.

6.2 ABA-Mediated Regulation of Homeobox Genes

The plant hormone ABA plays a key role in adaptive stress responses to environmental stimuli (Cutler et al. 2010; Fujita et al. 2011; Raghavendra et al. 2010). Much information is not available about the role of homeobox genes in ABA-dependent abiotic stress response pathways. Many of the *Arabidopsis* HD-ZIP class homeobox genes, including *ATHB6*, *ATHB7*, and *ATHB12* have been shown to be induced in response to exogenous ABA (Henriksson et al. 2005; Lee and Chun 1998; Soderman et al. 1996). However, no induction of *ATHB7* and *ATHB12* was observed in the ABA-deficient mutants (Olsson et al. 2004; Soderman et al. 1996). Further, the mutants of these genes exhibited reduced sensitivity and transgenic *Arabidopsis* plants overexpressing them were hypersensitive to exogenous ABA as compared to wild type, indicating their role in ABA-dependent abiotic stress response pathways (Olsson et al. 2004). Another homeobox gene, *ATHB5*, has been characterized as a positive regulator of ABA response in developing seedlings (Johannesson et al. 2003). ATHB6 protein has been identified as the interacting protein of protein phosphatase ABI1 and acts as negative regulator of ABA signaling pathway downstream of ABI1 (Himmelbach et al. 2002). The N-terminal domain of ATHB6 and protein phosphatase domain of ABI1 were found to be crucial for their interaction. It is known that the phosphorylation of transcription factors play a major role in DNA binding and in this case, the PP2C activity of ABI1 has been shown to be responsible for the interaction between ATHB6 and ABI1. A single point mutation is sufficient to prevent the interaction to take place, further suggesting the importance of these phosphorylation reactions for interacting proteins. The binding site of ATHB6 protein (CAATTATTA) has also been identified, which along with minimal promoter was sufficient to mediate ABA-dependent activation of gene expression (Himmelbach et al. 2002). We have also found the expression of some rice homeobox genes to be regulated by exogenous ABA [Bhattacharjee and Jain, unpublished]. Altogether, these studies suggest that HD TFs may be implicated mainly in ABA-dependent abiotic stress response pathways. However, their role in ABA-independent pathways cannot be ruled out.

6.3 Implication of Homeobox Genes in Abiotic Stress Tolerance

Only a few mutant and transgenic studies have been performed in plants, which implicate the homeobox genes in abiotic stress responses. In *Arabidopsis*, a few mutants have been isolated and analyzed, which validate the role of homeobox genes in abiotic stress responses. A homeobox gene mutant, *hos9,* hypersensitive to freezing before and after cold acclimation as compared to wild-type plants, was identified by large scale screening of *Arabidopsis* mutants (Zhu et al. 2004). The transcript levels of *RD29A* and other stress-inducible genes increased in the *hos9* mutant as compared to wild-type plants after cold treatment. However, there was no alteration in the expression level of CBF genes, which are otherwise believed to play an important role in cold acclimation. Further, none of the CBF family genes were differentially expressed in the *hos9* mutant, suggesting the role of HOS9 homeobox TF in mediating cold tolerance via CBF-independent pathway (Zhu et al. 2004).

More recently, one mutant with improved drought tolerance, *enhanced drought tolerance1* (*edt1*), was isolated from *Arabidopsis* in a gain-of-function genetic screen (Yu et al. 2008). The enhanced drought tolerance of the mutant was found due to the activated expression of a HD-ZIP IV class member, *HDG11*. The mutant exhibited more extensive root system, higher levels of superoxide dismutase activity and elevated levels of ABA and proline as compared to wild-type plants. Further, the overexpression of this homeobox gene conferred drought tolerance in transgenic tobacco plants, which was attributed to the improved root architecture and reduced leaf stomatal density. Unlike other TFs, such as DREB (Kasuga et al. 1999), the constitutive expression of *HDG11* did not cause growth retardation (Yu et al. 2008), which is agronomically very important for crop improvement. Several stress-responsive genes involved in ABA signaling and calcium signaling were induced in the mutant plants. It has been suggested that HDG11 may regulate a complex network of genes to impart stress tolerance. Further, the possibility of binding of START domain of HDG11 to a lipid ligand to provide drought tolerance had also not been ruled out. Interestingly, *HDG11* was found to be expressed specifically in flower buds, flowers and immature siliques, and did not respond to stress conditions (Yu et al. 2008). Although it is not clear that how exactly HDG11 mediates stress tolerance, it has been suggested that the expression pattern of this gene has allowed it to gain novel function in drought tolerance (Yu et al. 2008).

The transgenic plants overexpressing *ATHB7* and *ATHB12* showed altered phenotype mimicking stress conditions and suggested their involvement in the growth response to water-deficit in shoot and root (Olsson et al. 2004). In addition, *ATHB12* functionally complemented the NaCl-sensitive phenotype of a calcineurin-deficient yeast mutant and increased salt tolerance by regulating sodium ion homeostasis (Shin et al. 2004). In another study, the cDNA encoding for ZFHD1 TF was identified in a yeast one-hybrid screen using 62 bp promoter region of *ERD1* containing ZF-HD recognition sequence (Tran et al. 2006). The transgenic plants overexpressing *ZFHD1* revealed a significant improvement in drought tolerance and upregulation of several stress-inducible genes. Further, yeast two-hybrid analysis identified NAC proteins as interacting partners of ZFHD1. Interestingly, the *ERD1* transcripts accumulated at higher level in the plants overexpressing both ZFHD1 and NAC proteins simultaneously, but not in transgenics overexpressing ZFHD1 only (Tran et al. 2006). These results validated the earlier observation that induction of *ERD1* gene requires coordinated activity of ZFHDRS and NACRS *cis*-regulatory elements (Simpson et al. 2003) and provides evidence for regulation of abiotic stress responses by homeobox TFs in conjunction with other transcription factors.

In crop plants, the role of only one homeobox gene from rice, *OsBIHD1*, encoding a BELL-type TF, in stress responses has been analyzed in transgenics so far (Luo et al. 2005). The overexpression of *OsBIHD1* in transgenic tobacco plants elevated the levels of defense-related *PR-1A* gene expression. Various analyses revealed that these transgenic lines had developed enhanced disease resistance against tomato mosaic virus, tobacco mosaic virus, and *Phytopthora parasitica.* However, in contrast, these transgenics exhibited enhanced sensitivity to salt and oxidative stresses. It has been suggested that OsBIHD1 may be acting as a negative regulator of stress tolerance by suppressing the abiotic stress signaling cascade in overexpression transgenic tobacco plants. Further, these results suggested that *OsB1HD1* might be involved in different pathways to regulate abiotic and biotic stress responses (Luo et al. 2005).

The functional characterization of homeobox genes in abiotic stress responses has been carried out in other plants also. For example, *Hahb-4*, a sunflower HD-ZIP gene acts as a developmental regulator and has been shown to confer drought tolerance in *Arabidopsis* plants (Dezar et al. 2005). Additionally, in sunflower, the *Hahb-4* gene has been potentially shown to be involved in ABA-dependent responses to water stress (Gago et al. 2002). Moreover, in resurrection plant *Craterostigma plantagineum*, identification of five novel dehydration-responsive HD-ZIP proteins has been done (Deng et al. 2002). In *Brassica napus*, a HD-ZIP gene, *BnHB6*, has been reported to be involved in both biotic and abiotic stress responses (Yu et al. 2005). These evidences suggest crucial involvement of the homeobox genes in stress-responsive signaling in various plant species.

Very little is known about the identity of downstream target genes of HD TFs. Although a few studies have reported the identification of putative targets of homeobox TFs, convincing evidences are not available as of now. For example, a *late embryo-genesis abundant/dehydrin* gene, *CdeT6-19*, has been identified as the potential target of *CpHB-7* gene in *Craterostigma plantageneum* (Deng et al. 2006). In addition, based on the macroarray analysis and mining of HD-ZIP binding site containing genes, several known ABA-responsive genes were proposed to be the targets of CpHB-7. Further, in another study, microarray analysis identified the genes involved in ethylene signaling and synthesis as targets in transgenic *Arabidopsis* plants overexpressing *Hahb-4* gene (Manavella et al. 2006). The identification of downstream targets of homeobox genes will be very important as it would help in unraveling the exact role of these homeobox TFs in the gene regulatory network of abiotic stress responses.

7 Conclusion and Perspectives

Plants face tremendous challenge when they are exposed to various abiotic stresses. Hence, they develop several modes of adaptability to protect themselves against these stresses. The recognition of key components which can reduce the deleterious effects of abiotic stresses to plants is very important for understanding the molecular mechanisms responsible for stress response and tolerance. Several genes including TFs involved in stress responses have been identified, but their exact role has not been studied yet in crop plants. So, there is a need to identify the master regulators and their regulatory pathways involved in stress adaptation. Further, it is imperative to select suitable candidates for conferring stress tolerance in plants via genetic engineering. The homeobox genes represent a family of transcription factors, which regulate the expression of a plethora of target genes. It is already well established that these genes play central role in regulating various developmental processes. The recent studies suggest that homeobox genes are promising candidate targets for manipulating abiotic stress tolerance in plants and can be used for crop improvement. These genes have largely overlapping roles in development and stress responses and thus, may offer multiple advantages, when attempts are made to raise overexpression transgenic plants. As the evidences related to the role of homeobox genes in providing stress tolerance available as of now are preliminary, it will be important to carry out their detailed functional analysis. Probably, the overexpression of homeobox genes may be able to impart stress tolerance in transgenics without compromising the yield, growth, and development of plants. Further, the identification of other regulatory components and target genes of stress-responsive homeobox TFs may lead to identification of novel pathways and better understanding of underlying molecular mechanisms. In addition, it would be very interesting to study the role of small RNAs in HD TFs-mediated regulation of abiotic stress responses.

Acknowledgements The work is supported financially by the Department of Science and Technology (grant file number SR/S0/PS/07/2011), Government of India, New Delhi and core grant from NIPGR. AB acknowledges the award of research fellowship from the Council of Scientific and Industrial Research, New Delhi.

References

Agalou A, Purwantomo S, Overnas E et al (2008) A genome-wide survey of HD-Zip genes in rice and analysis of drought-responsive family members. Plant Mol Biol 66:87–103

Aoyama T, Dong CH, Yan WU et al (1995) Ectopic expression of the *Arabidopsis* transcriptional activator *ATHB-1* alters leaf cell fate in tobacco. Plant Cell 7:1773–1785

Arce AL, Raineri J, Capella M, Cabello JV, Chan RL (2011) Uncharacterized conserved motifs outside the HD-Zip domain in HD-Zip subfamily I transcription factors; a potential source of functional diversity. BMC Plant Biol 11:42

Ariel FD, Manavella PA, Dezar CA, Chan RL (2007) The true story of the HD-Zip family. Trends Plant Sci 12:419–426

Bartels D, Sunkar R (2005) Drought and salt tolerance in plants. Crit Rev Plant Sci 24:23–58

Bharathan G, Janssen BJ, Kellogg EA, Sinha N (1997) Did homeodomain proteins duplicate before the origin of angiosperms, fungi, and metazoa? Proc Natl Acad Sci USA 94: 13749–13753

Byrne ME, Groover AT, Fontana JR, Martienssen RA (2003) Phyllotactic pattern and stem cell fate are determined by the *Arabidopsis* homeobox gene *BELLRINGER*. Development 130:3941–3950

Chan RL, Gago GM, Palena CM, Gonzalez DH (1998) Homeoboxes in plant development. Biochim Biophys Acta 1442:1–19

Chen H, Rosin FM, Prat S, Hannapel DJ (2003) Interacting transcription factors from the three-amino acid loop extension superclass regulate tuber formation. Plant Physiol 132:1391–1404

Chinnusamy V, Schumaker K, Zhu JK (2004) Molecular genetic perspectives on cross-talk and specificity in abiotic stress signaling in plants. J Exp Bot 55:225–236

Choi H, Hong J, Ha J, Kang J, Kim SY (2000) ABFs, a family of ABA-responsive element binding factors. J Biol Chem 275:1723–1730

Cutler SR, Rodriguez PL, Finkelstein RR, Abrams SR (2010) Abscisic acid: Emergence of a core signaling network. Annu Rev Plant Biol 61:651–679

Deng X, Phillips J, Meijer AH, Salamini F, Bartels D (2002) Characterization of five novel dehydration responsive homeodomain leucine zipper genes from the resurrection plant *Craterostigma plantagineum*. Plant Mol Biol 49:601–610

Deng X, Phillips J, Bräutigam A et al (2006) A homeodomain leucine zipper gene from *Craterostigma plantagineum* regulates abscisic acid responsive gene expression and physiological responses. Plant Mol Biol 61:469–489

Deyhle F, Sarkar AK, Tucker EJ, Laux T (2007) WUSCHEL regulates cell differentiation during anther development. Dev Biol 302:154–159

Dezar CA, Gago GM, Gonzalez DH, Chan RL (2005) *Hahb-4*, a sunflower homeobox-leucine zipper gene, is a developmental regulator and confers drought tolerance to *Arabidopsis thaliana* plants. Transgenic Res 14:429–440

Dubouzet JG, Sakuma Y, Ito Y et al (2003) *OsDREB* genes in rice, *Oryza sativa* L, encode transcription activators that function in drought, high-salt and cold-responsive gene expression. Plant J 33:751–763

Frank W, Phillips J, Salamini F, Bartels D (1998) Two dehydration-inducible transcripts from the resurrection plant *Craterostigma plantagineum* encode interacting homeodomain leucine zipper proteins. Plant J 15:413–421

Fujii H, Verslues PE, Zhu JK (2007) Identification of two protein kinases required for abscisic acid regulation of seed germination, root growth and gene expression in *Arabidopsis*. Plant Cell 19:485–494

Fujita M, Fujita Y, Maruyama K et al (2004) A dehydration induced NAC protein, RD26, is involved in a novel ABA-dependent stress-signaling pathway. Plant J 39:863–876

Fujita Y, Fujita M, Satoh R et al (2005) AREB1 is a transcription activator of novel ABRE-dependent ABA-signaling that enhances drought stress tolerance in *Arabidopsis*. Plant Cell 17:3470–3488

Fujita M, Fujita Y, Noutoshi Y et al (2006) Crosstalk between abiotic and biotic stress responses: a current view from the points of convergence in the stress signaling networks. Curr Opin Plant Biol 9:436–442

Fujita Y, Fujita M, Shinozaki K, Yamaguchi-Shinozaki K (2011) ABA-mediated transcriptional regulation in response to osmotic stress in plants. J Plant Res 124:509–525. http://dx.doi.org/10.1007/s10265-011-0412-3

Furihata T, Maruyama K, Fujita Y et al (2006) ABA-dependent multisite phosphorylation regulates the activity of a transcription activator AREB1. Proc Natl Acad Sci USA 103:1988–1993

Gago GM, Almoguera C, Jordano J, Gonzalez DH, Chan RL (2002) *Hahb-4*, a homeobox-leucine zipper gene potentially involved in abscisic acid-dependent responses to water stress in sunflower. Plant Cell Environ 25:633–640

Grant MR, Jones JDG (2009) Hormone (dis)harmony molds plant health and disease. Science 324:750–752

Hadiarto T, Tran LS (2011) Progress studies of drought-responsive genes in rice. Plant Cell Rep 30:297–310

Hake S, Smith HM, Holtan H, Magnani E, Mele G, Ramirez J (2004) The role of *knox* genes in plant development. Annu Rev Cell Dev Biol 20:125–151

Hay A, Tsiantis M (2010) *KNOX* genes: Versatile regulators of plant development and diversity. Development 137:3153–3165

Hay A, Craft J, Tsiantis M (2004) Plant hormones and homeoboxes: bridging the gap? Bioessays 26:395–404

Henriksson E, Olsson A, Johannesson H, Hanson J, Engstrom P, Soderman E (2005) Homeodomain leucine zipper class I genes in *Arabidopsis* expression patterns and phylogenetic relationships. Plant Physiol 139:509–518

Himmelbach A, Hoffmann T, Leube M, Hohener B, Grill E (2002) Homeodomain protein ATHB6 is a target of the protein phosphatase ABI1 and regulates hormone responses in *Arabidopsis*. EMBO J 21:3029–3038

Hirayama T, Shinozaki K (2007) Perception and transduction of abscisic acid signals: keys to the function of the versatile plant hormone ABA. Trends Plant Sci 12:343–351

Hirayama T, Shinozaki K (2010) Research on plant abiotic stress responses in the post-genome era: past, present and future. Plant J 61:1041–1052

Hjellstrom M, Olsson A, Engstrom P, Soderman E (2003) Constitutive expression of the water deficit-inducible homeobox gene *ATHB7* in transgenic *Arabidopsis* causes a suppression of stem elongation growth. Plant Cell Environ 26:1127–1136

Jain M, Khurana JP (2008) Small RNA regulation of rice homeobox genes. Plant Signal Behav 3:1024–1025

Jain M, Khurana JP (2009) Transcript profiling reveals diverse roles of auxin-responsive genes during reproductive development and abiotic stress in rice. FEBS J 276:3148–3162

Jain M, Tyagi AK, Khurana JP (2008) Genome-wide identification, classification, evolutionary expansion and expression analyses of homeobox genes in rice. FEBS J 275:2845–2861

Johannesson H, Wang Y, Hanson J, Engstrom P (2003) The *Arabidopsis thaliana* homeobox gene *ATHB5* is a potential regulator of abscisic acid responsiveness in developing seedlings. Plant Mol Biol 51:719–729

Kagaya Y, Hobo T, Murata M, Ban A, Hattori T (2002) Abscisic acid induced transcription is mediated by phosphorylation of an abscisic acid response element binding factor, TRAB1. Plant Cell 14:3177–3189

Kasuga M, Liu Q, Miura S, Yamaguchi-Shinozaki K, Shinozaki K (1999) Improving plant drought, salt and freezing tolerance by gene transfer of a single stress-inducible transcription factor. Nat Biotechnol 17:287–291

Kobayashi Y, Murata M, Minami H et al (2005) Abscisic acid-activated SNRK2 protein kinases function in the gene-regulation pathway of ABA signal transduction by phosphorylating ABA response element-binding factors. Plant J 44:939–949

Lee YH, Chun JY (1998) A new homeodomain-leucine zipper gene from *Arabidopsis thaliana* induced by water stress and abscisic acid treatment. Plant Mol Biol 37:377–384

Luo H, Song F, Zheng Z (2005) Overexpression in transgenic tobacco reveals different roles for the rice homeodomain gene *OsBIHD1* in biotic and abiotic stress responses. J Exp Bot 56:2673–2682

Mahajan S, Tuteja N (2005) Cold, salinity and drought stresses: An overview. Arch Biochem Biophys 444:139–158
Manavella PA, Arce AL, Dezar CA (2006) Cross-talk between ethylene and drought signaling pathways is mediated by the sunflower *Hahb-4* transcription factor. Plant J 48:125–137
McConnell JR, Emery J, Eshed Y, Bao N, Bowman J, Barton MK (2001) Role of *PHABULOSA* and *PHAVOLUTA* in determining radial patterning in shoots. Nature 411:709–713
Mukherjee K, Brocchieri L, Burglin TR (2009) A comprehensive classification and evolutionary analysis of plant homeobox genes. Mol Biol Evol 26:2775–2794
Nakamura M, Katsumata H, Abe M et al (2006) Characterization of the class IV homeodomain-leucine zipper gene family in *Arabidopsis*. Plant Physiol 141:1363–1375
Nakashima K, Tran LS, Van Nguyen D et al (2007) Functional analysis of a NAC-type transcription factor *OsNAC6* involved in abiotic and biotic stress-responsive gene expression in rice. Plant J 51:617–630
Nakashima K, Ito Y, Yamaguchi-Shinozaki K (2009) Transcriptional regulatory networks in response to abiotic stresses in *Arabidopsis* and grasses. Plant Physiol 149:88–95
Ni Y, Wang X, Li D, Wu Y, Xu W, Li X (2008) Novel cotton homeobox gene and its expression profiling in root development and in response to stresses and phytohormones. Acta Biochim Biophy Sin 40:78–84
Olsson A, Engstrom P, Soderman E (2004) The homeobox genes *ATHB12* and *ATHB7* encode potential regulators of growth in response to water deficit in *Arabidopsis*. Plant Mol Biol 55:663–677
Pieterse CMJ, Leon-Reyes A, Van der Ent S, Van Wees SCM (2009) Networking by small-molecule hormones in plant immunity. Nat Chem Biol 5:308–316
Prigge MJ, Otsuga D, Alonso JM, Ecker JR, Drews GN, Clark SE (2005) Class III homeodomain-leucine zipper gene family members have overlapping, antagonistic and distinct roles in *Arabidopsis* development. Plant Cell 17:61–76
Qin F, Sakuma Y, Li J et al (2004) Cloning and functional analysis of a novel DREB1/CBF transcription factor involved in cold-responsive gene expression in *Zea mays* L. Plant Cell Physiol 45:1042–1052
Raghavendra AS, Gonugunta VK, Christmann A, Grill E (2010) ABA perception and signaling. Trends Plant Sci 15:395–401
Reiser L, Modrusan Z, Margossian L et al (1995) The *BELL1* gene encodes a homeodomain protein involved in pattern formation in the *Arabidopsis* ovule primordium. Cell 83:735–742
Rhoades MW, Reinhart BJ, Lim LP, Burge CB, Bartel B, Bartel DP (2002) Prediction of plant microRNA targets. Cell 110:513–520
Sessa G, Carabelli M, Sassi M et al (2005) A dynamic balance between gene activation and repression regulates the shade avoidance response in *Arabidopsis*. Genes Dev 19:2811–2815
Shin D, Koo YD, Lee J et al (2004) Athb-12, a homeobox-leucine zipper domain protein from *Arabidopsis thaliana*, increases salt tolerance in yeast by regulating sodium exclusion. Biochem Biophys Res Commun 323:534–540
Shinozaki K, Yamaguchi-Shinozaki K, Seki M (2003) Regulatory network of gene expression in the drought and cold stress responses. Curr Opin Plant Biol 6:410–417
Simpson SD, Nakashima K, Narusaka Y, Seki M, Shinozaki K, Yamaguchi-Shinozaki K (2003) Two different novel *cis*-acting elements of *erd1*, a *clpA* homologous *Arabidopsis* gene, function in induction by dehydration stress and dark-induced senescence. Plant J 33:259–270
Smith HM, Boschke I, Hake S (2002) Selective interaction of plant homeodomain proteins mediates high DNA- binding affinity. Proc Natl Acad Sci USA 99:9579–9584
Soderman E, Mattsson J, Engstrom P (1996) The *Arabidopsis* homeobox gene *ATHB-7* is induced by water deficit and by abscisic acid. Plant J 10:375–381
Soderman E, Hjellstrom M, Fahleson J, Engstrom P (1999) The HD-Zip gene *ATHB6* in *Arabidopsis* is expressed in developing leaves, roots and carpels and up-regulated by water deficit conditions. Plant Mol Biol 40:1073–1083
Suzuki N, Sejima H, Tam R, Schlauch K, Mittler R (2011) Identification of the MBF1 heat-response regulon of *Arabidopsis thaliana*. Plant J 66:844–851

Talbert PB, Adler HT, Parks DW, Comai L (1995) The *REVOLUTA* gene is necessary for apical meristem development and for limiting cell divisions in the leaves and stems of *Arabidopsis thaliana*. Development 121:2723–2735

Tan QK, Irish VF (2006) The *Arabidopsis* zinc finger-homeodomain genes encode proteins with unique biochemical properties that are coordinately expressed during floral development. Plant Physiol 140:1095–1108

Tang G, Reinhart BJ, Bartel DP, Zamore PD (2003) A biochemical framework for RNA silencing in plants. Genes Dev 17:49–63

Tran LS, Nakashima K, Sakuma Y et al (2004) Isolation and functional analysis of *Arabidopsis* stress-inducible NAC transcription factors that bind to a drought-responsive *cis*-element in the early responsive to dehydration stress 1 promoter. Plant Cell 16:2481–2498

Tran LS, Nakashima K, Sakuma Y et al (2006) Co-expression of the stress-inducible zinc finger homeodomain ZFHD1 and NAC transcription factors enhances expression of the *ERD1* gene in *Arabidopsis*. Plant J 49:46–63

Tron AE, Bertoncini CW, Chan RL, Gonzalez DH (2002) Redox regulation of plant homeodomain transcription factors. J Biol Chem 277:34800–34807

Uno Y, Furihata T, Abe H, Yoshida R, Shinozaki K, Yamaguchi-Shinozaki K (2000) *Arabidopsis* basic leucine zipper transcription factors involved in an abscisic acid-dependent signal transduction pathway under drought and high-salinity conditions. Proc Natl Acad Sci USA 97:11632–1163725

Urano K, Kurihara Y, Seki M, Shinozaki K (2010) 'Omics' analyses of regulatory networks in plant abiotic stress responses. Curr Opin Plant Biol 13:1–7

van der Graaff E, Laux T, Rensing SA (2009) The WUS homeobox-containing (WOX) protein family. Genome Biol 10:248.1–248.9

Vij S, Tyagi AK (2007) Emerging trends in the functional genomics of the abiotic stress response in crop plants. Plant Biotechnol J 5:361–380

Wang Y, Henriksson E, Soderman E, Henriksson KN, Sundberg E, Engstrom P (2003) The *Arabidopsis* homeobox gene, *ATHB16*, regulates leaf development and the sensitivity to photoperiod in *Arabidopsis*. Dev Biol 264:228–239

Yamaguchi-Shinozaki K, Shinozaki K (2006) Transcriptional regulatory networks in cellular responses and tolerance to dehydration and cold stresses. Annu Rev Plant Biol 57:781–803

Yu SW, Zhang LD, Zuo KJ, Tang DQ, Sun XF, Tang KX (2005) *Brassica napus* L. homeodomain leucine zipper gene *BnHB6* responds to abiotic and biotic stresses. J Integr Plant Biol 47:1236–1248

Yu H, Chen X, Hong YY et al (2008) Activated expression of an *Arabidopsis* HD-START protein confers drought tolerance with improved root system and reduced stomatal density. Plant Cell 20:1134–1151

Zhang Y, Wu R, Qin G, Chen Z, Gu H, Qu LJ (2011) Over-expression of WOX1 leads to defects in meristem development and polyamine homeostasis in Arabidopsis. J Integr Plant Biol 53:493–506

Zhu JK (2002) Salt and drought stress signal transduction in plants. Annu Rev Plant Biol 53:247–273

Zhu J, Shi H, Lee BH et al (2004) An *Arabidopsis* homeodomain transcription factor gene, *HOS9*, mediates cold tolerance through a CBF-independent pathway. Proc Natl Acad Sci U S A 101:9873–9878

Chapter 8
APETALA2 Gene Family: Potential for Crop Improvement Under Adverse Conditions

Sowmya Krishnaswamy, Shiv Verma, Muhammad H. Rahman, and Nat Kav

1 Introduction

Plants are affected by many unfavorable conditions including both biotic (e.g., bacteria, fungi, nematode, virus, weeds, parasitic plants, and insects) and abiotic stresses (e.g., drought, cold, salinity, freezing, heat, and water logging) that negatively affect their growth and productivity. It has been estimated that 90 % of total arable land experience one or more kind of environmental stresses (Dita et al. 2006). These conditions are worsening over time because of global climatic changes and developing stress-tolerant crops are becoming more important to minimize crop loss and to increase productivity (Agarwal et al. 2006; Vinocur and Altman 2005; http://www.ipcc.ch).

Unlike other organisms, plants are sessile, and in an attempt to overcome the imposed stresses, they trigger a cascade of molecular events when subjected to stress. These events include changes in gene expression which eventually lead to physiological and biological modifications necessary to enhance tolerance to adverse conditions. The advent of genomics and proteomics has been helpful in understanding these stress-signal transduction regulatory networks and studies have suggested that transcription factors (TFs) play very important roles in the expression of stress-responsive genes (Eulgem 2005; Fowler and Thomashow 2002; Yamaguchi-Shinozaki and Shinozaki 2006).

S. Krishnaswamy
Department Agricultural Food and Nutritional Science, University of Alberta, Edmonton, AB, Canada T6G 2P5

Southern Crop Protection and Food Research Centre, Agriculture and Agri-Food Canada London, ON, Canada, N5V 4T3

S. Verma • M.H. Rahman • N. Kav (✉)
Department Agricultural Food and Nutritional Science, University of Alberta, Edmonton, AB, Canada T6G 2P5
e-mail: Nat.Kav@ualberta.ca

N. Tuteja and S. Singh Gill (eds.), *Plant Acclimation to Environmental Stress*,
DOI 10.1007/978-1-4614-5001-6_8, © Springer Science+Business Media New York 2013

TFs are DNA-binding proteins that interact with specific *cis*-elements in the promoter regions of genes and regulate their expression by activating or repressing the recruitment of RNA polymerase (Karin 1990; Nikolov and Burley 1997). A single TF regulates the expression of several other genes including TFs themselves and are therefore considered to be important molecular targets for the genetic manipulation of cellular processes in plants (Hussain et al. 2011). Indeed, transcriptional regulators, considered a dominant class of the gene family, played a major role in selection and domestication along with morphological development in plants, which led to dramatic improvement in productivity of most extensively grown crops worldwide like rice, wheat, and maize (Doebley et al. 2006). Given the importance of TFs in regulation of metabolic pathways, it is not surprising that significant portion of plant genome encodes TFs. For example, approximately 5% of the *Arabidopsis* genome encodes for TFs (Riano-Pachon et al. 2007).

APETALA2/ethylene response element-binding protein (AP2/EREBP) TF family is the major group among the TF families in *Arabidopsis* with 147 genes comprising about 9 % of the total TFs (Feng et al. 2005). In higher plants, close to 200 AP2 TF genes have been reported (http://plntfdb.bio.uni-potsdam.de/v3.0/). For instance, the genomes of rice (Nakano et al. 2006), grapevine (Jaillon et al. 2007), and poplar (Zhuang et al. 2008) encode 139, 132, and 200 AP2/ERF-related proteins, respectively. The name AP2 arises from the protein APETALA that is involved in flower development (Jofuku et al. 1994).

AP2/EREBPs are characterized by the presence of a DNA-binding domain called AP2 domain, about 68 amino acids long (Hao et al. 1998; Riechmann and Meyerowitz 1998). Based on the presence of one or two AP2-DNA-binding domains, the family is further divided into four subfamilies, the AP2, DREB, ERF, RAV, and others (Sakuma et al. 2002). AP2 subfamily encodes proteins with two AP2 domains and these proteins are implicated in various growth events like meristem determinance, organ identity, and flower development (Saleh and Pagés 2003). Examples of proteins belonging to this class include AP2, baby boom (BBM), Glossy15 (GL15), and AINTEGUMENTA (ANT) (Krizek 2009; Moose and Sisco 1996; Passarinho et al. 2008). The DREB (dehydration-responsive element binding), ERF (ethylene-responsive factors), and RAV (related to ABI3/VP1) subfamily genes encode proteins with only one AP2 domain and members of these subfamilies have been implicated in stress signaling network (Guo et al. 2005; Saleh and Pagés 2003; Gutterson and Reuber 2004; Shinwari et al. 1988). The DREB groups were identified as genes encoding TFs involved in dehydration-responsive regulon (Liu et al. 1998; Stockinger et al. 1997), whereas ERF groups were identified as binding factors mediating the ethylene response (Fujimoto et al. 2000). The RAV groups were identified by Kagaya et al. (1999) as proteins with two DNA-binding domains, an AP2 and a B3 motif, and these proteins are involved in hormone and stress responses (Alonso et al. 2003; Hu et al. 2004; Sohn et al. 2006). Examples of proteins belonging to DREB, ERF, and RAV subfamilies include C-repeat/dehydration-responsive element-binding factors (CBFs/DREBs), ERFs, LePtis, TINY, abscisic acid insensitive (ABI4), and RAV proteins (Riechmann 2000; Sakuma et al. 2002).

2 Gene Regulation by AP2 TFs

As mentioned previously, ERF and DREB subfamily proteins are the major groups in AP2 family. For instance, in *Arabidopsis* out of 147 AP2 genes, 65 belong to ERF and 56 belong to DREB subfamily (Sakuma et al. 2002). The ERF subfamily proteins interact with ethylene response elements (ERE) or GCC box and regulate the expression of ethylene-inducible pathogenesis-related genes such as prb-1b, β-1, 3-glucanase, chitinase, and osmotin (Büttner and Singh 1997; Ohme-Takagi and Shinshi 1995; Xu et al. 1998, 2006). They can act as both activators and repressors of gene expression. For example, *Arabidopsis* AtERF1, AtERF2, and AtERF5 function as activators of GCC-dependent transcription, while AtERF3, AtERF4, and AtERF7 act as repressors of GCC-dependent transcription (McGrath et al. 2005; Xu et al. 2006).

The DREB subfamily proteins interact with C-repeat or dehydration response elements (DRE) and regulate the expression of low-temperature and/or water deficit responsive genes (Jaglo-Ottosen et al. 1998; Kasuga et al. 1999; Liu et al. 1998). The DRE (5′-TACCGACAT-3′) elements are found in the promoters of drought and cold-inducible genes like *rd29A* (Yamaguchi-Shinozaki and Shinozaki 1994). Similar to DRE, the C-repeat 5′-TGGCCGAC-3′ (containing the core 5′-CCGAC-3′) elements are found in the COR (cold-regulated) genes like *cor15a*, *rab18*, *kin1*, and *kin2* (Baker et al. 1994; Kurkela and Borg-Franck 1992; Kurkela and Franck 1990; Lang and Palva 1992). Liu et al. (1998) using reporter genes demonstrated, for the first time, that DREB proteins DREB1 and DREB2 act as activators of promoters harboring DRE elements. Overexpression studies have also demonstrated the role of DREB genes in DRE-dependent gene regulation. For example, overexpression of *DREB1*- and *DREB2*-induced expression of regulatory region rd29A which is involved in abiotic stress signaling (Liu et al. 1998). Moreover, *DREB1A* overexpression induced the expression of COR genes such as *rd29A/cor78/lti78*, *kin1*, *cor6.6/kin2*, *cor15a*, *cor47/rd17*, and *erd10* (Kasuga et al. 1999).

Earlier, it was thought that DREB-dependent regulation is involved only in abiotic stresses, whereas ERF genes are involved mostly in biotic stresses (Guo and Ecker 2004; Shinozaki and Yamaguchi-Shinozaki 2000). However, recent studies have indicated that DREB and ERF-type AP2 TFs are involved in multiple pathways activated by both kinds of stresses. For example, overexpression of tobacco ERF, *Tsi1*, enhances resistance to both *Pseudomonas syringae* as well as osmotic stress. In addition, the Tsi1 protein was shown to be capable of binding to DRE/CRT elements *in vitro* (Park et al. 2001). Furthermore, the overexpression of *CaERFLP1* resulted in enhanced expression of salt-inducible *LT145* which contains multiple DRE/CRT elements in its promoter (Lee et al. 2004). In addition, a series of ERF TFs were found to interact with the DRE/CRT motif *in vitro* (Yi et al., 2004; Li et al. 2005; Xu et al. 2007). Similarly, DREB-type AP2 TFs have been shown to regulate biotic stress signaling. For instance, *DREB2A*, a regulator of dehydration-responsive pathway was found to cross talk with *Adr1* (activated disease resistance 1) activated signaling network (Chini et al. 2004). Furthermore, a DREB-like factor,

TINY, demonstrated its ability to interact with both DRE and ERE elements with similar affinity and activated reporter genes containing these elements (Sun et al. 2008). Moreover, the overexpression of *TINY* in seedlings enhanced the expression of both DRE- and the ERE-containing genes in transgenic *Arabidopsis* (Sun et al. 2008). Similarly, it was demonstrated that RAP2.4 acts as a transactivator of both DRE- and ERE-mediated genes that are responsive to light, drought, and ethylene (Lin et al. 2008). Recently, *DEAR1* (DREB and EAR motif protein 1) gene whose expression is elevated in response to both biotic and abiotic stresses, when overexpressed, showed constitutive expression of PR genes and tolerance to *P. syringae* in transgenic *Arabidopsis* plants (Tsutsui et al. 2009). Similarly, the overexpression of *PgDREB2A* resulted in the upregulation of dehydrins and heat-shock protein genes as well as *NtERF5* that mediate expression of PR genes (Agarwal et al. 2010). Therefore it appears that some DREB and ERF transcription factors have a regulatory role in mediating cross talk between biotic and abiotic stress signaling pathways.

AP2 TFs also regulate the expression of members of the same family. For instance, it was demonstrated that CBF2/DREB1C acts as a negative regulator of *CBF1/DREB1B* and *CBF3/DREB1A* expression during cold acclimatization (Novillo et al. 2004). In addition, AP2 TFs are subjected to different temporal regulation to ensure transient and controlled expression of stress-related genes. For instance, in *Brassica napus,* it has been observed that *trans*-active Group I factors that bind with DREBs are expressed immediately on exposure to cold stress to turn on the DRE-mediated signaling pathway, whereas *trans*-inactive Group II proteins were expressed at later stages compete with the Group I to bind with the DRE and prevent the activation, and thus block the signal pathway (Zhao et al. 2006). AP2 TFs can also regulate their own expression like many other TFs. For example, the protein RAP2.1 possesses an AP2 domain that binds to DREs and regulates desiccation/cold-regulated (*RD/COR*) gene. Additionally, RAP2.1 can negatively regulate its own expression and keep the expression of stress response genes under tight control (Dong and Liu 2010).

Since AP2 TFs have been demonstrated to have important role in regulation of many genes in addition to their own expression, they have received much attention in recent time as ideal candidates for crop improvement. In addition, other proteins like inducer of CBF expression 1 (ICE1), calmodulin-binding transcription activator (CAMTA), ZAT12 (a zinc finger protein) that are involved in the regulation of AP2 family proteins, may also serve as good targets for manipulation (Chinnusamy et al. 2003; Doherty et al. 2009; Vogel et al. 2005).

3 Abiotic Stress Tolerance

Once the cis-regulatory elements of DREB/ERF TFs were identified as CRT/DRE and GCC elements, genetic and molecular approaches were used to investigate the potential utility of AP2/EREBP TFs from a wide variety of plants in order to enhance stress tolerance. Much of the data has come from overexpression and loss-of-function analysis and a list of characterized AP2 TFs from various species is presented in Table 8.1. AP2/EREBP members have demonstrated their crucial role in regulating different kinds of abiotic stress response, including drought, low temperature, salinity, and hypoxia (Haake et al. 2002; Hinz et al. 2010; Novillo et al. 2004; Oh et al. 2005, 2007; Yang et al. 2011).

DREB subfamily has been classified into six (A1–A6) groups (Sakuma et al. 2002), and *DREB1A* and *DREB2A* are the most studied genes among the DREBs. Group A1 contains *CBF*s and *DDF* (dwarf and delayed flowering) genes. The DREB1/CBF cold-response pathway is well characterized in *Arabidopsis* and rice (*Oryza sativa*) (Yamaguchi-Shinozaki and Shinozaki 2006). Three DREB1/CBF genes, namely *CBF1* (also called as *DREB1b*), *CBF2* (also called as *DREB1c*), and *CBF3* (also called as *DREB1a*) have been isolated from *Arabidopsis* (Gilmour et al. 1998; Liu et al. 1998; Stockinger et al. 1997). From rice, *DREB1/CBF* homologs such as *OsDREB1A*, *OsDREB1B*, *OsDREB1C*, and *OsDREB1D* have been isolated (Dubouzet et al. 2003). In response to low-temperature stress, these genes are quickly induced, and their products activate the CBF regulon to improve freezing tolerance (Agarwal et al. 2006; Nakashima and Yamaguchi-Shinozak 2006). Overexpression of *DREB1A* with constitutive and stress-inducible promoters in *Arabidopsis* has resulted in multiple abiotic stress tolerance including freezing stress tolerance (Kasuga et al. 2004; Liu et al. 1998). Additionally, overexpression of rice *OsDREB1A* in *Arabidopsis* resulted in expression of stress-related genes and consequent improved tolerance to abiotic stresses including drought, high salt, and freezing (Dubouzet et al. 2003).

Metabolome analysis of *DREB1A/CBF3* overexpressing *Arabidopsis* plants has demonstrated that monosaccharides, disaccharides, oligosaccharides, and sugar alcohol profiles were similar to the low-temperature-regulated metabolome (Cook et al. 2004; Maruyama et al. 2009) which suggests DREB1A may enhance tolerance by regulating genes involved in stress response. Indeed, the crucial role of group A1 DREBs in regulation of COR genes has been verified (Baker et al. 1994; Ouellet et al. 1998). Overexpression of *DREB1A/CBF3* in *Arabidopsis* resulted in elevated levels of P5CS transcripts and proline, in addition to elevated levels of total soluble sugars (sucrose, raffinose, glucose, and fructose), the components involved in cold acclimatization (Gilmour et al. 2000).

Similar to *DREB1A*, overexpression of *CBF1/DREB1B* in *Arabidopsis* induces COR genes and enhances freezing tolerance (Jaglo-Ottosen et al. 1998). CBF1 and CBF3 have an additive effect in inducing the whole CBF regulon (Novillo et al. 2007). On the other hand, *CBF2/DREB1C* acts as a negative regulator of both *CBF1* and *CBF3*, therefore, *cbf2* mutants show freezing, dehydration, and salt stress tolerance (Novillo et al. 2004, 2007). *CBF4* is a close *CBF/DREB1* homolog but its expression is rapidly induced only during drought stress and by ABA treatment, but not by cold (Haake et al. 2002). However, its overexpression has demonstrated increased tolerance to both drought and freezing stress (Haake et al. 2002). Overexpression of CBFs in crop plants has also demonstrated similar results with improved abiotic stress tolerance. For example, rice plants overexpressing

Table 8.1 List of genetically modified plant species with AP2 TFs that exhibited improved tolerance/resistance to various biotic and abiotic stresses

AP2 TF gene	Species	Biotic/abiotic stress tolerance/resistance	References
DREB1A	*Arabidopsis*	Freezing	Liu et al. (1998), Kasuga et al. (2004)
CBF1/DREB1B	*Arabidopsis*	Freezing	Jaglo-Ottosen et al. (1998)
CBF2/DREB1C	*Arabidopsis*	Freezing	Novillo et al. (2004, 2007)
CBF4	*Arabidopsis*	Drought, freezing	Haake et al. (2002)
CBF3/DREB1A	Rice	Drought, salinity, cold	Oh et al. (2005)
OsDREB1A, OsDREB1F	Rice	Cold, drought, salinity	Ito et al. (2006), Wang et al. (2008)
CBF1/DREB1b	Tomato	Chilling, oxidative	Hsieh et al. (2002)
BNCBF5 BNCBF17	*Brassica napus*	Freezing	Savitch et al. (2005)
TaDREB2, TaDREB3	Wheat and Barley	Drought, frost	Morran et al. (2011)
AtDREB2A ZmDREB2A	*Arabidopsis*	Heat	Sakuma et al. (2006a, b), Qin et al. (2007)
DREB2C	*Arabidopsis*	Heat	Lim et al. (2007), Chen et al. (2010)
AtDREB2A	*Arabidopsis*	Heat, drought	Sakuma et al. (2006a, b)
OsDREB2B	*Arabidiopsis*	Heat, drought	Matsukura et al. (2010)
ZmDREB2A	Maize	Drought	Qin et al. (2007)
SbDREB2	Rice	Drought	Bihani et al. (2011)
HARDY	Rice	Drought	Karaba et al. (2007)
GmERF3	Tobacco	Drought	Zhang et al. (2009)
SodERF3	Tobacco	Drought, osmotic stress	Trujillo et al. (2009)
RAP2.2	*Arabidopsis*	Hypoxia	Hinz et al. (2010)
AP37	Rice	Drought, high salinity, low temperature	Oh et al. (2009)
Sub1	Rice	Submergence	Xu et al. (2000, 2006), Jung et al. (2010)
WIN1	*Arabidopsis*	Drought	Aharoni et al. (2004), Broun (2004)
WXP1	*Medicago sativa*	Drought	Zhang et al. (2005)
HvRAF	*Arabidopsis*	Salinity, *Ralstonia solanacearum*	Jung et al. (2007)
TaERF1	*Arabidopsis*	*Botrytis cinerea, Pseudomonas syringae*	Xu et al. (2007)
GmERF3	Tobacco	Salinity, dehydration, *Ralstonia solan acearum, Alternaria alternata,* tobacco mosaic virus (TMV)	Zhang et al. (2009)
CaRAV1	*Arabidopsis*	*P. syringae*	Sohn et al. (2006)

(continued)

Table 8.1 (continued)

AP2 TF gene	Species	Biotic/abiotic stress tolerance/resistance	References
ERF1	*Arabidopsis*	*B. cinerea, Plectosphaerella cucumerina*	Berrocal-Lobo et al. (2002)
AtCBF1	Tomato	*Ralstonia solanacearum*	Li et al. (2011)
RAP2.6	Arabidopsis	Salt, drought	Krishnaswamy et al. (2011)
DREB19	Arabidopsis	Salt, drought	Krishnaswamy et al. (2011)
TSRF1	Rice	Drought	Quan et al. (2010)
TsCBF1	Maize	Drought	Zhang et al. (2010)
WXP1, WXP2	Arabidopsis	Drought, freezing	Zhang et al. (2007)
DREB1A	Rice	Drought, salt	Oh et al. (2005)
DREB1A	Tobacco	Drought, cold	Kasuga et al. (2004)
DREB1A	Wheat	Drought	Pellegrineschi et al. (2004)

HvCBF4 and *CBF3/DREB1A* exhibited enhanced tolerance to drought, high-salinity, and low-temperature stresses without stunting growth (Oh et al. 2005, 2007). Similarly, rice plants overexpressing *OsDREB1A* and *OsDREB1F* showed improved tolerance to low temperature, drought, and high-salinity conditions by expressing stress-inducible genes and by accumulating higher levels of osmoprotectants (Ito et al. 2006; Wang et al. 2008). Furthermore, transgenic tomato plants expressing *Arabidopsis CBF1/DREB1b* cDNA demonstrated increased chilling and oxidative stress tolerance (Hsieh et al. 2002). The overexpression of CBF/DREB1-like TFs (*BNCBF5* and *BNCBF17*) in *Brassica napus* also resulted in the constitutive expression of *COR* genes and those involved in photosynthesis and chloroplast development (Savitch et al. 2005). As a result, these transgenic *Brassica* plants were more tolerant to freezing stress and exhibited higher photosynthetic efficiency (Savitch et al. 2005). Furthermore, transgenic wheat and barley plants overexpressing *TaDREB2* and *TaDREB3* (close homologues of CBF factors) induced *LEA/COR/DHN* genes and improved drought and frost tolerance (Morran et al. 2011). Similarly, heterologous expression of CBFs in tobacco (Cong et al. 2008), potato (Behnam et al. 2006), and grasses (Zhao et al. 2007) resulted in enhanced tolerance to one or more abiotic stresses. From these studies, it appears that multiple mechanisms contribute to freezing tolerance through *CBF* regulon, and *CBF*s can be exploited to improve low temperature and other abiotic stress tolerance in crop plants.

Group A-2 genes are also well characterized and contain genes like *DREB2A* and *DREB2B*. These genes are induced in response to dehydration and induce the expression of various genes involved in dehydration tolerance (Liu et al. 1998; Sakuma et al. 2006a).

For instance, overexpression of *AtDREB2A* or corn *ZmDREB2A* in *Arabidopsis* induced the expression of LEA and heat stress-inducible genes (such as *HSPs* and *HsfA3*) leading to increased thermotolerance (Qin et al. 2007; Sakuma et al. 2006a, b). Similarly, overexpression of *DREB2C* in *Arabidopsis* increased the expression of heat stress-related genes and enhanced thermotolerance (Chen et al. 2007; Lim et al. 2007). It has also been demonstrated that sunflower (*Helianthus annuus*) *HaDREB2* enhances *Hahsp17.6G1* expression through a synergistic interaction with *HaHSFA9* (sunflower heat stress factor A9) (Díaz-Martín et al. 2005). In addition, Schramm et al. (2008) demonstrated that *HsfA3* is transcriptionally activated by DREB2A during heat stress, which in turn, regulates the expression of Hsp-encoding genes.

Some of the DREB2 group proteins can occur in both functional and nonfunctional forms, and require post-transcriptional or posttranslational modifications for their activation (Agarwal et al. 2007; Liu et al. 1998; Sakuma et al. 2006a, b). For instance, the deletion of negative regulatory domain from AtDREB2A transforms it to a constitutively active form (*DREB2A CA*). The overexpression of the constitutively active form of *AtDREB2A* results in significant drought stress and heat stress tolerance in *Arabidopsis* (Sakuma et al. 2006a, b). Similarly, it has been demonstrated that DREB2A from pearl millet (*Pennisetum glaucum*) is a phosphoprotein and phosphorylation negatively regulates its DRE-binding activity (Agarwal et al. 2007). Furthermore, it has been shown that alternative splicing of pre-mRNA is an important regulatory mechanism in rice *OsDREB2B* which led to enhanced expression of DREB2A target genes and improved drought and heat-shock stress tolerance in transgenic *Arabidiopsis* (Matsukura et al. 2010). The efficacy of DREB2 group proteins in improving abiotic stress tolerance has also been investigated in crop plants and similar results have been observed. For example, overexpression of *ZmDREB2A* enhanced drought tolerance of transgenic maize (Qin et al. 2007). Recently it has been shown that the expression of sorghum *SbDREB2* with *rd29A* promoter in transgenic rice plants improved seed set and tolerance to drought stress (Bihani et al. 2011). These results suggest that the DREB2 subgroup members play important roles in DRE/CRT-mediated drought and thermotolerance, and these could be exploited in genetic engineering for crop improvement.

Most of the studies about DREB-type TFs are focused on A-1 and A-2 groups. However, the proteins from other groups are also being characterized in an attempt towards finding valuable genes for stress tolerance. DREB genes from other groups like *TINY2* and *HARDY* (A-4), *RAP2.1* (A-5), *RAP2.4* (A-6) have also demonstrated their role in abiotic stress responses (Dong and Liu 2010; Karaba et al. 2007; Lin et al. 2008; Wei et al. 2005). A very good example is *HARDY* (*HRD*), a DREB TF encoding gene (from group A-4) from *Arabidopsis* which, when overexpressed in rice, improved water use efficiency by enhancing photosynthate assimilation and reducing transpiration and imparting overall enhanced drought tolerance (Karaba et al. 2007). In the last few years, several *DREB* genes have been cloned and characterized as stress-responsive genes in many crop plants, including *O. sativa* (Dubouzet et al. 2003; Tian et al. 2005), *Zea mays* (Kizis and Pagès 2002; Qin et al. 2004), *Triticum aestivum* (Xu et al., 2008; Andeani et al., 2009), *Hordeum vulgare* (Choi et al. 2002; Xu et al. 2009), *Glycine max* (Li et al. 2005; Chen et al. 2009; Chen et al., 2007), and *Gossypium hirsutum* (Huang and Liu 2006). Nevertheless, there are still many DREB genes that are yet to be characterized and their functional evaluation for abiotic stress tolerance has to be explored in order to obtain other potential stress regulating genes.

ERF subfamily is classified into six (B1–B6) groups and includes proteins like AtERF1 to 7, RAP2.2, RAP2.6, RAP2.11, and RAP2.612 (Sakuma et al. 2002). Members of ERF subfamily regulate diverse biological functions in plant growth and development, as well as participate in hormonal signaling (Boutilier et al. 2002; Elliott et al. 1996; Rashotte et al. 2006; Alonso et al., 2003). In addition, ERF TFs have also been shown to play critical roles in regulating stress-responsive genes that are required for plant survival under abiotic stress conditions. Overxpression of ERFs has resulted in improved tolerance to abiotic stresses. For example, an ERF-type TF gene from soybean (*GmERF3*) when overexpressed in tobacco, resulted in the accumulation of higher levels of free proline and soluble carbohydrates and demonstrated enhanced drought tolerance compared to wild-type plants (Zhang et al. 2009). The transgenic plants also exhibited enhanced salinity tolerance compared to controls (Zhang et al. 2009). Similarly, another ERF TF from sugarcane (*SodERF3*) imparted increased tolerance to drought and osmotic stress when overexpressed in tobacco (Trujillo et al. 2009). In addition, *RAP2.2 and RAP2.6L,* when overexpressed in *Arabidopsis* increased hypoxia and salinity tolerance (Hinz et al. 2010; Krishnaswamy et al. 2011). Furthermore, *RAP2.2* knock out plants had inferior survival rates than controls, demonstrating the importance of this AP2 TF in hypoxia stress tolerance (Hinz et al. 2010). Furthermore, overexpression of *AP37* in rice enhanced tolerance to multiple abiotic stresses including drought, high salinity, and low temperature under field conditions with a higher seed set over controls (Oh et al. 2009). These results demonstrate the potential of ERF-type AP2 TFs in improving crop plants for abiotic stress tolerance.

Submergence tolerance imparted by ERF is yet another example worth mentioning. Genetic analysis has demonstrated that *Submergence1* (*Sub1*) locus is the major source of submergence tolerance in rice (Xu et al. 2000, 2006). The *Sub1* locus is characterized by the presence of three ERF transcriptional regulators: *Sub1A*, *Sub1B*, and *Sub1C* (Fukao et al. 2006; Xu et al. 2006). The introgression of *Sub1* locus into submergence-intolerant rice using marker-assisted selection has led to the development of submergence tolerant near isogenic line (Fukao et al. 2006). Submergence tolerance is particularly important during monsoon flooding season in Southeast Asia, where water logging seriously limits rice production (Xu et al. 2006). Studies have demonstrated that *Sub1* locus modulates ethylene and gibberellin (GA) signaling during submergence to activate genes associated with acclimatization process (Fukao et al. 2006; Steffens and Sauter 2005). Transcriptome analysis has revealed that *Sub1* locus regulates another 12 ERF genes that are involved various process like anaerobic respiration and cytokinin-mediated delay in senescence via ethylene accumulation, negative regulation of ethylene-dependent gene expression and negative regulation of gibberellin mediated shoot elongation (Jung et al. 2010). This demonstrates the critical role of ERFs in plant adaptation to adverse conditions, and suggests that ERFs could be valuable targets to manipulate plants for enhanced stress tolerance.

Genes involved in leaf cutin/wax biosynthesis are expected to have great potential for crop improvement as composition of the leaf surface has a large influence on its ability to protect the plant from adverse conditions like inadequate water supply and pathogen attack (Kannangara et al. 2007; Zhang et al. 2005). An ERF family member from *Arabidopsis*, *WAX INDUCER1/SHINE1* (*WIN1/SHN1*) has been demonstrated to be involved in cutin biosynthesis (Kannangara et al. 2007). WIN1 modulates cuticle permeability in *Arabidopsis* by regulating genes encoding cutin biosynthetic enzymes, including a gene that encodes long-chain acyl-CoA synthetase (LACS2) (Kannangara et al. 2007). Indeed, *WIN1*, when overexpressed in *Arabidospis*, induced the production of epidermal waxes and improved drought tolerance (Aharoni et al. 2004; Broun 2004). It would be interesting to investigate whether *WIN1* performs a similar function when overexpressed in crop plants. Another distinct member of AP2 TF family called WXP1 from *Medicago truncatula* has been shown to be important in leaf wax synthesis (Zhang et al. 2005). *WXP1* and a closely related paralog (*WXP2*) enhanced drought and freezing tolerance in transgenic *Arabidopsis* (Zhang et al. 2005, 2007). In addition, overexpression of *WXP1* activated wax production and conferred drought tolerance in alfalfa (*Medicago sativa*) by reducing water loss and chlorophyll leaching (Zhang et al. 2005). These studies clearly imply that members of AP2 TF family are regulatory masters in multiple pathways whose products are essential for plant adaptation for adverse conditions. Further understanding the role and regulation of the TFs under abiotic stress could lead to the production of superior plant types that are able to withstand imposed stress leading to enhanced and sustained yield.

4 Biotic Stress Tolerance

AP2 TFs regulate both abiotic and biotic stress-related signaling, although DREB types are involved mostly in an ABA-independent abiotic stress responses (Lin et al. 2008; Sakuma et al. 2002), while ERFs family members are generally implicated in ethylene signaling and pathogen defense (Berrocal-Lobo et al. 2002; Nakano et al. 2006; Yang et al. 2005). Nevertheless, as discussed earlier, some genes regulate both DRE- and ERE-mediated signaling (Agarwal et al. 2010; Sun et al. 2008; Tsutsui et al. 2009). Many transgenic studies have demonstrated the potential usefulness of AP2 TFs in enhancing resistance/tolerance to biotic stresses. For example, an ERF-type TF from barley, *HvRAF* (*Hordeum vulgare root abundant factor*), that is homologous to *RAP2.2* (in *Arabidopsis*) and *AAK92635* (in rice), when overexpressed in *Arabidopsis,* induced the expression of many stress-responsive genes like *PDF1.2*, *JR3*, *PR1*, *PR5*, *KIN2*, and *GSH1* (Jung et al. 2007). This led to enhanced resistance to the pathogen *Ralstonia solanacearum,* in addition to imparting salinity tolerance in *HvRAF* transgenic *Arabidopsis* (Jung et al. 2007). Furthermore, *TaERF1,* an ERF-type TF from wheat, responded to abiotic stresses as well as to *Blumeria graminis* f. sp. *tritici* infection and salicylic acid treatment (Xu et al. 2007). Overexpression of *TaERF1* in *Arabidopsis* activated PR and COR/RD genes under normal growth conditions and improved resistance to pathogens (*Botrytis cinerea* and *P. syringae*) in addition to imparting abiotic stress tolerance (Xu et al. 2007). Furthermore, ectopic expression of soybean *GmERF3* gene in tobacco induced the expression of PR genes and enhanced resistance against infection by *R. solanacearum*, *Alternaria alternata*, and tobacco mosaic virus (TMV), in addition to imparting high salinity and dehydration tolerance (Zhang et al. 2009). Similarly, ectopic expression of the pathogen-induced transcription factor gene *CaRAV1* from pepper (*Capsicum annuum*) in *Arabidopsis* induced PR genes and enhanced resistance to *P. syringae* pv. *Tomato* (Sohn et al. 2006). In addition, overexpression of ERF1 that encodes an AP2 TF, was sufficient to confer resistance to *Arabidopsis* against necrotrophic fungi such as *B. cinerea* and *Plectosphaerella cucumerina* (Berrocal-Lobo et al. 2002). Although many studies demonstrate the importance of AP2 TFs against pathogen infection in *Arabidopsis*, only a few studies have confirmed their role in crop species. Recently, it was reported that overexpression of *AtCBF1* in tomato led to activation of many stress-related genes including RAV, ERF, and PR genes, and enhanced tolerance to bacterial wilt caused by *Ralstonia solanacearum* (Li et al. 2011). Furthermore transcriptome analysis of bacterial wilt tolerant *AtCBF1* tomato plants suggested that RAV protein (an AP2 TF) is a pivotal modulator involved in AP2-mediated defense pathway (Li et al. 2011). It is therefore evident that TFs are involved in the biotic stress signaling and their engineering have led to some progress in genetically engineering crop plants towards producing tolerant genotypes. However, there are still a number of factors that are as of yet unknown, and understanding the role of TFs can further our knowledge and ability to generate biotic stress-tolerant plants.

5 Conclusions

AP2 transcription factors are key regulators of multiple signaling pathways activated by plants to overcome adverse conditions, and therefore their modulation can potentially serve as a valuable tool towards achieving enhanced crop productivity. Manipulation of crops with genes encoding TFs has been purported to be a more promising approach in the development of abiotic- and biotic-stress-tolerant plants than engineering individual functional genes (Bartels and Hussain 2008). Transgenic expression of a single AP2 TF can lead to improved tolerance to different types of stresses like salinity, drought, and heat stress, in addition to various biotic stresses, through its effect on a number of structural genes. These genes are involved in various physiological functions related to growth and development, from flower to root, and their modulation has been shown to impart various beneficial agronomic characteristics. It is now essential to improve the agronomical crops with biotic and abiotic tolerance traits, as plants are exposed to variety of stresses under field conditions. Also, plants affected by drought and salinity will be more susceptible to biotic stresses due to their reduced vigor, and healthier plants may overcome stresses because of their increased fitness. Therefore, the use of TFs for enhancing crop productivity and fitness offers a novel and attractive way of genetically engineering crop plants for increased productivity. However, the overexpression of TFs sometimes might lead to pleiotropic effects such as stunted plant growth. This is due to the fact that many of RAV, DREB, ERFs are also involved in plant growth and development, and therefore they may have negative effect on plant growth and productivity when overexpressed (Ito et al. 2006; Kasuga et al. 1999, 2004; Liu et al. 1998). Therefore it is very important to choose right cloning strategies depending on the gene function. For instance, rice plants constitutively expressing *SbDREB2*, exhibited pleiotropic effects such as lower seed set, although they were drought tolerant. However, *rd29A*: *SbDREB2* transgenic rice plants (under the regulation of stress-induced promoter) showed enhanced drought resistance along with higher number of panicles (Bihani et al. 2011). Another aspect that needs to be considered is post-translational modifications of AP2 TFs. They undergo post-transcriptional/posttranslational modifications such as alternate splicing and phosphorylation (Sakuma et al. 2006a, b). These events would lead to the modulation of the activity of the regulatory sequences. AP2 TFs therefore have definite potential as molecular targets for genetic engineering of crop plants when the right approach is used for transgenic expression. However, there are still many AP2 TFs that remain uncharacterized, whose characterization might lead to the discovery of valuable additional candidate genes for crop improvement. To date, a number of AP2 TF family genes have been cloned and characterized. Recently discovered and characterized TFs (DREB/CBF) can possibly provide a more effective means of generating stress-tolerant, agronomically important crops, without compromising yield. As a matter of fact, the modulation of TFs is foreseen as an invaluable means of crop trait manipulation, and is likely to play a prominent role in the next generation of biotechnology-derived crops with desirable traits. With the advancement of molecular biology and bioinformatics techniques, it is envisaged that rapid gains will be made in characterizing additional TFs and their successful utilization in crop improvement.

References

Agarwal P, Agarwal PK, Joshi AJ, Sopory SK, Reddy MK (2010) Overexpression of *PgDREB2A* transcription factor enhances abiotic stress tolerance and activates downstream stress-responsive genes. Mol Biol Rep 37(2):1125–35

Agarwal P, Agarwal PK, Nair S, Sopory SK, Reddy MK (2007) Stress inducible DREB2A transcription factor from Pennisetum glaucum is a phosphoprotein and its phosphorylation negatively regulates its DNA binding activity. Mol Genet Genomics 277(2):189–198

Agarwal PK, Agarwal P, Reddy MK, Sopory SK (2006) Role of DREB transcription factors in abiotic and biotic stress tolerance in plants. Plant Cell Rep 25(12):1263–1274

Aharoni A, Dixit S, Jetter R, Thoenes E, Van Arkel G, Pereira A (2004) The SHINE clade of AP2 domain transcription factors activates wax biosynthesis, alters cuticle properties, and confers drought tolerance when overexpressed in *Arabidopsis*. Plant Cell 16(9):2463–2480

Alonso JM, Stepanova AN, Leisse TJ, et al. (2003) Genome-wide insertional mutagenesis of *Arabidopsis thaliana*. Science 301(5641):653–657

Andeani JK, Mohsenzadeh S, Mohabatkar H (2009) Isolation and characterization of partial DREB gene from four Iranian *Triticum aestivum* cultivars. World J Agri Sci 5(5):561–566

Baker SS, Wilhelm KS, Thomashow MF (1994) The 5'- region of *Arabidopsis thaliana* cor15a has cis-acting elements that confer cold-, drought- and ABA-regulated gene expression. Plant Mol Biol 24(5):701–713

Bartels D, Hussain SS (2008) Current status and implications of engineering drought tolerance in plants using transgenic approaches. CAB Rev: Perspect Agric Vet Sci Nutr Nat Resour 3(020):17

Behnam B, Kikuchi A, Celebi-Toprak F, Yamanaka S, Kasuga M, Yamaguchi-Shinozaki K, Watanabe KN (2006) The Arabidopsis *DREB1A* gene driven by the stress-inducible rd29A promoter increases salt-stress tolerance in proportion to its copy number in tetrasomic tetraploid potato (*Solanum tuberosum*). Plant Biotechnol 23(2):169–177

Berrocal-Lobo M, Molina A, Solano R (2002) Constitutive expression of ETHYLENE-RESPONSE-FACTOR1 in *Arabidopsis* confers resistance to several necrotrophic fungi. Plant J 29(1):23–32

Bihani P, Char B, Bhargava S (2011) Transgenic expression of sorghum *DREB2* in rice improves tolerance and yield under water limitation. J Agric Sci 149(1):95–101

Boutilier K, Offringa R, Sharma VK et al (2002) Ectopic expression of *BABY BOOM* triggers a conversion from vegetative to embryonic growth. Plant Cell 14(8):1737–1749

Broun P (2004) Transcription factors as tools for metabolic engineering in plants. Curr Opin Plant Biol 7(2):202–209

Büttner M, Singh KB (1997) *Arabidopsis thaliana* ethylene-responsive element binding protein (AtEBP), an ethylene-inducible, GCC box DNA-binding protein interacts with an ocs element binding protein. Proc Natl Acad Sci U S A 94(11):5961–5966

Chen H, Hwang JE, Lim CJ, Kim DY, Lee SY, Lim CO (2010) *Arabidopsis* DREB2C functions as a transcriptional activator of *HsfA3* during the heat stress response. Biochem Biophys Res Commun 401(2):238–44

Chen M, Wang QY, Cheng XG, Xu ZS, Li LC, Ye XG, Xia LQ, Ma YZ (2007) GmDREB2, a soybean DRE-binding transcription factor, conferred drought and high-salt tolerance in transgenic plants. Biochem Biophys Res Commun 353(2):299–305

Chen M, Xu Z, Xia L, Li L, Cheng X, Dong J, Wang Q, Ma Y (2009) Cold-induced modulation and functional analyses of the DRE-binding transcription factor gene, *GmDREB3*, in soybean (*Glycine max* L.). J Exp Bot 60(1):121–35

Chini A, Grant JJ, Seki M, Shinozaki K, Loake GJ (2004) Drought tolerance established by enhanced expression of the CC-NBSLRR gene, *ADR1*, requires salicylic acid, *EDS1* and *ABI1*. Plant J 38(5):810–822

Chinnusamy V, Ohta M, Kanrar S, Lee BH, Hong X, Agarwal M, Zhu JK (2003) ICE1: a regulator of cold-induced transcriptome and freezing tolerance in Arabidopsis. Genes Dev 17(8):1043–1054

Choi DW, Rodriguez EM, Close TJ (2002) Barley *Cbf3* gene identification, expression pattern, and map location. Plant Physiol 129(4):1781–1787

Cong L, Zheng HC, Zhang YX, Chai TY (2008) *Arabidopsis* DREB1A confers high salinity tolerance and regulates the expression of GA dioxygenases in Tobacco. Plant Sci 174(2):156–164

Cook D, Fowler S, Fiehn O, Thomashow M (2004) A prominent role for the CBF cold response pathway in configuring the low-temperature metabolome of Arabidopsis. Proc Natl Acad Sci U S A 101(42):15243–15248

Dita MA, Rispail N, Prats E, Rubiales D, Singh KB (2006) Biotechnology approaches to overcome biotic and abiotic stress constraints in legume. Euphytica 147(1–2):1–24

Díaz-Martín J, Almoguera C, Prieto-Dapena P, Espinosa JM, Jordano J (2005) Functional interaction between two transcription factors involved in the developmental regulation of a small heat stress protein gene promoter. Plant Physiol 139(3):1483–1494

Doebley JF, Gaut BS, Smith BD (2006) The molecular genetics of crop domestication. Cell 127(7):1309–1321

Doherty CJ, Van Buskirk HA, Myers SJ, Thomashow MF (2009) Roles for *Arabidopsis* CAMTA transcription factors in cold-regulated gene expression and freezing tolerance. Plant Cell 21(3):972–984

Dong CJ, Liu JY (2010) The Arabidopsis EAR-motif-containing protein RAP2.1 functions as an active transcriptional repressor to keep stress responses under tight control. BMC Plant Biol 10:47

Dubouzet JG, Sakuma Y, Ito Y, Kasuga M, Dubouzet EG, Miura S, Seki M, Shinozaki K, Yamaguchi-Shinozaki K (2003) *OsDREB* genes in rice, *Oryza sativa* L., encode transcription activators that function in drought-, high-salt- and cold-responsive gene expression. Plant J 33(4):751–763

Elliott RC, Betzner AS, Huttner E, Oakes MP, Tucker WQJ, Gerentes D, Perez P, Smyth DR (1996) *AINTEGUMENTA*, an APETALA2-like gene of *Arabidopsis* with pleiotropic roles in ovule development and floral organ growth. Plant Cell 8(2):155–168

Eulgem T (2005) Regulation of the *Arabidopsis* defense transcriptome. Trends Plant Sci 10(2):71–78

Feng JX, Liu D, Pan Y, Gong W, Ma LG, Luo JC, Deng XW, Zhu YX (2005) An annotation update via cDNA sequence analysis and comprehensive profiling of developmental, hormonal or environmental responsiveness of the *Arabidopsis* AP2/EREBP transcription factor gene family. Plant Mol Biol 59(6):853–868

Fowler S, Thomashow MF (2002) *Arabidopsis* transcriptome profiling indicates that multiple regulatory pathways are activated during cold acclimation in addition to the CBF cold response pathway. Plant Cell 14(8):1675–1690

Fujimoto SY, Ohta M, Usui A, Shinshi H, Ohme-Takagi M (2000) *Arabidopsis* ethylene-responsive element binding factors act as transcriptional activators or repressors of GCC box-mediated gene expression. Plant Cell 12(3):393–404

Fukao T, Xu K, Ronald PC, Bailey-Serres J (2006) A variable cluster of ethylene response factor-like genes regulates metabolic and developmental acclimation responses to submergence in rice. Plant Cell 18(8):2021–2034

Gilmour SJ, Sebolt AM, Salazar MP, Everard JD, Thomashow MF (2000) Overexpression of the *Arabidopsis CBF3* transcriptional activator mimics multiple biochemical changes associated with cold acclimation. Plant Physiol 124(4):1854–1865

Gilmour SJ, Zarka DG, Stockinger EJ, Salazar MP, Houghton JM, Thomashow MF (1998) Low temperature regulation of the *Arabidopsis* CBF family of AP2 transcriptional activators as an early step in cold-induced COR gene expression. Plant J 16(4):433–442

Guo A, He K, Liu D et al (2005) DATF: a database of Arabidopsis transcription factors. Bioinformatics 21(10):2568–2569

Guo HW, Ecker JR (2004) The ethylene signaling pathway: new insights. Curr Opin Plant Biol 7(1):40–49

Gutterson N, Reuber TL (2004) Regulation of disease resistance pathways by AP2/ERF transcription factors. Curr Opin Plant Biol 7(4):465–471

Haake V, Cook D, Riechmann JL, Pineda O, Thomashow MF, Zhang JZ (2002) Transcription factor CBF4 is a regulator of drought adaptation in *Arabidopsis*. Plant Physiol 130(2):639–648

Hao D, Ohme-Takagi M, Sarai A (1998) Unique mode of GCC box recognition by the DNA-binding domain of ethylene-responsive element-binding factor (ERF domain) in plant. J Biol Chem 273(31):26857–26861

Hinz M, Wilson IW, Yang J et al (2010) Arabidopsis RAP2.2: an ethylene response transcription factor that is important for hypoxia survival. Plant Physiol 153(2):757–772

Hsieh TH, Lee JT, Charng YY, Chan MT (2002) Tomato plants ectopically expressing *Arabidopsis CBF1* show enhanced resistance to water deficit stress. Plant Physiol 130(2):618–626

Hu YX, Wang YX, Liu XF, Li JY (2004) *Arabidopsis* RAVI is down-regulated by brassinosteroid and may act as a negative regulator during plant development. Cell Res 14(1):8–15

Huang B, Liu JY (2006) Cloning and functional analysis of the novel gene *GhDBP3* encoding a DRE-binding transcription factor from *Gossypium hirsutum*. Biochim Biophys Acta 1759(6):263–269

Hussain SS, Kayani MA, Amjad M (2011) Transcription factors as tools to engineer enhanced drought stress tolerance in plants. Biotechnol Progr 27(2):297–306

Ito Y, Katsura K, Maruyama K, Taji T et al (2006) Functional analysis of rice DREB1/CBF type transcription factors involved in cold responsive gene expression in transgenic rice. Plant Cell Physiol 47(1):141–153

Jaglo-Ottosen KR, Gilmour SJ, Zarka DG et al (1998) *Arabidopsis CBF1* overexpression induces cor genes and enhances freezing tolerance. Science 280(5360):104–106

Jaillon O, Aury JM, Noel B, Policriti A et al (2007) The grapevine genome sequence suggests ancestral hexaploidization in major angiosperm phyla. Nature 449(7161):463–467

Jofuku KD, Denboer BGW, van Montagu M, Okamuro JK (1994) Control of *Arabidopsis* flower and seed development by the homeotic gene apetala 2. Plant Cell 6(9):1211–1225

Jung J, Won SY, Suh SC et al (2007) The barley ERF-type transcription factor HvRAF confers enhanced pathogen resistance and salt tolerance in *Arabidopsis*. Planta 225(3):575–88

Jung KH, Seo YS, Walia H et al (2010) The submergence tolerance regulator *Sub1A* mediates stress-responsive expression of AP2/ERF transcription factors. Plant Physiol 152(3): 1674–1692

Kagaya Y, Ohmiya K, Hattori T (1999) RAV1, a novel DNA-binding protein, binds to bipartite recognition sequence through two distinct DNA-binding domains uniquely found in higher plants. Nucleic Acids Res 27(2):470–478

Kannangara R, Branigan C, Liu Y et al (2007) The transcription factor WIN1/ SHN1 regulates cutin biosynthesis in *Arabidopsis thaliana*. Plant Cell 19(4):1278–1294

Karaba A, Dixit S, Greco R et al (2007) Improvement of water use efficiency in rice by expression of *HARDY*, an *Arabidopsis* drought and salt tolerance gene. Proc Natl Acad Sci U S A 104(39):15270–15275

Karin M (1990) Too many transcription factors: positive and negative interactions. New Biol 2(2):126–131

Kasuga M, Liu Q, Miura S, Yamaguchi-Shinozaki K, Shinozaki K (1999) Improving plant drought, salt, and freezing tolerance by gene transfer of a single stress-inducible transcription factor. Nat Biotechnol 17(3):287–291

Kasuga M, Miura S, Shinozaki K, Yamaguchi-Shinozaki K (2004) A combination of the *Arabidopsis DREB1A* gene and stress inducible rd29A promoter improved drought- and low-temperature stress tolerance in tobacco by gene transfer. Plant Cell Physiol 45(3):346–350

Kizis D, Pagès M (2002) Maize DRE-binding proteins DBF1 and DBF2 are involved in rab17 regulation through the drought-responsive element in an ABA-dependent pathway. Plant J 30(6):679–89

Krishnaswamy S, Verma S, Rahman MH, Kav NN (2011) Functional characterization of four APETALA2-family genes (RAP2.6, *RAP2*.6 L, *DREB19* and *DREB26*) in *Arabidopsis*. Plant Mol Biol 75(1–2):107–127

Krizek B (2009) AINTEGUMENTA and AINTEGUMENTA-LIKE6 act redundantly to regulate Arabidopsis floral growth and patterning. Plant Physiol 150(4):1916–1929

Kurkela S, Borg-Franck M (1992) Structure and expression of kin 2, one of two cold- and ABA-induced genes of *Arabidopsis thaliana*. Plant Mol Biol 19(4):689–692

Kurkela S, Franck M (1990) Cloning and characterization of a cold-and ABA-inducible *Arabidopsis* gene. Plant Mol Biol 15(1):137–144

Lang V, Palva ET (1992) The expression of a rab-related gene rabl8 is induced by abscisic acid during the cold acclimation process of *Arabidopsis thaliana* (L) Heynh. Plant Mol Biol 20(5):951–962

Lee JH, Hong JP, Oh SK, Lee S, Choi D, Kim WT (2004) The ethylene-responsive factor like protein 1 (CaERFLP1) of hot pepper (*Capsicum annuum* L.) interacts in vitro with both GCC and DRE/CRT sequences with different binding affinities: possible biological roles of CaERFLP1 in response to pathogen infection and high salinity conditions in transgenic tobacco plants. Plant Mol Biol 55(1):61–81

Li C-W, Su R-C, Cheng C-P, Sanjaya, You S-J, Hsieh T-H, Chao T-C, Chan M-T (2011) Tomato RAV transcription factor is a pivotal modulator involved in the AP2/EREBP-mediated defense pathway. Plant Physiol 156(1):213–227

Li XP, Tian AG, Luo GZ, Gong ZZ, Zhang JS et al (2005) Soybean DRE-binding transcription factors that are responsive to abiotic stresses. Theor Appl Genet 110(8):1355–1362

Lim CJ, Hwang JE, Chen H et al (2007) Over-expression of the *Arabidopsis* DRE/CRT-binding transcription factor *DREB2C* enhances thermotolerance. Biochem Biophys Res Commun 362(2):431–436

Lin RC, Park HJ, Wang HY (2008) Role of *Arabidopsis RAP2.4* in regulating light- and ethylene-mediated developmental processes and drought stress tolerance. Mol Plant 1(1):42–57

Liu Q, Kasuga M, Sakuma Y et al (1998) Two transcription factors, DREB1 and DREB2, with an EREBP/AP2 DNA binding domain separate two cellular signal transduction pathways in drought- and low-temperature- responsive gene expression, respectively, in *Arabidopsis*. Plant Cell 10(8):1391–1406

Maruyama K, Takeda M, Kidokoro S et al (2009) Metabolic pathways involved in cold acclimation identified by integrated analysis of metabolites and transcripts regulated by DREB1A and DREB2A. Plant Physiol 150(4):1972–1980

Matsukura S, Mizoi J, Yoshida T (2010) Comprehensive analysis of rice DREB2-type genes that encode transcription factors involved in the expression of abiotic stress-responsive genes. Mol Genet Genomics 283(2):185–196

McGrath KC, Dombrecht B, Manners JM et al (2005) Repressor- and activator-type ethylene response factors functioning in jasmonate signaling and disease resistance identified via a genome-wide screen of *Arabidopsis* transcription factor gene expression. Plant Physiol 139(2):949–959

Morran S, Eini O, Pyvovarenko T et al (2011) Improvement of stress tolerance of wheat and barley by modulation of expression of DREB/CBF factors. Plant Biotechnol J 9(2):230–49

Moose SP, Sisco PH (1996) Glossy15, an APETALA2-like gene from maize that regulates leaf epidermal cell identity. Genes Dev 10(23):3018–27

Nakano T, Suzuki K, Fujimura T, Shinshi H (2006) Genome-wide analysis of the ERF gene family in *Arabidopsis* and rice. Plant Physiol 140(2):411–432

Nakashima K, Yamaguchi-Shinozak IK (2006) Regulons involved in osmotic stress-responsive and cold stress-responsive gene expression in plants. Physiol Plant 126(1):62–71

Nikolov DB, Burley SK (1997) RNA polymerase II transcription initiation: a structural view. Proc Natl Acad Sci U S A 94(1):15–22

Novillo F, Medina J, Salinas J (2007) Arabidopsis CBF1 and CBF3 have a different function than CBF2 in cold acclimation and define different gene classes in the CBF regulon. Proc Natl Acad Sci USA 104(52):21002–21007

Novillo F, Alonso JM, Ecker JR, Salinas J (2004) CBF2/DREB1C is a negative regulator of CBF1/DREB1B and CBF3/DREB1A expression and plays a central role in stress tolerance in *Arabidopsis*. Proc Nat Acad Sci U S A 101(11):3985–3990

Oh SJ, Kim YS, Kwon CW, Park HK, Jeong JS, Kim JK (2009) Overexpression of the transcription factor *AP37* in rice improves grain yield under drought conditions. Plant Physiol 150(3):1368–1379

Oh SJ, Kwon CW, Choi DW, Song SIK, Kim JK (2007) Expression of barley *HvCBF4* enhances tolerance to abiotic stress in transgenic rice. J Plant Biotechnol 5(5):646–656

Oh SJ, Song SI, Kim YS, Jang HJ, Kim M, Kim YK (2005) *Arabidopsis CBF3/DREB1A* and *ABF3* in transgenic rice increased tolerance to abiotic stress without stunting growth. Plant Physiol 138(2):341–351

Ohme-Takagi M, Shinshi H (1995) Ethylene-inducible DNA binding proteins that interact with an ethylene-responsive element. Plant Cell 7(2):173–182

Ouellet F, Vazquez-Tello A, Sarhan F (1998) The wheat wcs120 promoter is cold-inducible in both monocotyledonous and dicotyledonous species. FEBS Lett 423(3):324–328

Park JM, Park CJ, Lee SB, Ham BK, Shin R, Paek KH (2001) Overexpression of the tobacco *Tsi1* gene encoding an EREBP/AP2-type transcription factor enhances resistance against pathogen attack and osmotic stress in tobacco. Plant Cell 13(5):1035–1046

Passarinho P, Ketelaar T, Xing M, van Arkel J, Maliepaard C, Hendriks MW, Joosen R, Lammers M, Herdies L, den Boer B, van der Geest L, Boutilier K (2008) BABY BOOM target genes provide diverse entry points into cell proliferation and cell growth pathways. Plant Mol Biol 68(3):225–37

Pellegrineschi A, Reynolds M, Pacheco M, Brito RM, Almeraya R, Yamaguchi-Shinozaki K, Hoisington D (2004) Stress induced expression in wheat of the Arabidopsis thaliana DREB1A gene delays water stress symptoms under greenhouse conditions. Genome 47:493–500

Qin F, Sakuma Y, Liu Q, Li YQ, Shinozaki K, Yamaguchi-Shinozaki K (2004) Cloning and functional analysis of a novel DREB1/CBF transcription factor involved in cold-responsive gene expression in *Zea mays* L. Plant Cell Physiol 45(8):1042–1052

Qin QL, Liu JG, Zhang Z et al (2007) Isolation, optimization, and functional analysis of the cDNA encoding transcription factor *OsDREB1B* in *Oryza Sativa* L. Mol Breeding 19(4):329–340

Quan R, Hu S, Zhang Z, Zhang H, Zhang Z, Huang R (2010) Overexpression of an ERF transcription factor *TSRF1* improves rice drought tolerance. Plant Biotechnol J 8(4):476–488

Rashotte AM, Mason MG, Hutchison CE, Ferreira FJ, Schaller GE, Kieber JJ (2006) A subset of *Arabidopsis* AP2 transcription factors mediates cytokinin responses in concert with a two-component pathway. Proc Natl Acad Sci USA 103(29):11081–11085

Riano-Pachon DM, Ruzicic S, Dreyer I, Mueller-Roeber B (2007) PlnTFDB: an integrative plant transcription factor database. BMC Bioinform 8:42

Riechmann JL, Meyerowitz EM (1998) The AP2/EREBP family of plant transcription factors. Biol Chem 379(6):633–646

Riechmann JL (2000) *Arabidopsis* transcription factors: genome-wide comparative analysis among eukaryotes. Science 290(5499):2105–2110

Sakuma Y, Liu Q, Dubouzet JG, Abe H, Shinozaki K, Yamaguchi-Shinozaki K (2002) DNA-binding specificity of the ERF/AP2 domain of *Arabidopsis* DREBs, transcription factors involved in dehydration and cold-inducible gene expression. Biochem Biophys Res Commun 290(3):998–1009

Sakuma Y, Maruyama K, Osakabe Y, Qin F, Seki M, Shinozaki K, Yamaguchi-Shinozaki K (2006a) Functional analysis of an *Arabidopsis* transcription factor, DREB2A, involved in drought responsive gene expression. Plant Cell 18(5):1292–1309

Sakuma Y, Maruyama K, Qin F et al (2006b) Dual function of an *Arabidopsis* transcription factor DREB2A in water stress- responsive and heat-stress-responsive gene expression. Proc Nat Acad Sci USA 103(49):18828–18833

Saleh A, Pagés M (2003) Plant AP2/ERF transcription factors. Genetika 35(1):37–50

Savitch LV, Allard G, Seki M, Robert LS, Tinker NA, Huner NPA et al (2005) The effect of overexpression of two *Brassica* CBF/ DREB1-like transcription factors on photosynthetic capacity and freezing tolerance in *Brassica napus*. Plant Cell Physiol 46(9):1525–1539

Schramm F, Larkindale J, Kiehlmann E et al (2008) A cascade of transcription factor DREB2A and heat stress transcription factor HsfA3 regulates the heat stress response of Arabidopsis. Plant J 53(2):264–274

Shinozaki K, Yamaguchi-Shinozaki K (2000) Molecular responses to dehydration and low temperature: differences and cross-talk between two stress signalling pathways. Curr Opin Plant Biol 3(3):217–223

Shinwari ZK, Nakashima K, Miura S et al (1988) An Arabidopsis gene family encoding DRE/CRT binding proteins involved in low temperature responsive gene expression. Biochem Biophys Res Commun 250(1):161–170

Sohn KH, Lee SC, Jung HW, Hong JK, Hwang BK (2006) Expression and functional roles of the pepper pathogen-induced transcription factor *RAV1* in bacterial disease resistance, and drought and salt stress tolerance. Plant Mol Biol 61(6):897–915

Steffens B, Sauter M (2005) Epidermal cell death in rice is regulated by ethylene, gibberellin, and abscisic acid. Plant Physiol 139(2):713–721

Stockinger EJ, Gilmour SJ, Thomashow MF (1997) *Arabidopsis thaliana CBF1* encodes an AP2 domain-containing transcriptional activator that binds to the C-repeat/DRE, a cis-acting DNA regulatory element that stimulates transcription in response to low temperature and water deficit. Proc Natl Acad Sci USA 94(3):1035–1040

Sun S, Yu JP, Chen F et al (2008) TINY, a dehydrationresponsive element (DRE)-binding protein-like transcription factor connecting the DRE and ethylene-responsive element-mediated signaling pathways in *Arabidopsis*. J Biol Chem 283(10):6261–6271

Tian XH, Li XP, Zhou HL, Zhang JS, Gong ZZ, Chen SY (2005) *OsDREB4* genes in rice encode AP2-containing proteins that bind specifically to the dehydration- responsive element. J Integr Plant Biol 47(4):467–476

Trujillo L, Menendez C, Ochogavia ME et al (2009) Engineering drought and salt tolerance in plants using SodERF3, a novel sugarcane ethylene responsive factor. Biotechnol Apl 26(2):168–171

Tsutsui T, Kato W, Asada Y et al (2009) DEAR1, a transcriptional repressor of DREB protein that mediates plant defense and freezing stress responses in *Arabidopsis*. J Plant Res 122(6):633–643

Vinocur B, Altman A (2005) Recent advances in engineering plant tolerance to abiotic stress: achievements and limitations. Curr Opin Biotechnol 16(2):123–132

Vogel JT, Zarka DG, Van Buskirk HA, Fowler SG, Thomashow MF (2005) Roles of the CBF2 and ZAT12 transcription factors in configuring the low temperature transcriptome of *Arabidopsis*. Plant J 41(2):195–211

Wang ZL, An XM, Li B et al (2008) Identification and characterization of CBF/DREB1-related genes in *Populus hopeiensis*. Forest Stud China 10(3):143–148

Wei G, Pan Y, Lei J, Zhu YX (2005) Molecular cloning, phylogenetic analysis, expressional prowling and in vitro studies of *TINY2* from *Arabidopsis thaliana*. J Biochem Mol Biol 38(4):440–446

Xu K, Xu X, Fukao T et al (2006) *Sub1A* is an ethylene-response-factor-like gene that confers submergence tolerance to rice. Nature 442(7103):705–8

Xu K, Xu X, Ronald PC, Mackill DJ (2000) A high-resolution linkage map in the vicinity of the rice submergence tolerance locus Sub1. Mol Gen Genet 263(4):681–689

Xu ZS, Ni ZY, Li ZY (2009) Isolation and functional characterization of *HvDREB1*-a gene encoding a dehydration- responsive element binding protein in *Hordeum vulgare*. J Plant Res 122(1):121–130

Xu ZS, Ni ZY, Liu L, Nie LN, Li LC, Chen M, Ma YZ (2008) Characterization of the *TaAIDFa* gene encoding a CRT/DRE binding factor responsive to drought, high-salt, and cold stress in wheat. Mol Genet Genomics 280(6):497–508

Xu ZS, Xia LQ, Chen M et al (2007) Isolation and molecular characterization of the *Triticum aestivum* L. ethylene-responsive factor 1 (TaERF1) that increases multiple stress tolerance. Plant Mol Biol 65(6):719–732

Xu P, Narasimhan ML, Samson T et al (1998) A nitrilase-like protein interacts with GCC box DNA-binding proteins involved in ethylene and defense responses. Plant Physiol 118(3):867–874

Yamaguchi-Shinozaki K, Shinozaki K (1994) A novel cis-acting element in an *Arabidopsis* gene is involved in responsiveness to drought, low temperature, or high-salt stress. Plant Cell 6(2):251–264

Yamaguchi-Shinozaki K, Shinozaki K (2006) Transcriptional regulatory networks in cellular responses and tolerance to dehydration and cold stresses. Annu Rev Plant Biol 57:781–803

Yang Z, Tian L, Latoszek-Green M, Brown D, Wu K (2005) *Arabidopsis* ERF4 is a transcriptional repressor capable of modulating ethylene and abscisic acid responses. Plant Mol Biol 58(4):585–596

Yang CY, Hsu FC, Li JP, Wang NN, Shih MC (2011) The AP2/ERF transcription factor AtERF73/HRE1 modulates ethylene responses during hypoxia in *Arabidopsis*. Plant Physiol 56(1): 202–212

Yi SY, Kim JH, Joung YH, Lee S, Kim WT, Yu SH, Choi D (2004) The pepper transcription factor CaPFl confers pathogen and freezing tolerance in *Arabidopsis*. Plant Physiol 136(1): 2862–2874

Zhang G, Chen M, Li L et al (2009) Overexpression of the soybean *GmERF3* gene, an AP2/ERF type transcription factor for increased tolerances to salt, drought and diseases in transgenic tobacco. J Exp Bot 60(13):3781–3796

Zhang JY, Broeckling CD, Blancaflor EB, Sledge MK, Sumner LW, Wang ZY (2005) Overexpression of *WXP1*, a putative *Medicago truncatula* AP2 domain-containing transcription factor gene, increases cuticular wax accumulation and enhances drought tolerance in transgenic alfalfa (*Medicago sativa*). Plant J 42(5):689–707

Zhang JY, Broeckling CD, Sumner LW, Wang ZY (2007) Heterologous expression of two *Medicago truncatula* putative ERF transcription factor genes, *WXP1* and *WXP2*, in *Arabidopsis* led to increased leaf wax accumulation and improved drought tolerance, but differential response in freezing tolerance. Plant Mol Biol 64(3):265–278

Zhang SJ, Li N, Gao F, Yang AF, Zhang JR (2010) Over-expression of *TsCBF1* gene confers improved drought tolerance in transgenic maize. Mol Breeding 26(3):455–465

Zhao JS, Ren W, Zhi DY, Wang L, Xia GM (2007) *Arabidopsis DREB1A/CBF3* bestowed transgenic tall fescue increased tolerance to drought stress. Plant Cell Rep 26(9):1521–1528

Zhao TJ, Sun S, Liu Y et al (2006) Regulating the drought-responsive element (DRE)-mediated signaling pathway by synergic functions of trans-active and trans-inactive DRE binding factors in *Brassica napus*. J Biol Chem 281(16):10752–10759

Zhuang J, Cai B, Peng RH et al (2008) Genome-wide analysis of the AP2/ERF gene family in *Populus trichocarpa*. Biochem Biophys Res Commun 371(3):468–474

Chapter 9
Osmoprotectants: Potential for Crop Improvement Under Adverse Conditions

Saurabh C. Saxena, Harmeet Kaur, Pooja Verma, Bhanu P. Petla, Venkateswara R. Andugula, and Manoj Majee

Abbreviations

ROS	Reactive oxygen species
M6PI	Mannose-6-phosphate isomerase
M6PR	Mannose-6-phosphate reductase
M1PP	Mannose-1-phosphate phosphatase
MtlD	Mannitol-1-phosphate dehydrogenase
NAD	Nicotinamide adenine dinucleotide
GFOR	Glucose-fructose oxidoreductase
S6PDH	Sorbitol-6-phosphate dehydrogenase
NADP	Nicotinamide adenine dinucleotide phosphate
S6PP	Sorbitol-6-phosphate phosphatase
Stpd1	Gene encoding sorbitol-6-phosphate dehydrogenase
MIPS	*myo*-Inositol-1-phosphate synthase
IMP	Inositol monophosphatase
ABA	Abscisic acid
PINO1	*Porteresia coarctata* inositol-1-phosphate synthase
TPS	Trehalose-6-phosphate synthase
TPP	Trehalose-6-phosphate phosphatase
OtsA	*E. coli* gene encoding TPS
OtsB	*E. coli* gene encoding TPP
P5CS	L-Δ^1-pyrroline-5-carboxylate synthetase
P5CR	L-Δ^1-pyrroline-5-carboxylate reductase
ProDH	Proline dehydrogenase
P5C	L-Δ^1-pyrroline-5-carboxylate

S.C. Saxena • H. Kaur • P. Verma • B.P. Petla • V.R. Andugula • M. Majee (✉)
National Institute of Plant Genome Research, Aruna Asaf Ali Marg,
New Delhi, 110 067, India
e-mail: manoj_majee@nipgr.res.in

N. Tuteja and S. Singh Gill (eds.), *Plant Acclimation to Environmental Stress*,
DOI 10.1007/978-1-4614-5001-6_9,

Put	Putrescine
Spd	Spermidine
Spm	Spermine
ODC	Ornithine decarboxylase
ADC	Arginine decarboxylase
CPA	*N*-carbamoylputrescine amidohydrolase
SPDS	Spermidine synthases
SPMS	Spermine synthases
SAMDC	*S*-adenosylmethionine decarboxylase
SMCs	Small molecule chaperones
DABA	L-2,4-diaminobutyrate
EctB	L-2,4-diaminobutyric acid transaminase
EctA	L-2,4-diaminobutyric acid acetyltransferase
EctC	Ectoine synthase
CMO	Choline monooxygenase
BADH	Betaine aldehyde dehydrogenase
codA	Gene encoding choline oxidase

1 Introduction

The world human population is constantly rising and is expected to reach eight billion by 2025 and 8.9 billion by 2050. Hence, there is an urgent need to double the world food production to feed eight billion people by 2025 (FAO 2008). This is even more challenging to meet such huge demand in the current context of climate variability, particularly extreme temperature and unusual rainfall. It has been estimated that approximately 70 % of yield reduction is direct result of abiotic stresses alone (Acquaah 2007; Lobell and Field 2007).

One approach to increase crop production is to develop stress tolerant crops by transferring gene(s) for the adaptive traits from the tolerant species to the crops. However, through conventional breeding, this process has only been partially successful, partly because of poorly described traits and transfer of unavoidable genes during crossing (Yeo and Flowers 1989). Furthermore, complexity of stress tolerance trait, low genetic variance of yield component under stress and lack of efficient selection techniques make it more difficult to produce such stress resistant germplasms (Ribaut et al. 1996, 1997; Frova et al. 1999).

In contrast to traditional breeding, genetic engineering appears to be an attractive alternative with respect to the possibility of direct introduction of single or multiple genes into crops for betterment (Holmberg and Bülow 1998; Smirnoff 1998). Among various abiotic stresses, drought, salinity, and temperature (low and high) are the major factors that primarily limit plant growth and productivity and the common effect that all these factors impose on plant is osmotic stress.

In response to such stress, certain plants, marine algae, bacteria, and few other organisms synthesize and accumulate various low molecular weight organic compounds known as osmoprotectants or osmolytes or compatible solutes (Johnson et al. 1968; Yancey et al. 1982; Serraj and Sinclair 2002). Though, many crops lack the ability to synthesize some specific osmoprotectants found in stress tolerant organisms, ectopic expression of osmoprotectants is reported to be functional in several crop plants.

These osmoprotectants have been one of the favorite targets for genetic engineering for many years. Many crops are engineered using osmoprotectants like mannitol, glycine betaine, and trehalose, though the level of tolerance exhibited by these engineered crops varies greatly (Sheveleva et al. 1997; Huang et al. 2000). In this chapter, we elaborate the role of these osmoprotectants in stress tolerance including constraints and prospects of their use in metabolic engineering.

2 Osmoprotectants

Osmoprotectants are low molecular weight organic compounds primarily accumulated in response to osmotic stresses in diverse taxa including plants (Yancey et al. 1982). These are highly soluble compounds carrying no net charge at physiological pH and are nontoxic even at high concentrations. These molecules increase the osmotic pressure in the cytoplasm, thereby maintaining driving gradient for both water uptake and turgor pressure. Apart from osmotic adjustment, these compounds are reported to function as scavengers of reactive oxygen species (ROS), having chaperone-like activity and help in metabolic detoxification (Serraj and Sinclair 2002). In addition, osmoprotectants play an essential role in stabilizing proteins and membranes during oxidative damage by stress-induced ROS outburst (Yancey 1994; Bohnert and Jensen 1996).

Chemically they fall into three major groups viz. amino acids (e.g., Proline), quaternary ammonium compounds (e.g., glycine betaine), polyols and sugars (mannitol, D-ononitol, trehalose, fructans) (Yancey 1994). Among these osmoprotectants, proline, glycine betaine, and mannitol are commonly found in plants. In plant cells, osmoprotectants are primarily accumulated in cytosol and chloroplast but are also reported to be distributed in few other organelles.

2.1 Polyols

Polyols such as glycerol, mannitol, and sorbitol are straight chain metabolites and cyclic polyols like inositols, pinitol have been shown to accumulate in evolutionary diverse organisms in response to dehydration, salinity, and osmotic stress.

2.1.1 Mannitol and Sorbitol

Mannitol is a hexitol sugar alcohol and widely distributed in nature including more than 100 species of vascular plants. Mannitol is known to serve as a major carbon source in many organisms (Stoop et al. 1996). The mannitol biosynthetic pathway

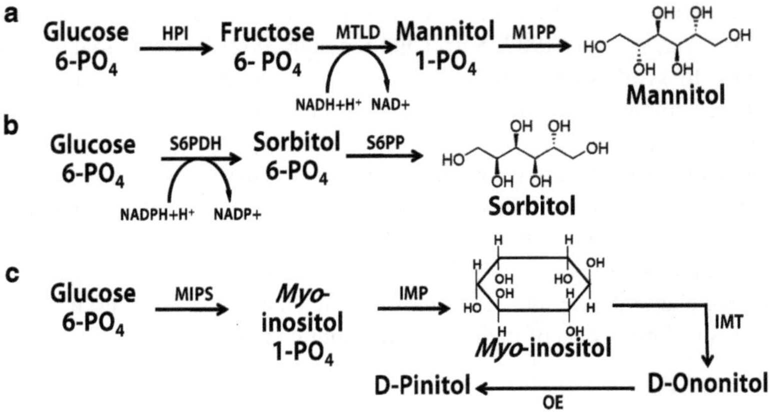

Fig. 9.1 Polyol biosynthetic pathways. [a] Mannitol biosynthesis: *HPI* hexose phosphate isomerase; *MTLD* mannitol-1-phosphate dehydrogenase; *M1PP* mannitol-1-phosphate phosphatase. [b] Sorbitol biosynthesis: *S6PDH* sorbitol-6-phosphate dehydrogenase; *S6PP* sorbitol-6-phosphate phosphatase. [c] *Myo*-Inositol biosynthesis: *MIPS myo*-inositol-1-phosphate synthase; *IMP* inositol monophosphatase; *IMT* inositol methyltransferase; *OE* ononitol epimerase

in higher plants starts with the isomerization of fructose-6-phosphate to mannose-6-phosphate by mannose-6-phosphate isomerase (M6PI, EC 5.3.1.8) which is then converted to mannitol-1-phosphate by mannose-6-phosphate reductase (M6PR, EC 1.1.1.224) (Loescher et al. 1992). In the final step, mannitol-1-phosphate is acted upon by mannose-1-phosphate phosphatase (M1PP, EC 3.1.3.22) to release free mannitol (Fig. 9.1a). In *E. coli*, mannitol is catabolized by the enzyme mannitol-1-phosphate dehydrogenase (MtlD, EC 1.1.1.17) in a reversible reaction whereas when expressed in transgenic tobacco it functions anabolically and synthesizes mannitol (Tarczynski et al. 1993).

Initially Tarczynski et al. (1993) demonstrated transgenic plants engineered for *MtlD* from *E. coli* in tobacco and *Arabidopsis* result in salinity tolerant phenotype. Targeted mannitol biosynthesis in chloroplasts with the help of an amino terminal transit peptide in tobacco resulted in increased tolerance to methyl viologen-induced oxidative stress and a better photosynthetic efficiency in transgenics, which was attributed to their increased ROS scavenging capacity (Shen et al. 1997b). The gene *MtlD* has also been engineered in economically important plants with substantial results, e.g., Sorghum transgenics overexpressing this gene were found to perform better under salt stress and demonstrated an overall better growth in comparison to control (Maheswari et al. 2010).

Another report of mannitol engineering in potato (*Solanum tuberosum* L.) revealed enhanced NaCl tolerance in both *in vitro* and in hydroponic culture, where transgenic plants were shown to retain more fresh weight than wild-type plants during salt stress (Rahnama et al. 2011). In addition to these, a series of experiments demonstrate transgenic eggplants expressing *mtlD* gene to be tolerant not only towards abiotic stress but biotic stress as well since they demonstrated increased resistance towards three fungal wilts caused by *Fusarium oxysporum*, *Verticillium dahlia* and *Rhizoctonia solani* under both *in vitro* and *in vivo* conditions (Prabhavathi et al. 2002; Prabhavathi and Rajam 2007).

Sorbitol is a sugar alcohol accumulated in higher plants especially in Rosaceae (Bieleski 1982). In microorganisms (*Zymomonas mobilis*), sorbitol biosynthesis requires a one step reaction catalyzed by the enzyme glucose-fructose oxidoreductase (GFOR, EC 1.1.99.28) from glucose and fructose. While in higher plants, NADP-dependent sorbitol-6-phosphate dehydrogenase (S6PDH, EC 1.1.1.200) catalyzes the key step conversion of glucose-6-phosphate to sorbitol-6-phosphate, which is later converted into sorbitol by sorbitol-6-phosphate phosphatase (S6PP, EC 3.1.3.50) (Fig. 9.1b). Many plants use it as a major photosynthetic product which is translocated from mature leaves to growing tissues such as fruits and young leaves (Webb and Burley 1962; Bieleski and Redgwell 1985). Studies show that transgenic tobacco plants over expressing *Stpd1* gene coding for S6PDH from apple accumulate higher amounts of sorbitol and were found to be phenotypically altered with necrotic lesions on the leaves. This was explained on the basis of higher concentration of sorbitol interfering with inositol biosynthesis and leading to osmotic imbalance (Sheveleva et al. 1998).

2.1.2 Inositol and Derivatives

Inositols and their derivatives are a functionally important class of compounds required for normal growth of cells. These inositols are cyclohexane hexitols and exist in nine isomeric forms, out of which *myo*-inositol is the most favored form in nature. The two step inositol biosynthetic pathway is the only *de novo* pathway for inositol synthesis and an out branch of the central glycolytic pathway. This inositol biosynthetic pathway is highly conserved throughout the biological kingdom where the rate limiting enzyme *myo*-inositol-1-phosphate synthase (MIPS, EC 5.5.1.4) catalyzes the conversion of glucose-6-phosphate to *myo*-inositol-1-phosphate and subsequently *myo*-inositol-1-phosphate is converted to free *myo*-inositol by the enzyme *myo*-inositol mono phosphatase (IMP, EC 3.1.3.25) (Fig. 9.1c). Free inositol can be further channelized to other physiologically significant pathways and produce various inositol derivatives (Loewus and Murthy 2000; Stevenson et al. 2000).

These inositols are required for normal growth and development, membrane biogenesis along with the roles of their phosphorylated derivatives as phosphorus store and as a secondary messenger in signal transduction pathways (Loewus and Murthy 2000). In addition to this, inositol and its derivatives such as pinitol, galactinol and other raffinose series oligosaccharides have been found to act as osmoprotectants and provide protection against abiotic stresses like salt and osmotic stress (Taji et al. 2002). Inositol is also utilized by the cell for the synthesis of molecules like stachyose and verbose which are carbohydrate stores for the cells and are stress induced in some species (Bohnert et al. 1995).

The very first plant gene for MIPS was isolated from *Spirodela polyrrhiza* and was shown to be spatially upregulated during ABA-induced morphogenic responses

(Smart and Fleming 1993). The gene was further overexpressed in *Arabidopsis* and the plants were shown to contain fourfold increase in *myo*-inositol content (Smart and Flores 1997). Paul and Cockburn (1989) demonstrated that *Mesembryanthemum crystallinum* (Ice plant) could tolerate upto 400 mM NaCl by accumulating an inositol derivative pinitol which accounts for around two third of the soluble carbohydrate content. The osmotic adjustment of this particular plant under such stress was thus attributed to its high level of pinitol. Further, coordinated induction of *myo*-inositol-1-phosphate synthase with inositol methyl transferase (*IMT1*) in ice plant was shown, resulting in tenfold accumulation of free inositol during salt stress condition. However, no such response was observed in *Arabidopsis thaliana* during similar stresses, which indicates a remarkable difference in the regulation of gene expression between halophytes and glycophytes (Ishitani et al. 1996). Tobacco plants expressing *McIMT1* gene accumulated increased amounts of D-ononitol and were shown to be less inhibited in growth and photosynthetic carbon fixation than wild-type plants in salt and drought stress condition (Sheveleva et al. 1997).

A novel salt tolerant MIPS (PINO1) from *Porteresia coarctata* has been reported and it's over expression in tobacco plant results in better growth and photosynthetic efficiency than control plants under high salinity stress (Majee et al. 2004). In a follow up study, it was shown that functional over expression of this gene could confer salt tolerance to a wide variety of organisms from bacteria to crop plants (Das Chatterjee et al. 2006). Later on, it has also been shown that co expression of *PINO1* and *McIMT1* allowed the transgenic tobacco plants to perform better under salt stress in comparison to expression of *PINO1* or *McIMT1* alone (Patra et al. 2010).

Recently, two divergent genes (*CaMIPS1* and *CaMIPS2*) encoding MIPS have been reported in chickpea and *CaMIPS2* has been shown to be highly induced under dehydration stress and provides better stress tolerance to transformed yeast under high salt and temperature stress (Kaur et al. 2008).

2.2 *Trehalose*

Trehalose is a nonreducing disaccharide (1,1 α-D glucopyranosyl, α-D-glucopyranoside) found in various organisms including bacteria, algae, fungi, yeast, insects, and some plants (Miranda et al. 2007; Elbein et al. 2003). Besides being a carbohydrate reserve, trehalose protects organisms against several physical and chemical stresses (Van Laere 1989; Wiemken 1990; Eleutherio et al. 1993). Trehalose is synthesized in a two step process in bacteria and yeast, first reaction catalyzed by trehalose-6-phosphate synthase (TPS, EC 2.4.1.15) forming trehalose-6-phosphate from UDP-glucose and glucose-6-phosphate; in second reaction trehalose-6-phosphate phosphatase (TPP, EC 3.1.1.12) converts trehalose-6-phosphate to trehalose (Goddijn and Van Dun 1999) (Fig. 9.2). In *E. coli*, these TPS and TPP enzymes have been shown to be encoded by genes *OtsA* and *OtsB*, where as *Saccharomyces cerevisiae* have evolved a trehalose synthase complex which includes a TPS (*Tps1*) and a TPP (*Tps2*) along with a regulatory subunit TSL (*Tps3*). In *Arabidopsis thaliana*,

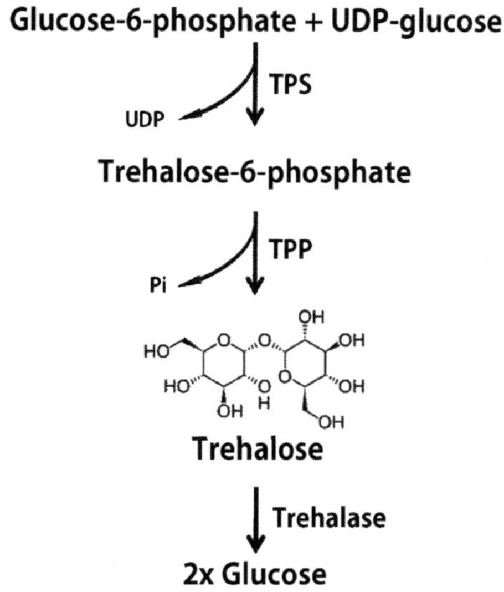

Fig. 9.2 Trehalose biosynthetic pathway in plants. *TPS* trehalose-6-phosphate synthase; *TPP* trehalose-6-phosphate phosphatase

a family of TPS genes with 11 members including trehalose-6-phosphate synthase exists with a subfamily of TPPs (Leyman et al. 2001).

Trehalose is having a unique water absorption capacity which protects the macromolecules from desiccation-induced damage (Rontein et al. 2002). During dehydration, trehalose has been thought to replace water molecules and thereby prevent protein denaturation and membrane fusion (Clegg 1985). It has been shown that trehalose along with other compounds like glycine betaine, proline, and mannitol is active in scavenging ROS (both hydrogen peroxide and superoxide anion) in a concentration-dependent manner (Zhu 2001; Luo et al. 2008). A significant amount of trehalose has been found in two resurrection plants *Myrothamnus flabellifolia* and *Sporobolus stapfianus* (Phillips et al. 2002) where trehalose is thought to prevent intracellular structural damage due to anhydrobiosis (Lunn 2007).

Trehalose metabolism and its engineering in plants for stress tolerance has been an area of immense interest. But studies in tobacco and potato plants (Holmström et al. 1996; Romero et al. 1997; Goddijn et al. 1997; Goddijn and Van Dun 1999; Paul et al. 2001) with a constitutive over expression of yeast or bacterial TPS and TPP genes have shown undesirable effects like stunted growth and abnormal metabolism. Later on, transgenic rice plants were generated using a fusion construct of coding regions of *OtsA* and *OtsB* (with TPS and TPP activity respectively) with either stress inducible (ABA) or tissue specific (rice rbcs) promoter. The phenotypically normal and fertile transgenic rice was achieved with an increased amount of trehalose with increased tolerance to a variety of stresses like salt, drought and low temperature. Transgenic plants also showed increased photosynthetic capacity (Garg et al. 2002). The over expression of trehalose-6 phosphate synthase (*AtTps1*) using 35S promoter in *Arabidopsis* led to significant dehydration tolerance without affecting its morphological traits (Avonce et al. 2004). The level of tolerance provided by these transgenic plants did not correlate well with amount of trehalose accumulated, signifying the other roles of trehalose apart from osmoprotection (Iordachescu and Imai 2008).

2.3 Proline

Proline, an imino acid, is one of the most common compatible osmolyte with high water solubility and stable conformation. It is an essential component of cellular and metabolic events and also responsible for osmotic adjustment in cell (Yancey 2005). Apart from plants, the accumulation of proline has been observed in bacteria, protozoa, algae, and marine invertebrates (McCue and Hanson 1990; Delauney and Verma 1993).

In plants, the biosynthesis of proline can occur via glutamate or ornithine pathway. Glutamate is the primary precursor for proline synthesis in osmotically stressed out and nitrogen deficient cells, while at higher levels of available nitrogen, the ornithine pathway is followed (Delauney et al. 1993). Biosynthetic pathway from glutamate to proline involves two important enzymes L-Δ^1-pyrroline-5-carboxylate synthetase (P5CS, EC 2.7.2.11) and L-Δ^1-pyrroline-5-carboxylate reductase (P5CR, EC 1.5.1.2). First glutamate is converted to glutamic-γ-semialdehyde (GSA) and L-Δ^1-pyrroline-5-carboxylate (P5C) by the action of P5CS, and then P5CR catalyzes the conversion of P5C to L-proline (Fig. 9.3). The level of proline in plants is controlled by degradation or metabolism of proline, where ProDH (proline dehydrogenase, EC 1.5.1.12) oxidizes proline to P5C in plant mitochondria and finally P5C dehydrogenase (P5CD, EC 1.5.1.12) converts P5C to L-glutamate (Boggess and Koeppe 1978; Elthon and Stewart 1981). In normal conditions, this oxidation pathway is followed whereas, under salt and water stress such proline degradation pathway is inhibited, as a result proline level increases (Delauney and Verma 1993; Peng et al. 1996).

Increased cellular proline content is reported to stabilize protein structure and protect cellular functions possibly by scavenging ROS under osmotic stress. Proline may also serve as a source of organic nitrogen, carbon, and energy during recovery from stress (Tyagi and Sairam 2004). This molecule is also involved in maintaining osmotic balance in the cell under dehydration conditions (Singh et al. 1972; Wyn Jones and Storeys 1978). During stress, higher proline content helps in maintaining the $NADP^+$/NADPH ratio in the cell (Hare and Cress 1997). In *E. coli*, proline has been shown to be a potent osmoprotectant as proline over-producing mutant of *E. coli* was found to possess increased osmotolerance and enhanced stability of proteins and membranes in low water and high temperature conditions (Csonka et al. 1988).

Transgenic plants or mutants raised in several studies demonstrate metabolism and accumulation of proline and its importance for development and survival of plants in various adverse environmental conditions (Hong et al. 2000; Mattioli et al. 2008; Szekely et al. 2008). Over expression of moth bean P5CS in rice, wheat and in carrot cell lines conferred enhanced tolerance to salt stress (Zhu et al. 1998; Sawahel and

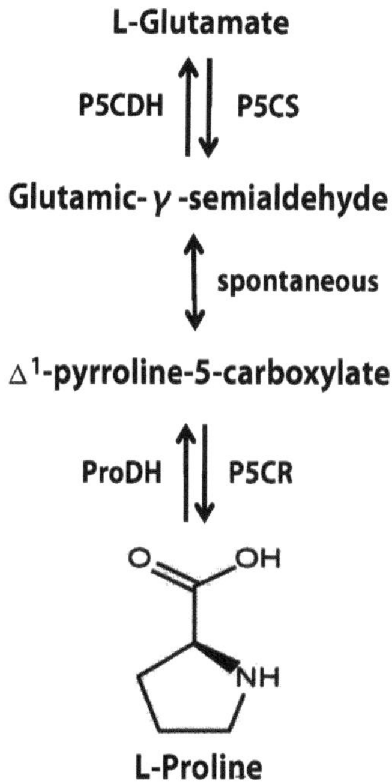

Fig. 9.3 Glutamate pathway for the biosynthesis and metabolism of proline in plants. *P5CS* Δ^1pyrroline-5-carboxylate synthetase; *ProDH* proline dehydrogenase; *P5CDH* P5C dehydrogenase; *P5CR* P5C reductase

Hassan 2002; Han and Hwang 2003). Various studies revealed upregulation of *P5CS* in *Oryza sativa* and *Arabidopsis thaliana* exposed to salt, dehydration, and ABA (Yoshiba et al. 1995; Igarashi et al. 1997). Tolerance to freezing and high salinity was established in antisense transgenic *Arabidopsis* plants carrying *AtProDH* encoding proline dehydrogenase, resulting in higher proline accumulation (Nanjo et al. 1999).

Studies have shown that P5CS is feedback inhibited by proline (Hu et al. 1992). A correlation between induction of P5CS gene and accumulation of proline has been found in *Arabidopsis thaliana* under abiotic stress (Savouré et al. 1995), but this feed back regulation of P5CS is relieved in plants under stress conditions, so as to accumulate more proline for combating disturbance in osmotic balance. In a study of transgenic tobacco plants, over expressing wild-type P5CS from *Vigna aconitifolia* and P5CSF1298 (a mutated P5CS, where feedback inhibition was removed through site directed mutagenesis) were used to compare proline level. Tobacco plant over expressing mutated P5CS accumulated almost twofold more proline than that of transgenic plants expressing wild-type P5CS (Kishor et al. 1995; Verma 1999).

2.4 Polyamines

Polyamines are small organic compounds with two or more primary amino groups, found in all eukaryotic cells. Putrescine (Put, a diamine), spermidine (Spd, a triamine), and spermine (Spm, a tetramine) are the major polyamines found in plants involved in various processes such as cell proliferation, growth, morphogenesis, differentiation, and programmed cell death (Yamaguchi et al. 2007; Alcázar et al. 2010a). In addition, several uncommon polyamines such as homospermidine, 1,3-diaminopropane, cadaverine, and canavalamine have been reported across the kingdoms of life (Minguet et al. 2008). Polyamines occur in free or conjugated forms either with phenolic compounds or macromolecules such as proteins and nucleic acids. Polycationic nature of polyamines at physiological pH is attributed for their biological activity (Gill and Tuteja 2010).

Polyamines play an important role in several plant developmental processes such as cell division, embryogenesis (Bastola and Minocha 1995), fruit ripening (Mehta et al. 1997, 2002), root growth (Watson et al. 1998), tuber development (Kumar et al. 1996; Rafart-Pedros et al. 1999), floral initiation, floral development, and stem elongation (Gerats et al. 1988; Masgrau et al. 1997; Hanzawa et al. 2000; Panicot et al. 2002).

Putrescine, spermidine, spermine, and cadaverine accumulation is well studied under abiotic stress conditions and has been reported in many plant species (Evans and Malmberg 1989; Alcázar et al. 2006, 2010b). Putrescine in plants is either directly synthesized from ornithine by ornithine decarboxylase (ODC, EC 4.1.1.17) or from arginine via *N*-carbamoylputrescine and agmatine. Arginine conversion requires the enzymes arginine decarboxylase (ADC, EC 4.1.1.19), *N*-carbamoylputrescine amidohydrolase (CPA, EC 3.5.1.53) and agmatine deiminase (ADI, EC 3.5.3.12) (Urano et al. 2003). Putrescine is further converted into spermidine and consequently to spermine by spermidine or spermine synthases (SPDS, EC 2.5.1.16; SPMS, EC 2.5.1.22) by the addition of an aminopropyl moiety from decarboxylated *S*-adenosylmethionine generated by *S*-adenosylmethionine decarboxylase (SAMDC, EC 4.1.1.50) (Fig. 9.4). *S*-adenosylmethionine is also the precursor of an important source of ethylene, aminocyclopropane carboxylic acid, thus metabolism of polyamine and ethylene is coupled together, which has significance in stress response (Zapata et al. 2004).

The less common polyamine cadaverine is the product of direct decarboxylation of lysine (Bakhanashvili et al. 1985). In *Arabidopsis*, genes involved in polyamine synthesis were identified as ADC, SAMDC, SPDS, SPMS (Urano et al. 2003), CPA (Piotrowski et al. 2003) and ADI (Janowitz et al. 2003). Beside their possible effects on the osmotic adjustment, polyamines are also involved in stomata closure by regulating voltage-dependent inward K^+ channels in the plasma membrane of guard cells (Liu et al. 2000). In addition polyamines are known to be components of the cellular antioxidant system and are usually regarded as scavengers of hydroxyl radicals. Cadaverine via hydroxyl radical-generating system inhibits DNA oxidative degradation in vitro (Kuznetsov et al. 2007). Putrescine, spermidine, and spermine act as hydroxyl radical scavengers in a dose-dependent manner. In addition spermine or spermidine was shown to quench singlet oxygen at higher concentrations (Das and Misra 2004). Transgenic approaches helped to generate plants expressing polyamine biosynthetic enzymes such as ADC, ODC, SAMDC, SPDS, ACC (1-amino cyclopropane-1-carboxylic acid) synthase and ACC oxidase, with enhanced environmental stress tolerance (Gill and Tuteja 2010; Rubén et al. 2010).

Fig. 9.4 Polyamine biosynthetic pathway in plants. *ADC* arginine decarboxylase; *ADI* agmatine deiminase; *CPA* *N*-carbamoylputrescine amidohydrolase; *ODC* ornithine decarboxylase; *SAMDC* *S*-adenosylmethionine decarboxylase; *SPDS* spermidine synthase; *SPMS* spermine synthase

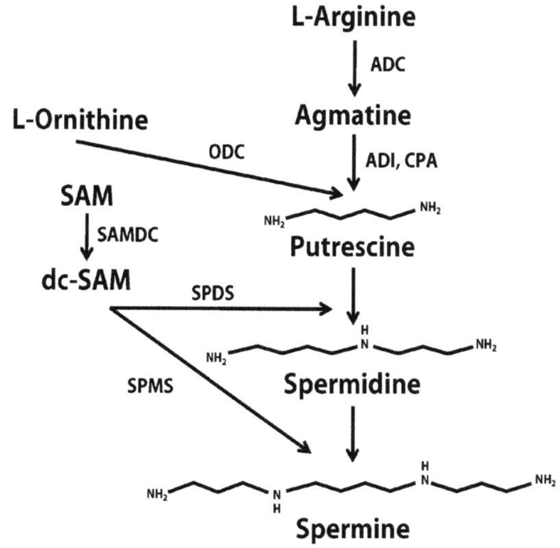

2.5 *Ectoine*

Ectoine (1,4,5,6-tetrahydro-2-methyl-4-pyrimidine carboxylic acid), a common solute of aerobic heterotrophic bacteria (Kempf and Bremer 1998; Galinski 1995; Severin et al. 1992; Kalyuzhnaya et al. 2001), was first discovered as an osmoprotectant in the halophilic bacterium *Ectothiorhodospira halochloris* (Galinski et al. 1985). Ectoines constitute a class of small molecule chaperones (SMCs), which accumulate to high intracellular concentrations without affecting the cellular functions and prevent the misfolding of proteins and other labile macromolecules from environmental stresses (Marina et al. 2008). This organic solute can either be synthesized *de novo* or taken up from the environment when available (Galinski and Trüper 1994; Kempf and Bremer 1998). The exact mechanisms of protein stabilization by ectoines are poorly understood, but they are believed to aid in hydration of proteins with solvent molecules (Kanapathipillai et al. 2005). Ectoine is synthesized from aspartate semialdehyde which is converted to L-2,4-diaminobutyrate (DABA) by L-2,4-diaminobutyric acid transaminase (EctB, EC 2.6.1.76). After that, DABA is acetylated to form Nγ-acetyl-L-2,4-diaminobutyrate (Nγ-acetyl-DABA) by L-2,4-diaminobutyric acid acetyltransferase (EctA, EC 2.3.1.178) (Fig. 9.5). The final step is the cyclization of Nγ-acetyl-DABA to form ectoine by the action of ectoine synthase (EctC, EC 4.2.1.108) (Reshetnikov et al. 2011).

The *ectABC* gene cluster involved in the biosynthesis of ectoine has been isolated from *Chromohalobacter salexigens* (Cánovas et al. 1997), *Marinococcus halophilus* (Louis and Galinski 1997), and *Halomonas elongata* (Göller et al. 1998). Functional expression of *Marinococcus halophilus* ectoine biosynthetic pathway genes in *E. coli* resulted in enhanced tolerance to salt (Louis and Galinski 1997).

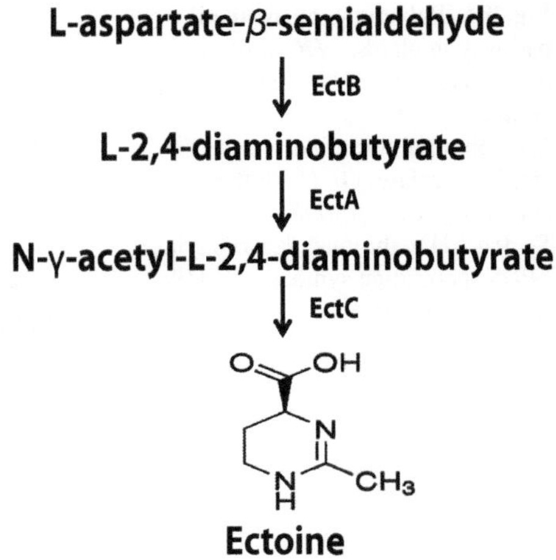

Fig. 9.5 Ectoine biosynthetic pathway in bacteria. *EctB* diaminobutyric acid (DABA) aminotransferase; *EctA* DABA acetyltransferase; *EctC* ectoine synthase

Plants transformed with ectoine biosynthesis genes from *Halomonas elongata* demonstrated enhanced tolerance to mannitol and NaCl (Nakayama et al. 2000; Moghaieb et al. 2006, 2011).

2.6 Glycine Betaine

Glycine betaine, a quaternary ammonium compound is widely distributed in microorganisms, higher plants and animals and one of the most common betaines found in plants (Rhodes and Hanson 1993). In many halotolerant plants, glycine betaine is reported to accumulate in plastids (Allard et al. 1998) and higher levels of glycine betaine correlates with higher level of stress tolerance (McNeil et al. 1999). Glycine betaine has diverse functions in plant cell such as stabilization of the quaternary structure of enzyme, proteins, and maintenance of membrane integrity under salt, cold, and heat stress (Sakamoto and Murata 2000).

The biosynthetic pathway in most plants follows the conversion of choline to glycine betaine in two oxidation steps via the intermediate betaine aldehyde. The first reaction is catalyzed by choline monooxygenase (CMO, EC 1.14.15.7) that converts choline to betaine aldehyde hydrate thus spontaneously forming betaine aldehyde which is acted upon by betaine aldehyde dehydrogenase (BADH, EC 1.2.1.8) to form glycine betaine, whereas in *Arthrobacter spp.* only one enzyme, choline oxidase (CO, EC 1.1.3.17) is required (Ikuta et al. 1977) (Fig. 9.6).

In higher plants, both these enzymes are localized in stroma of chloroplast (Lerma et al. 1988; Rathinasabapathi et al. 1997). Glycine betaine, when engineered in plants or exogenously applied provides sufficient tolerance to a variety of abiotic stresses (Sakamoto and Murata 2001, 2002). Transgenics generated in rice and tomato using choline oxidase (*codA*) targeting both chloroplast and cytosol have shown that the accumulation of glycine betaine in chloroplast is more efficient in providing stress

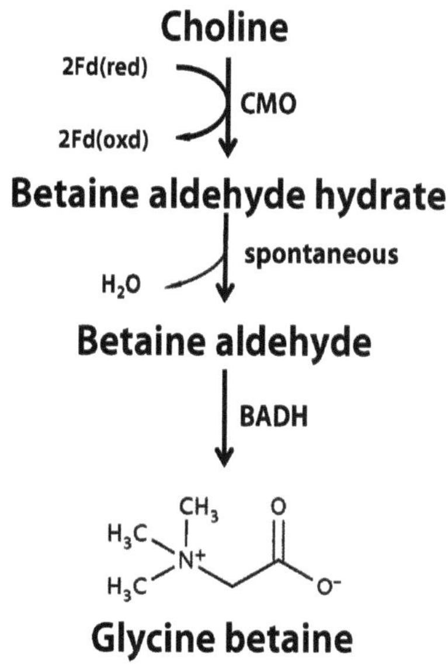

Fig. 9.6 Pathway for the biosynthesis of glycine betaine in plants. *CMO* choline monooxygenase; *BADH* betaine aldehyde dehydrogenase; *Fd(red)* and *Fd(ox)* ferredoxin in reduced and oxidized forms, respectively

tolerance than accumulation of glycine betaine in cytosol (Sakamoto et al. 1998; Chen and Murata 2002; Park et al. 2004, 2007). The photosynthetic machinery was found to be protected against salt and cold stresses in transgenic rice expressing *codA* with no negative effects on growth and development (Alia et al. 1998; Sakamoto et al. 1998). Interestingly, the *codA* over expressing *Arabidopsis* produced more flowers, siliques, and seeds than wild-type plant when grown under normal conditions (Park et al. 2004). Most of the plants are vulnerable to abiotic stress in their reproductive stage and it has been observed that accumulation of glycine betaine in reproductive organs can effectively protect the various organs from the damaging effect of stress and increase the crop yield (Park et al. 2004; Quan et al. 2004). Microarray studies in *Arabidopsis* reveals that exogenous application of glycine betaine also enhances the expression of other genes that are directly or indirectly involved in stress tolerance such as genes for ROS scavenging enzymes, transcription factors, ferric reductase, and membrane trafficking components (Einset et al. 2007).

3 Mechanism of Stress Tolerance

Osmoprotectants generally localize in cytoplasm following osmotic stress, though the mechanism by which these molecules provide tolerance under stress is not clearly understood (Ramanjulu and Bartels 2002). These osmoprotectants are thought to counteract osmotic imbalance by reducing cell's osmotic potential and thereby maintaining turgor pressure under conditions of low water potential and high ionic strength (Pathan et al. 2004). They also function to protect or replace the water shell around proteins (Yancey et al. 1982; Stoop et al. 1996) and stabilize

Table 9.1 Osmoprotectants and their role in stress tolerance

Osmoprotectant	Role in stress tolerance	Reference
Mannitol	Protects cellular structures from hydroxyl radical by reducing it	Shen et al. (1997a, b)
Glycine betaine	Salt and cold tolerance by protecting photosynthetic protein complex and reducing lipid peroxidation. Also works as chaperon in refolding of enzymes	Holmström et al. (2000), Chen et al. (2000), Sakamoto and Murata (2001)
Proline	Adjustment of cellular redox state	Shen et al. (1999), Kuznetsu and Shevyakova (1999)
Ectoine, trehalose, fructan	Mainly stabilize the membranes from oxidative damage	Romero et al. (1997), Nakayama et al. (2000)
Polyols (myo-inositol, D-ononitol, D-pinitol)	Dual functions—osmotic adjustment and supporting redox control	Shen et al. (1999)
Polyamines	Scavengers of hydroxyl radicals and stomata closure	Liu et al. (2000), Kuznetsov et al. (2007)

protein complexes and membranes (Murata et al. 1992; Papageorgiou and Murata 1995). The accumulation of these osmolytes in overexpressing transgenic plants is too low to provide protection by the way of osmotic mass action alone (Sheveleva et al. 1997; Sakamoto et al. 1998; Huang et al. 2000). Apart from this, investigators have also revealed some alternative modes of stress protection offered by these osmoprotectants like scavenging of ROS and chaperon like activities that protect protein structure (Shen et al. 1997b; Serraj and Sinclair 2002). Table 9.1 summarizes the specific functions of some common osmoprotectants under abiotic stress.

4 Metabolic Engineering for Osmoprotectant Synthesis

Genetic transformation technology enables us to achieve gene transfer in precise and predictable manner. Hence genetic engineering approaches would be useful to manipulate these osmoprotectants biosynthetic pathways for accumulating such molecules that act by scavenging ROS, reducing lipid peroxidation, maintaining protein structure and functions (Hare et al. 1998; McNeil et al. 1999; Diamant et al. 2001; Yamada et al. 2005). The physiological and agricultural implications of metabolic engineering of plants for osmoprotectant biosynthesis have been thoroughly reviewed and analyzed (Jain and Selvaraj 1997; Nelson et al. 1998; Bohnert and Sheveleva 1998; Yeo 1998). Table 9.2 summarizes different transgenics developed using genes involved in osmoprotectant biosynthesis for abiotic stress tolerance.

Table 9.2 List of transgenic plants engineered for stress tolerance using osmoprotectant genes

Osmolyte	Gene	Gene source	Plant species engineered	Stress tolerance	Reference
Mannitol					
Mannitol	*mtlD*, Mannitol-1-phosphate dehydrogenase	*E. coli*	Tobacco	No contribution to sustain growth under salinity and drought stress	Tarczynski et al. (1993)
Mannitol	*mtlD*, Mannitol-1-phosphate dehydrogenase	*E. coli*	*Arabidopsis*	Increased germination under salinity stress	Thomas et al. (1995)
Mannitol	*mtlD*, Mannitol-1-phosphate dehydrogenase	*E. coli*	Wheat	Drought and salinity tolerance of calli and plants	Abebe et al. (2003)
Mannitol	*M6PR*, Mannose-6-phosphate reductase	*Apium graveolens*	*Arabidopsis*	Mannitol accumulation under salt stress leading to salt tolerance	Zhifang and Loescher (2003)
Mannitol	*mtlD and GutD,* Mannitol-1-phosphate dehydroge nase and glucitol-6-phosphate dehydrogenase	*E. coli*	Loblolly pine	High salt tolerance due to mannitol and glucitol accumulation	Tang et al. (2005)
Mannitol	*mtlD,* Mannitol-1-phosphate dehydrogenase	*E. coli*	*Petunia*	Chilling tolerance	Chiang et al. (2005)
Mannitol	*mtlD,* Mannitol-1-phosphate dehydrogenase	*E. coli*	Tobacco	Increased plant height and fresh weight under salinity stress	Hu et al. (2005)
Mannitol	*M6PR*, Mannose-6-phosphate reductase	*A. graveolens*	*Arabidopsis*	Mannitol accumulation led to higher yield as well as chloroplast protection under salt stress	Sickler et al. (2007)

(continued)

Table 9.2 (continued)

Osmolyte	Gene	Gene source	Plant species engineered	Stress tolerance	Reference
Sugars	*AtGolS2*, Galactinol and Raffinose accumulation	*Arabidopsis*	*Arabidopsis*	Tolerance to cold, salinity, and drought stresses	Taji et al. (2002)
Sorbitol					
Sorbitol	*Stpd1*, Sorbitol-6-phosphate dehydrogenase	Apple	Tobacco	Not much stress tolerant; phenotypic alterations with necrotic lesions on leaves	Sheveleva et al. (1998)
Inositols					
Ononitol	*IMT1*, *Myo*-inositol-*o*-methyltransferase	Ice plant	Tobacco	Higher photosynthetic rate under salinity stress. Better recovery after drought stress	Sheveleva et al. (1997)
Myo-Inositol	*INO1*, *Myo*-inositol-1-phosphate synthase	*Porteresia coarctata* (wild rice)	Tobacco	Higher photosynthetic efficiency and better growth under salt stress	Majee et al. (2004)
Trehalose					
Trehalose	*TPS*, Trehalose-6-phosphate synthase	Yeast	Tobacco	Drought tolerance but decreased growth rate	Holmström et al. (1996)
Trehalose	*otsA*, *otsB*, Trehalose-6-phosphate synthase, Trehalose-6-phosphate phosphatase	*E. coli*	Tobacco	Improved growth under stress conditions, morphological alterations	Goddijn et al. (1997), Goddijn and Van Dun (1999)
Trehalose	*TPS1*, Trehalose-6-phosphate synthase	Yeast	Tobacco	Stunted growth, lancet shaped leaves, reduced sucrose content, and improved drought tolerance	Romero et al. (1997)
Trehalose	*otsA*, *otsB*, Trehalose-6-phosphate synthase, Trehalose-6-phosphate phosphatase	*E. coli*	Tobacco	Increased leaf dry weight and photosynthetic activity under drought	Pilon-Smits et al. (1998)

Trehalose	*TPS*, Trehalose-6-phosphate synthase	*E. coli*	Potato	Trehalose level was not increased	Goddijn et al. (1999)
Trehalose	*TPS*, *TPP*, Trehalose-6-phosphate synthase, Trehalose-6-phosphate phosphatase	*E. coli* and Yeast	Tobacco	Enhanced rate of photosynthesis (TPS), Reduced rate of photosynthesis (TPP)	Paul et al. (2001)
Trehalose	*otsA*, *otsB*, Trehalose-6-phosphate synthase, Trehalose-6-phosphate phosphatase	*E. coli*	Rice	Sustained plant growth, less photo oxidative damage, Increased stress tolerance	Garg et al. (2002)
Trehalose	*TPSP* (Fusion protein), Trehalose-6-phosphate synthase, Trehalose-6-phosphate phosphatase	*E. coli*	Rice	Drought, salt and cold tolerance	Jang et al. (2003)
Trehalose	*TPS1*, Trehalose-6-phosphate synthase	Yeast	Tomato	Apart from improved tolerance to drought, salinity, & oxidative stresses, transgenics had pleiotropic anatomical changes	Carolina and Francisco (2005)
Trehalose	*TPS1–TPS2*, Trehalose-6-phosphate synthase, Trehalose-6-phosphate phosphatase	Yeast	*Arabidopsis*	Improved drought, freezing, salt, and heat tolerance	Miranda et al. (2007)
Trehalose	*TPS1*, Trehalose-6-phosphate synthase	*Arabidopsis*	Tobacco	Drought resistance and sustained photosynthesis	Almeida et al. (2007)
Trehalose	*TPS1–TPS2*, Trehalose-6-phosphate synthase, Trehalose-6-phosphate phosphatase	Yeast	Tobacco	Maintenance of water status under drought stress	Karim et al. (2007)

(continued)

Table 9.2 (continued)

Osmolyte	Gene	Gene source	Plant species engineered	Stress tolerance	Reference
Proline					
Proline	*P5CR*, Pyrroline-5-carboxylate reductase	*Vigna aconitifolia*	Tobacco	Enhanced P5CR activity in transgenics did not yield significant increase in proline level	LaRosa et al. (1991)
Proline	*P5CS*, Pyrroline-5-carboxylate synthetase	*V. aconitifolia*	Tobacco	Increased biomass production and enhanced flower and seed development under salinity stress	Kishor et al. (1995)
Proline	*P5CS*, Pyrroline-5-carboxylate synthetase	*V. aconitifolia*	Rice	Increased biomass production under drought and salinity stress	Zhu et al. (1998)
Proline	*P5CS*, Pyrroline-5-carboxylate synthetase	*V. aconitifolia*	*Arabidopsis*	Antisense plants showed hypersensitivity to osmotic stress and showed morphological changes during non-stress condition	Nanjo et al. (1999b)
Proline	*ProDH*, Proline dehydrogenase	*V. aconitifolia*	*Arabidopsis*	Altered levels of proline dehydrogenase conferred salt and freezing tolerance	Nanjo et al. (1999b)
Proline	*P5CS*, Pyrroline-5-carboxylate synthetase	*V. aconitifolia*	Rice	Elevated proline and reduced free radical levels	Hong et al. (2000)
Proline	*P5CR*, Pyrroline-5-carboxylate reductase	*V. aconitifolia*	Soybean	Antisense plants failed to survive after 6 days of drought stress	De Ronde et al. (2000)
Proline	*P5CS*, Pyrroline-5-carboxylate synthetase	*V. aconitifolia*	Wheat	Wheat transgenic plants showed enhanced proline levels and conferred salt tolerance	Sawahel and Hassan (2002)

Proline	*ProDH*, Proline dehydrogenase	*V. aconitifolia*	*Arabidopsis*	Antisense plants showed hypersensitivity to exogenous proline	Mani et al. (2002)
Proline	*OAT*, Ornithine-D-aminotransferase	*Arabidopsis*	Tobacco	Overexpression increased proline biosynthesis and osmotolerance	Roosens et al. (2002)
Proline	*P5CS*, Pyrroline-5-carboxylate synthetase	*V. aconitifolia*	Rice	Transgenic rice plants showed better root growth and biomass development during NaCl treatment	Anoop and Gupta (2003)
Proline	*OAT*, Ornithine-D-aminotransferase	*Arabidopsis*	Rice	Overexpression led to increase in proline levels during osmotic stress and transgenic plants showed improved yield under stress	Wu et al. (2003)
Proline	*P5CS*, Pyrroline-5-carboxylate synthetase	*V. aconitifolia*	Carrot	Tolerance to salt stress	Han and Hwang (2003)
Proline	*P5CR*, Pyrroline-5-carboxylate reductase	Tomato	Soybean	Enhanced heat and drought stress tolerance	De Ronde et al. (2004)
Proline	*ProDH*, Proline dehydrogenase	*V. aconitifolia*	Tobacco	Antisense plants showed increased proline content and cytoplasmic osmotic pressure	Kochetov et al. (2004)
Proline	*P5CS*, Pyrroline-5-carboxylate synthetase	*V. aconitifolia*	Rice	Stress inducible expression of *P5CS* gene in rice seedlings showed significantly higher tolerance to drought and salt stress	Su and Wu (2004)

(continued)

Table 9.2 (continued)

Osmolyte	Gene	Gene source	Plant species engineered	Stress tolerance	Reference
Proline	*OsP5CS2*, Pyrroline-5-carboxylate synthetase	Rice	Rice	Enhanced salt and cold stress tolerance	Hur et al. (2004)
Proline	*P5CSF129*, Pyrroline-5-carboxylate synthetase	Tomato	Citrus	Drought tolerance	Molinari et al. (2004)
Proline	*P5CR*, Pyrroline-5-carboxylate reductase	*Arabidopsis*	Potato	Salinity tolerance	Hmida-Sayari et al. (2005)
Proline	*P5CS*, Pyrroline-5-carboxylate synthetase	Tomato	Sugarcane	Protection from drought stress and higher biomass yield	Molinari et al. (2007)
Proline	*P5CS*, Pyrroline-5-carboxylate synthetase	*V. aconitifolia*	*Medicago truncatula*	Enhanced tolerance to osmotic stress	Verdoy et al. (2006)
Polyamines					
Putrescine	*ADC*, Arginine decarboxylase	Oat	Rice	Salt tolerance	Roy and Wu (2001)
Putrescine	*ODC*, Ornithine decarboxylase	Mouse	Tobacco	Salt tolerance	Kumria and Rajam (2002)
Spermidine, spermine	*SAMDC*, *S*-Adenosylmethionine decarboxylase	*Tritordeum*	Rice	Salt tolerance	Roy and Wu (2002)
Spermidine	ACC (1-amino cyclopro pane-1-carboxylic acid) synthase, ACC oxidase, Arginine decarboxylase	Carnation	Tobacco	Tolerance to many stresses	Wi and Park (2002)
Spermidine, spermine	*SAMDC*, *S*-Adenosylmethionine decarboxylase	Human	Tobacco	Tolerance to osmotic stress	Waie and Rajam (2003)

Putrescine	*ADC*, Arginine decarboxylase	*Datura stramonium*	Rice	Drought tolerance	Capell et al. (2004)
Spermidine	*SPE*, Spermidine synthase	Yeast	*Arabidopsis*	Tolerance to chilling, freezing, salinity, drought, hyperosmosis	Kasukabe et al. (2004)
Spermidine	*SAMDC*, *S*-Adenosylmethionine decarboxylase	Carnation	Tobacco	Broad spectrum stress tolerance	Wi et al. (2006)
Putrescine	*ADC1*, Arginine decarboxylase	*Arabidopsis*	*Arabidopsis*	Freezing tolerance	Altabella et al. (2009)
Spermidine	*SAMDC*, *S*-Adenosylmethionine decarboxylase	Yeast	Tomato	Heat tolerance	Cheng et al. (2009)
Spermidine	*SPDS*, Spermidine synthase	Apple	Pear	Aluminum stress tolerance	Wen et al. (2009)
Putrescine	*ADC2*, Arginine decarboxylase	*Arabidopsis*	*Arabidopsis*	Drought tolerance	Alcázar et al. (2010b)
Ectoine					
Ectoine	*ectABC*, L-2,4-diaminobutyric acid acetyltransferase L-2,4-diaminobutyric acid transaminase Ectoine synthase	*Halomonas elongata*	Tobacco cell lines	Tolerance hyperosmotic stress	Hideki et al. (2000)
Ectoine	*ectABC* L-2,4-diaminobutyric acid acetyltransferase, L-2,4-diaminobutyric acid transaminase, Ectoine synthase	*H. elongata*	Tobacco	Tolerance to salt and osmotic stress	Moghaieb et al. (2006)

(continued)

Table 9.2 (continued)

Osmolyte	Gene	Gene source	Plant species engineered	Stress tolerance	Reference
Ectoine	*ectABC*	*H. elongata*	Tomato	Salt tolerance	Moghaieb et al. (2011)
	L-2,4-diaminobutyric acid acetyltransferase				
	L-2,4-diaminobutyric acid transaminase				
	Ectoine synthase				
Glycine betaine					
Betaine	*betA*, Choline dehydrogenase	*E. coli*	Tobacco	Increased tolerance to salinity stress	Lilius et al. (1996)
Betaine	*codA*, Choline oxidase	*Arthrobacter globiformis*	*Arabidopsis*	Seedlings tolerant to salinity stress and increased germination under cold	Hayashi et al. (1997), Alia et al. (1998)
Betaine	*codA*, Choline oxidase	*A. globiformis*	Rice	Increased tolerance to salinity and cold	Sakamoto et al. (1998)
Betaine	*GS2*, Chloroplastic glutamine synthetase	Rice	Rice	Increased salinity resistance and chilling tolerance	Hoshida et al. (2000)
Betaine	*codA*, Choline oxidase	*A. globiformis*	*Arabidopsis*, Tobacco	Increased stress tolerance to salt and cold	Huang et al. (2000)
Betaine	*BADH1*, Betaine aldehyde dehydrogenase	Sorghum	Tomato	Maintenance of osmotic potential	Moghaieb et al. (2000)
Betaine	*CMO*, Choline monooxygenase	*Atriplex hortensis*	Tobacco	Better in vitro growth under salinity and osmotic stress	Yi-Guo et al. (2002)
Betaine	*codA*, Choline oxidase	*A. globiformis*	Rice	Recovery from a week long salt stress	Mohanty et al. (2003)
Betaine	*codA*, Choline oxidase	*A. globiformis*	*Arabidopsis*	Salt tolerance in terms of flowering	Ronan et al. (2003)

Betaine	*BADH1*, Betaine aldehyde dehydrogenase	Carrot	Carrot	Salinity tolerance	Kumar et al. (2004)
Betaine	*betA*, Choline dehydrogenase	*E. coli*	Maize	Drought resistance at seedling stage and high yield after drought condition	Ruidang et al. (2004)
Betaine	*codA*, Choline oxidase	*A. globiformis*	*Brassica juncea*	Tolerance to stress-induced photoinhibition	Prasad and Saradhi (2004)
Betaine	*BADH1*, Betaine aldehyde dehydrogenase	*Atriplex hortensis*	Tobacco	Heat tolerance and increased photosynthetic efficiency	Yang et al. (2005)
Betaine	*COX*, Choline oxidase	*Arthrobacter pascens*	Rice	Salt stress tolerance	Su et al. (2006)

5 Constraints in Path of Metabolic Engineering

It has been observed that out of many transgenics developed for higher osmoprotectant accumulation, only a few succeeded due to metabolic constraints, a few are enlisted here:

1. Transgenes used for transforming a plant were of non-plant origin, mainly bacterial, while plants have their own genes for osmoprotectant synthesis. Use of plant origin genes can aid in overcoming this hurdle (Hanson et al. 1994).
2. Two major factors that generally limit the accumulation of osmoprotectants in transgenic plants are the availability of endogenous substrate and transport of osmolytes across the membranes (Nuccio et al. 1998, 2000; McNeil et al. 2000; Huang et al. 2000).
3. Some of the metabolic pathways are very rigid from flux point of view; they oppose the flux redistribution which arises due to over expression of transgene for metabolite biosynthesis (Stephanopoulos and Vallino 1991; Fernie et al. 2002).
4. Metabolic flux of the transgenics developed using constitutive promoter remains diverted all the time and there by affects plant's growth and development. Employing tissue specific and stress inducible promoters may support in balancing metabolic flux (Nelson et al. 1998; Russell et al. 1998; Garg et al. 2002).
5. Over expression of transgene may lead to diversion of metabolic flux from primary metabolism and therefore this can give rise to undesirable consequences (Sheveleva et al. 1998; Bohmert et al. 2000; Roessner et al. 2001; Garg et al. 2002) or it may lead to feedback effects on engineered pathway (Fernie et al. 2002; Regierer et al. 2002).
6. Cells may recognize the over expressed metabolite as non-self and may degrade it using endogenous machinery (Goddijn et al. 1997) or the host may lack regulatory control upon the over expressing enzyme (Trethewey 2004).
7. Over accumulation of various compatible solutes (mannitol, sorbitol, and trehalose) in transgenic plants have shown some harmful side effects (Karakas et al. 1997; Sheveleva et al. 1998; Yeo et al. 2000).
8. Studies show that osmolytes have minor impact on cellular water retention or osmotic adjustment in comparison to stabilization and protection of cellular components (Blum et al. 1996; Konstantinova et al. 2002; Turner et al. 2007).
9. Transgenic plants engineered for over expression of osmoprotectant synthesis gene could not be assessed rigorously for their stress tolerance potential (Bhatnagar-Mathur et al. 2008).

6 Conclusion

The avenues and possibilities of plants engineered for osmoprotectants has been an area of consistent research for plant scientists and have been reviewed extensively in Bohnert et al. 1995; Nuccio et al. 1999; Rathinasabapathi 2000; Chen and Murata

2002. Although the mode of action of these diverse categories of osmoprotectants might be overlapping, it is still a mystery as to what triggers the accumulation of different osmolytes under different stress conditions. Additionally, the protection offered by these molecules is still under speculation as whether it is a result of a better osmotic adjustment of the cell under stressful situations or they have some deeper impacts on the cellular system coping with stress. Among many attempts made at installing genes for osmoprotectant biosynthesis in plants, only a moderate level of stress tolerance has been achieved in controlled stress conditions and no significant performance has been reported from the field trials if any.

The past has nevertheless shown us that the way forward now is to first understand the comprehensive roles of these molecules in relieving stress in the cellular system along with the implications of over expressing these genes in terms of energy efficiency and channelization of metabolic flux away from physiologically important pathways.

7 Future Prospects

Considering the multigenicity of stress tolerance trait, transgenics developed through single gene insertions are inefficient in providing sustainable stress tolerance to crop plants. Therefore, it is important to carefully identify regulatory factors, which affect expression of key genes following any abiotic stress. Use of these regulatory factors like stress inducible transcription factor in transforming any crop plant may lead to regulation of many genes involved in stress response and thereby impart tolerance to multiple stresses. The overall functional analysis of transgenics made for different osmoprotectants may help us to select key regulatory genes for developing multiple stress tolerant crop varieties. So far, attempts for developing stress tolerant transgenics are restricted mostly to model plants, therefore focus on crop plants is the need of the hour.

Though in some cases, it has been reported that modification of compatible solute machinery could lead to no benefit in terms of yield under stress, therefore further research is necessary in order to genetically manipulate tissue specific and stress inducible osmoprotection in crop plants as these transgenics will be more efficient in abiotic stress tolerance without much affecting the metabolic flux. Broad stress tolerant genotypes may be generated by combining different strategies involved in enhancing stress tolerance, like stress-related genes, and their regulatory transcription factors.

Defining the exact mechanism of action of osmolyte and the specific macromolecules being targeted will lead to further improvement in metabolic engineering of osmoprotectants. Identification and characterization of novel osmoprotectants from stress tolerant crop varieties will also aid in achieving this objective.

Therefore, after analyzing these prospects it can be safely concluded that there exists a lot of scope in crop improvement using osmoprotectants but further developments will demand extensive evaluation of stress tolerance potential of these transgenic crops as there is much difference between controlled lab and field conditions.

References

Abebe T, Guenzi AC, Martin B, Cushman JC (2003) Tolerance of mannitol-accumulating transgenic wheat to water stress and salinity. Plant Physiol 131:1748–1755

Acquaah G (2007) Principles of plant genetics and breeding. Blackwell Publishing, Oxford, UK

Alcázar R, Marco F, Cuevas JC, Patrón M, Ferrando A, Carrasco P, Tiburcio AF, Altabella T (2006) Involvement of polyamines in plant response to abiotic stress. Biotechnol Lett 28:1867–1876

Alcázar R, Altabella T, Marco F, Bortolotti C, Reymond M, Koncz C, Carrasco P, Tiburcio AF (2010a) Polyamines: molecules with regulatory functions in plant abiotic stress tolerance. Planta 231:1237–1249

Alcázar R, Planas J, Saxena T, Zarza X, Bortolotti C, Cuevas JC, Bitrián M, Tiburcio AF, Altabella T (2010b) Putrescine accumulation confers drought tolerance in transgenic *Arabidopsis* plants over-expressing the homologous Arginine decarboxylase 2 gene. Plant Physiol Biochem 48(7):547–552

Alia HH, Chen THH, Murata N (1998) Transformation with a gene for choline oxidase enhances the cold tolerance of *Arabidopsis* during germination and early growth. Plant Cell Environ 21:232–239

Allard F, Houde M, Krol M, Ivanov A, Huner NPA, Sarhan F (1998) Betaine improves freezing tolerance in wheat. Plant Cell Physiol 39:1194–1202

Almeida AM, Silva AB, Arajo SS, Cardoso LA, Santos DM, Torn JM, Silva JM, Paul MJ, Fevereiro PS (2007) Responses to water withdrawal of tobacco plants genetically engineered with the *AtTPS1* gene: a special reference to photosynthetic parameters. Euphytica 154:113–126

Altabella T, Tiburcio AF, Ferrando A (2009) Plant with resistance to low temperature and method of production thereof. Spanish patent application WO2010/004070

Anoop N, Gupta AK (2003) Transgenic indica rice cv IR-50 over-expressing *Vigna aconitifolia* delta (1)-pyrroline-5-carboxylate synthetase cDNA shows tolerance to high salt. J Plant Biochem Biotechnol 12:109–116

Avonce N, Leyman B, Mascorro-Gallardo JO, Van Dijck P, Thevelein JM, Iturriaga G (2004) The *Arabidopsis* trehalose-6-P synthase *AtTPS1* gene is a regulator of glucose, abscisic acid, and stress signaling. Plant Physiol 136:3649–3659

Bakhanashvili M, Icekson I, Apelbaum A (1985) Cadaverine formation by specific lysine decarboxylation in *Pisum sativum* seedlings. Plant Cell Rep 4(6):297–299

Bastola DR, Minocha SC (1995) Increased putrescine biosynthesis through transfer of mouse ornithine decarboxylase cDNA in carrot promotes somatic embryogenesis. Plant Physiol 109:63–71

Bhatnagar-Mathur P, Vadez V, Sharma KK (2008) Transgenic approaches for abiotic stress tolerance in plants: retrospect and prospects. Plant Cell Rep 27:411–424

Bieleski RL (1982) Sugar alcohols. In: Loewus FA, Tanner W (eds) Plant carbohydrates I: intracellular carbohydrates, vol 13A, Encyclopedia of Plant physiology. Springer, New York, pp 158–192

Bieleski RL, Redgwell RJ (1985) Sorbitol versus sucrose as photosynthesis and translocation products in developing apricot leaves. Aust J Plant Physiol 12:657–668

Blum A, Munns R, Passioura JB, Turner NC (1996) Genetically engineered plants resistant to soil drying and salt stress: how to interpret osmotic relations? Plant Physiol 110:1051–1053

Boggess SF, Koeppe DE (1978) Oxidation of proline by plant mitochondria. Plant Physiol 62:22–25

Bohmert K, Balbo I, Kopka J, Mittendorf V, Nawrath C, Poirier Y, Tischendorf G, Trethewey RN, Willmitzer L (2000) Transgenic *Arabidopsis* plants can accumulate polyhydroxybutyrate to up to 4 % of their fresh weight. Planta 211:841–845

Bohnert HJ, Jensen RG (1996) Strategies for engineering water stress tolerance in plants. Trends Biotechnol 14:89–97

Bohnert HJ, Sheveleva E (1998) Plant stress adaptations- making metabolism move. Curr Opin Plant Biol 1:267–274

Bohnert HJ, Nelson DE, Jensen RG (1995) Adaptations to environmental stresses. Plant Cell 7:1099–1111

Cánovas D, Vargas C, Iglesias-Guerra F, Csonka LN, Rhodes D, Ventosa A, Nieto JJ (1997) Isolation and characterization of salt-sensitive mutants of the moderate halophile *Halomonas elongata* and cloning of the ectoine synthesis genes. J Biol Chem 272:25794–25801

Capell T, Bassie L, Christou P (2004) Modulation of the polyamine biosynthetic pathway in transgenic rice confers tolerance to drought stress. Proc Natl Acad Sci U S A 101:9909–9914

Carolina C, Francisco ACM (2005) Tomato abiotic stress enhanced tolerance by trehalose biosynthesis. Plant Sci 169:75–82

Chen THH, Murata N (2002) Enhancement of tolerance of abiotic stress by metabolic engineering of betaines and other compatible solutes. Curr Opin Plant Biol 5:250–257

Chen WP, Li PH, Chen THH (2000) Glycinebetaine increases chilling tolerance and reduces chilling-induced lipid peroxidation in *Zea mays*. Plant Cell Environ 23:609–618

Cheng L, Zou YJ, Ding SL, Zhang JJ, Yu XL, Cao JS, Lu G (2009) Polyamine accumulation in transgenic tomato enhances the tolerance to high temperature stress. J Integr Plant Biol 51:489–499

Chiang YC, Stushnoff C, McSay AE, Jones ML, Bohnert H (2005) Overexpression of Mannitol-1-phosphate dehydrogenase increases mannitol accumulation and adds protection against chilling injury in *Petunia*. J Am Soc Horti Sci 30:605–610

Clegg JS (1985) The physical properties and metabolic status of *Artemia* cysts at low water contents: the water replacement hypothesis. In: Leopold AC (ed) Membranes, metabolism and dry organisms. Cornell University Press, New York, pp 169–187

Csonka LN, Gelvin SB, Goodner BW, Orser CS, Siemieniak D, Slightom JL (1988) Nucleotide sequence of a mutation in the *proB* gene of *Escherichia coli* that confers proline overproduction and enhanced tolerance to osmotic stress. Gene 64:199–205

Das Chatterjee A, Goswami L, Maitra S, Dastidar KG, Ray S, Majumder AL (2006) Introgression of a novel salt tolerant L *myo*-inositol 1 phosphate synthase from *Porteresia coarctata* (Roxb.) Tateoka (*PcINO1*) confers salt tolerance to evolutionary diverse organisms. FEBS Lett 580:3980–3988

Das KC, Misra HP (2004) Hydroxyl radical scavenging and singlet oxygen quenching properties of polyamines. Mol Cell Biochem 262:127–133

De Ronde JA, Spreeth MH, Cress WA (2000) Effect of antisense-pyrroline-5-carboxylate reductase transgenic soybean plants subjected to osmotic and drought stress. Plant Growth Regul 32:13–26

De Ronde JA, Cress WA, Krüger GHJ, Strasser RJ, Van Staden J (2004) Photosynthetic response of transgenic soybean plants, containing an *Arabidopsis* P5CR gene, during heat and drought stress. J Plant Physiol 161:1211–1224

Delauney AJ, Verma DPS (1993) Proline biosynthesis and osmoregulation in plants. Plant J 4:215–223

Delauney AJ, Hu CA, Kishor BP, Verma DP (1993) Cloning of ornithine delta amino transferase cDNA from *Vigna aconitifolia* by transcomplementation in *Escherichia coli* and regulation of proline biosynthesis. J Biol Chem 268:18673–18678

Diamant S, Eliahu N, Rosenthal D, Goloubinoff P (2001) Chemical chaperones regulate molecular chaperones *in vitro* and in cells under combined salt and heat stresses. J Biol Chem 276:39586–39591

Einset J, Nielsen E, Connolly EL, Bones A, Sparstad T, Winge P, Zhu JK (2007) Membrane-trafficking RabA4c involved in the effect of glycinebetaine on recovery from chilling stress in *Arabidopsis*. Physiol Plant 130:511–518

Elbein AD, Pan YT, Pastuszak I, David Carroll D (2003) New insights on trehalose: a multifunctional molecule. Glycobiology 13:17–27

Eleutherio ECA, Araujo PS, Panek AD (1993) Protective role of trehalose during heat stress in *Saccharomyces cerevisiae*. Cryobiology 30:591–596

Elthon TE, Stewart CR (1981) Sub-mitochondrial location and electron transport characteristics of enzymes involved in proline oxidation. Plant Physiol 67:780–784

Evans PT, Malmberg RL (1989) Do polyamines have roles in plant development? Annu Rev Plant Physiol Plant Mol Biol 40:235–269

FAO (2008) Soaring food prices: facts, perspectives, impacts and actions required. Document HLC/08/INF/1 prepared for the High-level conference on world food security: the challenges of climate change and bioenergy, June 3–5, Rome, pp. 3–5

Fernie A, Willmitzer L, Trethewey R (2002) Sucrose to starch: a transition in molecular plant physiology. Trends Plant Sci 7:35–41

Frova C, Krajewski P, Di Fonzo N, Villa M, Sari-Gorla M (1999) Genetic analysis of drought tolerance in maize by molecular markers. I. Yield components. Theor Appl Genet 99:280–288

Galinski EA (1995) Osmoadaptation in bacteria. Adv Microb Physiol 37:273–328

Galinski EA, Trüper HG (1994) Microbial behavior in salt-stressed ecosystems. FEMS Microbiol Rev 15:95–108

Galinski EA, Pfeiffer HP, Trüper HG (1985) 1,4,5,6,-Tetrahydro-2-methyl-4-pyrimidinecarboxylic acid, a novel cyclic acid from halophilic phototrophic bacteria of the genus *Ectothiorhodospira*. Eur J Biochem 149:135–139

Garg AK, Kim JK, Owens TG, Ranwala AP, Choi YC, Kochian LV, Wu RJ (2002) Trehalose accumulation in rice plants confers high tolerance levels to different abiotic stresses. Proc Natl Acad Sci U S A 99:15898–15903

Gerats AGM, Kaye C, Collins C, Malmberg ML (1988) Polyamine levels in *Petunia* genotypes with normal and abnormal floral morphologies. Plant Physiol 86:390–393

Gill SS, Tuteja N (2010) Polyamines and abiotic stress tolerance in plants. Plant Signal Behav 51:26–33

Goddijn OJM, Van Dun K (1999) Trehalose metabolism in plants. Trends Plant Sci 4:315–319

Goddijn OJM, Verwoerd TC, Voogd E, Krutwagen RWHH, De Graaf PTHM, Poels J, Van Dun K, Ponstein AS, Damm B, Pen J (1997) Inhibition of trehalase activity enhances trehalose accumulation in transgenic plants. Plant Physiol 113:181–190

Göller K, Ofer A, Galinski EA (1998) Construction and characterization of an NaCl-sensitive mutant of *Halomonas elongata* impaired in ectoine biosynthesis. FEMS Microbiol Lett 161:293–300

Han KH, Hwang CH (2003) Salt tolerance enhanced by transformation of a *P5CS* gene in carrot. J Plant Biotechnol 5:149–153

Hanson AD, Rathinasabapathi B, Rivoal J, Burnet M, Dillon DO, Gage DA (1994) Osmoprotective compounds of the Plumbaginaceae: a natural experiment in metabolic engineering of stress tolerance. Proc Natl Acad Sci U S A 91:306–310

Hanzawa Y, Takahashi T, Michael AJ, Burtin D, Long D, Pineiro M, Coupland G, Komeda Y (2000) *ACAULIS5*, an *Arabidopsis* gene required for stem elongation, encodes a spermine synthase. EMBO J 19:4248–4256

Hare PD, Cress WA (1997) Metabolic implications of stress-induced proline accumulation in plants. Plant Growth Regul 21:79–102

Hare PD, Cress WA, Van Staden J (1998) Dissecting the roles of osmolyte accumulation during stress. Plant Cell Environ 21:535–553

Hayashi H, Alia Mustardy L, Deshnium P, Ida M, Murata N (1997) Transformation of *Arabidopsis thaliana* with the *codA* gene for choline oxidase: accumulation of glycinebetaine and enhanced tolerance to salt and cold stress. Plant J 12:133–142

Hideki N, Kazuya Y, Hisayo O, Yoshikatsu M, Atsuhiko S (2000) Ectoine, the compatible solute of *Halomonas elongata*, confers hyperosmotic tolerance in cultured tobacco cells. Plant Physiol 122:1239–1247

Hmida-Sayari A, Gargouri-Bouzid R, Bidani A, Jaoua L, Savouré A, Jaou S (2005) Over expression of 1-pyrroline-5-carboxylate synthetase increases proline production and confers salt tolerance in transgenic potato plants. Plant Sci 169:746–752

Holmberg N, Bülow L (1998) Improving stress tolerance in plants by gene transfer. Trends Plant Sci 3:61–66

Holmström KO, Mantyala E, Welin B, Mandal A, Palva TE, Tunnela O, Londsborough J (1996) Drought tolerance in tobacco. Nature 379:683–684

Holmström KO, Somersalo S, Mandal A, Palva TE, Welin B (2000) Improved tolerance to salinity and low temperature in transgenic tobacco producing glycinebetaine. J Exp Bot 51:177–185
Hong Z, Lakkineni K, Zhang Z, Verma DPS (2000) Removal of feedback inhibition of delta1-pyrroline-5-carboxylate synthetase results in increased proline accumulation and protection of plants from osmotic stress. Plant Physiol 122:1129–1136
Hoshida H, Tanaka Y, Hibino T, Hayashi Y, Tanaka A, Takabe T (2000) Enhanced tolerance to salt stress in transgenic rice that over expresses chloroplast glutamine synthetase. Plant Mol Biol 43:103–111
Hu CA, Delauney AJ, Verma DP (1992) A bifunctional enzyme (delta^1-pyrroline-5-carboxylate synthetase) catalyzes the first two steps in proline biosynthesis in plants. Proc Natl Acad Sci U S A 89:9354–9358
Hu L, Lu H, Liu Q, Chen X, Jiang X (2005) Over expression of *mtlD* gene in transgenic *Populus tomentosa* improves salt tolerance through accumulation of mannitol. Tree Physiol 25:1273–1281
Huang J, Hirji R, Adam L, Rozwadowski KL, Hammerlindl JK, Keller WA, Selvaraj G (2000) Genetic engineering of glycinebetaine production toward enhancing stress tolerance in plants: metabolic limitations. Plant Physiol 122:747–756
Hur J, Hong Jong K, Lee C-H, An G (2004) Stress-inducible *OsP5CS2* gene is essential for salt and cold tolerance in rice. Plant Sci 167:417–426
Igarashi Y, Yoshiba Y, Sanada Y, Wada K, Yamaguchi-Shinozaki K, Shinozaki K (1997) Characterization of the gene for delta1-pyrroline-5-carboxylate synthetase and correlation between the expression of the gene and salt tolerance in *Oryza sativa* L. Plant Mol Biol 33:857–865
Ikuta S, Mamura S, Misaki H, Horiuti Y (1977) Purification and characterization of choline oxidase from *Arthrobacter globiformis*. J Biochem 82:1741–1749
Iordachescu M, Imai R (2008) Trehalose biosynthesis in response to abiotic stresses. J Integr Plant Biol 50(10):1223–1229
Ishitani M, Majumder AL, Bornhouser A, Michalowski CB, Jensen RG, Bohnert HJ (1996) Coordinate transcriptional induction of *myo*-inositol metabolism during environmental stress. Plant J 9:537–548
Jain RK, Selvaraj G (1997) Molecular genetic improvement of salt tolerance in plants. Biotechnol Annu Rev 3:245–267
Jang I, Oh S, Seo J, Choi WB, Song SI, Kim CH, Kim YS, Seo HS, Choi YD, Nahm BH, Kim JK (2003) Expression of a bifunctional fusion of the *Escherichia coli* genes for Trehalose-6-Phosphate synthase and Trehalose-6-Phosphate phosphatase in transgenic rice plants increases trehalose accumulation and abiotic stress tolerance without stunting growth. Plant Physiol 131:516–524
Janowitz T, Kneifel H, Piotrowski M (2003) Identification and characterization of plant agmatine iminohydrolase, the last missing link in polyamine biosynthesis of plants. FEBS Lett 544:258–261
Johnson MK, Johnson EJ, MacElroy RD, Speer HL, Bruff BS (1968) Effects of salts on the halophilic alga *Dunaliella viridis*. J Bacteriol 95:1461–1468
Kalyuzhnaya MG, Khmelenina VN, Eshinimaev BT, Suzina NE, Nikitin D, Solonin A, Lin JL, McDonald IR, Murrell JC, Trotsenko YA (2001) Taxonomic characterization of new alkaliphilic and alkali tolerant methanotrophs from soda lakes of the Southeastern transbaikal region and description of *Methylomicrobium buryatense* sp. Syst Appl Microbiol 24:166–176
Kanapathipillai M, Lentzen G, Sierks M, Park CB (2005) Ectoine and hydroxyectoine inhibit aggregation and neurotoxicity of Alzheimer's β-amyloid. FEBS Lett 579:4775–4780
Karakas B, Ozias-Akins P, Stushnoff C, Suefferheld M, Rieger M (1997) Salinity and drought tolerance of mannitol-accumulating transgenic tobacco. Plant Cell Environ 20:609–616
Karim S, Aronsson H, Ericson H, Pirhonen M, Leyman B, Welin B, Palva ET, Patrick VD, Kjell-Ove H (2007) Improved drought tolerance without undesired side effects in transgenic plants producing trehalose. Plant Mol Biol 64:371–386

Kasukabe Y, He LX, Nada K, Misawa S, Ihara I, Tachibana S (2004) Over expression of spermidine synthase enhances tolerance to multiple environmental stresses and up-regulates the expression of various stress regulated genes in transgenic *Arabidopsis thaliana*. Plant Cell Physiol 45:712–722

Kaur H, Shukla RK, Yadav G, Chattopadhyay D, Majee M (2008) Two divergent genes encoding L-*myo*-inositol 1-phosphate synthase1 (*CaMIPS1*) and 2 (*CaMIPS2*) are differentially expressed in chickpea. Plant Cell Environ 31:1701–1716

Kempf B, Bremer E (1998) Uptake and synthesis of compatible solutes as microbial stress responses to high-osmolality environments. Arch Microbiol 170:319–330

Kishor PBK, Hong Z, Miao G-H, Hu CAA, Verma DPS (1995) Over expression of delta-1-pyrroline-5-carboxylate synthetase increases proline production and confers osmotolerance in transgenic plants. Plant Physiol 108:1387–1394

Kochetov AV, Titov SE, Kolodyazhnaya YS, Komarova ML, Koval VS, Makarova NN, Il'yinskyi YY, Trifonova EA, Shumny VK (2004) Tobacco transformants bearing antisense suppressor of proline dehydrogenase gene are characterized by higher proline content and cytoplasm osmotic pressure. Russ J Genet 40:216–218

Konstantinova T, Pavanova D, Atanassov A, Djilianov D (2002) Freezing tolerant tobacco, transformed to accumulate osmoprotectants. Plant Sci 163:157–164

Kumar A, Taylor MA, Mad Arif SA, Davies HV (1996) Potato plants expressing antisense and sense SAMDC transgenes show altered levels of polyamines and ethylene: antisense plants display abnormal phenotypes. Plant J 9:147–158

Kumar S, Dhingra A, Daniell H (2004) Plastid-expressed Betaine aldehyde dehydrogenase gene in carrot cultured cells, roots and leaves confers enhanced salt tolerance. Plant Physiol 135:2843–2854

Kumria R, Rajam MV (2002) Ornithine decarboxylase transgene in tobacco affects polyamines, in vitro morphogenesis and response to salt stress. J Plant Physiol 159:983–990

Kuznetsov Vl V, Shevyakova NI (1999) Proline under stress conditions: biological role, metabolism, and regulation. Russ J Plant Physiol 46:321–336

Kuznetsov V, Shorina M, Aronova E, Stetsenko L, Rakitin V, Shevyakova N (2007) NaCl and ethylene-dependent cadaverine accumulation and its possible protective role in the adaptation of the common ice plant to salt stress. Plant Sci 172:363–370

LaRosa PC, Rhodes D, Rhodes JC, Bressan RA, Csonka LN (1991) Elevated accumulation of proline in NaCl-adapted tobacco cells is not due to altered D1-pyrroline-5-carboxylate reductase. Plant Physiol 96:245–250

Lerma C, Hanson AD, Rhodes D (1988) Oxygen-18 and deuterium labeling studies of choline oxidation by spinach and sugar beet. Plant Physiol 88:695–702

Leyman B, Dijck PV, Thevelein JM (2001) An unexpected plethora of trehalose biosynthetic genes in *Arabidopsis thaliana*. Trends Plant Sci 6:510–513

Lilius G, Holmberg N, Bülow L (1996) Enhanced NaCl stress tolerance in transgenic tobacco expressing bacterial choline dehydrogenase. Biotechnology 14:177–180

Liu Q, Zhang Y, Shouyi C (2000) Plant protein kinase genes induced by drought, high salt and cold stresses. Chin Sci Bull 45:1153–1157

Lobell DB, Field CB (2007) Global scale climate-crop yield relationships and the impacts of recent warming. Environ Res Lett 2:014002

Loescher WH, Tyson RH, Everard JD, Redgwell RJ, Bieleski RL (1992) Mannitol synthesis in higher plants: evidence for the role and characterization of a NADPH-dependent mannose-6-phosphate reductase. Plant Physiol 98(4):1396–1402

Loewus FA, Murthy PN (2000) *Myo*-inositol metabolism in plants. Plant Sci 150:1–19

Louis P, Galinski EA (1997) Characterization of genes for the biosynthesis of the compatible solute ectoine from *Marinococcus halophilus* and osmoregulated expression in *Escherichia coli*. Microbiology 143:1141–1149

Lunn JE (2007) Gene families and evolution of trehalose metabolism in plants. Funct Plant Biol 34:550–563

Luo Y, Li WM, Wang W (2008) Trehalose: protector of antioxidant enzymes or reactive oxygen species scavenger under heat stress? Environ Exp Botany 63:378–384
Maheswari M, Varalaxmi Y, Vijayalakshmi A, Yadav SK, Sharmila P, Venkateswarlu B, Vanaja M, PardhaSardhi P (2010) Metabolic engineering using *mtlD* gene enhances tolerance to water deficit and salinity in sorghum. Biol Plant 54:647–652
Majee M, Maitra S, Dastidar KG, Pattnaik S, Chatterjee A, Hait NC, Das KP, Majumder AL (2004) A novel salt tolerant L *myo*-inositol-1-phosphate synthase from *Porteresia coarctata* (Roxb) Tateoka, a halophytic wild rice. J Biol Chem 279:28539–28552
Mani S, Van de Cotte B, Van Montagu M, Verbruggen N (2002) Altered levels of proline dehydrogenase cause hypersensitivity to proline and its analogs in *Arabidopsis*. Plant Physiol 128:73–83
Marina M, Maiale SJ, Rossi FR, Romero MF, Rivas EI, Gárriz A, Ruiz OA, Pieckenstain FL (2008) Apoplastic polyamine oxidation plays different roles in local responses of tobacco to infection by the necrotrophic fungus *Sclerotinia sclerotiorum* and the biotrophic bacterium *Pseudomonas viridiflava*. Plant Physiol 147(4):2164–2178
Masgrau C, Altabella T, Farrás R, Flores D, Thompson AJ, Besford RT, Tiburcio AF (1997) Inducible over expression of oat arginine decarboxylase in transgenic tobacco plants. Plant J 11:465–473
Mattioli R, Marchese D, D'Angeli S, Altamura MM, Costantino P, Trovato M (2008) Modulation of intracellular proline levels affects flowering time and inflorescence architecture in *Arabidopsis*. Plant Mol Biol 66:277–288
McCue KF, Hanson AD (1990) Drought and salt tolerance: towards understanding and application. Trends Biotechnol 8:358–362
McNeil SD, Nuccio ML, Hanson AD (1999) Betaines and related osmoprotectants. Targets for metabolic engineering of stress resistance. Plant Physiol 120:945–949
McNeil SD, Rhodes D, Russell BL, Nuccio ML, Shachar-Hill Y, Andrew D, Hanson AD (2000) Metabolic model identifies key constraints on an engineered glycinebetaine synthetic pathway in tobacco. Plant Physiol 124:153–162
Mehta RA, Handa A, Li N, Mattoo AK (1997) Ripening activated expression of S-adenosylmethionine decarboxylase increases polyamine levels and influences ripening in transgenic tomato fruits (abstract no.134). Plant Physiol 114:S-44
Mehta RA, Cassol T, Li N, Ali N, Handa AK, Mattoo AK (2002) Engineered polyamine accumulation in tomato enhances phytonutrient content, juice quality, and vine life. Nat Biotechnol 20:613–618
Minguet EG, Vera-Sirera F, Marina A, Carbonell J, Blazquez MA (2008) Evolutionary diversification in polyamine biosynthesis. Mol Biol Evol 25:2119–2128
Miranda JA, Avonce N, Suarez R, Thevelein JM, Dijck PV, Iturriaga GA (2007) A bifunctional TPS-TPP enzyme from yeast confers tolerance to multiple and extreme abiotic-stress conditions in transgenic *Arabidopsis*. Planta 226:1411–1421
Moghaieb REA, Tanaka N, Saneoka H, Hussein HA, Yousef SS, Ewada MAF, Aly MAM, Fujika K (2000) Expression of betaine aldehyde dehydrogenase gene in transgenic tomato hairy roots leads to the accumulation of glycinebetaine and contributes to the maintenance of the osmotic potential under salt stress. Soil Sci Plant Nutr 46:873–883
Moghaieb REA, Tanaka N, Sanoeka H, Murooka Y, Ono H, Morikawa H, Nakamura A, Nguyen NT, Suwa R, Fujita K (2006) Characterization of salt tolerance in ectoine-transformed tobacco plants (*Nicotiana tabaccum*): photosynthesis, osmotic adjustment, and nitrogen partitioning. Plant Cell Environ 29:173–182
Moghaieb REA, Nakamura A, Saneoka H, Fujita K (2011) Evaluation of salt tolerance in ectoine-transgenic tomato plants (*Lycopersicon esculentum*) in terms of photosynthesis, osmotic adjustment, and carbon partitioning. GM Crops 2–1:58–65
Mohanty C, Kathuria H, Ferjani A, Sakamoto A, Mohanty P, Murata N, Tyagi AK (2003) Transgenics of an elite indica rice variety Pusa Basmati 1 harbouring the codA gene are highly tolerant to salt stress. Theor Appl Genet 106:51–57
Molinari HBC, Marur CJ, Filho JCB, Kobayashi AK, Pileggi M, Junior RPL, Pereira LFP, Vieira LGE (2004) Osmotic adjustment in transgenic citrus rootstock Carrizo citrange (Citrus sinensis Osb. Poncirus trifoliate L. Raf.) overproducing proline. Plant Sci 167:1375–1381

Molinari HBC, Marur CJ, Daros E, Freitas de Campos MK, Portela de Carvalho JFR, Filho JCB, Pereira LFP, Vieira LGE (2007) Evaluation of the stress-inducible production of proline in transgenic sugarcane (*Saccharum* spp.): osmotic adjustment, chlorophyll fluorescence and oxidative stress. Physiol Planta 130:218–229

Murata N, Mohanty PS, Hayashi H, Papageorgiou GC (1992) Glycinebetaine stabilizes the association of extrinsic proteins with the photosynthetic oxygen-evolving complex. FEBS Lett 296:187–189

Nakayama H, Yoshida K, Ono H, Murooka Y, Shinmyo A (2000) Ectoine, the compatible solute of *Halomonas elongata*, confers hyperosmotic tolerance in cultured tobacco cells. Plant Physiol 122:1239–1247

Nanjo T, Kobayashi M, Yoshiba Y, Kakubari Y, Yamaguchi-Shinozaki K, Shinozaki K (1999) Antisense suppression of proline degradation improves tolerance to freezing and salinity in *Arabidopsis thaliana*. FEBS Lett 461:205–210

Nelson DE, Shen B, Bohnert HJ (1998) Salinity tolerance-mechanisms, models and the metabolic engineering of complex traits. Genet Eng 20:153–176

Nuccio ML, Russell BL, Nolte KD, Rathinasabapathi B, Gage DA, Hanson AD (1998) The endogenous choline supply limits glycinebetaine synthesis in transgenic tobacco expressing choline monooxygenase. Plant J 16:487–496

Nuccio ML, Rhodes D, McNeil SD, Hanson AD (1999) Metabolic engineering of plants for osmotic stress resistance. Curr Opin Plant Biol 2:128–134

Nuccio ML, McNeil SD, Ziemak MJ, Hanson AD, Jain RK, Selvaraj G (2000) Choline import into chloroplasts limits glycine betaine synthesis in tobacco: analysis of plants engineered with a chloroplastic or a cytosolic pathway. Metab Eng 2:300–311

Panicot M, Masgrau C, Borrell A, Cordeiro A, Tiburcio AF, Altabella T (2002) Effects of putrescine accumulation in tobacco transgenic plants with different expression of oat arginine decarboxylases. Physiol Planta 114:281–287

Papageorgiou GC, Murata N (1995) The unusually strong stabilizing effects of glycinebetaine on the structure and function in the oxygen-evolving photosystem II complex. Photosyn Res 44:243–252

Park EJ, Jeknic Z, Sakamoto A, DeNoma J, Yuwansiri R, Murata N, Chen THH (2004) Genetic engineering of glycinebetaine synthesis in tomato protects seeds, plants, and flowers from chilling damage. Plant J 40:474–487

Park EJ, Jeknic Z, Pino MT, Murata N, Chen THH (2007) Glycinebetaine accumulation in chloroplasts is more effective than that in cytosol in protecting transgenic tomato plants against abiotic stress. Plant Cell Environ 30:994–1005

Pathan MS, Subudhi PK, Courtois B, Nguyen HT (2004) Molecular dissection of abiotic stress tolerance in sorghum and rice. In: Nguyen HT, Blum A (eds) Physiology and biotechnology integration for plant breeding. Marcel Dekker, New York, pp 525–569

Patra B, Ray S, Richter A, Majumder AL (2010) Enhanced salt tolerance of transgenic tobacco plants by co-expression of PcINO1 and McIMT1 is accompanied by increased level of *myo*-inositol and methylated inositol. Protoplasma 245(1–4):143–152

Paul MJ, Cockburn W (1989) Pinitol, a compatible solute in *Mesembryanthemum crystallinum* L. J Exp Bot 40:1093–1098

Paul M, Pellny T, Goddijn OJM (2001) Enhancing photosynthesis with sugar signals. Trends Plant Sci 6:197–200

Peng Z, Lu Q, Verma DPS (1996) Reciprocal regulation of D1-pyrroline-5-carboxylate synthetase and proline dehydrogenase genes control levels during and after osmotic stress in plants. Mol Gen Genet 253:334–341

Phillips JR, Oliver MJ, Bartels D (2002) Molecular genetics of desiccation and tolerant systems. In: Black M, Pritchard H (eds) Desiccation and survival in plants: drying without dying. CAB International, Wallingford, pp 319–341

Pilon-Smits EAH, Terry N, Sears T, Kim H, Zayed A, Hwang S, van Dun K, Voogd E, Verwoerd TC, Krutwagen RWHH, Goddijn OJM (1998) Trehalose-producing transgenic tobacco plants show improved growth performance under drought stress. J Plant Physiol 152:525–532

Piotrowski M, Janowitz T, Kneifel H (2003) Plant C-N Hydrolases and the identification of a plant N-Carbamoylputrescine amidohydrolase involved in polyamine biosynthesis. J Biol Chem 278(3):1708–1712

Prabhavathi V, Rajam MV (2007) Mannitol-accumulating transgenic eggplants exhibit enhanced resistance to fungal wilts. Plant Sci 173:50–54

Prabhavathi S, Yadav JS, Kumar PA, Rajam MV (2002) Abiotic stress tolerance in transgenic eggplant (*Solanum melongena* L.) by introduction of bacterial mannitol phosphodehydrogenase gene. Molecular Breed 9:137–147

Prasad KVSK, Saradhi PP (2004) Enhanced tolerance to photoinhibition in transgenic plants through targeting of glycinebetaine biosynthesis into the chloroplasts. Plant Sci 166:1197–1212

Quan R, Shang M, Zhang H, Zhao Y, Zhang J (2004) Improved chilling tolerance by transformation with betA gene for the enhancement of glycinebetaine synthesis in maize. Plant Sci 166:141–149

Rafart-Pedros A, Mac Leod MR, Ross HA, McRae D, Tiburcio AF, Davies HD, Taylor M (1999) Manipulation of the S-adenosylmethionine decarboxylase transcript level in potato tubers. Over-expression leads to an increase in tuber number and a change in tuber size distribution. Planta 209:153–160

Rahnama H, Vakilian H, Fahimi H, Ghareyazie B (2011) Enhanced salt stress tolerance in transgenic potato plants (*Solanum tuberosum* L.) expressing a bacterial *mtlD* gene. Acta Physiol Plant 33:1521–1532

Ramanjulu S, Bartels D (2002) Drought and dessication-induced modulation of gene expression in plants. Plant Cell Environ 25:141–151

Rathinasabapathi B (2000) Metabolic engineering for stress tolerance: installing osmoprotectant synthesis pathways. Ann Bot 86:709–716

Rathinasabapathi B, Burnet M, Russell BL, Gage DA, Liao PC, Nye GJ, Scott P, Golbeck JH, Hanson AD (1997) Choline monooxygenase, an unusual iron-sulphur enzyme catalyzing the first step of glycinebetaine synthesis in plants: prosthetic group characterization and cDNA cloning. Proc Natl Acad Sci U S A 94:3454–3458

Regierer B, Fernie AR, Springer F, Perez-Melis A, Leisse A, Koehl K, Willmitzer L, Giegenberger P, Kossmann J (2002) Starch content and yield increase as a result of altering adenylate pools in transgenic plants. Nat Biotechnol 20:1256–1260

Reshetnikov AS, Khmelenina VN, Mustakhimov II, Trotsenko YA (2011) Genes and enzymes of ectoine biosynthesis in halotolerant methanotrophs. Methods Enzymol 495:15–30

Rhodes D, Hanson AD (1993) Quaternary ammonium and tertiary sulfonium compounds in higher plants. Annu Rev Plant Physiol Plant Mol Biol 44:357–384

Ribaut JM, Hosington DA, Deitsch JA, Jiang C, Gonzalez-de-Leon D (1996) Identification of quantitative trait loci under drought conditions in tropical maize. 1. Flowering parameters and the anthesis-silking interval. Theor Appl Genet 92:905–914

Ribaut JM, Jiang C, Gonzalez-de-Leon D, Edmeades GO, Hosington DA (1997) Identification of quantitative trait loci under drought conditions in tropical maize, 2. Yield components and marker assisted selection strategies. Theor Appl Genet 94:887–896

Roessner U, Luedemann A, Brust D, Fiehn O, Linke T, Willmitzer L, Fernie AR (2001) Metabolic profiling allows comprehensive phenotyping of genetically or environmentally modified plant systems. Plant Cell 13:11–29

Romero C, Belles JM, Vaya JL, Serrano R, Culianez-Macia FA (1997) Expression of the yeast trehalose-6-phosphate synthase gene in transgenic tobacco plants: pleiotropic phenotypes include drought tolerance. Planta 201:293–297

Ronan S, Hirokazu T, Hideko N, Laszlo M, Chen THH, Norio M (2003) Enhanced formation of flowers in salt-stressed *Arabidopsis* after genetic engineering of the synthesis of glycinebetaine. Plant J 36:165–176

Rontein D, Basset G, Hanson AD (2002) Metabolic engineering of osmoprotectant accumulation in plants. Metab Eng 4:49–56

Roosens NH, Bitar FA, Loenders K, Angenon G, Jacobs M (2002) Over expression of ornthine-d-aminotransferase increases proline biosynthesis and confers osmotolerance in transgenic plants. Mol Breed 9:73–80

Roy M, Wu R (2001) Arginine decarboxylase transgene expression and analysis of environmental stress tolerance in transgenic rice. Plant Sci 160:869–875

Roy M, Wu R (2002) Overexpression of S-adenosylmethionine decarboxylase gene in rice increases polyamine level and enhances sodium chloride-stress tolerance. Plant Sci 163:987–992

Rubén A, Teresa A, Francisco M, Cristina B, Matthieu R, Csaba K, Pedro C, Antonio FT (2010) Polyamines: molecules with regulatory functions in plant abiotic stress tolerance. Planta 231:1237–1249

Ruidang Q, Shang M, Zhang H, Zhao Y, Zhang J (2004) Improved chilling tolerance by transformation with betA gene for the enhancement of glycinebetaine synthesis in maize. Plant Sci 166:141–149

Russell BL, Rathinasabapathi B, Hanson AD (1998) Osmotic stress induces expression of choline monooxygenase in sugar beet and amaranth. Plant Physiol 116:859–865

Sakamoto A, Murata N (2000) Genetic engineering of glycinebetaine synthesis in plants: current status and implications for enhancement of stress tolerance. J Exp Bot 51:81–88

Sakamoto A, Murata N (2001) The use of bacterial choline oxidase, a glycinebetaine-synthesizing enzyme, to create stress-resistant transgenic plants. Plant Physiol 125:180–188

Sakamoto A, Murata N (2002) The role of glycinebetaine in the protection of plants from stress: clues from transgenic plants. Plant Cell Environ 25:163–171

Sakamoto A, Murata A, Murata N (1998) Metabolic engineering of rice leading to biosynthesis of glycinebetaine and tolerance to salt and cold. Plant Mol Biol 38:1011–1019

Savouré A, Jaoua S, Hua XJ, Ardiles W, Van MM, Verbruggen N (1995) Isolation, characterization and chromosomal location of a gene encoding the delta 1-pyrroline-5-carboxylate synthetase in *Arabidopsis thaliana*. FEBS Lett 372:13–19

Sawahel WA, Hassan AH (2002) Generation of transgenic wheat plants producing high levels of the osmoprotectant proline. Biotechnol Lett 24:721–725

Serraj R, Sinclair TR (2002) Osmolyte accumulation: can it really help increase crop yield under drought conditions? Plant Cell Environ 25:333–341

Severin J, Wohlfarth A, Galinski EA (1992) The predominant role of recently discovered tetrahydropyrimidines for the osmoadaptation of halophilic eubacteria. J Gen Microbiol 138:1629–1638

Shen B, Jensen RG, Bohnert HJ (1997a) Increased resistance to oxidative stress in transgenic plants by targeting mannitol biosynthesis to chloroplasts. Plant Physiol 113:1177–1183

Shen B, Jensen RG, Bohnert HJ (1997b) Mannitol protects against oxidation by hydroxyl radicals. Plant Physiol 115:527–532

Shen B, Hohmann S, Jensen RG, Bohnert HJ (1999) Roles of sugar alcohols in osmotic stress adaptation. Replacement of glycerol by mannitol and sorbitol in yeast. Plant Physiol 121:45–52

Sheveleva EV, Chmara W, Bohnert HJ, Jensen RG (1997) Increased salt and drought tolerance by D-ononitol production in transgenic *Nicotiana tabaccum* L. Plant Physiol 115:1211–1219

Sheveleva EV, Marquez S, Chmara W, Zegeer A, Jensen RG, Bohnert HJ (1998) Sorbitol-6-phosphate dehydrogenase expression in transgenic tobacco: high amounts of sorbitol lead to necrotic lesions. Plant Physiol 117:831–839

Sickler CM, Edwards GE, Kiirats O, Gao Z, Loescher W (2007) Response of mannitol-producing *Arabidopsis thaliana* to abiotic stress. Funct Plant Biol 34:382–391

Singh TN, Aspinall D, Paleg LG (1972) Proline accumulation and varietal adaptability to drought in barley: a potential metabolic measure of drought resistance. Nat New Biol 236:188–190

Smart CC, Fleming AJ (1993) A plant gene with homology to D-myo-inositol-3 phosphate synthase is rapidly and spatially upregulated during an abscisic acid induced morphogenic response in *Spirodela polyrrhiza*. Plant J 4:279–293

Smart CC, Flores S (1997) Overexpression of d-myo-inositol-3-phosphate synthase leads to elevated levels of inositol in *Arabidopsis*. Plant Mol Biol 33:811–820

Smirnoff N (1998) Plant resistance to environmental stress. Curr Opin Biotechnol 9:214–219

Stephanopoulos G, Vallino JJ (1991) Network rigidity and metabolic engineering in metabolite overproduction. Science 252:1675–1681

Stevenson JM, Perera IY, Heilmann I, Persson S, Boss WF (2000) Inositol signaling and plant growth. Trends Plant Sci 5:252–258

Stoop JMH, Williamson JD, Pharr DM (1996) Mannitol metabolism in plants: a method for coping with stress. Trends Plant Sci 1:139–144

Su J, Wu R (2004) Stress-inducible synthesis of proline in transgenic rice confers faster growth under stress conditions than that with constitutive synthesis. Plant Sci 166:941–948

Su J, Hirji R, Zhang L, He C, Selvaraj G, Wu R (2006) Evaluation of the stress-inducible production of choline oxidase in transgenic rice as a strategy for producing the stress-protectant glycinebetaine. J Exp Bot 57:1129–1135

Szekely G, Abraham E, Cseplo A, Rigo G, Zsigmond L, Csiszar J, Ayaydin F, Strizhov N, Jasik J, Schmelzer E, Koncz C, Szabados L (2008) Duplicated *P5CS* genes of *Arabidopsis* play distinct roles in stress regulation and developmental control of proline biosynthesis. Plant J 53:11–28

Taji T, Ohsumi C, Iuchi S, Seki M, Kasuga M, Kobayashi M, Yamaguchi-Shinozaki K, Shinozaki K (2002) Important roles of drought and cold inducible genes for galactinol synthase in stress tolerance in *Arabidopsis thaliana*. Plant J 29:417–426

Tang W, Peng X, Newton RJ (2005) Enhanced tolerance to salt stress in transgenic loblolly pine simultaneously expressing two genes encoding mannitol-1-phosphate dehydrogenase and lucitol-6-phosphate dehydrogenase. Plant Physiol Biochem 43:139–146

Tarczynski MC, Jensen RG, Bohnert HJ (1993) Stress protection of transgenic tobacco by production of the osmolyte mannitol. Science 259:508–510

Thomas JC, Sepahi M, Arendall B, Bohnert HJ (1995) Enhancement of seed germination in high salinity by engineering mannitol expression in *Arabidopsis thaliana*. Plant Cell Environ 18:801–806

Trethewey RN (2004) Metabolite profiling as an aid to metabolic engineering in plants. Curr Opin Plant Biol 7:196–201

Turner NC, Shahal A, Berger JD, Chaturvedi SK, French RJ, Ludwig C, Mannur DM, Singh SJ, Yadava HS (2007) Osmotic adjustment in chickpea (*Cicer arietinum* L.) results in no yield benefit under terminal drought. J Exp Bot 58:187–194

Tyagi A, Sairam RK (2004) Physiology and molecular biology of salinity stress tolerance in plants. Curr Sci 86(3):407–420

Urano K, Yoshiba Y, Nanjo T, Igarashi A, Seki M, Sekiguchi F, Yamaguchi-Shinozaki K, Shinozaki K (2003) Characterization of *Arabidopsis* genes involved in biosynthesis of polyamines in abiotic stress responses and developmental stages. Plant Cell Environ 26:1917–1926

Van Laere A (1989) Trehalose, reserve and/or stress metabolite? FEMS Microbiol Rev 63(3):201–210

Verdoy D, Coba De La Pena T, Redondo FJ, Lucas MM, Pueyo JJ (2006) Transgenic *Medicago truncatula* plants that accumulate proline display nitrogen-fixing activity with enhanced tolerance to osmotic stress. Plant Cell Environ 29:1913–1923

Verma DPS (1999) Osmotic stress tolerance in plants: role of proline and sulfur metabolisms. In: Shinozaki K, Yamaguchi-Shinozaki K (eds) Molecular responses to cold, drought, heat and salt stress in higher plants. Austin, RG Landes, pp 153–168

Waie B, Rajam MV (2003) Effect of increased polyamine biosynthesis on stress responses in transgenic tobacco by introduction of human S-adenosylmethionine gene. Plant Sci 164:727–734

Watson MB, Emory KK, Piatak RM, Malmberg RL (1998) Arginine decarboxylase (polyamine synthesis) mutants of *Arabidopsis thaliana* exhibit altered root growth. Plant J 13:231–239

Webb KL, Burley JWA (1962) Sorbitol translocation in apple. Science 137:766

Wen XP, Ban Y, Inoue H, Matsuda N, Moriguchi T (2009) Aluminum tolerance in a spermidine synthase-overexpressing transgenic European pear is correlated with the enhanced level of spermidine via alleviating oxidative status. Environ Exp Bot 66:471–478

Wi SJ, Park KY (2002) Antisense expression of carnation cDNA encoding ACC synthase or ACC oxidase enhances polyamine content and abiotic stress tolerance in transgenic tobacco plants. Mol Cells 13:209–220

Wi SJ, Kim WT, Park KY (2006) Over expression of carnation S-adenosylmethionine decarboxylase gene generates a broad-spectrum tolerance to abiotic stresses in transgenic tobacco plants. Plant Cell Rep 25:1111–1121

Wiemken A (1990) Trehalose in yeast, stress protectant rather than reserve carbohydrate. J Gen Microbiol 58:209–217

Wu LQ, Fan ZM, Guo L, Li YQ, Zhang WJ, Qu LJ, Chen ZL (2003) Over-expression of an *Arabidopsis* delta-OAT gene enhances salt and drought tolerance in transgenic rice. Chin Sci Bull 48:2594–2600

Wyn Jones RG, Storeys R (1978) Salt stress and comparative physiology in the *Gramineaell.* Glycine betaine and proline accumulation in two salt and water stressed barley cultivar. Aust J Plant Physiol 5:817–829

Yamada M, Morishita H, Urano K, Shiozaki N, Yamaguchi-Shinozaki K, Shinozaki K, Yoshiba Y (2005) Effects of free proline accumulation in petunias under drought stress. J Exp Bot 56:1975–1981

Yamaguchi K, Takahashi Y, Berberich T, Imai A, Takahashi T, Michael AJ, Kusano T (2007) A protective role for the polyamine spermine against drought stress in *Arabidopsis*. Biochem Biophys Res Commun 352:486–490

Yancey PH (1994) Compatible and counteracting solutes. In: Strange K (ed) Cellular and molecular physiology of cell volume regulation. CRC, Boca Raton, pp 81–109

Yancey PH (2005) Organic osmolytes as compatible, metabolic and counteracting cytoprotectants in high osmolarity and other stresses. J Exp Biol 208:2819–2830

Yancey PH, Clark ME, Hand SC, Bowlus RD, Somero GN (1982) Living with water stress: evolution of osmolyte systems. Science 217:1214–1222

Yang X, Liang Z, Lu C (2005) Genetic engineering of the biosynthesis of glycinebetaine enhances photosynthesis against high temperature stress in transgenic tobacco plants. Plant Physiol 138:2299–2309

Yeo AR (1998) Molecular biology of salt tolerance in the context of whole-plant physiology. J Exp Bot 49:915–929

Yeo AR, Flowers TJ (1989) Selection for physiological characters—examples from breeding for salt tolerance. In: Flowers TJ, Jackson MB, Jones HG (eds) Plants under stress: biochemistry, physiology and ecology and their application to plant improvement. Cambridge University Press, Cambridge, pp 217–234

Yeo ET, Kwon HB, Han SE, Lee JT, Ryu JC, Byun MO (2000) Genetic engineering of drought-resistant potato plants by introduction of the trehalose-6-phosphate synthase (TPS1) gene from *Saccharomyces cerevisiae*. Mol Cells 10:263–268

Yi-Guo S, Bao-Xing D, Wan-Ke Z, Jin-Song Z, Shou-Yi C (2002) AhCMO, regulated by stresses in *Atriplex hortensis*, can improve drought tolerance in transgenic tobacco. Theor Appl Genet 105:815–821

Yoshiba Y, Kiyosue T, Katagiri T, Ueda H, Mizoguchi T, Yamaguchi-Shinozaki K, Wada K, Harada Y, Shinozaki K (1995) Correlation between the induction of a gene for delta 1-pyrroline-5-carboxylate synthetase and the accumulation of proline in *Arabidopsis thaliana* under osmotic stress. Plant J 7:751–760

Zapata PJ, Serrano M, Pretel MT, Amoros A, Botella MA (2004) Polyamines and ethylene changes during germination of different plant species under salinity. Plant Sci 167:781–788

Zhifang G, Loescher WH (2003) Expression of a celery mannose-6-phosphate reductase in *Arabidopsis thaliana* enhances salt tolerance and induces biosynthesis of both mannitol and a glucosyl-mannitol dimer. Plant Cell Environ 26:275–283

Zhu JK (2001) Plant salt tolerance. Trends Plant Sci 6:66–71

Zhu B, Su J, Chang M, Verma DPS, Fan YL, Wu R (1998) Overexpression of a D1-pyrroline-5-carboxylate synthetase gene and analysis of tolerance to water-and salt-stress in transgenic rice. Plant Sci 139:41–48

Chapter 10
Epigenetic Modifications in Plants Under Adverse Conditions: Agricultural Applications

Alex Boyko and Igor Kovalchuk

1 Introduction

Global climate change and a continuous decrease in the amount of agricultural land together with the growing demand for food production present a significant challenge to plant breeding and crop improvement programs. Unlike the classical breading approaches, plant transgenesis permits the fast improvement of economically important crops. Since plant transgenesis often involves the introduction of foreign DNA into the plant genome, its safety is frequently debated. Consequently, the commercial application of transgenic crops is still severely restricted in many countries.

In very general terms, the ability to improve plant fitness and increase crop yield often depends on altering genome transcription to mediate specific agricultural characteristics or traits in the field. While conventional breeding utilizes random mutagenesis and breeding-mediated transfer of desired traits from related species, plant transgenesis relies on relatively site-specific and controlled gene integration events. Furthermore, it allows the introduction of genes from distant and even non-plant species. The use of both classical plant breeding and genetic engineering help achieve the desired transcriptional output of the genome by irreversibly changing the genetic composition of the plant.

During the past decades, the attention was drawn toward elucidating the mechanisms that could allow genetically identical cells or even whole organisms to achieve and maintain different terminal phenotypes. This was accomplished by using different non-genetic or epigenetic determinants that could modify gene expression

A. Boyko
Institute of Plant Biology, University of Zürich, Zollikerstrasse 107, 8008
Zürich, Switzerland

I. Kovalchuk (✉)
Department of Biological Sciences, University of Lethbridge, Lethbridge,
AB, Canada T1K 3M4
e-mail: igor.kovalchuk@uleth.ca

N. Tuteja and S. Singh Gill (eds.), *Plant Acclimation to Environmental Stress*,
DOI 10.1007/978-1-4614-5001-6_10,

heritably (mitotically and/or meiotically) and reversibly (without changing the gene sequence encoded in DNA). These epigenetic determinants/marks are enzyme-mediated chemical modifications of DNA and DNA-associated proteins. They include DNA methylation, histone modifications, nucleosome positioning, and small (sm) RNAs. Epigenetic marks modify the properties of chromatin and change gene transcriptional states on the scale from the entire genome to a single specific gene. These marks allow for greater genome plasticity which results in better adaptation of plants to changing environmental conditions. Understanding a role of epigenetic modifications in plant responses to stress and adaptation and uncovering the mechanisms which mediate and guide the deposition of these modifications throughout the genome may provide us with new insights into possible strategies to improve economically important crops. It would be of vital importance to incorporate newly acquired knowledge of epigenetic mechanisms mediating stress responses and plant adaptation into an already existing system of genetic-based knowledge and tools which would provide crop improvement.

2 Epigenetic Modifications and Gene Expression Control

In order to be considered as an epigenetic signal, any molecular signal should satisfy three main criteria, including the presence of a mechanism for its propagation, evidence of transmission, and the effect on gene expression (Bonasio et al. 2010). Depending on how they act, epigenetic signals can be further divided into the *cis*- (e.g., DNA methylation and histone modifications) and *trans*-acting (e.g., smRNAs) groups. Perhaps, DNA methylation is the best and most studied example of the molecular signal which satisfies all three epigenetic criteria. The ample experimental evidence supports its involvement in gene expression control, mitotic and transgenerational epigenetic inheritance. DNA methylation is maintained by a highly complex network of molecular mechanisms which are very sensitive to various developmental and environmental cues. On the contrary, the current knowledge about histone modifications and smRNAs is still far from complete. It is still debatable whether the posttranslational modification of histone tails and regulation of gene expression by smRNAs are true epigenetic marks. Nevertheless, in this review, we will consider and discuss both these marks as the epigenetic ones (Table 10.1).

2.1 DNA Methylation

DNA methylation is a *cis*-acting repressive epigenetic signal. It involves a covalent modification of cytosine by a methyl group for producing 5-methyl-cytosine. Cytosine methylation can occur in several nucleotide sequence contexts, including symmetric CG and CHG and asymmetric CHH (where H is C, T, or A) sites. Methylation of CG, CHG, and CHH occurs with the frequency of 24, 6.7, and 1.7%

Table 10.1 Selected examples of the epigenetic pathways that control agriculturally important characteristics and traits

Epigenetic pathway	Effector protein	Transcription	Phenotype	Reference
DNA Hypermethylation	DNA methyltransferases (MET1; CMT3; DRM2)	Repression	A two-fold increase in CHG methylation under high salinity conditions was associated with switching from C3-photosynthesis to CAM metabolism in *M. crystallinum* plants	Dyachenko et al. (2006)
			An increase in DNA methylation in the genomic regions involved in pathogen defense in response to virus infection in tomato	Mason et al. (2008)
			An age-dependent increase in methylation-mediated resistance to blight pathogen *X. oryzae* in rice	Sha et al. (2005)
			Insertion and methylation of the *Mutator*-like transposon in the first intron of the *FLC* gene established a phenotype of early flowering in Landsberg *erecta* (*Ler*) accession plants	Liu et al. (2004)
DNA Hypomethylation	Passive or active DNA demethylation by DNA glycosylases (ROS1; DME; DML2; DML3)	Activation	Temperature-induced demethylation activated *Tam3* transposition from the *nivea* gene promoter resulting in the red-flower phenotype	Hashida et al. (2003, 2006)
			Transposon immobilization may result in the formation of a new stress-inducible allele (e.g., an acquired transcriptional respon siveness to high temperature at the *Onsen* neo-insertion sites)	Ito et al. (2011)
			Stably-inherited DNA hypomethylation at several MS-AFLP marker loci correlated with low respiration rates, high EUE and improved seed yield under normal physiological and drought conditions in canola plants	Hauben et al. (2009)
			Demethylation of the *Xa21G* resistance gene promoter by a pulse treatment in rice seeds with 5-azadeoxycytidine resulted in the acquisition of a stably-inherited trait of disease resistance	Akimoto et al. (2009)
			Loci-specific hypomethylation in the globally hypermethylated genome mediated the increased expression of DNA transcription and repair genes in the progeny of plants exposed to salt; Enhanced tolerance to high salt concentrations and MMS acquired by the progeny of plants exposed to salt	Boyko et al. (2010a; 2010b)

(continued)

Table 10.1 (continued)

Epigenetic pathway	Effector protein	Transcription	Phenotype	Reference
Histone deacetylation	HDA6, Histone deacetylase	Repression	Activity of HDA6 is induced by jasmonic acid and ethylene	Zhou et al. (2005)
	HDA19, Histone deacetylase		The increased expression of the *ERF1* and *PR* genes and enhanced resistance to fungal pathogen in transgenic plants that overex press *HDA19*	Zhou et al. (2005)
			Down-regulation of light-responsive and photomorphogenesis genes	Benhamed et al. (2006)
	AtHD2C, Histone deacetylase		Increased tolerance to salinity and drought conditions in transgenic plants that overexpress *AtHD2C*	Sridha and Wu (2006)
	HOS15, WD-40 domain protein; interacts with histone H4		Freezing stress-hypersensitive phenotype in *hos15* mutants	Zhu et al. (2007b)
Histone acetylation	GCN5, Histone acetyltransferase protein	Activation	Up-regulation of light-responsive and photomorphogenesis genes	Benhamed et al. (2006)
			Activation of cold-responsive gene transcription	Stockinger et al. (2001)
	HAF2, Histone acetyltransferase protein		Up-regulation of light-responsive and photomorphogenesis genes	Benhamed et al. (2006)
	HAC1, Histone acetyltransferase		Transcriptional up-regulation of the heat-shock gene *HSP17*	Bharti et al. (2004)

Histone methylation	*Polycomb* group (*PcG*) proteins, Histone methyltransferase	Repression	Transcriptional up-regulation of cold-responsive genes *COR15A* and *ATGOLS3* mediated by a decrease in H3K27me3	Kwon et al. (2009)
	MSI1, WD-40 domain protein; subunit of PRC2 complex		Drought stress tolerance in *msi1-cs* transgenic plants	Alexandre et al. (2009)
	Trithorax group (trxG) proteins, Histone methyltransferase	Activation	Transcriptional activation of drought stress-inducible genes mediated by the enrichment in H3K4me3	Kim et al. (2008)
			Hypoxia stress tolerance in the submerged rice seedlings via the enhanced expression of the *ADH1* and *PDC1* genes mediated by the enrichment in H3K4me3	Tsuji et al. (2006)
Histone variants	HIS1-S, Histone 1 variant	N/A	Alleviates negative effects of drought stress by reducing transpira tion rates when present in chromatin of wilted tomato leaves	Scippa et al. (2004)
	HIS1-3, Histone 1 variant		Improved drought tolerance and ABA-hypersensitivity phenotypes in transgenic plants that overexpress an active form of AREB1, a positive regulator of *HIS1-3*	Fujita et al. (2005)
	H2A.Z, Histone H2A variant	Depends on the inter-acting transcrip-tion factors	Enriched within transcription start sites of high temperature stress-responsive genes	Kumar et al. (2010)
			Down-regulation of the expression of phosphate-starvation response genes	Smith et al. (2010)
			Down-regulation of genes mediating systemic acquired resistance in *Arabidopsis*	March-Díaz et al. (2008)
			Up-regulated *FLC* transcription and repression of flowering upon the enrichment in the H2A.Z variant at the *FLC* gene loci	Deal et al. (2007)

of all genome-wide sites available, respectively. The DNA methylation landscapes are largely responsible for the transcriptional genome output and can direct the deposition of other epigenetic marks and chromatin remodeling (Zilberman et al. 2007). While CG methylation represents a very stable epigenetic mark that can be faithfully preserved throughout multiple rounds of cell division and transmitted to the next generation via the gametes (Mathieu et al. 2007), asymmetric CHH methylation is quickly lost during a DNA replication process, and its maintenance requires persistent *de novo* methylation activity (Huettel et al. 2007).

Maintaining proper DNA methylation is critical for regulating gene transcription and transposon silencing during plant development and response to stress. Overall, approximately 20% of *Arabidopsis* genome is methylated, with transposons and DNA repeats comprising the largest fraction of methylated DNA sequences. Whereas transposons are heavily methylated throughout their whole sequence, transcriptional gene repression is usually associated with DNA methylation localized in the gene promoter regions (Zhang et al. 2006). Furthermore, methylation of transcribed regions does not result in gene silencing, and about one-third of *Arabidopsis* genes contain methylated cytosines in their coding regions (Zhang et al. 2006; Cokus et al. 2008). Methylation of transcribed regions occurs exclusively at CG sites and may be involved in fine tuning of gene transcription (Zilberman et al. 2007). Genes methylated within the coding sequence display moderate expression levels and are less likely to have tissue-specific expression, as compared to unmethylated genes (Zhang et al. 2006; Vaughn et al. 2007; Zilberman et al. 2007).

In *Arabidopsis*, CG methylation is maintained throughout the DNA replication process by the DNA METHYLTRANSFERASE 1 (MET1) enzyme (Vongs et al. 1993) and its cofactor, VARIATION IN METHYLATION 1 (VIM1) (Woo et al. 2007). VIM1 mediates the recognition of hemimethylated DNA sequences at the replication foci and insures the correct transfer of epigenetic information to the newly synthesized DNA strand. Therefore, CG methylation is highly resistant to reprogramming and displays robust transgenerational inheritance (Mathieu et al. 2007). The maintenance of DNA methylation at CG sites also requires the activity of a chromatin-remodeling factor decreased DNA methylation 1 (DDM1). DDM1 may facilitate localization of methyl-CG-binding domain proteins (MBDs) in specific nuclear domains (Zemach et al. 2005), thus promoting heterochromatin formation and gene silencing (Ben-Porath and Cedar 2001; Zemach and Grafi 2007). Mutations in DDM1 result in a severe genome-wide loss of DNA methylation (Vongs et al. 1993; Jeddeloh et al. 1999) and a decrease in H3K9 methylation (Gendrel et al. 2002), suggesting yet another cross talk between DNA methylation, histone and chromatin modifications.

CHG methylation is maintained by a feed-forward loop formed by a histone methyltransferase, KRYPTONITE (KYP) (Jackson et al. 2002) and a plant-specific DNA methyltransferase, cromomethylase 3 (CMT3) (Lindroth et al. 2001). The chromodomain of CMT3 binds directly to dimethylated histone H3K9me2 and methylates DNA at the CHG sites (Lindroth et al. 2004). Similar to the recruitment of CMT3 to the H3K9me2 sites, the SRA domain of KYP interacts with the methylated CHG sites leading to H3K9 dimethylation at these loci (Johnson et al. 2007).

Additionally, two other H3K9 histone methyltransferases, SU(VAR)3-9 homolog 5 (SUVH5) and SUVH6, may contribute to the maintenance of CHG methylation (Ebbs et al. 2005; Ebbs and Bender 2006). The presence of a self-reinforcing DNA-histone methylation loop can explain a strong genome-wide correlation between the distribution of H3K9me2 and CHG methylation marks. Likewise, the recruitment of KYP to the methylated CG sites via its SRA domain followed by histone H3K9 dimethylation (Johnson et al. 2007) can facilitate the recruitment of CMT3 to DNA. This lends further supports to the idea that CG methylation can be used as a platform for depositing other epigenetic marks.

In contrast to symmetric DNA methylation, asymmetric CHH methylation is quickly lost during DNA replication. Hence, maintaining DNA methylation at the CHH sites requires constant *de novo* DNA methylation activity. This is achieved by the domains rearranged methyltransferase 2 (DRM2) *de novo* methyltransferase (Cao and Jacobsen 2002). DRM2 is actively targeted to the CHH methylation sites using the RNA-dependent DNA methylation (RdDM) pathway (Matzke et al. 2009). The RdDM pathway utilizes 24-nt-long short interfering (si)RNAs which are generated by a dicer-like 3 (DCL3) protein and delivered to their target sites in chromatin by the argonaute 4 (AGO4) effector complex to guide methylation of homologous DNA sequences. DRM2 activity may also depend on surrounding epigenetic marks such as DNA methylation and histone modifications. Two histone methyltransferases, SUVH9 and SUVH2, can preferentially bind the methylated CHH and CG sites, respectively, and help recruit DRM2 to the DNA loci targeted by the RdDM pathway (Johnson et al. 2008). Additionally, at some genomic loci, CMT3 may act redundantly with DRM2 to control CHH methylation (Cao et al. 2003).

Whereas asymmetric DNA methylation is quickly lost in the absence of a siRNA trigger, the removal of symmetric DNA methylation may require mechanisms of active DNA demethylation. Also, DNA methylation can be passively lost in the absence of active maintenance of symmetric methylation during DNA replication. A passive demethylation mechanism, however, is not suitable for nondividing cells; it is also not applicable when fast modification of the DNA methylation landscape is required in response to environmental stimuli. Active DNA methylation in *Arabidopsis* is achieved by using specific DNA glycosylases that remove methylated cytosines from DNA (Zhu 2009). Four DNA glycosylases have been described in *Arabidopsis,* including repressor of silencing 1 (ROS1) (Gong et al. 2002), DEMETER (DME) (Choi et al. 2002), demeter-like 2 (DML2), and DML3 (Ortega-Galisteo et al. 2008). DME plays a critical role during gametogenesis by mediating a genome-wide decrease in CG methylation and establishing gene imprinting in endosperm (Huh et al. 2008). In contrast, ROS1, DME2 and DME3 DNA glycosylases are active in vegetative tissues in which they prevent transcriptional gene silencing and counteract DNA methylation introduced by the RdDM pathway (Kapoor et al. 2005; Penterman et al. 2007a, b; Zhu et al. 2007a). These enzymes insure a tight transcriptional control over transposons and other normally silenced loci, and they act on the genes residing in heterochromatin or close to a heterochromatic environment (Penterman et al. 2007b; Zhu et al. 2007a). The recent identification of ROS3 protein that contains an RNA-binding motif and is a member of the ROS1 DNA demethylation pathway indicates sequence-specific targeting of DNA demethylating enzymes (Zheng et al. 2008). ROS3 binds smRNAs in vitro and in vivo and co-localizes with ROS1 in discrete foci dispersed throughout the plant nucleus. It can be hypothesized that ROS1 can be targeted to the specific genome loci using smRNAs bound to ROS3. Therefore, ROS3 can be considered to be an important functional link between smRNA biogenesis and DNA demethylation pathways.

2.2 Histone Modifications and Chromatin Remodeling

Histone modifications form a layer of epigenetic information in the plant genome which is highly interactive and responsive to the developmental and environmental cues. The high complexity of information carried by this epigenetic mark results from a large number of possible histone modifications combined with possible combinatorial effects when new epigenetic information can arise by combining certain histone marks together. The fast reversibility of histone modifications and multiple cross talks between histone-modifying pathways, DNA methylation and chromatin remodeling make histone modifications an ideal choice for regulating genome transcription under changeable growth conditions.

There are several distinct molecular levels at which epigenetic information can be recorded using histones. As histones form nucleosomes, the exchange of canonical histones with specialized histone variants can alter the transcriptional properties of chromatin (Talbert and Henikoff 2010). Similarly, ATP-dependent chromatin remodelers can change nucleosome position by moving the histone core with respect to DNA sequence. This may allow an easier access of the general transcriptional machinery to the targeted gene (Rando and Ahmad 2007). Indeed, a number of constitutively expressed genes contain the nucleosome-depleted regions in their promoters (Zhang et al. 2007). Next, numerous posttranslational modifications in the N-terminal tails of histones alter their physical properties, and thus change histone–DNA and protein–protein interactions in chromatin (Berger 2007). The most common histone tail modifications include acetylation, methylation, phosphorylation, ubiquitination, biotinylation, and sumoylation. Whereas histone acetylation acts directly by loosening the histone association with DNA, which leads to transcriptional activation, histone methylation helps recruit other effector proteins and their complexes and thus can have either repression or activation properties. Finally, the type of modified amino acids (e.g., lysine (K) or arginine (R)) and the degree of modification (e.g., mono-, di- or trimethylation) further specify the effect on transcription and define chromatin localisation of a given histone mark. For example, while the repressive marks H3K9me and H3K9me2 localize to heterochromatin, transcriptional repression of euchromatin is mediated by H3K9me3 (Berger 2007).

To date, the effects of histone H3 acetylation and methylation on gene expression are probably the best-studied. General acetylation of lysines in histone H3 is mediated by the histone acetyl transferase (HAT) enzymes; it also promotes up-regulated transcription. Histone acetylation is usually associated with gene promoters and the 5′-end of the transcribed sequences. Conversely, deacetylation of lysines in histone H3 by the histone deacetylase (HDAC) enzymes leads to transcriptional repression of the targeted genes (Chen and Tian 2007).

Trimethylated histones H3K4me3 and H3K27me3 are associated with transcriptional activation and silencing, respectively. Trimethylation of H3K4me3 is mediated by the trithorax group (trxG) protein complexes at the 5' end of the actively transcribed genes (Santos-Rosa et al. 2002). The presence of this mark can recruit histone acetyltransferases and chromatin remodeling complexes such as NURF by mediating a decrease in the nucleosome density at the target loci (Ruthenburg et al. 2007). The positive effect of H3K4me3 on transcription from the target genes can be reversed by the H3K4-specific histone demethylase enzyme (Jiang et al. 2007). Similarly, the dimethylated histone H3K4me2 is associated with active genes in euchromatin, and it is depleted in transposons (Lippman et al. 2004). On the contrary, trimethylation of H3K27me3 mediated by the *Polycomb* Repressive 2 (PRC2) complex leads to transcriptional repression of developmentally important genes and transcription factors (Margueron and Reinberg 2011). Interestingly, in contrast to animals, plants contain at least three distinct PRC2 complexes that are active during different stages of plant development and may control different target genes (Pien and Grossniklaus 2007). Overall, over 18% of *Arabidopsis* genes contain H3K27me3 in their promoters (Zhang et al. 2007) where it is believed to mediate the tissue-specific gene expression patterns. H3K27me3 serves as a binding site for the like heterochromatin protein 1 (LHP1) protein that further reinforces transcriptional repression in euchromatin (Libault et al. 2005).

The presence of H3K27me3 and H3K9me3 in the regions of euchromatin is mutually exclusive, which indicates that different repressive mechanisms may be involved in regulating the expression of different groups of genes (Turck et al. 2007). H3K9me2 is found only in heterochromatin where it exclusively overlaps with transposons and DNA repeats (Fuchs et al. 2006). The methylated histone H3K9me2 is an important link between different epigenetic pathways. It facilitates the recruitment of CMT3 to DNA. Additionally, histone H3K9 and DNA methylation at heterochromatic loci are maintained by DDM1 chromatin remodeling factor (Lippman et al. 2004).

Transcriptional properties of chromatin can be also modified by incorporating histone H2A and H3 variants (Talbert and Henikoff 2010). There are two histone H2A variants found in plants. The H2A.Z variant occurs throughout the plant genome, and it is mainly found in nucleosomes close to the transcriptional start sites. It mediates transcription regulation and the formation of heterochromatin boundaries possibly by preventing DNA methylation (Raisner and Madhani 2006; Zilberman et al. 2008). The function of H2AX in pants is less clear; in animals, it is phosphorylated near the sites of DNA strand breaks and is involved in DNA damage repair (van Attikum and Gasser 2009). Similarly, there are two histone H3 variants found in plants. The CenH3 variant is incorporated at centromeres where it is involved in chromosome segregation (Zhang et al. 2008). The incorporation of the H3.3 variant occurs within chromatin regions where active nucleosome remodeling takes place.

H3.3 is deposited into promoters, gene regulatory elements, and transcribed genomic regions (Mito et al. 2005; Deal et al. 2010).

2.3 Small RNAs

In the recent years, an ample progress has been made in understanding the role of smRNAs in establishing and maintaining epigenetic landscapes of genomes. smRNAs are *trans*-acting epigenetic signals that can reversibly and in a sequence-specific manner modify gene expression at transcriptional (the RdDM pathway) and posttranslational (micro (mi) RNAs) levels (Carthew and Sontheimer 2009; Malone and Hannon 2009; Voinnet 2009). The high sensitivity of smRNAs to developmental and environmental cues enables them to orchestrate multifaceted transcriptional responses. Being an important part of interactive epigenetic network, smRNAs can influence DNA methylation, histone modifications, and help recruit chromatin modifiers to their genome targets (Saze 2008; Hammoud et al. 2009; Bourc'his and Voinnet 2010). smRNAs serve as mobile signals during plant development (Chitwood et al. 2009; Dunoyer et al. 2010; Molnar et al. 2010). They also can mediate the inheritance of epigenetic information during mitosis and meiosis via maternal/paternal pools of smRNAs (Saze 2008; Grant-Downton et al. 2009; Mosher et al. 2009; Slotkin et al. 2009). It is plausible to suggest that smRNAs can serve as a molecular basis for the transgenerational inheritance of environmental memories in plants (Boyko and Kovalchuk 2011a, b; Hauser et al. 2011).

Four main groups of smRNA can be distinguished based on their biogenesis pathways, structure and biological functions (Vazquez 2006). These include miRNA, trans-acting short interfering RNAs (ta-siRNA), natural-antisense siRNA (nat-siRNA) and repeat-associated siRNA (ra-siRNA). While the first three groups function predominantly at the posttranscriptional level through messenger RNA degradation or translation inhibition, ra-siRNAs mediate transcriptional gene silencing via the RdDM pathway by directing *de novo* DNA methylation to heterochromatin and genomic loci which contain transposons and repetitive sequences. The latter mechanism is responsible for directing approximately 30% of cytosine methylation in the *Arabidopsis* genome (Cokus et al. 2008; Lister et al. 2008). smRNAs are generated by dicer-like proteins, they are delivered to their RNA/DNA targets by ARGONAUTE proteins that can direct posttranslational and transcriptional silencing of the targeted genes.

The siRNA biogenesis machinery includes two DNA polymerases, Pol IV and Pol V, with Pol V acting downstream of Pol IV. Pol IV acts in a complex with a SNF2-like chromatin remodeling factor CLASSY 1 (CLSY) and an RNA-dependent RNA polymerase 2 (RDR2) to copy single-stranded (ss)RNAs transcribed by Pol IV into double-stranded (ds) RNAs. These dsRNAs are later cleaved by DCL3 to produce 24-nt-long siRNAs that are recruited by an effector complex containing either AGO4 or AGO6 to guide chromatin modifications to homologous DNA sequences. The amplification and reinforcement of siRNA production together with *de novo* DNA methylation at the siRNA-targeted site is mediated by Pol V and the DRD complex. The DDR complex is composed of a SNF2-like chromatin remodeling factor defective in RNA-directed DNA methylation 1 (DRD1), a structural-maintenance-of-chromosomes hinge domain-containing protein defective in meristem silencing 3 (DMS3), and a novel protein with single-stranded methyl-DNA-binding activities, RNA-directed DNA methylation 1 (RDM1). At the DNA loci targeted by siRNAs, both the transcripts produced by Pol V from genomic sequences or perhaps, the transcribed DNA sequence itself may interact with the AGO4-bound siRNA complex to facilitate the recruitment of the *de novo* DNA methylation machinery and histone-modifying complexes to chromatin (Simon and Meyers 2011 and references within).

It is not surprising that there is a strong correlation between siRNAs and DNA methylation. The balanced activity of the siRNA-directed DNA methylation and ROS1 DNA demethylation pathways may be required to reversibly modulate gene expression in nondividing cells (Penterman et al. 2007a; Lister et al. 2008). The DNA demethylation pathways are necessary to maintain a proper composition of smRNA populations. The *ros1 dml2 dml3* triple mutant displays the altered composition of smRNA populations due to *de novo* methylation of previously active DNA regions located in the proximity of the ta-siRNA generating loci (Lister et al. 2008). The RdDM pathway is also involved in control of chromatin organization since *pol V* and *drd1* mutants exhibit decondensation of pericentromeric repeats and depletion of the repressive mark, H3K9me2, at centromeres (Pontes et al. 2009). Altogether, this supports the role of siRNAs as a core element of the signalling network that mediates epigenetic modifications in plant cells.

3 The Role of Epigenetic Networks in Plant Stress Responses

A complex interplay between DNA methylation, histone modification, chromatin remodeling and smRNAs biogenesis pathways enables the fine-tuned adjustment of gene expression in plant cells. It allows the integration of input from various developmental and environmental cues into the system of gene transcription control, which permits an accurate orchestration of complex developmental events such as the induction of flowering in response to season changes (Dennis and Peacock 2007). Moreover, the high sensitivity of epigenetic modifications provides a fast genome-wide adjustment of gene expression in response to rapidly changing growth conditions. The transient nature of epigenetic modifications can be advantageous when dealing with frequent and short-lasting environmental stresses. As some stresses may persist at the given location for the duration of several plant generations, the transmission of epigenetic information (i.e., environmental memories) and the gene expression patterns associated with it to the progeny can be an important strategy for plant survival and adaptation. The robustness of the process of CG methylation makes it an ideal mechanism for the stable transgenerational maintenance of epigenetic traits. In contrast to genetic alleles where the acquisition of new gene expression properties is achieved by permanent changes in DNA sequences, no such changes are needed for epialleles. The newly acquired information is encoded using reversible yet stable epigenetic marks such as DNA methylation. This allows for the greater plasticity of gene expression and can offer a superior and adaptive advantage to sessile organisms like plants. In fact, many of the known plant epialleles control the important environmentally regulated developmental traits like flowering time (e.g., *FLC* (Sheldon et al. 2000) and *FWA* (Soppe et al. 2000)) and flower morphology (e.g., *LCYC*) (Cubas et al. 1999).

3.1 AChanges in the DNA Methylation Landscape in Response to Stress

DNA methylation controls gene expression by restricting an access of the general transcriptional machinery to the target genes. DNA methylation serves as a substrate for the recruitment of higher order protein complexes involved in histone and chromatin modification. Altogether, it makes chromatin less accessible to the processes like transcription, transposition, DNA damage and DNA repair. Generally, responses to stress involve numerous changes in plant transcriptome that may require active modification of the existing DNA methylation landscape.

Up-regulation of stress-specific genes frequently correlates with a decrease in DNA methylation. Two independent studies demonstrated that in tobacco, the accumulation of specific abiotic and biotic stress-induced transcripts was associated with an active demethylation process (Wada et al. 2004; Choi and Sano 2007). Similar epigenetic responses to stress were documented for hemp and clover plants subjected to heavy metal stress when DNA hypomethylation at several marker loci was observed (Aina et al. 2004). Cold stress was reported to reduce DNA replication and trigger demethylation in DNA of the nucleosome core of the *ZmMI1* gene in root tissues of maize seedlings (Steward et al. 2002). Also, Choi and Sano (2007) reported stress-induced DNA demethylation of the *NtGPDL* gene under cold, salt and aluminum stress. In these two studies, DNA demethylation resulted in the induced expression of Z*mMI1* and *NtGPDL* genes. Infection of *Arabidopsis* plants with *Pseudomonas syringae* led to DNA hypomethylation at centromeric repeats (Pavet et al. 2006). Using tomato plants infected with a virus, Mason et al. (2008) demonstrated that pathogen attacks triggered changes in DNA methylation at several marker loci. The majority of detected polymorphisms were associated with the genomic regions involved in defense and stress responses (Mason et al. 2008).

Stress-induced changes in the DNA methylation landscape may include both DNA hyper- and hypomethylation. While some genomic regions can become hypomethylated, others may display a significant increase in the presence of methyl-cytosine. Consistently, exposure of apomictic dandelion populations to various abiotic and biotic stresses triggered significant changes in the methylation pattern in exposed plants (Verhoeven et al. 2010). Exposure of pea plants to water-deficit stress resulted in hypermethylation as a response (Labra et al. 2002).

Similarly, transient hypermethylation at heterochromatic loci was observed in tobacco cell-suspension culture under osmotic stress (Kovarik et al. 1997). The possible adaptive nature of stress-induced changes in DNA methylation is supported by several independent studies. Using *M. crystallinum* plants exposed to high salinity conditions, Dyachenko et al. (2006) demonstrated that a twofold increase in CHG methylation was associated with switching from C3-photosynthesis to CAM metabolism. An age-dependent increase in methylation was sufficient to mediate resistance to the blight pathogen *X. oryzae* in rice (Sha et al. 2005).

DNA methylation plays a key role in restricting transposon movement. The activation of transposons in response to stress is a common phenomenon in plants. The stress-mediated transposon induction was previously reported for *Tos17* (rice) (Hirochika et al. 1996), *Tto1* (tobacco) (Takeda et al. 1999), *Tnt1* (tobacco) (Beguiristain et al. 2001), and *BARE-1* retrotransposons (barley) (Kalendar et al. 2000); and it was often associated with the decreased DNA methylation at the transposon loci. The temperature-dependent activation of the *Tam3* transposon in *Antirrhinum majus* plants is a prominent example of how stress-dependent transposon activation can affect plant phenotype. Shifting *A. majus* plants to low temperature conditions decreased methylation and increased the excision rate of the *Tam3* transposon (Hashida et al. 2003, 2006). Since *Tam3* is located in the promoter region of the *nivea* gene controlling flower pigmentation, high levels of methylation in the transposon result in transcriptional silencing of the *nivea* gene and white flowers. Demethylation and transposition of *Tam3* led to the activation of the *nivea* gene and appearance of red flowers (Hashida et al. 2003, 2006). This example suggests that the presence of transposons may influence transcription of the neighboring genes.

The genome-wide distribution and high sensitivity of transposons to stress add another degree of complexity to transcriptional regulatory networks. The high mobility of activated transposons offers a plausible mechanism for the diversification of genomic sequences and formation of new stress-responsive alleles. Indeed, Ito et al. (2011) studied the *Onsen* transposon activated by elevated temperature and showed that genes neighboring the *Onsen* neo-insertion sites could acquire the transcriptional response to high temperature. Similarly, transcriptional silencing via the siRNA pathway triggered by recently relocated transposons can result in the appearance of new phenotypes. *Arabidopsis* accession Landsberg *erecta* (L*er*) plants display early flowering due to the insertion of the *Mutator*-like transposon into the first intron of the *FLC* gene which transcription is required for repression of flowering (Liu et al. 2004). A broad spectrum of siRNAs originating from transposons can target various stress-tolerance genes (Hilbricht et al. 2008). Furthermore, due to siRNA mobility (Chitwood et al. 2009; Dunoyer et al. 2010; Molnar et al. 2010), the transposon-derived siRNAs could regulate gene expression in distant non-effected plant organs, thus mediating systemic stress and possibly acclimation responses.

Noteworthy, changing DNA methylation is not an absolute prerequisite for the transcriptional response to stress. The activation of several repetitive elements in *Arabidopsis* upon prolonged heat stress occurs without loss of DNA methylation and is mainly accompanied by nucleosome loss and heterochromatin decondensation (Pecinka et al. 2010). In some cases, histone modifications can be sufficient to remove transcriptional silencing. Heat, freezing and UVB treatments can release transgene silencing without loss of DNA methylation via altering histone occupancy and inducing histone H3 acetylation (Lang-Mladek et al. 2010). Finally, the contribution of yet unknown molecular mechanisms to heterochromatic transcription repression cannot be excluded. The presence of such unidentified mechanisms is supported by the studies by Tittel-Elmer et al. (2010) indicating that a transient genome-wide release of heterochromatin-associated silencing in response to temperature treatment can be achieved without modifying repressive epigenetic marks.

3.2 Histone Modification Changes Under Stress Conditions

Chromatin structure is another important epigenetic element in global gene regulatory networks. Transcription regulation of many stress-responsive genes depends on the activity of histone-modifying enzymes and chromatin remodeling complexes (Chinnusamy and Zhu 2009). In addition to having direct effects on gene expression, some of the stress-induced histone modifications can also affect DNA methylation. Knockout mutants of *HDA6* in *Arabidopsis* result in increased histone acetylation and loss of DNA methylation at the rRNA gene promoters leading to their transcription derepression (Earley et al. 2006).

Histone deacetylases are highly responsive to various environmental signals and mediate transcriptional repression by reducing levels of histone acetylation at the targeted genes. The *Arabidopsis* histone deacetylase, HDA6, is an important component of the transcriptional gene silencing and RdDM pathway (Aufsatz et al. 2002; Probst et al. 2004). The activity of HDA6 is induced by two important plant stress-response hormones, jasmonic acid and ethylene (Zhou et al. 2005). The sensitivity to hormonal signals enables histone deacetylation to be directed by various biotic and abiotic stresses. Down-regulation of *HDA6* and *HDA19* in *Arabidopsis* by RNAi was reported to result in transcription repression of the *ETHYLENE RESPONSE FACTOR-1* (*ERF1*) and *PATHOGENESIS-RELATED* (*PR*) stress-response genes. Consistently, transgenic *Arabidopsis* lines that overexpress *HDA19* display an increased expression of the *ERF1* and *PR* genes (Zhou et al. 2005). Also, photomorphogenesis and expression of light-responsive genes were reported to be under the negative control of HDA19 (Benhamed et al. 2006). Another important cross talk exists between abscisic acid (ABA) signalling and histone deacetylation. Overexpression of the *AtHD2C* gene, a member of the HDAC family, which is normally down-regulated by ABA, resulted in an enhanced expression of some ABA-responsive genes such as LEA class genes and increased tolerance to salinity and drought conditions (Sridha and Wu 2006). The importance of histone deacetylation in stress response is further supported by hypersensitivity of the *hos15* mutants to freezing stress. The *HOS15* gene encodes a WD-40 domain protein that interacts with histone H4 and is important for H4 deacetylation (Zhu et al. 2007b).

Transcriptional activation of stress-tolerance genes can be eased by the activity of HATs that interact directly with transcriptional factors activating stress-responsive genes (Stockinger et al. 2001). Indeed, light-dependent transcriptional induction of the pea plastocyanin gene correlates with increased histone acetylation in the promoter and 5′ gene coding region (Chua et al. 2003). Similarly, HAF2 and GCN5 histone acetyltransferase proteins were shown to promote the expression of light-responsive genes and photomorphogenesis (Benhamed et al. 2006). Whereas histone acetyltransferase HAC1 is required for transcriptional up-regulation of the heat-shock gene *HSP17* (Bharti et al. 2004), GCN5 interacts with the CBF1 transcriptional factor and activates transcription of cold-responsive genes (Stockinger et al. 2001). Exposure to cold also results in the progressive enrichment in H3K27me3 at the *FLC* gene locus and is a molecular key for the induction of flowering (Crevillén and Dean 2011). Consistently with the repressive role of H3K27me3, transcriptional up-regulation of two *Arabidopsis* cold-responsive genes, *COR15A* and *ATGOLS3,* was found to be associated with a decrease in H3K27me3 at the gene loci (Kwon et al. 2009). Recently, transcriptional profiling of *Arabidopsis* co-suppression lines (*msi1-cs*) of the *MSI1* gene encoding a subunit of the PRC2 complex revealed up-regulation of a subset of the ABA-responsive genes eliciting the response to drought and salt stress (Alexandre et al. 2009).

The adaptive response to stress is frequently mediated by the simultaneous deposition of several distinct activating or repressive histone marks. The activation of stress-responsive genes in *Arabidopsis* by drought is associated with the enrichment in H3K4me3 and H3K9ac in four drought stress-inducible genes (Kim et al. 2008). Similarly, Tsuji et al. (2006) reported that the enhanced expression of alcohol dehydrogenase 1 (ADH1) and pyruvate decarboxylase (PDC1) enzymes in the submerged rice seedlings correlated with the enrichment in H3K4me3 and H3 acetylation in the *ADH1* and *PDC1* genes. Importantly, these epigenetic modifications were dynamic, and the basal expression level of the *ADH1* and *PDC1* genes was restored upon reaeration. Consistently, Sokol et al. (2007) reported that a rapid and transient increase in histone H3 Ser-10 phosphorylation, H3 phosphoacetylation, and H4 acetylation in response to salt, cold and ABA treatments correlated with the stress-type-specific gene expression.

Deploying different histone variants to nucleosomes may serve as another strategy for regulating gene transcription in response to stress. Deposition of the linker histone variant HIS1-S was shown to alleviate negative effects resulted from drought stress in tomato. When present in chromatin of wilted tomato leaves, HIS1-S reduced transpiration rates by negatively regulating stomatal conductance (Scippa et al. 2004). Similarly, the expression of the *HIS1-3* gene, the *Arabidopsis* linker histone H1 homolog, is induced in root meristem and elongation zone in response to drought and ABA treatment (Ascenzi and Gantt 1997, 1999). Transgenic *Arabidopsis* plants overexpressing an active form of a positive regulator of the expression of the *HIS1-3* gene, the AREB1 transcription factor, display ABA hypersensitivity and improved drought tolerance (Fujita et al. 2005). The H2A.Z variant of histone H2A is another example of a histone variant sensitive to environmental cues. Genes which expression is altered by high temperature stress are usually enriched in the H2A.Z variant within their transcription start sites (Kumar and Wigge 2010). Due to the thermal instability of H2A.Z-containing nucleosomes, high temperature results in exposure of the gene promoter, thereby facilitating an access of transcriptional activators or repressors. Consequently, the final transcriptional output of this epigenetic switch depends on the type of a transcriptional regulator recruited to the target. In addition to the temperature response genes, the H2A.Z variant down-regulates the expression of phosphate-starvation response genes (Smith et al. 2010) and genes mediating systemic acquired resistance in *Arabidopsis* (March-Díaz et al. 2008). In contrast to these repressive effects on gene transcription, the enrichment in the H2A.Z variant at *FLC* loci results in up-regulated transcription and repression of flowering (Deal et al. 2007).

3.3 Small RNAs: A Stress-Sensitive Signal that Shapes the Plant Epigenome

The RdDM pathway is an important component of the gene regulatory network that can use siRNA-derived signals to modify transcription of the targeted genes. The sensitivity of siRNAs to various environmental stimuli combined with the ability to guide DNA methylation in a sequence-specific manner makes siRNAs an ideal choice for directing dynamic changes in plant transcriptome under the ever-changing growth conditions. The siRNA pathway controls at least one-third of all methylated loci in the genome of *Arabidopsis* (Lister et al. 2008). A recent study by Zheng et al. (2008) suggested that siRNAs can also direct sequence-specific DNA demethylation via the ROS1 pathway. The expression of smRNAs can change in response to a variety of different stresses including pathogen attacks, mechanical stress, dehydration, salinity, cold, ABA, and nutrient deprivation (Sunkar and Zhu 2004; Borsani et al. 2005; Lu et al. 2005; Katiyar-Agarwal et al. 2007; Sunkar et al. 2007; Ben Amor et al. 2009). Low temperatures can promote virus-induced gene silencing, while high temperatures delay it (Tuttle et al. 2008). The expression of siRNA may display tissue- and organ-specific patterns (Sunkar and Zhu 2004; Lu et al. 2005; Ben Amor et al. 2009), suggesting their importance in the development and morphogenesis (Swiezewski et al. 2007). The hypersensitivity of siRNA biogenesis mutants to genotoxic stress (Yao et al. 2010) supports the contribution of epigenetic control toward maintaining genome stability.

Recent studies showed that siRNAs can act as a mobile signal and have an effect on transposon and DNA methylation in distant tissues (Chitwood et al. 2009; Dunoyer et al. 2010; Molnar et al. 2010). These observations add an additional degree of complexity to the system of epigenetically mediated transcriptional control and may serve as a key component in understanding molecular mechanisms behind stress-induced systemic responses and transgenerational epigenetic inheritance of gene expression patters in plants. In grafting experiments with *dcl2*, *dcl3*, and *dcl4* mutants, Molnar et al. (2010) demonstrated that mobile siRNAs could direct DNA methylation in the recipient cells. Consistently, using labeled siRNA duplexes, Dunoyer et al. (2010) demonstrated the physical cell-to-cell movement of siRNA duplexes supporting their role as mobile silencing signals between plant cells. Finally, the activation of transposons and the production of 24-nt siRNAs in the vegetative nucleus of pollen and in a central cell during female gametogenesis may reinforce silencing at transposons in the sperm, egg cell and developing embryo (Mosher et al. 2009; Hsieh et al. 2009; Slotkin et al. 2009).

The identification of nat-siRNAs that originate from natural-antisense transcripts (Borsani, et al. 2005; Zhou et al. 2009) allowed better understanding of molecular mechanisms mediating resistance to pathogen attacks (Katiyar-Agarwal et al. 2007), salt and drought stress tolerance (Borsani, et al. 2005; Zhou et al. 2009). These smRNAs are involved in the stress-mediated regulation of genes located in antisense overlapping pairs that are capable of generating complimentary transcripts. The studies by Borsani, et al. (2005) demonstrated that the induction of one of the genes by stress in such antisense pair results in the production of nat-siRNA which guides the cleavage of other gene transcript, thus down-regulating the gene expression level. A similar molecular response can be observed in plants challenged with bacterial pathogens. Katiyar-Agarwal et al. (2007) established that resistance to *Pseudomonas syringae* in *Arabidopsis* is mediated by the induction of the specific nat-siRNA in response to the infection. Importantly, the genome of *Arabidopsis* contains over 2,000 genes that are found in convergent overlapping pairs and respond to various environmental stimuli (Borsani, et al. 2005). Therefore, the nat-siRNA-mediated regulation of gene expression in response to environmental stimuli can be an important molecular pathway mediating stress tolerance.

miRNAs represent another stress-sensitive trigger that may modulate gene expression at the posttranscriptional level. The work by Sunkar and Zhu (2004) supported the high abundance of stress-inducible miRNAs in *Arabidopsis*. Among the most interesting miRNAs uncovered by this study were miRNA402 and miRNA407 regulated by dehydration, salinity, cold, and ABA. Whereas miRNA402 targets the ROS-like DNA glycosylase, miR407 targets a SET domain protein functioning in histone lysine methylation (Sunkar and Zhu 2004). This suggests yet another stress-sensitive cross talk between DNA methylation and histone modification pathways. The hypersensitivity of *hen1-1* and *dcl1-9* plants that are partially deficient in miRNAs biogenesis to abiotic stresses further supports an important role of miRNAs in mediating tolerance to stress (Sunkar and Zhu 2004). The presence of multiple miRNA target sites within the gene transcript may imply that different levels of gene repression might be achieved through a various number of miRNAs bound to the target (Doench et al. 2003). The stress-induced miRNAs can also display a tissue-specific expression pattern. This may reflect the need in the organ-specific metabolic differences during response to stress. Indeed, miR393 down-regulates TIR1, a positive regulator of auxin signalling, and has the strongest expression in the inflorescence under physiological conditions. Thus, the inhibition of plant growth by stress is consistent with the strong induction of miR393 expression (Sunkar and Zhu 2004).

4 Epigenetic Mechanisms of Crop Improvement: Present and Future Challenges

Today, plant breeding and transgenesis mainly focus on genetic modifications of economically important crops trying to achieve higher yields, better field performance and increased resistance to stress. At the same time, the epigenetic component of plant response to environment is often overlooked, and epigenetic effects on gene expression in plants and their progeny are often disregarded. To date, an increasing amount of experimental evidence has accumulated indicating that epigenetic modifications acquired by the ancestral generation can be transmitted to the progeny. The inherited epigenetic landscape may not only change the progeny's transcriptome but also affect genome stability and facilitate acclimation to stress (reviewed in Boyko and Kovalchuk 2011a). We believe that uncovering the molecular mechanisms mediating transgenerational epigenetic changes and environmental memories associated with them would help create better crops for the field application. Below, we describe the recent advancements in plant transgenesis and selection of epigenetic traits. We also discuss several prominent examples which illustrate how parental environmental memories can improve performance and stress tolerance in the immediate progeny. In future, manipulating growth conditions of plants used for seed production may become an important method for improving their progeny performance in the field.

4.1 *The Contribution of Plant Transgenesis to Crop Improvement*

The genetic engineering of stress-tolerant plants requires modifications of the expression of genes involved in stress-related response. In contrast to plant breeding, plant transgenesis provides a faster alternative for crop improvement. The ability to introduce non-plant genes into the plant genome is among great benefits offered by genetic engineering, and it has found an extensive practical application. The initial efforts in genetic engineering of stress-tolerant crops were mainly directed on the use of the so-called single-action genes. As the name suggests, these genes were responsible for the production of a single protein or metabolite that could mediate stress resistance. Successful examples of the single-action gene strategy are a large number of insect-resistant and herbicide-tolerant plants dominating nowadays in the field. The single-action gene method was also used to improve crop tolerance to drought and salt stress. Here, the main emphasis was made on the expression of key enzymes of the osmolyte biosynthesis pathway, transport proteins, detoxifying enzymes, etc. Using this strategy, transgenic rice with the improved survival under submergence conditions was developed by overexpressing the *PYRUVATE DECARBOXYLASE 1* (*PDC1*) gene (Quimlo et al. 2000). Similarly, the engineered increase in the accumulation of compatible solutes achieved by transgenic modifications of their biosynthetic pathway improved stress tolerance and provided better protection against the secondary oxidative stress (Diamant et al. 2001). The overexpression of superoxide dismutase, glutathione peroxidase, ascorbate peroxidase, and glutathione reductase enzymes that mediate deactivation of reactive oxygen species was successfully applied for alleviating the effects of chilling stress on photosynthetic performance in tobacco (Gupta et al. 1993) and potato plants (Perl et al. 1993). The expression of HVA1, a LEA group protein, that normally accumulates during water-deficit stress was sufficient to improve stress tolerance and growth characteristics in transgenic rice and wheat (Sivamani et al. 2000; Rohila et al. 2002). Genetically engineered enhanced biosynthesis of the heat-shock protein HSP101 significantly improved thermotolerance in basmati rice (Katiyar-Agarwal et al. 2003). Despite the considerable success in engineering stress-tolerant crops, a wide application of the single-action gene strategy remains rather limited since to achieve tolerance to multiple abiotic stresses usually requires the expression of multiple genes at a time.

This conceptual problem was overcome by engineering transgenic plants that expressed transcription factors and signal transduction genes. This strategy proved to be particularly beneficial as a number of abiotic stresses, such as drought, salinity, and chilling, induce transcription from a similar set of the target genes, display significant cross talks in their signalling pathways and can co-occur in nature (Seki et al. 2001; Chen and Murata 2002). Consequently, engineering a single transcription factor or a signalling molecule could be sufficient to activate hundreds of stress-related genes mimicking gene expression profiles corresponding to physiological stress responses. Furthermore, the use of this strategy could help develop transgenic plants with the enhanced tolerance to multiple abiotic stresses. Indeed, the constitutive expression of the cold-inducible transcription factor *CBF1* engineered in tomato significantly improved chilling, drought, and salt stress tolerance (Hsieh et al. 2002). Similarly, the overexpression of the mitogen-activated protein kinase kinase kinase (MAPKKK) NPK1 that activates an oxidative signal cascade significantly improved tolerance to cold, heat, salinity, and drought stresses in transgenic tobacco and maize (Kovtun et al. 2000; Shou et al. 2004).

Among the frequent drawbacks resulting from the constitutive overexpression of transcription factors and permanently activated signalling pathways were reduced crop yield, a dwarf phenotype, and early senescence responses (Hsieh et al. 2002). Hence, the recent efforts in plant genetic engineering were directed on isolation, characterization, and application of the inducible and stress-inducible gene promoters for transgene expression. The application of inducible-gene promoters permitted the targeted time- and tissue-specific expression transgenes and made it possible to exert a tighter control over levels of transgene expression for its optimal function. By focusing tolerance-mediating transcriptional and metabolic responses in time, it was possible to decrease the total energy required by plant to continuously maintain defense gene expression. Indeed, a combination of the drought-inducible *rd29a* promoter and a sequence of the *CBF* (*DREB1A*) transcription factor enhanced drought, salt and low temperature tolerance in tobacco, wheat, and potato plants (Kasuga et al. 2004; Pellegrineschi et al. 2004; Behnam et al. 2006, 2007). Importantly, the phenotype and growth performance of transgenic plants expressing transcription factors from stress-inducible promoters usually surpass those of plants in which transgene expression is driven by a constitutive promoter.

Recently, the knowledge of smRNA pathways mediating gene regulation was successfully deployed for transgenic crop improvement. Sunkar et al. (2006) engineered transgenic plans expressing a modified miRNA-resistant form of Cu/Zn superoxide dismutase (SOD). In wild-type plants, the expression level of SOD is regulated by miRNA398 that targets both cytosolic and chloroplast-localized forms of SOD, thereby reducing the total level of SOD in a cell. In contrast to wild-type plants, transgenic plants accumulated high levels of SOD mRNA that enhanced tolerance to high light levels, heavy metals, and other oxidative stresses (Sunkar et al. 2006). Another strategy for crop improvement that uses smRNAs is engineering *in-planta* expression of smRNAs that can trigger silencing of essential genes in pathogens, insects, and nematodes. A prominent example is engineering root-knot nematode, *Meloidogyne incognita,* resistance in tobacco plants (Huang et al. 2006). In this work, tobacco plants were modified to express dsRNAs that target two *Meloidogyne* genes required for interactions between plants and parasitic nematodes. Due to a high degree of conservation of these target genes among *Meloidogyne* species, tobacco plants acquired resistance to four economically important nematode species.

Undoubtedly, the future advancement in the field of plant genetic engineering will be focused on the use of inducible-gene expression systems exploiting transcriptional regulators and signalling proteins. Additionally, an increasing interest can be anticipated toward the ability to manipulate epigenetic regulators and modify the epigenetic landscapes while attempting to improve stress tolerance. The great potential of deploying epigenetic mechanisms for crop improvement has been already indicated by several experimental studies. Below, we give a few insights on how our knowledge of epigenetic responses to stress can be used for crop improvement.

4.2 Transgenerational Inheritance: Is it a Road to Transgenerational Hardening?

To isolate and integrate the desired traits into economically important crops, plant breeding relies on rare genetic mutation events or chemically accelerated genome-wide mutagenesis followed by selection of plants with the desired traits. Overall, this process is based on the Mendelian laws and on the notion of hard inheritance and can be quite time consuming. The recent progress made in plant genetic transformation has significantly accelerated the production of new plant cultivars and permitted more precise and controlled modifications of plant genomes with plant and non-plant DNA. Since gene targeting techniques are still far from being well-established in plants, to obtain transgenic plants with the desired transgene integration sites is still a very laborious process. Furthermore, the need for selection markers to improve the efficiency of the whole transformation process has raised a number of public concerns regarding environmental safety of genetically modified organisms, which sometimes impedes the use of newly developed transgenic plant lines in the field.

Soft inheritance (Youngson and Whitelaw 2008) can provide a valuable compliment to the existing techniques for crop improvement that are based on hard inheritance. Since the soft inheritance relies on epigenetic mechanisms, it could serve as a fast and flexible system allowing improving plant performance in the next generation without altering its genetic makeup. Soft inheritance can be advantageous for natural plant populations located in dynamic environments where growth conditions may fluctuate within the frequency period for only several generations. Using epigenetic marks as carriers of environmental memories of gene expression would enable descendant plants to establish new phenotypes that provide a greater fitness in their growth environment. Importantly, this can be achieved on a reduced time-scale and without changing genetic information encoded in the genome. Indeed, two populations of mangrove tree species that inhabit two different environments and consequently display distinctly different phenotypes, differ in epigenetic rather than genetic changes in the genome (Lira-Medeiros et al. 2010). Similar extensive epigenetic differences were also documented between clonal off-shoot offspring and mother plants in date palms (Fang and Chao 2007).

Plants are able to adapt their transcriptomes to the upcoming severe stresses by using environmental cues to predict upcoming changes. This phenomenon in plants is well known and is often described as seasonal hardening. Exposure to mild stress serves as a necessary indicator of a more severe type of upcoming stress and triggers the development of increased stress tolerance (Beck et al. 2004; Turunen and Latola 2005). Sensing and preparing for environmental changes are two necessary adaptations used by plants as sessile organisms. Strikingly, within 1 month of seasonal hardening, some pine trees can establish tolerance to temperatures as low as −70 °C (Beck et al. 2004). Undoubtedly, such phenotypic plasticity can only be achieved by well-orchestrated changes in gene expression and metabolome composition within a short period of time. Overall, it is not by changing genetic information but by manipulating the existing gene pool, we can make plants to survive extreme growth conditions.

The importance of the phenomenon of hardening for crop improvement is that it may also occur on a transgenerational scale. In contrast to animals, in plants, the germline is separated from the soma late in development. This will allow the incorporation of the acquired genetic and epigenetic changes into the gametes and then their transmission to the progeny. Indeed, rearrangements that occur within a transgene sequence upon exposure to UVC or a virus can be inherited by the plant progeny (Ries et al. 2000; Kovalchuk et al. 2003a). Passing the memory of preexisted or newly appeared environmental conditions to the progeny is advantageous for plant survival as it prepares the progeny to the changed growth conditions or a new stress. Experimental evidence supports the existence of environmental memories in plants. In *Campanulastrum americanum* species, the maternal light environment can influence offspring life history (annual vs. biennial) in an adaptive manner (Galloway and Etterson 2007). Specifically, growing *C. americanum* in the light environment was similar to the maternal environment led to enhanced seed survival and germination. Temperature treatments can also result in transgenerational hardening of the progeny. In *Arabidopsis*, exposure ancestors to elevated growth temperatures for two consecutive generations (P and F1 generations) resulted in increased fitness in the F3 generation when exposed to heat stress (Whittle et al. 2009). Persistence of adaptive stress memories over one unexposed generation (the F2 generation) supports the involvement of heritable epigenetic effects in this phenomenon. In another study, exposure of *Arabidopsis* plants to cold during bolting and seed maturation was sufficient to improve the recovery of photosynthetic yield under chilling and freezing conditions in their immediate progeny (Blödner et al. 2007). Adaptive transgenerational responses are not restricted to grasses. In fact, the maternal photoperiod and temperature were shown to have a positive adaptive effect on phenology and frost hardiness in progeny of *Picea abies* (Norway spruce) (Johnsen et al. 2005). This transgenerational adaptive plasticity indicates that plants can sense changes in growth conditions and modify gene expression in their progeny to better fit the new growth environment.

It is possible that differential genome methylation is one of the mechanisms of transgenerational stress response since it may maintain the patterns of gene expression required to mediate acclimation to stress. Indeed, a correlation between changes in global DNA methylation, genome stability and stress tolerance was previously documented for *Pinus silvestris* grown under the conditions of radioactive contamination (Kovalchuk et al. 2003b, 2004). Consistently, a correlation between changes in transgenerational methylation and stress tolerance was also reported for the progeny of plants exposed to different abiotic and biotic stresses (Boyko et al. 2010a; Kathiria et al. 2010). The study by Boyko et al. (2010a) suggested that transgenerational response to salt stress includes hypermethylation of repetitive elements and differential changes in methylation patterns elsewhere in the genome. Importantly, changes in DNA methylation were accompanied by a significant increase in the expression of genes involved in DNA transcription and repair as well as by elevated tolerance to salt stress (Boyko et al. 2010a, b). Similarly, a delay in the appearance of viral infection symptoms was observed in the immediate progeny of tobacco plants challenged with the virus (Kathiria et al. 2010). These findings support the hypothesis that changes in DNA methylation acquired by progeny could have an adaptive effect. The heritability of stress-induced changes in DNA methylation received further support by Verhoeven et al. (2010) who showed that many of DNA methylation changes that occurred in genetically identical apomictic dandelion plants exposed to various abiotic and biotic stresses could be faithfully transmitted to the offspring.

A heritable increase in the frequency of somatic homologous recombination (HR) events in plants exposed to stress and in their immediate progeny can be another important mechanism involved in transgenerational acclimation to stress. A number of biotic and abiotic stresses were found to alter genome stability by changing the frequency of HR events in somatic and meiotic cells. These stimuli include pathogen attacks, bacterial elicitor flagellin, high and low temperatures, day length, UVB, and UVC, drought and flood, salt, osmotic, and oxidative stresses as well as drugs that modify chromatin and change DNA methylation patterns (Lucht et al. 2002; Kovalchuk et al. 2003a; Boyko et al. 2005, 2006a, b, 2010a, b; Molinier et al. 2006; Pecinka et al. 2009). The stress-induced increase in the frequency of HR can be inherited and persist for at least one generation following stress exposure (Molinier et al. 2006; Boyko et al. 2007, 2010a; Kathiria et al. 2010; Yao and Kovalchuk 2011). An intriguing hypothesis is that stress can guide plant genome evolution using repair pathways, particularly HR, to trigger locus-specific genome rearrangements, thereby allowing a rapid evolution of targeted sequences and associated phenotypes (reviewed in Boyko and Kovalchuk 2011b).

In the light of this hypothesis, HR may provide a molecular mechanism for a rapid diversification of genomic sequences in response to stress (Boyko et al. 2007; DeBolt 2010). Unfortunately, the experimental evidence supporting this hypothesis is still scarce. Nevertheless, two independent studies provide some interesting insights regarding this matter. In the first study, Boyko et al. (2007) showed that challenging tobacco plants with a compatible virus is sufficient to decrease DNA methylation and increase the frequency of rearrangements in the leucine-rich repeat (LRR) regions of the resistance (*R*) gene loci in the noninfected progeny. These findings are very intriguing since the evolution of plant *R*-genes involved multiple gene duplication and recombination events (Meyers et al. 2005). The second study was focused on the effects of stress-induced genome rearrangements on plant genome diversification and evolution. In his work, DeBolt (2010) compared gene copy number variation (CNV) among sibling individuals in plant populations that were exposed to biotic or abiotic stresses and selected for fecundity for five consecutive generations. A high number of the repetitive CNVs observed among sibling individuals exposed to the same stress for multiple generations supported a nonrandom occurrence of rearrangements (DeBolt 2010). In agreement with the study by Boyko et al. (2007), the initiation sites of CNVs were most frequently located within the stress response genes including multiple LRR-containing disease resistance proteins and transposons. Finally, the presence of similar types of repetitive CNVs in plants exposed to the temperature of 16 °C and salicylic acid treatments indicated that certain genome regions were generally more prone to rearrangements in response to stress.

5 Using Epigenetics for Crop Improvement: Current Advances and Future Prospective

Epigenetic changes can generate multiple epigenetic polymorphisms within a plant population (Fang and Chao 2007; Lira-Medeiros et al. 2010). Since epigenetic modifications are sensitive to environmental stimuli, it is reasonable to suggest that some of the newly acquired epigenetic polymorphisms could be beneficial for plant survival and could provide a source for new heritable variations among the plants in a population. In contrast to classical plant breeding that is performed only at the population level, breeding epigenetic traits should be done at the level of a single plant organism since the population of genetically identical individuals is expected to have multiple heritable epigenetic variations.

Naturally occurring and artificially induced epigenetic variation between the plants in a population can be a promising source of economically important traits for modern breeding and crop improvement programs. Unfortunately, exploiting epigenetic variation in crop breeding is still far away from being well-established. Nevertheless, the first successful steps in this direction have been made. The work by Hauben et al. (2009) demonstrated that quantitative traits can be recursively selected through recurrent selfing in isogenic lines. The authors isolated several stable canola lines with an increase in seed yield from a genetically isogenic population of *Brassica napus* plants. In this selection experiment, two seedlings with the highest and lowest cellular respiration rates were chosen from the population of a double haploid canola line. These two plants were self-fertilized to generate two populations. These two populations were further selected for four additional rounds to isolate the sublines with the highest and lowest respiration rates, respectively, as compared to the control line from which the selection was initiated. Increased seed yield correlated with the energy-use efficiency (EUE). Strikingly, during field trials, the selected lines with low respiration rates and high EUE demonstrated up to an 8% increase in seed yield when compared to the control line. In contrast, in the selected lines with high respiration rates and low EUE, a 10% reduction in seed yield was observed (Hauben et al. 2009). Furthermore, when grown in the field under drought conditions, the line with low respiration rates and the highest EUE showed seed yield that was 20% higher than that of the control line. Not surprisingly, the stable phenotypic variation between the lines was associated with the difference in DNA methylation as it was determined by using the Methylation-Sensitive (MS-)AFLP analysis. Differential DNA methylation was always localized in the coding sequences. These MS-AFLP patterns were line-specific and stable for at least eight generations. The lines with low respiration rates that displayed the improved seed yield under field conditions were characterized by pronounced DNA hypomethylation and showed stable line-specific changes in the levels of histone 3 and histone 4 acetylation. At the same time, no significant genetic differences were detected between the lines that used the AFLP analysis. Hence, physiological and agronomical differences between the selected canola lines were solely mediated by distinct heritable epigenetic states that were isolated by the artificial selection for increased cellular respiration rates and EUE.

To better understand the association between epigenetic modifications and stable plant phenotypes, methods that enable genome-wide generation of epialleles are required. To date, a reasonable success has been achieved by using various chemical treatments and epigenetic mutants that trigger the genome-wide epigenetic changes. A rapid loss of genome-wide DNA methylation can be mediated by using plants deficient in the factors that control global DNA methylation, such as MET1 and DDM1. Since the *met1* and *ddm1* mutants display progressive accumulation of severe developmental defects over generations, this approach is less likely to be widely used for plant breeding. Nonetheless, it may still be possible to segregate some interesting phenotypes independently from mutations in the *DDM1* or *MET1* genes. The accumulation of negative effects associated with progressive loss of DNA methylation can be avoided by using constitutive or RNAi-mediated down-regulation of MET1 and DDM1 enzymes. Fujimoto et al. (2008) reported stable inheritance (independent of the RNAi construct) of hypomethylated states in transgenic canola plants in which DDM1 was down-regulated.

The application of chemical inhibitors of DNA methylation may offers a better alternative to the use of epigenetic mutants. In fact, global genome hypomethylation, similar to one observed in the *met1* or *ddm1* plants, can be triggered by the application of 5-azacytidine (5-AzaC) (Christman 2002; Akimoto et al. 2007) and zebularine (Baubec et al. 2009). Both chemicals inhibit DNA methylation by covalently binding the MET1 protein and limiting its catalytic activity. The application of zebularine, however, offers several important advantages compared to 5-AzaC. While both chemicals transiently decrease DNA methylation in a sequence-independent manner, zebularine has much longer half-life under physiological conditions, which makes it more suitable for applications in plant tissue culture and growth media (Cheng et al. 2003; Baubec et al. 2009). Also, the effects of zebularine on DNA methylation loss are clearly dose-dependent. Therefore, the investigation of a correlation between DNA methylation and gene transcription can be easier (Baubec et al. 2009). Both chemicals can be used to produce "epimutagenized" plant populations for the reverse genetic screens and high-throughput bisulphite sequencing in a way similar to conventional chemical mutagenesis. Such epimutagenesis will greatly accelerate the identification of epialleles corresponding to the desired agricultural traits. In the recent study, Marfil et al. (2009) were able to reproduce naturally occurring abnormal floral phenotypes associated with DNA methylation polymorphisms in potato. Using 5-AzaC, the authors generated hypomethylated potato plants and showed that new phenotypes segregated with similar DNA methylation patterns but not with DNA sequences (Marfil et al. 2009).

There is great potential in the use of DNA methylation inhibitors for engineering epigenetic traits that are beneficial to agriculture. The recent study of epigenetic inheritance in rice led to the identification of a novel stably-inherited epiallele allowing for disease resistance (Akimoto et al. 2009). Specifically, by treating *Oryza sativa* seeds with 5-azadeoxycytidine, a chemical similar to 5-AzaC, Akimoto et al. (2009) had been cultivating the progeny in the field for 10 years. The lines that demonstrated stable inheritance of phenotypic traits (height and the day of ear emergence) were selected for further analysis of DNA methylation by MSAP. One of these lines, designated as Line-2, displayed the phenotype of dwarfism in the parental generation and in nine successive generations of the progeny. Among six fragments identified by MSAP in Line-2 plants, one fragment corresponding to the gene encoding the Xa21-like protein, *Xa21G*, was identified (Akimoto et al. 2009). The expression of the *Xa21* gene family members in rice is known to mediate resistance to *Xanthomonas oryzae* pv. *oryzae* infection in a gene-for-gene manner. From seven members of this gene family in rice, only one, the *Xa21* gene, is actively expressed, while others are silenced by DNA methylation (Wang et al. 1996, 1998).

The wild-type cultivar *Oryza sativa* ssp. *japonica* used in this study did not contain the *Xa21* gene, and thus it was susceptible to *X. oryzae* infection. In contrast, Line-2 plants showed resistance to *X. oryzae* infection. The trait of acquired disease resistance was mediated by the constitutive expression of the *Xa21G* gene due to complete erasure of DNA methylation at the gene promoter in Line-2 plants (Akimoto et al. 2009). No *Xa21G* transcripts were found in wild-type plants, which was consistent with the highly methylated sequence of the *Xa21G* promoter and disease susceptibility. Hence, the acquisition of the stably-inherited disease resistance trait was mediated by demethylation of the resistance gene promoter caused by a pulse treatment of germinated rice seeds with 5-azadeoxycytidine.

The establishment of the heritable epigenetic landscapes corresponding to gene expression patterns that are advantageous at the time of stress can be a possible alternative to isolation of single epialleles via 5-AzaC treatment and laborious genetic screens. Perhaps, this can be achieved by exposing plants used for the production of seed stock to a stress that is likely to occur at the future field sites. While being technically simple, this pretreatment procedure could significantly boost tolerance to stress in plant progeny in the field. Initial studies offered proof of concept indicating that it is indeed possible to increase stress tolerance in the immediate progeny by exposing ancestral plants to mild and/or short-term stress signals (Johnsen et al. 2005; Blödner et al. 2007; Galloway and Etterson 2007; Whittle et al. 2009; Boyko et al. 2010a, b; Kathiria et al. 2010). This strategy is likely to be mediated by the same molecular mechanisms as plant hardening in which exposure to mild stress serves as a necessary indicator of more severe stress (Beck et al. 2004; Turunen and Latola 2005). During plant hardening, the first mild stress triggers adaptive changes in the plant transcriptome, thereby enhancing tolerance to a more severe stress. It is plausible that the memory of growth environment, including stress, can be transmitted to the next generation, thus preparing the progeny to the changed growth conditions. Transgenerational transmission of stress memory and the associated patterns of gene expression could be mediated by epigenetic signals such as DNA methylation and would result in the progeny having gene expression that is best suitable under existing stress conditions (reviewed in Boyko and Kovalchuk 2011a). In fact, Boyko et al. (2010a, b) reported that the first progeny (the G1 generation) of *Arabidopsis* plants (the P generation) exposed to mild salt stress (25 mM NaCl) displayed improved germination rates and higher biomass accumulation when grown under high salt (125 and 150 mM NaCl) and genotoxic (MMS) conditions. In parallel, Kathiria et al. (2010) described delayed signs of viral infection symptoms in the immediate progeny (the G1 generation) of tobacco plants (the P generation) challenged with the virus. In both studies, enhanced tolerance of the progeny to stress correlated with changes in DNA methylation, thereby supporting the role of epigenetic mechanisms in this process. Furthermore, the deficiency in establishing transgenerational changes in DNA methylation and stress tolerance in the *dcl2 Arabidopsis* mutants indicated that smRNAs could play a role of a *trans*-acting signal of epigenetic stress memory (Boyko et al. 2010a).

References

Aina R, Sgorbati S, Santagostino A, Labra M, Ghiani A, Citterio S (2004) Specific hypomethylation of DNA is induced by heavy metals in white clover and industrial hemp. Physiol Planta 121:472–480

Akimoto K, Katakami H, Kim HJ, Ogawa E, Sano CM, Wada Y, Sano H (2007) Epigenetic inheritance in rice plants. Ann Bot 100:205–217

Alexandre C, Möller-Steinbach Y, Schönrock N, Gruissem W, Hennig L (2009) Arabidopsis MSI1 is required for negative regulation of the response to drought stress. Mol Plant 2:675–687

Ascenzi R, Gantt JS (1997) A drought-stress-inducible histone gene in *Arabidopsis thaliana* is a member of a distinct class of plant linker histone variants. Plant Mol Biol 34:629–641

Ascenzi R, Gantt JS (1999) Molecular genetic analysis of the drought-inducible linker histone variant in *Arabidopsis thaliana*. Plant Mol Biol 41:159–169

Aufsatz W, Mette MF, Van Der Winden J, Matzke M, Matzke AJ (2002) HDA6, a putative histone deacetylase needed to enhance DNA methylation induced by double-stranded RNA. EMBO J 21:6832–6841

Baubec T, Pecinka A, Rozhon W, Mittelsten Scheid O (2009) Effective, homogeneous and transient interference with cytosine methylation in plant genomic DNA by zebularine. Plant J 57:542–554

Beck EH, Heim R, Hansen J (2004) Plant resistance to cold stress: mechanisms and environmental signals triggering frost hardening and dehardening. J Biosci 29:449–459

Beguiristain T, Grandbastien MA, Puigdomènech P, Casacuberta JM (2001) Three *Tnt1* subfamilies show different stress-associated patterns of expression in tobacco. Consequences for retrotransposon control and evolution in plants. Plant Physiol 127(1):212–221

Behnam B, Kikuchi A, Celebi-Toprak F, Kasuga M, Yamaguchi-Shinozaki K, Watanabe KN (2007) *Arabidopsis rd29A::DREB1A* enhances freezing tolerance in transgenic potato. Plant Cell Rep 26:1275–1282

Behnam B, Kikuchi A, Celebi-Toprak F, Yamanaka S, Kasuga M, Yamaguchi-Shinozaki K, Watanabe KN (2006) The *Arabidopsis DREB1A* gene driven by the stress-inducible *rd29A* promoter increases salt-stress tolerance in proportion to its copy number in tetrasomic tetraploid potato (*Solanum tuberosum*). Plant Biotechnol 23:169–177

Ben Amor B, Wirth S, Merchan F, Lapirte P, d'Aubenton-Carafa Y, Hirsch J, Maizel A, Mallory A, Lucas A, Deragon JM, Vaucheret H, Thermes C, Crespi M (2009) Novel long non-protein coding RNAs involved in *Arabidopsis* differentiation and stress responses. Genome Res 19:57–69

Benhamed M, Bertrand C, Servet C, Zhou DX (2006) *Arabidopsis* GCN5, HD1, and TAF1/HAF2 interact to regulate histone acetylation required for light-responsive gene expression. Plant Cell 18:2893–2903

Ben-Porath I, Cedar H (2001) Epigenetic crosstalk. Mol Cell 8:933–935

Berger SL (2007) The complex language of chromatin regulation during transcription. Nature 447:407–412

Bharti K, Von Koskull-Doring P, Bharti S, Kumar P, Tintschl-Korbitzer A, Treuter E, Nover L (2004) Tomato heat stress transcription factor HsfB1 represents a novel type of general transcription coactivator with a histone-like motif interacting with the plant CREB binding protein ortholog HAC1. Plant Cell 16:1521–1535

Blödner C, Goebel C, Feussner I, Gatz C, Polle A (2007) Warm and cold parental reproductive environments affect seed properties, fitness, and cold responsiveness in *Arabidopsis thaliana* progenies. Plant Cell Environ 30:165–175

Bonasio R, Tu S, Reinberg D (2010) Molecular signals of epigenetic states. Science 330:612–616

Borsani O, Zhu J, Verslues PE, Sunkar R, Zhu JK (2005) Endogenous siRNAs derived from a pair of natural *cis*-antisense transcripts regulate salt tolerance in *Arabidopsis*. Cell 123:1279–1291

Bourc'his D, Voinnet O (2010) A small-RNA perspective on gametogenesis, fertilization, and early zygotic development. Science 330:617–622

Boyko A, Blevins T, Yao Y, Golubov A, Bilichak A, Ilnytskyy Y, Hollander J, Meins F Jr, Kovalchuk I (2010a) Transgenerational adaptation of *Arabidopsis* to stress requires DNA methylation and the function of *Dicer*-like proteins. PLoS One 5:e9514

Boyko A, Filkowski J, Kovalchuk I (2005) Homologous recombination in plants is temperature and day-length dependent. Mutat Res 572:73–83

Boyko A, Golubov A, Bilichak A, Kovalchuk I (2010b) Chlorine ions but not sodium ions alter genome stability of *Arabidopsis thaliana*. Plant Cell Physiol 51:1066–1078

Boyko A, Greer M, Kovalchuk I (2006a) Acute exposure to UVB has a more profound effect on plant genome stability than chronic exposure. Mutat Res 602:100–109

Boyko A, Hudson D, Bhomkar P, Kathiria P, Kovalchuk I (2006b) Increase of homologous recombination frequency in vascular tissue of *Arabidopsis* plants exposed to salt stress. Plant Cell Physiol 47:736–742

Boyko A, Kathiria P, Zemp FJ, Yao Y, Pogribny I, Kovalchuk I (2007) Transgenerational changes in the genome stability and methylation in pathogen-infected plants: (virus-induced plant genome instability). Nucleic Acids Res 35:1714–1725

Boyko A, Kovalchuk I (2011a) Genome instability and epigenetic modification-heritable responses to environmental stress? Curr Opin Plant Biol 14(3):260–266 [Epub ahead of print]

Boyko A, Kovalchuk I (2011b) Genetic and epigenetic effects of plant-pathogen interactions: an evolutionary perspective. Mol Plant 4(6):1014–1023 [Epub ahead of print]

Cao X, Aufsatz W, Zilberman D, Mette MF, Huang MS, Matzke M, Jacobsen SE (2003) Role of the DRM and CMT3 methyltransferases in RNA-directed DNA methylation. Curr Biol 13:2212–2217

Cao X, Jacobsen SE (2002) Role of the *Arabidopsis* DRM methyltransferases in *de novo* DNA methylation and gene silencing. Curr Biol 12:1138–1144

Carthew RW, Sontheimer EJ (2009) Origins and Mechanisms of miRNAs and siRNAs. Cell 136:642–655

Chen TH, Murata N (2002) Enhancement of tolerance of abiotic stress by metabolic engineering of betaines and other compatible solutes. Curr Opin Plant Biol 5:250–257

Chen ZJ, Tian L (2007) Roles of dynamic and reversible histone acetylation in plant development and polyploidy. Biochim Biophys Acta 1769:295–307

Cheng JC, Matsen CB, Gonzales FA, Ye W, Greer S, Marquez VE, Jones PA, Selker EU (2003) Inhibition of DNA methylation and reactivation of silenced genes by zebularine. J Natl Cancer Inst 95:399–409

Chinnusamy V, Zhu JK (2009) Epigenetic regulation of stress responses in plants. Curr Opin Plant Biol 12:133–139

Chitwood DH, Nogueira FT, Howell MD, Montgomery TA, Carrington JC, Timmermans MC (2009) Pattern formation via small RNA mobility. Genes Dev 23:549–554

Choi CS, Sano H (2007) Abiotic-stress induced demethylation and transcriptional activation of a gene encoding a glycerophosphodiesterase-like protein in tobacco plants. Mol Genet Genomics 277(5):589–600

Choi Y, Gehring M, Johnson L, Hannon M, Harada JJ, Goldberg RB, Jacobsen SE, Fischer RL (2002) DEMETER, a DNA glycosylase domain protein, is required for endosperm gene imprinting and seed viability in *Arabidopsis*. Cell 110:33–42

Christman JK (2002) 5-Azacytidine and 5-aza-2-deoxycytidine as inhibitors of DNA methylation: mechanistic studies and their implications for cancer therapy. Oncogene 21:5483–5495

Chua YL, Watson LA, Gray JC (2003) The transcriptional enhancer of the pea plastocyanin gene associates with the nuclear matrix and regulates gene expression through histone acetylation. Plant Cell 15(6):1468–1479

Cokus SJ, Feng S, Zhang X, Chen Z, Merriman B, Haudenschild CD, Pradhan S, Nelson SF, Pellegrini M, Jacobsen SE (2008) Shotgun bisulphite sequencing of the *Arabidopsis* genome reveals DNA methylation patterning. Nature 452:215–219

Crevillén P, Dean C (2011) Regulation of the floral repressor gene *FLC*: the complexity of transcription in a chromatin context. Curr Opin Plant Biol 14(1):38–44

Cubas P, Vincent C, Coen E (1999) An epigenetic mutation responsible for natural variation in floral symmetry. Nature 401:157–161

Deal RB, Henikoff JG, Henikoff S (2010) Genome-wide kinetics of nucleosome turnover determined by metabolic labeling of histones. Science 328:1161–1164

Deal RB, Topp CN, McKinney EC, Meagher RB (2007) Repression of flowering in *Arabidopsis* requires activation of *FLOWERING LOCUS C* expression by the histone variant H2A.Z. Plant Cell 19:74–83

DeBolt S (2010) Copy number variation shapes genome diversity in *Arabidopsis* over immediate family generational scales. Genome Biol Evol 2:441–453

Dennis ES, Peacock WJ (2007) Epigenetic regulation of flowering. Curr Opin Plant Biol 10(5):520–527

Diamant S, Eliahu N, Rosenthal D, Goloubinoff P (2001) Chemical chaperones regulate molecular chaperones *in vitro* and in cells under combined salt and heat stresses. J Biol Chem 276:39586–39591

Doench JG, Petersen CP, Sharp PA (2003) siRNAs can function as miRNAs. Genes Dev 17(4):438–442

Dunoyer P, Schott G, Himber C, Meyer D, Takeda A, Carrington JC, Voinnet O (2010) Small RNA duplexes function as mobile silencing signals between plant cells. Science 328:912–916

Dyachenko OV, Zakharchenko NS, Shevchuk TV, Bohnert HJ, Cushman JC, Buryanov YI (2006) Effect of hypermethylation of CCWGG sequences in DNA of *Mesembryanthemum crystallinum* plants on their adaptation to salt stress. Biochemistry (Mosc) 71:461–465

Earley K, Lawrence RJ, Pontes O, Reuther R, Enciso AJ, Silva M, Neves N, Gross M, Viegas W, Pikaard CS (2006) Erasure of histone acetylation by *Arabidopsis* HDA6 mediates large-scale gene silencing in nucleolar dominance. Genes Dev 20:1283–1293

Ebbs ML, Bartee L, Bender J (2005) H3 lysine 9 methylation is maintained on a transcribed inverted repeat by combined action of SUVH6 and SUVH4 methyltransferases. Mol Cell Biol 25:10507–10515

Ebbs ML, Bender J (2006) Locus-specific control of DNA methylation by the *Arabidopsis* SUVH5 histone methyltransferase. Plant Cell 18:1166–1176

Fang J-G, Chao CT (2007) Methylation-sensitive amplification polymorphism in date palms (*Phoenix dactylifera L.*) and their off-shoots. Plant Biol 9:526–533

Fuchs J, Demidov D, Houben A, Schubert I (2006) Chromosomal histone modification patterns — from conservation to diversity. Trends Plant Sci 11:199–208

Fujimoto R, Kinoshita Y, Kawabe A, Kinoshita T, Takashima K, Nordborg M, Nasrallah ME, Shimizu KK, Kudoh H, Kakutani T (2008) Evolution and control of imprinted *FWA* genes in the genus *Arabidopsis*. PLoS Genet 4:1–11

Fujita Y, Fujita M, Satoh R, Maruyama K, Parvez MM, Seki M, Hiratsu K, Ohme-Takagi M, Shinozaki K, Yamaguchi-Shinozaki K (2005) AREB1 is a transcription activator of novel ABRE-dependent ABA signaling that enhances drought stress tolerance in *Arabidopsis*. Plant Cell 17:3470–3488

Galloway LF, Etterson JR (2007) Transgenerational plasticity is adaptive in the wild. Science 318:1134–1136

Gendrel AV, Lippman Z, Yordan C, Colot V, Martienssen RA (2002) Dependence of heterochromatic histone H3 methylation patterns on the *Arabidopsis* gene *DDM1*. Science 297:1871–1873

Gong Z, Morales-Ruiz T, Ariza RR, Roldán-Arjona T, David L, Zhu JK (2002) ROS1, a repressor of transcriptional gene silencing in *Arabidopsis*, encodes a DNA glycosylase/lyase. Cell 111:803–814

Grant-Downton R, Le Trionnaire G, Schmid R, Rodriguez-Enriquez J, Hafidh S, Mehdi S, Twell D, Dickinson H (2009) MicroRNA and tasiRNA diversity in mature pollen of *Arabidopsis thaliana*. BMC Genomics 10:643

Gupta AS, Heinen JL, Holady AS, Burke JJ, Allen RD (1993) Increased resistance to oxidative stress in transgenic plants that over-express chloroplastic Cu/Zn superoxide dismutase. Proc Natl Acad Sci U S A 90:1629–1633

Hammoud SS, Nix DA, Zhang H, Purwar J, Carrell DT, Cairns BR (2009) Distinctive chromatin in human sperm packages genes for embryo development. Nature 460:473–478

Hashida SN, Kitamura K, Mikami T, Kishima Y (2003) Temperature shift co-ordinately changes the activity and methylation state of transposon *Tam3* in *Antirrhinum majus*. Plant Physiol 132:1207–1216

Hashida SN, Uchiyama T, Martin C, Kishima Y, Sano Y, Mikami T (2006) The temperature-dependent change in methylation of the *Antirrhinum* transposon *Tam3* is controlled by the activity of its transposase. Plant Cell 18(1):104–118

Hauben M, Haesendonckx B, Standaert E, Van Der Kelen K, Azmi A, Akpo H, Van Breusegem F, Guisez Y, Bots M, Lambert B, Laga B, De Block M (2009) Energy use efficiency is characterized by an epigenetic component that can be directed through artificial selection to increase yield. Proc Natl Acad Sci U S A 106:20109–20114

Hauser MT, Aufsatz W, Jonak C, Luschnig C (2011) Transgenerational epigenetic inheritance in plants. Biochim Biophys Acta 1809(8):459–468 [Epub ahead of print]

Hilbricht T, Varotto S, Sgaramella V, Bartels D, Salamini F, Furini A (2008) Retrotransposons and siRNA have a role in the evolution of desiccation tolerance leading to resurrection of the plant *Craterostigma plantagineum*. New Phytol 179:877–887

Hirochika H, Sugimoto K, Otsuki Y, Tsugawa H, Kanda M (1996) Retrotransposons of rice involved in mutations induced by tissue culture. Proc Natl Acad Sci U S A 93:7783–7788

Hsieh TF, Ibarra CA, Silva P, Zemach A, Eshed-Williams L, Fischer RL, Zilberman D (2009) Genome-wide demethylation of *Arabidopsis* endosperm. Science 324(5933):1451–1454

Hsieh TH, Lee JT, Charng YY, Chan MT (2002) Tomato plants ectopically expressing *Arabidopsis CBF1* show enhanced resistance to water deficit stress. Plant Physiol 130:618–626

Huang G, Allen R, Davis EL, Baum TJ, Hussey RS (2006) Engineering broad root-knot resistance in transgenic plants by RNAi silencing of a conserved and essential root-knot nematode parasitism gene. Proc Natl Acad Sci U S A 103:14302–14306

Huettel B, Kanno T, Daxinger L, Bucher E, van der Winden J, Matzke AJ, Matzke M (2007) RNA-directed DNA methylation mediated by DRD1 and Pol IVb: a versatile pathway for transcriptional gene silencing in plants. Biochim Biophys Acta 1769:358–374

Huh JH, Bauer MJ, Hsieh TF, Fischer RL (2008) Cellular programming of plant gene imprinting. Cell 132:735–744

Ito H, Gaubert H, Bucher E, Mirouze M, Vaillant I, Paszkowski J (2011) An siRNA pathway prevents transgenerational retrotranspositionin plants subjected to stress. Nature 472(7341):115–119

Jackson JP, Lindroth AM, Cao X, Jacobsen SE (2002) Control of CpNpG DNA methylation by the KRYPTONITE histone H3 methyltransferase. Nature 416:556–560

Jeddeloh JA, Stokes TL, Richards EJ (1999) Maintenance of genomic methylation requires a SWI2/SNF2-like protein. Nat Genet 22:94–97

Jiang D, Yang W, He Y, Amasino RM (2007) Arabidopsis relatives of the human lysine-specific *Demethylase1* repress the expression of *FWA* and *FLOWERING LOCUS C* and thus promote the floral transition. Plant Cell 19:2975–2987

Johnsen O, Daehlen OG, Ostreng G, Skrøppa T (2005) Day length and temperature during seed production interactively affect adaptive performance of *Picea abies* progenies. New Phytol 168:589–596

Johnson LM, Bostick M, Zhang X, Kraft E, Henderson I, Callis J, Jacobsen SE (2007) The SRA methyl-cytosine-binding domain links DNA and histone methylation. Curr Biol 17:379–384

Johnson LM, Law JA, Khattar A, Henderson IR, Jacobsen SE (2008) SRA-domain proteins required for DRM2-mediated *de novo* DNA methylation. PLoS Genet 4:e1000280

Kalendar R, Tanskanen J, Immonen S, Nevo E, Schulman AH (2000) Genome evolution of wild barley (*Hordeum spontaneum*) by *BARE-1* retrotransposon dynamics in response to sharp microclimatic divergence. Proc Natl Acad Sci U S A 97(12):6603–6607

Kapoor A, Agius F, Zhu JK (2005) Preventing transcriptional gene silencing by active DNA demethylation. FEBS Lett 579:5889–5898

Kasuga M, Miura S, Shinozaki K, Yamaguchi-Shinozaki K (2004) A combination of the *Arabidopsis DREB1A* gene and stress-inducible *rd29A* promoter improved drought- and low-temperature stress tolerance in tobacco by gene transfer. Plant Cell Physiol 45:346–350

Kathiria P, Sidler C, Golubov A, Kalischuk M, Kawchuk LM, Kovalchuk I (2010) Tobacco mosaic virus infection results in an increase in recombination frequency and resistance to viral, bacterial, and fungal pathogens in the progeny of infected tobacco plants. Plant Physiol 153:1859–1870

Katiyar-Agarwal S, Agarwal M, Grover A (2003) Heat-tolerant basmati rice engineered by over-expression of *hsp101*. Plant Mol Biol 51:677–686

Katiyar-Agarwal S, Gao S, Vivian-Smith A, Jin H (2007) A novel class of bacteria-induced small RNAs in *Arabidopsis*. Genes Dev 21(23):3123–3134

Kim JM, To TK, Ishida J, Morosawa T, Kawashima M, Matsui A, Toyoda T, Kimura H, Shinozaki K, Seki M (2008) Alterations of lysine modifications on histone H3 N-tail under drought stress conditions in *Arabidopsis thaliana*. Plant Cell Physiol 49:1580–1588

Kovalchuk I, Abramov V, Pogribny I, Kovalchuk O (2004) Molecular aspects of plant adaptation to life in the Chernobyl zone. Plant Physiol 135:357–363

Kovalchuk I, Kovalchuk O, Kalck V, Boyko V, Filkowski J, Heinlein M, Hohn B (2003a) Pathogen-induced systemic plant signal triggers DNA rearrangements. Nature 423:760–762

Kovalchuk O, Burke P, Arkhipov A, Kuchma N, James SJ, Kovalchuk I, Pogribny I (2003b) Genome hypermethylation in *Pinus silvestris* of Chernobyl – a mechanism for radiation adaptation? Mutat Res 529:13–20

Kovarik A, Koukalova B, Bezdek M, Opatrn Z (1997) Hypermethylation of tobacco heterochromatic loci in response to osmotic stress. Theor Appl Genet 95:301–306

Kovtun Y, Chiu WL, Tena G, Sheen J (2000) Functional analysis of oxidative stress-activated mitogen-activated protein kinase cascade in plants. Proc Natl Acad Sci U S A 97:2940–2945

Kumar SV, Wigge PA (2010) H2A.Z-containing nucleosomes mediate the thermosensory response in *Arabidopsis*. Cell 140:136–147

Kwon CS, Lee D, Choi G, Chung W (2009) Histone occupancy dependent and -independent removal of H3K27 trimethylation at cold-responsive genes in *Arabidopsis*. Plant J 60:112–121

Labra M, Ghiani A, Citterio S, Sgorbati S, Sala F, Vannini C, Ruffini-Castiglione M, Bracale M (2002) Analysis of cytosine methylation pattern in response to water deficit in pea root tips. Plant Biol 4:694–699

Lang-Mladek C, Popova O, Kiok K, Berlinger M, Rakic B, Aufsatz W, Jonak C, Hauser MT, Luschnig C (2010) Transgenerational inheritance and resetting of stress-induced loss of epigenetic gene silencing in *Arabidopsis*. Mol Plant 3(3):594–602

Libault M, Tessadori F, Germann S, Snijder B, Fransz P, Gaudin V (2005) The *Arabidopsis* LHP1 protein is a component of euchromatin. Planta 222:910–925

Lindroth AM, Cao X, Jackson JP, Zilberman D, McCallum CM, Henikoff S, Jacobsen SE (2001) Requirement of CHROMOMETHYLASE3 for maintenance of CpXpG methylation. Science 292:2077–2080

Lindroth AM, Shultis D, Jasencakova Z, Fuchs J, Johnson L, Schubert D, Patnaik D, Pradhan S, Goodrich J, Schubert I, Jenuwein T, Khorasanizadeh S, Jacobsen SE (2004) Dual histone H3 methylation marks at lysines 9 and 27 required for interaction with CHROMOMETHYLASE3. EMBO J 23:4286–4296

Lippman Z, Gendrel AV, Black M, Vaughn MW, Dedhia N, McCombie WR, Lavine K, Mittal V, May B, Kasschau KD, Carrington JC, Doerge RW, Colot V, Martienssen R (2004) Role of transposable elements in heterochromatin and epigenetic control. Nature 430:471–476

Lira-Medeiros CF, Parisod C, Fernandes RA, Mata CS, Cardoso MA, Ferreira PC (2010) Epigenetic variation in mangrove plants occurring in contrasting natural environment. PLoS One 5(4):e10326

Lister R, O'Malley RC, Tonti-Filippini J, Gregory BD, Berry CC, Millar AH, Ecker JR (2008) Highly integrated single-base resolution maps of the epigenome in *Arabidopsis*. Cell 133:523–536

Liu J, He Y, Amasino R, Chen X (2004) siRNAs targeting an intronic transposon in the regulation of natural flowering behaviour in *Arabidopsis*. Genes Dev 18:2873–2878

Lu S, Sun YH, Shi R, Clark C, Li L, Chiang VL (2005) Novel and mechanical stress-responsive microRNAs in *Populus trichocarpa* that are absent from *Arabidopsis*. Plant Cell 17:2186–2203

Lucht JM, Mauch-Mani B, Steiner H-Y, Metraux J-P, Ryals J, Hohn B (2002) Pathogen stress increase somatic recombination frequency in *Arabidopsis*. Nat Genet 30:311–314

Malone CD, Hannon GJ (2009) Small RNAs as guardians of the genome. Cell 136:656–668
March-Díaz R, García-Domínguez M, Lozano-Juste J, León J, Florencio FJ, Reyes JC (2008) Histone H2A.Z and homologues of components of the SWR1 complex are required to control immunity in *Arabidopsis*. Plant J 53:475–487
Marfil CF, Camadro EL, Masuelli RW (2009) Phenotypic instability and epigenetic variability in a diploid potato of hybrid origin, Solanum ruiz-lealii. BMC Plant Biol 9:21
Margueron R, Reinberg D (2011) The Polycomb complex PRC2 and its mark in life. Nature 469:343–349
Mason G, Noris E, Lanteri S, Acquadro A, Accotto GP, Portis E (2008) Potentiality of methylation-sensitive amplification polymorphism (MSAP) in identifying genes involved in tomato response to tomato yellow leaf curl Sardinia virus. Plant Mol Biol Reporter 26:156–173
Mathieu O, Reinders J, Caikovski M, Smathajitt C, Paszkowski J (2007) Transgenerational stability of the *Arabidopsis* epigenome is coordinated by CG methylation. Cell 130:851–862
Matzke M, Kanno T, Daxinger L, Huettel B, Matzke AJ (2009) RNA-mediated chromatin-based silencing in plants. Curr Opin Cell Biol 21:367–376
Meyers BC, Kaushik S, Nandety RS (2005) Evolving disease resistance genes. Curr Opin Plant Biol 8:129–134
Mito Y, Henikoff JG, Henikoff S (2005) Genome-scale profiling of histone H3.3 replacement patterns. Nat Genet 37:1090–1097
Molinier J, Ries G, Zipfel C, Hohn B (2006) Transgeneration memory of stress in plants. Nature 442:1046–1049
Molnar A, Melnyk CW, Bassett A, Hardcastle TJ, Dunn R, Baulcombe DC (2010) Small silencing RNAs in plants are mobile and direct epigenetic modification in recipient cells. Science 328:872–875
Mosher RA, Melnyk CW, Kelly KA, Dunn RM, Studholme DJ, Baulcombe DC (2009) Uniparental expression of PolIV-dependent siRNAs in developing endosperm of *Arabidopsis*. Nature 460:283–286
Ortega-Galisteo AP, Morales-Ruiz T, Ariza RR, Roldan-Arjona T (2008) *Arabidopsis* DEMETER-LIKE proteins DML2 and DML3 are required for appropriate distribution of DNA methylation marks. Plant Mol Biol 67:671–681
Pavet V, Quintero C, Cecchini NM, Rosa AL, Alvarez ME (2006) *Arabidopsis* displays centromeric DNA hypomethylation and cytological alterations of heterochromatin upon attack by *Pseudomonas syringae*. Mol Plant Microbe Interact 19:577–587
Pecinka A, Dinh HQ, Baubec T, Rosa M, Lettner N, Scheid OM (2010) Epigenetic regulation of repetitive elements is attenuated by prolonged heat stress in *Arabidopsis*. Plant Cell 22(9):3118–3129
Pecinka A, Rosa M, Schikora A, Berlinger M, Hirt H, Luschnig C, Mittelsten SO (2009) Transgenerational stress memory is not a general response in *Arabidopsis*. PLoS One 4(4):e5202
Pellegrineschi A, Reynolds M, Pacheco M, Brito RM, Almeraya R, Yamaguchi-Shinozaki K, Hoisington D (2004) Stress-induced expression in wheat of the *Arabidopsis thaliana DREB1A* gene delays water stress symptoms under greenhouse conditions. Genome 47:493–500
Penterman J, Uzawa R, Fischer RL (2007a) Genetic interactions between DNA demethylation and methylation in *Arabidopsis*. Plant Physiol 145:1549–1557
Penterman J, Zilberman D, Huh JH, Ballinger T, Henikoff S, Fischer RL (2007b) DNA demethylation in the *Arabidopsis* genome. Proc Natl Acad Sci U S A 104:6752–6757
Perl A, Perl-Treves R, Galili S, Aviv D, Shalgi E, Malkin S, Galun E (1993) Enhanced oxidative stress defense in transgenic potato expressing tomato Cu, Zn superoxide dismutases. Theor Appl Genet 85:568–576
Pien S, Grossniklaus U (2007) Polycomb group and trithorax group proteins in *Arabidopsis*. Biochim Biophys Acta 1769:375–382
Pontes O, Costa-Nunes P, Vithayathil P, Pikaard CS (2009) RNA polymerase V functions in *Arabidopsis* interphase heterochromatin organization independently of the 24-nt siRNA-directed DNA methylation pathway. Mol Plant 2:700–710

Probst AV, Fagard M, Proux F, Mourrain P, Boutet S, Earley K, Lawrence RJ, Pikaard CS, Murfett J, Furner I, Vaucheret H, Mittelsten Scheid O O (2004) *Arabidopsis* histone deacetylase HDA6 is required for maintenance of transcriptional gene silencing and determines nuclear organization of rDNA repeats. Plant Cell 16:1021–1034

Quimlo CA, Torrizo LB, Setter TL, Ellis M, Grover A, Abrigo EM, Oliva NP, Ella ES, Carpena AL, Ito O, Peacock WJ, Dennis E, Datta SK (2000) Enhancement of submergence tolerance in transgenic rice plants overproducing pyruvate decarboxylase. J Plant Physiol 156:516–521

Raisner RM, Madhani HD (2006) Patterning chromatin: form and function for H2A.Z variant nucleosomes. Curr Opin Genet Dev 16:119–124

Rando OJ, Ahmad K (2007) Rules and regulation in the primary structure of chromatin. Curr Opin Cell Biol 19:250–256

Ries G, Buchholz G, Frohnmeyer H, Hohn B (2000) UV-damage-mediated induction of homologous recombination in *Arabidopsis* is dependent on photosynthetically active radiation. Proc Natl Acad Sci U S A 97:13425–13429

Rohila JS, Jain RK, Wu R (2002) Genetic improvement of Basmati rice for salt and drought tolerance by regulated expression of a barley *Hva1* cDNA. Plant Sci 163:525–532

Ruthenburg AJ, Allis CD, Wysocka J (2007) Methylation of lysine 4 on histone H3: intricacy of writing and reading a single epigenetic mark. Mol Cell 25:15–30

Santos-Rosa H, Schneider R, Bannister AJ, Sherriff J, Bernstein BE, Emre NC, Schreiber SL, Mellor J, Kouzarides T (2002) Active genes are tri-methylated at K4 of histone H3. Nature 419:407–411

Saze H (2008) Epigenetic memory transmission through mitosis and meiosis in plants. Semin Cell Dev Biol 19:527–536

Scippa GS, Di Michele M, Onelli E, Patrignani G, Chiatante D, Bray EA (2004) The histone-like protein H1-S and the response of tomato leaves to water deficit. J Exp Bot 55:99–109

Seki M, Narusaka M, Abe H, Kasuga M, Yamaguchi-Shinozaki K, Carninci P, Hayashizaki Y, Shinozaki K (2001) Monitoring the expression pattern of 1300 *Arabidopsis* genes under drought and cold stresses by using a full-length cDNA microarray. Plant Cell 13:61–72

Sha AH, Lin XH, Huang JB, Zhang DP (2005) Analysis of DNA methylation related to rice adult plant resistance to bacterial blight based on methylation-sensitive AFLP (MSAP) analysis. Mol Genet Genomics 273:484–490

Sheldon CC, Rouse DT, Finnegan EJ, Peacock WJ, Dennis ES (2000) The molecular basis of vernalization: the central role of *FLOWERING LOCUS C* (*FLC*). Proc Natl Acad Sci U S A 97:3753–3758

Shou H, Bordallo P, Wang K (2004) Expression of the *Nicotiana* protein kinase (NPK1) enhanced drought tolerance in transgenic maize. J Exp Bot 55:1013–1019

Simon SA, Meyers BC (2011) Small RNA-mediated epigenetic modifications in plants. Curr Opin Plant Biol 14:148–155

Sivamani E, Bahieldin A, Wraith JM, Al-Niemi T, Dyer WE, Ho THD, Qu R (2000) Improved biomass productivity and water use efficiency under water deficit conditions in transgenic wheat constitutively expressing the barley *HVA1* gene. Plant Sci 155:1–9

Slotkin RK, Vaughn M, Borges F, Tanurdzic M, Becker JD, Feijo JA, Martienssen RA (2009) Epigenetic reprogramming and small RNA silencing of transposable elements in pollen. Cell 136:461–472

Smith AP, Jain A, Deal RB, Nagarajan VK, Poling MD, Raghothama KG, Meagher RB (2010) Histone H2A.Z regulates the expression of several classes of phosphate starvation response genes but not as a transcriptional activator. Plant Physiol 152:217–225

Sokol A, Kwiatkowska A, Jerzmanowski A, Prymakowska-Bosak M (2007) Up-regulation of stress-inducible genes in tobacco and *Arabidopsis* cells in response to abiotic stresses and ABA treatment correlates with dynamic changes in histone H3 and H4 modifications. Planta 227:245–254

Soppe WJJ, Jacobsen SE, Alonso-Blanco C, Jackson JP, Kakutani T, Koornneef M, Peeters AJM (2000) The late flowering phenotype of *fwa* mutants is caused by gain-of-function epigenetic alleles of a homeodomain gene. Mol Cell 6:791–802

Sridha S, Wu K (2006) Identification of *AtHD2C* as a novel regulator of abscisic acid responses in *Arabidopsis*. Plant J 46:124–133

Steward N, Ito M, Yamaguchi Y, Koizumi N, Sano H (2002) Periodic DNA methylation in maize nucleosomes and demethylation by environmental stress. J Biol Chem 277:37741–37746

Stockinger EJ, Mao Y, Regier MK, Triezenberg SJ, Thomashow MF (2001) Transcriptional adaptor and histone acetyltransferase proteins in *Arabidopsis* and their interactions with CBF1, a transcriptional activator involved in cold regulated gene expression. Nucleic Acids Res 29:1524–1533

Sunkar R, Chinnusamy V, Zhu J, Zhu JK (2007) Small RNAs as big players in plant abiotic stress responses and nutrient deprivation. Trends Plant Sci 12(7):301–309

Sunkar R, Kapoor A, Zhu JK (2006) Posttranscriptional induction of two cu/zn superoxide dismutase genes in *Arabidopsis* is mediated by downregulation of miR398 and important for oxidative stress tolerance. Plant Cell 18:2051–2065

Sunkar R, Zhu JK (2004) Novel and stress-regulated microRNAs and other small RNAs from *Arabidopsis*. Plant Cell 16(8):2001–2019

Swiezewski S, Crevillen P, Liu F, Ecker JR, Jerzmanowski A, Dean C (2007) Small RNA-mediated chromatin silencing directed to the 3' region of the *Arabidopsis* gene encoding the developmental regulator, FLC. Proc Natl Acad Sci U S A 104:3633–3638

Takeda S, Sugimoto K, Otsuki H, Hirochika H (1999) A 13-bp *cis*-regulatory element in the LTR promoter of the tobacco retrotransposon *Tto1* is involved in responsiveness to tissue culture, wounding, methyl jasmonate and fungal elicitors. Plant J 18(4):383–393

Talbert PB, Henikoff S (2010) Histone variants – ancient wrap artists of the epigenome. Nat Rev Mol Cell Biol 11:264–275

Tittel-Elmer M, Bucher E, Broger L, Mathieu O, Paszkowski J, Vaillant I (2010) Stress-induced activation of heterochromatic transcription. PLoS Genet 6(10):e1001175

Tsuji H, Saika H, Tsutsumi N, Hirai A, Nakazono M (2006) Dynamic and reversible changes in histone H3-Lys4 methylation and H3 acetylation occurring at submergence-inducible genes in rice. Plant Cell Physiol 47:995–1003

Turck F, Roudier F, Farrona S, Martin-Magniette ML, Guillaume E, Buisine N, Gagnot S, Martienssen RA, Coupland G, Colot V (2007) *Arabidopsis* TFL2/LHP1 specifically associates with genes marked by trimethylation of histone H3 lysine 27. PLoS Genet 3:e86

Turunen M, Latola K (2005) UV-B radiation and acclimation in timberline plants. Environ Pollut 137:390–403

Tuttle JR, Idris AM, Brown JK, Haigler CH, Robertson D (2008) Geminivirus-mediated gene silencing from cotton leaf crumple virus is enhanced by low temperature in cotton. Plant Physiol 148:41–50

van Attikum H, Gasser SM (2009) Crosstalk between histone modifications during the DNA damage response. Trends Cell Biol 19:207–217

Vaughn MW, Tanurdzić M, Lippman Z, Jiang H, Carrasquillo R, Rabinowicz PD, Dedhia N, McCombie WR, Agier N, Bulski A, Colot V, Doerge RW, Martienssen RA (2007) Epigenetic natural variation in *Arabidopsis thaliana*. PLoS Biol 5:e174

Vazquez F (2006) *Arabidopsis* endogenous small RNAs: highways and byways. Trends Plant Sci 11:460–468

Verhoeven KJ, Jansen JJ, van Dijk PJ, Biere A (2010) Stress-induced DNA methylation changes and their heritability in asexual dandelions. New Phytol 185(4):1108–1118

Voinnet O (2009) Origin, biogenesis, and activity of plant microRNAs. Cell 136:669–687

Vongs A, Kakutani T, Martienssen RA, Richards EJ (1993) *Arabidopsis thaliana* DNA methylation mutants. Science 260:1926–1928

Wada Y, Miyamoto K, Kusano T, Sano H (2004) Association between up-regulation of stress-responsive genes and hypomethylation of genomic DNA in tobacco plants. Mol Genet Genomics 271(6):658–666

Wang GL, Ruan DL, Song WY, Sideris S, Chen L, Pi LY, Zhang S, Zhang Z, Fauquet C, Gaut BS, Whalen MC, Ronald PC (1998) *Xa21D* encodes a receptor-like molecule with a leucine-rich repeat domain that determines race-specific recognition and is subject to adaptive evolution. Plant Cell 10:765–779

Wang G-L, Song W-Y, Ruan D-L, Sideris S, Ronald PC (1996) The cloned gene, *Xa21*, confers resistance to multiple *Xanthomonas oryzae* pv. *oryzae* isolates in transgenic plants. Mol Plant Microbe Interact 9:850–855

Whittle CA, Otto SP, Johnston MO, Krochko JE (2009) Adaptive epigenetic memory of ancestral temperature regime in *Arabidopsis thaliana*. Botany 87:650–657

Woo HR, Pontes O, Pikaard CS, Richards EJ (2007) VIM1, a methyl-cytosine binding protein required for centromeric heterochromatinization. Genes Dev 21:267–277

Yao Y, Bilichak A, Golubov A, Blevins T, Kovalchuk I (2010) Differential sensitivity of *Arabidopsis* siRNA biogenesis mutants to genotoxic stress. Plant Cell Rep 29:1401–1410

Yao Y, Kovalchuk I (2011) Abiotic stress leads to somatic and heritable changes in homologous recombination frequency, point mutation frequency and microsatellite stability in *Arabidopsis* plants. Mutat Res 707:61–66

Youngson NA, Whitelaw E (2008) Transgenerational epigenetic effects. Annu Rev Genomics Hum Genet 9:233–257

Zemach A, Grafi G (2007) Methyl-CpG-binding domain proteins in plants: interpreters of DNA methylation. Trends Plant Sci 12(2):80–88

Zemach A, Li Y, Wayburn B, Ben-Meir H, Kiss V, Avivi Y, Kalchenko V, Jacobsen SE, Grafi G (2005) DDM1 binds *Arabidopsis* methyl-CpG binding domain proteins and affects their subnuclear localization. Plant Cell 17:1549–1558

Zhang W, Lee HR, Koo DH, Jiang J (2008) Epigenetic modification of centromeric chromatin: hypomethylation of DNA sequences in the CENH3-associated chromatin in *Arabidopsis thaliana* and maize. Plant Cell 20:25–34

Zhang X, Clarenz O, Cokus S, Bernatavichute YV, Pellegrini M, Goodrich J, Jacobsen SE (2007) Whole-genome analysis of histone H3 lysine 27 trimethylation in *Arabidopsis*. PLoS Biol 5:e129

Zhang X, Yazaki J, Sundaresan A, Cokus S, Chan SW, Chen H, Henderson IR, Shinn P, Pellegrini M, Jacobsen SE, Ecker JR (2006) Genome-wide high-resolution mapping and functional analysis of DNA methylation in *Arabidopsis*. Cell 126:1189–1201

Zheng X, Pontes O, Zhu J, Miki D, Zhang F, Li WX, Iida K, Kapoor A, Pikaard CS, Zhu JK (2008) ROS3 is an RNA-binding protein required for DNA demethylation in *Arabidopsis*. Nature 455:1259–1262

Zhou C, Zhang L, Duan J, Miki B, Wu K (2005) HISTONE DEACETYLASE19 is involved in jasmonic acid and ethylene signaling of pathogen response in *Arabidopsis*. Plant Cell 17:1196–1204

Zhou X, Sunkar R, Jin H, Zhu JK, Zhang W (2009) Genome-wide identification and analysis of small RNAs originated from natural antisense transcripts in *Oryza sativa*. Genome Res 19:70–78

Zhu J, Jeong JC, Zhu Y, Sokolchik I, Miyazaki S, Zhu JK, Hasegawa PM, Bohnert HJ, Shi H, Yun DJ, Bressan RA (2007a) Involvement of *Arabidopsis HOS15* in histone deacetylation and cold tolerance. Proc Natl Acad Sci U S A 105:4945–4950

Zhu J, Kapoor A, Sridhar VV, Agius F, Zhu JK (2007b) The DNA glycosylase/lyase ROS1 functions in pruning DNA methylation patterns in *Arabidopsis*. Curr Biol 17:54–59

Zhu JK (2009) Active DNA demethylation mediated by DNA glycosylases. Annu Rev Genet 43:143–166

Zilberman D, Coleman-Derr D, Ballinger T, Henikoff S (2008) Histone H2A.Z and DNA methylation are mutually antagonistic chromatin marks. Nature 456:125–129

Zilberman D, Gehring M, Tran RK, Ballinger T, Henikoff S (2007) Genome-wide analysis of *Arabidopsis thaliana* DNA methylation uncovers an interdependence between methylation and transcription. Nat Genet 39:61–69

Chapter 11
Physiological Role of Nitric Oxide in Plants Grown Under Adverse Environmental Conditions

Mirza Hasanuzzaman, Sarvajeet Singh Gill, and Masayuki Fujita

Abbreviations

ABA	Abscisic acid
APX	Ascorbate peroxidase
AsA	Ascorbic acid
ATP	Adenosine triphosphate
CAT	Catalase
CDPK	Calcium-dependent protein kinase
cGMP	Cyclic guanosine monophosphate
cPTIO	2-(4-Carboxyphenyl)-4,4,5,5-tetramethylimidazoline-l-oxyl-3-oxide
COX	Cytochrome c oxidase
DHA	Dehydroascorbate
DHAR	Dehydroascorbate reductase
ETC	Electron transport chain
GAP	Glycerinaldehyde-3-phosphate
GPX	Glutathione peroxidase
GR	Glutathione reductase

M. Hasanuzzaman
Laboratory of Plant Stress Responses, Department of Applied Biological Science, Kagawa University, Miki-cho, Kita-gun, Kagawa 761-0795, Japan

Department of Agronomy, Sher-e-Bangla Agricultural University, Dhaka 1207, Bangladesh
e-mail: mhzsauag@yahoo.com

S. Singh Gill
Stress Physiology and Molecular Biology Lab, Centre for Biotechnology, MD University, Rohtak 124 001, India

M. Fujita (✉)
Laboratory of Plant Stress Responses, Department of Applied Biological Science, Kagawa University, Miki-cho, Kita-gun, Kagawa 761-0795, Japan
e-mail: fujita@ag.kagawa-u.ac.jp

N. Tuteja and S. Singh Gill (eds.), *Plant Acclimation to Environmental Stress*,
DOI 10.1007/978-1-4614-5001-6_11,

GS	Glutathione synthase
GSH	Reduced glutathione
GSNO	*S*-nitrosoglutathione
GSSG	Oxidized glutathione
GST	Glutathione *S*-transferase
IAA	Indole-3-acetic acid
JA	Jasmonic acid
LNNA	N^{ω}-nitro L-arginine
LOOH	Lipid hydroperoxides
MAPK	Mitogen-activated protein kinase
MDA	Malondialdehyde
MDHA	Monodehydroascorbate
MDHAR	Monodehydroascorbate reductase
NADH	Nicotinamide adenine dinucleotide
NADPH	Nicotinamide adenine dinucleotide phosphate
$NADPH_{ox}$	NADPH oxidases
NiR	Nitrite reductase
NO	Nitric oxide
NOHA	N-hydroxyarginine
NOS	Nitric oxide synthase
NR	Nitrate reductase
PA	Polyamine
PAL	Phenylalanine ammonia-lyase
PCD	Programmed cell death
PEG	Polyethylene glycol
POX	Peroxidases
RNS	Reactive nitrogen species
ROOH	Organic hydroperoxides
ROS	Reactive oxygen species
RWC	Relative water content
SA	Salicylic acid
SAM	*S*-adenosyl methionine
SNAP	*S*-nitroso-*N*-acetylpenicillamine
SNP	Sodium nitroprusside
TFBS	Transcription factor binding sites
XDH	Xanthine dehydrogenase
XOR	Xanthine oxidoreductase
γ-ecs	γ-Glutamylcysteine synthetase

1 Introduction

By 2050, the world's population will have increased by a third and demand for agricultural products will rise by 70% (Noble and Ruaysoongnern 2010). In meeting future food production demands without consuming more land, it is necessary to boost up the yield of crop. However, due to rapid climate changes crop plants are suffering from different adverse conditions, termed as abiotic stress. Abiotic stresses, particularly salinity, drought, temperature extremes, flooding, toxic metals, high-light intensity, UV-radiation, herbicides, and ozone, are the major causes of yield loss in cultivated crops worldwide and pose major threats to agriculture and food security (Rodríguez et al. 2005; Acquaah 2007). Abiotic stress leads to a series of morphological, physiological, biochemical, and molecular changes that adversely affect plant growth and productivity (Wang et al. 2001). However, the rapidity and efficiency of these responses may be decisive for the viability of the given species. Plants are only able to survive under such stressful conditions if they are able to perceive the stimulus, generate and transmit signals, and initiate various physiological and biochemical changes (Bohnert and Jensen 1996). Abiotic stresses can also lead to oxidative stress through the increase in reactive oxygen species (ROS), including singlet oxygen ($^1O_2$), superoxide (O_2^-), hydrogen peroxide (H_2O_2), and hydroxyl radicals (OH·), all of which are highly reactive and may cause cellular damage through oxidation of lipids, proteins, and nucleic acids (Apel and Hirt 2004; Gill and Tuteja 2010).

Exploring suitable crop improvements or ways to alleviate stress is one of the tasks of plant biologists. Nitric oxide (NO) is a highly reactive, membrane-permeable free radical which was previously considered to be a highly toxic compound (Gordge 1998). The discovery and elucidation of its biological functions in the 1980s came as a surprise. NO was named "Molecule of the Year" in 1992 by the journal "Science," a NO Society was founded, and a scientific journal, "Nitric Oxide," devoted entirely to NO, was created (Delledonne 2005). Its emission from plants has been reported several years ago in soybean plants (Klepper 1979). Later, *in vivo* and *in vitro* nitrate reductase (NR)-dependent NO production has been found in other plants such as sunflower and maize (Rockel et al. 2002). However, the discovery of NO's signaling role in cardiovascular system regulation has changed the paradigm concerning its cytotoxicity (Korhonen et al. 2005). The biological functions of NO have gradually been elucidated. NO can provoke both beneficial and harmful effects in plant cells (Hasanuzzaman et al. 2010). This dual role probably depends on the local concentration of NO as an effect of the rate of synthesis, translocation, effectiveness of removal of this reactive nitrogen species, as well as its ability to directly interact with other molecules and signals (Arasimowicz and Floryszak-Wieczorek 2007).

In plant system, many possible sources work together for the production or synthesis of NO which depends on the plant species, plant organs, environmental conditions, and the signal pathway in the plant (Neill et al. 2002a). Recently, different groups reviewed the sources of NO in plant (Popova and Tuan 2010; Baudouin 2011; Misra et al. 2011a). It can be produced non-enzymatically or enzymatically through cytosolic nitrate reductase (NR), plasma membrane nitrite reductase (NiR), nitric oxide synthase (NOS) and xanthine dehydrogenase (XDH), etc.

Research on NO in plants has gained considerable attention in recent years mainly due to its function in plant growth and development and as a key signaling molecule in different intracellular processes. Nitric oxide now can be designated as a "jack-of-all-trades" molecule which regulates plant cell responses under physiological conditions throughout the life cycle of plants (Yemets et al. 2011). As reviewed in several recent reports (Besson-Bard et al. 2008; Wilson et al. 2008; Leitner et al. 2009; Hao and Zhang 2010; Corpas et al. 2011; Mazid et al. 2011a, b; Siddiqui et al. 2011; Wimalasekera et al. 2011), NO production has been associated with a number of physiological situations in plants. These cover the entire lifespan of the plant and include germination (Šírová et al. 2011), root development (Yemets et al. 2011), nodulation (del Giudice et al. 2011; Meilhoc et al. 2011), control of stomatal movements (Hancock et al. 2011; He et al. 2011), flowering (Khurana et al. 2011), pollen tube growth (Šírová et al. 2011), and leaf senescence (Procházková and Wilhelmová 2011).

Recently, NO has emerged as an important signaling molecule and antioxidant. NO triggers many kinds of redox-regulated (defense-related) gene expressions, directly or indirectly, to establish plant stress tolerance (Polverari et al. 2003; Sung and Hong 2010). Several recent reports indicated that the application of exogenous NO donors confers tolerance to various abiotic stresses like salinity (Hasanuzzaman et al. 2011a), drought (Bai et al. 2011), high temperature (Hossain et al. 2010b), chilling (Liu et al. 2011), toxic metals (Xiong et al. 2010), flooding (Gupta et al. 2011), high-light intensity (Xu et al. 2010c), UV-B radiation (Kim et al. 2010), and elevated ozone (Ahlfors et al. 2009). It was also suggested that NO, itself, possesses antioxidant properties and might act as a signal in activating ROS-scavenging enzyme activities under various abiotic stresses (Palavan-Unsal and Arisan 2009; Hao and Zhang 2010; Mazid et al. 2011a; Siddiqui et al. 2011). Several lines of study have shown that the protective effect of NO against abiotic stress is closely related to the NO-mediated reduction of ROS in plants (Beligni and Lamattina 1999a; Wang and Yang 2005; Hao and Zhang 2010; Corpas et al. 2011).

In this chapter, we discuss recent progress in understanding the function of NO in plant responses and tolerance to abiotic stresses and in plant development. The physiological and biochemical mechanisms of NO-induced abiotic stress tolerance and the translation of signal transduction into cellular responses towards stress tolerance are the foci of this review.

2 Nitric Oxide Synthesis/Production in Plants

In plant system, many possible sources work together for the production or synthesis of NO which depends on the plant species, plant organs, environmental conditions, and the signal pathway in the plant (Neill et al. 2002a). Recently, different groups reviewed the sources of NO in plant (Popova and Tuan 2010; Baudouin 2011; Misra et al. 2011a); Fig. 11.1.

Higher plants can react both to the atmospheric or soil NO and they are also able to emit substantial amounts of NO (Durner and Klessig 1999). In the atmosphere, nitrification/denitrification cycles provide NO as a by-product of N_2O oxidation into the atmosphere. Nitrification of NH_4^+ is the primary source of N_2 emitted to the atmosphere, where it oxidizes to NO and NO_2^- (Wojtaszek 2000). In plant, NO can be formed both enzymatically and non-enzymatically (Fig. 11.1).

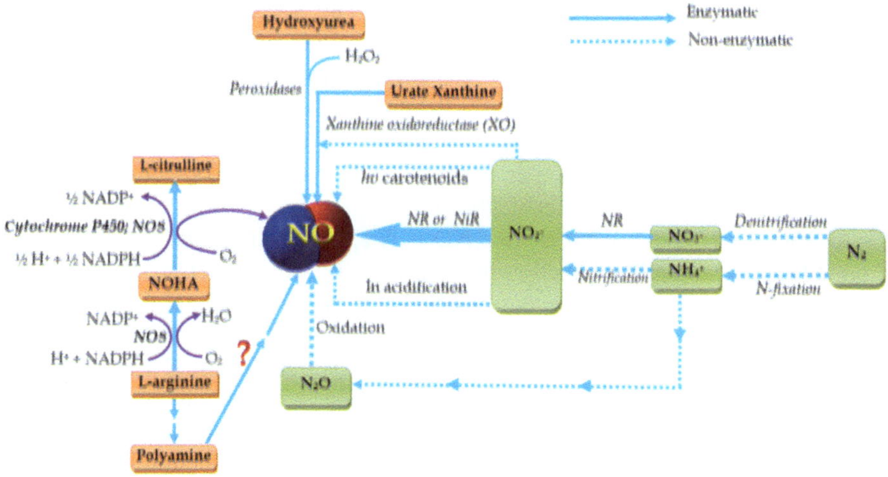

Fig. 11.1 Different mechanisms of NO synthesis/production in plant

Production of NO from NO_2^- is a common non-enzymatic phenomenon which occurs at low pH compartments (Igamberdiev et al. 2010). In this case, NO_2^- can dismutate to NO and NO_3^- (Stöhr and Ullrich 2002). The generation *in vitro* of NO by the reaction of H_2O_2 (10–50 mM) and L-arginine (10–20 mM) at pH 7.4 and 37 °C has been reported by Nagase et al. (Nagase et al. 1997). The non-enzymatic synthesis of NO has also been demonstrated by Gotte et al. (2002), with short-time kinetics, by shock waves treatment of solutions containing 1 mMH_2O_2 and 10 mM L-arginine. Beligni et al. (Beligni et al. 2002) obtained the NO synthesis in barley aleurone cells as reduction of NO_2^- by AsA at acidic pH. Light-mediated reduction of NO_2^- by carotenoids was also proposed as another non-enzymatic mechanism of NO formation (Cooney et al. 1994).

There are several enzymes in plants that may produce NO. The key enzymes involved in the production of NO in plants are: cytosolic nitrate reductase (NR; EC 1.6.6.1), plasma membrane nitrite reductase (NiR, EC. 1.6.6.4), nitric oxide synthase (NOS; EC 1.14.13.39), and xanthine dehydrogenase (XDH; EC 1.1.1.204). One of the major origin of NO production in plants, however, is probably through the action of NADPH-dependent NR which provided the first known mechanism to make NO in plants. This enzyme can generate NO from NO_2^- with NAD(P)H as electron donor and the catalysis site is probably the molybdenum cofactor (Moco) (Yamasaki et al. 1999; Rockel et al. 2002; Crawford 2006; Ferreira and Cataneo 2010); Fig 11.1). This is the only enzyme whose NO-producing activity has been confirmed both *in vivo* and *in vitro* (Courtois et al. 2008; Kaiser et al. 2002). In plant cells, NO_2^- can be accumulated when the photosynthetic activity is inhibited or under anaerobic conditions (Lamattina et al. 2003; Rockel et al. 2002). Production of NO, dependent on NR activity, was recorded in many cultivated plants such as in *Ccucumis sativus* (de la Haba et al. 2001), *Helianthus annuus*, *Zea mays* (Rockel et al. 2002), *Triticum aestivum* (Xu and Zhao 2003), *Nicotiana tabacum* (Planchet et al. 2005), and *Medicago truncatula* (Horchani et al. 2011). Recently, a number of plant studies provided evidence for the role of NR in NO synthesis in plant (Moreau et al. 2010). It has been reported that NR is responsible for NO production during stomatal closure (Desikan et al. 2002; Bright et al. 2006; Neill et al. 2008), in response to defense elicitors (Shi and Li 2008; Srivastava et al. 2009; Wu et al. 2009), under abiotic stress (Sang et al. 2008), and during developmental processes (Kolbert et al. 2008; Seligman et al. 2008).

Another enzyme that can generate NO is NiR by which plants synthesize NO from NO_2^-. Nitric oxide production in plants by NiR has been observed in several plant species, viz., *Helianthus annuus* (Rockel et al. 2002), *Glycine max* (Delledonne et al. 1998), and *Chlamydomonas reinhardtii* (Sakihama et al. 2002). A plasma membrane-bound, root-specific enzyme, NO_2^--NO oxidoreductase (Ni-NOR), using cytochrome c as an electron donor in vivo and having a comparatively reduced pH optimum is reported by Stöhr and Stremlau (Stöhr and Stremlau 2006). Recently, Gupta and Kaiser (2010) showed the NO_2^--dependent NO production in plant cells under anoxic condition, which is localized in and mediated by the electron transport chain in the mitochondrial membranes.

Nitric oxide synthase is another enzyme for NO synthesis in plants, whose activity in higher plants was first reported by Cueto et al. (1996) as well as Ninnemann and Maier (1996) by using the method of conversion of arginine, the substrate of NOS, into citrulline. Since last 20 years, there have been an increasing number of reports showing the presence of NOS activity in plants similar, to a certain extent, to mammalian NOS (del Río et al. 2004). Later, NOS-like activity in plants has been detected widely. Corpas et al. (2006) showed arginine-dependent NOS activity, which was dependent on the plant organ and its developmental stage. The enzymatic oxidation of L-arginine to yield NO and L-citrulline has been reported in extracts from *Pisum sativum* (Leshem and Haramaty 1996), *Glycine max* (Delledonne et al. 1998), *Nicotiana tabacum* (Durner et al. 1998), and *Zea mays* (Ribeiro et al. 1999), which implicated NOS activity. NOS (Moncada et al. 1991) catalyzes the two-step oxidation of L-arginine to NO and citrulline (L-arginine + NADPH + H + $O_2 \rightarrow N^{\omega}$ hydroxyarginine + O_2 + $NADP^+$ + H_2O and thereafter N^{ω} hydroxyarginine + ½ NADP H + ½ $H^+ \rightarrow$ Citrulline + NO^+ + ½ $NADP^+$ H_2O), a reaction that might also be catalyzed by a cytochrome P450 (Boucher et al. 1992; Wojtaszek 2000); Fig. 11.1). Zemojtel et al. (2004) postulated the discovery of a novel conserved family of NOS and showed significant homology in NOS sequence in as divergent organisms as plants, snails, and mammals. In fact, the discovery of a new class of NOS in *Arabidopsis thaliana* is a real breakthrough in the studies on NO occurrence and function in plants. Recently, Gas et al. (2009) reported that plant NOS provides new evidence of a central role for plastids in NO metabolism.

In addition to these enzymes, xanthine oxidase/dehydrogenase (XDH) also been rarely suggested as a source for NO using NO_2^- and xanthine as substrates (Millar et al. 1998). Xanthine oxidoreductase (XOR) is another Moco-containing enzyme which has been recently demonstrated to produce NO (Harrison 2002). It occurs into two interconvertible forms: the O_2^--producing XO (form O; EC1.1.3.22) and xanthine dehydrogenase (form D; EC1.1.1.204) (Palma et al. 2002). XOR has been found present in pea leaf peroxisomes where the preponderant form of the enzyme is xanthine oxidase (XO) and only a 30% is present as xanthine dehydrogenase (XD) (Corpas et al. 1997; Sandalio et al. 1988). More recently, horseradish peroxidase was also demonstrated to generate NO from hydroxyurea and H_2O_2 (Huang et al. 2002; Veitch 2004). Other heme proteins that have been proposed as good candidates for the enzymatic generation of NO are cytochromes P450. These proteins have been shown to catalyze the oxidation of NOHA (N-hydroxyarginine) by NADPH and O_2 with the generation of NO (Boucher et al. 1992; Mansuy and Boucher 2002); Fig. 11.1). Hemoglobin and catalase (CAT) were also reported to produce NO and other nitrogen oxides by catalyzing the oxidation of NOHA (Boucher et al. 1992).

Because of this rapid response and having direct correlation between polyamines (PAs) and NO, a number of studies reported that PAs like spermine and spermidine trigger NO production in planta (Tun et al. 2006; Gaupels et al. 2008; Groppa et al. 2008); Fig. 11.1). The discovery that hydroxylamines (R-NHOH) can be oxidized to NO by $O_2^{\cdot-}$- or H_2O_2-generating systems, as well as by tobacco cells, has led to the recent proposal of another oxidative pathway for NO synthesis (Rumer et al. 2009). Gao et al. (2009) found that PA levels correlate with NO because L-arginine is a common precursor in their biosynthesis. However, the efficiency of this oxidative process is low and the existence of hydroxylamines in plants has not been confirmed (Moreau et al. 2010).

3 Signaling Mechanisms of NO

In plants NO regulates several physiological processes such as germination, growth, nodulation, stomatal closure, flowering, orientation of pollen tubes, adaptation to abiotic and biotic stresses, and cell death (Delledonne 2005; Krasylenko et al. 2010; Misra et al. 2011a, b). Although the underlying mechanisms by which this is achieved are still unrevealed, different plant studies through the application of NO donor provided the evidence supporting the signaling role of NO (Wendehenne et al. 2006). To play the signaling function, a molecule has to possess certain properties facilitating its direct influence on second messengers. Properties of a signaling molecule, such as a simple structure, small dimensions, and high diffusivity, are obviously found in a molecule of NO (Arasimowicz and Floryszak-Wieczorek 2007). Nitric oxide is highly reactive due to the presence of an unpaired electron, which explains its existence in a cell as three interchangeable species such as NO^- (nitroxyl anion), $NO\cdot$(free radical), and NO^+ (nitrosonium cation) usually referred to as RNS (Stamler et al. 1992; Neill et al. 2003). In response, the main question to be answered is how NO regulates these diverse biological processes. Some studies do shed some light on the subject. Different experimental results indicated that NO is an endogenous signal in plants that mediates responses to several stimuli which is outlined in Fig. 11.2.

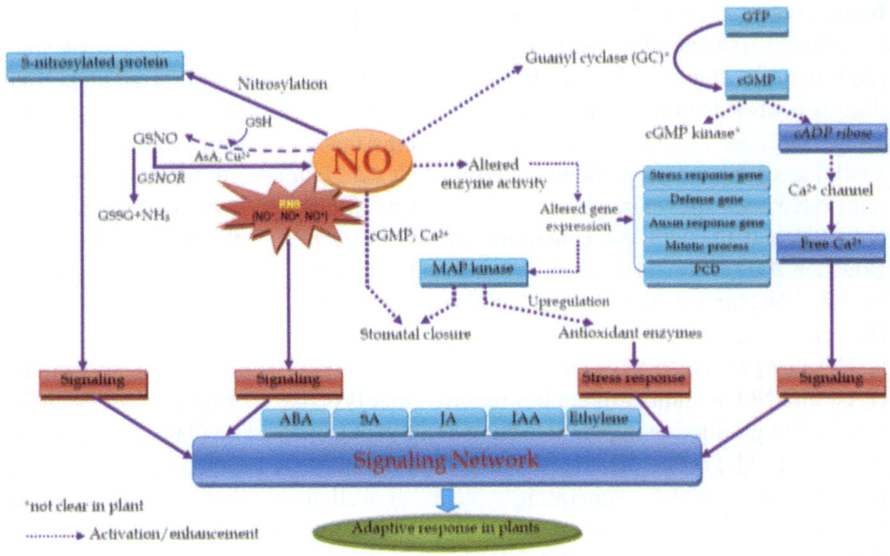

Fig. 11.2 NO signaling network in plant

The signaling function of NO mediated by direct and indirect interactions can be accomplished in individual cells and even in microcompartments, which is consistent with a recently suggested notion on the role of Ca^{2+}, H_2O_2, and cyclic nucleotides (Krasylenko et al. 2010). The modulating effect of NO on signal transduction in plant cells might be mediated by its influence on cyclic guanosine monophosphate (cGMP), cADP-ribose, and Ca^{2+} levels (Correa-Aragunde et al. 2006; Pagnussat et al. 2004), as well as on mitogen-activated protein kinase (MAPK, Leitner et al. 2009) and on gene expression profiles (Besson-Bard et al. 2009a, b). In the signaling network, NO is also interrelated with other signaling molecules (Fig 11.2). The cross talk between NO, protein kinases, the second messengers (Ca^{2+}, cGMP and cADPR, phosphatidic acid, ROS), and also phytohormones, provides the molecular basis for many physiological processes indirectly regulated by NO in plant cell (Lamotte et al. 2006; Besson-Bard et al. 2008; Courtois et al. 2008; Erdei and Colbert 2008; Wilson et al. 2008; Lanteri et al. 2008; Leitner et al. 2009).

The cGMP was first detected in *Zea mays* by Janistyn (1983) and then in *Phaseolus vulgaris* by Newton et al. (1999). Later, the evidence that cGMP is an NO signaling intermediate has been reported in several systems (Neill et al. 2003; Delledonne 2005). This signaling pathway showed increases in cytosolic Ca^{2+} either by a release from intracellular sources or by influx from the extracellular environment. The other main procedure in signaling pathway is reversible protein phosphorylation (Palavan-Unsal and Arisan 2009). Several experimental results indicated the necessity for cGMP synthesis and its action for plant responses to NO. The necessity of cGMP for abscisic acid (ABA)- and NO-induced stomatal closure has been identified in *Pisum sativum* and *Arabidopsis* (Neill et al. 2002b).

Palavan-Unsal and Arisan (2009) concluded that cGMP is an intracellular mediator for some signaling pathways, but for others additional signals are necessary for this process. Donaldson et al. (2004) reported that stress-induced enhanced ABA synthesis caused a rapid increase in the cGMP content of *Arabidopsis* seedlings. It would appear that although an elevated level of cGMP is required for effective ABA-induced stomatal closure, additional signaling pathways stimulated by ABA must operate in concert for such an increase to mediate its effects (Misra et al. 2011b; Neill et al. 2008). Delledonne et al. (1998) introduced an animal NOS to tobacco leaves and treated tobacco cell suspension with an NO donor (*S*-Nitrosoglutathione, GSNO) and observed a prompt increase in cGMP level. Synthesis of cGMP also correlated with NO-induced cell death in *Arabidopsis* (Clarke et al. 2000). It was also reported that NO may act through cGMP and cADPR to modulate intracellular Ca^{2+}- permeable channels in order to elevate free cytosolic Ca^{2+} levels in cells (Arasimowicz and Floryszak-Wieczorek 2007). In *Arabidopsis*, cGMP synthesis is also required during NO-induced PCD (Clarke et al. 2000; Neill et al. 2002a).

Nitric oxide or its RNS relatives may modify proteins on cysteine residues through *S*-nitrosylation or on tyrosine residues through nitration. Nitric oxide also nitrosylates metals, especially within the heme moiety. Much information has been produced by recent studies on protein *S*-nitrosylation (Besson-Bard et al. 2008; Lindermayr and Durner 2009; Moreau et al. 2010). This process leads to the formation of nitrosylated cysteine residues, either by the transfer of NO from nitrosothiols to the cysteine sulfhydryl group or by direct reaction with RNS. So far, many proteins have been identified which were nitrosylated upon treatment with GSNO in culture cell and leaf protein extracts (Abat et al. 2008; Baudouin 2011). These lead to a direct impact on plant response through metabolic adjustments as well as related to downstream signaling (Baudouin 2011). In their recent study, Holzmeister et al. (2011) postulated that the concentration of GSNO and the level of *S*-nitrosylated proteins are regulated by GSNO reductase, which seems to play a major role in NO signaling. In their study, Chaki et al. (2011) observed that mechanical wounding induces a nitrosative stress by down-regulation of GSNO reductase and an increase in *S*-nitrosothiols in *Helianthus annuus* seedlings and thus SNOs constitute a new signal in plants

Calcium ion is a well-known intracellular secondary messenger in signaling processes (Courtois et al. 2008), which is also functionally interconnected with NO signaling activity (Courtois et al. 2008; Krasylenko et al. 2010). For instance, concurrent increases of NO concentration and cytosolic level of free Ca^{2+} were found to occur during signal transduction initiated by abiotic and biotic stressors (Arasimowicz and Floryszak-Wieczorek 2007). It has been observed that cytosolic Ca^{2+} mediates the effects of NO leading to stomatal closure (Neill et al. 2002a; Garcia-Mata et al. 2003; Neill et al. 2003). In addition, treatment of NO stimulates an increase of intracellular Ca^{2+} in *Vicia faba* guard and *Nicotiana tabacum* cells (Garcia-Mata et al. 2003; Lamotte et al. 2004). Increase of cytosolic-free Ca^{2+} induced by osmotic stress and by the elicitor cryptogein in tobacco cells is also influenced by NO (Gould et al. 2003; Lamotte et al. 2004). These data clearly suggested that NO functions as a Ca^{2+}-activating intracellular compound in plant cells leading to cell signaling (Palavan-Unsal and Arisan 2009). Courtois et al. (2008) reported that Ca^{2+} also interact with NOS-like enzymes in plants.

Similar to that in mammals, NO is also known to activate MAPK signaling pathways in plant cells (Kumar and Klessig 2000; Pagnussat et al. 2004; Palavan-Unsal and Arisan 2009; Baudouin 2011). The primary targets of NO in plant cells might include MAPK. In plants, MAPKs can be activated in response to extracellular signals such as drought, cold, phytohormones, pathogen attack and osmotic stress that cause the activation of signal transduction pathways resulting in altered gene expression (Hirt 1997; Misra et al. 2011a; Palavan-Unsal and Arisan 2009). It has been reported that H_2O_2 stimulates the activation of a MAPK in *Arabidopsis* suspension cultures (Desikan et al. 1999) and H_2O_2 have been determined to activate two MAPKs in *Arabidopsis* plants, at least one of which is activated independently of salicylic acid (SA) and jasmonic acid (JA) and ethylene signaling pathways (Grant et al. 2000). In another report, the NO-activated MAPK in tobacco can also be activated by other signals such as SA (Kumar and Klessig 2000) and H_2O_2 (Samuel et al. 2000). Thus, activation of a central MAPK cascade could be a focal point of convergence of both H_2O_2 and NO signaling pathways activated in response to various stresses. However, it is still not clear whether MAPK activation by NO occurs directly or via other messengers (Lamotte et al. 2004). In order to explain signal transduction mechanisms that operate during IAA- and NO-induced adventitious root formation, Pagnussat et al. (2004) investigated the involvement of a MAPK cascade in this process. In this study, cucumber explants were treated with SNP or with SNP plus the specific NO scavenger (cPTIO) and it was observed that a MAPK signaling cascade is activated during the adventitious rooting process induced by IAA in a NO-mediated but cGMP-independent pathway. Later on, Zhang et al. (2007) also observed that MAPK activation is targeted by H_2O_2 and NO in mesophyll cells same way, which is required for downstream signaling to enhance antioxidant gene expression and enzyme activity. In their study, both ABA and H_2O_2 activate an MAPK enzyme in *Zea mays* leaves (or at least an enzyme with properties characteristic of MAPKs), but this activation is largely prevented by removal of NO with the NO scavenger cPTIO. Moreover, as with enhancement of antioxidant activity, the MAPK is activated by treatment with NO (Zhang et al. 2007). Hao and Zhang (2010) reported that there may be a causal and interdependent relationship between MAPKK/CDPK and NO in darkness-induced stomatal closure, and in the process this cross talk may lead to the formation of a self-amplification loop about them. One of the most studied interactions in plants is NO–ROS cooperation during the hypersensitive reaction, which is characterized by programmed cell death that contributes to plant resistance to stress (Kovacic and Somanathan 2011).

3.1 Interactions Between NO with Other Signaling Molecules

It is generally observed that NO and ROS are generated in response to similar stimuli and with similar kinetics; however, NO and ROS interact in various ways. In several situations, such as during pathogen attack and stomatal closure induced by the hormone ABA, both H_2O_2 and NO appear to be generated and function in parallel (Desikan et al. 2004). Moreover, all these signals can induce the generation of antioxidant activity that ameliorates oxidative stress (Neill 2007). Several defense responses are activated by stress, where one of the most important one is stomatal closure induced by ABA redistribution and synthesis (Hao and Zhang 2010). Zhang et al. (2007) also demonstrated the connection between ABA and H_2O_2 and NO in *Zea mays* leaves, where endogenous ABA synthesized in response to dehydration induces H_2O_2 production that in turn accelerates NO synthesis and subsequent up-regulation of antioxidant enzymes' activities. ABA synthesis and action are essential for plant survival during water stress. In fact, ABA signaling in guard cells is especially complex, with H_2O_2, NO, and MAPKs all playing roles (Neill 2007). Bright et al. (2006) reported that ABA-induced NO production in guard cells depends on H_2O_2 generation. Hao and Zhang (2010) presented a key "ABA–H_2O_2–NO–MAPK–antioxidant survival Cycle" and suggest that during water stress ABA has several ameliorative functions that involve NO as a key signaling intermediate and which include the rapid induction of stomatal closure to reduce transpirational water loss and the activation of antioxidant defenses to combat oxidative damage.

Nitric oxide biosynthesis has also been established to be induced by auxin in cucumber roots (Pagnussat et al. 2002; Guo et al. 2003), which was needed for root growth and the formation of lateral roots. Recently, it has been indicated that NO can stimulate cell division and embryogenic cell formation in leaf protoplast-derived cells of alfalfa in the presence of auxin (Ötvös et al. 2005). It was found that various NO-releasing compounds promoted auxin-dependent division of leaf protoplast-derived alfalfa cells. In contrast, application of NO scavenger or NO synthesis inhibitor inhibited the same process (Palavan-Unsal and Arisan 2009). The role of gibberellic acid (GA) related with NO in seed germination was also reported (Palavan-Unsal and Arisan 2009). It was observed that NO donor, SNP and *S*-nitroso-*N*-acetylpenicillamine, delayed GA-induced programmed cell death in *Hordeum vulgare* aleurone layers (Beligni et al. 2002). Tun et al. (2006) reported a linkage between PA and NO and showed that PAs induce the production of NO in various tissues within seedlings of *Arabidopsis thaliana* (Palavan-Unsal and Arisan 2009).

It was also reported that low concentrations of NO either endogenously produced or exogenously applied in the 1 μM range exert significant growth promoting and ethylene inhibiting effects, which are reversed by higher NO concentrations or equimolar applications of NOS inhibitor N_6-methyl-arginine or NO-releasing compounds (Leshem 1996; Palavan-Unsal and Arisan 2009). The alternative oxidase 1, a gene (*AOX1a*), was used as a molecular probe to investigate its regulation by signal molecules such as H_2O_2, NO, ethylene, SA, and JA, all of them reported to be involved in the O_3 response (Ederly et al. 2006; Palavan-Unsal and Arisan 2009). Ethylene biosynthesis also found to be influenced by NO in the maturation and senescence of plant tissue (Arasimowicz and Floryszak-Wieczorek 2007). It was observed that the application of exogenous NO to plants modulates the generation of ethylene (Zhu and Zhou 2007). Lindermayr et al. (2006) observed that NO directly acts by down-regulating ethylene synthesis through *S*-nitrosylation of methionine adenosyl transferase (*MAT1*) in *Arabidopsis* plants. The improvement of NO leads to the inhibition of *MAT1* activity and results in the reduction of the pool of ethylene precursor *S*-adenosyl methionine (SAM).

3.2 NO and Gene Expression

The physiological effects of NO signaling are actively involved in the modification of gene expression. Transcriptomic analyses have recently provided the identity of many NO-regulated genes (Ahlfors et al. 2008; Badri et al. 2008; Ferrarini et al. 2008; Palmieri et al. 2008; Besson-Bard et al. 2009b). A high proportion (~30%) biological effect of NO-mediated functional gene expression is associated to the plant stress response (Besson-Bard et al. 2009a). However, a major question raised by the transcriptomic data available comes from the extremely low quantity of genes commonly regulated when comparing different studies using similar experimental approaches (i.e., exogenous treatments of plant material with NO gas, NO-releasing chemicals or mammalian NOS inhibitors). The particularities of chemicals, plant material, and growing conditions used could afford these differences. However, further studies using standardized conditions are therefore required to identify and compare NO-dependent gene expression controlled by endogenous NO in particular physiological conditions. Some answers may also come from unraveling how NO triggers specific gene expression (Baudouin 2011). No transcriptional regulators have been identified yet to find out the *S*-nitrosylated or nitrated proteins. Palmieri et al. (2008) analyzed the promoter of 28 NO-regulated genes and identified eight families of transcription factor binding sites (TFBS) that are markedly over-represented. These correspond to the binding sites of stress-related transcription factors, which is in good accordance with the function of NO-responsive genes. Whether an over-representation of these TFBS is also found in promoters of other NO-responsive genes previously identified is currently unknown.

4 Protective Role of NO Under Abiotic Stress Condition

It is well-established that NO is a signaling molecule involved in many physiological processes in plants. Many authors reported that NO plays a crucial role in plant growth and development, starting from germination to flowering, ripening of fruit and senescence of organs, respiratory metabolism (Siddiqui et al. 2011; Wimalasekera et al. 2011). In recent years, NO has been found to be involved in plants response to different abiotic stresses like salinity, drought, high or low temperature, toxic metals, flooding, high light, UV-B radiation, and ozone (Ahlfors et al. 2009; Hossain et al. 2010b; Kim et al. 2010; Xiong et al. 2010; Xu et al. 2010c; Bai et al. 2011; Gupta et al. 2011; Hasanuzzaman et al. 2011a; Liu et al. 2011); Table 11.1). It was also suggested that NO, itself, possesses antioxidant properties and might act as a signal in activating ROS-scavenging enzyme activities under various abiotic stresses

(Palavan-Unsal and Arisan 2009; Hao and Zhang 2010; Mazid et al. 2011a; Siddiqui et al. 2011; Table 11.2). However, there has been lack of clarity about the mechanism(s) by which NO reduces abiotic stresses.

4.1 Salinity

Soil salinity, one of the most severe abiotic stresses, limits the production of nearly over 6% of the world's land and 20% of irrigated land (15% of total cultivated areas) and negatively affects crop production worldwide. On the other hand, increased salinity of agricultural land is expected to have destructive global effects, resulting in up to 50% land loss by the middle of the twenty-first century (Mahajan and Tuteja 2005). Osmotic stress due to salinity leads to a slow growth rate and developmental characteristics such as vegetative development, net assimilation capacity, leaf expansion rate, and leaf area index (Zheng et al. 2008; Hasanuzzaman et al. 2009). A reduction in photosynthesis is also one of the most conspicuous effects of salinity stress (Leisner et al. 2010; Raziuddin et al. 2011). In plants, salt stress can lead to the reduction of CO_2 availability and inhibit carbon fixation, exposing chloroplasts to excessive excitation energy which in turn could increase the generation of ROS (Gill and Tuteja 2010). Enhanced ROS production under salt stress induces phytotoxic reactions such as lipid peroxidation, protein degradation, and DNA mutations (Tanou et al. 2009c). Several reports showed the overproduction of ROS in plants under saline conditions and ROS-induced membrane damage is a major cause of cellular toxicity by salinity (Mittova et al. 2004; Hasanuzzaman et al. 2011a, b; Hossain et al. 2011). Salt stress tolerance is a complex trait which involves the coordinated action of many gene families that perform diverse roles such as ion sequestration, control of water loss through stomata, osmotic adjustment, other metabolic adjustments, and antioxidative defense (Abogadallah 2010).

Several reports indicated the protective role of NO on salt stress tolerance in various plant species. Under saline conditions, tolerant plants typically maintain high K^+ and low Na^+ in the cytosol of cells. These processes appear to be mediated by several transport systems, such as H^+-ATPase, carriers, and channels associated with plasma membranes (Kovacic and Somanathan 2011). In this regard, NO serves as a signal in inducing salt resistance by increasing the K^+:Na^+ ratio, which is dependent on the increased plasma membrane H^+-ATPase activity (Zhao et al. 2004). Zhang et al. (2006) reported that NO signaling enhanced salt tolerance in *Zea mays* seedlings through increased activity of proton pump and Na^+/H^+ antiport in the tonoplast. Uchida et al. (2002) observed an enhanced tolerance to salt stress (100 mM NaCl, 8 days) in rice seedlings when pretreated NO (1 μM SNP, 2 days). This pretreatment induced the activity of antioxidant enzymes, viz., superoxide dismutase (SOD), catalase (CAT) and ascorbate peroxidase (APX) as well some stress-related genes (sucrose-phosphatesynthase, Δ'-pyrroline-5-carboxylate synthase, and small heat-shock protein 26). Enhanced seed germination and root growth of

Table 11.1 Nitric oxide-mediated physiological changes in plants under major abiotic stresses

Type of stress	Plants	Stress treatments and duration	NO treatment	Effects	References
Salinity	*Oryza sativa* L. cv. Nipponbare	100 mM NaCl, 8 days	1 μM SNP, 2 days	Enhanced seedling growth	Uchida et al. (2002)
	Triticum aestivum L., cv. Huaimai 17	300 mM NaCl, 1–5 days	100 μM SNP, 20 h	Increased seed germination	Zheng et al. (2009)
				Enhanced seed respiration rate and ATP synthesis	
	Cucumis sativus L. cv. Jinchun 2	50 mM NaCl, 8 days	100 μM SNP, 8 days	Increased seedling growth, photosynthetic pigment content, proline accumulation, net photosynthetic rate, stomatal conductance, and transpiration rate	Fan et al. (2007)
	Kosteletzkya virginica	200–400 mM NaCl, 5 days	600 μM SNP, 5 days	Increased dry weight, proline accumulation	Guo et al. (2009)
				Maintained a lower ratio of $[Na^+]/[K^+]$	
	Cucumis sativus L. cv. Jinchun 2	50 mM NaCl, 8 days	100 μM SNP, 8 days	Increased plant height, stem thickness, fresh weight and increased dry matter accumulation	Fan et al. (2010)
				Increased polyamines biosynthesis	
	Cicer arietinm L. cv HC-3	25 mM NaCl, 2, 4 and 6 days	0.2 and 1 mM SNP, 2, 4 and 6 days	Increased RWC	Sheokand et al. (2010)
				Decreased relative membrane injury	
	Lycopersicom esculentum Mill. cv. Hufan1480 and Hufan2496	100 mM, 8 days	100 μM SNP, 8 days	Increased shoot and root dry weight	Wu et al. (2011)
	Oryza sativa L.	80 mM NaCl, 5 days	100 and 200 μM SNP, 16 h	Increased germinability of seeds	Habib et al. (2010)

Drought	*Triticum aestivum* L. var. Yunong949	15% PEG-6000, 24 h	300 μM SNP, 24 h	Maintained higher RWC (RWC) and lower leaf water loss	Tan et al. (2008)
				Increased proline accumulation	
	Antiaris toxicaria seed	Dessication, 12 days	30 μM SNP, 12 h	Improved seed germination	Bai et al. (2011)
	Triticum aestivum L.	15% PEG-6000, 12–72 h	100 μM SNP	Stabilized the structure and function of biomembrane, increased the activities of H^+-adnosinetriphosphatase and Ca^{2+}-ATP	Hui et al. (2009)
High temperature	*Oryza sativa* L. cv. Nipponbare	50 °C, 5 h	1 μM SNP, 2 days	Improved survival rate of seedlings	Uchida et al. (2002)
				Improved quantum yield for photosystem II	
	Phaseolus radiatus	45 °C, 90 min	150 μM SNP, 60 min	Increased chlorophyll *a* fluorescence parameters, membrane integrity, and maximal quantum yield of photosystem II (PSII) (measured as *F*v/*F*m)	Yang et al. (2006)
				Decreased electrolyte leakage	
	Phragmites communis Trin. callus	45 °C, 2 h	100 μM SNP and SNAP, 24 h	Decreased relative ion leakage	Song et al. (2006)
				Increased relative growth rate and cell viability	
Low temperature	*Cucumis sativus* L. cv. ZND407	4 °C, 72 h	1 mM SNP, 72 h	Increased soluble sugar and chlorophyll content	Liu et al. (2011)
	Cucumis sativus L. cv. Deltastar	2 ± 1 °C, 15 days	25 μM NO, 12 h	Increases in membrane permeability	Yang et al. (2011)
				Reduced chilling injury index	

(continued)

Table 11.1 (continued)

Type of stress	Plants	Stress treatments and duration	NO treatment	Effects	References
Toxic metals	*Hordeum vulgare* L. cv. Weisuobuzhi and Dong 17	5 μM $CdCl_2$, 1–25 days	0.25 mM SNP, 1–25 days	Increased chlorophyll content and photosynthesis Improved the ultrastructure of root cells (increased starch grains and reduced osmiphilic plastoglobuli)	Chen et al. (2010)
	Triticum aestivum L.	0.1 mM $CdCl_2$	SNP 0.01 or 0.1 mM	Enhanced root growth	Groppa et al. (2008)
	Oryza sativa L. cv. Zhonghua 11	0.2 mM $CdCl_2$, 10 days	100 μM SNP, 10 days	Increased root and shoot length as well as total biomass Increased chlorophyll content and photosynthesis Increased pectin and hemicellulose content	Xiong et al. (2009)
	Arabidopsis thaliana L. Heyn	100 μM $Pb(NO_3)_2$, 7 days	0.5 mM SNP, 3 h	Increased root length	Phang et al. (2011)
	Lycopersicon esculentum Mill. cv. No. 4 Zhongshu	1 μM $CuSO_4$, 24 h	100 μM SNP, 24 h	Increased chlorophyll content and biomass of fresh/dry leaves	Wang et al. (2010)
	Triticum aestivum L cv. Yangmai 158	5 mM $CuCl_2$, 3 days	100 μM SNP, 3 h	Improved seeds germination	Hu et al. (2007)
	Festuca arundinacea cv. Arid3	25 μM AsO_4^{3-}, 4 and 8 days	100 μM SNP	Decreased ion leakage Increase dry mass of leaves	Jin et al. (2010)
	Hibiscus moscheutos	100 μM $AlCl_3$, 12 h	100 μM SNP, 12 h	Decreased inhibition of root elongation Growth enhancement of root	Tian et al. (2007)
	Triticum aestivum L. cv. Yangmai 158	0.2 mM $AlCl_3$, 2–8 days	100 μM SNP, 2–8 days	Increased chlorophyll content Increased proline accumulation and soluble protein	Zhang et al. (2008)

High light	*Festuca arundinacea* (Schreb.) cvs. Arid3 and Houndog5	500 μmol/m²/s	1 mM SNP	Reduced light-induced electrolyte leakage	Xu et al. (2010b)
UV-B radiation	*Glycine max* L.	30 kJ/m², 100 min	0.8 mM SNP, 12 h	Increased chlorophyll content and decrease ion leakage.	Santa-Cruz et al. (2010)
	Zea mays L. cv. Yuyu No. 22,	4.8 kJ/m²/day	100 μM SNP	Increased leaf area and biomass of plants	An et al. (2005)
	Pisum sativum L. No. 8711-2	4.8 kJ/m²/day	300 μM SNP	Increased stem length	Qu et al. (2006)
	Zea mays L.	UV-B radiation	SNP	Prevented chlorophyll content reduction and of higher quantum yield for photosystem II	Kim et al. (2010)
				Increased flavonoids and anthocyanin, UV-B absorbing compounds	
Ozone	*Arabidopsis thaliana*	300 or 350 nL/L, 6–8 h	0.5 mM SNP, 1–2 h	Decreased cell death	Ahlfors et al. (2009)
				Increased hormone biosynthesis	

Table 11.2 NO-induced regulation of antioxidant capacity in plants under major abiotic stresses

Types of stress	Plant	Stress treatment and duration	NO treatment	Effects	References
Salinity	*Triticum aestivum* L.	300 mM NaCl, 72 h	SNP 1 mM, 24 h	Increased AsA, GSH levels and enhanced the activities of MDHAR, DHAR, GR, GST, GPX, and Cat activities	Hasanuzzaman et al. (2011a)
	Kosteletzkya virginica	200–400 mM NaCl, 5 days	600 μM SNP, 5 days	Increased activities of CAT, POD, and SOD	Guo et al. (2009)
				Decrease MDA contents	
	Cucumis sativus L. cv. Jinchun 2	50 mM NaCl, 8 days	100 μM SNP, 8 days	Increased activity of SOD, POD, CAT, and APX	Fan et al. (2007)
	Oryza sativa L. cv. Nipponbare	100 mM NaCl, 8 days	1 μM SNP, 2 days	Enhanced the activity of SOD, CAT, and APX	Uchida et al. (2002)
	Triticum aestivum L., cv. Huaimai 17	300 mM NaCl, 1–5 days	0.1 mM SNP, 20 h	Increased SOD and CAT activities	Zheng et al. (2009)
				Decreased the contents of MDA and H_2O_2, and $O_2^{\cdot-}$ release rate	
	Lycopersicom esculentum Mill. cv. Hufan1480 and Hufan2496	100 mM NaCl, 8 days	100 μM SNP, 8 days	Increased activities of SOD, POD, CAT, and APX	Wu et al. (2011)
				Increased the levels of AsA and GSH	
				Reduced MDA level and $O_2^{\cdot-}$ production	
	Cicer arietinm L. cv HC-3	25 mM NaCl, 2, 4 and 6 days	0.2 and 1 mM SNP, 2, 4 and 6 days	Increased activities of SOD, CAT, APX, GR, and DHAR	Sheokand et al. (2010)
				Increased the GSH/GSSG and ASC/DHA ratio	
				Partially decreased MDA and H_2O_2 content	

Drought	*Triticum aestivum* L. var Yunong949	15% PEG-6000, 24 h	300 μM SNP, 24 h	Increased activities of SOD and CAT	Tan et al. (2008)
	Antiaris toxicaria seed	Dessication, 12 days	30 μM SNP, 12 h	Increased activity of antioxidant AsA-GSH pathway enzymes (APX, MDHAR, DHAR, and GR) and metabolites (AsA: DHA and GSH:GSSG ratio) Decreased the production of H_2O_2	Bai et al. (2011)
	Triticum aestivum L.	15% PEG-6000, 12–72 h	0.1 mM SNP	Increased SOD, POD, and CAT activities Decreased $O_2{\cdot}^-$ generation and H_2O_2 production	Hui et al. (2009)
High temperature	*Phaseolus radiatus*	45 °C, 90 min	150 μM SNP, 60 min	Increased the activities of CAT, SOD, and POD	Yang et al. (2006)
	Phragmites communis Trin.	45 °C, 2 h	100 μM SNP and SNAP, 24 h	Decreased H_2O_2 and MDA contents. Increased activities of SOD, CAT, APX, and POD	Song et al. (2006)
Low temperature	*Cucumis sativus* cv. ZND407	4 °C, 72 h	1 mM SNP, 48 h	Increased SOD, GR, POD, and CAT Decrease in MDA content	Liu et al. (2011)
	Cucumis sativus L. cv. Deltastar	2±1 °C, 15 days	25 μM NO, 12 h	Delayed the increases in both $O_2{\cdot}^-$ production rate and H_2O_2 Increased activities of SOD, CAT, APX, and POD and higher DPPH-radical scavenging activity	Yang et al. (2011)

(continued)

Table 11.2 (continued)

Types of stress	Plant	Stress treatment and duration	NO treatment	Effects	References
Toxic metals	*Hordeum vulgare* L. cvs. Weisuobuzhi and Dong 17	5 μM $CdCl_2$, 1–25 day	0.25 mM SNP, 1–25 days	Increased SOD, APX, and CAT activities; cAPX activity and gene expression of root/leaf *cAPX* and leaf *CAT1*	Chen et al. (2010)
	Triticum aestivum L.	0.1 mM $CdCl_2$	SNP 0.1 mM	Increased GSH content Decreased MDA content	Groppa et al. (2008)
	Lycopersicon esculentum Mill. cv. No. 4 Zhongshu	1 μM $CuSO_4$, 24 h	100 μM SNP, 24 h	Increased CAT, POD, SOD and APX Reduction in H_2O_2 accumulation	Wang et al. (2010)
	Triticum aestivum L. cv. Yangmai 158	5 mM $CuCl_2$, 3 days	100 μM SNP, 3 h	Stimulated activities of SOD and CAT and decreased the activities LOX Sustained a lower level MDA and H_2O_2	Hu et al. (2007)
	Arabidopsis thaliana L. Heyn	100 mM $Pb(NO_3)_2$, 7 days	0.5 mM SNP, 3 h	Reversed activities of SOD, CAT, GR, GPX, and POD	Phang et al. (2011)
	Triticum aestivum L. cv. Yangmai 158	0.2 mM $AlCl_3$, 2–8 days	0.1 mM SNP, 2–8 days	Decreased MDA and H_2O_2 levels Increased SOD, CAT, and APX activities	Zhang et al. (2008)
	Festuca arundinacea cv. Arid3	25 μM AsO_4^{3-}, 4 and 8 days	100 μM SNP	Increased SOD, CAT, and APX activities Decreased MDA and H_2O_2 content	Jin et al. (2010)

High light	*Festuca arundinacea* (Schreb.) cvs. Arid3 and Houndog5	500 μmol/m²/s	1 mM SNP	Increased the activities of SOD, CAT, APX, and GR Reduced contents of MDA, H_2O_2, and $O_2^{\cdot-}$. Decreased LOX activity	Xu et al. (2010b)
UV-B radiation	*Glycine max* L.	30 kJ/m², 100 min	0.8 mM SNP, 12 h	Increased CAT and APX activities Prevented H_2O_2 and $O_2^{\cdot-}$ accumulation	Santa-Cruz et al. (2010)
	Zea mays L.	UV-B radiation	SNP	Increased the activities of CAT and APX Decreased MDA and H_2O_2 content	Kim et al. (2010)

Lupinus luteus seedlings (Kopyra and Gwóźdź 2003) and increased growth and dry weight of *Zea mays* seedlings (Zhang et al. 2006) were also observed with the treatment of NO donor under stressed condition. Treating *Hordeum vulgare* leaves with exogenous NO (50 μM SNP), Li et al. (2008) observed that it could alleviate the damage of salt stress (50 mM NaCl) which was reflected by decreased ion leakage, malondialdehyde (MDA), carbonyl, and H_2O_2 content. Additionally, the presence of the NO donor enhanced the activities of SOD, APX, and CAT. In our recent study, we observed that exogenous NO modulated the ROS detoxification systems in *Triticum aestivum* seedlings (Hasanuzzaman et al. 2011a). The seedlings pretreated with NO donor (1 mM SNP, 24 h) when exposed to salt (150 and 300 mM NaCl, 4 days) showed an increase in the ascorbate (AsA) and glutathione (GSH) contents and the GSH:GSSG ratio as well as the activities of monodehydroascorbate reductase (MDHAR), dehydroascorbate reductase (DHAR), glutathione reductase (GR), glutathione *S*-transferase (GST), and glutathione peroxidase (GPX) as compared to the seedlings without NO pretreatment, which ultimately decreased the contents of MDA and H_2O_2.

Liu et al. (2007) found that salt tolerance of *Phaseolus vulgaris* root was enhanced by the NR-dependant NO production where glucose-6-phosphate dehydrogenase enzyme played an important role. NO interacts with other salt-dependent signaling molecules in establishing systemic defense response. ROS, phytohormones, and MAPKs play important roles in plant responses to salt stress. Protein post-translational modifications like *S*-nitrosylation could also contribute to NO signaling during salt stress (Tanou et al. 2009a). In another study they (Tanou et al. 2009b) observed that preexposure to SNP, prior to salinity, resulted in higher GSH redox compared to NaCl-treated citrus plants providing a link between GSH and NO during the establishment of salt tolerance. Fan et al. (2007) showed that exogenous NO (100 μM SNP) significantly alleviated the salt (50 mM NaCl) injury to cucumber seedlings and increased seedling growth. In addition, photosynthetic pigment content, proline, as well as the activity of SOD, POD, CAT, and APX were also increased. Similarly, net photosynthetic rate, stomatal conductance, and transpiration rate also increased significantly. However, exogenous NO application markedly decreased membrane permeability, rate of $O_2^{\cdot-}$ production, the contents of MDA and H_2O_2, and intercellular CO_2 concentration. Song et al. (2009) observed enhanced seedlings growth in *Suaeda salsa*. An increase of the dry weight, proline accumulation, and lower ratio of $[Na^+]/[K^+]$ were observed in salt-stressed *Kodtetzkya virginica* seedlings when treated with SNP (Guo et al. 2009). In *Triticum aestivum*, Zheng et al. (2008) investigated the protective roles of NO (presoaking with 0.1 mM SNP) on seed germination under salt stress (300 mM). They observed the positive effects of exogenous NO on wheat seeds exposed to salinity included an increased germination rate, enhanced respiration rate, and ATP synthesis and maintained balance of Na^+ and K^+ ions. Similarly, SNP triggered an increase in the activities of antioxidant enzymes, SOD and CAT, whereas decreased the contents of MDA, H_2O_2, and $O_2^{\cdot-}$ release rate in the mitochondria leading to a decrease in ROS accumulation (Zheng et al. 2009).

Qiao and Fan (2008) observed the expression of a rice gene *OsNOA1* homologous to *Arabidopsis AtNOA1* that can re-establish diminished NO synthesis in *Atnoa1* and induced the expression of plasma membrane Na^+/H^+ antiporter gene *AtSOS1* and H^+-ATPase gene *AtAHA2*, resulting in the restoration of *Atnoa1* in terms of Na^+/K^+ ratio and salt tolerance phenotypes. They also suggested that this phenomenon can be mimicked by exogenous application of NO donor. Studies using *Arabidopsis* mutant *Atnoa1* with an *in vivo* NOS activity and a reduced endogenous NO level were more sensitive to NaCl stress than wild type (Zhao et al. 2007). However, treatment of *Atnos1* plants with exogenous SNP alleviated the oxidative damage caused by NaCl stress. *Atnoa1* mutants displayed a greater Na^+/K^+ ratio in shoots than wild type when exposed to NaCl, but SNP treatment led to a decrease of Na^+/K^+ ratio back to the levels observed in the wild type (Zhao et al. 2007). In *Arabidopsis*, the wild-type plants exhibited higher survival rates under salt stress than *Atnoa1* plants which have a reduced level of endogenous NO (Guo et al. 2003; Zhao et al. 2007). More importantly, exogenous NO application to *Atnoa1* mutants alleviated the salt-induced oxidative damage. More recently, Zhang et al. (2010a) reported that the transgenic *Arabidopsis* line TL9 had higher proline, soluble protein, and chlorophyll contents as well as lower MDA content compared to its receptor, *Atnoa1* mutant, under salt stress condition. Root elongation and survival rate in TL9 were significantly higher than those in *Atnoa1* seedlings under salt stress. present study proved that *StNOA1* participated in *Arabidopsis thaliana* salt stress responses and increased its salinity tolerance. They concluded that present study proved that *StNOA1* participated in *Arabidopsis thaliana* salt stress responses and increased its salinity tolerance.

Recently, a number of studies have been carried out to observe the effect of exogenous NO on salt stress tolerance. David et al. (2010) reported that NO enhanced biochemical adaptation during the seedling growth of *Helianthus annuus* under salinity conditions (40–120 mM NaCl). They found an increased Na^+/K^+ ratio (fourfold) in roots, and Na^+ was rapidly transported to the cotyledons, which registered a concomitant increase in this ratio. They also concluded that the origin of this endogenous generation of NO appears to be mediated by NOS activity (David et al. 2010). In *Cucumis sativus* seedlings, Fan et al. (2010) observed that exogenous SNP increased the salt tolerance by adjusting the biosynthesis of PAs and the ratio of three different PAs. Their results showed that treatment with 100 μM SNP significantly improved the growth of cucumber seedlings under NaCl stress for 8 days, as indicated by increased, plant height, stem thickness, fresh weight, and increased dry matter accumulation. Zheng et al. (2010) reported that pretreatment of NO donor significantly maintained the balance between C and N metabolism through increasing total soluble protein and by up-regulating the endopeptidase and carboxypeptidase activities in plants grown under salt stress. Exogenous NO supplementation as SNP has significant ameliorating effect against NaCl-induced oxidative damage in chickpea leaves as observed by Sheokand et al. (2010). They exposed 5-day-old *Cicer arietinum* plants to NaCl treatment (250 mM) alone and in combination with two concentrations of SNP (0.2 and 1 mM) for 2, 4, and 6 days. Both the SNP treatments had a positive effect on antioxidant enzymes SOD, CAT,

APX, GR, and DHAR under salt stress. NaCl treatment resulted in a decline in the GSH/GSSG and AsA/DHA ratio; however, SNP treatments increased the reduced form of both the metabolites thus elevating the ratio of GSH/GSSG and AsA/DHA. Exogenous NO partially decreased MDA and H_2O_2 content. Habib et al. (2010) demonstrated that the application of lower concentrations of NO (0.1 and 0.2 mM) as presowing seed treatment (for 16 h) showed a significant improvement of seed germinability of rice seed under salt stress (80 mM, 5 days). However, higher concentration of NO showed no significant effects; rather it caused negative effect on the germinability. When exposed to NO donors, NO-associated salt priming action was evident in halophytes in tolerating high salinity during germination and early growth stages (Molassiotis and Fotopoulos 2011) which was due to the better induction of antioxidant enzyme activity in response to high salinity conditions. Under salt stress, NO-mediated signaling mechanisms involve in the family of protein kinases. Very recently, Corpas et al. (2011) reported that tobacco-cell suspensions exposed to salt stress, the osmotic stress-activated protein kinase (NtOSAK) is activated by NO and confer stress signals. While studying with *Lycopersicom esculentum* cv. Hufan1480 and Hufan2496, Wu et al. (2011) observed notable improvement of growth and enhanced antioxidant defense in salt-stressed (100 mM NaCl) plants when treated with exogenous NO (100 μM SNP). They observed that in the presence of 100 μM SNP under salt stress, the reduction in shoot and root dry mass declined to 16 and 3%, respectively in Hufan1480, and to 21 and 6%, respectively in Hufan2496. The MDA content of Hufan1480 and Hufan2496 decreased significantly by 22 and 12% over the salt treatment, respectively. The rate of $O_2^{\cdot-}$ production in Hufan1480 and Hufan2496 decreased significantly by 20 and 17%, respectively, over the salt stress. A remarkable increase in the activities of SOD, POD, CAT, and APX as the levels of non-enzymatic antioxidants, AsA and GSH, was also obtained by NO treatments under stress condition.

4.2 Drought

Drought is one of the most devastating environmental stresses that affect the growth and development of plants. The effects of drought stress are expected to increase with climate change and a growing water crisis (Harb et al. 2010). A plant suffers from drought stress due to the unavailability of water to the root zone or excessive transpiration rate. In general, drought stress affects the growth, dry matter production, and economic yield of plants. Drought stress is characterized by a reduction of water content, decreased leaf water potential, turgor loss, stomatal closure, and decrease in cell elongation and expansion (Jaleel et al. 2009; Mingchi et al. 2010; Din et al. 2011). Drought stress may lead to stomatal closure, which reduces CO_2 availability in the leaves and inhibits carbon fixation, exposing chloroplasts to excessive excitation energy, which in turn could increase the generation of ROS and induce oxidative stress (Mittler 2002; de Carvalho 2008) by generating free radicals like $O_2^{\cdot-}$, $^1O_2$, H_2O_2, and OH·, which are potentially dangerous under drought stress

(Li et al. 2010a; Faize et al. 2011; Hasanuzzaman and Fujita 2011). Thus, the enhancement of antioxidant defense mechanisms is considered to be an adaptive mechanism of plants to drought stress and the strengthening of these defense mechanisms, through the enhanced functions of antioxidant components (enzymatic and non-enzymatic), may reduce or prevent oxidative damage and improve the drought resistance of plants (Sharma and Dubey 2005; de Carvalho 2008; Jaleel et al. 2009).

Different plant studies provided the evidence that NO could protect the drought-induced damage in plants. Among the different mechanisms to avoid water deficit, stomatal closure is important which is a response triggered by a signal that originates in the root system. Neill et al. (2008) reported that stomatal closure, initiated by ABA, is affected through a complex intracellular signaling in which NO appears to be one component. It was indicated that drought induces NO generation, which activates cellular processes that afford some protection against the stress (Kovacic and Somanathan 2011). Previously, NO induced stomatal closure and enhanced adaptive plant response to drought stress has also been observed by Garcia-Mata and Lamattina (2001). Later, Desikan et al. (2004) hypothesized that involvement of NR-mediated NO synthesis in *Arabidopsis* guard cells responsive to ABA and was shown to be required for ABA-induced stomatal closure. Both NO and ROS were reported to participate in the osmotic tolerance of wheat seedlings by stimulating ABA biosynthesis (Xing et al. 2004). According to Tian and Lei (2006), *Triticum aestivum* leaves exogenous NO treatment (2 mM SNP) enhanced drought tolerance by up-regulating the activities of SOD, CAT, and phenylalanine ammonia-lyase (PAL). As a result, the NO-treated plants showed lower levels of MDA and H_2O_2 as well as enhanced growth. Exogenous NO (SNP)-treated reed (*Phragmites communis*) suspension cultures exposed to stressful action of PEG-6000 was accompanied by deceleration of ion leakage, lowering of H_2O_2 and O_2^- content, and by activation of antioxidant defense enzymes (Zhao et al. 2008). Tan et al. (2008) reported that exogenous NO (300 μM SNP) alleviated oxidative damage, accelerated protein synthesis and enhanced photosynthesis rate, and increased the activities of SOD and CAT and also maintained higher relative water content (RWC) and lower leaf water loss in leaves of wheat seedlings exposed to drought stress (15% PEG). Interestingly, addition of NO scavenger (c-PTIO) reversed such effects of NO, which suggested that application of NO might confer an enhanced resistance to drought stress in plants.

Hao et al. (2008) suggested that NO participated in the signaling of drought-induced protective responses in *Zea mays* seedlings which is dependent on NOS-like activity. They also observed that both NOS activity and the NO production markedly increased under dehydration stress. After NO pretreatment and subsequent dehydration stress, detached leaves maintained more water content by decreasing transportation rate which was due to the prevention of membrane permeability by exogenous application of NO donor (SNP). In tomato plants, Nasibi and Kalantari (2009) observed that the seedling sprayed with 100 μM prevented drought-induced decrease in RWC and membrane stability index and reduced lipid peroxidation and H_2O_2 content, while NO scavenger (200 μM PTIO) reversed the protective effects of SNP suggesting that protective effect by SNP is attributable to NO release. They also found that the activity of APX and GR increased under SNP pretreatment which indicated that the reduction of drought-induced oxidative damages by NO in tomato leaves is most likely mediated through either NO ability to scavenge ROS or stimulation of antioxidant enzymes. In a recent study, Bai et al. (2011) demonstrated that pretreatment with NO increases the activities of antioxidant AsA-GSH cycle enzymes (APX, MDHAR, DHAR, and GR) and the efficiencies of the metabolites (AsA: DHA and GSH:GSSG ratio), decreases H_2O_2 production and minimizes the inhibitory effects of desiccation on seed germination. Desiccation stress also increases the protein carbonylation levels and reduces protein *S*-nitrosylation of these antioxidant enzymes which was reversed by NO treatment. The results by Xiong et al. (2011) showed that the increase of endogenous NO is dispensable for proline accumulation in the leaves of rice under drought stress. More importantly, exogenous application of NO alleviates drought-induced water loss and ion leakage by decreasing transpiration rate of rice leaves.

4.3 High Temperature

High temperature or heat stress results from temperatures high enough to damage plant tissues, substantially influencing the growth and metabolism of plants (Balla et al. 2009). Now-a-days, one of the serious challenges for plant growth and productivity is to cope with the abrupt and often unpredictable temperature fluctuations. Different global circulation models predict that greenhouse gases will gradually increase the world's average ambient temperature and lead to global warming (Meehl et al. 2007). Therefore, plants' responses and adaptation to elevated temperature and the mechanisms to develop heat-tolerant cultivars should be examined. High temperatures caused cell injury or death, inhibited growth, reduced ion flux, scorching of leaves and twigs, sunburn on plant organs, leaf senescence and abscission, delay in seed germination and a loss of vigor, reduction of photosynthesis and respiration, reduction in shoot dry mass, relative growth rate and net assimilation rate, fruit discoloration and damage, and reduced yield significantly (Egli et al. 2005; Howarth 2005; Ismail and Hall 1999; Wahid et al. 2007). Extreme temperature stress accelerates the generation and reactions of ROS including $^1O_2$, $O_2^{\cdot-}$, H_2O_2, and OH·, thereby inducing oxidative stress (Mittler 2002; Yin et al. 2008).

Results suggest that NO might act as a signal and extreme temperature tolerance might be through decreasing the ROS level (Neill et al. 2002a). NO is involved in signal transduction of JA-induced stomatal closure of *Vicia faba* (Xin et al. 2005). They observed that NO exposure effectively protects calluses from two ecotypes of reed when exposed to heat stress. Increased NO production was observed in response to heat stress in tobacco, rice, and alfalfa (Qiao and Fan 2008). NO donor treatment in rice and *Triticum aestivum* reported to be effective in reducing damages caused by high temperatures (Qiao and Fan 2008; Uchida et al. 2002). Yang et al. (2006) showed that NO (SNP 150 μM, 60 min) presoaked leaf discs of *Phaseolus radiatus* when exposed to a heat shock (45 °C, 90 min) significantly improved the chlorophyll a fluorescence parameters, membrane integrity, and activities of CAT, POD, and SOD as compared to unsoaked heat-shocked leaf discs. The maximal quantum yield of photosystem II (PSII) (measured as *Fv*/*Fm*) was significantly increased. Moreover, the electrolyte leakage due to heat shock was reduced by 48%, lipid peroxidation and H_2O_2 content were kept at control level by SNP presoaking. In *Aarabidopsis*, several mutants have been identified by Lee et al. (2008) which impair the *GSNOR1* gene, showing the involvement of this gene in the mechanism of response against high temperature. Thus, the mutant HOT5 (sensitive to hot temperatures) showed that GSNOR modulates the intracellular level of SNOs, enabling thermo tolerance as well the regulation of plant growth and development (Lee et al. 2008). Song et al. (2006) pretreated callus of *Phragmites communis* (reed) with two different NO donors, viz. SNP and *S*-nitroso-*N*-acetylpenicillamine (SNAP), for 24 h and then exposed to high temperature (45 °C) for 2 h. They observed that exogenous NO caused dramatic alleviation of high temperature-induced ion leakage increase, growth suppression, and cell viability as well as H_2O_2 and MDA contents. However, the activities of SOD, CAT, APX, and POD increased in both calluses in the presence of NO donors under heat stress. On the other hand, NO scavenger (cPTIO) arrested NO donors-mediated protective effects. They concluded that it provided a good indication that NO can effectively overcome oxidative stress induced by heat stress and that NO might act as a signal in activating ROS-scavenging enzymes under heat stress and thus confer thermotolerance (Song et al. 2006). In a recent study, it was reported that excessive NO production under high temperature might be involved in the thermoinhibition of seed germination in *Arabidopsis thaliana* (Hossain et al. 2010b).

4.4 Low Temperature

In plants both chilling and freezing stresses are together termed as low temperature or cold stress. Chilling stress results from temperatures cool enough to produce injury without the formation of ice in plant tissues, whereas in freezing stress ice formed in plant tissues. Chilling stress usually occurs at temperature between 0 and 10 °C, but a few tropical species such as rice and sugarcane are exceptionally sensitive to chilling and show injury signs up to 15 °C (Thomashow 1999). Low temperature stress affects seedlings more than mature plants with noticeable symptoms on plants including surface lesions, a water-soaked appearance, desiccation, discoloration, tissue breakdown, accelerated senescence, and faster decay due to leakage of plant metabolites (Sharma et al. 2005; Solanke and Sharma 2008). Another major negative effect of low temperature stress is that it induces severe membrane damage which is largely due to acute dehydration associated with freezing (Yadav 2010). Low temperature stress also severely hampers the reproductive development of plants which may cause floral sterility (Nahar et al. 2009; Yadav 2010). Chilling stress also affects the root growth of plants (Einset et al. 2007; Farooq et al. 2009). These changes limit the roots' capacity for water and mineral uptake and ultimately overall plant growth (Ercoli et al. 2004; Farooq et al. 2009). Low temperature reduces dry matter production and partitioning in crop plants (Verheul et al. 1996). With decreasing temperature, the solubility of a gas increases, which leads to a higher concentration of O_2 and thus enhances the risk of oxidative stress at low temperature which leads to the increased production of $O_2^{\cdot -}$, H_2O_2, $^1O_2$, and OH· (Guo et al. 2006).

Exogenously applied NO was found to enhance low temperature tolerance in many plant species like *Lycopersicon esculentum*, *Triticum aestivum*, and *Zea mays* (Neill et al. 2003). Experimental evidence indicates NOS-like enzymes are sources of NO in response to low temperature (Corpas et al. 2008). Similarly, in *Arabidopsis*, freezing tolerance was shown to be achieved by NR-dependent NO production by modulating proline accumulation (Zhao et al. 2009). A slightly enhanced NO synthesis in the cells of root tips and in the surrounding elongation zone has been observed of cucumber seedlings by Arasimowicz-Jelonek et al. (2009). However, this NO production was reduced by pretreatment with NOS and NR inhibitors. Additionally, exogenous NO also reduced lipid peroxidation by diminishing the LOX activity (Arasimowicz-Jelonek et al. 2009). In another study, Zhang et al. (2010b) reported that up-regulation of arginase activity and gene expression may be a chilling tolerance strategy in *Lycopersicon esculentum* fruit. Inhibition of chilling-induced arginase activity could aggravate chilling injury and oxidation damage. Arginase appears to play an important role in the chilling resistance process of cherry tomato fruit induced by L-Arginine which has contribution to NO synthesis.

In a recent study, Liu et al. (2011) pretreated *Cucumis sativus* seedlings with 1 mM SNP (NO donor) and exposed to 4 °C temperature. They observed that SNP-treated MDA content was significantly decreased (27%) in SNP-pretreated chilling-stressed seedlings as compared to stress alone. In addition, soluble sugar and chlorophyll content increased with NO pretreatment. Further investigations revealed that treatment with NO donor stimulated the activities of various enzymes such as SOD, GR, POD, and CAT, which indicated that exogenous NO at 1.0 mM SNP enhanced chilling stress tolerance. However, higher dose of NO (2 mM SNP) did not show any protective effect, rather they somewhat showed negative toxicity to plants. Cantrel et al. (2011) demonstrated that NO content increased in *Arabidopsis thaliana* plants in response to low temperature (4 °C, 1–4 h) which is dependent upon NR activity. They also suggested a new function for NO as an intermediate in gene regulation and lipid-based signaling during cold transduction. Very recently, Cui et al. (2011) observed that scavenging or inhibition of NO production inhibited brassinosteroids-induced tolerance to photooxidative and cold stress and partly blocked brassinosteroids-induced expression and activities of several antioxidant enzymes. Pretreatment of the exogenous NO precursor, on the other hand, led to both increased stress tolerance and increased expression of antioxidant enzymes. They concluded that NO plays an important role in plant stress tolerance by brassinosteroids. Yang et al. (2011) pretreated *Cucumis sativus* fruit with 25 μM NO for 12 h and then stored at low temperature (2 ± 1 °C) and observed that NO at 25 μM was most effective in reducing chilling injury index (CI) in cucumber fruit, reduced the increases in membrane permeability and MDA, and delayed the increases in both $O_2^{\cdot -}$ production rate and H_2O_2 content. The NO-treated fruit also exhibited significantly higher activities of SOD, CAT, APX, and POD and higher DPPH-radical scavenging activity than control fruit during the storage which suggest that NO enhanced chilling tolerance in cucumber fruit by improving the antioxidative defense system.

4.5 Toxic Metals

In recent years, substantial amounts of toxic metals (especially heavy metals) have been released by geological activities or by accelerated anthropogenic impacts causing serious environmental problems (Sun et al. 2008). Since these metals are often found both in soil and water as contaminants, studies on complex metal toxicity in different plant species have come into focus. Making a generalization about the effect of metals on plants is difficult due to the multidimensional variations in parameters under different concentrations, types of metals, duration of exposure, target organs of plants, plant age, etc. Several physio-biochemical processes in plants cells are affected by toxic metals (Dubey 2011). Direct phytotoxic effects of metals include their direct interactions with proteins, enzymes, displacement of essential cations from specific binding sites, causing altered metabolism, inhibiting the activities of enzymes, etc. (Sharma and Dubey 2007; Sharma and Dietz 2008; Hossain et al. 2010a). Toxic metals influence homeostatic events, including water uptake, transport, and transpiration and thus symptoms start to develop and become visible, eventually leading to the death of plant cells (Fodor 2002; Poschenrieder and Barceló 2004). The most obvious plant reaction under metal toxicity is the inhibition of growth rate (Sharma and Dubey 2007). Heavy metals also cause chlorosis, necrosis, leaf rolling, inhibition of root growth, stunted plant growth, altered stomatal action, decreased water potential, efflux of cations, alterations in membrane functions, inhibition of photosynthesis, altered metabolism, altered activities of several key enzymes, etc. (Sharma and Dubey 2007; Dubey 2011). There is enough evidence that exposure of plants to excess concentrations of redox active metals results in oxidative injury.

A number of reports have revealed that exogenous NO treatment helps the plants to protect against the adverse effects of metal toxicity, starting from a decrease of metal accumulation (Xiong et al. 2009) and ending with the decrease of metal-induced oxidative stress (Kopyra and Gwóźdź 2003; Hsu and Kao 2005; Singh et al. 2008; Tewari et al. 2008; Chen et al. 2010; Xu et al. 2010a; Arasimowicz-Jelonek et al. 2011). Bartha et al. (2005) investigated the protective role of NO in *Brassica juncea* and *Pisum sativum* in response to heavy metals (100 μM Cd, Cu, or Zn). Different NO levels with different heavy metal loads were observed; the most effective metals were Cu and Cd, where the NO production doubled after 1 week of treatment. In the case of Cu treatment, two-phase kinetics was found, that is, a rapid NO burst in the first 6 h was followed by a slower and gradual increase. The fast appearance of NO in the presence of Cu^{2+} suggests that this can be a novel reaction hitherto not studied in plants under heavy metal stress. After a long-term treatment, NO levels were inversely related to NO_2^- concentrations that originated from NR activity, suggesting conversion of NO_2^- to NO.

Several reports have provided the indication regarding the contribution of NO to Cd toxicity by promoting Cd uptake and subsequent metal-induced reduction of root growth (Besson-Bard et al. 2009b). Hsu and Kao (2004) showed the protective effect of NO in preventing Cd-induced accumulation of NH_4^+, decrease in the activity of glutathione synthase (GS), and increase in the specific activity of PAL. Laspina et al. (2005) observed that *Helianthus annuus* leaves exposed to a 10-day Cd stress showed a decrease in GSH level, but NO was able to efficiently counteract GSH depletion. In *Brassica juncea* and *Pisum sativum* roots exposed to 100 μM Cd, NO accumulation began after 24 h and an enhanced production was observed also after long-term (5 days) Cd exposure (Bartha et al. 2005). In *Helianthus annuum* leaves, NO pretreatment alleviated the toxic effect of Cd^{2+} by preventing the oxidative stress development (Groppa et al. 2008). Exogenous NO was reported to alleviate toxicity of arsenic, whose application suppressed elongation of rice roots and coleoptiles. In rice plants, Xiong et al. (2009) observed that exogenous application of NO enhances Cd tolerance by increasing pectin and hemicelluloses content in the cell wall of roots. In another report, Singh et al. (2008) concluded that exogenous NO ameliorates Cd toxicity in wheat roots, increases the ROS-scavenging activity, and reverses Cd-induced increases in the activities of antioxidant enzymes. In following year, same authors (Singh et al. 2009) observed that NO restored growth of roots and coleoptiles, by serving as ROS scavenger which resulted in decreased MDA content and lower levels of $O_2^{\cdot-}$ and H_2O_2. In *Triticum aestivum* roots growing for 4 weeks at a low Cd concentration (1 μM) ca., 2.4-fold increase in NO emission was recorded, thus confirming the stimulatory effect of Cd stress on NO production in roots (Mahmood et al. 2009). In contrary, Rodríguez-Serrano et al. (2009) reported that a long-term (14-day period) Cd exposure resulted in the significant reduction of NO content in leaves in the *Pisum sativum*. Innocenti et al. (2007) observed that γ-glutamylcysteine synthetase (γ-ecs) and GSH synthetase (*gshs*) genes were upregulated by NO treatment, suggesting that NO is involved in the regulation of GSH synthesis-related genes expression.

Cross talk between ROS and NO has been also proposed for the defense responses of *Pisum sativum* plants exposed to Cd (Rodríguez-Serrano et al. 2009). Chen et al. (2010) reported the Cd-induced NO synthesis stimulated by NR and NOS-like enzymes in roots/leaves which might partly contribute to its Cd tolerance in barley roots. In their study, exogenous NO dramatically alleviated Cd toxicity, markedly diminished Cd-induced ROS and MDA accumulation, ameliorated Cd-induced damage to leaf/root ultrastructure, and increased chlorophyll content and photosynthesis. Exogenous NO significantly elevated the depressed SOD, APX, and CAT activities in the Cd-sensitive *Hordeum vulgure* genotype after 10- and 15-day treatments. Moreover, NO treatment significantly increased stromal APX and Mn-SOD activities and upregulated Cd-induced decrease in cAPX activity and gene expression of root/leaf cAPX and leaf *CAT1* in the Cd-sensitive genotype. They finally concluded that NO, as a potent antioxidant, protects barley seedlings against oxidative damage under Cd stress, by directly and indirectly scavenging ROS, and helps to maintain stability and integrity of the subcellular structure (increased starch grains and reduced osmiophilic plastoglobuli). Overall, exogenous NO donors in various plants and following protective role makes it possible to monitor the effects of NO on a broad cellular antioxidant machinery upon Cd exposure (Xiong et al. 2010). In their recent study, Xu et al. (2010a) showed that NO may participate in maintaining the auxin equilibrium by reducing IAA oxidase activity in roots of *Medicago truncatula* subjected to Cd stress, thus alleviating the negative effect of Cd on root growth inhibition. There is a scarcity of information related to the role of internal NO content in plants grown under heavy metals stress. It was reported that Cd is able to enhance NO synthesis in plant roots within the first several hours of stress duration. In another reports, a 48 h exposure to Cd of *Medicago truncatula* roots showed marked decrease in endogenous NO accumulation and GSH level. More importantly, exogenous NO also recovered the Cd-diminished GSH pool (Xu et al. 2010a) which was attributed to the enhanced expression of GSH synthesis-related genes. Xiong et al. (2010) indicated that application of exogenous NO decreases both ROS accumulation in roots and H_2O_2 accumulation in leaves of *Oryza sativa* under Cd stress. The formation of NO has been demonstrated in various plant tissues exposed to Cd stress. However, the time and intensity of NO generation relatively frequently show conflicting data (Arasimowicz-Jelonek et al. 2011).

Hu et al. (2007) reported that pretreatment with NO (100 μM SNP, 3 h) could significantly improve wheat seed germination and alleviate oxidative stress against Cu toxicity (5 mM $CuCl_2$, 24 h). Pretreatment with NO donor also upregulated the activities of SOD and CAT and decreased the LOX activity. As a result, it sustained a lower level of MDA and interfered with H_2O_2 excessive accumulation compared with the control, thereby enhancing the antioxidative capacity. Tewari et al. (2008) concluded that NO is most likely to mediate Cu toxicity in *Panax ginseng* roots through the modulation in the activities of antioxidant enzymes (CAT, POD, APX, and GR) involved in H_2O_2 detoxification and in the maintenance of cellular redox couples and contents of molecular antioxidants such as non-protein thiol, AsA, and its redox status. Recently, Wang et al. (2010) suggest that application of the NO donor (SNP) efficiently alleviated the toxic effects of Cu, as shown by increases in chlorophyll content and the biomass of fresh/dry leaves in *Lycopersicon esculentum*. Exogenous NO treatment also induced the transcription and increased activities of antioxidant enzymes, including CAT, POD, SOD, and APX, led to reduction in H_2O_2 accumulation in the leaves. However, NO inhibitors or scavengers reverse the effect of NO on Cu toxicity, suggesting that the protective effect of SNP is attributable to NO released. In wheat leaves, Tian et al. (2007) showed that exogenous NO decreased the Al^{3+} toxicity in root elongation of *Hibiscus moschetuos*. They suggested that both NO scavenger and inhibitor were correlated with endogenous NO levels in root cells and reduction of endogenous NO concentrations resulting from inhibition of NOS activity. Zhang et al. (2008) reported the enhancement of antioxidant capacity by exogenous NO under Al stress was due to the increased activities of SOD, CAT, and APX and increasing the proline content, whereas it decreases H_2O_2 and MDA concentrations and maintains the level of soluble protein, compared with water controls.

In *Sorghum bicolor*, the application of NO donors increased Fe bioavailability, which was associated with the promotion of oxidative stress and ROS formation. In parallel, NO donors protected the seed from Fe toxicity by decreasing the protein and lipid oxidative modifications (Jasid et al. 2008). In *Festuca arundinacea* (tall fescue) leaves, Jin et al. (2010) observed that application of NO donor (100 μM SNP) before As stress (25 μM As) alleviated arsenic-induced electrolyte leakage, lipid peroxidation, and the levels of H_2O_2 and $O_2{\cdot}^-$. Moreover, the activities of SOD, CAT, and APX increased in presence of SNP under As stress. However, this effect was altered by application of NO scavenger (PTIO) before As treatment. Most recently, Phang et al. (2011) reported the protective role of exogenous NO on Pb toxicity in *Arabidopsis thaliana* seedlings. Pretreatment of seeds with SNP counteracted Pb toxicity by reducing the H_2O_2 and lipid hydroperoxide contents of Pb-exposed seedlings. Moreover, Pb-induced rises in the activities of antioxidant enzymes, viz. SOD, CAT, GR, GPX, and POD, were reversed by SNP pretreatment of seeds.

4.6 High-Light Intensity

Although light is a requisite for photosynthesis, when the amount of absorbed light exceeds the amount required for photosynthesis, the excess light can be harmful. Above a certain threshold, carbon fixation becomes saturated and photosynthesis is incapable of using all of the energy absorbed by the plants. Under these conditions of excess light absorption, the chloroplast lumen becomes acidic in nature, reduces the electron transport chain, and excitation energy accumulates within chloroplast. Excess excitation energy (EEE) could result in increases in the triplet form of chlorophyll and in the singlet oxygen, which are toxic in nature (Ali et al. 2005).

Under high light, NO and Ca^{2+} are active components of signaling events in ABA inhibition of light-induced stomatal opening. Garcia-Mata and Lamattina (2007) showed that both endogenous and exogenous NO inhibited the light-induced stomatal opening in *Vicia faba* epidermal strips. In another study, second messenger Ca^{2+} as well as protein kinases including MAPK and SnRK2 are very plausible mediators of the NO signals (Besson-Bard et al. 2008). Recently, Xu et al. (2010b) postulated that high-light stress-induced NOS activity leading to elevated NO which might act as a signaling molecule triggering enhanced activities of antioxidant enzymes, further protecting against injuries caused by high intensity light. In their experiment with *Festuca arundinacea* (tall fescue), pretreatment with SNP prior to exposure to high-light stress reduced light-induced electrolyte leakage and contents of MDA, H_2O_2, and $O_2{\cdot}^-$. Additionally, the activities of SOD, CAT, APX, and GR increased in presence of SNP under high-light stress, but LOX activity was inhibited. Application of NO scavenger (PTIO), however, reversed these effects of NO. Later, same researchers have reported that the treatment of tall fescue leaves with

100 μM SNP before high-light stress alleviated light-induced electrolyte leakage, MDA, and carbonyl contents (Xu et al. 2010c). The levels of H_2O_2 and $O_2^{\cdot-}$ were reduced as well. Moreover, the activities of SOD, CAT, and APX increased in tall fescue in presence of SNP under high-light stress (Xu et al. 2010c).

4.7 Flooding

Due to the increased frequency of extreme climate events, flooding or waterlogging has become an important constraint to crop production globally, causing a significant reduction in yield (Wollenweber et al. 2003). Flooding induces the progressive reduction in soil O_2 concentration and redox potential (Ruiz-Sánchez et al. 1996), which contribute to the appearance of several reduced compounds of either chemical or biochemical origin (Kozlowski 1997). Alarming changes in the earth's average temperature, erratic rainfall, and rise in sea level due to increasing melting glaciers could exaggerate flooding problems in the near future. One of the initial responses to flooding stress appears to involve the closing of stomata to avoid water loss, with a subsequent down-regulation of the photosynthetic machinery (García-Sánchez et al. 2007). Under submerged conditions, there is a decrease in total chlorophyll content in plants (Damanik et al. 2010), which sometimes respond to flooding by reducing leaf water potential, stomatal conductance, gas exchange, and plant growth (Arbona et al. 2008). Waterlogging, like other abiotic stresses, also leads to oxidative stress through an increase in ROS, such as $O_2^{\cdot-}$, $^1O_2$, H_2O_2, and OH· (Arbona et al. 2008). ROS are produced at the transition when a plant or any of its parts either enters to hypoxia/anoxia from normoxic conditions or returns to an aerobic environment (Irfan et al. 2010). Kumutha et al. (2009) and Sairam et al. (2009) showed that hypoxia-induced ROS are due to induction of membrane-linked NADPH oxidase. Higher accumulation of H_2O_2 and increased lipid peroxidation under anaerobic conditions have been reported by several groups (Hossain et al. 2009; Kumutha et al. 2009; Sairam et al. 2011).

In *Pisum sativum*, germinating seeds treated with NO could regulate the respiratory O_2 consumption; as a result, the seeds maintained some O_2 in order to prevent themselves from encountering complete anoxia (Borisjuk et al. 2007). Benamar et al. (2008) also suggested a NO_2^-–NO cycle to occur under hypoxia. Under hypoxic condition, *Medicago truncatula* leaves were found to release substantial amounts of NO (Dordas et al. 2003). More importantly, it is also known that NO is engaged in plant adaptation to hypoxia, as well as in the formation of aerenchyma during hypoxia and anoxia (Hebelstrup et al. 2007). Another important function of NO_2^- reduction under hypoxa is to contribute to ATP generation. Stoimenova et al. (2007) reported that under hypoxia, the accumulated NAD(P)H (via inhibition of glycolysis and lipid breakdown) can be oxidized by the externally facing mitochondrial NAD(P)H dehydrogenases, transferring electrons to the ubiquinone pool. When oxygen concentration decreases below the K_m of cytochrome c oxidase (COX), NO_2^- acts as an alternative electron acceptor and concomitant reduction of NO_2^- to

NO leads to a limited ATP production. Finally, the NO produced in mitochondria is oxidized by non-symbiotic cystolic hemoglobins, and the resulting NO_3^- becomes available as substrate for nitrate reductase. This cyclic process helps to generate ATP during oxygen-deprived conditions (Igamberdiev et al. 2010; Igamberdiev and Hill 2009; Gupta et al. 2011).

4.8 Ultraviolet Radiation

Plants use solar radiation for photosynthesis and accordingly are also exposed to UV-B radiation. Under exposure to UV-B radiation, different kinds of morphological, biochemical, and physiological responses of plants have been reported. UV-B radiation has detrimental effects such as reduced photosynthesis, biomass reduction, decreased protein synthesis, impaired chloroplast function, damage to DNA, etc. (He et al. 2003; Zhang et al. 2003). Enhanced UV-B radiation significantly decreases plant height and leaf area and increases leaf thickness (Ren et al. 2007). Increased leaf thickness suggests the possibility of a lower penetration of UV-B radiation into the deeper mesophyll layer (Bornman and Vogelmann 1991). Exposure to UV-B leads to the generation of ROS such as $^1O_2$, $O_2^{\cdot -}$, H_2O_2, and OH· (Moldau 1999). An increase in ROS by UV-B radiation has been observed in several plant species (Agrawal and Rathore 2007; Du et al. 2011; Singh et al. 2011), leading to the oxidative destruction of cell components through oxidative damage of nucleic acids, membrane lipids, proteins, and enzymes (Roleda et al. 2006a, b).

Protective role of NO under UV-B-induced damages in plants has been studied by several researchers. Nitric oxide plays a dual role in plant responses to UV-B irradiation. After pretreatment of *Zea mays* seedlings with NO donors, the deleterious effect of UV-B irradiation was mitigated in parallel with activation of NOS in microsomes and cytosol (An et al. 2005). In addition, UV-B induced stomatal closure, which was mediated by NO generation which was due to the NOS-like activity (He et al. 2005). Although exogenous NO mitigated the inhibitory effect of UV-B irradiation, the endogenous NO was found to be the main factor responsible for inhibition of mesocotyle growth upon UV-B irradiation (Hu et al. 2005). Wang et al. (2006) reported that NO generated from NOS-like activity appeared to act in the same direction or synergistically with ROS to induce ethylene synthesis in defense response under UV-B radiation in *Zea mays* leaves. In *Vicia faba* leaves, exogenous NO donor alleviated the injurious effect of UV-B, leading to the increased chlorophyll content and to the increase in potential and effective quantum yields of electron flow in photosystem II; the oxidative damage to thylakoid membranes was reduced to minimum owing to activation of SOD, APX, and CAT (Shi et al. 2005). They also reported that addition of NO donor can partially alleviate UV-B-induced decrease of chlorophyll content, PSII photochemistry (F_v/Fm) and quantum yield of PSII electron transport ($\O_{PS\text{-}I}$), and oxidative damage to the thylakoid membrane in bean leaves. Exogenous NO also decreased H_2O_2 by up-regulating the activities of CAT and APX. Later, Qu et al. (2006) proposed the role of NO as a signal in UV-B induced inhibition of *Pisum sativum* stems elongation. In *Zea mays* leaves, UV-B radiation accelerate ABA production, which activated $NADPH_{ox}$ and H_2O_2 generation, and that an NOS-like-dependent mechanism increased NO production to maintain cell homeostasis and attenuate UV-B-derived cell damage (Tossi et al. 2009).

In recent study, Santa-Cruz et al. (2010) demonstrated that NO protects against oxidative damage. Pretreatments with SNP, a NO donor, prevented chlorophyll loss, H_2O_2 and $O_2^{\cdot-}$ accumulations, and ion leakage in UV-B-treated plants. NOS-like activity is also required for heme oxygenase gene (*HO-1*) induction under UV-B radiation. Application of SNP was also found to alleviate UV-B stress-induced growth suppression of *Zea mays* (Kim et al. 2010). In this study, NO donor enhanced the survival of more green leaf tissue preventing chlorophyll content reduction and of higher quantum yield for photosystem II than in non-treated controls under UV-B stress. Moreover, the increase of flavonoids and anthocyanin, UV-B absorbing compounds, was observed in the NO-treated seedlings. Application of NO donor also prevented UV-B-induced increase in the contents of MDA and H_2O_2 which were accompanied by the enhancement of the activities of CAT and APX enzymes. However, it was also observed that using NO scavenger (PTIO) to the maize leaves arrested NO-induced protective effect. The inhibitor of NOS (LNNA), in addition, significantly increased H_2O_2 and MDA accumulation and decreased antioxidant enzyme activities in maize leaves under UV-B stress. These results concluded that NO might act as a signal in up-regulating ROS-scavenging system that protects plants from oxidative stress induced by UV-B radiation and thus confer UV-B tolerance (Kim et al. 2010).

4.9 Ozone

It is predicted that significant crop losses due to O_3 damage will increase 25% in background O_3 concentration over the next 30–50 years (Meehl et al. 2007). In many industrialized countries, tropospheric ozone (O_3) reaches to such high concentration which is harmful for the plant species (Schraudner et al. 1997). Therefore, considering the predicted effect of O_3, it is necessary to explore the multifarious responses of plants and their adaptation under elevated O_3. Many reports indicate that O_3 leads to a general reduction of growth and competitive fitness of plants (Gillespie et al. 2011) in which elevated O_3 concentrations cause oxidative injury in living tissues and may result in negative long-term effects on the vitality of plants, leaf damage, biomass reduction, altered metabolism, and accelerated senescence, which lead to losses in yield (Ashmore 2005; Li et al. 2010b; Feng et al. 2011). Being a strong oxidant, O_3 can interact with constituents of the apoplast to generate ROS such as H_2O_2, $O_2^{\cdot-}$, OH,· and HOO· (Yan et al. 2010a, b).

In *Arabidopsis* plants, elevated O_3 induced NOS activity that preceded accumulation of SA and cell death (Rao and Davis 2001). In tobacco, NO was found to induce SA synthesis (Durner et al. 1998). Ahlfors et al. (2009) suggested that NO can modify signaling, hormone biosynthesis, and gene expression in plants during

O_3 exposure, which modulates ozone-induced cell death of *Arabidopsis thaliana*. In their study, the NO donor (SNP) and O_3 individually induced a large set of defense-related genes; however, in a combined treatment SNP accelerated the O_3-induced SA biosynthesis and other defense-related genes. Moreover, exogenous NO also decreased O_3-induced SA accumulation. The O_3-sensitive mutant *rcd1* was found to be a NO overproducer; in contrast, *Atnoa1/rif1* (*Arabidopsis* NO-associated 1/resistant to inhibition by *FSM1*), a mutant with decreased production of NO, was also O_3-sensitive.

4.10 Role of NO under Oxidative Stress

Under adverse environmental conditions like salinity, drought, temperature extremes, heavy metal toxicity, high-light intensity, nutrient deficiency, UV-B radiation, ozone, etc. oxidative stress is occurred through accelerating the production of ROS such as $^1O_2$, $O_2{\cdot}^-$, H_2O_2, and OH·. ROS are extremely reactive in nature because they can interact with a number of cellular molecules and metabolites, thereby leading to irreparable metabolic dysfunction and death. In general, plant cells are adequately equipped to keep ROS within the limits that are generated as a consequence of normal cellular metabolic activities. Under different stress conditions, however, ROS generation often exceeds the overall cellular antioxidative potential leading to stress-induced adverse effects on plant growth and physiology. A steady state balance is required to protect plant cells from oxidative damage. Plants possess an efficient non-enzymatic (AsA, GSH, α-tocopherol, phenolic compounds, alkaloids, and non-protein amino acids) and enzymatic (SOD, CAT, APX, MDHAR, DHAR, GR, GPX, GST, POD) antioxidant defense systems which work in concert to control the cascades of uncontrolled oxidation and protect plant cells from oxidative damage by scavenging ROS (Mittler et al. 2004; Gill and Tuteja 2010). These antioxidant defense systems are found in almost all cellular compartments, demonstrating the importance of ROS detoxification for cellular survival (Mittler et al. 2004). These defenses are not restricted to the intracellular compartment, but are also found in the apoplast to a limited extent (Mittler 2002; Gill and Tuteja 2010). Different plant studies indicated that endogenous NO is a key factor in the tolerance of cells to oxidative stress induced by a range of abiotic conditions, and this probably involves the enhanced expression of genes encoding antioxidant enzymes (Hao and Zhang 2010). Several studies have also shown that exogenous NO ameliorates the oxidative stress induced by a range of abiotic stress conditions (Bai et al. 2011; Hasanuzzaman et al. 2011a; Liu et al. 2011; Phang et al. 2011; Wu et al. 2011).

NO exerts a protective function against oxidative stress mediated by (1) reaction with lipid radicals, which stops the propagation of lipid oxidation; (2) scavenging the O_2^- and formation of peroxynitrite ($ONOO^-$) that can be neutralized by other cellular processes; (3) activation of antioxidant enzymes (SOD, CAT, APX, GPX, GR, POX, etc.); and (4) functioning as a signaling molecule in the cascade of events leading to changes of gene expression. These mechanisms together confer enhanced antioxidant protection against oxidative stress (Hao and Zhang 2010; Hasanuzzaman et al. 2010; Misra et al. 2011a); Fig. 11.3). However, whether or not endogenous NO has an antioxidant function is debatable. The presence of an unpaired electron within the NO molecule makes it a reactive species and is also the origin of its duality. As mentioned above, NO readily reacts with O_2^- to form peroxynitrite $ONOO^-$. Peroxynitrite can provoke the nitration of tyrosine residues both in vitro and in vivo, and this reaction has been proposed as a regulatory mechanism for protein activity. Nitric oxide is generally toxic and in these conditions, when combined with low amounts of O_2^-, the formation of $ONOO^-$ was reported to be deleterious to lipids, proteins, and DNA (Wink et al. 1993). However, ROS-induced toxicity is minimized as NO acts as chain breaker and hence enhance protection. In these situations, peroxides have proven to be much more toxic than NO and $ONOO^-$, and NO is considered to have a protective function (Wink et al. 1993). Several studies have reported the involvement of nitrated proteins in plants (Cecconi et al. 2009; Chaki et al. 2009; Baudouin 2011). Moreover, modification of the nitrated protein pattern occurs in response to several stresses (Corpas et al. 2008; Cecconi et al. 2009; Chaki et al. 2009). In addition, the reaction of NO with lipid alcoxyl (LO·) and peroxyl (LOO·) radicals is rapid, giving rise to the expectation that NO could also stop the propagation of radical-mediated lipid oxidation (Baudouin 2011). Beligni and Lamattina (1999b) showed that NO is able to prevent the chlorophyll decay produced by two ROS-generating compounds and that this effect is mimicked by OH· and iron scavengers. NO is capable of producing complexes with metal-containing proteins, namely, with hemoglobins, cytosolic and mitochondrial aconitase, CAT, APX, and cytochrome coxidase (Besson-Bard et al. 2008). Furthermore, a great deal of attention is paid to covalent post-translational protein modifications caused by synergistic action of NO and other reactive forms of nitrogen and oxygen.

In plants $O_2^{\cdot -}$ can arise from several sources, such as mitochondria, chloroplasts, or $NADPH_{ox}$. Superoxide is readily dismutased to H_2O_2 at low pH or in a reaction catalyzed by SOD. Both $O_2^{\cdot -}$ and H_2O_2 have been suggested as signaling molecules in plants (Neill et al. 2002a). Nitric oxide reacting with $O_2^{\cdot -}$ or H_2O_2 could potentially disrupt $O_2^{\cdot -}/H_2O_2$ signaling. According to Dubovskaya et al. (2007), during H_2O_2-induced oxidative stress, low concentrations of NO inhibit lipid peroxidation, counteract the fragmentation of DNA, and prevent accumulation of soluble proteins in tobacco cells, while at high concentrations it promoted degradation of DNA and soluble proteins and reduced ATP synthesis. The results are consistent with the hypothesis that NO performs a dual role in plants, acting as antioxidant and signaling messenger as well. Cui et al. (2011) found that pretreatment of the exogenous NO precursor led to both increased stress tolerance and increased expression of antioxidant enzymes in *Cucumis sativus* plants. Zhang et al. (2006) had shown that osmotic stress, ABA, and H_2O_2 enhance the expression of several antioxidant genes such as *CAT1*, cytosolic ascorbate peroxidase (*cAPX*), and plastidial glutathione reductase 1 (*GR1*), and the total enzyme activities of CAT, APX, GR, and SOD. Later, Zhang et al. (2007) demonstrated that NO is an essential intermediate in these ABA and H_2O_2 enhancements. Pretreatment with

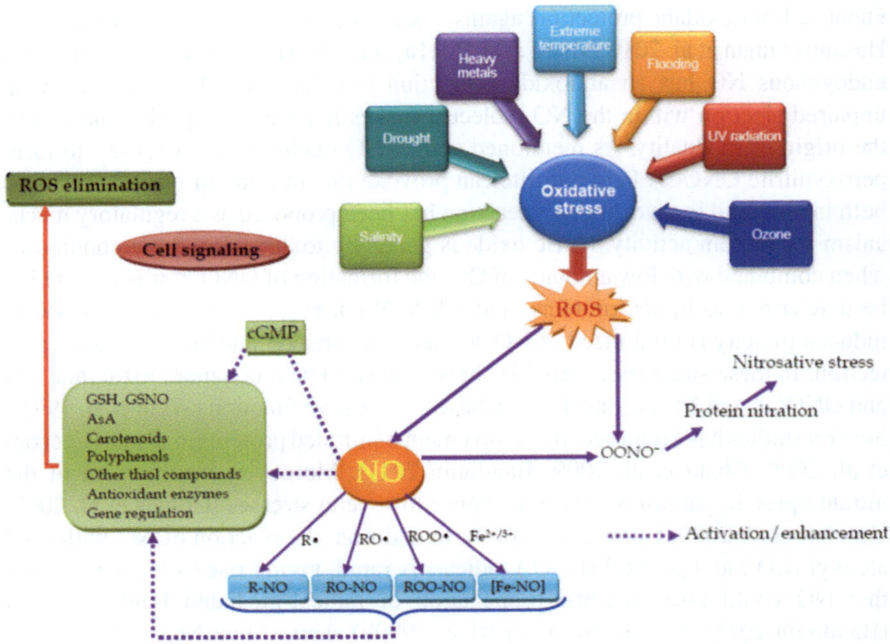

Fig. 11.3 Protection of NO under oxidative stress condition

NO scavenger, c-PTIO substantially prevented increases in gene expression and enzyme activity. Moreover, treatment with the NO donor (SNP) essentially reproduced the effects of ABA or H_2O_2. Importantly, the removal of the NO released from SNP with c-PTIO prevented the increases, and treatment with $Na_3Fe(CN)_6$ (a molecule similar to SNP but does not release NO) had no effect. A number of studies have already shown that exogenously applied NO can impart protective antioxidant properties. In previous study, it was believed that NO is involved in two respiratory electron transport pathways in mitochondria (Yamasaki et al. 2001; Zottini et al. 2002) where it detoxifies ROS and enhances antioxidant defense systems in plants under abiotic stresses. Shi et al. (2007) reported that the exogenous NO treatment protects plant from damage by eliminating the ($O_2^{\cdot -}$) and lipid radical and upregulates the antioxidant enzymes activities especially SOD. Some recent work has indicated that endogenous NO induces antioxidant defenses, potentially via ABA signaling (Song et al. 2006; Zhou et al. 2005). Recently, Hao and Zhang (2010) indicated a key "ABA–H_2O_2–NO–MAPK–antioxidant survival Cycle" and proposed that during water stress ABA have several protective functions that involve NO as a key signaling intermediate through the induction of stomatal closure to reduce water loss and the activation of antioxidant defenses during oxidative stress.

An additional route for NO removal implies its reaction with the antioxidant thiol GSH to form GSNO and the subsequent reduction of GSNO to GSSG and NO_3^- (Feechan et al. 2005; Rusterucci et al. 2007; Lee et al. 2008). It has also been proposed that increase of NO production and corresponding decrease of NO removal through the repression of GSNO reductase gene expression could take place to form the bioactive NO signal (Díaz et al. 2003; Rusterucci et al. 2007). However, NO production and removal spatially integrated to generate the operating NO signal remains unclear.

The fact that NO is intricately linked to generation and detoxification of ROS suggests that a fine-tuned system exists in plants for ensuring maintenance of basal levels of NO, coupled with interaction between NO signals and ROS signals, required for signaling purpose that ensures cellular homeostasis (Leach et al. 2010). In conclusion, NO is generated as pivotal role in alleviating oxidative stress that is consequent to abiotic stress (de Gara et al. 2010).

5 Conclusion

The roles of NO in plant responses to abiotic stresses are studied through investigating the effects on plant physiological and biochemical changes under stress. NO has been found to play a crucial role in plant growth and development, starting from cell cycle regulation, differentiation, and morphogenesis, including flowering and root formation. However, the most important and best documented function of NO is the up-regulation of antioxidant defense or directly functions as an antioxidant. Although several NO synthetic pathways in plants have been suggested, biochemical and molecular details of each pathway remain obscure; and it is unclear how these identified pathways cooperate with each other in plants, and which pathway operates in each particular tissue or organ or at a specific time. Regarding NO biosynthesis, future studies should focus on how NO is produced in a particular tissue or organ (and in which pathway), at what time scale NO production is elicited by a developmental or environmental stimulus, and how the above described pathways work in concert when/if they all work in the same tissue or organ at the same time scale. Rapidly increasing evidences indicate that NO is actively involved in several physiological processes; however, there has been much disagreement regarding the mechanism(s) by which NO reduces abiotic stress. Therefore, most of the research work has still to be done to elucidate the functions of NO as a signaling molecule in interaction with plant hormones, nutrients, and metals; functions of endogenous NO in plants; actual biosynthesis pathways of NO in plants and its regulation to environmental stimulus and cellular redox homeostasis regulation; and NO-mediated defense gene regulation in plants. In the last few years NO and H_2O_2 have emerged to be central players in the world of plant cell signaling, particularly under various stressful situations. The full range of biological functions for these two signaling molecules remains to be catalogued, and determining the ways in which they interact, both together and with the ever-increasing array of signals known to be recognized by plants, will need to be elucidated (Neill et al. 2002a). Other research priorities must include full characterization of the enzymes through which the intracellular concentrations of H_2O_2 and NO are regulated, and where these enzymes are located in different cells and tissues. The intracellular signaling cascades that transduce H_2O_2 and NO perceptions into cellular responses have so far been characterized only superficially. Finally, there arises the question how H_2O_2 and NO are detected by cells. Such perception could conceivably involve direct interaction of H_2O_2 and NO with various cellular proteins, such as transcription factors, ion channels, or enzymes. $H_2O_2^-$ and NO-sensitive enzymes could include signaling enzymes such as protein kinases and phosphatases (Neill et al. 2002a). NOS-deficient mutant and/or gene knock-out mutant are now available. Genomics tools are accelerating the discovery of NO-producing genes on a global scale and are expanding our understanding of the oxidative stress response and the pleiotropic roles of NO in signaling, gene expression, and plant stress tolerance.

References

Abat JK, Mattoo AK, Deswal R (2008) *S*-nitrosylated proteins of a medicinal CAM plant *Kalanchoe pinnata* – ribulose-1,5-bis-phosphate carboxylase/oxygenase activity targeted for inhibition. FEBS J 275:2862–2872

Abogadallah GM (2010) Antioxidative defense under salt stress. Plant Signal Behav 5:369–374

Acquaah G (2007) Principles of plant genetics and breeding. Blackwell, Oxford, UK, p 385

Agrawal SB, Rathore D (2007) Changes in oxidative stress defense in wheat (*Triticum aestivum* L.) and mung bean (*Vigna radiata* L.) cultivars grown with or without mineral nutrients and irradiated by supplemental ultraviolet-B. Environ Exp Bot 59:21–33

Ahlfors R, Brosché M, Kollist H, Kangasjärvi J (2008) Nitric oxide modulates ozone-induced cell death, hormone biosynthesis and gene expression in *Arabidopsis thaliana*. Plant J 58:1–12

Ahlfors R, Brosché M, Kollist H, Kangasjärvi J (2009) Nitric oxide modulates ozone-induced cell death, hormone biosynthesis and gene expression in *Arabidopsis thaliana*. Plant J 58:1–12

Ali MB, Hahn E-J, Paek K-Y (2005) Effects of light intensities on antioxidant enzymes and malondialdehyde content during short-term acclimatization on micropropagated *Phalaenopsis* plantlet. Environ Exp Bot 54:109–120

An L, Liu Y, Zhang M, Chen T, Wang X (2005) Effects of nitric oxide on growth of maize seedling leaves in the presence or absence of ultraviolet radiation. J Plant Physiol 162:317–326

Apel K, Hirt H (2004) Reactive oxygen species: metabolism, oxidative stress, and signal transduction. Annu Rev Plant Biol 55:373–399

Arasimowicz M, Floryszak-Wieczorek J (2007) Nitric oxide as a bioactive signalling molecule in plant stress responses. Plant Sci 172:876–887

Arasimowicz-Jelonek M, Floryszak-Wieczorek J, Kubis J (2009) Involvement of nitric oxide in water stress-induced responses of cucumber roots. Plant Sci 177:682–690

Arasimowicz-Jelonek M, Floryszak-Wieczorek J, Gwóźdźa EA (2011) The message of nitric oxide in cadmium challenged plants. Plant Sci. doi:10.1016/j.plantsci.2011.03.019

Arbona V, Hossain Z, López-Climent MF, Pérez-Clemente RM, Gómez-Cadenas A (2008) Antioxidant enzymatic activity is linked to waterlogging stress tolerance in citrus. Physiol Plant 132:452–466

Ashmore MR (2005) Assessing the future global impacts of ozone on vegetation. Plant Cell Environ 28:949–964

Badri DV, Loyola-Vargas VM, Du J, Stermitz FR, Broeckling CD, Iglesias-Andreu L, Vicanco JM (2008) Transcriptome analysis of *Arabidopsis* roots treated with signalling compounds: a focus on signal transduction, metabolic regulation and secretion. New Phytol 179:209–223

Bai X, Yang L, Tian M, Chen J, Shi J, Yang Y, Hu X (2011) Nitric oxide enhances desiccation tolerance of recalcitrant *Antiaris toxicaria* seeds via protein *S*-nitrosylation and carbonylation. PLoS One 6:e20714. doi:10.1371/journal.pone.0020714

Balla K, Bencze S, Janda T, Veisz O (2009) Analysis of heat stress tolerance in winter wheat. Acta Agron Hung 57:437–444
Bartha B, Kolbert Z, Erdei L (2005) Nitric oxide production induced by heavy metals in *Brassica juncea* L. Czern. and *Pisum sativum* L. Acta Biol Szeged 49:9–12
Baudouin E (2011) The language of nitric oxide signaling. Plant Biol 13:233–242
Beligni MV, Lamattina L (1999a) Is nitric oxide toxic or protective? Trends Plant Sci 4:299–300
Beligni MV, Lamattina L (1999b) Nitric oxide counteracts cytotoxic processes mediated by reactive oxygen species in plant tissues. Planta 208:337–344
Beligni MV, Fath A, Bethke PC, Lamattina L, Jones RL (2002) Nitric oxide acts as an antioxidant and delays programmed cell death in barley aleurone layers. Plant Physiol 129:1642–1650
Benamar A, Rolletschek H, Borisjuk L, Avelange-Macherel MH, Curien G, Mostefai HA, Andriantsitohaina R, Macherel D (2008) Nitrite-nitric oxide control of mitochondrial respiration at the frontier of anoxia. Biochim Biophys Acta 1777:1268–1275
Besson-Bard A, Pugin A, Wendehenne D (2008) New insights into nitric oxide signalling in plants. Annu Rev Plant Biol 59:21–39
Besson-Bard A, Astier J, Rasul S, Wawer I, Dubreuil-Maurizi C, Jeandroz S, Wendehenne D (2009a) Current view of nitric oxide-responsive genes in plants. Plant Sci 177:302–309
Besson-Bard A, Gravot A, Richaud P, Auroy P, Duc C, Gaymard F, Taconnat L, Renou JP, Pugin A, Wendehenne D (2009b) Nitric oxide contributes to cadmium toxicity in *Arabidopsis* by promoting cadmium accumulation in roots and by upregulating genes related to iron uptake. Plant Physiol 149:1302–1315
Bohnert HJ, Jensen RG (1996) Strategies for engineering water stress tolerance in plants. Trends Biotechnol 14:89–97
Borisjuk L, Macherel D, Benamar A, Wobus U, Rolletschek H (2007) Low oxygen sensing and balancing in plant seeds: a role for nitric oxide. New Phytol 176:813–823
Bornman JF, Vogelmann TC (1991) Effect of UV-B radiation on leaf optical properties measured with fiber optics. J Exp Bot 42:547–554
Boucher JL, Genet A, Valdon S, Delaforge M, Henry Y, Mansuy D (1992) Cytochrome P450 catalyzes the oxidation of Nω-hydroxy-L-arginine by NADPH and O_2 to nitric oxide and citrulline. Biochem Biophys Res Commun 187:880–886
Bright J, Desikan R, Hancock JT, Weir IS, Neill SJ (2006) ABA induced NO generation and stomatal closure in *Arabidopsis* are dependent on H_2O_2 synthesis. Plant J 45:113–122
Cantrel C, Vazquez T, Puyaubert J, Rezé N, Lesch M, Kaiser WM, Dutilleul C, Guillas I, Zachowski A, Baudouin E (2011) Nitric oxide participates in cold-responsive phosphosphingolipid formation and gene expression in *Arabidopsis thaliana*. New Phytol 189:415–427
Cecconi D, Orzetti S, Vandelle E, Rinalducci S, Zolla L, Delledonne D (2009) Protein nitration during defence response in *Arabidopsis thaliana*. Electrophor 30:2460–2468
Chaki M, Fernandez-Ocana AM, Valderrama R, Carreras A, Esteban FJ, Luque F, Gomez-Rodriguez MV, Begara-Morales JC, Corpas FJ, Barroso JB (2009) Involvement of reactive nitrogen and oxygen species (RNS and ROS) in sunflower–mildew interaction. Plant Cell Physiol 50:265–279
Chaki M, Valderrama R, Fernández-Ocaña AM, Carreras A, Gómez-Rodríguez MV, Pedrajas JR, Begara-Morales JC, Sánchez-Calvo B, Luque F, Leterrier M, Corpas FJ, Barroso JB (2011) Mechanical wounding induces a nitrosative stress by down-regulation of GSNO reductase and an increase in *S*-nitrosothiols in sunflower (*Helianthus annuus*) seedlings. J Exp Bot 62:1803–1813
Chen F, Wang F, Sun H, Cai Y, Mao W, Zhang G, Vincze E, Wu F (2010) Genotype-dependent effect of exogenous nitric oxide on Cd-induced changes in antioxidative metabolism, ultrastructure, and photosynthetic performance in barley seedlings (*Hordeum vulgare*). J Plant Growth Regul 29:394–408
Clarke A, Desikan R, Hurst RD, Hancock JT, Neill ST (2000) NO way back: nitric oxide and programmed cell death in *Arabidopsis thaliana* suspension cultures. Plant J 24:667–677
Cooney RV, Harwood PJ, Custer LJ, Franke AA (1994) Light-mediated conversion of nitrogen dioxide to nitric oxide by carotenoids. Environ Health Perspect 102:460–462
Corpas FJ, de la Colina C, Sanchez-Rasero F, del Río LA (1997) A role for leaf peroxisomes in the catabolism of purines. J Plant Physiol 151:246–250

Corpas F, Barroso J, Carreras A, Valderrama R, Palma J, León A, Sandalio L, del Río L (2006) Constitutive arginine-dependent nitric oxide synthase activity in different organs of pea seedlings during plant development. Planta 224:246–254

Corpas FJ, Chaki M, Fernandez-Ocana AM, Valderrama R, Palma JM, Carreras A, Begara-Morales JC, Airaki M, del Rio LA, Barroso JB (2008) Metabolism of reactive nitrogen species in pea plants under abiotic stress conditions. Plant Cell Physiol 49:1711–1722

Corpas FJ, Leterrier M, Valderrama R, Airaki M, Chaki M, Palma JM, Barroso JB (2011) Nitric oxide imbalance provokes a nitrosative response in plants under abiotic stress. Plant Sci. doi:10.1016/j.plantsci.2011.04.005

Correa-Aragunde N, Graziano M, Chevalier C, Lamattina L (2006) Nitric Oxide mediates the expression of cell-cycle regulatory genes during lateral root formation in tomato. J Exp Bot 57:581–588

Courtois C, Besson A, Dahan J, Bourque S, Dobrowolska G, Pugin A, Wendehenne D (2008) Nitric oxide signalling in plants: Interplays with Ca^{2+} and protein kinases. J Exp Bot 59:155–163

Crawford NM (2006) Mechanisms for nitric oxide synthesis in plants. J Exp Bot 57:471–478

Cueto M, Hernandez-Perera O, Martin R, Bentura ML, Rodrigo J, Lamas S, Golvano MP (1996) Presence of nitric oxide synthase activity in roots and nodules of *Lupinus albus*. FEBS Lett 398:159–164

Cui J-X, Zhou Y-H, Ding J-G, Xia X-J, Shi K, Chen S-C, Asam T, Chen Z, Yu J-Q (2011) Role of nitric oxide in hydrogen peroxide-dependent induction of abiotic stress tolerance by brassinosteroids in cucumber. Plant Cell Environ 34:347–358

Damanik RI, Maziah M, Ismail MR, Ahmad S, Zain AM (2010) Responses of the antioxidative enzymes in Malaysian rice (*Oryza sativa* L.) cultivars under submergence condition. Acta Physiol Plant 32:739–747

David A, Yadav S, Bhatla SC (2010) Sodium chloride stress induces nitric oxide accumulation in root tips and oil body surface accompanying slower oleosin degradation in sunflower seedlings. Physiol Plant 140:342–354

de Carvalho MHC (2008) Drought stress and reactive oxygen species. Plant Signal Behav 3:156–165

de Gara L, Locato V, Dipierro S, De Pinto MC (2010) Redox homeostasis in plants: The Challenge of living with endogenous oxygen production. Resp Physiol Neurobiol 173:13–19

de la Haba P, Agüera E, Benítez L, Maldonado JM (2001) Modulation of nitrate reductase activity in cucumber (*Cucumis sativus*) roots. Plant Sci 161:231–237

del Giudice J, Cam Y, Damiani I, Fung-Chat F, Meilhoc E, Bruand C, Brouquisse R, Puppo A, Boscari A (2011) Nitric oxide is required for an optimal establishment of the *Medicago truncatula–Sinorhizobium meliloti* symbiosis. New Phytol 191:405–717

del Río LA, Corpas FJ, Barroso JB (2004) Nitric oxide and nitric oxide synthase activity in plants. Phytochemistry 65:783–792

Delledonne M (2005) NO news is good news for plants. Curr Opin Plant Biol 8:390–396

Delledonne M, Xia Y, Dixon RA, Lamb C (1998) Nitric oxide functions as a signal in plant disease resistance. Nature 394:585–588

Desikan R, Clarke A, Hancock JT, Neill SJ (1999) H_2O_2 activates a MAP kinase-like enzyme in *Arabidopsis thaliana* suspension cultures. J Exp Bot 50:1863–1866

Desikan R, Griffiths R, Hancock J, Neill S (2002) A new role for an old enzyme: nitrate reductase-mediated nitric oxide generation is required for abscisic acid-induced stomatal closure in *Arabidopsis thaliana*. Proc Natl Acad Sci USA 99:16314–16318

Desikan R, Cheung M-K, Bright J, Henson D, Hancock JT, Neill SJ (2004) ABA, hydrogen peroxide and nitric oxide signalling in stomatal guard cells. J Exp Bot 55:205–212

Díaz M, Achkor H, Titarenko E, Martínez MC (2003) The gene encoding glutathione-dependent formaldehyde dehydrogenase/GSNO reductase is responsive to wounding, jasmonic acid and salicylic acid. FEBS Lett 543:136–139

Din J, Khan SU, Ali I, Gurmani AR (2011) Physiological and agronomic response of canola varieties to drought stress. J Anim Plant Sci 21:78–82

Donaldson L, Ludidi N, Knight MR, Gehring C, Denby K (2004) Salt and osmotic stress cause rapid increases in *Arabidopsis thaliana* cGMP levels. FEBS Lett 569:317–320
Dordas C, Rivoal J, Hill RD (2003) Plant hemoglobins, nitric oxide and hypoxic stress. Ann Bot 91:173–178
Du H, Liang Y, Pei K, Ma K (2011) UV Radiation-responsive proteins in rice leaves: A proteomic analysis. Plant Cell Physiol 52:306–316
Dubey RS (2011) Metal toxicity, oxidative stress and antioxidative defense system in plants. In: Gupta SD (ed) Reactive oxygen species and antioxidants in higher plants. CRC Press, Boca Raton, FL, pp 177–203
Dubovskaya LV, Kolesneva EV, Knyazev DM, Voltovskii ID (2007) Protective role of nitric oxide during hydrogen peroxide-induced oxidative stress in tobacco plants. Russ J Plant Physiol 54:755–762
Durner J, Klessig DF (1999) Nitric oxide as a signal in plants. Curr Opin Plant Biol 2:369–374
Durner J, Wendehenne D, Klessig DF (1998) Defense gene induction in tobacco by nitric oxide, cyclic GMP and cyclic ADP-ribose. Proc Natl Acad Sci USA 95:10328–10333
Ederly L, Morettini R, Borgogni A, Wasternack C, Miersch O, Reale L, Ferranti L, Tosti N, Pasqualini S (2006) Interaction between nitric oxide and ethylene in the induction of alternative oxidase in ozone-treated tobacco plants. Plant Physiol 142:595–608
Egli DB, Tekrony DM, Heitholt JJ, Rupe J (2005) Air temperature during seed filling and soybean seed germination and vigor. Crop Sci 45:1329–1335
Einset J, Winge P, Bones A (2007) ROS signaling pathways in chilling stress. Plant Signal Behav 2:365–367
Ercoli L, Mariotti M, Masoni A, Arduini I (2004) Growth responses of sorghum plants to chilling temperature and duration of exposure. Eur J Agron 21:93–103
Erdei L, Colbert Z (2008) Nitric oxide as a potent signalling molecule in plants. Acta Biol Szeged 52:1–5
Faize M, Burgos L, Faize L, Piqueras A, Nicolas E, Barba-Espin G, Clemente-Moreno MJ, Alcobendas R, Artlip T, Hernández JA (2011) Involvement of cytosolic ascorbate peroxidase and Cu/Zn-superoxide dismutase for improved tolerance against drought stress. J Exp Bot. doi:10.1093/jxb/erq432
Fan H, Guo S, Jiao Y, Zhang R, Li J (2007) Effects of exogenous nitric oxide on growth, active oxygen species metabolism, and photosynthetic characteristics in cucumber seedlings under NaCl stress. Front Agric China 1:308–314
Fan H-F, Du C-X, Guo S-R (2010) Nitric oxide enhances salt tolerance in cucumber seedlings by regulating free polyamine content. Environ Exp Bot. doi:10.1016/j.envexpbot.2010.09.007
Farooq M, Aziz T, Wahid A, Lee DJ, Siddique KHM (2009) Chilling tolerance in maize: agronomic and physiological approaches. Crop Past Sci 60:501–516
Feechan A, Kwon E, Yun BW, Wang Y, Pallas JA, Loake GJ (2005) A central role for *S*-nitrosothiols in plant disease resistance. Proc Natl Acad Sci USA 102:8054–8059
Feng Z, Pang J, Kobayashi K, Zhu ZN, Ort DR (2011) Differential responses in two varieties of winter wheat to elevated ozone concentration under fully open-air field conditions. Global Change Biol 17:580–591
Ferrarini A, de Stefano M, Baudouin E, Pucciariello C, Polverari A, Puppo A, Delledonne M (2008) Expression of *Medicago truncatula* genes responsive to nitric oxide in pathogenic and symbiotic conditions. Mol Plant Microbe Interact 21:781–790
Ferreira LC, Cataneo AC (2010) Nitric oxide in plants: a brief discussion on this multifunctional molecule. Sci Agric 67:236–243
Fodor F (2002) Physiological responses of vascular plants to heavy metals. In: Prasad MNV, Strzalka K (eds) Physiology and biochemistry of metal toxicity and tolerance in plants. Kluwer, Dortrech, pp 149–177
Gao HJ, Yang HQ, Wang JX (2009) Arginine metabolism in roots and leaves of apple (*Malus domestica* Borkh.): the tissue-specific formation of both nitric oxide and polyamines. Sci Hortic 119:147–152

Garcia-Mata C, Lamattina L (2001) Nitric oxide induces stomatal closure and enhances the adaptive plant responses against drought stress. Plant Physiol 126:1196–1204

Garcia-Mata C, Lamattina L (2007) Abscisic acid (ABA) inhibits light-induced stomatal opening through calcium- and nitric oxide-mediated signaling pathways. Nitric Oxide 17:143–151

Garcia-Mata C, Gay R, Sokolovski S, Hills A, Lamattina L, Blatt MR (2003) Nitric oxide regulates K^+ and Cl^- channels in guard cells through a subset of abscisic acid-evoked signaling pathways. Proc Natl Acad Sci USA 100:11116–11121

García-Sánchez F, Syvertsen JP, Gimeno V, Botia P, Pérez-Pérez JG (2007) Responses to flooding and drought stress by two citrus rootstock seedlings with different water-use efficiency. Physiol Plant 130:532–542

Gas E, Flores-Pérez U, Sauret-Güeto S, Rodríguez-Concepción M (2009) Hunting for plant nitric oxide synthase provides new evidence of a central role for plastids in nitric oxide metabolism. Plant Cell 21:18–23

Gaupels F, Furch ACU, Will T, Mur LAJ, Kogel KH, van Bel AJE (2008) Nitric oxide generation in *Vicia faba* phloem cells reveals them to be sensitive detectors as well as possible systemic transducers of stress signals. New Phytol 178:634–646

Gill SS, Tuteja N (2010) Reactive oxygen species and antioxidant machinery in abiotic stress tolerance in crop plants. Plant Physiol Biochem 48:909–930

Gillespie KM, Rogers A, Ainsworth EA (2011) Growth at elevated ozone or elevated carbon dioxide concentration alters antioxidant capacity and response to acute oxidative stress in soybean (*Glycine max*). J Exp Bot. doi:10.1093/jxb/erq435

Gordge MP (1998) How cytotoxic is nitric oxide? Exp Nephrol 6:12–16

Gotte G, Amelio E, Russo S, Marlinghaus E, Musci G, Suzuki H (2002) Short-time non-enzymatic nitric oxide synthesis from L-arginine and hydrogen peroxide induced by shock waves treatment. FEBS Lett 520:153–155

Gould KS, Lamotte O, Klinguer A, Pugin A, Wendehenne D (2003) Nitric oxide production in tobacco leaf cells: a generalized stress response? Plant Cell Environ 26:1851–1862

Grant M, Brown I, Adams S, Knight M, Ainslie A, Mansfield J (2000) The *RPM1* plant disease resistance gene facilitates a rapid and sustained increase in cytosolic calcium that is necessary for the oxidative burst and hypersensitive cell death. Plant J 23:441–450

Groppa D, Rosales EP, Iannone MF, Benavides MP (2008) Nitric oxide, polyamines and Cd-induced phytotoxicity in wheat roots. Phytochemistry 69:2609–2615

Guo FQ, Okamoto M, Crawford NM (2003) Identification of a plant nitric oxide synthase involved in hormonal signaling. Science 302:100–103

Guo Z, Ou W, Lu S, Zhong Q (2006) Differential responses of antioxidant system to chilling and drought in four rice cultivars differing in sensitivity. Plant Physiol Biochem 44:828–836

Guo Y, Tian Z, Yan D, Zhang J, Qin P (2009) Effects of nitric oxide on salt stress tolerance in *Kosteletzkya virginica*. Life Sci J 6:67–75

Gupta KJ, Kaiser WM (2010) Production and scavenging of Nitric oxide by barley root mitochondria. Plant Cell Physiol 51:576–584

Gupta KJ, Igamberdiev AU, Manjunatha G, Segu S, Moran JF, Neelawarne B, Bauwe H, Kaiser WM (2011) The emerging roles of nitric oxide (NO) in plant mitochondria. Plant Sci. doi:10.1016/j.plantsci.2011.03.018

Habib N, Ashraf M, Ahmad MSA (2010) Enhancement in seed germinability of rice (*Oryza sativa* L.) by pre-sowing seed treatment with nitric oxide (NO) under salt stress. Pak J Bot 42:4071–4078

Hancock JT, Neill SJ, Wilson ID (2011) Nitric oxide and ABA in the control of plant function. Plant Sci. doi:10.1016/j.plantsci.2011.03.017

Hao GP, Zhang JH (2010) The role of nitric oxide as a bioactive signaling molecule in plants under abiotic stress. In: Hayat S, Mori M, Pichtel J, Ahmad A (eds) Nitric oxide in plant physiology. Weinheim, Wiley-VCH Verlag, pp 115–138

Hao GP, Xing Y, Zhang JH (2008) Role of nitric oxide dependence on nitric oxide synthase-like activity in the water stress signaling of maize seedling. J Integr Plant Biol 50:435–442

Harb A, Krishnan A, Ambavaram MMR, Pereira A (2010) Molecular and physiological analysis of drought stress in *Arabidopsis* reveals early responses leading to acclimation in plant growth. Plant Physiol 154:1254–1271

Harrison R (2002) Structure and function of xanthine oxidoreductase: where are we now? Free Radic Biol Med 33:774–797

Hasanuzzaman M, Fujita M (2011) Selenium pretreatment up-regulates the antioxidant defense and methylglyoxal detoxification system and confers enhanced tolerance to drought stress in rapeseed seedlings. Biol Trace Elem Res 143:1758–1776

Hasanuzzaman M, Fujita M, Islam MN, Ahamed KU, Nahar K (2009) Performance of four irrigated rice varieties under different levels of salinity stress. Int J Integr Biol 6:85–90

Hasanuzzaman M, Hossain MA, Fujita M (2010) Physiological and biochemical mechanisms of nitric oxide induced abiotic stress tolerance in plants. Am J Plant Physiol 5:295–324

Hasanuzzaman M, Hossain MA, Fujita M (2011a) Nitric oxide modulates antioxidant defense and methylglyoxal detoxification system and reduces salinity induced damage in wheat seedling. Plant Biotechnol Rep 5:353–365

Hasanuzzaman M, Hossain MA, Fujita M (2011b) Selenium-induced upregulation of the antioxidant defense and methylglyoxal detoxification system reduces salinity-induced damage in rapeseed seedlings. Biol Trace Elem Res 143:1704–1721

He DL, Wong CH, He YH (2003) The effect of reduction of ultraviolet–B radiance on the content of flavonoid in leaves of wheat. Chinese J Agromet 24:32

He JM, Xu H, She XP, Song XG, Zhao WM (2005) The role and the interrelationship of hydrogen peroxide and nitric oxide in the UV-B-induced stomatal closure in broad bean. Funct Plant Biol 32:237–247

He J-M, Zhang Z, Wang R-B, Chen Y-P (2011) UV-B-induced stomatal closure occurs via ethylene-dependent NO generation in *Vicia faba*. Funct Plant Biol 38:293–302

Hebelstrup KH, Igamberdiev AU, Hill RD (2007) Metabolic effects of hemoglobin gene expression in plants. Gene 398:86–93

Hirt H (1997) Multiple roles of MAP kinases in plant signal transduction. Trends Plant Sci 2:11–15

Holzmeister C, Fröhlich A, Sarioglu H, Bauer N, Durner J, Lindermayr C (2011) Proteomic analysis of defense response of wildtype *Arabidopsis thaliana* and plants with impaired NO homeostasis. Proteomics 11:1664–1683

Horchani F, Prévot M, Boscari A, Evangelisti E, Meilhoc E, Bruand C, Raymond P, Boncompagni E, Aschi-Smiti S, Puppo A, Brouquisse R (2011) Both plant and bacterial nitrate reductases contribute to nitric oxide production in *Medicago truncatula* nitrogen-fixing nodules. Plant Physiol 155:1023–1036

Hossain Z, López-Climent MF, Arbona V, Pérez-Clemente RM, Gómez-Cadenas A (2009) Modulation of the antioxidant system in citrus under waterlogging and subsequent drainage. J Plant Physiol 166:1391–1404

Hossain KK, Itoh RD, Yoshimura G, Tokuda G, Oku H, Cohen MF, Yamasaki H (2010a) Effects of nitric oxide scavengers on thermoinhibition of seed germination in *Arabidopsis thaliana*. Russ J Plant Physiol 57:222–232

Hossain MA, Hasanuzzaman M, Fujita M (2010b) Up-regulation of antioxidant and glyoxalase systems by exogenous glycinebetaine and proline in mung bean confer tolerance to cadmium stress. Physiol Mol Biol Plants 16:259–272

Hossain MA, Hasanuzzaman M, Fujita M (2011) Coordinate induction of antioxidant defense and glyoxalase system by exogenous proline and glycinebetaine is correlated with salt tolerance in mung bean. Front Agric China 5:1–14

Howarth CJ (2005) Genetic improvements of tolerance to high temperature. In: Ashraf M, Harris PJC (eds) Abiotic stresses: Plant resistance through breeding and molecular approaches. Howarth Press Inc., New York, pp 277–300

Hsu YT, Kao CH (2004) Cadmium toxicity is reduced by nitric oxide in rice leaves. Plant Growth Regul 42:227–238

Hsu YT, Kao CH (2005) Abscisic acid accumulation and cadmium tolerance in rice seedlings. Physiol Plant 124:71–80

Hu X, Neill S, Tang Z, Cai W (2005) Nitric oxide mediates gravitropic bending in soybean roots. Plant Physiol 137:663–670

Hu KD, Hu LY, Li YH, Zhang FQ, Zhang H (2007) Protective roles of nitric oxide on germination and antioxidant metabolism in wheat seeds under copper stress. Plant Growth Regul 53:173–183

Huang J, Sommers EM, Kim-Shapiro DB, King SB (2002) Horseradish peroxidase catalyzed nitric oxide formation from hydroxyurea. J Am Chem Soc 124:3473–3480

Hui L, Cai ZW, Jie ZH, Lai HY, Fang TJ (2009) Effects of exogenous nitric oxide donor sodium nitroprusside on adnosinetriphosphatase activity and membrane lipid peroxidation in wheat seedling leaves under drought stress. Plant Physiol Commun 45:455–458

Igamberdiev AU, Hill RD (2009) Plant mitochondrial function during anaerobiosis. Ann Bot 103:259–268

Igamberdiev AU, Bykova NV, Shah JK, Hill RD (2010) Anoxic nitric oxide cycling in plants: participating reactions and possible mechanisms. Physiol Plant 138:393–404

Innocenti G, Pucciariello C, LeGleuher M, Hopkins J, De Stefano M, Delledonne M, Puppo A, Baudouin E, Frendo P (2007) Glutathione synthesis is regulated by nitric oxide in Medicago truncatula roots. Planta 225:1597–1602

Irfan M, Hayat S, Hayat O, Afroz S, Ahmad A (2010) Physiological and biochemical changes in plants under waterlogging. Protoplasma 241:3–17

Ismail AM, Hall AE (1999) Reproductive-stage, heat tolerance, leaf membrane thermostability and plant morphology in cowpea. Crop Sci 39:1762–1768

Jaleel CA, Manivannan P, Wahid A, Farooq M, Somasundaram R, Panneerselvam R (2009) Drought stress in plants: a review on morphological characteristics and pigments composition. Int J Agric Biol 11:100–105

Janistyn B (1983) Gas chromatographic mass spectrometric identification and quantification of cyclic guanosine 3',5'-cyclic monophosphate in maize seedlings. Planta 159:382–388

Jasid S, Simontacchi M, Puntarulo S (2008) Exposure to nitric oxide protects against oxidative damage but increases the labile iron pool in sorghum embryonic axes. J Exp Bot 59:3953–3962

Jin J-W, Xu Y-F, Huang Y-F (2010) Protective effect of nitric oxide against arsenic-induced oxidative damage in tall fescue leaves. Afr J Biotechnol 9:1619–1627

Kaiser WM, Weiner H, Kandlbinder A, Tsai C-B, Rockel P, Sonoda M, Planchet E (2002) Modulation of nitrate reductase: some new insights, an unusual case and a potentially important side reaction. J Exp Bot 53:875–882

Khurana A, Khurana JP, Babbar SB (2011) Nitric oxide induces flowering in the duckweed *Lemna aequinoctialis* Welw. (Syn. *L. paucicostata* Hegelm.) under noninductive conditions. J Plant Growth Regul. doi:10.1007/s00344-011-9199-7

Kim T-Y, Jo M-H, Hong J-H (2010) Protective effect of nitric oxide against oxidative stress under UV-B radiation in maize leaves. J Environ Sci 19:1323–1334

Klepper LA (1979) Nitric oxide (NO) and nitrogen dioxide (NO_2) emissions from herbicide-treated soybean plants. Atmosph Environ 13:537–541

Kolbert Z, Bartha B, Erdei L (2008) Exogenous auxin-induced NO synthesis is nitrate reductase-associated in *Arabidopsis thaliana* root primordia. J Plant Physiol 165:967–975

Kopyra M, Gwóźdź EA (2003) Nitric oxide stimulates seed germination and counteracts the inhibitory effect of heavy metals and salinity on root growth of *Lupinus luteus*. Plant Physiol Biochem 41:1011–1017

Korhonen R, Lahti A, Kankaanranta H, Moilanen E (2005) Nitric oxide production and signaling in inflammation. Curr Drug Targets Inflamm Allergy 4:471–479

Kovacic P, Somanathan R (2011) Integrated approach to nitric oxide in animals and plants (mechanism and bioactivity): Cell signaling and radicals. J Recept Signal Transduct Res 31:111–120

Kozlowski TT (1997) Responses of woody plants to flooding and salinity. Tree Physiol Monogr 1:1–29

Krasylenko YA, Yemets AI, Blume YB (2010) Functional role of nitric oxide in plants. Russ J Plant Physiol 57:451–461

Kumar D, Klessig DF (2000) Differential induction of tobacco MAP kinases by the defence signals nitric oxide, salicylic acid, ethylene and jasmonic acid. Mol Plant Microbe Interact 13:347–351

Kumutha D, Ezhilmathi K, Sairam RK, Srivastava GC, Deshmukh PS, Meena RC (2009) Waterlogging induced oxidative stress and antioxidant activity in pigeon pea genotypes. Biol Plant 53:75–84

Lamattina L, Garcia-Mata C, Graziano M, Pagnussat G (2003) Nitric oxide: the versatility of an extensive signal molecule. Annu Rev Plant Biol 54:109–136

Lamotte O, Gould K, Lecourieux D, Sequeira-Legrand A, Lebrun-Garcia A, Durner J, Pugin A, Wendehenne D (2004) Analysis of nitric oxide signaling functions in tobacco cells challenged by the elicitor cryptogein. Plant Physiol 135:516–529

Lamotte O, Courtois C, Dobrowolska G, Besson A, Pugin A, Wendehenne D (2006) Mechanisms of nitric oxide-induced increase of free cytosolic Ca^{2+} concentration in *Nicotiana plumbaginifolia* cells. Free Rad Biol Med 40:1369–1376

Lanteri ML, Laxalt AM, Lamattina L (2008) Nitric oxide triggers phosphatidic acid accumulation via phospholipase D during auxin-induced adventitious root formation in cucumber. Plant Physiol 147:188–198

Laspina NV, Groppa MD, Tomaro ML, Benavides MP (2005) Nitric oxide protects sunflower leaves against Cd-induced oxidative stress. Plant Sci 169:323–330

Leach J, Keyster M, Du Plessis M, Ludidi N (2010) Nitric oxide synthase activity is required for development of functional nodules in soybean. J Plant Physiol 167:1584–1591

Lee U, Wie C, Fernández BO, Feelisch M, Vierling E (2008) Modulation of nitrosative stress by *S*-nitrosoglutathione reductase is critical for thermo- tolerance and plant growth in *Arabidopsis*. Plant Cell 20:786–802

Leisner CP, Cousins AB, Offermann S, Okita TW, Edwards GE (2010) The effects of salinity on photosynthesis and growth of the single-cell C4 species *Bienertia sinuspersici* (Chenopodiaceae). Photosynth Res 106:201–214

Leitner M, Vandelle E, Gaupels F, Bellin D, Delledonne M (2009) NO signals in the haze: Nitric oxide signalling in plant defence. Curr Opin Plant Biol 12:451–458

Leshem YY (1996) Nitric oxide in biological systems. Plant Growth Regul 18:155–159

Leshem YY, Haramaty E (1996) The characterization and contrasting effects of the nitric oxide free radicals in vegetative stress and senescence of *Pisum sativum*. J Plant Physiol 148:258–263

Li Q-Y, Niud H-B, Yind J, Wanga M-B, Shao HB, Deng DZ, Chen X-X, Ren J-P, Li Y-C (2008) Protective role of exogenous nitric oxide against oxidative-stress induced by salt stress in barley (*Hordeum vulgare*). Colloids Surf B: Biointer 65:220–225

Li CH, Li Y, Wuyun TN, Wu GL, Jiang GM (2010a) Effects of high concentration ozone on soybean growth and grain yield. Ying Yong Sheng Tai Xue Bao 21:2347–2352

Li X, Shen X, Li J, Eneji AE, Li Z, Tian X, Duan L (2010b) Coronatine alleviates water deficiency stress on winter wheat seedlings. J Integr Plant Biol 52:616–625

Lindermayr C, Durner J (2009) S-nitrosylation in plants: pattern and function. J Proteomics 73:1–9

Lindermayr C, Saalbach G, Bahnweg G, Durner J (2006) Differential inhibition of *Arabidopsis* methionine adenosyltransferases by protein *S*-nitrosylation. J Biol Chem 281:4285–4291

Liu Y, Wu R, Wan Q, Xie G, Bi Y (2007) Glucose-6-phosphate dehydrogenase plays a pivotal role in nitric oxide-involved defense against oxidative stress under salt stress in red kidney bean roots. Plant Cell Physiol 48:511–522

Liu X, Wang L, Liu L, Guo Y, Ren H (2011) Alleviating effect of exogenous nitric oxide in cucumber seedling against chilling stress. Afr J Biotechnol 10:4380–4386

Mahajan S, Tuteja N (2005) Cold, salinity and drought stresses: An overview. Arch Biochem Biophys 444:139–158

Mahmood T, Gupta KJ, Kaiser WM (2009) Cadmium stress stimulates nitric oxide production by wheat roots. Pak J Bot 41:1285–1290

Mansuy D, Boucher JL (2002) Oxidation of *N*-hydroxyguanidines by cytochromes P450 and NO-synthases and formation of nitric oxide. Drug Metab Rev 34:593–606

Mazid M, Khan TA, Mohammad F (2011a) Role of Nitric oxide in regulation of H_2O_2 mediating tolerance of plants to abiotic stress: A synergistic signalling approach. J Stress Physiol Biochem 7:34–74
Mazid M, Khan TA, Mohammad F (2011b) Potential of NO and H_2O_2 as signaling molecules in tolerance to abiotic stress in plants. J Ind Res Tech 1:56–68
Meehl GA, Stocker TF, Collins WD, Friedlingstein P, Gaye AT, Gregory JM, Kitoh A, Knutti R, Murphy JM, Noda A, Raper SCB, Watterson IG, Weaver AJ, Zhao ZC (2007) Global Climate Projections. In: Solomon S, Qin D, Manning M, Chen Z, Marquis M, Averyt KB, Tignor M, Miller HL (eds) Climate change 2007: The physical science basis. Cambridge University Press, Cambridge, New York, Contribution of working group I to the fourth assessment report of the Intergovernmental Panel on Climate Change, pp 749–845
Meilhoc E, Boscari A, Bruand C, Puppo A, Brouquisse R (2011) Nitric oxide in legume–rhizobium symbiosis. Plant Sci. doi:10.1016/j.plantsci.2011.04.007
Millar TM, Stevens CR, Benjamin N, Eisenthal R, Harrison R, Blake DR (1998) Xanthine oxidoreductase catalyses the reduction of nitrate and nitrite to nitric oxide under hypoxic conditions. FEBS Lett 427:225–228
Mingchi L, Xiangli L, Jing H, Lihong G (2010) Effect of simulated drought stress on plant growth, yield and fruit properties of tomato. Acta Hort 856:193–202
Misra AN, Misra M, Singh R (2011a) Nitric oxide: A ubiquitous signaling molecule with diverse role in plants. Afr J Plant Sci 5:57–74
Misra AN, Misra M, Singh R (2011b) Nitric oxide biochemistry, mode of action and signaling in plants. J Med Plants Res 4:2729–2739
Mittler R (2002) Oxidative stress, antioxidants and stress tolerance. Trends Plant Sci 7:405–410
Mittler R, Vanderauwera S, Gollery M, Van Breusegem F (2004) Reactive oxygen gene network of plants. Trends Plant Sci 9:490–498
Mittova V, Guy M, Tal M, Volokita M (2004) Salinity upä regulates the antioxidative system in root mitochondria and peroxisomes of the wild saltä tolerant tomato species *Lycopersicon pennellii*. J Exp Bot 55:1105–1113
Molassiotis A, Fotopoulos V (2011) Oxidative and nitrosative signaling in plants: two branches in the same tree? Plant Signal Behav 6:210–214
Moldau H (1999) Ozone detoxification in the mesophyll cell wall during a simulated oxidative burst. Free Radic Res 31:19–24
Moncada S, Palmer RMJ, Higgs EA (1991) Nitric oxide: physiology, pathophysiology and pharmacology. Pharmacol Rev 43:109–142
Moreau M, Lindermayr C, Durner J, Klessig DF (2010) NO synthesis and signalling in plants: where do we stand? Physiol Plant 138:372–383
Nagase S, Takemura K, Ueda A, Hirayama A (1997) A novel nonenzymatic pathway for the generation of nitric oxide by the reaction of hydrogen peroxide and D-or L-arginine. FEBS Lett 233:150–153
Nahar K, Biswas JK, Shamsuzzaman AMM, Hasanuzzaman M, Barman HN (2009) Screening of indica rice (*Oryza sativa* L.) genotypes against low temperature stress. Bot Res Int 2:295–303
Nasibi F, Kalantari KM (2009) Influence of nitric oxide in protection of tomato seedling against oxidative stress induced by osmotic stress. Acta Physiol Plant 31:1037–1044
Neill S (2007) Interactions between abscisic acid, hydrogen peroxide and nitric oxide mediate survival responses during water stress. New Phytol 175:4–6
Neill SJ, Desikan R, Clarke A, Hancock JT (2002a) Nitric oxide is a novel component of abscisic acid signalling in stomatal guard cells. Plant Physiol 128:13–16
Neill SJ, Desikan R, Clarke A, Hurst RD, Hancock JT (2002b) Hydrogen peroxide and nitric oxide as signalling molecules in plants. J Exp Bot 53:1237–1247
Neill SJ, Desikan R, Hancock JT (2003) Nitric oxide signalling in plants. New Phytol 159:11–35
Neill S, Barros R, Bright J, Desikan R, Hancock J, Harrison J, Morris P, Ribeiro D, Wilson I (2008) Nitric oxide, stomatal closure, and abiotic stress. J Exp Bot 59:165–176
Newton RP, Roef L, Witters E, van Onckelen H (1999) Cyclic nucleotides in higher plants: the enduring paradox. New Phytol 143:427–455

Ninnemann H, Maier J (1996) Indications for the occurrence of nitric oxide synthases in fungi and plants and the involvement in photoconidiation of *Neurospora crassa*. Photochem Photobiol 64:393–398

Noble AD, Ruaysoongnern S (2010) The nature of sustainable agriculture. In: Dixon GR, Tilston EL (eds) Soil microbiology and sustainable crop production. Springer, Dordrecht, pp 1–25

Ötvös K, Pasternak T, Miskolczi P, Domoki M, Dorjgotov D, Szücs A, Bottka S, Dudits S, Fehér A (2005) Nitric oxide is involved in the activation of cell division and somatic embryo formation in alfalfa. Plant J 43:849–860

Pagnussat GC, Simontacchi M, Puntarulo S, Lamattina L (2002) Nitric oxide is required for root organogenesis. Plant Physiol 129:954–956

Pagnussat GC, Lanteri ML, Lombardo MC, Lamattina L (2004) Nitric oxide mediates the indole-acetic acid activation of a mitogen-activated protein kinase cascade involved in adventitious root formation. Plant Physiol 135:279–286

Palavan-Unsal N, Arisan D (2009) Nitric oxide signaling in plants. Bot Rev 75:203–229

Palma JM, Sandalio LM, Corpas FJ, Romero-Puertas MC, McCarthy I, del Rıo LA (2002) Plant proteases, protein degradation, and oxidative stress: role of peroxisomes. Plant Physiol Biochem 40:521–530

Palmieri MC, Sell S, Huang X, Scherf M, Werner T, Durner J, Lindermayr C (2008) Nitric oxide-responsive genes and promoters in *Arabidopsis thaliana*: a bioinformatics approach. J Exp Bot 59:177–186

Phang IC, Leung DWM, Taylor HH, Burritt DJ (2011) The protective effect of sodium nitroprusside (SNP) treatment on Arabidopsis thaliana seedlings exposed to toxic level of Pb is not linked to avoidance of Pb uptake. Ecotoxicol Environ Safety 74:1310–1315

Planchet E, Gupta KJ, Sonoda M, Kaiser WM (2005) Nitric oxide emission from tobacco leaves and cell suspensions: rate limiting factors and evidence for the involvement of mitochondrial electron transport. Plant J 41:732–743

Polverari A, Molesini B, Pezzotti M, Buonaurio R, Marte M, Delledonne M (2003) Nitric oxide-mediated transcriptional changes in *Arabidopsis thaliana*. Mol Plant Microbe Interact 16:1094–1105

Popova L, Tuan T (2010) Nitric oxide in plants: properties, biosynthesis and physiological functions. Iranian J Sci Technol Trans A 34:173–183

Poschenrieder C, Barceló J (2004) Water relations in heavy metal stressed plants. In: Prasad MNV (ed) Heavy metal stress in plants, 3rd edn. Springer, Berlin, pp 249–270

Procházková D, Wilhelmová N (2011) Nitric oxide, reactive nitrogen species and associated enzymes during plant senescence. Nitric Oxide 24:61–65

Qiao W, Fan L-M (2008) Nitric oxide signalling in plant responses to abiotic stresses. J Integrative Plant Biol 50:1238–1246

Qu Y, Feng H, Wang Y, Zhang M, Cheng J (2006) Nitric oxide functions as a signal in ultraviolet-B induced inhibition of pea stems elongation. Plant Sci 170:994–1000

Rao MV, Davis KR (2001) The physiology of ozone induced cell death. Planta 213:682–690

Raziuddin, Farhatullah, Hassan G, Akmal M, Shah SS1, Mohammad F, Shafi M, Bakht J, Zhou W (2011) Effects of cadmium and salinity on growth and photosynthesis parameters of *Brassica* species. Pak J Bot 43:333–340

Ren J, Dai W, Xuan Z, Yao Y, Korpelainen H, Li C (2007) The effect of drought and enhanced UV-B radiation on the growth and physiological traits of two contrasting poplar species. For Ecol Manage 239:112–119

Ribeiro EA, Cunha FQ, Tamashiro WMSC, Martins IS (1999) Growth phase-dependent subcellular localization of nitric oxide synthase in maize cells. FEBS Lett 445:283–286

Rockel P, Strube F, Rockel A, Wildt J, Kaiser WM (2002) Regulation of nitric oxide (NO) production by plant nitrate reductase *in vivo* and *in vitro*. J Exp Bot 53:103–110

Rodríguez M, Canales E, Borrás-Hidalgo O (2005) Molecular aspects of abiotic stress in plants. Biotecnol Aplic 22:1–10

Rodríguez-Serrano M, Romero-Puertas MC, Pazmiño DM, Testillano PS, Risueño MC, del Río LA, Sandalio LM (2009) Cellular response of pea plants to cadmium toxicity: cross talk between reactive oxygen species, nitric oxide, and calcium. Plant Physiol 150:229–243

Roleda MY, Hanelt D, Wiencke C (2006a) Growth and DNA damage in young *Laminaria sporophytes* exposed to ultraviolet radiation: implication for depth zonation of kelps on Helgoland (North Sea). Marine Biol 148:1201–1211

Roleda MY, Wiencke C, Lüder UH (2006b) Impact of Ultraviolet radiation on cell structure, UV-absorbing compounds, photosynthesis, DNA damage, and germination in zoospores of Arctic *Saccorhiza dermatodea*. J Exp Bot 57:3847–3856

Ruiz-Sánchez MC, Domingo R, Morales D, Torrecillas A (1996) Water relations of Fino lemon plants on two rootstocks under flooded conditions. Plant Sci 120:119–125

Rumer S, Gupta KJ, Kaiser WM (2009) Plant cells oxidize hydroxylamines to NO. J Exp Bot 60:2065–2072

Rusterucci C, Espunya MC, Diaz M, Chabannes M, Martinez MC (2007) *S*-Nitrosoglutathione reductase affords protection against pathogens in *Arabidopsis*, both locally and systemically. Plant Physiol 143:1282–1292

Sairam RK, Kumutha D, Ezhilmathi K, Chinnusamy V, Meena RC (2009) Water-logging induced oxidative stress and antioxidant enzymes activity in pigeon pea. Biol Plant 53:493–504

Sairam RK, Dharmar K, Lekshmy S, Chinnusam V (2011) Expression of antioxidant defense genes in mung bean (*Vigna radiata* L.) roots under water-logging is associated with hypoxia tolerance. Acta Physiol Plant 33:735–744

Sakihama Y, Nakamura S, Yamasaki H (2002) Nitric oxide production mediated by nitrate reductase in the green alaga *Chlamydomonas reinhardtii*. An alternative NO production pathway in photosynthetic organisms. Plant Cell Physiol 43:290–297

Samuel MA, Miles GP, Ellis BE (2000) Ozone treatment rapidly activates MAP kinase signalling in plants. Plant J 22:367–376

Sandalio LM, Fernandez VM, Ruperez FL, del Rıo LA (1988) Superoxide free radicals are produced in glyoxysomes. Plant Physiol 87:1–4

Sang J, Jiang M, Lin F, Xu S, Zhang A, Tan M (2008) Nitric oxide reduces hydrogen peroxide accumulation involved in water stress-induced subcellular anti-oxidant defense in maize plants. J Integr Plant Biol 50:231–243

Santa-Cruz DM, Pacienza NA, Polizio AH, Balestrasse KB, Tomaro ML, Yannarelli GG (2010) Nitric oxide synthase-like dependent NO production enhances heme oxygenase up-regulation in ultraviolet-B-irradiated soybean plants. Phytochemistry 71:1700–1707

Schraudner M, Langebartels C, Sandermann H (1997) Changes in the biochemical status of plant cells induced by the environmental pollutant ozone. Physiol Plant 100:274–280

Seligman K, Saviani EE, Oliveira HC, Pinto-Maglio CA, Salgado I (2008) Floral transition and nitric oxide emission during flower development in Arabidopsis thaliana is affected in nitrate reductase-deficient plants. Plant Cell Physiol 49:1112–1121

Sharma SS, Dietz KJ (2008) The relationship between metal toxicity and cellular redox imbalance. Trends Plant Sci 14:43–50

Sharma P, Dubey RS (2005) Drought induces oxidative stress and enhances the activities of antioxidant enzymes in growing rice seedlings. Plant Growth Regul 46:209–221

Sharma P, Dubey RS (2007) Involvement of oxidative stress and role of antioxidative defense system in growing rice seedlings exposed to toxic levels of aluminium. Plant Cell Rep 26:2027–2038

Sharma P, Sharma N, Deswal R (2005) The molecular biology of the low-temperature response in plants. Bioessays 27:1048–1059

Sheokand S, Bhankar V, Sawhney V (2010) Ameliorative effect of exogenous nitric oxide on oxidative metabolism in NaCl treated chickpea plants. Braz J Plant Physiol 22:81–90

Shi FM, Li YZ (2008) *Verticillium dahliae* toxins-induced nitric oxide production in *Arabidopsis* is major dependent on nitrate reductase. BMB Rep 41:79–85

Shi S, Wang G, Wang Y, Zhang L, Zhang L (2005) Protective effect of nitric oxide against oxidative stress under ultraviolet-B radiation. Nitric Oxide 13:1–9

Shi Q, Ding F, Wang X, Wei M (2007) Exogenous nitric oxide protect cucumber roots against oxidative stress induced by salt stress. Plant Physiol Biochem 45:542–550

Siddiqui MH, Al-Whaibi MH, Basalah MO (2011) Role of nitric oxide in tolerance of plants to abiotic stress. Protoplasma 248:447–455

Singh HP, Batish DR, Kaur G, Arora K, Kohli RK (2008) Nitric oxide (as sodium nitroprusside) supplementation ameliorates Cd toxicity in hydroponically grown wheat roots. Environ Exp Bot 63:158–167

Singh HP, Kaur S, Batish DR, Sharma VP, Sharma N, Kohli RK (2009) Nitric oxide alleviates arsenic toxicity by reducing oxidative damage in the roots of rice. Nitric Oxide 20:289–297

Singh R, Singh S, Tripathi R, Agrawal SB (2011) Supplemental UV-B radiation induced changes in growth, pigments and antioxidant pool of bean (*Dolichos lablab*) under field conditions. J Environ Biol 32:139–145

Šírová J, Sedlářová M, Piterková J, Luhová L, Petřivalský M (2011) The role of nitric oxide in the germination of plant seeds and pollen. Plant Sci. doi:10.1016/j.plantsci.2011.03.014

Solanke AU, Sharma AK (2008) Signal transduction during cold stress in plants. Physiol Mol Biol Plants 14:69–79

Song L, Ding W, Zhao M, Sun B, Zhang L (2006) Nitric oxide protects against oxidative stress under heat stress in the calluses from two ecotypes of reed. Plant Sci 171:449–458

Song J, Shi G, Xing S, Chen M, Wang B (2009) Effects of nitric oxide and nitrogen on seedling emergence, ion accumulation, and seedling growth under salinity in the euhalophyte *Suaeda salsa*. J Plant Nutr Soil Sci 172:544–549

Srivastava N, Gonugunta VK, Puli MR, Raghavendra AS (2009) Nitric oxide production occurs downstream of reactive oxygen species in guard cells during stomatal closure induced by chitosan in abaxial epidermis of *Pisum sativum*. Planta 229:757–765

Stamler JS, Singel DJ, Loscalzo J (1992) Biochemistry of nitric oxide and its redox-activated forms. Science 258:1898–1902

Stöhr C, Stremlau S (2006) Formation and possible roles of nitric oxide in plant roots. J Exp Bot 57:463–470

Stöhr C, Ullrich WR (2002) Generation and possible roles of NO in plant roots and their apoplastic space. J Exp Bot 53:2293–2303

Stoimenova M, Igamberdiev AU, Gupta KJ, Hill RD (2007) Nitrite-driven anaerobic ATP synthesis in barley and rice root mitochondria. Planta 226:465–474

Sun Y, Li Z, Guoc B, Chud G, Wei C, Liang Y (2008) Arsenic mitigates cadmium toxicity in rice seedlings. Environ Exp Bot 64:264–270

Sung CH, Hong JK (2010) Sodium nitroprusside mediates seedling development and attenuation of oxidative stresses in Chinese cabbage. Plant Biotechnol Rep 4:243–251

Tan J, Zhao H, Hong J, Han Y, Li H, Zhao W (2008) Effects of exogenous nitric oxide on photosynthesis, antioxidant capacity and proline accumulation in wheat seedlings subjected to osmotic stress. World J Agric Sci 4:307–313

Tanou G, Job C, Rajjou L, Arc E, Belghzi M, Diamantidis G, Molassiotis A, Job D (2009a) Proteomics reveal the overlapping roles of hydrogen peroxide and nitric oxide in the acclimation of citrus plants to salinity. Plant J 60:795–804

Tanou G, Molassiotis A, Diamantidis G (2009b) Induction of reactive oxygen species and necrotic death-like destruction in strawberry leaves by salinity. Environ Exp Bot 65:270–281

Tanou G, Molassiotis A, Gr D (2009c) Hydrogen peroxide- and nitric oxide-induced systemic antioxidant prime-like activity under NaCl-stress and stress-free conditions in citrus plants. J Plant Physiol 166:1904–1913

Tewari RK, Hahn EJ, Paek KY (2008) Modulation of copper toxicityinduced oxidative damage by nitric oxide supply in the adventitious roots of *Panax ginseng*. Plant Cell Rep 27:171–181

Thomashow MF (1999) Plant cold acclimation: Freezing tolerance genes and regulatory mechanisms. Annu Rev Plant Physiol Plant Mol Biol 50:571–599

Tian X, Lei Y (2006) Nitric oxide treatment alleviates drought stress in wheat seedlings Biol Plant. Biol Plant 50:775–778

Tian QY, Sun DH, Zhao MG, Zhang WH (2007) Inhibition of nitric oxide synthase (NOS) underlies aluminum-induced inhibition of root elongation in *Hibiscus moscheutos*. New Phytol 174:322–331

Tossi V, Lamattina L, Cassia R (2009) An increase in the concentration of abscisic acid is critical for nitric oxide-mediated plant adaptive responses to UV-B irradiation. New Phytol 181:871–879

Tun NN, Santa-Catarina C, Begum T, Silveira V, Handro W, Floh EIS, Scherer GFE (2006) Polyamines induce rapid biosynthesis of nitric oxide (NO) in *Arabidopsis thaliana* seedlings. Plant Cell Physiol 47:346–354

Uchida A, Jagendorf AT, Hibino T, Takabe T (2002) Effects of hydrogen peroxide and nitric oxide on both salt and heat stress tolerance in rice. Plant Sci 163:515–523

Veitch NC (2004) Horseradish peroxidase: a modern review of a classic enzyme. Phytochemistry 65:249–259

Verheul MJ, Picatto C, Stamp P (1996) Growth and development of maize (*Zea mays* L.) seedlings under chilling conditions in the field. Eur J Agron 5:31–43

Wahid A, Gelani S, Ashraf M, Foolad MR (2007) Heat tolerance in plants: An overview. Environ Exp Bot 61:199–223

Wang YS, Yang ZM (2005) Nitric oxide reduces aluminium toxicity by preventing oxidative stressing the roots of *Cassia tora* L. Plant Cell Physiol 46:1915–1923

Wang WX, Vinocur B, Shoseyov O, Altman A (2001) Biotechnology of plant osmotic stress tolerance: physiological and molecular considerations. Acta Hort 560:285–292

Wang Y, Feng H, Qu Y, Cheng J, Zhao Z, Zhang M, Wang X, An L (2006) The relationship between reactive oxygen species and nitric oxide in ultraviolet-B–induced ethylene production in leaves of maize seedlings. Environ Exp Bot 57:51–61

Wang L, Yang L, Yang F, Li X, Song Y, Wang X, Hu X (2010) Involvements of H_2O_2 and metallothionein in NO-mediated tomato tolerance to copper toxicity. J Plant Physiol 167:1298–1306

Wendehenne D, Courtois C, Besson A, Gravot A, Buchwalter A, Pugin A, Lamotte O (2006) NO-based signaling in plants. In: Lamattina L, Polacco JC (eds) Nitric oxide in plant growth, development and stress physiology. Springer, Berlin, pp 35–51

Wilson ID, Neill SJ, Hancock JT (2008) Nitric oxide synthesis and signalling in plants. Plant Cell Environ 31:622–631

Wimalasekera R, Tebartz F, Scherer GFE (2011) Polyamines, polyamine oxidases and nitric oxide in development, abiotic and biotic stresses. Plant Sci. doi:10.1016/j.plantsci.2011.04.002

Wink DA, Hanbauer I, Krishna MC, De Graff W, Gamson J, Mitchel JB (1993) Nitric oxide protects against cellular damage and cytotoxicity form reactive oxygen species. Proc Natl Acad Sci USA 90:9813–9817

Wojtaszek P (2000) Nitric oxide in plants: To NO or not to NO. Phytochemistry 54:1–4

Wollenweber B, Porter JR, Schellberg J (2003) Lack of interaction between extreme high temperature events at vegetative and reproductive growth stages in wheat. J Agron Crop Sci 189:142–150

Wu SJ, Qi JL, Zhang WJ, Liu SH, Xiao FH, Zhang MS, Xu GH, Zhao WG, Shi MW, Pang YJ, Shen HG, Yang YH (2009) Nitric oxide regulates shikonin formation in suspension-cultured *Onosma paniculatum* cells. Plant Cell Physiol 50:118–128

Wu X, Zhu W, Zhang H, Ding H, Zhang HJ (2011) Exogenous nitric oxide protects against salt-induced oxidative stress in the leaves from two genotypes of tomato (*Lycopersicom esculentum* Mill.). Acta Physiol Plant 33:1199–1209

Xin L, Wuliang S, Shuqiu Z, Chenghou Z (2005) Nitric oxide involved in signal transduction of jasmonic acid-induced stomatal closure of *Vicia faba* L. Chinese Sci Bull 50:520–525

Xing HL, Tan L, Zhao An Z, Wang S, Zhang C (2004) Evidence for the involvement of nitric oxide and reactive oxygen species in osmotic stress tolerance of wheat seedlings: Inverse correlation between leaf abscisic acid accumulation leaf water loss. Plant Growth Regul 42:61–68

Xiong J, An L, Lu H, Zhu C (2009) Exogenous nitric oxide enhances cadmium tolerance of rice by increasing pectin and hemicelluloses contents in root cell wall. Planta 230:755–765

Xiong J, Fu G, Tao L, Zhu C (2010) Roles of nitric oxide in alleviating heavy metal toxicity in plants. Arch Biochem Biophys 497:13–20

Xiong J, Zhang L, Fu G, Yang Y, Zhu C (2011) Longxing Tao (2011) Drought-induced proline accumulation is uninvolved with increased nitric oxide, which alleviates drought stress by decreasing transpiration in rice. J Plant Res. doi:10.1007/s10265-011-0417-y
Xu YC, Zhao BL (2003) The main origin of endogenous NO in higher non-leguminous plants. Plant Physiol Biochem 41:833–838
Xu J, Wang W, Yin H, Liu X, Sun H, Mi Q (2010a) Exogenous nitric oxide improves antioxidative capacity and reduces auxin degradation in roots of *Medicago truncatula* seedlings under cadmium stress. Plant Soil 326:321–330
Xu Y, Sun X, Jin J, Zhou H (2010b) Protective effect of nitric oxide on light-induced oxidative damage in leaves of tall fescue. J Plant Physiol 167:512–518
Xu Y-F, Sun X-L, Jin J-W, Zhou H (2010c) Protective roles of nitric oxide on antioxidant systems in tall fescue leaves under high-light stress. Afr J Biotechnol 9:300–306
Yadav SK (2010) Cold stress tolerance mechanisms in plants: A review. Agron Sustain Dev 30:515–527
Yamasaki H, Sakihama Y, Takahashi S (1999) An alternative pathway for nitric oxide production in plants: new features of an old enzyme. Trends Plant Sci 4:128–129
Yamasaki H, Shimoji H, Ohshiro Y, Sakihama Y (2001) Inhibitory effects of nitric oxide on oxidative phosphorylation in plant mitochondria. Nitric Oxide 5:261–270
Yan K, Chen W, He X, Zhang G, Xu S, Wang L (2010a) Responses of photosynthesis, lipid peroxidation and antioxidant system in leaves of *Quercus mongolica* to elevated O_3. Environ Exp Bot 69:198–204
Yan K, Chen W, Zhang GY, He XY, Li X, Xu S (2010b) Effects of elevated CO_2 and O_3 on active oxygen metabolism of *Quercus mongolica* leaves. Ying Yong Sheng Tai Xue Bao 21:557–562
Yang J-D, Yun J-Y, Zhang T-H, Zhao H-L (2006) Presoaking with nitric oxide donor SNP alleviates heat shock damages in mung bean leaf discs. Bot Stud 47:129–136
Yang H, Wu F, Cheng J (2011) Reduced chilling injury in cucumber by nitric oxide and the antioxidant response. Food Chem 127:1237–1242
Yemets AI, Krasylenko YA, Lytvyn DI, Sheremet YA, Blume YB (2011) Nitric oxide signalling via cytoskeleton in plants. Plant Sci. doi:10.1016/j.plantsci.2011.04.017
Yin H, Chen QM, Yi MF (2008) Effects of short-term heat stress on oxidative damage and responses of antioxidant system in *Lilium longiflorum*. Plant Growth Regul 54:45–54
Zemojtel T, Penzkofer T, Dandekar T, Schultz J (2004) A novel conserved family of nitric oxide synthase? Trends Biochem Sci 29:224–226
Zhang M, An L, Feng H, Chen T, Chen K, Liu Y, Tang H, Wang CH (2003) The cascade mechanisms of nitric oxide as second messenger of ultraviolet-B in inhibiting mesocotyl elongation. Photochem Photobiol 77:219–225
Zhang YY, Wang LL, Liu YL, Zhang Q, Wei QP, Zhang WH (2006) Nitric oxide enhances salt tolerance in maize seedlings through increasing activities of proton-pump and Na^+/H^+ antiport in the tonoplast. Planta 224:545–555
Zhang A, Jiang M, Zhang J, Ding H, Xu S, Hu X, Tan M (2007) Nitric oxide induced by hydrogen peroxide mediates abscisic acid-induced activation of the mitogen-activated protein kinase cascade involved in antioxidant defense in maize leaves. New Phytol 175:36–50
Zhang H, Li YH, Hu LY, Wang SH, Zhang FQ, Hu KD (2008) Effects of exogenous nitric oxide donor on antioxidant metabolism in wheat leaves under aluminum stress. Russ J Plant Physiol 55:469–474
Zhang B, Wang HQ, Liu BL, Liu J, Wang X, Liu Q, Zhang HG (2010a) A potato NOA gene increased salinity tolerance in *Arabidopsis thaliana*. Afr J Biotechnol 9:5869–5878
Zhang X, Shen L, Li F, Zhang Y, Meng D, Sheng J (2010b) Up-regulating arginase contributes to amelioration of chilling stress and the antioxidant system in cherry tomato fruits. J Sci Food Agric 90:2195–2202
Zhao L, Zhang F, Guo J, Yang Y, Li B, Zhang L (2004) Nitric oxide functions as a signal in salt resistance in the calluses from two ecotypes of reed. Plant Physiol 134:848–857
Zhao M, Zhao X, Wu Y, Zhang L (2007) Enhanced sensitivity to oxidative stress in an *Arabidopsis* nitric oxide synthase mutant. J Plant Physiol 164:737–745

Zhao L, He JX, Wang XM, Zhang LX (2008) Nitric oxide protects against polyethylene glycol-induced oxidative damage in two ecotypes of reed suspension cultures. J Plant Physiol 165:182–191

Zhao MG, Chen L, Zhang LL, Zhang WH (2009) Nitric reductase-dependent nitric oxide production is involved in cold acclimation and freezing tolerance in Arabidopsis. Plant Physiol 151:755–767

Zheng YH, Jia AJ, Ning TY, Xu JL, Li ZJ, Jiang GM (2008) Potassium nitrate application alleviates sodium chloride stress in winter wheat cultivars differing in salt tolerance. J Plant Physiol 165:1455–1465

Zheng C, Jiang D, Liu F, Dai T, Liu W, Jing Q, Cao W (2009) Exogenous nitric oxide improves seed germination in wheat against mitochondrial oxidative damage induced by high salinity. Environ Exp Bot 67:222–227

Zheng C, Jiang D, Dai T, Jing Q, Cao W (2010) Effects nitroprusside, a nitric oxide donor, on carbon and nitrogen metabolism and the activity of the antioxidant system in wheat seedlings under salt stress. Acta Ecol Sin 30:1174–1183

Zhou B, Guo Z, Xing J, Huang B (2005) Nitric oxide is involved in abscisic acid-induced antioxidant activities in *Stylosanthes guianensis*. J Exp Bot 56:3223–3228

Zhu SH, Zhou J (2007) Effect of nitric oxide on ethylene production in strawberry fruit during storage. Food Chem 100:1517–1522

Zottini M, Formentin E, Scattolin M, Carimi F, Schiavo FL, Terzi M (2002) Nitric oxide affects plant mitochondrial functionality *in vivo*. FEBS Lett 515:75–78

Chapter 12
Weeds as a Source of Genetic Material for Crop Improvement Under Adverse Conditions

Bhumesh Kumar, Meenal Rathore, and A.R.G. Ranganatha

1 Introduction

World population is expanding at an alarming pace with more than 90 % in developing countries; however, on the other hand, resources are shrinking. Despite continuous efforts in crop breeding, stagnancy is observed in production resulting in an ever-widening gap between demand and supply of food grains. An increase in crop production is indispensable to minimize this ever-increasing gap and to feed the exploding population. In short we are already in need of a second green revolution as emphasized by Hon'ble Prime Minister of India in the lecture delivered on Foundation Day of Indian Council of Agriculture Research on July 16, 2011 (www.icar.org.in). Access to an array of genetic diversity is critical to the success of breeding programs. The introgression of *Rht1* and *Rht2* genes from "Norin 10," (a cultivar from Japan) that reduced plant height and increased disease resistance in wheat provided the foundation for the "Green Revolution" and demonstrated that genetic resources have tremendous potential for crop improvement. Plant breeding for crop improvement has its limitations too and introgression of several unwanted genes can result in inferior phenotypes. Another constraint to conventional plant breeding is greater time requirement, labor intensive, and tediousness which sometimes make the output less profitable. Hybrids and synthetics may hold an answer and provide the yield increase needed in the future. It is worthwhile to mention here that enhanced yield is dependent on the genes of a crop for their ability to tolerate biotic or abiotic stress(es), efficiency to uptake nutrients and assimilate them, water use efficiency along with the agronomic practices. Genes are already available in nature, some known, others untapped, e.g., a wild relative of maize, *Tripsacum*, represents an

B. Kumar (✉) • M. Rathore • A.R.G. Ranganatha
Directorate of Weed Science Research, Jabalpur, MP, India
e-mail: kumarbhumesh@yahoo.com

N. Tuteja and S. Singh Gill (eds.), *Plant Acclimation to Environmental Stress*,
DOI 10.1007/978-1-4614-5001-6_12,

untapped genetic resource for abiotic and biotic stress resistance that could provide developing world farmers access to hybrid technology (Hoisington et al. 1999).

India has 52 % of cultivable land with different climates and soil types. With 20 agro-climatic regions, 10 biodiversity zones with over 45,000 known species representing 11 % of earth's flora, India has an upper hand in available genetic diversity that can be utilized for crop breeding. But factors like incompatibility within and among species, varying reproductive phase timings, etc., limit the utility of these available genetic resources in conventional plant breeding. Recent advancements in molecular and genetic engineering technologies provide us the liberty to transfer desirable genetic material across the border of species and genera, thus expanding the horizon for exploring genetic diversity for crop improvement in shorter duration and better efficacy. Despite the availability of gene pool and advanced technologies to transfer genetic material, success is far away from its potential. One of the many reasons responsible for this status may be unrelatedness between the donor and receptor species at the molecular level which will be discussed in following sections.

Various abiotic and biotic factors pose a serious threat to crop production which sometime leads to complete failure of crops. Genetic engineering of crop plants can play a vital role to boost economy of developing countries like India where a large part of the GDP comes from agriculture. Progress has been noticed in minimizing the crop damage through the use of disease and pest resistant varieties, reducing the use of chemicals and enhancing stress tolerance in crop plants enabling utilization of unproductive lands. Still, concerted efforts need to be continued for utilizing the complete genetic potential. Crop production is also limited due to other parameters like day length, efficacy to utilize CO_2 and nutrients, water usage, weed competitiveness, etc. Emphasis can be laid on extending the growing season of crops, increasing crop density, minimizing losses due to environmental and biotic factors, reducing losses due to food spoilage by means of enhancing the shelf life of fruits and vegetables, and reducing preharvest sprouting in food grains. Biotechnology offers an array of opportunities for crop improvement, management of biotic and abiotic stresses, bioremediation, and undoubtedly has the potential to bring the "Evergreen Revolution."

India is the hub of agricultural biodiversity and its gene wealth can greatly complement the developments in biotechnology. Wild species have been utilized for crop improvement through conventional breeding as well as molecular marker-assisted breeding. But, in genetic engineering work the source of transgenes, by and large, have either been model plant species like *Arabidopsis thaliana* or microorganism(s) like *Escherichia coli, Bacillus*, *Agrobacterium*, etc. Transgenic crops have also been developed and commercialized using the aforesaid gene sources, e.g., Bt-cotton, glyphosate-resistant soybean and maize, etc. Though commercialized, each of these transgenic had to face the concern related to source of gene(s), here microorganisms. Consumers, on large, are reluctant to consume food having genetic material from bacteria, be it candidate gene, marker genes, or reporter genes. The hue and cry made way to cisgenesis which uses all genetic components from plant species to develop trans/cisgenic plants. This requires intensive basic work to discover, isolate, and then utilize genetic components from plants to construct vectors for transformation. Mother Nature has it all... our job is to hunt it out and use it, this time from plants alone.

Gene transfer from a related species to crop plants with a similarity in habitat can be a good model system to achieve genetic engineering of the crop plants. Weeds, recognized to be hardier than crop plants in terms of abiotic and biotic stress can be an alternate and potential source of genes for the production of transgenic plants. Competitiveness of weeds and their better ability to survive under adverse conditions, and more important, their coexistence with crop plants makes them a potential source for trait-based gene transfer into crop plants. An effort is made to highlight the potential of weed plants as a source for genetic material for crop improvement with special emphasis on abiotic and biotic tolerance.

2 Crop Improvement

Genetic improvement of crop plants has been underway since long back. During the last few decades, a better understanding of basic genetics and crop physiology has paved the way to dramatic increases in the production of agriculturally important crop via improved breeding and agronomic practices (hybrid seeds, use of improved fertilizers and pesticides, and increase in the irrigation facilities). These approaches are being further modified as per the need of the hour. However, reduced availability of species-specific gene pool is hindering the expansion of the horizon of these practices. It is now felt that an alteration in the genetic makeup of crops to enhance competitiveness and resistance to environmental and biotic stresses (e.g., salt, heat, drought, pollution, insects, and diseases) is an indispensable option. For the molecular biologist, the current need to alter physiological traits in crop genome is an interesting avenue. To give a success story we need to identify genes for desirable traits from coexisting natural bio resources, clone, and then transfer into host crops. Keeping in mind the macromolecular interactions, codon biasness, and the inevitable ethical concerns, etc., it would be definitely desirable to hunt for genes from similar habitat ecotypes, and weeds perfectly fulfill this requirement.

Hon'ble former President of India, Dr APJ Abdul Kalam, at the concluding session of "ANA World Conference" on "Animal Nutrition-Preparedness to combat Challenges-2009" spoke on how we can fill a gap between demand and supply of the food materials without increasing the land and other fixed resources. In his speech, it was emphasized to explore the mechanism of competitiveness of weeds in an effort to develop more competitive crop plant. A big question arises as to what makes a plant a weed and which traits enable them to over-compete crop plants despite the man-made hindrances, i.e., weed control measures. Few characteristics can be highlighted here in a summarized form as to have an idea about the unique traits of weeds which can be explored for crop improvement. The first of such traits is undoubtedly competitiveness. Another trait is their ability to cope with adverse environments—biotic factors like diseases and insects. A shorter life cycle leading to variation in maturity time and dormancy of seeds which help in establishment of seed bank for future generations is also worth mentioning here. All these factors together make weeds more competitive and hardier as compared to crop plants. For example, weedy rice has vigorous growth and by virtue of that competes with cultivated rice and reduces the crop yield. Farmers cannot harvest the grains of weedy rice as it matures earlier and shatters the grains into soil for the next generation. Above all, unlike the cultivated rice, grains of weedy rice become dormant till next season, thus establishing an effective seed bank. All these characters together provide an edge over cultivated rice and provide competitiveness. Attempts have been made to know what is responsible for speedy and vigorous growth of weedy rice. Research was conducted to compare: (1) the relative efficiencies of rice and weedy rice under a competitive condition, (2) the accumulation and partitioning of N by rice and weedy rice, and (3) the N use efficiency of rice and weedy rice (Burgos et al. 2006). They reported that nitrogen uptake and nitrogen use efficiency for biomass production was higher in weedy rice as compared to that in cultivated rice which gave weedy rice an edge in terms of vigorous growth and biomass accumulation. In initial stage of establishment of seedling, the fast and vigorous growth can be seen in most of the weed species in agricultural fields including *Phalaris minor* in wheat crop and *Echinocloa* species in rice crop. The strategy behind this seems to be simple but very effective. In initial stage, weeds use the soil resources, i.e., soil nitrogen more effectively than crop plants in order to achieve a fast and vigorous growth which results in greater biomass and once they are successful in doing so, then they can easily capture the other resources like space and photosynthetically active radiation (PAR); thus provide a tough competition to the crop plants.

It is said that if you know your enemies and know yourself, you will not be imperiled in a hundred battles. This fact holds equally true in the battle between crop and weeds. According to Norris et al. (2002), by knowing a weed's strengths and weaknesses, the weeds can be better managed. As weeds have challenged our efforts to manage them and continue to do so, research needs to be focused on identification of the traits that make weeds strong competitors of crop plants. With identification of traits, genes involved in the mechanism may be discovered, isolated, and cloned for transfer into crop plants. Expression of a competitive trait in crop plant will help to negate the advantages of a weed plant over the crop plants. Though the strategy may appear simple, it involves massive efforts, extensive basic work along with knowledge of the molecular interactions occurring within a crop and/or weed which can provide deep insights into the weedy traits. Another point to be kept in mind is fitness of engineered plants and the yield penalty, if there is any.

In *Trifolium pratense* (a lawn weed), postharvest oxidation of o-diphenols to o-quinones by endogenous polyphenol oxidases prevents breakdown of forage protein during storage. Forage crops like alfalfa (*Medicago sativa*) lack both polyphenol oxidase and o-diphenols, hence, breakdown of their storage proteins upon harvest and storage results in economic losses and release of excess nitrogen into the environment. A possible pathway for phaselic acid biosynthesis involves a hydroxycinnamoyl transferase (HCT) capable of forming caffeoyl and/or p-coumaroyl esters with malate. Genes encoding two distinct HCTs were identified in red clover. Understanding how *T. pratense* synthesizes o-diphenols such as phaselic acid and then transfer of genes like HCTs and polyphenol oxidase can help in the development of forage crops utilizing this natural system of protein protection (Sullivan 2009). This study further points out towards the potential of weeds in crop improvement especially for forage crops like *M. sativa*.

The molecular and genetic changes underlying the transformation of wild plants into agricultural weeds are poorly understood. In a study, Lai et al. (2008) used a sunflower cDNA microarray to detect variation in gene expression between two wild (non-weedy) populations and four weedy populations of *Helianthus annuus*. When grown in a common growth chamber environment, populations differed substantially in their gene expression patterns, indicating extensive gene expression pattern especially those which involved in abiotic and biotic tolerance. Another strategic point emerged from the study is down regulation of costly genes and pathways which provide an extra resource allocation for the element which is essential for vigorous and competitive growth rate in weedy sunflowers, however, further efforts are being made to verify the hypothesis.

New sources of dwarfing genes were identified from accessions of *Avena fatua* in Japan and Korea. The dwarfing genes were transferred from backcrossed and self-pollinated relatives to the cultivated oat. One of the lines "L169" segregated into two different recessive dwarf lines in BC_1, which were selected as semidwarf (L169a) and extreme-dwarf (L169b) lines. L169a was identified as a good genotype with a high grain yield. Another line, L288 is a semidwarf line conditioned by a semidominant dwarfing gene, with a unilateral panicle, large florets, and good grain quality due to strong resistance to lodging (Morikawa et al. 2007). The above study again indicates the potential of the coexisting weeds in crop improvement.

Use of molecular techniques now allows us to identify the gene segments responsible for improved performance and to focus on well-directed crosses in the future or production of hybrids and transgenics. A recent example is the role of the *Lr19*-containing segment which is now being investigated at CIMMYT in detail for further insights. This gene segment originally came from *Agropyron elongatum* (a weed) and was first incorporated into the wheat variety "Agatha." The yield trial data indicates that varieties containing the *Lr19* gene yield at least 10 % more than counterparts without *Lr19* (Ravi Singh, personal communication). Interestingly, *Lr19* gene was originally transferred for its possible role in conferring leaf rust resistance, but its potential to increase yield may become a more important factor for wheat breeders, again demonstrating the unanticipated potential of the weeds for useful transfers. Garg et al. (2009) screened 177 disomic addition lines (DAL) of wheat (*Triticum aestivum*) containing chromosomes from different alien species, and reported that the chromosome 1E addition line of weed *A. elongatum*, that is also known to be a potential genetic resource for drought and salinity tolerance, showed potential for improvement of bread making quality of wheat. Analysis of the high molecular weight glutenin subunit sequence of *A. elongatum* indicated that it closely resembled the sequence of the A and B genome of wheat. From these observations, it was inferred that 1E-encoded seed storage proteins which have considerable potential for improvement of wheat end product quality if transferred to specific chromosomes such as 1A of wheat, may enhance the overall effect on bread making quality.

3 Stress Tolerance

A number of adverse environmental factors such as drought, salinity, cold, freezing, high temperature, anoxia, high light intensity, nutrient imbalances, and herbicide toxicity are collectively termed as abiotic stresses. Abiotic stresses lead to disturbance in metabolic processes and internal homeostasis and damage to macromolecules which are otherwise vital to the normal cellular functions and maintenance. Stomatal closure, reduced supply of CO_2, hence, slows down the rate of biochemical reactions during prolonged periods of dehydration. During this period, high light intensity, associated with high temperature leads to generation of reactive oxygen species (ROS) in the chloroplasts thus causing irreversible cellular damages and photo inhibition. Since the understanding of the molecular pathways induced in response to one or more of the abiotic stresses, efforts have been made to raise the transgenic plants with the insertion of desirable traits for abiotic tolerance but only limited success has been achieved so far. Reasons for the scenario can be ascribed mainly to the abnormal phenotypes obtained, loss of vigor, or yield penalties in addition to insufficient transgene expression and hence reduced protection against the stress factor(s). The use of multiple tolerance mechanisms for one or more of the abiotic stresses through co-transformation or via gene pyramiding provides alternative strategies to engineer the plants for desired traits; however, transgene expression to its maximum potential and a boom in the area of abiotic stress plants is yet awaited. It is high time we reorient our strategies in the context of genetic engineering. A major bottleneck observed in plant genetic engineering experiments is the unrelatedness between donor and receptor species and the other involves the complex cross talking between abiotic and/or biotic stresses. Molecular understanding of the stress perception, signal transduction, and transcriptional regulation of abiotic stress-responsive genes may help to engineer tolerance for multiple stresses and further such an understanding can be more effective if we identify the desired traits and their molecular basis in a related species instead of unrelated ones (e.g., microorganisms or model plants). Weeds are potential donor species for desired traits for abiotic tolerance traits for the reasons that they are closely related to crop plants and spend time in the same habitat as crop plants ensuring a good chance of success. In following sections, attempts have been made to point out the useful and potential traits in weed species for various abiotic stresses tolerance which can be transferred into crop plants.

3.1 Salt Stress Tolerance

Soil salinity affects plant growth and development by way of osmotic stress, injurious effects of toxic ions like Na^+ Cl^-, SO_4^{2-}, and Mg^{2+}, and nutrient imbalance caused by these excess of ions. Salinity stress response is multigenic, as a number of processes involved in the tolerance mechanism such as various compatible solutes/osmolytes, polyamines, ROS and antioxidant defense mechanism, ion transport, and compartmentalization of injurious ions (Sairam and Tyagi 2004). Various genes/cDNAs encoding proteins involved in the above mentioned processes have been identified and isolated and their role in regulation of metabolic and signal transduction processes involving salicylic acid and polyamines, and strategies to improve salinity stress tolerance have also been discussed.

Pokkali rice is a unique variety of rice that is cultivated in an organic way in the waterlogged coastal regions of Kerala and has been reckoned as highly tolerant to salinity. Pokkali rice cultivation has not changed much over the centuries and still being done without fertilizers or manure in purely natural way of cultivation that relies on the monsoon and the sea tides. Despite shrinking farming of Pokkali rice, it has emerged a potential source as a genetic material for rice breeding for abiotic stress tolerance and being used for such programs worldwide. Root cDNA libraries and ESTs and gene expression profile from salt stressed young Pokkali rice is established by a pioneer research group in the USA (Kawasaki et al. 2000). A high number of salt-specific genes (approximately 800) were identified which were either novel genes (≈300) or their expression was altered under salt stress. Such a big number of novel genes in Pokkali in response to salt stress make the Pokkali rice a potential source for genetic material for rice breeding for salt stress tolerance. However, another important point is the coexistence of a number of weeds in Pokkali fields which can be used for improvement of other crops.

A survey was conducted in Kerala, India, covering the area where Pokkali rice cultivation is being practiced, to identify the major weeds and their adaptive potential in the saline ecosystem (Vidya et al. 2004). A total of 18 weeds were observed which include *Diplachne fusca*, *Echinochloa crusgalli*, *Panicum repens*, *Fimbristylis miliacea*, *Eleocharis dulcis*, *Cyperus difformis*, *Eichhornia crassipes*, *Lemna polyrrhiza*, *Spirodela polyrrhiza*, *Pistia stratiotes*, *Monochoria vaginalis*, *Alternanthera sessilis*, *Nymphaea nouchali*, *Sphenoclea zeylanica*, *Ludwigia parviflora*, *Sphaeranthus africanus*, *Salvinia molesta*, *Azolla pinnata*, and *Ceratopteris thalictroides*. From the above weed species, *D. fusca* was the most dominant weed species, occurring in approximately 85 % of the sites surveyed indicating its immense potential of adaptation under salt stress. The anatomical analysis of *D. fusca* and *E. crusgalli* revealed the presence of Kranz anatomy, a typical characteristic of the C_4 plants. Another adaptive feature observed in *D. fusca* is the robust and efficient growth and the presence of microhairs on the leaves which function as salt glands and secrete salt on the leaf surfaces.

The response to increasing salinity (in the range 0–420 mol/m^3 NaCl) of *D. fusca* has been studied in terms of growth, water relations, ion uptake, and leaf photosynthesis in Australia (Myers et al. 1990). It was reported that at salinity up to 200 mol/m^3 NaCl there was no significant effect on growth, water status, and most of the photosynthesis-related parameters. Even at higher salinity, initial inhibition was observed which recovered up to a significant level at later stage. Authors further confirmed that *D. fusca* possessed a C_4 mode of leaf photosynthesis and the optimum leaf temperature for photosynthesis was approximately 45 °C suggesting adaptability of this weed species at high temperature. Results from this study strongly indicate the potential of *D. fusca* as a candidate species for molecular manipulation of crop plants for salinity tolerance as well as to high temperature. In another study, Sandhu et al. (2006) studied various aspects of response of *D. fusca* to root salinity. In pot trials, a 50 % yield (dry matter) was obtained at ECe 22.3. While salt stress led to Na^+ and Cl^- uptake, most of these ions appeared to be secreted out selectively from the leaves. The shoot K^+ content on a tissue water basis remained unaffected by salt stress and the shoot tissue had a high K^+ selectivity. Osmotic adaptation was mainly brought about by tissue dehydration and by accumulation of the compatible solute glycinebetaine in fairly high concentrations. High salt tolerance of this species along with high dehydration tolerance and its potential of osmotic adjustment on account of high glycinebetaine accumulation make this species a highly promising as a source of genetic material for the breeding of salt tolerant crops by means of genetic engineering.

Another important weed species which showed a high degree of salt tolerance appears to be a gramineous weed *E. crusgalli*. Salt tolerance and physiochemical aspects contributing to the tolerance were studied in rice and *E. crusgalli* (Yamamoto et al. 2003). Growth inhibition and decrease in relative water content under salt stress was more severe in rice than in *E. crusgalli*. Proline accumulation in leaves was significantly higher in salt stressed *E. crusgalli* than in rice, suggesting the significance of proline production in the salt tolerance of this weed. The content of MDA (a measure of lipid peroxidation and membrane damage) of the rice increased more greatly with NaCl treatment than that in *E. crusgalli*. NaCl treatment also affected polyamine metabolism of both plant species; however, the response of each plant to salt stress was somewhat different, especially in the leaves. Putrescine and spermidine contents in leaves were high in non-stressed plants in rice, although rather lower in *E. crusgalli* in response to NaCl concentrations. These results indicate that an increase in proline and changes in polyamines relates to the salt tolerance of *E. crusgalli*.

Sawada et al. (2008) studied the comparative performance of rice and *E. crusgalli* to find out the factors responsible for salt tolerance of *E. crusgallii*. Accumulation of salicylic acid was observed in salt stressed rice seedlings, and they hypothesized that the accumulation of salicylic acid might potentiate oxidative injury in rice seedlings since the inhibition of salicylic acid synthesis alleviated the growth inhibition under high salinity. In *E. crusgalli* seedlings exposed to high salinity, neither free nor conjugated salicylic acid content showed any increase. Additionally, foliar application of salicylic acid led to the growth inhibition of *E. crusgalli* seedlings under salt stress concomitantly with striking reduction of the efficiency of PSII. Catalase and superoxide dismutase activities of *E. crusgalli* seedlings were induced by the salt treatment which further provides protection against salt-induced ROS. However, salicylic acid pretreatment suppressed such an induction of catalase activity but promoted superoxide dismutase activity, in turn, led to the buildup of the leaf hydrogen peroxide. Accumulation of the cellular salicylic acid facilitates the generation of H_2O_2 through the suppression of catalase activity and concomitantly promotion of superoxide dismutase activity, thus enhances the oxidative injury caused by salt stress. However, *E. crusgalli* seedlings have the adaptive mechanism to check the accumulation of salicylic acid and hence build up of hydrogen peroxide which otherwise seems to be damaging. Above study strongly suggest the inherent adaptive mechanism of *E. crusgalli*

Fig. 12.1 Effect of NaCl (300 mM) on growth of *Echinochloa crus-galli*

under salt stress which can be explored in terms of genetic material for salt tolerance engineering; however, a detailed molecular study is indispensable to pinpoint the potential gene(s) or transcription factor(s) involved in the process.

Work has been initiated at DWSR to find out potential weed species that are tolerant to abiotic stresses. Preliminary experiments indicate that *E. crusgalli* can tolerate up to 300 mM NaCl without considerable growth depression (Fig. 12.1). The molecular basis of salt tolerance in *E. crusgalli* is being studied to identify candidate gene(s) for genetic engineering of crops for salt tolerance.

3.2 Water Deficit Tolerance

Water deficit (drought) is an ever-increasing problem that puts the large population at risk globally in every crop season. Conventional breeding offers an opportunity for significant and predictable incremental improvements in the drought tolerance of new cultivars of crops (Banziger and Araus 2007). In this regard, significant progress in grain yield under drought stress has been made through selection in multi-environment trials (Campos et al. 2004). Likewise, the genetic dissection of crop performance in drought prone environments has greatly benefited from the use of DNA markers (Ribaut and Ragot 2007). Attempts have been made to identify donor species for genetic material for drought tolerance, yet nothing is a final word and only limited success has been achieved so far. Resurrection plants which survive extreme drought definitely provide a potential alternate for molecular engineering for drought tolerance of the crop plants. Weeds can provide some extra degree of compatibility as a donor for the genetic material for crop improvement. In the following section, an effort has been made to pinpoint the case studies as well as to highlight the potential species which can be considered as source for the genetic material for drought tolerance engineering.

Water deficit or drought represents one of the most significant constraints to agricultural productivity. Transgenic approaches can be used in combination with conventional breeding strategies to develop crops with enhanced drought tolerance, and one way in which this can be achieved is through the manipulation of polyamine metabolism. Modulation of the polyamine biosynthetic pathway in transgenic rice conferred tolerance to drought stress (Capell et al. 2004). Polyamines are small, ubiquitous, nitrogenous compounds that have been implicated in a variety of stress responses in plants. It has been difficult to establish a direct cause-and-effect relationship between increased putrescine levels in plants and abiotic stress. Elevated putrescine might be the cause of stress-induced injury or, alternatively, a protective response resulting from stress. Transgenic rice plants expressing the *arginine decarboxylase* (*adc*) gene from *Datura stramonium* were generated and their response to drought stress was investigated. Wild-type plants responded to the onset of drought stress by increasing endogenous putrescine levels, but this was insufficient to trigger the conversion of putrescine into spermidine and spermine (the agents that are believed to protect plants under stress). In contrast, transgenic plants expressing *Datura adc* produced much higher levels of putrescine under stress, promoting spermidine and spermine synthesis and ultimately protecting the plants from drought (Capell et al. 2004). From the results, it is obvious that manipulation of polyamine biosynthesis in plants can produce drought-tolerant genotypes. In this study, point worth to note is the source of *adc* gene is *D. stramonium*, a common weed, which itself is considered as a drought-tolerant species.

Flora of desert may provide some inroads into the mechanism by which they tolerate the extreme drought conditions. Weed species in *Thar* Desert of India, which experience high temperature and high drought conditions during summer and extreme low temperature in winter, might be explored for the potential for crop improvement specially the abiotic stress tolerance. In such category, resurrection plants come first as they can even tolerate complete dehydration and have the capacity to revive when the conditions improve. *Oropetium thomaeum* is abundant in the area around Kailana near Jodhpur, and sometimes it covers the whole ground between scattered vegetation (Gaff and Bole 1986). Physiological and molecular features during dehydration and rehydration classify it as a resurrection plant. Metabolic adaptation of this species can be characterized as high content of sucrose, raffinose, and stachyose and low content of monosaccharides in dehydrated leaf tissues along with the presence of two prominent water stress proteins, namely the LEA dehydrin and aldose reductase (Bartels and Mattar 2002). A comparison in induction pattern of these proteins in dehydration-sensitive maize and *O. thomaeum* indicates that aldose reductase protein was only expressed in desiccation-tolerant embryos in maize but never in leaves; however, it was present in desiccated leaves of *O. thomaeum*. These results together suggest that the presence of aldose reductase protein in the leaves can be considered as a marker for desiccation tolerance at least in monocotyledonous plants and *O. thomaeum* can be a suitable source for molecular manipulation of crop plants for drought tolerance. *Physalis minima*, another common weed tolerate high degree of water deficit. In an ongoing pot experiment at authors' institute, *P. minima*

Fig. 12.2 Effect of water deficit (13 days) on growth of *Physalis minima*

showed tolerance up to 13 days of water deficit and recovered within 24 h without showing any injury symptoms (Fig. 12.2).

3.3 Submergence Tolerance

Flooding causes immense damage to agricultural production globally. Submergence creates the anaerobic condition in which aerobic respiration ceases and glycolysis dominates leading to the production of ethanol. Ethanol above the threshold is toxic to most of the crop species. Tolerance to ethanol and the ability to metabolize key intermediary substrates under anaerobiosis were studied in *E. crusgalli* (L.) Beauv. var *oryzicola* seeds to further characterize the mechanism which enable it to germinate and grow without O_2 (Mary and Robert 1983). Results indicate that *E. crusgalli* possesses high tolerance to ethanol and is able to metabolize ethanol (45-fold greater than endogenous levels) in the absence of O_2. Five day anaerobically grown seedlings of *E. crusgalli* metabolized added [^{14}C] sucrose primarily to CO_2 and ethanol. Of the soluble compounds labeled, the phosphorylated intermediates of glycolysis and the oxidative pentose phosphate pathway predominated more under anaerobiosis than in air. In addition, organic acids and lipids were labeled from [^{14}C] sucrose, the latter indicating that metabolism of carbohydrate via acetyl-CoA occurred in the absence of O_2. Lipids were also labeled when seeds were supplied with [^{14}C] ethanol or [^{14}C] acetate. Further, labeling experiments using the above compounds showed labeling of organic acids, i.e., succinate and citrate being labeled under anaerobic conditions, while fumarate was formed under aerobic conditions only. The above metabolic characteristics would allow maintenance of an active alcoholic fermentation system during which high alcohol dehydrogenase activity helps recycle NAD and continued energy production without O_2. In addition, the ability of *E. crusgalli* to metabolize carbohydrate intermediates and to synthesize lipids indicate that mechanisms exist for providing the carbon intermediates for biosynthesis, particularly membrane synthesis for growth, even in the absence of O_2. For submergence tolerance, crop plant can be engineered to enhance ethanol tolerance by incorporation of a gene(s) like alcohol dehydrogenase, and for this purpose *E. crusgalli* emerges as a potential species for genetic material.

3.4 Temperature Stress Tolerance

Low and high temperatures are always associated with oxidative stress which adversely affects the crop performance and yield. The effect of thermal stress on the antioxidant system was investigated in two invasive plants species, *Eupatorium adenophorum* Spreng and *E. odoratum* L. (Lu et al. 2008) with the aim to explore the relationship between the response of antioxidant enzymes and temperature tolerance in the these invasive weeds. For the heat treatments, temperature was increased stepwise to 30, 35, 38, and finally to 42 °C. For the cold treatments, temperature was decreased stepwise to 20, 15, 10, and finally to 5 °C. In *E. adenophorum*, the coordinated increase of the activities of antioxidant enzymes was effective in protecting the plant from the accumulation of ROS at low temperature, however, no such protection was observed during the heat treatment as evident from the higher level of lipid peroxidation in *E. adenophorum* than under cold stress. In *E. odoratum*, a reverse trend was observed in terms of membrane damage, as indicated by low monodehydroascorbate content, and the coordinated increase antioxidant enzymes was observed in heat-treated plants, but the antioxidant enzymes were unable to operate in cold stress. This indicates that *E. odoratum* plants have a higher capacity for scavenging oxygen radicals in heat stress than in cold stress. Differential responses of antioxidant enzymes may be one of the possible mechanisms of the differing temperature tolerance between the two weed species, and this differential ability may be exploited for transferring the responsible genes into crop plants. However, detailed molecular studies are a prerequisite before proceeding for crop improvement studies.

Scientists have made a breakthrough in overcoming corn's intolerance to cold, which results extended length of the growing season and better yield of corn (Wang et al. 2008). Corn is C_4 plant and until recently, the higher productivity achieved by C_4 species was thought to be possible only in warm environments. Low temperature at either end of the growing season limits the growing period, thus limiting yield also. Recently, a C_4 grass related to corn, *Miscanthus* × *giganteus*, was found to be exceptionally productive in cold climates. To exactly find what is responsible for cold tolerance of *Miscanthus* × *giganteus*, researchers focused on four extra chemical reactions that separate C_4 from C_3 plants. The enzyme for one of these steps, pyruvate phosphate dikinase (PPDK) is made up of two subunits. At low temperature, these subunits have been observed to fall apart. The researchers examined the expression of the gene coding for this enzyme and found that when leaves of corn were placed in cold, PPDK slowly disappeared in parallel with the decline in rate of photosynthesis, however, when *Miscanthus* leaves were placed in the cold, they produced PPDK at greater rate so that the leaf became able to maintain photosynthesis in the cold conditions (Wang et al. 2008). Cloning the PPDK gene from both corn and *Miscanthus* into *E. coli* facilitated the production of large quantities of recombinant protein and it was observed that as the enzyme was concentrated, it could resistant to cold. Thus, the difference in PPDK expression amongst the two plants was not due to a change in structure of protein component but rather the amount of protein present. The finding suggests that modifying corn to synthesize more PPDK during cold climate may allow corn to be cultivated in colder climate and be productive for more months of the year in its current locations, and thus may boost corn production. From the above study, it can be inferred that enzyme PPDK from the two species are not different structurally; however, the ability of the *Miscanthus* PPDK to get concentrated in cold makes it novel and hence this may be utilized for the genetic engineering of corn for cold tolerance.

3.5 Heavy Metal Tolerance

Screening of plant species which can tolerate and accumulate higher amounts of heavy metal has been in focus to unravel the mechanism underlying heavy metal tolerance, as well as hunting for a potential species for phytoremediation of polluted soils. A study was conducted on the native flora growing on fly ash affected areas near a thermal power plant, Tanda, Uttar Pradesh, India for accumulation of toxic heavy metals (Dwivedi et al. 2008). Among nine aquatic plants *Hydrilla verticillata* (an aquatic weed) was found to be the most efficient metal accumulator and it was further emphasized that this weed can be utilized for phytoremediation of water bodies. Recently, Liu et al. (2009) studied the potential of grasses for copper tolerance and found that elephant grass (*Pennisetum purpureum* Schumach) is a potential candidate for growth on Cu contaminated soils.

The ability of seven hyper accumulator weeds which grow naturally in heavy metal contaminated channels of three different industries (Hindustan Aeronautical Ltd., Eveready Ltd., and Scooter India Ltd.) to accumulate heavy metals was studied (Sahu et al. 2007). In general, accumulation of heavy metals depends upon the plant species and the metal concentration in media. It was observed that *Eichhornnia crassipes* accumulated high level of metals (Fe –4052.44 μg/g, Mn-788.42 μg/g, and Cu-315.50 μg/g), while *S. polyrrhiza* accumulated Cd (12.75 μg/g), Pb (20.25 μg/g), and Cr (128.27 μg/g) even when the metal concentrations were not high in the effluent. In summary, these two plants were found to be the best accumulators at different contaminated sites. The results will be helpful in the selection of plant species which can be used as bioaccumulators. In addition, other weed species like *Typha angustata* (Bose et al. 2008), *Alternanathrea philoxeroides*, and *Mollugo verticillata* (Khankhane and Varshney 2008) also have been reported to have the potential of metal accumulation. The capability of these weeds to tolerate and accumulate heavy metals can be utilized for the transformation of some crops with a partitioning into waste materials (not edible parts) enabling them to grow in heavy metal polluted areas. Phytochelatin synthase (PCS) catalyzes synthesis of phytochelatin from its immediate precursor "GSH" in the presence of metal ions. Phytochelatin plays a pivotal role in sequestering heavy metal ions by binding them and transporting them into the vacuole and thus avoiding toxic effects on metabolism (Kumar et al. 2009). *PCS* gene can be a potential candidate for transfer into crop plants; however, it needs to be studied in detail in the above mentioned weed species for its involvement in heavy metal tolerance and translocation.

3.6 Herbicide Tolerance

Herbicide tolerance is the inherent ability of a species to survive and reproduce after herbicide treatment. A herbicide which is effective for a particular weed may not be effective for other weed species or even for the same species after prolonged use of the same herbicide. To date, resistant weeds can be seen against almost all herbicide groups figuring a total of 332 resistant biotypes globally (www.weedscience.org) and the problem is almost analogous to well-known insecticide resistance as houseflies became resistant to DDT within a few years after this insecticide's introduction. The most dramatic example is inhibition of acetolactate synthase (ALS) group with 102 resistant biotypes followed by photosystem II inhibitors group with 68 biotypes. Recently, multiple herbicide resistance has been reported in *P. minor* which poses a serious threat to the wheat production in India (Chhokar and Sharma 2008). Other cases of resistance to glyphosate have been reported in *Chenopodium album* (Hite et al. 2008) and *Lolium multiflorum* (Sherwood and Jasieniuk 2009). The appearance of herbicide-resistant weeds must be taken as a serious matter because if weeds become resistant to more chemicals, our options for weed control via herbicides will become more limited. In the meantime, however, the molecular biologist can see all these adversaries from a different angle making a silver line in the cloud—if we can establish how weeds became resistant to triazines, isoproturan, and glyphosate and other group of herbicides, perhaps we can do the same for our crops by means of genetic engineering. How weeds become resistant to herbicide(s)? What are the molecular changes underlying in resistance mechanism and their associated changes, these are few questions we need to answer at molecular level. Once we get an answer, weeds could be a potential source of genetic material for making herbicide-resistant crop plants. For example, to find out why atrazine did not kill weeds of resistant biotypes, researchers first identified the protein that contained the herbicide-binding site and then the gene which codes for this protein. It was firmly established that the gene for this protein resided on the chloroplast DNA. This led to the successful isolation of the gene encoding the herbicide receptor protein. When the pair of genes from susceptible and resistant chloroplasts was analyzed to determine their nucleotide sequence, a single change was observed which led to the substitution of a serine to a glycine in the protein of the resistant chloroplasts (Salava et al. 2006). Since the gene involved in triazine-binding site is altered in triazine-resistant weeds, the next question is: how can we make the same change in crop plants which are also sensitive to triazine? In this context, we have one major success story which can signify the potential of weeds in making crop plants resistant to herbicides. The work is the outcome of a classical breeding program conducted at the University of Guelph, Ontario. The Guelph scientists identified a new triazine-resistant biotype of the weed *Brassica campestris* (Beversdorf and Kott 1987). The Guelph scientists took pollen from the flowers of *Brassica napus* (an important oil crop in Canada and a close relative of the weed) and fertilized the egg of the weed flower. This means that the progeny of this cross will have half of the genes in their nucleus coming from the crop plant and half from the weed. The novelty of these experiments lies in the fact that all the chloroplasts came from the weed, since it was the female parent. This is a special case where we can document the genetic contribution of a weed species to a crop plant's physiological properties. Guelph researchers continued back crossing, always taking pollen from the crop plant, for several generations. Since the genes of the weed were diluted out by 50 % at each cross, they eventually ended up with a crop plant with cells containing the chloroplasts of the weed (which, of course, determine atrazine resistance). Scientists registered the triazine-resistant seed of oilseed rape in 1980 which currently known as Canola. Further, they have accomplished the same thing with summer turnip rape, another important crop in Canada. This novel crop breeding program provides new horizons for weed control in these two important crops. Unfortunately, this approach is not applicable to many other crop genetic systems due to inability of cross fertilization between resistant weed and sensitive crop. However, with the advancement of the biotechnological techniques, we can transfer the desirable genes at our will even in a non-compatible pair of weed and crop.

3.7 Climate Resilience

A hard fact is that weeds are tougher than crop plants and despite a number of control measures, they keep coming back. Based on this ability, scientists are now thinking to harness the resilience of weeds to modify food crops against global warming and its consequences (Fox 2009). Studies have shown that many weeds survive far better in the high heat and high atmospheric carbon dioxide environment. Thus, as rise in CO_2 level is almost sure, many crops such as rice and corn face the dual problem of increased competition from weeds and reduced yield in a harsh environment due to global warming. Lessons can be learnt from the weed species which have shown greater potential to cope up with the global warming effects and then either breed or genetically engineer the genes that make weeds a "tough plant" into the food crops. For example, rice does not produce seeds when temperatures are high (above 35 °C) and at the same time cannot compete with closely related weedy rice in a high CO_2 environment as suggested by Ziska and Mcclung (2008). Options are to study first the potential of weedy type of rice to compete with yield in high CO_2 concentrations and bring the spotlight to a particular gene(s) that may be exploited for further rice improvement.

A study was conducted in "Free Air CO_2 enrichment" (FACE) to observe effects of elevated CO_2 on physiological, biochemical, and antioxidant defense systems in mungbean and associated weeds (*Euphorbia geniculata* and *Commelina diffusa*). Results suggest that elevated CO_2 affects the overall growth of plants and crop-weed interaction changed in favor of weeds on account of higher benefit to weed species in terms of growth, development, and photosynthesis. A unique finding of this study was the adaptive potential of weed species *E. geniculata* to high atmospheric CO_2 which showed higher transpiration and stomatal conductance to cope up with high temperature associated with rising CO_2, however, such adaptation was never evident in mungbean (Awasthi 2010). Other species which provides tough competition to legume crops due to dominant growth at elevated CO_2 is *C. diffusa* L (Awasthi 2010). Both these weed species together resulted in considerable yield reduction in high CO_2 environment as compared to ambient CO_2 concentration. However, potential genes responsible for such dominant behavior of these weed species yet to be identified.

It has been reported that ambient ozone (O_3), either alone or in concurrence with acid rain precursors, accounts for a huge crop losses in crop production (Tong et al. 2007). Along with other greenhouse gases, O_3 exposure is of particular concern due to its high potential to damage the crop plants. Plant competition may be altered by ongoing climate changes including rising tropospheric O_3. To explore this possibility, growth and gas exchange responses of a crop weed system in partial additive competition in open top chambers (OTCs) was studied in cotton (*Gossypium barbadense* L.) by Grantz and Shrestha (2006). Cotton was found to be more sensitive to O_3 than *Cyperus esculentus* L. and hence the competitiveness of *C. esculentus* L. over cotton was observed.

4 Disease and Insect Resistance

Weeds can be a potential source for the development of disease resistance in crop plants. Several attempts have been made to harness the potential of weeds as genetic material for improving disease resistance. Hoisington et al. (1999) emphasized the significance of weeds or weed-like wild relatives of crop plants in conferring the disease resistance into important crops. A good number of genes have been identified and already incorporated into modern wheat and other crops for rust and virus resistance through conventional breeding programs.

Incorporation of specific rust resistance genes in wheat and their confirmation through biochemical and molecular markers tested by Reddy et al. (1996). Leaf rust resistance lines were developed by incorporating *Lr9* gene from *Aegilops umbellulata* and *Lr19* and *Lr24* from *A. elongatum* into four rust susceptible Indian wheat cultivars LOK-1, HUW234, J28, and K68 through a backcross breeding program and the presence of LR19 was confirmed in four lines. The leaf rust-resistant line with *Lr19* was crossed with the leaf rust-susceptible wheat cultivar "Agra Local" and F_2 plants were tested and found to be resistant. Further, efforts are being made to confirm the presence of other leaf rust-resistant genes (*Lr9* and *Lr24*) with the help of RAPD/STS markers.

From the annual weed *E. crusgalli* (L.) two novel defensin proteins Ec-AMP-D1 and Ec-AMP-D2 that differ by a single amino acid substitution were isolated by a combination of different chromatographic procedures (Tatyana et al. 2008). Both defensins were active against several phytopathogenic fungi and the oomycete *Phytophthora infestans* at micromolar concentrations. The Ec-AMP-D1 showed higher activity against the oomycete than Ec-AMP-D2. Novelty of the antifungal *E. crusgalli* defensin proteins was the presence of a C-terminal region of the molecule as the main determinant of the antifungal activity.

Identification of genes which confer resistance to disease is the key for the breeding of disease resistance. A full length cDNA (*Hv-GR*) whose transcript increased in response to infection by *Blumeria graminis* sp. *tritici* (*Bgt*) was isolated from *Haynaldia villosa* (Chen et al. 2007). This gene encodes a glutathione reductase (GR) with high similarity to chloroplastic GRs from other plant species. Chloroplastic localization of Hv-GR was confirmed by targeting of the green fluorescent protein (GFP)-Hv-GR fusion protein to chloroplasts of epidermal guard cells. Following inoculation with *Bgt*, transcript accumulation of *Hv-GR* increased in a resistant line of wheat, but no significant change was observed in a susceptible line. In vivo function of *Hv-GR* in converting oxidized glutathione (GSSG) to the reduced form (GSH) was verified through heterologous expression of *Hv-GR* in a yeast GR-deficient mutant. As expected, overexpression of this gene resulted in increased resistance of the mutant to H_2O_2, indicating a critical role for *Hv-GR* in protecting cells against oxidative stress. Further, overexpression of *Hv-GR* in a susceptible wheat variety, *T. aestivum* cv. Yangmai 158, enhanced resistance to powdery mildew and induced transcript accumulation of other pathogenesis-related genes, *PR-1a* and *PR-5*, through increasing the foliar GSH/GSSG ratio. From the study, it can be inferred that a high ratio of GSH to GSSG is required for wheat defense against *Bgt*, and that chloroplastic GR enzymes might serve as a redox mediator for the activation of pathogenesis-related genes.

Transgenic tobacco plants were produced by overexpressing a serine proteinase inhibitor gene, *SaPIN2a*, from the American black nightshade *Solanum americanum* under the control of the *CaMV35S* promoter using *Agrobacterium tumefaciens*-mediated transformation (Luo et al. 2009). Bioassays for insect resistance showed that *SaPIN2a* overexpressing transgenic tobacco plants were more resistant to *Helicoverpa armigera* and *Spodoptera litura* larvae, two devastating pests of crop plants. Interestingly, overexpression of *SaPIN2a* in transgenic tobacco plants resulted in a significant increase in glandular trichome density and promotion of trichome branching, which could also provide an additional resistance mechanism in transgenic plants against insect pests. Therefore, SaPIN2a could be used as an alternative proteinase inhibitor for the production of insect-resistant transgenic plants.

5 Conclusions

Weeds are a reservoir of naturally available germplasm that may provide potential alternative source of important genes for generating abiotic and biotic stress-tolerant crop plants, as well as other characters for crop improvement. Utilization of weeds as a source of genetic material has certain advantages over the wild species, microorganisms or model plant species. Coexistence of weeds and crop plant ensures better chance of success and provides better control over other co-regulated processes. Another benefit of weeds as a source of genetic material can be noticed in terms of improved fitness and less yield penalty. With only a few successful examples, most of the transgenics have not come out of containment into the natural environment. Evidences discussed in this chapter strongly suggest that weeds have the potential and to be exploited as a source of genetic material to be utilized for crop improvement.

References

Awasthi JP (2010) Effect of elevated CO_2 on physiological and biochemical aspects in mungbean and associated weeds (*Euphorbia geniculata* and *Commelina diffusa*). M.Sc. Thesis, Rani Durgawati Vushwa Vidhyalaya, Jabalpur (M.P.), India

Banziger M, Araus JL (2007) Recent advances in breeding maize for drought and salinity stress tolerance. In: Jenks MA (ed) Advances in molecular breeding toward drought and salt tolerant crops. Springer, Berlin, pp 587–601

Bartels D, Mattar MZM (2002) *Oropetium thomaeum*: a resurrection grass with a diploid genome. Maydica 47:185–192

Beversdorf WD, Kott LS (1987) Development of triazine resistance in crops by classical plant breeding. Weed Sci 35:9–11

Bose S, Vedamati J, Rai V et al (2008) Metal uptake and transport by *Typha angustata* L. grown on metal contaminated waste amended soil: An implication of phytoremediation. Geoderma 145:136–142

Burgos NR, Norman RJ, Gealy DR et al (2006) Competitive N uptake between rice and weedy rice. Field Crops Res 99:96–105

Campos H, Cooper M, Habben JE et al (2004) Improving drought tolerance in maize: a view from industry. Field Crops Res 90:19–34

Capell T, Bassie L, Christou P (2004) Modulation of the polyamine biosynthetic pathway in transgenic rice confers tolerance to drought stress. Proc Natl Acad Sci U S A 101:9909–9914

Chen YP, Xing LP, Wu GJ et al (2007) Plastidial glutathione reductase from *Haynaldia villosa* is an enhancer of powdery mildew resistance in wheat (*Triticum aestivum*). Plant Cell Physiol 48:1702–1712

Chhokar RS, Sharma RK (2008) Multiple herbicide resistance in littleseed canarygrass (*Phalaris minor*): A threat to wheat production in India. Weed Biol Manage 8:112–123

Dwivedi S, Srivastava S, Mishra S et al (2008) Screening of native plants and algae growing on fly-ash affected areas near National Thermal Power Corporation, Tanda, Uttar Pradesh, India for accumulation of toxic heavy metals. J Hazard Mater 158:359–365

Fox S (2009) Weeds can be a tool to fight against global warming: Inserting weed genes to protect crops from global warming. http://www.popsci.com/environment/article/2009-06/weed-genes-protect-crops-global-warming

Gaff DF, Bole PV (1986) Resurrection grasses in India. Oecologia 71:159–160

Garg M, Tanaka H, Ishikawa N et al (2009) *Agropyron elongatum* HMW-glutenins have a potential to improve wheat end-product quality through targeted chromosome introgression. J Cereal Sci 50:358–363
Grantz DA, Shrestha A (2006) Tropospheric ozone and interspecific competition between yellow nutsedge and Pima cotton. Crop Sci 46:1879–1889
Hite GA, King SR, Hagood ES et al (2008) Differential response of a Virginia common lambsquarters (*Chenopodium Album*) collection to glyphosate. Weeds Sci 56:203–209
Hoisington D, Khairallah M, Reeves T et al (1999) Plant genetic resources: What can they contribute toward increased crop productivity? Proc Natl Acad Sci U S A 96:5937–5943
Kawasaki S, Borchert C, Deyholos M (2000) Gene expression profiles during the Initial phase of salt stress in rice. Plant Cell 13:889–906
Khankhane PJ, Varshney JG (2008) Accumulation of heavy metals by weeds grown along drains of Jabalpur. Indian J Weed Sci 40:55–59
Kumar B, Singla-Pareek SL, Sopory SK (2009) Glutathione homeostasis: crucial for abiotic stress tolerance in plants. In: Pareek A et al (eds) Abiotic stress adaptation in plants: physiological, molecular and genomic foundation. Springer Science, Dordrecht
Lai Z, Kane NC, Zou Y et al (2008) Natural variation in gene expression between wild and weedy populations of *Helianthus annuus*. Genetics 179:1881–1890
Liu X, Shen Y, Lou L et al (2009) Copper tolerance of the biomass crops elephant grass (*Pennisetum purpureum* Schumach), Vetiver grass (*Vetiveria zizanioides*) and the upland reed (*Phragmites australis*) in soil culture. Biotechnol Adv 27:633–640
Lu P, Sang WG, Ma KP (2008) Differential responses of the activities of antioxidant enzymes to thermal stresses between two invasive eupatorium species in China. J Integr Plant Biol 50:393–401
Luo M, Wang Z, Li H et al (2009) Overexpression of a weed (*Solanum americanum*) proteinase inhibitor in transgenic tobacco results in increased glandular trichome density and enhanced resistance to *Helicoverpa armigera* and *Spodoptera litura*. Int J Mol Sci 10:1896–1910
Mary ER, Robert A (1983) Anaerobiosis in *Echinochloa crus-galli* (Barnyard Grass) Seedlings: Intermediary metabolism and ethanol tolerance. Plant Physiol 72:44–49
Morikawa T, Sumiya M, Kuriyama S (2007) Transfer of new dwarfing genes from the weed species *Avena fatua* into cultivated oat *A. byzantina*. Plant Breed 126:30–35
Myers BA, Neales TF, Jones MBBA (1990) The Influence of salinity on growth, water relations and photosynthesis in *Diplachne fusca* (L.) P. Beauv. Ex Roemer & Schultes. Aust J Plant Physiol 17:675–691
Norris RF, Caswell-Chen EP, Kogan M (2002) Concepts in integrated pest management. Prentice Hall, Upper Saddle River
Reddy VRK, Damodaran S, Asir R et al (1996) Development of disease rust resistance in hexaploid wheat—an overview. In: Siddiqui BA, Khan S (eds) Plant breeding advances and in-vitro culture. CBS, New Delhi, pp 179–194
Ribaut JM, Ragot M (2007) Marker-assisted selection to improve drought adaptation in maize: the backcross approach, perspectives, limitations, and alternatives. J Exp Bot 58:351–360
Sahu RK, Naraian R, Chandra V (2007) Accumulation of metals in naturally grown weeds (*Aquatic Macrophytes*) grown on an industrial effluent channel. Clean Soil Air Water 35:261–265
Sairam RK, Tyagi A (2004) Physiology and molecular biology of salinity stress tolerance in plants. Current Sci 86:10–11
Salava J, Chodová D, Kočová M et al. (2006). Molecular basis of atrazine resistance in Czech biotypes of *Digitaria sanguinalis* (L.) Scop. The international survey of herbicide resistant weeds. www.weedscience.com
Sandhu GR, Aslam Z, Salim M et al (2006) The effect of salinity on the yield and composition of *Diplachne fusca* (Kallar grass). Plant Cell Environ 4:177–181
Sawada H, Ie-S S, Usui K et al (2008) Adaptive mechanism of *Echinochloa crus-galli* Beauv. var. *formosensis* Ohwi under salt stress: Effect of salicylic acid on salt sensitivity. Plant Sci 174:583–589

Sherwood AM, Jasieniuk M (2009) Molecular identification of weedy glyphosate-resistant *Lolium* (Poaceae) in California. Weed Res 49:354–364
Sullivan ML (2009) Phenylalanine ammonia lyase genes in red clover: expression in whole plants and in response to yeast fungal elicitor. Biol Plant 53:301–306
Tatyana IO, Rogozhin EA, Baranov Y et al (2008) Seed defensins of barnyard grass *Echinochloa crusgalli* (L.) Beauv. Biochimie 90:1667–1673
Tong D, Mathur R, Schere K et al (2007) The use of air quality forecasts to assess impacts of air pollution on crops: Methodology and case study. Atmos Environ 41:8772–8784
Vidya AS, Abraham CT, Girija T (2004) Weed spectrum of Pokkali lands: the salt marsh rice ecosystem of Kerala. Ind J Weed Sci 34:157–159
Wang D, Portis A, Moose S et al (2008) Add one enzyme, and corn can stand the cold. http://www.thaindian.com/newsportal/uncategorized/add-one-enzyme-and-corn-can-stand-the-cold_10090512.html
Yamamoto A, Ie-S S, Fujihara S et al (2003) Physiochemical factors affecting the salt tolerance of *Echinochloa crusgalli* Beauv. var. formosensis Ohwi. Weed Biol Manage 3:98–104
Ziska LH, Mcclung AM (2008) Differential response of cultivated and weedy (red) rice to recent and projected increases in atmospheric carbon dioxide. Agron J 100:1259–1263

Chapter 13
Sustainable Agriculture Practices for Food and Nutritional Security

Vibha Dhawan

1 Introduction

The domestication of plants started over 10,000 years ago and led to the birth of agriculture. Since then, the main emphasis has been on increasing productivity, sometimes compromising on long-term sustainability. During initial years, the farming community observed that if the same crop was cultivated over years on the same soil, there was gradual decline in productivity. Thus, without any scientific knowledge, the farmers realized that there was need to leave the soil fellow for a while so that it regains its vigour, a process which was later called shifting/jhum cultivation. Even today, in some regions of the North east India, the practice is still prevalent. Unfortunately, due to population pressures, the land is no more available in abundance which has led to shortening of Jhum cycles, thus resulting in insufficient time for soil to rejuvenate fully. Further, the process is not efficient as burning biomass to clear the land is a wasteful process. Other methods adopted for restoring soil fertility are the cereal-legume rotation and crop-livestock integrated farming. Legumes are known to fix atmospheric nitrogen and thus improve the nutrient status of the soil. Similarly, animal waste adds to the organic nutrient status and improves texture of the soil. The practice of ploughing back all agricultural residues into the soil is also being adopted to improve the physical structure and the organic matter status of the soil. Thus, a sustainable system of soil maintenance and enhancement was developed through experience and experiments ever since the agriculture started.

During early years of agriculture, crop health management was a serious challenge as not many agrochemcials were discovered. However, the farmers were quite successful in keeping pest control under control through maintenance of rich

V. Dhawan (✉)
The Energy and Resources Institute (TERI),
Habitat Place, Lodhi Road, New Delhi 110 003, India
e-mail: vassilis.fotopoulos@cut.ac.cy

N. Tuteja and S. Singh Gill (eds.), *Plant Acclimation to Environmental Stress*,
DOI 10.1007/978-1-4614-5001-6_13, © Springer Science+Business Media New York 2013

agrobiodiversity, planting of varieties resistant to major insects, use of botanical pesticides and intercropping and planting of different crops at any given time to spread the risk.

2 Green Revolution and Sustainable Agriculture

Unfortunately over the years, the traditional practices of conservation and sustainable use gradually gave way to high-input agriculture leading to monoculture, widespread planting of few selected varieties, unsustainable usage of water and excessive use of fertilizers and pest-control chemicals. Inspite of the fact that legumes and millets can be grown in difficult production environments (such as drought-prone areas), their cultivation has reduced drastically mainly due to lower yields and poor price realization. Green revolution marked the beginning of expansion of production through productivity improvement, became a blessing in terms of saving land and forests, but it led to overexploitation of land and water, and excessive use of chemical pesticides and mineral fertilizers resulting in environmental pollution. Prof. MS Swaminathan, known as "Father of Green Revolution in India" for introducing and developing high-yielding varieties of wheat in India, recognized the probable negative impacts of the technology. During his address in Indian Science Congress held in Varanasi in January 1968, he cautioned the world "that exploitive agriculture offers great dangers if carried out with only an immediate profit or production motive. The emerging exploitative farming community in India should become aware of this. Intensive cultivation of land without conservation of soil fertility and soil structure would lead, ultimately, to the springing up of deserts. Irrigation without arrangements for drainage would result in soils getting alkaline or saline. Indiscriminate use of pesticides, fungicides and herbicides could cause adverse changes in biological balance as well as lead to an increase in the incidence of cancer and other diseases, through the toxic residues present in the grains or other edible parts. Unscientific tapping of underground water will lead to the rapid exhaustion of this wonderful capital resource left to us through ages of natural farming. The rapid replacement of numerous locally adapted varieties with one or two high yielding strains in large contiguous areas would result in the spread of serious diseases capable of wiping out entire crops, as happened prior to the Irish potato famine of 1854 and the Bengal rice famine in 1942. Therefore, the initiation of exploitative agriculture without a proper understanding of the various consequences of every one of the changes introduced into traditional agriculture, and without first building up a proper scientific and training base on sustain it, may only lead us, in the long run, into an era of agricultural rather than one of agricultural prosperity" (Swaminathan 2010).

The Green Revolution of late 1960s transformed the agriculture in most parts of the world. The technological package of improved seeds of cereals, chemical fertilizers, irrigation and other pest-control measures transformed agriculture leading to rapid-productivity gains and widespread acceptability of the technology among the farmers (Timsina and Connor 2001; Gupta et al. 2003; Gupta and Seth

2007). The benefits of green revolution were not just restricted to food security, but changed the entire social fabric of rural communities and became the main driver of economic growth in rural areas. The associated services, such as sale of seed and other farm inputs, lead to entrepreneurship development. Also, with surplus income, farmers started investing in education and products that were earlier restricted to elite class, thus boosting the rural economy.

Unfortunately, in India the productivity gains since the beginning of this century have more or less stagnated (Duxbury 2001; Kataki et al. 2001; Kumar et al. 2002; Ladha et al. 2003a, b; Prasad 2005; Dhawan 2008) leading to concerns over national food security and lagging economic growth in the rural areas. An analysis of green revolution has further raised concerns about its sustainability as it has led to wide-spread salinity in many areas, overexploitation of ground water resulting in receding water tables, and ground water pollution with fertilizers and pesticides and even heavy metals such as arsenic in ground water. Dogra (1986) estimated that nearly 4.5 million ha of irrigated land was affected by salination and another 6 million ha due to water logging. These concerns, sometimes overstated, demand technological interventions to conserve resources, reduce production cost and improve productivity while sustaining environmental quality (Hobbs and Gupta 2003; Gupta and Sayre 2007; Gupta and Seth 2007; Erenstein and Laxmi 2008), leading to evergreen revolution. Also, the conventional agricultural practices must be revisited and evaluated in the light of R&D in other related fields for meeting the objective of long-term sustainability.

3 Future Challenges

Today agriculture in many parts of the world is challenged by the changing pattern of temperature and precipitation (climate change) and growing scarcity of water. The rivers are fully exploited, and at least during part of the year, there is more or less a crisis situation. The massive expansion of canals and tubewells has further led to serious overdrawing of groundwater. The virtual flow of water as grains or other agricultural produce must be studied, especially for crops such as rice, and adequate policies be formulated keeping in long-term vision. Water-efficient varieties of the crops that are grown over large acreage must be developed and promoted. Temperature extremes and change in the pattern of temperature regimes are also affecting the grain yield for major crops, drawing the attention of agricultural scientists to develop either early-maturing varieties or those which can tolerate wide range of temperatures.

Inspite of progress made in chemicals for controlling pests, plant diseases still remain to be the major problem. For many crops (for example, cabbage in India), inspite of impressive pest control measures being developed through R & D, the percent losses today are greater than what they were at the time of independence in 1947. In most cases, pest develops resistance to a particular chemical, resulting in continuous search for developing new formulation. The problem is further compounded by high cropping intensities, mono-cropping and high fertilizer use which creates dense lush green canopies in which pests can thrive well. The large-scale plantation of similar varieties leads to similar susceptibility causing pest problems. Initially, control was based on prophylactic chemical applications, driven by calendar rather than pests and diseases. This approach disrupted the natural pest predator balance and led to resurgence of pest problems that required even higher doses of pesticide applications for proper control. The excessive use of pesticides has led to environmental problems and has even affected health of the farm workers. Some of the pesticides are even carcinogenic and the leachate, which gets into the ground water has serious health implications to the entire community.

4 Organic Farming

Organic farming has been advocated in many parts of the world but most often the alternative farming approaches fail to match high productivity levels achieved by farming methods of green revolution (Pretty et al. 2007). The green manure production is also gradually losing ground as the land needed to grow the crop is not available readily (Hazell and Wood 2008).

Organic farming has enormous potential to improve yields of small and marginal farmers, who either cannot afford to invest in fertilizers/chemicals or are in remote areas with limited access to markets. The biofertilizers/biopesticides can also replace part of the chemical requirements, thus restoring soil health. For example, in most wheat-growing areas of India, upto 40% chemical fertilizers can be replaced through application of mycorrhizal biofertilizer (Mäder et al. 2011). Organic cultivation is also practiced for high-value crops for the elite customers and certified organic produce are being sold at premium price.

While benefits of organic cultivation have been documented and practiced for a long time, it has not become popular. More research interventions are required so as to develop reliable commercial products with adequate shelf life. Subsidies on synthetic chemicals further act as a disincentive for biofertilizers and biopesticides. Certification of inputs (e.g. biopesticides and biofertilizers) giving details of active ingredients and product (certified organic vegetables/fruit/grains) is essential so as to develop customers' confidence, be it at the producer level (farmer) or at the consumer level (public) to commercialize this technology further.

5 Technologies for Efficient Resource Utilization

Other efforts to improve sustainability involve precise matching of nutrients, switching to slow-release fertilizers and use of drip irrigation. At present, these are largely adopted for high-value horticultural species, but the success achieved definitely calls for policy interventions such as providing subsidies. While subsidies are available on electricity cost for drawing water for agricultural practices, they may be extended to put drip irrigation facilities leading to efficient water utilization.

In recent years, fortunately the environmental concerns have attracted attention, leading to enhanced research efforts towards developing sustainable technologies and farming practices. Apart from technological innovations, one must relook at the old technologies and supplement them with newer tools. Many of the technologies have failed in the past as they were labour intensive and thus were not viable. However, with invention of newer tools, it has now become practical to adopt them at wider scale. Some of the old technologies that are proving to be extremely promising are as follows.

5.1 Zero-Tillage Planting of Wheat and Rice

Zero tillage typically saves energy, helps in preventing soil and land degradation (such as decline of soil organic matter, soil structural breakdown and soil erosion) and leads to more efficient use of water and other inputs. The interest in zero tillage originated due to the time conflicts between rice harvesting and wheat planting (Harrington et al. 1993). Wheat is grown in cool dry winter (November-April in Northern India) while rice is grown during the warm monsoon season (May–November in Northern India). The technology involves tractor-drawn seed drill with 6–11 inverted T lines to seed wheat directly into unploughed fields with a single pass of the tractor. This specialized agriculture machinery was not originally available, thus demanding a lot of manpower which was actually not feasible. The technology was introduced in India by CIMMYT in 1989 and, thereafter, in 1991 a first prototype of Indian Zero tillage was developed in GB Pant University of Agriculture and Technology in Pant Nagar in the State of Uttarakhand. A collaborative program for further development and commercialization of zero-tillage drills by small-scale farm machinery manufacturers was initiated by the national agricultural research system in collaboration with CIMMYT and subsequently with the Rice-Wheat Consortium of the Indo-Gangetic Plains, leading to the adaptation of the technology by large number of farmers.

It is interesting to note that introduction of genetically modified (GM) herbicide (glyphosate tolerant) soybean in 1996 in Argentina was specially adopted for use with zero-tillage technologies that facilitated the wheat-soybean double cropping scheme (Trigo et al. 2010). This has been the major technological event in the agricultural history of the country. The zero-tillage systems expanded from about 3 million ha in 1990–1991 to more than 22 million hectares in 2007–2008 in Argentina, thus demonstrating acceptability of the technology and its gelling with the most advanced transgenic seed technology. Zero tillage has positive effect on environment although further research is needed for putting values to the environmental impacts (Akhtar 2006; Sarwar and Goheer 2007; Erenstein and Laxmi 2008; Hobbs and Govaerts 2009; Pathak 2009). The immediate impacts are in terms of saving of diesel cost (≅8%) and indirect in terms of greenhouse gas emissions.

5.2 Shallow Tubewells

Shallow tubewells were first adopted in Bangladesh and thereafter expanded in North eastern part of India. Since its independence in 1971, Bangladesh has tripled its production of cereal grains, despite a continuous decline in arable land. The progress can largely be attributed to expansion of ground water irrigation which was triggered by change in government policy in favour of liberalization in procuring and marketing of minor irrigation equipment such as low-lift power pumps and shallow and deep tubewells (Palmer-Jones 1999; Ahmed 2001; Dorosh 2006). The change in policy has radically changed dry season cultivation of *Boro* rice (Singh et al. 2003) and increased contribution to productivity by 56% in 2008 (while it was barely 9% in 1966–1967). Further, expansion of STWs has contributed to the development of rural entrepreneurship which has led to the growth of other agribusiness services (Mandal 2000).

The farmers are always looking for new technologies and if they find the technology to be beneficial, it is adopted on large scale. Nearly 22% of farm households in Bangladesh owns approximately 1.3 million STWs that provide irrigation services to 10.2 million out of 15 million farm households (Hossain et al. 2003) and has resulted in over 58% increase in average rice yield. This has helped in keeping the rice prices affordable and ensuring food availability to the poor.

The expansion of STWs for *boro* rice cultivation has also attracted criticism especially on its negative environmental impact. Some of the concerns are similar to the Green Revolution such as overemphasis on rice production thus pushing out pulses and oil seeds. It has resulted in nutritional security, negative impact on soil fertility as two crops of rice are raised, and application of heavy doses of pesticides having adverse impacts on the quality of water runoff thus affecting fish habitats, arsenic contamination of ground water and deteriorating quality of drinking water.

However, one must give credit to government policy in Bangladesh that encouraged the procurement and marketing of minor irrigation equipment for adopting new technologies thus increasing rice production (Hussain 2010). At the same, reducing the negative impacts of technology is the need of the hour.

In last few years, tremendous progress has been made in the field of plant tissue culture and molecular biology. The new tools have potential of supplementing conventional breeding technologies thus designing crops for the future. Some of these technologies are as follows:

5.3 Micropropagation

Micropropagation is a technique of producing identical plants from somatic cells. Enhanced axillary proliferation is preferred method of producing year-round pathogen-free plants. The technique is exploited on commercial-scale world for large number of ornamental and horticultural species. The technology has a potential of improving production of major horticultural species through cloning of elite genotype, production of hybrids through cloning of mother lines, multiplication of plants of desired sex, etc. Further, species that vegetatively propagated (such as citrus, apple and banana) harbour large population of microbes including viruses and results in gradual decline in productivity. Through identification of elites (chosen for desirable traits, such as fruit yield, and its quality), the selected tree can be tested for known viruses and, if found to be infected, freed of viruses through shoot-tip culture and heat therapy techniques. The virus-free stock can then be further multiplied in large numbers under aseptic conditions applying tissue culture techniques. The virus-free stock is suitable for raising plantation even in virgin areas without any danger of introducing new diseases and can be taken across international boundaries. The technology has been already commercially exploited by the farmers all over the globe and in India for number of horticultural species such as banana, strawberry, large cardamom, vanilla and black pepper. At global scale, it has been exploited for a large number of ornamental and other horticultural species such as kiwi, apple and citrus.

The micropropagation technology was accidentally discovered by George Mendel in 1960, while he was attempting to get virus-free plants of *Cymbidium* orchid. Realizing the potential of the technology, by early 1970s the technology was commercialized by the quality-conscious developed part of the world mainly for ornamental plants. Since it is a labour-intensive process, the developed countries shifted the production base to the developing counties and large number of tissue-culture companies were set up in India in late 1980s. With the experience of producing quality plants, the tissue-culture-propagated plants have found their way and acceptability among Indian farmers. To ensure that quality parameters are not compromised, the Ministry of Science and Technology, Government of India, through Department of Biotechnology, Government of India, has established a unique National Certification System for Tissue Culture raised plants (NCS-TCP) up to laboratory level and to regulate its genetic fidelity. Biotech Consortium India Limited (BCIL) has been identified as the Accreditation Unit and Project Management Unit for implementation of NCS-TCP. As the accreditation unit, BCIL is assisting in Recognition of Tissue Culture production facilities and Accreditation of Test Laboratories for virus testing and establishing clonal fidelity. The National Certification System has helped in _ gaining growers' confidence through distribution of quality and certified planting material among the farmers (for other details visit www.bcil.nic.in/certification-services.htm). In recent years, the micropropagated plants are gaining popularity and their impact is visible in few crops such as banana where significant improvement has been observed both in quality and quantity of produce.

5.4 Production of Virus-Free Plants

One of the most important applications of tissue culture propagation is the production of certified virus-free plants. One of the earliest certification programmes had its roots in the 1930s when the first graft transmissible agent, *Citrus psorosis,* was discovered at the Citrus Experimental Station in Riverside, California, United States. In 1956, following a string of advances in citrus indexing that greatly increased the number of diseases that could be induced, the Psoriasis Free programme evolved, into the California Valley Improvement Programme (CVIP), which is now known as Citrus Clonal Protection Program or (CCPP).

The Citrus Variety Improvement Program of Spain (CVIPS) was launched in 1975 based on the CCPP model. The pathogen-free plants of local cultivars were established by shoot-tip grafting (STG) technique and genotypes imported from other countries subjected to STG-based quarantine procedure. The healthy genotypes were maintained in germplasm bank, and health budwoods were supplied following a certification program. The CVIPS has had a positive impact on the Spanish Citrus Industry. Since its inception, about 92 million certified nursery trees originating from the germplasm bank have been planted in the fields, which represent more than 75% of planting in the country. Similar programmes are being followed by other countries such as France, Italy, Cyprus, Argentina, Brazil, Chile and Cuba and legislatively established health certification programme for citrus (Vapnek 2009).

Such programmes are being extended to other crops by many other countries. Replacement of infected trees with virus-free stock has not only helped in increasing land productivity but also lowered the application of chemicals leading to sustainability.

5.5 Double Haploidy

A double haploid (DH) is a genotype formed through chromosome doubling of haploid cells. It is known that gametic cells typically have half (n) the number of chromosomes of the somatic cells ($2n$). The sexual reproduction involves formation of gametes through a special cell division (meiosis) resulting in halving the number of chromosomes (n) in the gametes. Through fusion of male and female gametes, a diploid cell is formed which is heterozygous with each gamete contributing half of each of the chromosome.

Haploid production especially through plant tissue culture methods has enabled the production of completely homozygous lines from gametic cells in a single generation, thus shortening the time frame required for conventional plant breeding. The plants raised from the diploid cells give rise to heterozygous plants each representing different sets of traits inherited from each of the parent. The lines which exhibit desired characteristics are then chosen for large-scale field trials as well as for further breeding experiments.

Blakelsee et al. (1922) have reported natural occurrence of haploid plants of *Datura stramonium.* The revolution came when Guha and Maheshwari (1964) developed anther culture technique, giving promise that haploids can be produced under in vitro conditions in species where they do not occur naturally. Studies were thus extended to many other crops, and protocols have been perfected for most of the commercially important species. World over, large number of varieties have been released through diploidization of haploids raised through anther/pollen culture.

Other technologies such as parthenogenesis, pseudogamy and chromosome elimination after wide crossing have also been introduced for production of haploids.

The discovery of DNA as double helical structure and its role in inheritance of traits through genes has opened up new avenues for research and possibility of designing crops in much shorter time (molecular breeding) and even introducing which were not earlier known in the species of interest (transgenic technology). The double haploid technology found much wider application through integration of molecular tools such as the following:

6 Mapping Quantitative Trait Loci

Most of the economic traits are controlled by more than one gene and each gene having small but cumulative effect. As the quantitative trait loci (QTL) effects are small and are influenced by environmental factors, accurate phenotyping with replicated trials is required. This is possible with double haploid plants because of their homozygous nature and also possibility of producing them in large numbers.

6.1 Marker-Assisted Breeding

Traditionally, plant breeders select plants based on their visible or measurable traits called phenotypes. The process, however, is slow and is influenced by the environment and takes many years of back-crossing and testing in the field before a variety can be developed. The discovery of DNA and its role in carrying genetic material has led to development of another important tool called molecular-assisted selection (MAS). To help identify specific genes, scientists use what are called molecular or genetic markers. The markers are string sequence of nucleic acids which makes up a segment of DNA. The markers are located near DNA sequence of the desired gene and are transmitted by the standard lines of inheritance from one gene to the next. Since the markers and genes are close together on the same chromosome, they tend to set back as such and are collectively passed on from one generation to the next. This is called genetic linkage. Through molecular tools, genes can be detected in the laboratory and technologies have been developed (chip technology) whereby a part of the seed is subjected to molecular analysis. Only seeds with desired gene combination are taken to the field for further testing. Thus, it saves labour, time and space which otherwise is required for field evaluation. The technology has potential of testing huge collections of germplasm that are available in germplasm banks for future applications. Also, parents for the future breeding can be identified through these tools.

Several marker systems have been developed and are applied to a range of crop species. These are the restriction fragment length polymorphisms (RFLPs), random amplification of polymorphic DNAs (RAPDs), sequence tagged sites (STS), amplified fragment length polymorphisms (AFLPs), simple sequence repeats (SSRs) or microsatellites, and single nucleotide polymorphism (SNPs). Each technique has its own merits and demerits (Korzun 2003).

Major seed companies are investing heavily in molecular markers. One of the leading companies, Monsanto, has invested over 100 million dollars in this with firm platform and has invested over 75 million in proprietary software tools. It has capability of analysing millions of samples annually. Similarly, DuPont is using MAS for corn and soya bean to develop drought tolerant varieties. Syngenta has also developed water optimization technology called "agrisure artesian" and a limited quantity of hybrids with this technology was planted in 2011.

Molecular breeding through MAS has somehow limited scope as one can only incorporate traits that are already present in the crop. It cannot be used effectively to breed crops with long generation times and with crops that are vegetatively propagated. For such crops transgenics is a more powerful tool.

7 Production of Transgenics

The transgene organisms are the ones where genetic material has been transferred naturally or artificially (through genetic engineering techniques) from one organism to another. To be more precise, the term describes presence of segment of DNA containing gene sequence that has been isolated from one organism and is introduced into a different organism.

While in plant breeding, to incorporate a trait of interest, one has to do series of back-crosses but in case of genetic engineering, it is possible to identify a gene of interest, clone it (make multiple copies) and insert into the genome of otherwise desirable plant. The gene has no boundaries and, therefore, it is possible to incorporate genes of even microbial origin to higher plant. It is important to understand that gene is nothing but a small portion of DNA consisting of four nucleotides, ATGC (adenine, thymine, guanine, and cytosine) organized in a particular fashion. Thus, there is no difference in a gene of animal, plant and microbial origin.

The DNA so incorporated may retain its normal metabolic function including production of proteins in transgene organism or may even alter the metabolic pathways of the organism in which it has been introduced. Thus, the product of interest must be carefully evaluated and compared with the original product (substantial equivalence).

The plants produced by genetic engineering are regulated through a set of rules and regulations by the countries which are either importing them as food (both in grain and product form) or are growing them for food or fuel. In India, even to initiate research, the institutions must convene regular meetings of the Institutional Biosafety Committee (IBSC). At the next level, when plants are to be tested in containment, a much larger regulatory body Review Committee on Genetic Manipulation (RCGM) under the Department of Biotechnology reviews the application, ensuring that all aspects of biosafety are duly considered. Further, it is ensured that nothing is leaked out of the premises and even remains of the test plants are carefully destroyed. The applicant also develops procedures for testing the product, meeting the criteria of substantial equivalence, other food safety concerns, and with proper test protocols for any possible allergenicity and toxicity. The approving body for commercial release is GEAC (Genetic Engineering Approval Committee) under the Ministry of Environment and Forests and apart from concerns related to food safety, environmental implications are carefully evaluated. The committees have representatives from various interest groups, such as researchers, policy-makers, government representatives from various ministries, and medical doctors.

In India, approval has so far been granted only for one GM crop and that is cotton. The trait is for conferring resistance (through *Bacillus thuringiensis* (Bt) protein) to insect pests of cotton. The technology was granted release for commercial plantation in 2002. During 2010, technology was adopted by 6.3 million farmers covering an area of 9.4 million hectares representing an unprecedented increase of 188-fold in 9 years. According to the recent ISAAA Brief (No. 42 of 2010), yield gains are approximately 31%, and a significant 39% reduction in insecticide sprays, leading to 88% increase in profitability (□US$250/ha). It is interesting to see that a total of 780 Bt cotton introduction (779 hybrids and one variety) were approved for planting in 2010, indicating that the perceived risk of loss of biodiversity was non-existent. The deployment of Bt cotton has helped India attain number one position as exporter of cotton and only next to China in terms of producer. While due to increased production, there has been some decline in international prices, but due to productivity gains, the individual farmer's income has still gone up. Further, while the measurable benefit is in terms of savings from reduced applications of pesticides, there are hidden additional benefits of cleaner environment.

We must realize that there is an opportunity cost—what would have been the plight of the cotton farmer in India if the technology was not adopted.

The next crop which was expected to get approval for commercial release was Bt Brinjal. GEAC in October 2009 allowed the release of Bt eggplant but immediately had put a moratorium. The then Minister of State for Environment and Forests, Shri Jairam Ramesh, has called public consultations in different parts of the country and based on the public opinion has put a hold on its commercial release. Public has voiced number of concerns such as lack of long-term studies on the safety of the crop, biosafety concerns due to large number of existing varieties of Brinjal, and authenticity of food safety tests as they were largely done by the developer of the technology and difficulty in labelling as consumer has right to know what they are consuming.

The issue is not restricted to the introduction of Bt Brinjal alone, but is a much larger one. Today, we are living in a global village, and thus cannot expect to be untouched by the developments in other parts of the world. With 15.5 million farmers growing biotech crops in the year 2010 in 29 countries and over 30 countries importing products of biotech crops thus making total of 59 countries using biotech crops, we can afford to keep on debating on the issues of perceived risks and ignore the existing risk of food insecurity, malnutrition coupled with environmental damage due to excessive applications of chemicals. Globally, we are also witnessing rise in food prices at a rate which is consistently higher than the consumer price index. This is definitely a departure from past many years when the increase in food prices was in line with CPI. It is quite depressing for the developing part of the world, where people are spending substantial part of their income on food and large percentage of our population today is forced to make hard decision of relying largely on cereals resulting in nutritional insecurity. The prices of pulses, vegetables and fruits have gone skyrocketing, making them a luxury item for the poor rather than a part of their routine food. Thus, if globally any technology is adapted which improves productivity without compromising on food or environmental safety, the products of that will reach other parts of the globe. To ensure that India does not reach to a situation where importing food will become cheaper than producing its own, we must carefully evaluate all technological options that are available for meeting food and nutritional security. The global leaders and policy makers also have responsibility in terms of creating enabling environment for adaptation of new technologies leading to long-term sustainability of agriculture.

The technology is evolving and, as we move, we will be learning new lessons be it extension or its regulation. We do not have to rush and proper risk assessment and management practices must be evolved. But at the same time we should not be doing analysis to an extent that it kills the technology. Similarly, regulations are definitely required but not at a prohibitory high cost. We should have faith in our farmer who is the best judge to decide how to increase land productivity and also in our government regulations as they are made by experts keeping long-term vision in the horizon.

8 Nutritional Security

A critical but yet often overlooked component in food security is nutritional security. Worldwide, while the number of people with insufficient availability of food or access to calories is declining, these figures do not capture widespread problem of hidden hunger or nutrient deficiency. A large number of children are suffering from Vitamin A deficiency and a large number of pregnant women suffer from iron and zinc deficiency. The nutrient deficiency raises the risk of mortalities among infants as their immune system is weakened resulting in increased susceptibility to diseases such as pneumonia, diarrhoea, measles, malaria, dengue and even viral infections. Several strategies to combat micronutrient deficiency have been undertaken in the recent past ranging from food supplementation, fortification and promotion of dietary diversification both at the national level by the governments and by international organization through aids. Through proper education and awareness programmes, people are being made conscious of the benefits of balanced food, and role of fruits and vegetables. Strategy of homestead food production by incorporating horticultural species in the farms has been widely accepted in the Asian countries (Katake 2002). However, it is worthwhile to mention here that socioeconomic conditions play an extremely important role as there is unequal distribution of food even at household level because of the social fabric. Traditionally, the head of the family and men folks get the best food, followed by boy child, and typically women diet consists of only cereals. The declining of millet and other coarse grains is a matter of great concern as it might have serious health implications on the economically disadvantageous section of the society.

Unfortunately, research investments in these crops have been very little and are thus called orphan crops. In the changed scenario, where private investments are increasing in agricultural research and development, these crops are still not gaining any attention (Hazell and Haddad 2001; Rosegrant and Hazell 2000). Perhaps, they are the most suitable candidates for public–private partnership as technologies developed for major crops can be shared with public sector institutions to bring about improvements in the orphan crops. However, there are still unresolved issues related to Intellectual Property Rights, stewardship and liability, especially associated with new technologies (such as transgenic) that are hindering sharing of technologies on humanitarian ground.

9 Conclusion

There is enormous potential for improving land productivity by supplementing conventional technologies with biotechnologies. These applications include conservation agriculture, tissue culture for cloning of disease-free elite genotypes, locally produced bio-pesticides and bio-fertilizers for small and marginal farmers, part replacement of inorganic fertilizers with bio-fertilizers for improving soil health, bio-pesticides safe for environment and humans, and applications of di-haploids, molecular markers and transgenics. While the benefits of these technologies have been proved beyond doubt, they are yet to be adopted on commercial scale. The reasons are many, such as bringing awareness through proper extension; adequate micro-financing avenues; crop insurance, entrepreneurship development for marketing of agro-inputs; value addition of farm produce so that farmer is compensated adequately; certification of bio-fertilizers and bio-pesticides; and organic certification of food. While, on one hand, our productivity gains have plateaued and are unable to keep pace with growing population, we are further challenged due to climate change and reduced availability of water. The increasing prices of food are further raising concerns on food security. The advancement in molecular tools has opened new avenues of designing future crops, be it in terms of molecular analysis for desirable trait (MAS) or introducing trait which was non-existent in the species of interest through transgenic technology. Agriculture requires much serious investments in research and ensuring that research results percolates down to the farmer thus ensuring that every citizen on this planet has access to adequate and safe food that is produced in sustainable way.

References

Ahmed RA (2001) Prospects of the rice economy in Bangladesh. University Press, Dhaka

Akhtar MR (2006) Impact of resource conservation technologies for sustainability of irrigated agriculture in Punjab-Pakistan. Pak J Agric Res 44(3):239–255

Blakelsee AF, Belling J, Farhnam ME, Bergner AD (1922) A haploid mutant in the Jimson weed, Datura stramonium. Science 55:646–647

Dhawan V (ed) (2008) Agriculture for food security and rural growth. TERI, New Delhi

Dogra B (1986) The Indian experience with large dams. In: Goldsmith E, Hildyard N (eds) The social and environmental effects of large dams, vol 2. Wadebridge Ecological Centre, London

Dorosh P (2006) Accelerating income growth in rural Bangladesh. World Bank, Washington DC

Duxbury JM (2001) Long-term yield trends in the rice-wheat cropping system: results from experiments in Northwest India. J Crop Prod 3(2):27–52

Erenstein O, Laxmi V (2008) Zero tillage impacts in India's rice-wheat systems: a review. Soil Tillage Res 100(1–2):1–14

Fan S (2007) Agricultural research and urban poverty in China and India. In: Adato M, Meinzen-Dick R (eds) Agricultural research livelihoods, and poverty: studies of economic and social impacts in six countries. Johns Hopkins University Press, Baltimore

Guha S and Maheshwari SC (1964). In vitro production of embryos from anthers of Datura. Nature, 204:497

Gupta R, Sayre K (2007) Conservation agriculture in South Asia. J Agric Sci 145(3):207–214

Gupta R, Seth A (2007) A review of resource-conserving technologies for sustainable management of the rice-wheat cropping systems of the Indo-Gangetic plaints (IGP). Crop Protect 26(3):436–447

Gupta RK, Naresh RK, Hobbs PR, Jiaguo Z, Ladha JK (2003) Sustainability of post-green revolution agriculture: the rice-wheat cropping systems of the Indo-Gangetic Plains and China. In: Ladha JK, Hill JE, Duxbury JM, Gupta RK, Buresh RJ (eds) Improving the productivity and sustainability of rice-wheat systems: issues and impacts. American Society of Agronomy, Crop Science Society of America and Soil Science Society of America, Madison

Harrington LW, Fujisaka S, Hobbs PR, Sharma HC, Singh RP, Chaudhary MK, Dhiman SD (1993) Wheat and rice in Karnal and Kurukshetra Districts, Haryana, India: Farmers' practices, problems and an agenda for action. Haryana Agricultural University, Indian Council of Agricultural Research, International Maize and Wheat Improvement Center and International Rice Research Centre, Mexico City

Hazell PBR, Haddad L (2001) Agricultural research and poverty reduction. In: 2020 Vision for food, agriculture and the environment. Discussion Paper 34. International Food Policy Research Institute, Washington, DC

Hazell PBR, Wood S (2008) Drivers of changes in global agriculture. Philos Trans R Soc Lond B Biol Sci 363(1491):495–515

Hobbs PR, Govaerts B (2009) How conservation agriculture can contribute to buffering climate change. In: Reynolds M (ed) Climate change and crop production. CAB International, Wallingford

Hobbs PR, Gupta RK (2003) Resource-conserving technologies for wheat and rice-wheat system. In: Ladha JK, Hill JE, Duxbury JM, Gupta RK, Buresh RJ (eds) Improving the productivity and sustainability of rice-wheat systems: issues and impacts. American Society of Agronomy, Crop Science Society of America, and Soil Science Society of America, Madison

Hussain M (2010) Shallow tube wells, boro rice, and their impact on food security in Bangladesh. In Spielman D j and Pandya-Lorch R (eds), Proven Successes in Agricultural Development. A technical compendium to millions fed. International Food Policy Research Institute, Washington, DC

Hossain M, Gollin D, Cabanilla V, Cabrera E, Johnson N, Khush GS, McLaten G (2003) International research and genetic improvement in rice: evidence from Asia and Latin America.

In: Evenson R, Golin D (eds) Crop variety improvement and its effect on productivity: the impact of international rice research. CAB International, Wallingford

Kataki PK (2002) Shifts in cropping system and its effect on human nutrition: case study. Indian J Crop Prod 6(1/2):119–144

Kataki PK, Hobbs P, Adhikary B (2001) The rice-wheat cropping system of South Asia: trends, constraints and productivity—a prologue. J Crop Prod 3(2):1–26

Korzun V (2003). Molecular markers and their applications in cereals breeding. A paper presented during the FAO international workshop on "Marker assisted selection: A fast track to increase genetic gain in plant and animal breeding?" 17–18 October 2003, Turin, Italy

Kumar P, Jha D, Kumar A, Chaudhary MK, Grover RK, Singh RK, Singh RKP, Mitra A, Joshi PK, Singh A, Badal PS, Mittal S, Ali J (2002) Economic analysis of total factor productivity of crop sector in Indo-Gangetic Plain of India by district and region, vol 2, Agricultural Economics Research Report. Indian Agricultural Research Institute, New Delhi

Ladha JK, Hill JE, Duxbury JM, Gupta RK, Buresh RJ (eds) (2003a) Improving the productivity and sustainability of rice-wheat systems: Issues and impacts. Madison: American Society of Agronomy, Crop Science Society of America, Soil Science Society of America

Ladha JK, Dawe D, Pathak H, Padre AT, Yadav RL, Singh B, Singh Y, Singh Y, Sinh P, Singh A, Kundu L, Sakal R, Ram N, Regmi AP, Gami SK, Bhandari AL, Amin R, Yadav CR, Bhattarai EM, Das S, Aggarwal HP, Gupta RK, Hobbe PR (2003b) How extensive are yield declines in long-term rice: Wheat experiments in Asia? Field Crop Res 81:159–180

Mäder P, Kaiser F, Adholeya A, Singh R, Uppal HS, Sharma AK, Srivastava R, Sahai V, Aragno M, Wiemken A, Johri BN, Fried PM (2011) Inoculation of root microorganisms for sustainable wheat-rice and wheat-black gram rotations in India. Soil Biol Biochem 43:609–619

Mandal MAS (2000) Dynamics of irrigation water market in Bangladesh. In: Mandal MAS (ed) Changing rural economy of Bangladesh. Bangladesh Economic Association, Dhaka

Palmer-Jones RW (1999) Slowdown in agricultural growth in Bangladesh: neither a good description nor a description good to give. In: Rogali B, Harriss-White BB, Bose S (eds) Sonar Bangla? Agricultural growth and agrarian change in West Bengal and Bangladesh Sonar Bangla? Agricultural growth and agrarian change in West Bengal and Bangladesh. Sage, New Delhi

Pathak H (2009) Greenhouse gas mitigation in rice-wheat system with resource conserving technologies. In: Proceedings of the 4th World congress in conservation agriculture, February 4–7, 2009. New Delhi, India. New Delhi: World Congress on Conservation Agriculture

Prasad R (2005) Rice-wheat cropping system. Adv Agron 86:255–339

Pretty JN, Noble AD, Bossio D, Dixon J, Hine RE, Penning de Vries FWT, Morrison JIL (2007) Resource conserving agriculture increases yields in developing countries. Environ Sci Technol 40(4):1114–1119

Rosegrant MW, Hazell PBR (2000) Transforming the rural Asia economy: the unfinished revolution. Oxford University Press, Hong Kong

Sarwar MN, Goheer MA (2007) Adoption and impact of zero tillage technology for wheat in rice-wheat system—water and cost saving technology. A case study from Pakistan (Punjab). Paper presented at the International Forum on Water Environmental Governance in Asia, March 14–15, Bangkok, Thailand

Singh RK, Hossain M, Thakur R (2003) Boro rice. International Rice Research Institute, New Delhi

Swaminathan MS (2010) From green to evergreen revolution Indian agriculture: performance and challenges. Academic Foundation, New Delhi

Timsina J, Connor DJ (2001) Productivity and management of rice-wheat cropping systems issues and challenges. Field Crop Res 69(2):93–132

Trigo EJ, Cap EJ, Malach VN, Villarreal F (2010) The case of Zero-tillage technology in Argentina. In: Spielman D J, Pandya-Lorch R (eds) Proven success in agricultural development: a technical compendium to millions fed: an IFPRI 2020 book, pp. 191–213

Vapnek J (2009) A health certification programme for citrus. FAO legal papers. Online # 81

Chapter 14
Arbuscular Mycorrhiza: Approaches for Abiotic Stress Tolerance in Crop Plants for Sustainable Agriculture

Rupam Kapoor, Heikham Evelin, Piyush Mathur, and Bhoopander Giri

1 Introduction

Plant growth and productivity is severely affected by various environmental stresses, such as, drought, salinity, and heavy metals worldwide. Global effects on desertification, soil salinization, scarcity of water resources, soil heavy metal contamination, and effects of other unfavorable factors are predicted to cause dramatic changes in the climatic conditions of arable lands, and together represent the primary cause of crop loss by more than 50% (Alcázar et al. 2006). Improving plant tolerance and maintaining crop productivity against such abiotic stresses is a major challenge for sustainable agriculture. Besides the intrinsic capacity of plants to tolerate abiotic stresses, most plants in nature establish a symbiosis with arbuscular mycorrhizal fungi (AMF) by which plants increase their tolerance to several stressful conditions thus, maintain growth and productivity (Ruiz-Lozano et al. 2006).

Arbuscular mycorrhiza is the symbiotic association between plant roots and fungi of the phylum Glomeromycota. Both of them derive benefits from the interaction; mycorrhizal fungi improve the nutrient status, water absorption, growth and enhance the host plant's resistance to biotic and abiotic stresses, while the host plant is necessary for fungal growth and reproduction by providing carbon in the form of photosynthates (Smith and Gianinazzi 1988; Smith and Read 1997). This interaction plays very important functions in plant ecosystem because more than 80% plant species depend on it for mineral uptake and their normal functioning (Remy et al. 1994; Smith and Read 1997).

R. Kapoor (✉) • H. Evelin • P. Mathur
Department of Botany, University of Delhi, Delhi 110 007, India
e-mail: kapoor_rupam@yahoo.com

B. Giri
Department of Botany, Swami Shraddhanand College, University of Delhi, Delhi 110 036, India

N. Tuteja and S. Singh Gill (eds.), *Plant Acclimation to Environmental Stress*, DOI 10.1007/978-1-4614-5001-6_14,

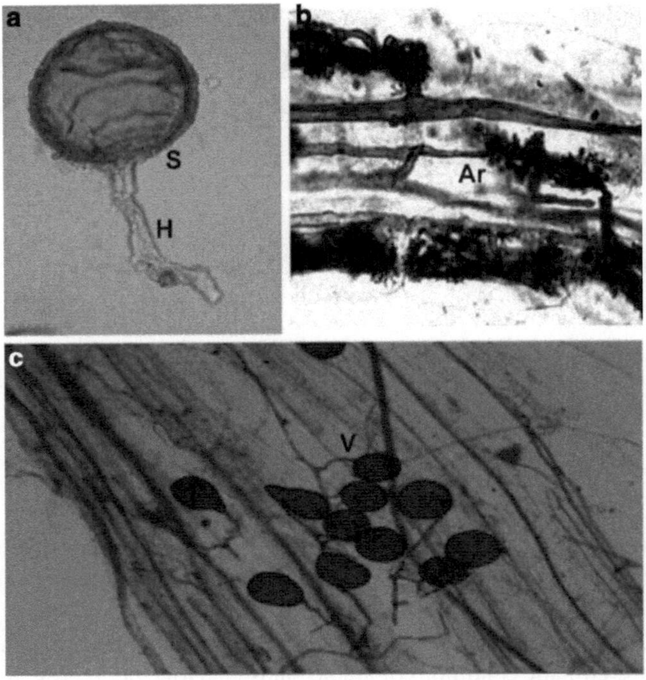

Fig. 14.1 AMF colonization in *Trigonella foenum-graecum*. The figure shows the different structures of AMF colonization: (**a**) a germinating spore (**b**) the arbuscules and (**c**) vesicles. The germinating spore colonizes the plant roots and forms arbuscules and vesicles in the cortical cell. The arbuscules are dichotomously branched tree like structures originating from branches of the intraradical hyphae after the branch hypha penetrates through the cortical cell wall. An arbuscule forms between the cell wall and plasma membrane and are the sites for exchange between the two symbiotic partners. Vesicles are thin-walled lipid-containing bodies produced terminally or intercalarily from hyphae in the root cortex and serve as the storage organ of the fungus

The three important components of mycorrhizal symbiosis are the root, the fungal intraradical structures in the root and an extraradical mycelium (ERM) in the soil. The intraradical structures formed by AMF in the host root cortical cells are intracellular hyphal coils and/or dichotomously branched structures—the arbuscules, hypertrophied hyphae—the vesicles (Fig. 14.1). Arbuscules are the sites of nutrient exchange between the symbionts and vesicles function as the fungal organs of storage (Smith and Read 1997). ERM surrounding the root is profusely branched absorbing hyphae, which form a network extending into the soil. ERM increases the total absorptive surface area of root and help in acquiring nutrients even beyond the depletion zone which develops around plant roots (George et al. 1992). The effect of AMF on phosphorus uptake is particularly large because of poor mobility of phosphate ions in soil (Hooker and Black 1995). AMF produces copious amounts of a stable hydrophobic glycoprotein called glomalin, which is deposited on the outer hyphal walls of the ERM and on adjacent soil particles and play an important role in soil structure stabilization (Wright et al. 1996; Wright and Upadhyaya 1998).

This involvement of fungal mycelium in soil conservation has a great relevance in sustainable systems (Barea and Jeffries 1995).

The development of a functional AM symbiosis is a multi-step process which is dependent on molecular cross talk between the two symbionts. It involves a sequence of recognition events leading to the morphological and physiological integration of these symbionts (Gianinazzi et al. 1995; Giovannetti and Sbrana 1998) (Fig. 14.1). The development of AM symbiosis involves induction of spore germination and hyphal growth (Becard and Piche 1990) contact between hyphae and root surfaces which includes recognition and formation of appresoria on the root epidermal cells (Giovannetti and Citernesi 1993; Bonfante and Bianciotto 1995); hyphal penetration into roots, which is characterized by localized production of wall-degrading hydrolytic enzymes by the fungus and by the exertion of hydrostatic pressure by the hyphal tip (Perotto et al. 1994; Giovannetti and Sbrana 1998); formation of arbuscules and establishment of a functional symbiosis (Samra et al. 1997).

Though, it is well established that AMF play very significant role in plant growth and productivity under abiotic stress conditions, however, due to their obligate nature the main obstacle in using AMF at large scale in agriculture is the lack of reliable and easily implementable methods for mass inoculum production. The most widely used standard and conventional method of maintaining and multiplying AMF is soil-based pot culture method. Besides, a few other methods such as inoculum rich soil pellets, aeroponic culture nutrient film technique, polymer-based inoculum, and root organ culture have also been introduced.

Under abiotic stress, colonization of plants with AMF assist plants to withstand these stresses. AMF enhance plant growth, productivity, and nutrient uptake under stress conditions. They also enhance osmolyte production, influence plant–water relation, and rate of photosynthesis, alter leaf water potential, ionic balance, antioxidant production and other physiological and biological parameters and thus improve plant's capacity to tolerate abiotic stresses. This chapter provides an overview of the mechanisms evolved by AMF to aid plants survive in these stressful conditions (drought, salinity, and heavy metals).

2 Salinity

Huge losses in arable land due to salinity are a major concern for sustainable agriculture. Salinity is a soil condition characterized by a high concentration of soluble salts. Soils are classified as saline when the electrical conductivity of the soil is 4 dS/m or more (Richards 1954), which is equivalent to approximately 40 mM NaCl and generates an osmotic pressure of approximately 0.2 MPa (Munns and Tester 2008). Salinity arises due to deposition of salts via two natural processes—weathering of rocks containing soluble salts of various types, mainly chlorides of sodium, calcium, and magnesium, and to a lesser extent, sulfates and carbonates; and deposition of oceanic salt (mainly NaCl) carried by inland wind and rain (Munns and Tester 2008). Thus, the major ions contributing to soil salinity include cations

(Na^+, Ca^{2+}, Mg^{2+}, and K^+), anions (Cl^-, SO_4^{2-}, HCO_3^-, CO_3^{2-}, and NO_3^-). Other constituents contributing to salinity in hyper saline soils and water include B, Sr^{2+}, SiO_2, Mo, Ba^{2+}, and Al^{3+} (Hu and Schmidhalter 2002). Besides, irrigation water and fertilizers used in agriculture, low precipitation, and over-exploitation of available water resources also contribute significantly to soil salinity (Canterll and Linderman 2001; Al-Karaki 2006). Of all the salts, sodium chloride is the most soluble and abundant salt released (Munns and Tester 2008). At present, cultivated land affected by salt amounts to 77 million hectares and constitutes 5% of 1.5 billion hectares cultivated land around the world. It is projected that increased salinization of arable land will result in to 50% land loss by the middle of the twenty-first century (Wang et al. 2003). The alarming rate of increase in soil salinity in agricultural land creates a distress to agriculture as most of the economically important crop species are very sensitive to soil salinity (Mahajan and Tuteja 2005) and have resulted in decreased crop production of more than 20% irrigated land worldwide (Porcel et al. 2012).

Excessive salts in soil, in particular, Na^+ ions alter the basic structure of the soil (Mahajan and Tuteja 2005). The presence of Na^+ ions in the cation exchange complex render the soil compact and subsequently reduce soil porosity and hamper soil aeration (Manchanda and Garg 2008). Low soil aeration due to high salt concentration has a direct relation with all major living processes, such as reduction in growth, photosynthesis, protein and lipid metabolism (due to salt-induced osmotic imbalance), nutritional disorder, and ion toxicity in plants (Evelin et al. 2012). Osmotic imbalance in a salt stressed plant is often manifested as retardation in growth of the plants with the leaves and stems appearing stunted (Singh and Charath 2001). This effect of salt is primarily due to (1) decrease in the plant's ability to take up water and nutrients as a result of osmotic or water-deficit (physiological drought) effect of salt (Evelin et al. 2009) and; (2) uptake of salt by plants from the soil through transpiration stream which injure cells in the transpiring leaves thereby inhibiting cell division and enlargement in plant's growing point (Manchanda and Garg 2008). Diversion of energy to counteract the accumulation of salts in the cells may also contribute to the stunted growth of the plants grown in saline soils (Evelin et al. 2009). Continued uptake of salt by plants and subsequent significant increase in the concentration of salts decreases the size of the leaves (Singh and Charath 2001) and induce leaf senescence by affecting the structure of chlorophyll molecules and photosynthesis. Specific effects of salt stress on leaf senescence have been related to accumulation of toxic ions (Na^+ and Cl^-) or to K^+ and Ca^{2+} depletion (Yeo et al. 1991). Salt-induced ionic imbalance and toxicity is perceptible in the form of disruption in the plant mineral relations. This may be explicated by the effects of salinity on nutrient availability, competitive uptake, transport or partitioning within the plant or may be caused by physiological inactivation of a given nutrient resulting in an increase in the plant's internal requirement for that essential element thereby causing ionic imbalance in the cell (Grattan and Grieve 1999). At the whole plant level, salinity frequently induces an increase in Na^+ and Cl^- ions as well as a decrease in K^+, Ca^{2+}, NO_3^-, and Pi concentrations (Shokri and Maadi 2009). Therefore, high concentrations of Na^+ and Cl^- ions in the soil solution may depress nutrient-ion activities and produce extreme ratios of Na^+: K^+, Na^+: Ca^{2+}, Ca^{2+}: Mg^{2+}, and Cl^-: NO_3^- (Evelin et al. 2012).

3 Arbuscular Mycorrhiza in Mitigation of Salt Stress

The role of AMF in mitigation of the effects of salt stress in plants is widely recognized. Arbuscular mycorrhizal fungi stimulated alleviation of salt stress effects and tolerance of host plants is brought about by combining nutritional (enhancing/selective uptake of nutrients and prevention of nutritional disorder), biochemical (accumulation of osmoregulators, control of reactive oxygen species (ROS) and enhanced activities of antioxidant enzymes and molecules), physiological (photosynthetic efficiency) and structural adaptations. Recent molecular studies have indicated that the regulation of AM-induced plant salt tolerance may involve mechanisms at the molecular level. Figure 14.2 illustrates the mechanisms of AM-mediated improved tolerance of host plants in salt and drought stress.

3.1 Enhance Growth and Biomass

Salinity-induced retardation of growth and fitness of plants has been shown to be alleviated by AM colonization. Under salt stress, plant growth and biomass, measured as indicators of plant fitness are higher in AM than the non-AM plants (Giri et al. 2007; Shokri and Maadi 2009; Evelin et al. 2012; Latef and Chaoxing 2011). The increased dependency of host plants on AMF under salt stress is an affirmation that AMF enhance plant growth (Borde et al. 2011; Kumar et al. 2011). Better growth of AM plants as compared to non-AM plants under salt stress is attributed to improved nutrient uptake, especially phosphorus (P), and other nutrients, mycorrhiza-mediated effects on water absorption and increased photosynthetic system (Giri et al. 2003; Garg and Manchanda 2009; Abdel-Fattah and Asrar 2011; Evelin et al. 2012; Kumar et al. 2011).

3.2 Prevention of Nutrient Deficiency and Ion Toxicity

Plants, for their growth and survival require 14 essential nutrients of which the macro-nutrients nitrogen (N), phosphorus (P), potassium (K), calcium (Ca), magnesium (Mg) and sulfur (S) are present in large amounts in tissues. The micronutrients (copper (Cu), iron (Fe), molybdenum (Mo), manganese (Mn), zinc (Zn)), though required in very small amount by the plants, are indispensable for plant growth and survival. Under salt stress, the plants are deprived of these nutrients due to the effects of salinity on availability, uptake, transport or physiological inactivation of a given nutrient (Grattan and Grieve 1999). Several studies have shown that the effects of salt-induced ion toxicity and nutrient deficiency can be ameliorated if the plants are colonized by AMF. The ERM of AMF has the ability to explore and exploit the growth medium thereby facilitating the uptake of nutrients.

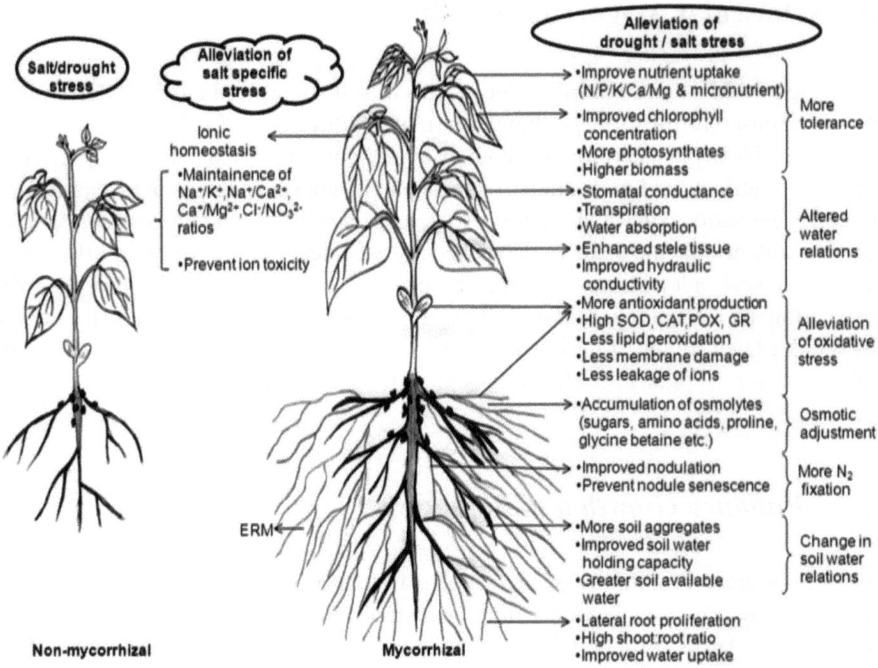

Fig. 14.2 Role of AMF colonization in alleviation of salt and drought stress. Under salt and drought stresses, the plants experience similar physiological effects. The plant on the *left* side depicts a salt/drought-stressed plant and shows lesser growth and biomass due to lower photosynthetic efficiency, greater degree of oxidative damage, lesser uptake water and nutrients and osmotic imbalance. On the *right* side is depicted an AMF-colonized plant under salt and drought stress. The presence of AMF in the plant roots and rhizosphere improves the quality and water holding capacity of the soil thereby altering soil-water relations. AMF colonization also induces lateral root proliferation and aids in uptake of water and nutrients. This is also facilitated by sturdier vascular system in AM plants which possess a higher rate of transpiration and stomatal conductance; thus improving water use efficiency. Better nutrient uptake in AM plants prevents nutrient deficiency. In plants under salt stress, better nutrient uptake also resulted in prevention of ion toxicity and maintenance favorable ionic ratios. There is higher accumulation of osmolytes in AM plants and to counteract the osmotic imbalance generated due to salt and drought stress. The oxidative damage under stress is also better prevented in AM plants by activation of antioxidant enzymes and molecules thereby preventing membrane and cellular damage preventing the leakage of ions. Lower damage in photosynthetic apparatus results in a higher rate of photosynthesis which leads to higher biomass, shoot: root ratio and leaf area, which in turn is responsible for more photosynthates. In legumes, AMF colonization improved nodulation and prevent nodule senescence thereby improving N_2 fixation

Furthermore, the ERM can also detect and show physiological plasticity in response to the nutrient status of their environment and host (Hodge et al. 2010). It is by virtue of this property of ERM that AMF pursue improved and/or selective uptake of nutrients as one key mechanism to prevent nutritional disorder and ionic imbalance in host plants in saline soils. The runner hypha radially extends the AMF colony and upon perception of signals from the nutrient ions, it produces branched absorbing structures (BAS) or spores which absorb the nutrients and translocates them to the plants (Hodge et al. 2010). Mycorrhizal plants can therefore potentially access nutrients from a larger area than the non-mycorrhizal (NM) equivalents So, under salt stress, when the availability of the nutrients to plants are limited due to physiological drought, AM colonization offers huge benefit to host plants by improving the uptake of essential nutrients.

Arbuscular mycorrhizal fungi facilitate the uptake of N, P, K, Ca, Mg, Cu, Fe, Zn while successfully limiting the uptake of Na^+ and Cl^-. This buffering activity of AMF aids in preventing salt-induced ion toxicity while alleviating nutrient deficiency and other related cellular effects. For example, improved N and Mg uptake prevents degradation of chlorophyll and protein by Na^+. Higher P and Ca relieve the membrane system of the plant from the possible attack by ROS that cause peroxidation of the membrane lipids. While the maintenance of membrane integrity in the tonoplast imparts successful compartmentalization of excess Na^+ and Cl^- ions in the vacuole, the integrity of plasma membrane prevents leakage of cellular contents. In this way, ion toxicity is contained and cellular damage is prevented. The compartmentalization of Na^+ and Cl^- ensures higher K^+:Na^+, Ca^{2+}:Na^+, and NO_3^-:Cl^- ratios in the plant indicating smooth metabolic processes in the plant. Higher P-concentration in tissues also induces uptake and translocation of micro-nutrients by increasing the sink size in plants (Liu et al. 2000b). Understanding and elucidation of the above mechanism is a result of extensive investigation.

At the moment, the explanation of the molecular regulation of the mechanisms of AM-regulated uptake of nutrients is not complete. An encouraging point is that significant breakthroughs have been made in this aspect in the last few years. For instance, it has been proved that N absorbed through AMF constitutes a fifth of plant N (Leigh et al. 2009). Nitrogen is taken up by ERM as inorganic N from the soil in the form of nitrate and assimilated via nitrate reductase located in the arbuscule-containing cells (Kaldorf et al. 1998), and converted to arginine via the GS-GOGAT (glutamine synthetase-glutamine: 2-oxoglutarate amidotransferase) cycle. Arginine in the hyphae is broken down to urea and ultimately transferred to the plant as NH_4^+ with the resulting C skeletons from arginine breakdown re-incorporated in to the fungal C pools (Bago et al. 2002; Govindarajulu et al. 2005). The identification of ammonium transporter gene in ERM of *Glomus intraradices* (López-Pedrosa et al. 2006), a mycorrhiza-specific plant ammonium transporter in *Lotus japonicus* that is expressed in arbusculated cells (Guether et al. 2009), and up-regulation of an ammonium transporter in *Medicago truncatula* (Gomez et al. 2009) makes the mechanism clearer.

Similarly, in the case of P, high affinity phosphate transporters in ERM take up P and transport it within the fungus as polyphosphate, and once in the intraradical hyphae the long chains are hydrolyzed, facilitating transfer to the host plant via a phosphate transporter (Harrison and van Buuren 1995; Harrison 1999; Bago et al. 2002; Ohotomo and Saito 2005). Adding up, higher affinity for phosphate ions, lower threshold concentration for P, ability to store larger amounts of absorbed P than the plant roots also aids the continued movement of P by AM symbiosis into the hyphae (Bolan 1991).

AM-mediated maintenance of higher K^+:Na^+ ratio in host plants is accomplished by regulating the expression and activity of K^+ and Na^+ transporters and of H^+ pumps that generate the driving force for transport of ions (Parida and Das 2005). The Na^+/H^+ antiporter catalyze the transfer of Na^+ out of the cytoplasm into either vacuole or apoplasm (Ouziad et al. 2006). Though Ca^{2+}:Na^+ ratio has been shown to be influenced by AM colonization, the molecular mechanism is yet to be deciphered.

Compared to macro-nutrients, studies on the regulation of the mechanism of micro-nutrients uptake is rather rare. Till date, there is only a single report for a putative zinc transporter (González-Guerrero et al. 2005) in AMF. This lack of knowledge masks the AM-mediated uptake of micro-nutrients, therefore deserves extensive investigations.

3.3 Osmotic Adjustment

One of the mechanisms on how AMF impart tolerance to host plants is the adjustment of salt-induced osmotic imbalance. Osmotic adjustment is brought about by the accumulation of metabolites that act as compatible solutes. The compatible solutes, also called osmolytes, are so named because they do not interfere with normal biochemical reactions; rather they replace water in biochemical reactions. Frequently investigated osmolytes include proline, glycinebetaine, sugars, and polyols (Hasegawa et al. 2000; Parida and Das 2005).

The primary role of these osmolytes in plant cell is to adjust the osmotic potential of the soil with respect to the surroundings. However, these osmolytes are also bestowed with various other functions in plants. For instance, proline can also act as a signaling molecule and influence defense pathways, regulate complex metabolic and developmental processes, offers additional opportunities for plant improvement (Szabados and Savoure 2009).

Glycinebetaine, the non-toxic cellular osmolyte enhances tolerance to salt stress by various ways: (1) stabilize the structures and activities of enzymes and protein complexes; (2) maintain the integrity of membranes against the damaging effects of excessive salt; (3) protection of the photosynthetic machinery; (4) induction of specific genes whose products are involved in stress tolerance; (5) reductions in levels of ROS under stress; and (6) regulation of the activities of ion-channel proteins either directly or via protection of the plasma membrane (Gorham 1995; Chen and Murata 2008).

In AM plants, often the accumulation of proline, glycinebetaine, sugars, organic acids, and polyols is higher as compared to their non-AM equivalents under salt stress (Al-Garni 2006; Sannazzaro et al. 2007; Garg and Manchanda 2009; Sheng et al. 2011). Higher accumulation of osmolytes leads to better osmo-regulation. However, a few studies have reported lower proline accumulation in AM plants as compared to non-AM plants and may be explained due to stress avoidance (exclusion of salt). The variable effects of mycorrhizal colonization on proline levels of plants under salt stress may be related to the differences between plant species and effects of different composition of the inoculum.

The increase in total sugars in mycorrhizal plants under salt stress has also been related to improved osmotic adjustment. Sugars can prevent structural changes in soluble protein, maintain the osmotic equilibrium in plant cells, and protect membrane integrity (Abd-El Baki et al. 2000). The enhanced sugar accumulation in AM plants may be explained by the sink effect of fungus demanding sugars from the shoot tissues (Augé 2001), increased rates of photosynthesis and of carbon compounds to the root system, hydrolysis of starch, higher concentration of organic acid in AM plants (Sheng et al. 2011).

While modulation of polyamine pools may be projected as one of the mechanisms used by AMF to improve plant adaptation to saline soils (Sannazzaro et al. 2007), organic acids besides osmotic adjustement are credited with balancing cation excess and pH homeostasis (Hatzig et al. 2010). Moreover, high amount of organic acids, especially malic acid, can enhance sugar synthesis through facilitation of CO_2 delivery to the Calvin cycle (Chollet et al. 1996).

3.4 *Maintenance of Reactive Oxygen Species Level*

AM-mediated salt tolerance by host plants has also been attributed to its ability to detoxify ROS and maintain the delicate balance between ROS and antioxidants. ROS is an inevitable byproduct of plant metabolism produced essentially from photosynthesis, photorespiration and respiration. However, salt stress encourages the production of ROS in excess. While at low concentrations, ROS are required for signaling, growth and behavior, high concentrations of ROS is a threat to cells as it causes membrane lipid peroxidation, protein oxidation, enzyme inhibition and DNA and RNA damage (Miller et al. 2010; Singh et al. 2011). Enhanced generation of ROS under salt stress is accomplished in four ways: first, plants responds to salt stress by decreasing stomatal conductance to avoid excessive water loss. This in turn decreases the internal CO_2 concentrations (Ci) and slows the reduction of CO_2 by Calvin cycle. This response leads to depletion of oxidized $NADP^+$, which acts as the final acceptor of electrons in PSI, and alternatingly increases the leakage of electrons to O_2 forming O_2^- (Hsu and Kao 2003). Furthermore, Na^+ or Cl^- toxicity resulting from salt stress could disrupt the photosynthetic electron transport and provoke electron leakage to O_2 (Borsani et al. 2001; Slesak et al. 2002). Second: the decrease in C_i slows down the reactions of Calvin cycle and induces photorespiration particularly in C_3 plants, resulting in generation of more H_2O_2 in the peroxisome (Wingler et al. 2000; Ghannoum 2009). Third: the cell membrane-bound NADPH oxidase and the apoplastic diamine oxidase get activated during salt stress and therefore contribute to generation of ROS (Hernandez et al. 2001; Mazel et al. 2004; Tsai et al. 2005). Fourth: salt stress increases the rate of respiration resulting in higher respiratory electron leakage to O_2 (Fry et al. 1986; Moser et al. 1991; Jeanjean et al. 1993).

AM colonization restricts the excessive generation of ROS by enhancing the activities of antioxidant enzymes (superoxide dismutase (SOD), catalase, peroxidase

(POX), ascorbate peroxidase, glutathione reductase) and antioxidant molecules (carotenoids, ascorbic acid, glutathione, tocopherols). The benefits of AM symbiosis in containing the levels of ROS with concomitant enhanced activities of antioxidant enzymes and antioxidant molecules have been shown by many researchers (Wu et al. 2010b; Hajiboland et al. 2010; Borde et al. 2011; Singh et al. 2011). Mycorrhizal plants have lower concentration of malondialdehyde concentration while maintaining higher activities of SOD, catalase (CAT), POX, ascorbate peroxidase (APOX) activities than in non-mycorrhizal plants (Wu et al. 2010b; Latef and Chaoxing 2011). The increased SOD will help detoxify super oxide ($O_2^{\cdot-}$) to hydrogen peroxide (H_2O_2) (Smirnoff 1993). This H_2O_2 generated are scavenged by CAT and POD. APOX is also reported to involve in detoxification on hydrogen peroxide produced in the chloroplasts of the stressed host plants (Benavides et al. 2000; Lopez et al. 1996). The elevated levels of GR activity may serve to ensure the availability of $NADP^+$ to accept electrons derived from photosynthetic electron transport, thereby directing electrons away from oxygen and minimizing the production of $O_2^{\cdot-}$ (Gamble and Burke 1984; Menconi et al. 1995). Thus, AM-mediated rapid removal of excess ROS helps in maintaining the optimum concentration of ROS to perform its physiological role; at the same time, preventing the shift towards destructive mode.

3.5 *Prevention of Membrane Damage*

It is now becoming clearer that the presence of AMF in roots of plants help in maintaining the integrity and stability of plasma membrane under saline stress. Under NaCl stress, lower lipid peroxidation and concurrent lower electrolyte leakage were reported in mycorrhizal maize, pigeon pea, and fenugreek as compared to their non-AM counterparts (Feng et al. 2002; Garg and Manchanda 2009; Evelin et al. 2012). It is suggested that AM-conferred resistance to peroxidation of membrane lipids and hence leakage of electrolytes from the cell, mediated through improved nutrition, especially P and maintenance of higher Ca^{2+}:Na^+, are the key factors contributing to this beneficial effect of AM colonization on membrane integrity (Evelin et al. 2012). Other mechanisms include maintenance of higher antioxidant capacity and containment of ROS concentration in mycorrhizal plants as against non-AM plants.

3.6 *Higher Photosynthetic Efficiency*

Salinity-induced degradation of chlorophyll molecule and subsequent decrease in photosynthesis is major process driving the plants to die prematurely. On the positive side, these salt-induced toxic effects on photosynthetic machinery of the plant may be restricted or prevented by inoculating plant with AMF. In this regard, many authors have demonstrated that AM plants had higher chlorophyll concentration and non-photochemical quenching capacity as compared to their non-AM equivalents (Sheng et al. 2008; Hajiboland et al. 2010; Kumar et al. 2011; Evelin et al. 2012). Arbuscular mycorrhiza-facilitated improved uptake of nutrients (higher K^+, Ca^{2+}, and Mg^{2+} ions) helps in avoiding the specific effects of salt stress on chlorophyll degradation and leaf senescence (Evelin et al. 2012). Furthermore, the ability of AM plants to regulate the energy bifurcation between photochemical and non-photochemical events also helps in maintaining photosynthesis (Sheng et al. 2008).

3.7 Improved Water Status

Arbuscular mycorrhizal colonization and the effective development of external mycelium is an important means for uptake of water in plants grown in saline soils. Under salt stress, mycorrhizal colonization have shown to provide better water status by maintaining higher relative water content over the non-AM plants (Aroca et al. 2006; Colla et al. 2008; Jahromi et al. 2008; Sheng et al. 2008). Improved water status due to AM colonization may also be attributed to its role in ensuring liquid continuity, high hydraulic conductivity of roots and hence water uptake (Smith et al. 2010). Since hydraulic conductivity is dependent on P-concentration (Carvajal et al. 1996), it is likely that water uptake would be more strongly expressed in P-sufficient AM plant than in P-deficient non-AM plants. Mycorrhizal plants are also shown to accumulate solutes and maintain the osmotic balance. Lower water saturation deficit and higher turgor potential in AM plants also improves the water status of the plant (Al-Garni 2006; Sheng et al. 2008).

Various studies have shown better water status in AM plants than their corresponding non-AM plants, however, it is still unclear of how water from the AM fungus is translocated to the plant system. The possibility of direct water transfer to plants via fungal hyphae have been put forward, however the idea remains controversial to be established (Smith et al. 2010). This gap in our knowledge is surprising, given the importance of AM-mediated uptake of water in plants under salt stress. However, a positive step towards bringing down this gap is the recent discovery of an aquaporin gene *GintAQP1* from *Glomus intraradices* (Aroca et al. 2009), the expression of *GintAQP1* is a compensatory alternative to plant aquaporins (Aroca et al. 2009). Aquaporins belong to the major intrinsic protein (MIP) family of transmembrane channels, which permit the selective membrane passage of water (and other few compounds) but not of H^+ and other ions (Chen et al. 2001; Hill et al. 2004) through the plasma lemma (by plasma lemma intrinsic proteins (PIPs)) and the tonoplast (by tonoplast intrinsic protein (TIPs)). Though a few studies have shown AMF influence on plant aquaporins (Ouziad et al. 2006; Aroca et al. 2007; Jahromi et al. 2008), these findings are not persuasive enough to establish and characterize the role of aquaporins and point to the possibility that AMF differentially exert control on each PIP (plasma membrane intrinsic protein) gene and each *PIP* gene analysed may have a different function and regulation in AM-mediated alleviation of water stress.

3.8 Abscissic Acid Concentrations

Plant growth and response to a stress condition is largely under the control of hormones (Mahajan and Tuteja 2005). Under salt stress, the increase in transpiration results in increase in pH of leaf and accumulation of ABA. ABA, in turn promotes the efflux of K^+ ions from the guard cells, resulting in loss of turgor leading to stomatal closure. On the other hand, the analogous findings that AM colonization can alter the ABA levels in host plant, sponsor for an AM-induced regulatory mechanism for ABA accumulation (Duan et al. 1996; Ludwig-Müller 2000; Estrada-Luna and Davies 2003). AM-mediated higher ABA levels regulate free polyamine pools in the plant (Sannazzaro et al. 2007). However, in contrast to this report, Jahromi et al. (2008) reported lower ABA levels in *Glomus intraradices* colonized lettuce plants than the non-AM plants indicating that AM plants were less strained than non-AM plants by salinity stress imposed; hence, they accumulated less ABA.

3.9 Nodulation and Nitrogen Fixation

Plant inoculation with AMF has been shown to promote nodulation and prevent premature nodule senescence during salt stress (Garg and Manchanda 2009; Manchanda and Garg 2011; Evelin et al. 2012). Pre-mature nodule senescence during salt stress is induced due to acceleration of lytic activities, formation of green pigments from leghaemoglobin (Sarath et al. 1986) and loss of nitrogen fixation (Delgardo et al. 1994). In AM plants, higher leghaemoglobin concentration delays the change of colour in nodule from pink to brownish pink due to synthesis of green pigments from leghaemoglobin. Mycorrhizal plants also possess a higher nitrogenase activity. All these parameters contribute to a higher nitrogen fixing ability of AM plants. The increased in nitrogenase activity and nitrogen fixation in AM plants than non-AM plants has been attributed to relief from P-stress, which is beneficial for the functioning of nitrogenase enzyme of the bacterial symbionts and possibly due to uptake of some essential micro-nutrients which results both in improved growth of plants (Founoune et al. 2002) or vice versa (Rabie and Almadini 2005). Therefore it may be suggested that mycorrhizal and nodule symbioses often act synergistically on infection rate, mineral nutrition, and plant growth (Patreze and Cordeiro 2004), which supports the need for both N and P and increased tolerance of plants to salinity stress (Rabie and Almadini 2005).

3.10 Molecular Mechanism

The plant function is ultimately explained by operation of genes in cells and tissues to regulate its growth with a symbiont in coordination with the environmental stress. Notable absence of knowledge regarding the molecular basis for AM-induced tolerance to salt stress in plants indicates that study aimed at molecular levels is at nascent stage. Only a handful of reports in molecular studies (Harrison and van Buuren 1995; González-Guerrero et al. 2005; Ouziad et al. 2006; Aroca et al. 2007; Jahromi et al. 2008; Aroca et al. 2009; Guether et al. 2009), are a testimony to the declaration. Though molecular studies are racing at a fast pace in other plant biology/microbiology studies, the AM plant molecular studies are often impeded due to the obligate and heterokaryotic nature of AMF. So far, the scientists have concerted their efforts in unraveling the mechanisms for nutrient and water uptake in plants. In this regard, expression analyses of plant aquaporins have been studied in tomato (Ouziad et al. 2006), *Phaseolus vulgaris* (Aroca et al. 2007) and lettuce (Jahromi et al. 2008) (as described in the section on improved water status). Recently, an aquaporins gene (*GintAQP1*) from *Glomus intraradices* was reported (Aroca et al. 2009). This is a step nearer to elucidation of water uptake mechanism in AM symbioses; however, it deserves more research to reach a conclusion.

Molecular studies on AM plant mineral uptake have revealed a phosphate transporter (Harrison and van Buuren 1995), an ammonium transporter (López-Pedrosa et al. 2006; Guether et al. 2009), a putative zinc transporter (González-Guerrero et al. 2005) in AM fungus. The role of AMF in nutrient uptake at the molecular level is only beginning to be understood. Expression studies have been conducted for Na^+/H^+ antiporters—*LeNHX1* and *LeNHX2* in tomato (Ouziad et al. 2006). The authors reported that salt and mycorrhizal colonization had no significant effects on these two antiporters.

In another study, Jahromi et al. (2008) analysed the expression pattern of genes encoding Δ1-pyrroline-5-carboxylate synthetase (*LsP5CS*), late embryogenesis abundant protein (*LsLea*) and ABA (*Lsnced*) in mycorrhizal and non-mycorrhizal lettuce plants subjected to varied salt treatments (0–100 mM NaCl). The *PC5S* enzyme catalyzes the rate-limiting step in the biosynthesis of proline (Kishor et al. 1995), an osmoregulator in plants. Late embryogenesis abundant protein acts as stress markers. They also possess chaperone-like activity, thus having a protective role during osmotic stress. *Lsnced* encodes for 9-cis-epoxycarotenoid dioxygenase, a key enzyme for the biosynthesis of stress hormone ABA. ABA promotes stomatal closure to minimize transpirational water loss. It also mitigates stress damage through the activation of many stress-responsive genes, which collectively increase plant stress tolerance (Bray 2002). The authors reported a higher expression of genes *LsP5CS* and *Lsnced* in non-AM plants than AM plants at 50 mM NaCl, though at 100 mM, the levels were similar. *LsLea* gene was found to express under conditions of salt stress and the induction of this gene was found to be lower in AM plants than non-AM plants. The lower expression of this gene suggests that AM plants suffer less stress than non-AM plants, which may be likely due to primary salt avoidance mechanism.

4 Drought

Available water resources for successful crop production have been decreasing in recent years rendering the lands more drought-prone. Furthermore, in view of various climate change models scientists suggest that in many regions of world, crop losses due to increasing water shortage will further aggravate its impacts (Anjum et al. 2011). Drought is the most important environmental stress and affects plant performance more than any other environmental factor. It severely impairs growth and development and limit plant production (Shao et al. 2009).

Plant experiences drought stress either when the water supply to root becomes deficient or transpiration rate becomes very high. Drought affects growth, yield, membrane integrity, pigment content, osmotic balance, water relations, and photosynthetic activity (Benjamin and Nielsen 2006; Praba et al. 2009). The susceptibility of plants to drought stress varies depending on stress degree, different accompanying stress factors, plant species, and their developmental stages (Demirevska et al. 2009). Acclimation of plants to water deficit is the result of different events, which lead to adaptive changes in plant growth and physio-biochemical processes, such as changes in plant structure, maintenance of growth rate, tissue osmotic potential, and antioxidant defenses (Duan et al. 2007).

5 Arbuscular Mycorrhiza in Mitigation of Drought Stress

AM symbiosis is known to complement plant's innate ability to tolerate drought stress and alleviate stress symptoms (Fig. 14.2). A number of studies have demonstrated that AM symbiosis can protect the host plants against the detrimental effects of drought stress (Rodriguez and Redman 2005; Cho et al. 2006; Subramanian et al. 2006; Augé et al. 2007; Wu et al. 2007; Aroca et al. 2008; Wu et al. 2008; Manoharan et al. 2010; Fan and Liu 2011).

5.1 Enhanced Growth and Nutrient Uptake

The most obvious explanation for AM-induced changes in plant water balance and drought resistance is indirect effects associated with changes in plant size and phenology that occur via increased acquisition of phosphorous and sometimes other nutrients (Augé et al. 2004). AM symbiosis generally leads to greater plant biomass and altered root-shoot; root length-leaf area ratios, enhanced plant growth, and nutrient uptake ratios (Al-Karaki et al. 2004). The size of a plant can affect its water relations; larger plants with larger root system may have access to more extensive soil water reserves. The favorable contribution of AMF inoculation on plant growth may be attributed to the following facts (1) AMF inoculation leads to formation of AM, which might have enhanced water uptake from the soil in the host plant (Augé 2001). It has been well documented that AM formation results in an ecological niche for roots to be more accessible to water resources, as the fungal hyphae can penetrate soil pores inaccessible to the root hairs (Ruiz-Lozano 2003) (2) Enhanced plant growth might also result from soil-borne pathogen protection (St-Arnaud and Vujanovic 2007), improved soil structural development, aggregate stabilization (Rillig 2004; Wright 2005) (3) coupled with the water absorption, nutrient uptake in the host plants may be facilitated in a better manner than in the non-mycorrhizal one (Ruiz-Lozano 2003).

5.2 Prevention of Nutrient Deficiency

AM symbiosis has been shown to improve the acquisition of nutrients including phosphate, nitrogen, sulfur or even more trace elements like copper and zinc (Subramanian et al. 2006; Cavagnaro et al. 2010; Tian et al. 2010; Latef and Chaoxing 2011). Since nutrient mobility is limited under drought conditions, AMF may have a larger impact on overall plant growth and development in dry relative to well-watered conditions (Sánchez-Díaz and Honrubia 1994). AM hyphae have the potential to access nutrients from drier areas. For instance, phosphorous becomes less mobile in arid soils, and an enhanced P acquisition by AM would hence become more important in improving water relations of host plants (Augé et al. 2004; Allen 2006, 2007). Soil P is usually in the form of orthophosphate that may be directly absorbed at soil–root interface through root epidermis and hairs, and indirectly at the fungal-root interface through external AM hyphae (Garg et al. 2006; Requena et al. 2007). In addition, AM colonization has been found to be related with the increase in activities of certain enzymes that help in hydrolysis and mobilization of nutrients. Often, P forms complexes with Ca and Mg rendering P unavailable for uptake. Higher acid phosphatase in mycorrhizosphere as compared to rhizosphere of non-AM plants enables the hydrolysis and mobilization of P_i releasing P for uptake. Higher soil acid phosphatase activities in mycorrhizosphere were also found to be positively correlated to soil water content (Chethan Kumar et al. 2008; Sardans et al. 2008). Thus, increase of soil acid phosphatase mediated by AMF partially alleviates plant drought stress (Wu et al. 2011).

During periods of drought, N availability is reduced resulting in decreased N uptake and lower rates of N assimilation. In plants N assimilation begins with the reduction of (nitrate) NO_3^- to (nitrite) NO_2^- by nitrate reductase (NR). This step often serves as the rate-limiting step in the assimilation process and is drastically slowed by water stress (Ruiz-Lozano 2003). Colonization by AMF improves both the nutritional status and N-assimilation rate of drought-stressed plant (Subramanian and

Charest 1998; Boomsma and Vyn 2008). Increase in N assimilation may result from the direct uptake of NO_3^- or NH_4^+ by hyphae (Cardoso and Kuyper 2006). This may contribute to greater protein concentration in AM plants over non-AM plants under drought stress. Stress also affects the rate of catalysis of downstream NH_4^+-assimilation enzyme glutamine synthetase (GS) and glutamine synthase (GOGAT). Drought-stressed AM shoots and roots of maize cultivars exhibited higher (NR), GS and GOGAT activities than water-stressed non-AM shoots and roots (Subramanian and Charest 1998). More robust uptake of water and nutrients might provide sufficient substance necessary to maintain a well defined growth in AM plants.

5.3 Osmotic Adjustment

One of the significant responses to drought is the accumulation of compatible solutes, also known as osmolytes, which function for osmotic adjustment, so as to maintain favorable gradient for water flow from soil into roots (Ruiz-Lozano 2003). Osmotic adjustment in plants specifically involves the active accumulation of various ions, amino acids, and sugars. It is primarily necessary for the maintenance of turgor, cellular expansion and growth, stomatal opening, photosynthesis, and water influx during water stress (Chaves et al. 2003; Ruiz-Lozano 2003). Changes in amino acid concentrations in response to drought in AM colonized plants are variable with some studies reporting increased levels (Ogawa and Yamauchi 2006b) and others reporting the opposite (Augé 2001). The greater concentration of amino acids in AM roots and shoots indicates a greater capability of osmotic adjustment through amino acid accumulation in these plants. Proline has been regarded as a key osmolytes participating in osmotic adjustment (Molinari et al. 2007; Hassine et al. 2008). In plants exposed to water stress, proline often serves as an osmoprotectant, as a solute for the protection of proteins and enzymes from denaturation, as a hydroxyl radical scavenger, as an alleviator of cell acidity and as a sink for energy to control redox potential (Chaves et al. 2003; Ruiz-Lozano 2003). AM plants have been reported to accumulate significantly higher level of proline upon exposure to drought stress when compared with non-mycorrhizal plants. Through osmotic adjustment via sugar accumulation, plant typically maintains turgor better in roots than in leaves in response to water deficits. Higher root turgor helps maintain root growth, elongation, and nutrient water uptake (Ogawa and Yamauchi 2006a; Studer et al. 2007).

The enhanced sugar content in AM roots under well-watered conditions may be due to the sink effect of the mycorrhizal fungus demanding sugars from shoot tissues. Under drought, Porcel and Ruiz-Lozano (2004) observed that the sugar content in roots was similar in both AM and non-AM treatments, suggesting that osmotic adjustment occurred. In contrast, in shoots the sugar content of droughted AM plants was considerably lower than in non-AM plants. The authors proposed two explanations for lower hexose accumulation in leaves of mycorrhizal plants. One, lower availability of photosynthates for storage in these tissues; and another explanation, that AM shoots are less strained by drought than non-AM ones. The lower accumulation of compatible solutes may indicate that the AM plants more successfully avoided drought stress (Augé 2001). In fact proline and other osmoregulator also accumulated less in shoots of AM plants than in non-AM plants suggest drought avoidances.

5.4 Maintenance of ROS Level

In plants, metabolism of ROS such as superoxide (O_2^-), hydrogen peroxide (H_2O_2), and hydroxyl radical ($OH^·$) is kept in dynamic balance. Under water stress condition, this balance is broken leading to enhanced production of ROS, which is deleterious to cells and cause oxidative damage. AM symbiosis provokes a more powerful ROS scavenging system in host plants and reduces destruction of biomolecules at the cellular level. AM plants subjected to drought shows lower lipid peroxidation than non-mycorrhizal plants (Porcel and Ruiz-Lozano 2004). SOD and POX are two important enzymes involved in elimination of superoxide anion and H_2O_2 respectively. Ruiz-Lozano et al. (1996) and Porcel et al. (2003) proposed that AMF protects host plants against oxidative damage by increments of enzymatic and non enzymatic antioxidants. Higher activities of SOD and POX in AM plants have been reported by many authors indicating that the activity of antioxidant enzymes is induced by AMF inoculation (Wu and Xia 2006; Bressano et al. 2010; Latef and Chaoxing 2011; Fan and Liu 2011). In addition, Fan and Liu (2011) observed that mRNA levels of four genes in the AM plants were apparently higher than in non-mycorrhizal plants under drought stress. Genes were annotated as homologues to *CSD1* (Copper/zinc SOD), *MIOX1* (myco-inositol oxygenase), *GLX1* (glyoxalase) and TTC5 (Transparent Testa 5) gene of *Arabidopsis*. These genes encode enzymes responsible for elimination of ROS, alleviating oxidative stress and detoxification of cytotoxic compounds. All these studies demonstrate that AMF inoculation results in a well-established defense mechanism against the drought, in which mitigation of oxidative stress might be a crucial part (Fan and Liu 2011).

5.5 Higher Photosynthetic Efficiency

AM plants often display higher rate of photosynthesis than their non-mycorrhizal counter parts do, which is consistent with the AMF effects on stomatal conductance. Most of the studies suggest that AM symbiosis increases the units of photosynthesis (Ruiz-Sánchez et al. 2010) so as to increase the rates of photosynthetic storage and export at the same time (Augé 2001). It has been shown that concentration of chlorophyll in AM plants was higher than their control non-mycorrhizal plants (Fan and Liu 2011). Higher concentration of chlorophyll is associated with higher photosynthetic rate (Davies et al. 1993). The higher photosynthetic rates associated with mycorrhization can result in higher concentration of soluble sugars and photosynthetic byproducts in the leaf symplasm, which can manifest itself as an increased cytoplast osmolarity in AM plants as against non-AM plants (Porcel and Ruiz-Lozano 2004).

5.6 Improved Water Status

AMF have the ability to affect plant water relation in both water-limiting and well-watered conditions. Augé (2001) has provided an extensive review on the effects of AM symbiosis on plant water relations in numerous host species colonized by various fungal symbionts, with a particular emphasis on these effects under drought conditions. Early studies examining the effects of AM symbiosis on plant water relations generally concluded that improved drought tolerance results from enhanced P nutrition. Further studies revealed that existence of other mechanisms either only partially correlated with or unrelated to plant nutrition or size. One proposed mechanism primarily focuses upon the impact of AM colonization on water absorption rates, which further involves alleviation of plant gas exchange parameters and subsequently, overall leaf hydration (Boomsma and Vyn 2008). Other mechanisms involve changes in plant hydraulic conductance (e.g. enhanced stele tissue size), soil water relations (e.g. increased aggregate stability, greater soil available water), soil–root water potential gradient (e.g. enhanced soil drying), plant water potential components (stomatal conductance-leaf water potential relationship alterations), and non-hydraulic root signals (differing cytokinin and auxin (AA) concentrations) (Boomsma and Vyn 2008). As suggested by Sánchez-Díaz and Honrubia (1994), AM-induced changes in water relations may involve complex interactions among multiple mechanisms. Primary impact of AM symbiosis involves changes in stomatal conductance (g_s) and transpiration (T), with T typically higher and g_s frequently unaffected or greater during drought stress in AM relative to non-AM plants. At times, AMF also postpone reductions in leaf water potential during periods of drought and hasten returns to control levels upon the alleviation of water-limiting conditions (Augé 2001).

5.7 Molecular Mechanism

The beneficial effect of AM symbiosis under drought stress conditions has been studied largely at the physiological level including the regulation of transpiration rate or increasing root water absorption (Augé et al. 2004). In last decade, it has also been noted that, under drought conditions, AM and non-AM plants differently regulate the expression of several stress genes in root tissue (Ruiz-Lozano et al. 2006). Among the genes regulated by the AM symbiosis during drought, aquaporins genes have been described (Porcel et al. 2006; Ruiz-Lozano et al. 2006; Aroca et al. 2007, 2008). Aquaporins are membrane intrinsic proteins that facilitate water and small neutral solutes flow, always, following an osmotic gradient. Modulation of aquaporin gene in AM symbiosis under osmotic stress is discussed under salt stress.

Another important mechanism involved in adaptation to water deficit is the induction of specific genes encoding an important component of endoplasmic reticulum—the luminal binding protein (BiP). The protein BiP is a molecule present in all kingdoms. The role of BiP in the ER is to transiently bind to unfolded proteins and to prevent intramolecular and intermolecular interactions that can result in permanent misfolding or aggregation, with the subsequent loss of their function (Gething and Sambrook 1992). A BiP encoding gene from *Glomus intraradices* has been identified after differential hybridization of cDNA library constructed from the fungus growing in vitro and subjected to drought stress (Porcel et al. 2007). Its expression was up-regulated by drought stress not only during in vitro conditions (AM monoxenic cultures) but also ex vitro, when forming natural symbiosis with plants. The contribution of AMF to the enhanced drought tolerance of the host plant can be mediated by proteins with chaperone-like activity, such as that of BiP (Porcel et al. 2007).

5.8 Contributions of Extraradical AM Mycelia (Drought Avoidance)

Arbuscular mycorrhizal fungi also colonize soils, changing chemical and physical soil properties (Jastrow et al. 1998). These properties can affect plant response to drought (Augé 2001). Therefore, in addition to influencing plants directly by colonizing plant tissue, AM symbiosis has the potential to affect drought response by changing the soils in which plant is growing. In fact, merely growing plants in a soil that had previously been mycorrhizal resulted in higher stomatal conductance of non-AM bean plants, under drought conditions (Augé et al. 2004). The colonization of soil by hyphae may have a greater influence on host behavior during drought than colonization of roots. The ability to survive to lower soil hydration was associated with more soil hyphae implies that soil hyphae may somehow aid root systems in more thoroughly extracting water from drying soils. Others have suggested that, at similar bulk soil water potential or bulk water content in AM and non-AM soils, soil water potential might actually be slightly higher in the rhizosphere of AM plants, if mycorrhizae more effectively ramify and dry out a particular volume of soil than do non-AM roots (Hardie and Leyton 1981; Gupta 1991; Duan et al. 1996). On several occasions, AM plants have been observed to deplete soil water more thoroughly than non-AM plants before achieving a similar shoot response. AM plants developed lower soil water potential before wilting (Hardie and Leyton 1981) or at the permanent wilting point (Bethlenfalvay et al. 1988a, b), relative to non-AM plants. Soil of AM cowpeas had to lose more water than soils of similarly sized non-AM plants, before evoking similar stomatal conductance, shoot water potential, transpiration and ABA in xylem near stomatal closure (Duan et al. 1996). AM sorghum was also able to maintain leaf water potential to lower soil water potential than similarly sized non-AM plants (Osonubi 1994). Dakessian et al. (1986), Bethlenfalvay et al. (1988a) and Franson et al. (1991) have provided evidence that AM plants apparently have access to soil water below the permanent wilting water potential of non-AM plants.

The extraradical AM mycelia increase the efficacy of root water absorption in dry soil (e.g. Reid 1979; Fitter 1985; Davies et al. 1992), soil hyphae may increase soil-to-root contact in drying soils. Perirhizal resistance—resistance to water flow across the soil–root interface—results from draw-down resistance, diurnally imposed by the rapid loss of water from the soil immediately adjacent to the root, and from contact resistance, which increases as the surface of the root has less contact with rhizosphere water (Tinker 1976; Klepper 1990). Contact resistance increases as water retreats from large pores into smaller and smaller capillary areas in the soil and decreases the amount of root length actually wetted (Herkelrath et al. 1977). Root and soil shrinkage creates gaps between the root and the soil, which can decrease water absorption (e.g. Nobel and Cui 1992). Root hairs can help prevent air gaps at the soil–root interface, as they grow into very small pores and effectively "glue" themselves to soil particles with exuded mucilage (Klepper 1990). AMF soil hyphae might serve this same function, perhaps even more effectively than root hairs, because most hyphae can enter finer pores than can root hairs (Tisdall 1991). Further, extraradical hyphal development and soil aggregation by AM plants have been greater under drought conditions (Davies et al. 1992).

Soil structure refers to pore space as well as to aggregates. Soil aggregate stability is a crucial soil property affecting soil sustainability, crop production, biological activity, soil carbon storage, and the movement and storage of water (Amezketa 1999). AMF and roots counteract as factors that affect soil aggregate stability, although the mechanism is still not known (Piotrowski et al. 2004). Glomalin, a glycoprotein produced by AMF and first reported by Wright and Upadhyaya (1996), is long-lived in soils (Rillig et al. 2001) and is tightly correlated with soil aggregate stability (Wright and Upadhyay 1998). Because soil aggregates regulate soil water flow (Prove et al. 1990), it seems logical to suspect that AMF colonization may improve the water relations of plants. Improved soil structure generally has positive impacts on soil moisture retention properties (Hamblin 1985). Colonization of soil by AMF has been shown to change soil moisture retention properties, in concert with changes in soil hyphal density and associated soil characters (Bearden 2001). Glomalin could influence soil carbon storage indirectly by stabilizing soil aggregates. Mycorrhizal soils maintain better soil structure, especially soil water-stable aggregates and Bradford reactive soil protein, which are important for (1) maintaining soil porosity; (2) increasing stability against wind and water erosion and (3) storing C by protecting organic matter from microbial decomposition (Augé et al. 2004). Mycorrhizal soils have more water-stable aggregates and consequently higher soil moisture (Augé et al. 2001). AMF colonization enhances plant growth under drought stress indirectly through affecting soil moisture retention via glomalin's effect on soil water-stable aggregates (Wu et al. 2008).

6 Heavy Metal

Heavy metals occur naturally in the environment and constitute a potential hazard for water, soils, plants, and sediments. Agro-ecosystem receive inputs of heavy metals from the increased use of agro chemicals, the application of metal containing wastes such as sewage sludge, coal, and wood ashes to soils and from atmospheric deposition (Mhatre and Pankhurst 1997). Heavy metals are grouped into one category of elements with specific weight higher than 5 g cm^{-3} (Göhre and Paszkowski 2006). Some of these metals are essential plant micro-nutrients such as Cu, Fe, Mn, Ni, and Zn and are required for beneficial plant growth and development, while others have no known biological function such as Cd, Pd, and Hg. High contents of heavy metals in soils are generally considered a matter of concern as they may adversely affect the quality of soil water and compromise sustainable food production.

At high concentration, heavy metal influence the structure of enzymes and hence their functionality by affecting the protein structure or substituting a necessary element. As the structure of plasma membrane, protein such as H^+-ATPase is sensitive to alteration by heavy metals; the toxic effects of heavy metals can influence the permeability and function of plasma membrane. In addition, metals cause oxidative stress (production of ROS) adversely affecting cellular components and hence plant tissues (Sajedi et al. 2010).

Remediation of metal compounds presents a different set of problems when compared to organics. Organic compounds can be degraded while metals normally need to be physically removed or immobilized (Kroopnick 1994). Contaminated soil can be remediated by chemical, physical, or biological techniques. The most common remediation technique is off-site management; the metal contaminated soil is taken for burial at landfill sites. This method of remediation merely shifts the contamination problem elsewhere (Vidali 2001). Moreover, physico-chemical technologies used for soil remediation render the land useless as a medium for plant growth, as they remove all the biological activities. Therefore, sustainable on-site techniques for remediation of heavy metal contaminated sites need to be developed. In recent years, attention has been paid to the remediation of polluted soil, by use of plants (phytoremediation) and microbes (micro-remediation). These two approaches are preferred to chemical or physical remediation, because of their cost effectiveness, environment friendliness, and fewer side effects (Karimi et al. 2011).

In plants, the root is typically the organ, which is in continuous contact with metal ions in soil. Therefore, interaction between the microorganism in the rhizosphere and the plant activity related to soil remediation is inevitable (Compant et al. 2010). Combination of a hyper accumulating plant with beneficial rhizo- and/or endospheric microorganisms holds great promise for low cost cleaning of contaminated sites (Karimi et al. 2011).

7 Arbuscular Mycorrhiza in Mitigation of Heavy Metal Stress

Arbuscular mycorrhizal fungi have been observed in soils containing heavy metals (Wulf et al. 2003; Göhre and Paszkowski 2006; Hildebrandt et al. 2007). AMF can both positively and adversely affect the uptake of heavy metals by plants. Although considerable variability in plant responses to AMF inoculation has been observed in contaminated soils, the potential of AMF to buffer heavy metal stress has been demonstrated in a number of studies (e.g. Hildebrandt et al. 1999; Janoušková et al. 2006; Chen et al. 2007). Similar to stresses such as soil compaction and salinity, the alleviating effects of AMF on plant growth may intensify with increasing heavy metal concentration (Hildebrandt et al. 1999; Audet and Charest 2006), indicating a significant interaction between AM and stress level. The role of AMF in enhancing plant tolerance to heavy metals is very much dependent on AMF species, plant genotype and type of element in the soil (Sudová et al. 2008). For example, according to Khan et al. (2000), Zn is absorbed and crystallized in AMF hyphae and cortical cells of mycorrhizal root, Zn transfer to shoot decreases. Although AMF enable enhanced uptake of Fe and Mn in plants (Miransari et al. 2006), at high concentrations, AM are able to decrease the translocation of Mn in shoots and retain Fe in roots (Leyval et al. 2002).

Similar to plants, AMF from contaminated soils have been reported to cope better with heavy metal toxicity than those not exposed to such long-term selection pressure (Weissenhorn et al. 1993; Malcova et al. 2003). In spite of increasing knowledge in AMF-plant interaction under heavy metal stress, little is known about whether there is a synergism between plant and fungal heavy metal tolerance. It can be hypothesized that tolerant AMF may confer additional heavy metal tolerance on its host thus leading to their higher survival rate and reproductive success on contaminated sites. Alternatively, carbon invested into maintenance of AM symbiosis (4–20% of total plant photosynthates) may represent high costs for the tolerant plants that can cope well with heavy metal contamination without being mycorrhizal, but not for the heavy metal sensitive ones.

7.1 Heavy Metal Sequestration in Root/Soil

Heavy metals in soil are associated with a number of soil components which determine their behavior in the soil and influence their bio availability (Boruvka and Drabek 2004).

7.2 Fungal Cell Wall

The AM-mediated alleviation of the effects of heavy metal stress cannot be only attributed to nutrition effects, but also to the impact of AMF on metal distribution at

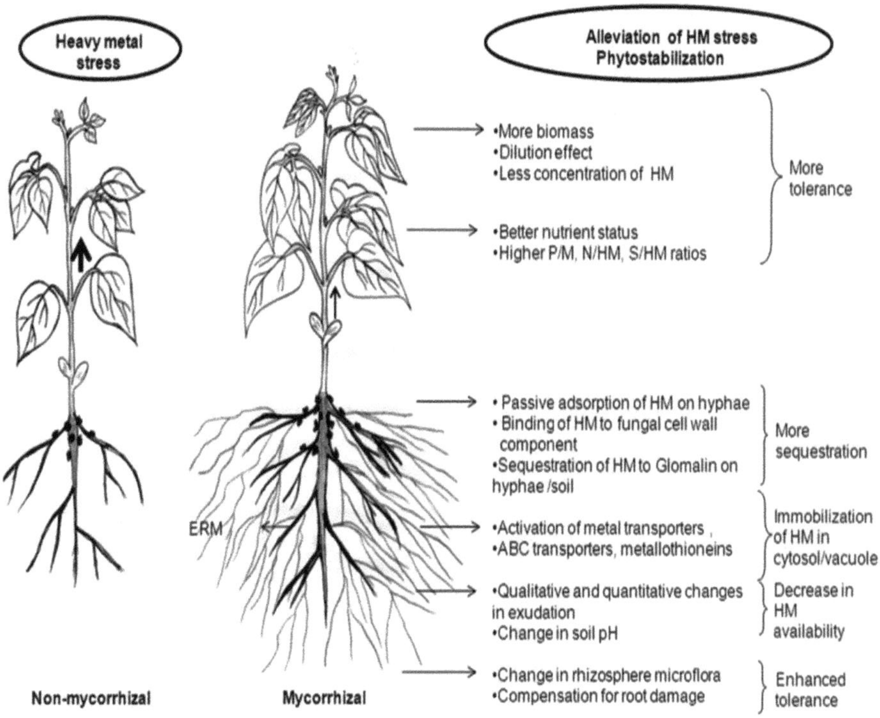

Fig. 14.3 Contribution of arbuscular mycorrhiza fungi to phytostabilization of heavy metals (HM). *Left*: Non-mycorrhizal plant in HM polluted soil showing enhanced uptake and transfer of HM to shoot. The plants are smaller as compared to the mycorrhizal plants on the *right* due to the effects of HM. *Right*: Mycorrhizal plant in HM polluted soil showing higher biomass and more tolerance to HM stress. In the rhizosphere, AMF colonization induces change in pH and microflora thereby decreasing HM availability. AMF in the roots also immobilize HM on its hyphae and sequester HM inside the cell; thereby lessening its transfer to shoot. On the other hand, AMF colonization also compensates for damaged root and enhances uptake of nutrients and water thereby maintaining better nutrient: HM ratios. These mechanisms facilitates for increased growth and higher biomass, which also complements to tolerance mechanism by diluting the effect of HM. Thickness of *arrows* indicates uptake level of HM

the soil-fungus-plant interfaces. AMF play an important ecological role in phytostabilization of toxic trace elements in soil by sequestration and, in turn help mycorrhizal plants survive in polluted soils (Fig. 14.3). The fungal cell wall components such as free amino, hydroxyl and carboxyl can bind to potentially toxic elements such as Cu, Pb and Cd (Kapoor and Viraraghavan 1995). Binding of heavy metal to chitin in the fungal cell wall reduces its local concentration in the soil. Joner et al. (2000) noted binding of up to 0.5 mg Cd per mg dry biomass, as a consequence of passive adsorption to the hyphae. Large surface area of hyphae in the soil is an important sink for heavy metals. Moreover, heavy metal tolerant fungi show greater affinity for heavy metals than roots (2–4 times more) and are thus suitable for immobilizing heavy metal in the soil (Joner et al. 2000).

Immobilization of metals on both extra and intraradical fungal tissue has been shown (Kaldorf et al. 1999; Joner et al. 2000), thus providing a plausible explanation of the barrier for metal translocation from the roots to the shoots of inoculated plants. Reduced transfer, as indicated by enhanced root:shoot Cd ratios in AM plants has been suggested (Tullio et al. 2003; Joner et al. 2000). This may be due to intracellular precipitation of metallic cations with phosphates.

7.3 Glomalin

Recently, glomalin, a glycoprotein produced by AMF has been suggested to have a metal chelating function influencing the metal availability for plants (Wright and Upadhyay 1998; Wright et al. 1998; Gonzalez-Chavez et al. 2004). One gram of glomalin could extract up to 4.3 mg Cu, 0.08 mg Cd and 1.12 mg Pb from the polluted sites (Gonzalez-Chavez et al. 2004). This protein may be one of the first cellular components in fungi coming in contact with ions from the surrounding environment; however, the exact mechanism of heavy metal binding by glomalin remains unclear (Volesky 1990). The glomalin stabilizes, reduces availability and decreases toxicity risk of heavy metals to other soil microorganisms and plants growing in these sites. The copious production and the recalcitrant nature of this molecule in the soil further enhance the potential usefulness of this compound in the soil (Rillig et al. 2001; Gonzalez-Chavez et al. 2004).

7.4 Change in Rhizosphere pH and Microflora

The rhizosphere of mycorrhizal plants (mycorrhizosphere) has lower heavy metal concentrations in soil solution as compared to that of non-mycorrhizal plants (Shen et al. 2006; Redon et al. 2008). This is because AMF reduces the availability of heavy metals to the host plant (Audet and Charest 2007). The lower soil solution concentration of heavy metals in the mycorrhizosphere has been often associated with high pH (Li and Christie 2001; Bi et al. 2003; Shen et al. 2006). In this regard, significant contributions from plants through AMF that is systemic effect of AM may also be responsible for modifications of soil pH and bacterial communities in the mycorrhizosphere or for the exudation of specific compounds (Marschner and Baumann 2003; Vierheilig et al. 2003). Root exudation is also an important factor influencing heavy metal mobility and bioavailability in soil (Hinsinger 2001). Thus, it may be suggested that AM-induced reduction in heavy metal availability in mycorrhizosphere is a combined endeavor of both plant-AMF. However, Janoušková and Pavliková (2010) showed cadmium immobilization in the rhizosphere of AM-*Nicotiana tobaccum* by the fungal ERM-induced alkalinization of substrate, while no indication was found for a significant involvement of plant-mediated effects of AM.

7.5 Enhance Growth and Biomass

Mycorrhizal associations increase the absorptive surface area of plant due to extra matrical fungal hyphae exploring rhizosphere beyond the root-hair zone, which in turn enhances water use efficiency and mineral uptake. The enhanced capability of uptake of minerals results in greater biomass production, a prerequisite for successful remediation. The stimulatory effect of AMF inoculation on the development of metal-treated plants was observed for maize, soybean, pea, and sunflower plants (Rivera-Becerril et al. 2002; Andrade et al. 2004, 2008; Jurkiewicz et al. 2004). Another possible mechanism of metal tolerance includes dilution of metal concentrations in plant tissues due to promotion of plant growth by AMF. For example Arsenate, As (V) is an analog of Pi and mycorrhizas would enhance uptake of both (Smith et al. 2010). Chen et al. (2007) pointed out that though total As (V) content increased concentration of As are frequently lower in AM plants, possibly reflecting tissue dilutions of As in the larger plants, rather than reduction in uptake per plant. AMF confers As tolerance in plants as the result of compensation by the fungal pathway for poor root growth. However, recent evidence indicate that this is an over simplified explanation and is not adequate (Smith et al. 2010).

8 Prevention of Nutrient Deficiency and Heavy Metal Toxicity

Excessive heavy metal concentrations (such as Cd) can induce deficiencies and imbalance of plant mineral nutrients (Greger and Lindberg 1987). AMF inoculation alleviates this effect and mycorrhizal plants show higher N, P, Ca, Mg, and S concentrations in shoot than non-mycorrhizal plants. Mycorrhizal maize plants showed higher P/Cd, N/Cd, and S/Cd ratios in both shoots and roots than non-mycorrhizal plants (Andrade et al. 2008). Higher ratios of P/metal in mycorrhizal plant species have been observed for several plant species (Andrade et al. 2004), suggesting that the higher P status of these plants may alleviate metal stress by phosphate complexation with metal ions inside the cells. The higher shoot N and S uptake in mycorrhizal plants lead to higher production of thiol rich proteins which, in addition to P complexation play an important role in heavy metal detoxification in vascular plants.

9 Phytoextraction

Phytoextraction is a rather recent technology and represents the most effective and attractive strategy on clean up contaminated soils (Kramer 2005). As phytoextraction relies on the capacity of plants to accumulate and tolerate heavy metal in their shoots, its efficiency depends both on the ability of metal to be translocated in aerial

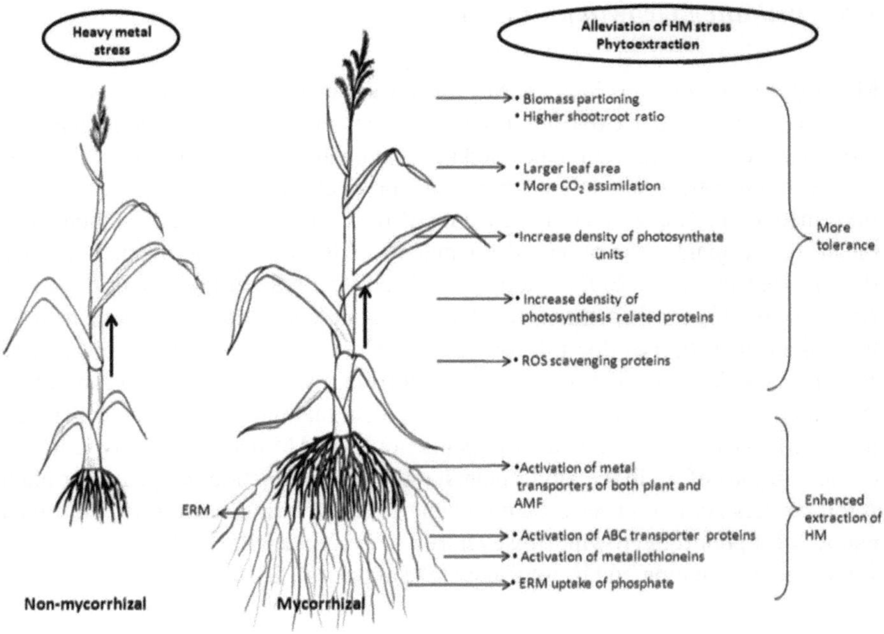

Fig. 14.4 Contribution of AMF to phytoextraction of HM. On the *left* is the non-mycorrhizal plant in HM polluted soil. The presence of HM in the rhizosphere reduced the growth of the plant; therefore the plant is smaller as compared to the mycorrhizal plants on the *right*. The mycorrhizal plant is able to tolerate HM stress due to the following reasons: (1) AMF colonization resulted in enhanced extraction of HM by activation of ABC transporter proteins and metallothioneins. At the same time, the extraradical mycelium (ERM) of extends the root system and enhances the uptake of P. (2) AMF also enhances the activities of ROS scavenging proteins and prevent the plants from oxidative stress. (3) AM plants maintain higher rate of photosynthesis thereby leading to higher biomass, shoot: root ratio and leaf area. Higher leaf area, in turn facilitates higher CO_2 assimilation. In this way, AMF colonization diminishes the phytotoxic effects of HM and imparts tolerance to the plant

plant organs and on shoot biomass production (Salt et al. 1998; Shi and Cai 2009; Wu et al. 2010a). Three main indicators have been used to measure plant effectiveness in extracting heavy metal from soil: the tolerance index expressed as the ratio of shoot growth parameters for plant grown in polluted soil to plants grown in metal-free soil (Shi and Cai 2009; Wu et al. 2010a); the transport factor calculated as the ratio of the heavy metal in shoot to that in roots (Wu et al. 2010a; Wang et al. 2007); and heavy metal partitioning that corresponds to the metal quantity present in plant organs (Redon et al. 2008). Figure 14.4 shows the mechanisms in mycorrhizal plants in alleviation of heavy metal stress through phytoextraction.

10 AM and Hyper Accumulators

There are only a limited number of plants (the metallophytes) that can grow under heavy metal stress (Miransari 2010; Tonin et al. 2001; Hildebrandt et al. 2006). In addition to the development of some special physiological processes, symbiosis with AM also enables metallophytes to grow under heavy metal stress by substantially reducing plant uptake of heavy metals (Berreck and Haselwandter 2001). Most metallophytes belong to the families Brassicaceae and Caryophyllaceae, which are non-mycorrhizal plants (de Mars and Boerner 1996). However, some species such as *Biscutella laerigata* and *Thlaspi* spp. are able to develop symbiosis with AM species such as *Glomus intraradices* (Hildebrandt et al. 2007).

It was discovered that addition of AMF further enhances the uptake and accumulation of As in *Pteris riltata* (Leung et al. 2006). At highest As concentration tested (100 mg/kg soil), non-mycorrhizal plants accumulated 60.4 mg As/kg soil while AM plants accumulated 88.1 mg As/kg soil. This was accompanied by enhanced growth, possibly due to improved phosphate (Pi) nutrition. Both effects combined allow for higher recovery of heavy metal. Similarly, *Berkheya coddii* belonging to Asteraceae family, used for phytomining Ni accumulated 30% more Ni on AMF inoculation by "adapted" AMF than in non-mycorrhizal controls (Turnau and Mesjasz-Przybylowicz 2003).

11 AM and Non-hyper Accumulators

Non-hyper accumulators can also be used for phytoextraction if they are sufficiently tolerant to heavy metal and produce high biomass. Research regarding shoot tolerance mechanism upon heavy metal phytoextraction has been essentially conducted in hyperacccumulator plant species; however, there is little evidence regarding processes by which mycorrhiza allow plant shoots to cope with metal stress (Davies et al. 2001). This is probably because roots are considered as the main site of metal toxicity exposure; therefore the cellular and molecular basis of heavy metal tolerance of mycorrhizal plants have been essentially grasped at the below ground level (Joner et al. 2000; Ouziad et al. 2005; Janoušková et al. 2006; Hildebrandt et al. 2007; Aloui et al. 2009; González-Guerrero et al. 2010).

12 Allocation Plasticity

Recently, Aloui et al. (2011) proposed that enhanced metal extracting capacity of mycorrhizal plant is not related to increase in root/shoot translocation rate, but to a high level of allocation plasticity. They observed that *Medicago truncatula* plants inoculated with *Glomus irregulare* displayed a significant increase in shoot tolerance to Cd relative to those non-mycorrhizal. In spite of reduced root to shoot translocation rate of Cd, shoots of mycorrhizal plants contained the highest metal quantity relative to shoots of non-mycorrhizal plants (Kapoor and Bhatnagar 2007). From these observations they concluded that shoots of mycorrhizal *M. truncatula* have a capacity for extracting Cd, which is not related to an increased root to shoot transport factor. It has been proposed that a significant shift in root to shoot biomass partitioning permitted some plants to reduce the incidence of metal-induced stress in photosynthetic organs, a process referred to as allocation plasticity (Audet and Charest 2008).

The allocation plasticity of heavy metals in mycorrhizal plants may also be related to improved photosynthesis. AMF promotes photosynthesis by increasing the plant's ability to use light energy, maximize the area available for CO_2 assimilation per unit of carbon invested, facilitate the electron transport, prevent inhibition of aminolevulenic acid synthesis and protochlorophyllide photoreduction and by increasing the density of photosynthetic units (Stobart et al. 1985; Wright et al. 1998; Schoefs 2005; Aloui et al. 2011). A more tolerant photosynthetic system would allow plants maintaining high transpiration efficiency thus creating a water flux that can drive metals from the roots into the stem and leaves where the metal can be compartmentalized (Visioli et al. 2010). Taken together, both biomass partitioning and photosynthesis-related indicators support the idea that mycorrhizal plants extract and tolerate heavy metal by displaying a high level of allocation plasticity (Aloui et al. 2011) and do not invest in an intrinsic tolerance mechanism typical of *Arabidopsis thaliana*, which involves for example phytochelatin production (Audet and Charest 2007).

Although, several studies have demonstrated that mycorrhizal legumes can accumulate and tolerate heavy metal in their above-ground organs (Rivera-Becerril et al. 2002; Göhre and Paszkowski 2006; Aloui et al. 2009) information regarding molecular mechanisms by which shoots of mycorrhizal plants can escape heavy metal toxicity is lacking. In a pioneering work, Aloui et al. (2011) gave a first picture of shoot proteome modifications upon AM symbiosis and/or heavy metal stress in a legume plant. On performing 2-DE/MALDI/TOF-based comparative proteomic analysis of *Medicago truncatula* shoot responses upon mycorrhization and Cd exposure, they observed that in non-mycorrhizal plants, metal-responsive shoot proteins impaired CO_2 assimilation; the mycorrhiza responsive shoot proteome was characterized by an increase in photosynthesis-related proteins coupled to a reduction in gluconeogenesis/glycolysis and antioxidant processes. By contrast, Cd was found to trigger the opposite response coupled with the up-accumulation of molecular chaperones in shoot of mycorrhizal plants relative to those metal-free.

13 Molecular Mechanism

Although macroscopic symptoms and physiological effects of heavy metal stress are well documented in higher plants, especially those of agricultural importance like legumes and cereals (Das et al. 1997; Sanità di Toppi and Gabbrielli 1999) there is lack of information concerning the molecular basis of such responses as well as on how they may be modulated by AM symbiosis. In this regard, targeted approaches have been developed within the last decade to monitor changes in plant gene expression and protein accumulation for understanding the molecular mechanism of heavy metal tolerance in mycorrhizal plants (Repetto et al. 2003; Rivera-Becerril et al. 2005).

When plant is subjected to high levels of heavy metal, these are translocated and accumulated in the parenchyma cells of the inner root, which is the place of colonization with different fungal structures including arbuscules, vesicles and hyphae (Kaldorf et al. 1999). The mechanisms directing heavy metal movement to plant roots by AMF have been proposed. These are (1) heavy metal may be deposited in the cellular wall or in the fungal vacuoles; (2) sequestration of heavy metals by siderophores may deposit heavy metal in root apoplasm or in soil; (3) metallothioneins or phytochelatins may result in the deposition of heavy metal in fungal or plant cell; and (4) the allocation of heavy metal from cytoplasm is performed by metal transporters located at the plasmalemma or tonoplast of both symbionts (Miransari 2011). Heavy metal content in roots of mycorrhizal plants is highly altered indicating that the related genes are expressed at transcriptional and translational levels by AMF (Ouziad et al. 2005).

At the genetic level, very few genes involved in heavy metal homeostasis have been analysed in detail in AMF: the Zn transporter *GintZnT1* from *Glomus intraradices* involved in Zn compartmentalization (González-Guerrero et al. 2005) and metallothionein gene *GmarMT1* and *GintMT1* from *Gigaspora margarita* and *Glomus intraradices* respectively that may provide protection against Cu (Lanfranco et al. 2002; González-Guerrero et al. 2007). Metallothioneins (MTs) constitute an extensive and diverse family of small cysteine-rich protein that bind metals via the thiol groups of their cysteine residues and may play a role in the intracellular sequestration of heavy metal. Although, the most widely accepted role for MTs is metal detoxification, several studies have indicated that MTs play a role in the protection against the effect of ROS. In fact, the MT protein itself acts as an antioxidant as it is a potent scavenger of hydroxyl radicals (Andrews and Geiser 1999). González-Guerrero et al. (2007) suggested that *GintMT1* might play a key role in the regulation of the redox status of the extraradical mycelia of *G. intraradices* through either its metal chelation activity or its thiol groups which might contribute to the pool of cytosolic thiols that regulate fungal redox status. In their other experiment, González-Guerrero et al. (2010) indicated that Cd and Cu result in expression of the related transporters *Gint ABC1* might play a key role in Cd and Cu detoxification. Given that Cu is an active redox metal that induces oxidative stress in *G. intraradices* (Benabdellah et al. 2009) and that transcription of *Gint ABC1* is induced by oxidative stress, up-regulation of *Gint ABC1* expression in the presence of Cu might be also due to the oxidative response elicited by Cu (González-Guerrero et al. 2010).

On the basis of 2D-based proteomic approach used to compare the proteomes of *Medicago truncatula* roots either colonized or not with the AM fungus *Glomus intraradices* in Cd-free and Cd-contaminated substrates, Aloui et al. (2009) reported on the protective effect conferred by AM symbiosis. Their results indicated that at the proteome level, nine out of the 15 Cd-induced changes in non-mycorrhizal roots were absent or reversed in those Cd-treated and colonized by *G. intraradices*—including the *G. intraradices*—dependent down accumulation of Cd stress-responsive proteins. Out of the 26 mycorrhizal-related proteins that were identified only six displayed changes in abundance upon Cd exposure, suggesting that part of the symbiotic program, which displays low sensitivity to Cd, may be recruited to counteract Cd toxicity through the mycorrhiza-dependent synthesis of proteins having functions putatively involved in alleviating oxidative damages, including a cyclophilin, a guanine nucleotide-binding protein, an ubiquitin carboxy terminal hydrolase, a thiazole biosynthetic enzyme, an annexin, a glutathione S-transferase (GST) like protein and a S-adenosylmethionine (SAM) synthase.

14 Research Concerns

Although AMF usually can enhance plant potential to grow in heavy metal polluted soil and in case of hyperaccumulators may improve their metal uptake, the selection of appropriate mycobiont for the right plant is very important. This would result in more efficient bioaugmentation strategy, which would bioremediate the soil more efficiently (Miransari 2011). As mentioned previously, selection of AM species from contaminated areas can be more beneficial, because such species are adapted to high concentrations of heavy metals The colonization by AMF can lead to increased uptake and subsequent accumulation of heavy metal in above-ground tissues of plants—while this trait is required in remediation of land by phytoextraction, it is not desirable in agriculture crops grown for food. On the other hand, in several cases mycorrhizal colonization leads to accumulation of heavy metal in roots as described earlier within this review. Although this scenario could be desirable for enhanced plant heavy metal tolerance, it may interfere with efficient phytoextraction (Göhre and Paszkowski 2006).

15 Future Directions

Present-day industrial agricultural practices place several constraints on the use of services provided by mycorrhiza; however in order to manipulate AMF and to achieve their efficient use for long-term agricultural stability and productivity, we have to increase our knowledge on the impact of different production strategies on both the diversity of AM fungal communities and its relation to production quantity and quality (Gianinazzi et al. 2010).

The effects of AM in amelioration of abiotic stress on a variety of crop plants have typically been studied using pot-cultures in greenhouse or growth chamber environments in which interaction between the two symbionts were studied in a controlled manner (Wright 2005). However, in field setting AM symbiosis is affected by factors not present in controlled greenhouse or laboratory conditions. In order to fully understand AMF effects on abiotic stress tolerance, future experiments should be conducted in field. These studies could be conducted at varying levels of stress and nutrient availability and at multiple locations so that the stress x soil fertility interaction could be interrogated across a number of environments (Boomsma and Vyn 2008). Cropping practices that mimic those used by growers should also be improved in future field research, including the use of various nutrient amendment, tillage systems, plant densities, crop rotations, and pesticide applications appropriate for optimum yield in a given environment. The findings of these studies could potentially serve as guides for examining the effects of cropping practices on the AM-mediated alleviation of abiotic stress.

The proportion of modern genotypes of crop plants capable of forming an effective symbiosis in a abiotic stress environment is generally unknown. A thorough analysis of the variation in colonization and responsiveness of susceptible and tolerant genotypes is necessary. Although future research should focus on identifying crop varieties capable of forming an effective symbiosis, it should also determine fungal isolates, species and communities that demonstrate the greatest enhancement of crop productivity in stress environment. Morphologically identical isolates of a particular fungal species can have varying effects on host growth, with the effect of a single isolate varying by such environmental factors as soil temperature, pH, moisture, nutrient status, and salinity (Picone 2003). Although an AM isolate or species can benefit a particular genotype it may not do so for another (Liu et al. 2000a, b). These conditional phenomena suggest that beneficial AM species may need to be determined both for specific genotypes and local agro-ecosystem environments. Such species need to be capable of surviving and spreading among native AM communities (when introduced) and in typical environmental and agronomic conditions.

Further understanding of plant metabolic and physiological responses to AM infection during a particular stress could be enhanced through the use of molecular and genomic techniques. There are a far and few studies on AM symbiosis under abiotic stress investigated on a molecular level (Porcel et al. 2005; Ruiz-Lozano 2003). Further studies could involve nontargeted screening of cDNA libraries from both AMF and AM colonized plants. Such techniques could enable the identification of stress-regulated genes that permit enhanced stress tolerance in AM colonized hosts. Using microarrays, stress-tolerance mechanisms (morpho-physiological and molecular processes) associated with AM symbiosis could be either identified or more fully understood through the comparison of AM and non-AM plants of same genotype.

In some cases, improving plant productivity through the use of AM inoculums instead of indigenous AM populations may be preferable or even necessary (e.g. in areas that received extensive use of fungicide or on severely degraded land) (Douds et al., 2005). Unfortunately, due to the obligate biotrophic nature of AM, the cost of inoculums production is presently quiet high. Thus, advances in technologies are imperative towards formulation of large scale, uniform, economically viable inoculums.

From the large number of studies extensively examined in this chapter, it is evident that AM symbiosis can potentially improve abiotic stress tolerance.

However, more research is required, particularly in field settings across multiple environments. Further endeavors seeking to elucidate, validate, and implement improvements in stress tolerance through AM symbiosis will likely require extensive collaboration among crop physiologists, plant breeders, genomicists, molecular and soil biologists.

References

Abd-El Baki GK, Siefritz F, Man HM et al (2000) Nitrate reductase in *Zea mays* L under salinity. Plant Cell Environ 23:515–521

Abdel-Fattah GM, Asrar AWA (2011) Arbuscular mycorrhizal fungal application to improve growth and tolerance of wheat (*Triticum aestivum* L.) plants grown in saline soil. Acta Physiol Plant. doi:10.1007/s11738-011-0825-6

Alcázar R, Marco F, Cuevas JC et al (2006) Involvement of polyamines in plant response to abiotic stress. Biotech Lett 28:1867–1876

Al-Garni SMS (2006) Increasing NaCl-salt tolerance of a halophytic plant *Phragmites australis* by mycorrhizal symbiosis. Am Eur J Agric Environ Sci 1:119–126

Al-Karaki G, McMichael B, Zak J (2004) Field response of wheat to arbuscular mycorrhizal fungi and drought stress. Mycorrhiza 14:263–269

Al-Karaki GN (2006) Nursery inoculation of tomato with arbuscular mycorrhizal fungi and subsequent performance under irrigation with saline water. Sci Hort 109:1–7

Allen MF (2006) Water dynamics of mycorrhizas in arid soils. In: Gadd GM (ed) Fungi in biogeochemical cycles. Cambridge University Press, New York, pp 74–97

Allen MF (2007) Mycorrhizal fungi: highways for water and nutrients in arid soils. Vadose Zone J 6:291–297

Aloui A, Recorbet G, Gollotte A et al (2009) On the mechanisms of cadmium stress alleviation in *Medicago truncatula* by arbuscular mycorrhizal symbiosis: a root proteomic study. Proteomics 9:420–433

Aloui A, Recorbet G, Robert F et al (2011) Arbuscular mycorrhizal symbiosis elicits shoot proteome changes that are modified during cadmium stress alleviation in *Medicago truncatula*. BMC Plant Pathol 11:75

Amezketa E (1999) Soil aggregate stability: a review. J Sustain Agric 14:83–151

Andrade SAL, Abreu CA, de Abreu MF et al (2004) Influence of lead addition on arbuscular mycorrhiza and *Rhizobium* symbioses under soybean plants. Appl Soil Ecol 26:123–131

Andrade SAL, Silviera APD, Jorge RA et al (2008) Cadmium accumulation in sunflower plants influenced by arbuscular mycorrhiza. Inter J Phytorem 10:1–13

Andrews GK, Geiser J (1999) Expression of the mouse metallothionein- I and -II genes provides a reproductive advantage during maternal dietary zinc deficiency. J Nutr 129:1643–1648

Anjum SA, Xie XY, Wang LC et al (2011) Morphological, physiological and biochemical responses of plants to drought stress. Afr J Agric Res 6:2026–2032

Aroca R, Bago A, Sutka M et al (2009) Expression analysis of the first arbuscular mycorrhizal fungi aquaporins described reveals concerted gene expression between salt stressed and non stressed mycelium. MPMI 22:1169–1178

Aroca R, Ferrante A, Vernieri P et al (2006) Drought, abscisic acid, and transpiration rate effects on the regulation of PIP aquaporin gene expression and abundance in *Phaseolus vulgaris* plants. Ann Bot 98:1301–1310

Aroca R, Porcel R, Ruiz-Lozano JM (2007) How does arbuscular mycorrhizal symbiosis regulate root hydraulic properties and plasma membrane aquaporins in *Phaseolus vulgaris* under drought, cold, or salinity stresses? New Phytol 173:808–816

Aroca R, Vernieri P, Ruiz-Lozano JM (2008) Mycorrhizal and non-mycorrhizal *Lactuca sativa* plants exhibit contrasting responses to exogenous ABA during drought stress and recovery. J Exp Bot 59:2029–2041
Audet P, Charest C (2006) Effects of AM colonization on "wild tobacco" plants grown in zinc-contaminated soil. Mycorrhiza 16:277–283
Audet P, Charest C (2007) Dynamics of arbuscular mycorrhizal symbiosis in heavy metal phytoremediation: meta-analytical and conceptual perspectives. Environ Pollut 147:609–614
Audet P, Charest C (2008) Allocation plasticity and plant-metal partitioning: meta-analytical perspectives in phytoremediation. Environ Pollut 156:290–296
Augé RM, Toler HD, Moore JL, Cho K, Saxton AM (2007) Comparing contributions of soil versus root colonization to variations in stomatal behaviour and soil drying in mycorrhizal *Sorghum bicolor* and *Cucurbita pepo*. J Plant Physiol 164:1289–1299
Augé RM (2001) Water relations, drought and vesicular mycorrhizal fungi symbiosis. Mycorrhiza 11:3–42
Augé RM, Sylvia DM et al (2004) Partitioning mycorrhizal influence on water relations of *Phaseolus vulgaris* into soil and root components. Can J Bot 82:503–514
Augé RM, Stodola AJW, Tims JE et al (2001) Moisture retention properties of a mycorrhizal soil. Plant Soil 230:87–97
Bago B, Pfeffer PE, Zipfel W et al (2002) Tracking metabolism and imaging transport in arbuscular mycorrhizal fungi. Metabolism and transport in AM fungi. Plant Soil 244:189–197
Barea JM, Jeffries P (1995) Arbuscular mycorrhizas in sustainable soil-plant systems. In: Varma A, Hock B (eds) Mycorrhiza: structure, function, molecular biology and biotechnology. Berlin, Springer, pp 521–560
Bearden BN (2001) Influence of arbuscular mycorrhizal fungi on soil structure and soil water characteristics of vertisols. Plant Soil 229:245–258
Becard G, Piche Y (1990) Physiological factors determining vesicular arbuscular mycorrhizal formation in host and non host Ri T-DNA transformed roots. Can J Bot 68:1260–1264
Benabdellah K, Merlos MA, Azco´n-Aguilar C et al (2009) *GintGRX1*, the first characterized glomeromycotan glutaredoxin, is a multifunctional enzyme that responds to oxidative stress. Fungal Genet Biol 46:94–103
Benavides MP, Marconi PL, Gallego SM et al (2000) Relationship between antioxidant defence systems and salt tolerance in *Solanum tuberosum*. Aust J Plant Physiol 27:273–278
Benjamin JG, Nielsen DC (2006) Water deficit effects on root distribution of soybean, field pea and chickpea. Field Crops Res 97:248–253
Berreck M, Haselwandter K (2001) Effect of the arbuscular mycorrhizal symbiosis upon uptake of cesium and other cations by plants. Mycorrhiza 10:275–280
Bethlenfalvay GJ, Brown MS, Ames RN et al (1988a) Effects of drought on host and endophyte development in mycorrhizal soybeans in relation to water use and phosphate uptake. Physiol Plant 72:565–571
Bethlenfalvay GJ, Thomas RS, Dakessian S et al (1988b) Mycorrhizae in stressed environments: effects on plant growth, endophyte development, soil stability and soil water. In: Hutchinson CF, Timmermann BN (eds) Arid Lands: Today and Tomorrow. Westview Press Inc, Boulder CO, pp 1015–1029
Bi YL, Li XL, Christie P (2003) Influence of early stages of arbuscular mycorrhiza on uptake of zinc and phosphorus by red clover from a low-phosphorus soil amended with zinc and phosphorus. Chemosphere 50:831–837
Bolan N (1991) A critical review on the role of mycorrhizal fungi in the uptake of phosphorus by plants. Plant Soil 134:189–207
Bonfante P, Bianciotto N (1995) Saprotrophic versus symbiotic phase in endomycorrhizal fungi: morphology and cytology. In: Varma A, Hock B (eds) Mycorrhizas: structure, function, molecular biology and biotechnology. Springer, Berlin, pp 229–247
Boomsma CR, Vyn TJ (2008) Maize drought tolerance: Potential improvements through arbuscular mycorrhizal symbiosis? Field Crops Res 108:14–31

Borde M, Dudhane M, Jite P (2011) Growth photosynthetic activity and antioxidant responses of mycorrhizal and non-mycorrhizal bajra (*Pennisetum glauca*) crop under salinity stress condition. Crop Prot 30:265–271

Borsani O, Valpuesta V, Botella MA (2001) Evidence for a role of salicylic acid in the oxidative damage generated by NaCl and osmotic stress in Arabidopsis seedlings. Plant Physiol 126:1024–1030

Boruvka L, Drabek O (2004) Heavy metal distribution between fractions of humic substances in heavy polluted soils. Plant Soil Environ 50:339–345

Bray EA (2002) Abscissic acid regulation of gene expression during water-deficit stress in the era of *Arabidopsis* genome. Plant Cell Environ 25:153–161

Bressano M, Curetti M, Giacheroa L et al (2010) Mycorrhizal fungi symbiosis as a strategy against oxidative stress in soybean plants. J Plant Physiol 167:1622–1626

Canterll IC, Linderman RG (2001) Preinoculation of lettuce and onion with VA mycorrhizal fungi reduces deleterious effects of soil salinity. Plant Soil 233:269–281

Cardoso IM, Kuyper TW (2006) Mycorrhizas and tropical soil fertility. Agric Ecosyst Environ 116:72–84

Carvajal M, Cooke DT, Clarkson DT (1996) Responses of wheat plants to nutrient deprivation may involve the regulation of water-channel function. Planta 199:372–381

Cavagnaro TR, Dickson S, Smith FA (2010) Arbuscular mycorrhizas modify plant responses to soil zinc addition. Plant Soil 329:307–313

Chaves MM, Maroco JP, Pereira JS (2003) Understanding plant responses to drought—from genes to the whole plant. Funct Plant Biol 30:239–264

Chen BD, Xiao X, Zhu YG et al (2007) The arbuscular mycorrhizal fungus *Glomus mosseae* gives contradictory effects on phosphorus and arsenic acquisition by *Medicago sativa*. Linn Sci Total Environ 379:226–234

Chen GP, Wilson ID, Kim SH et al (2001) Inhibiting expression of a tomato-ripening associated membrane protein increases organic acids and reduces sugar levels of fruit. Planta 212:799–807

Chen TTH, Murata N (2008) Glycinebetaine: an effective protectant against abiotic stress in plants. Trends Plant Sci 13:1360–1385

Chethan Kumar KV, Chandrashekar KR, Lakshmipathy R (2008) Variation in arbuscular mycorrhizal fungi and phosphatase activity associated with Sida cardifolia in Karnataka. World J Agr Sci 4:770–774

Cho K, Toler HD, Lee J et al (2006) Mycorrhizal symbiosis and response of sorghum plants to combined drought and salinity stresses. J Plant Physiol 163:517–528

Chollet R, Vidal J, O'Leary MH (1996) Phosphoenolpyruvate carboxylase: a ubiquitous, highly regulated enzyme in plants. Annu Rev Plant Physiol Plant Mol Biol 47:273–298

Colla G, Rouphael Y, Cardarelli M et al (2008) Alleviation of salt stress by arbuscular mycorrhizal in zucchini plants grown at low and high phosphorus concentration. Biol Fert Soils 44:501–509

Compant S, Clément B, Sessitsch A (2010) Plant growth-promoting bacteria in the rhizo- and endosphere of plants: their role, colonization, mechanisms involved and prospects for utilization. Soil Biol Biochem 42:669–678

Dakessian S, Brown MS, Bethlenfalvay GJ (1986) Relationship of mycorrhizal growth enhancement and plant growth with soil water and texture. Plant Soil 94:439–443

Das P, Samantaray S, Rout GR (1997) Studies on cadmium toxicity in plants: a review. Environ Pollut 98:29–36

Davies FT Jr, Puryear JD, Newton RJ et al (2001) Mycorrhizal fungi enhance accumulation of chromium in sunflower (*Helianthus annuus*). J Plant Physiol 158:777–786

Davies FT, Potter JR, Linderman RG (1992) Mycorrhiza and repeated drought exposure affect drought resistance and extraradical hyphae development of pepper plants independent of plant size and nutrient content. J Plant Physiol 139:289–294

Davies FT, Potter JR, Linderman RG (1993) Drought resistance of mycorrhizal pepper plants independent of leaf P concentration response in gas exchange and water relations. Physiol Plant 87:45–53

De Mars BG, Boerner REJ (1996) Vesicular arbuscular mycorrhizal development in the Brassicaceae in relation to plant life span. Flora 191:179–189
Delgardo MJ, Ligero F, Lluch C (1994) Effects of salt stress on growth and nitrogen fixation by pea, faba-bean, common bean, and soybean plants. Soil Biol Biochem 26:371–376
Demirevska K, Zasheva D, Dimitrov R et al (2009) Drought stress effects on Rubisco in wheat: changes in the Rubisco large subunit. Acta Physiol Plant 31:1129–1138
Duan B, Yang Y, Lu Y et al (2007) Interactions between drought stress, ABA and genotypes in *Picea asperata*. J Exp Bot 58:3025–3036
Duan X, Neuman DS, Reiber JM et al (1996) Mycorrhizal influence on hydraulic and hormonal factors implicated in the control of stomatal conductance during drought. J Exp Bot 47:1541–1550
Estrada-Luna AA, Davies FT (2003) Arbuscular mycorrhizal fungi influence water relations, gas exchange, abscissic acid and growth of micropropagated Chile ancho pepper (*Capsicum annuum*) plantlets during acclimatization and post-acclimatization. J Plant Physiol 160:1073–1083
Evelin H, Giri B, Kapoor R (2012) Contribution of *Glomus intraradices* inoculation to nutrient acquisition and mitigation of ionic imbalance in NaCl-stressed *Trigonella foenum-graecum*. Mycorrhiza 22:203–217
Evelin H, Kapoor R, Giri B (2009) Arbuscular mycorrhizal fungi in alleviation of salt stress: a review. Ann Bot 104:1263–1281
Fan QJ, Liu JH (2011) Colonization with arbuscular mycorrhizal fungus affects growth, drought tolerance and expression of stress-responsive genes in *Poncirus trifoliate*. Acta Physiol Plant 33:1533–1542
Feng G, Zhang FS, Xl L et al (2002) Improved tolerance of maize plants to salt stress by arbuscular mycorrhiza is related to higher accumulation of soluble sugars in roots. Mycorrhiza 12:185–190
Fitter AH (1985) Functioning of vesicular-arbuscular mycorrhizas under field conditions. New Phytol 99:257–265
Founoune H, Duponnis R, Ba AM et al (2002) Influence of the dual arbuscular endomycorrhizal/ ectomycorrhizal symbiosis on the growth of *Acacia holosericea* (A. Cunn.ex G. Don) in glasshouse conditions. Ann Forest Sci 59:93–98
Franson RL, Milford SB, Bethlenfalvay GJ (1991) The Glycine-Glomus Bradyrhizobium symbiosis, XI. Nodule gas exchange and efficiency as a function of soil and root water status in mycorrhizal soybean. Physiol Plant 83:476–482
Fry IV, Huflejt M, Erber WWA et al (1986) The role of respiration during adaptation of the freshwater cyanobacterium *Synechococcus* 6311 to salinity. Arch Biochem Biophys 244:686–691
Gamble P, Burke JJ (1984) Effect of water stress on the chloroplast antioxidant system. I. Alteration in glutathione reductase activity. Plant Physiol 76:615–621
Garg N, Jali G, Kaur A (2006) Arbuscular mycorrhiza: nutritional aspects. Arch Agron Soil Sci 52:593–606
Garg N, Manchanda G (2009) Role of arbuscular mycorrhizae in the alleviation of ionic, osmotic and oxidative stresses induced by salinity in *Cajanus cajan* (L.) Millsp. (pigeonpea). J Agron Crop Sci 195:110–123
George E, Haussler K, Vetterlein D et al (1992) Water and nutrient translocation by hyphae of *Glomus mosseae*. Can J Bot 70:2130–2137
Gething MJ, Sambrook J (1992) Protein folding in the cell. Nature 355:33–45
Ghannoum O (2009) C4 photosynthesis and water stress. Ann Bot 103:635–644
Gianinazzi S, Gianinazzi-Pearson V, Franken P et al (1995) Molecules and genes involved in mycorrhizal functioning. In: Stoechi V, Bonfante P, Nuti M (eds) Biotechnologies of Ectomycorrhizae. Plenum, New York, pp 67–76
Gianinazzi S, Gollotte A, Binet MN et al (2010) Agroecology: the key role of arbuscular mycorrhizas in ecosystem services. Mycorrhiza 20:519–530

Giovannetti M, Citernesi AS (1993) Time courase of appresorium formation on host plants by arbuscular mycorrhizal fungi. Mycol Res 97:1140–1142

Giovannetti M, Sbrana C (1998) Meeting a non host, the behavior of AM fungi. Mycorrhiza 8:123–130

Giri B, Kapoor R, Mukerji KG (2003) Influence of arbuscular mycorrhizal fungi and salinity on growth, biomass and mineral nutrition of *Acacia auriculiformis*. Biol Fert Soils 38:170–175

Giri B, Kapoor R, Mukerji KG (2007) Improved tolerance of *Acacia nilotica* to salt stress by arbuscular mycorrhiza, *Glomus fasciculatum* may be partly related to elevated K^+/Na^+ ratios in root and shoot tissues. Microb Ecol 54:753–760

Göhre V, Paszkowski U (2006) Contribution of the arbuscular mycorrhizal symbiosis to heavy metal phytoremediation. Planta 223:1115–1122

Gomez SK, Javot H, Deewatthanawong P et al (2009) *Medicago truncatula* and *Glomus intraradices* gene expression in cortical cells harboring arbuscules in the arbuscular mycorrhizal symbiosis. BMC Plant Biol 9:10

Gonzalez-Chavez MC, Carillo-Gonzalez R, Wright SE et al (2004) The role of glomalin, A protein produced by arbuscular mycorrhizal fungi, in sequestering potentially toxic elements. Environ Pollut 130:317–323

González-Guerrero M, Azccón-Aguilar C, Mooney M et al (2005) Characterization of a *Glomus intraradices* gene encoding a putative Zn transporter of the cation diffusion facilitator family. Fungal Genet Biol 42:130–140

González-Guerrero M, Benabdellah K, Valderas A et al (2010) *GintABC1* encodes a putative ABC transprter of the MRP subfamily induced by Cu, Cd, and oxidative stress in *Glomus intraradices*. Mycorrhiza 20:137–146

González-Guerrero M, Cano C, Azcon-Aguilar C et al (2007) *GintMT1* encodes a functional metallothionein in *Glomus intraradices* that responds to oxidative stress. Mycorrhiza 17:327–335

Gorham J (1995) Betaines in higher plants—biosynthesis and role in stress metabolism. In: Wallgrove RM (ed) Amino acids and their derivatives in higher plants. Cambridge University Press, Cambridge, pp 171–203

Govindarajulu M, Pfeffer PE, Jin HR et al (2005) Nitrogen transfer in the arbuscular mycorrhizal symbiosis. Nature 435:819–823

Grattan SR, Grieve CM (1999) Salinity, mineral nutrient relations in horticultural crops. Sci Hort 78:127–157

Greger M, Lindberg S (1987) Effects of Cd^{2+} and EDTA on young sugar beets (*Beta vulgaris*). II. Net uptake and distribution of Mg^{2+}, Ca^{2+} and Fe^{2+}/Fe^{3+}. Physiol Plant 69:81–86

Guether M, Neuhaüser B, Balestrini R et al (2009) A mycorrhizal-specific ammonium transporter from *Lotus japonicas* acquires nitrogen released by arbuscular mycorrhizal fungi. Plant Physiol 150:73–83

Gupta RK (1991) Drought response in fungi and mycorrhizal plants. In: Arora DK, Rai B, Mukerji KG, Knudsen GR (eds) Handbook of applied mycology, vol 1, Soil and plants. Marcel Dekker, New York, pp 55–75

Hajiboland R, Aliasgharzadeh N, Laiegh SF et al (2010) Colonization with arbuscular mycorrhizal fungi improves salinity tolerance of tomato (*Solanum lycopersicum* L.) plants. Plant Soil 331:313–327

Hamblin AP (1985) The influence of soil structure on water movement, crop root growth, and water uptake. Adv Agron 38:95–158

Hardie K, Leyton L (1981) The influence of vesicular-arbuscular mycorrhiza on growth and water relations of red clover. I. In phosphate deficient soil. New Phytol 89:599–608

Harrison MJ (1999) Molecular and cellular aspects of the arbuscular mycorrhizal symbiosis. Ann Rev Plant Physiol Plant Mol Biol 50:361–389

Harrison MJ, van Buuren ML (1995) A phosphate transporter from the mycorrhizal fungus *Glomus versiforme*. Nature 378:626–629

Hasegawa PM, Bressan RA, Zhu JK et al (2000) Plant cellular and molecular responses to high salinity. Ann Rev Plant Physiol Plant Mol Biol 51:463–499

Hassine AB, Ghanem MES, Bouzid Lutts S (2008) An inland and a coastal population of the Mediterranean xero-halophyte species Atriplex halimus L. differ in their ability to accumulate proline and glycinebetaine in response to salinity and water stress. J Exp Bot 59:1315–1326
Hatzig S, Kumar A, Neubert A et al (2010) PEP-carboxylase activity: a comparison of its role in a C4 and a C3 species under salt stress. J Agron Crop Sci 196:185–192
Herkelrath WN, Miller EE, Gardner WR (1977) Water uptake by plants: II. The roots contact model. Soil Sci Soc Amer J 41:1039–1043
Hernandez JA, Ferrer MA, Jimenez A et al (2001) Antioxidant systems and $O_2^{\cdot-}/H_2O_2$ production in the apoplast of pea leaves. Its relation with salt-induced necrotic lesions in minor veins. Plant Physiol 127:817–831
Hildebrandt U, Hoef-Emden K, Backhausen S et al (2006) The rare, endemic zinc violets of Central Europe originate from *Viola lutea* Huds. Plant Systematics Evol 257:205–222
Hildebrandt U, Kaldorf M, Bothe H (1999) The zinc violet and its colonization by arbuscular mycorrhizal fungi. J Plant Physiol 154:709–711
Hildebrandt U, Regvar M, Bothe H (2007) Arbuscular mycorrhiza and heavy metal tolerance. Phytochem 68:139–146
Hill AE, Shachar-Hill B, Shachar-Hill Y (2004) What are aquaporins for? J Memb Biol 197:1–32
Hinsinger P (2001) Bioavailability of soil inorganic P in the rhizosphere as affected by root-induced chemical changes: a review. Plant Soil 237:173–195
Hodge A, Helgasson T, Fitter AH (2010) Nutritional ecology of arbuscular mycorrhizal fungi. Fungal Ecol 3:267–273
Hooker JE, Black KE (1995) Arbuscular mycorrhizal fungi as components of sustainable soil-plant systems. Crit Rev Biotech 15:201–212
Hsu SY, Kao CK (2003) Ammonium ion, ethylene, and abscissic acid in polyethylene glycol-treated rice leaves. Biol Plant 46:617–619
Hu Y, Schmidhalter U (2002) Limitation of salt stress to plant growth. In: Hock B, Elstner CF (eds) Plant Toxicology. Marcel Dekker Inc, New York, pp 91–224
Jahromi F, Aroca R, Porcel R, Ruiz-Lozano JM (2008) Influence of salinity on the *in vitro* development of *Glomus intraradices* and on the *in vivo* physiological and molecular responses of mycorrhizal lettuce plants. Microb Ecol 55:45–53
Janoušková M, Pavliková D (2010) Cadmium immobilization in the rhizosphere of arbuscular mycorrhizal plants by the fungal extraradical mycelium. Plant Soil 332:511–520
Janoušková M, Pavliková D, Vosátka M (2006) Potential contribution of arbuscular mycorrhiza to cadmium immobilisation in soil. Chemosphere 65:1959–1965
Jastrow JD, Miller RM, Lussenhop J (1998) Contributions of interacting biological mechanisms to soil aggregate stabilization in restored prairie. Soil Biol Biochem 30:905–916
Jeanjean R, Matthijs HCP, Onana B et al (1993) Exposure of the cyanobacterium *Synechocystis* PCC6803 to salt stress induces concerted changes in respiration and photosynthesis. Plant Cell Physiol 34:1073–1079
Joner EJ, Briones R, Leyval C (2000) Metal-binding capacity of arbuscular mycorrhizal mycelium. Plant Soil 226:227–234
Jurkiewicz A, Orlowska E, Anielska T et al (2004) The influence of mycorrhiza and EDTA application on heavy metal uptake by different maize varieties. Acta Biol Cracov 46:7–18
Kaldorf M, Kuhn AJ, Schroder WH et al (1999) Selective element deposits in maize colonized by a heavy metal tolerance conferring arbuscular mycorrhizal fungus. J Plant Physiol 154:718–728
Kaldorf M, Schemelzer E, Bothe H (1998) Expression of maize and fungal nitrate reductase in arbuscular mycorhiza. MPMI 11:439–448
Kapoor R, Bhatnagar AK (2007) Attenuation of cadmium toxicity in mycorrhizal celery (*Apium graveolens* L.). World J Microbiol Biotech 23:1083–1089
Kapoor A, Viraraghavan T (1995) Fungal biosorption—an alternative treatment option for heavy metal bearing wastewater: a review. Biores Technol 53:195–206
Karimi A, Khodaverdillo H, Sepehri M et al (2011) Arbuscular mycorrhizal fungi and heavy metal contaminated soils. Afr J Microbiol 5:1571–1576

Khan AG, Kuek C, Chaudhary TM et al (2000) Role of plants, mycorrhizae and phytochelators in heavy metal contaminated land remediation. Chemosphere 41:197–207

Kishor PB, Hong Z, Miao GH et al (1995) Over expression of Δ'-pyrroline-5-carboxilate synthetase increases proline production and confers osmotolerance in transgenic plants. Plant Physiol 108:1387–1394

Klepper B (1990) Root growth and water uptake. In: Stewart BA, Nielsen DR (eds) Irrigation of agricultural crops, vol 30, Agronomy Series. ASA-CSSA-SSSA, Madison, pp 281–322

Kramer U (2005) Phytoremediation: novel approaches to cleaning up polluted soils. Curr Opin Biotech 16:133–141

Kroopnick PM (1994) Vapor abatement cost analysis methodology for calculating life cycle costs for hydrocarbon vapor extracted during soil venting. In: Wise DL, Trantolo DJ (eds) Remediation of hazardous waste. Marcel Dekker, New York, pp 779–790

Kumar A, Sharma S, Mishra S (2011) Influence of arbuscular mycorrhizal (AM) fungi and salinity on seedling growth, solute accumulation and mycorrhizal dependency of *Jatropha curcas* L. J Plant Growth Regul 29:297–306

Lanfranco L, Bolchi A, Cesale RE et al (2002) Differential expression of a metallothionein gene during the presymbiotic versus the symbiotic phase of an arbuscular mycorrhizal fungus. Plant Physiol 130:58–67

Latef AAHA, Chaoxing H (2011) Effect of arbuscular mycorrhizal fungi on growth, mineral nutrition, antioxidant enzymes activity and fruit yield of tomato grown under salinity stress. Sci Hort 127:228–233

Leigh J, Hodge A, Fitter AH (2009) Arbuscular mycorrhizal fungi can transfer substantial amounts of nitrogen to their host plant from organic material. New Phytol 181:199–207

Leung H, Ye Z, Wong M (2006) Interactions of mycorrhizal fungi with *Pteris vittata* (As hyperaccumulator) in As-contaminated soils. Environ Pollut 139:1–8

Leyval C, Joner EJ, del Val C et al (2002) Potential of arbuscular mycorrhizal fungi for bioremediation. In: Gianinazzi S, Schüepp H, Barea JM, Haselwandter K (eds) Mycorrhizal Technology in Agriculture. Birkhäuser Verlag, Basel, pp 175–186

Li XL, Christie P (2001) Changes in soil solution Zn and pH and uptake of Zn by arbuscular mycorrhizal red clover in Zn-contaminated soil. Chemosphere 42:201–207

Liu A, Hamel C, Hamilton RI et al (2000a) Mycorrhizae formation and nutrient uptake of new corn (*Zea mays* L.) hybrids with extreme canopy and leaf architecture as influenced by soil N and P levels. Plant Soil 221:157–166

Liu A, Hamel C, Hamilton RI et al (2000b) Acquisition of Cu, Zn, Mn, and Fe by mycorrhizal maize (*Zea mays* L.) grown in soil at different P and micronutrient levels. Mycorrhiza 9:331–336

Lopez F, Vansuyt G, Casse-Delbart F et al (1996) Ascorbate peroxidase activity, not the mRNA level, is enhanced in salt stressed *Raphanus sativus* plants. Physiol Plant 97:13–20

López-Pedrosa A, González-Guerrero M, Valderas A et al (2006) *GintAMT1* encodes a functional high-affinity ammonium transporter that is expressed in the extraradical mycelium of *Glomus intraradices*. Fungal Genet Biol 43:102–110

Ludwig-Müller J (2000) Indole-3-butyric acid in plant growth and development. Plant Growth Regul 32:219–230

Mahajan S, Tuteja N (2005) Cold, salinity and drought stresses: An overview. Arch Biochem Biophy 444:139–157

Malcova R, Rydlova J, Vosatka M (2003) Metal-free cultivation of Glomus sp. BEG 140 isolated from Mn-contaminated soil reduces tolerance to Mn. Mycorrhiza 13:151–157

Manchanda G, Garg N (2008) Salinity and its effects on the functional biology of legumes. Acta Physiol Plant 30:595–618

Manchanda G, Garg N (2011) Alleviation of salt-induced ionic, osmotic and oxidative stresses in *Cajanus cajan* nodules by AM inoculation. Plant Biosys 145:88–97

Manoharan P, Shanmugaiah V, Balasubramanian N et al (2010) Influence of AM fungi on the growth and physiological status of *Erythrina variegata* Linn. Grown under different water stress conditions. Eur J Soil Biol 46:151–156

Marschner P, Baumann K (2003) Changes in bacterial community structure induced by mycorrhizal colonization in split-root maize. Plant Soil 251:279–289

Mazel A, Leshem Y, Tiwari BS et al (2004) Induction of salt and osmotic stress tolerance by overexpression of an intracellular vesicle trafficking protein AtRab7 (AtRabG3e). Plant Physiol 134:118–128

Menconi M, Sgherri CLM, Pinzino C et al (1995) Activated oxygen species production and detoxification in wheat plants subjected to a water deficit programme. J Exp Bot 46:1123–1130

Mhatre GN, Pankhurst CE (1997) Bioindicators to detect contamination of soils with special reference to heavy metals. In: Pankhurst CE, Doube BM, Gupta VVSR (eds) Biological indicators of soil health. CAB International, New York, pp 349–369

Miller G, Suzuki N, Ciftci-Yilmaz S et al (2010) Reactive oxygen species homeostasis and signaling during drought and salinity stresses. Plant Cell Environ 33:453–457

Miransari M (2010) Contribution of arbuscular mycorrhizal symbiosis to plant growth under different types of soil stresses. Plant Biol 12:563–569

Miransari M (2011) Interactions between arbuscular mycorrhizal fungi and soil bacteria. Appl Microbiol Biotech 89:917–30

Miransari M, Bahrami HA, Rejali F et al (2006) Evaluating the effects of arbuscular mycorrhizae on corn (*Zea mays* L.) yield and nutrient uptake in compacted soils. Soil Water 1:106–122

Molinari HB, Marur CJ, Daros E et al (2007) Evaluation of the stress inducible production of proline in transgenic sugarcane (*Saccharum* spp.): osmotic adjustment, chlorophyll fluorescence and oxidative stress. Physiol Plant 130:218–229

Moser D, Nicholls P, Wastyn M et al (1991) Acidic cytochrome c_6 of unicellular cyanobacteria is an indispensable and kinetically competent electron donor to cytochrome oxidase in plasma and thylakoid membranes. Biochem Int 24:757–768

Munns R, Tester M (2008) Mechanisms of salinity tolerance. Ann Rev Plant Biol 59:651–681

Nobel PS, Cui M (1992) Hydraulic conductances of the soil, the rootsoil air gap, and the root: changes for desert succulents in drying soil. J Exp Bot 43:319–326

Ogawa A, Yamauchi A (2006a) Root osmotic adjustment under osmotic stress in maize seedlings. 1. Transient change of growth and water relations in roots in response to osmotic stress. Plant Prod Sci 9:27–38

Ogawa A, Yamauchi A (2006b) Root osmotic adjustment under osmotic stress in maize seedlings. 2. Mode of accumulation of several solutes for osmotic adjustment in the root. Plant Prod Sci 9:39–46

Ohotomo R, Saito M (2005) Polyphosphate dynamics in mycorrhizal roots during colonization of as arbuscular mycorrhizal fungus. New Phytol 167:571–578

Osonubi O (1994) Comparative effects of vesicular-arbuscular mycorrhizal inoculation and phosphorus fertilization on growth and phosphorus uptake of maize (*Zea mays* L.) and sorghum (*Sorghum bicolor* L.) plants under drought-stressed conditions. Biol Fert Soils 18:55–59

Ouziad F, Hildebrandt U, Schmelzer E et al (2005) Differential gene expressions in arbuscular mycorrhizal-colonized tomato grown under heavy metal stress. J Plant Physiol 162:634–649

Ouziad F, Wilde P, Schmelzer E et al (2006) Analysis of expression of aquaporins and Na^+/H^+ transporters in tomato colonized by arbuscular mycorrhizal fungi and affected by salt stress. Environ Exp Bot 57:177–186

Parida SK, Das AB (2005) Salt tolerance and salinity effects on plants. Ecotoxicol Environ Safety 60:324–349

Patreze CM, Cordeiro L (2004) Nitrogen fixing and vesicular arbuscular mycorrhizal symbioses in some tropical legume trees of tribe Mimosaceae. Forest Ecol Manag 196:275–285

Perotto S, Brewin N, Bonfante P (1994) Colonization of pea roots by the mycorrhizal fungus *Glomus versiformae* and by *Rhizobium* bacteria: Immunological comparison using monoclonal antibodies as probes for plant cell surface components. Mol Plant-Microbe Interactions 7:91–98

Picone C (2003) Managing mycorrhizae for sustainable agriculture in the tropics. In: Vandermeer JH (ed) Tropical agroecosystems. CRC, Boca Raton, pp 95–132
Piotrowski JS, Denich T, Klironomos JN et al (2004) The effects of arbuscular mycorrhizas on soil aggregation depend on the interaction between plant and fungal species. New Phytol 164:365–373
Porcel R, Aroca R, Azcon R et al (2006) PIP aquaporin gene expression in arbuscular mycorrhizal *Glycine max* and *Lactuca sativa* plants in relation to drought stress tolerance. Plant Mol Biol 60:389–404
Porcel R, Aroca R, Cano C et al (2007) A gene from the arbuscular mycorrhizal fungus *Glomus intraradices* encoding a binding protein is up-regulated by drought stress in some mycorrhizal plants. Environ Exp Bot 60:251–256
Porcel R, Aroca R, Ruiz-Lozano JM (2012) Salinity stress alleviation using arbuscular mycorrhizal fungi. A review. Agronomy Sust Dev 32:181–200
Porcel R, Azco´n R, Ruiz-Lozano JM (2005) Evaluation of the role of genes encoding for dehydrin proteins (LEA D-11) during drought stress in arbuscular mycorrhizal *Glycine max* and *Lactuca sativa* plants. J Exp Bot 56:1933–1942
Porcel R, Barea JM, Ruiz-Lozano JM (2003) Antioxidant activities in mycorrhizal soybean plants under drought stress and their possible relationship to the process of nodule senescence. New Phytol 157:135–143
Porcel R, Ruiz-Lozano JM (2004) Arbuscular mycorrhizal influence on leaf water potential, solute accumulation and oxidative stress in soybean plants subjected to drought stress. J Exp Bot 55:1743–1750
Praba ML, Cairns JE, Babu RC (2009) Identification of physiological traits underlying cultivar differences in drought tolerance in rice and wheat. J Agron Crop Sci 195:30–46
Prove BG, Loch RJ, Foley JH et al (1990) Improvements in aggregation and infiltration characteristics of a krasnozem under maize with direct drill and stubble retention. Aust J Soil Res 28:577–590
Rabie GH, Almadini AM (2005) Role of bioinoculants in development of salt-tolerance of *Vicia faba* plants under salinity stress. Afr J Biotech 4:210–222
Redon PO, Béguiristain T, Leyval C (2008) Influence of *Glomus intraradices* on Cd partitioning in a pot experiment with *Medicago truncatula* in four contaminated soils. Soil Biol Biochem 40:2710–2712
Reid CPP (1979) Mycorrhizae and water stress. In: Riedacher A, Gagnaire-Michard J (eds) Root physiology and symbiosis, IUFRO Symposium Proceedings, Nancy France, pp. 392–408.
Remy W, Taylor TN, Hass H et al (1994) Four hundred million year old vesicular arbuscular mycorrhizae. PNAS (USA) 91:11841–11843
Repetto O, Bestel-Corre G, Dumas-Gaudot E et al (2003) Targeted proteomics to identify cadmium-induced protein modifications in *Glomus mosseae* inoculated pea roots. New Phytol 157:555–567
Requena N, Serrano E, Ocón A et al (2007) Plant signals and fungal perception during arbuscular mycorrhiza establishment. Phytochem 68:33–40
Richards LA (ed) (1954) Diagnosis and improvement of saline and alkali soils. United States Department of Agriculture, Handbook no 60, Washington DC, pp. 4–18
Rillig MC (2004) Arbuscular mycorrhizae, glomalin and soil quality. Can J Soil Sci 84:355–363
Rillig MC, Wright SF, Nichols KA et al (2001) Large contribution of arbuscular mycorrhizal fungi to soil carbon pools in tropical forest soils. Plant Soil 233:167–177
Rivera-Becerril F, Calantzis C, Turnau K et al (2002) Cadmium accumulation and buffering of cadmium-induced stress by arbuscular mycorrhiza in three *Pisum sativum* L. genotypes. J Exp Bot 53:1177–1185
Rivera-Becerril F, Van Tuinen D, Martin-Laurent F et al (2005) Molecular changes in *Pisum sativum* L. roots during arbuscular mycorrhiza buffering of cadmium stress. Mycorrhiza 16:51–60
Rodriguez RJ, Redman RS (2005) Henson JM Symbiotic lifestyle expression by fungal endophytes and the adaptation of plants to stress: unraveling the complexities of intimacy.

In: Dighton J, Oudemans P, White J (eds) The Fungal Community: Its Organization and Role in the Ecosystem. Taylor & Francis/CRC Press, Boca Raton, pp 683–696

Ruiz-Lozano JM (2003) Arbuscular mycorrhizal symbiosis and alleviation of osmotic stress. New perspectives for molecular studies. Mycorrhiza 13:309–317

Ruiz-Lozano JM, Azcon R, Palma JM (1996) Suiperoxide dismutase activity in arbuscular-my corrhizal *Lactuca sativa* L. plants subjected to drought stress. New Phytol 134:327–333

Ruiz-Lozano JM, Porcel R, Aroca R (2006) Does the enhanced tolerance of arbuscular mycorrhizal plants to water deficit involve modulation of drought-induced plant genes? New Phytol 171:693–698

Ruiz-Sánchez M, Aroca R, Muñoz Y et al (2010) Arbuscular mycorrhizal symbiosis enhances the photosynthetic efficiency and the antioxidative response of rice plants subjected to drought stress. J Plant Physiol 167:862–869

Sajedi NA, Ardakani MR, Rejali F et al (2010) Yield and yield components of hybrid corn (*Zea mays* L.) as affected by mycorrhizal symbiosis and zinc sulfate under drought stress. Physiol Mol Biol Plants 16:343–51

Salt DE, Smith RD, Raskin I (1998) Phytoremediation. Annu Rev Plant Physiol Plant Mol Biol 49:643–668

Samra A, Dumas-Gaudot E, Gianinazzi S (1997) Detection of symbiosis related polypeptides during the early stages of the establishment of arbuscular mycorrhizal between *Glomus mosseae* and *Pisum sativum* roots. New Phytol 135:711–722

Sánchez-Díaz M, Honrubia M (1994) Water relations and alleviation of drought stress in mycorrhizal plants. In: Gianinazzi S, Schüepp H (eds) Impact of arbuscular mycorrhizas on Sustainable Agriculture and Natural Ecosystems. MA Birkhäuser, Boston, pp 167–178

Sanità di Toppi L, Gabbrielli R (1999) Response to cadmium in higher plants. J Exp Bot 41:105–130

Sannazzaro AI, Echeverria M, Alberto EO et al (2007) Modulation of polyamine balance in *Lotus glaber* by salinity and arbuscular mycorrhiza. Plant Physiol Biochem 45:39–46

Sarath G, Cohen HC, Wagner FW (1986) High performance liquid chromatographic separation of leghemoglobins from soybean root nodules. Anal Biochem 154:224–231

Sardans J, Penuelas J, Ogaya R (2008) Experimental drought reduced acid and alkaline phosphatase activity and increased organic extractable P in soil in a *Quercus ilex* Mediterranean forest. Eur J Soil Biol 44:509–520

Schoefs B (2005) Protochlorophyllide reduction □ what is new in 2005? Photosynthetica 43:329–343

Shao HB, Chu LY, Jaleel CA et al (2009) Understanding water deficit stress-induced changes in the basic metabolism of higher plants – biotechnologically and sustainably improving agriculture and the ecoenvironment in arid regions of the globe. Critic Rev Biotech 29:131–151

Shen H, Christie P, Li X (2006) Uptake of zinc, cadmium and phosphorus by arbuscular mycorrhizal maize (*Zea mays* L.) from a low available phosphorus calcareous soil spiked with zinc and cadmium. Environ Geochem Health 28:111–119

Sheng M, Tang M, Chan H et al (2008) Influence of arbuscular mycorrhizae on photosynthesis and water status of maize plants under salt stress. Mycorrhiza 8:287–296

Sheng M, Tang M, Zhang F et al (2011) Influence of arbuscular mycorrhiza on organic solutes in maize leaves under salt stress. Mycorrhiza 21:423–430

Shi G, Cai Q (2009) Cadmium tolerance and accumulation in eight potential energy crops. Biotechnol Adv 27:555–561

Shokri S, Maadi B (2009) Effect of arbuscular mycorrhizal fungus on the mineral nutrition and yield of *Trifolium alexandrinum* plants under salinity stress. J Agron 8:79–83

Singh KN, Charath R (2001) Salinity Tolerance. In: Reynolds MP, Monasteiro JIO, McNab A (eds) Application of Physiology in Wheat Breeding. DF, CIMMYT, Mexico, pp 101–110

Singh LP, Gill SS, Tuteja N (2011) Unraveling the role of fungal symbionts in plant abiotic stress tolerance. Plant Signal Behav 6:175–191

Slesak I, Miszalski Z, Karpinska B et al (2002) Redox control of oxidative stress responses in the C3-CAM intermediate plant *Mesembryanthemum crystallinum*. Plant Physiol Biochem 40:669–677

Smirnoff N (1993) The role of active oxygen in the response of plants to water deficit and dessication. New Phytol 125:27–58
Smith SE, Facelli E, Pope S et al (2010) Plant performance in stressful environments: interpreting new and established knowledge of the roles of arbuscular mycorrhizas. Plant Soil 326:3–20
Smith SE, Gianinazzi V (1988) Physiological interactions between symbionts in vesicular arbuscular mycorrhizal plants. Annu Rev Plant Physiol Plant Mol Biol 39:221–224
Smith SE, Read DJ (1997) Mycorrhizal Symbiosis. Academic, San Diego, California, USA
St-Arnaud M, Vujanovic V (2007) Effects of the arbuscular mycorrhizal symbiosis on plant diseases and pests. In: Hamel C, Plenchette C (eds) Mycorrhizae in Crop Production. Haworth Food & Agricultural Products Press, Binghamton, NY, pp 67–122
Stobart A, Griffiths W, Ameen-Bukhari I et al (1985) The effect of Cd^{2+} on the biosynthesis of chlorophyll in leaves of barley. Physiol Plant 63:293–298
Studer C, Hu Y, Schmidhalter U (2007) Evaluation of the differential osmotic adjustments between roots and leaves of maize seedlings with single or combined NPK-nutrient supply. Funct Plant Biol 34:228–236
Subramanian KS, Charest C (1998) Arbuscular mycorrhizae and nitrogen assimilation in maize after drought and recovery. Physiol Plant 102:285–296
Subramanian KS, Santhanakrishnan P, Balasubramanian P (2006) Responses of field grown tomato plants to arbuscular mycorrhizal fungal colonization under varying intensities of drought stress. Sci Hort 107:245–53
Sudová R, Doubkova P, Vosatka M (2008) Mycorrhizal association of *Agrostis capillaries* and *Glomus intraradices* under heavy metal stress: combination of plant clones and fungal isolates from contaminated and uncontaminated substrates. Appl Soil Ecol 40:19–29
Szabados L, Savoure A (2009) Proline: a multifunctional amino acid. Trends Plant Sci 15:89–98
Tian CJ, Kasiborski B, Koul R et al (2010) Regulation of the nitrogen transfer pathway in the arbuscular mycorrhizal symbiosis:gene characterization and the coordination of expression with nitrogen flux. Plant Physiol 153:1175–1187
Tinker PB (1976) Transport of water to plant roots in soil. Philos Trans Royal Soc London Ser B 273:445–461
Tisdall JM (1991) Fungal hyphae and structural stability of soil. Aust J Soil Res 29:729–743
Tonin C, Vandenkoornhuyse P, Joner EJ et al (2001) Assessment of arbuscular mycorrhizal fungi diversity in the rhizosphere of *Viola calaminaria* and effect of these fungi on heavy metal uptake by clover. Mycorrhiza 10:161–168
Tsai YC, Hong CY, Liu LF et al (2005) Expression of ascorbate peroxidase and glutathione reductase in roots of rice seedlings in response to NaCl and H_2O_2. J Plant Physiol 162:291–299
Tullio M, Pierandrei F, Salerno A et al (2003) Tolerance to cadmium of vesicular arbuscular mycorrhizae spores isolated from a cadmium-polluted and unpolluted soil. Biol Fertil Soils 37:211–214
Turnau K, Mesjasz-Przybylowicz J (2003) Arbuscular mycorrhizal of *Berkheya coddii* and other Ni-hyperaccumulating members of Asteraceae from ultramafic soils in South Africa. Mycorrhiza 13:185–190
Vidali M (2001) Bioremediation: an overview. J Appl Chem 73:1163–1172
Vierheilig H, Lerat S, Piché Y (2003) Systemic inhibition of arbuscular mycorrhiza development by root exudates of cucumber plants colonized by *Glomus mosseae*. Mycorrhiza 13:167–170
Visioli G, Marmiroli M, Marmiroli N (2010) Two-dimensional liquid chromatography technique coupled with mass spectrometry analysis to compare the proteomic response to cadmium stress in plants. J Biomed Biotechnol 2010:1–10
Volesky B (1990) Biosorption by fungal biomass. In: Volesky B (ed) Biosorption of heavy metals. CRC, Boca Raton, pp 140–171
Wang J, Fang W, Yang Z et al (2007) Inter- and intra-specific variations of cadmium accumulation of 13 leafy vegetable species in a greenhouse experiment. J Agric Food Chem 55:9118–9123
Wang W, Vinocur B, Altman A (2003) Plant responses to drought, salinity and extreme temperatures: toward genetic engineering for stress tolerance. Planta 218:1–14

Weissenhorn I, Leyval C, Berthelin J (1993) Cd-tolerant arbuscular mycorrhizal (AM) fungi from heavy-metal polluted soils. Plant Soil 157:247–256
Wingler A, Lea PJ, Quick WP et al (2000) Photorespiration: metabolic pathways and their role in stress protection. Philos Trans R Soc Lond B Biol Sci 355:1517–1529
Wright SE, Franke-Snder M, Morton JB et al (1996) Time course study and partial characterization of a protein on hyphae of arbuscular mycorrhizal fungi during active colonization of roots. Plant Soil 181:193–203
Wright SF Management of arbuscular mycorrhizal fungi. In: Zobel RW, Wright SF. eds. Roots and Soil Management: Interactions between Roots and the Soil. American Society of Agronomy, Crop Science Society of America, Soil Science Society of America, Madison, WI; 2005:183–197.
Wright SF, Upadhyay A (1998) A survey of soils for aggregrate stability and glomalin, a glycoprotein produced by hyphae of arbuscular mycorrhizal fungi. Plant Soil 198:97–107
Wright SF, Upadhyay A, Buyer JS (1998) Comparison of N-linked oligosaccharides of glomalin from arbuscular mycorrhizal fungi and soils by capillary electrophoresis. Soil Biol Biochem 30:1853–1857
Wright SF, Upadhyaya A (1996) Extraction of an abundant and unusual protein from soil and comparison with hyphal protein of arbuscular mycorrhizal fungi. Soil Sci 161:575–586
Wu F, Yang W, Zhang J et al (2010a) Cadmium accumulation and growth responses of a poplar (*Populus deltoids*×*Populus nigra*) in cadmium contaminated purple soil and alluvial soil. J Hazard Mater 177:268–273
Wu QS, Xia RX (2006) Arbuscular mycorrhizal fungi influence growth, osmotic adjustment and photosynthesis of citrus under well-watered and water stress conditions. J Plant Physiol 163:417–25
Wu QS, Zou YN, Xia RX et al (2007) Five *Glomus* species affect water relations of Citrus tangerine during drought stress. Bot Stud 48:147–154
Wu QS, Xia RX, Zou YN (2008) Improved soil structure and citrus growth after inoculation with three arbuscular mycorrhizal fungi under drought stress. Eur J Soil Biol 44:122–128
Wu QS, Zou YN, He XH (2011) Differences of hyphal and soil phosphatase activities in drought-stressed mycorrhizal trifoliate orange (*Poncirus trifoliata*) seedlings. Sci Horti 129:294–298
Wu QS, Zou YN, Liu W et al (2010b) Alleviation of salt stress in citrus seedlings inoculated with mycorrhiza: changes in leaf antioxidant systems. Plant Soil Environ 56:470–475
Wulf A, Manthey K, Doll J et al (2003) Transcriptional changes in response to arbuscular mycorrhiza development in the model plant *Medicago truncatula*. Mol Plant-Microb Interact 16:306–314
Yeo AR, Lee KS, Izard P et al (1991) Short- and long-term effects of salinity on leaf growth in rice (*Oryza sativa* L.). J Exp Bot 42:881–889

Chapter 15
Biofertilizers: A Sustainable Eco-Friendly Agricultural Approach to Crop Improvement

Ranjan Kumar Sahoo, Deepak Bhardwaj, and Narendra Tuteja

1 Introduction

Both the quality and quantum of agricultural production is dependent on the quality of soil. Soil is not only an important resource for the farmer but it also provides niche to the various organisms from microbes to mammals. Globally, the loss in crop productivity has been caused due to a poor management of top soil. Excessive exploitation of arable land without sufficient addition of useful nutrients as well as the detriment caused by salinization and drought has been known to be the chief reason for the poor quality of soil (Yuan et al. 2007). Hence the proper maintenance of soil is a key to crop productivity. Chemical fertilizers serve as the best and often a too convenient method to enrich soil with useful nutrients and to meet the global demands of food production; however, they are economically very expensive and hazardous to health (Righi et al. 2005). Fortunately, nature has provided innate machinery consisting of various microbes and useful flora of the soil to answer this challenge. This machinery not only maintains the richness of the soil but it also works in tandem with plants as part of an ecosystem. This machinery is what constitutes "biofertlizers" and is a central part of what we know as green agriculture.

Biofertlizers are the preparations containing efficient strains of micro-organisms, organic products and dead tissues of plants which give nutrients to the soil as well as plants. It gradually enhances soil fertility and increases crop yield. Biofertlizers convert unavailable form of nutrients to available form by increasing the microbial population in the rhizosphere. Microbial populations are responsible for the supply of soluble nutrients to the plants. They are useful in various ways that includes

R.K. Sahoo • D. Bhardwaj
International Centre for Genetic Engineering and Biotechnology (ICGEB), Aruna Asaf Ali Marg, New Delhi 110 067, India

N. Tuteja
Plant Molecular Biology, International Centre for Genetic Engineering and Biotechnology (ICGEB), Aruna Asaf Ali Marg, New Delhi, India

N. Tuteja and S. Singh Gill (eds.), *Plant Acclimation to Environmental Stress*,
DOI 10.1007/978-1-4614-5001-6_15,

fixing of atmospheric nitrogen and solubilization of plant nutrients like phosphorus and sulphur. Microbiota also stimulates plant health by suppression of disease, degradation of contaminants and promotion of plant growth through synthesis of growth-promoting substances, like auxins and cytokinins, and also provides protection against biotic and abiotic stresses (Pedraza 2008; Sturz and Christie 2000; Kader et al. 2002). Preparations containing these can also be considered as fertilizers under a broad term, microbial fertilizers.

The commercial history of biofertilizers began with the launch of "'Nitrogin" by Nobbe and Hiltner, bacterial inoculants for legumes in 1895. Timonin (1948) prepared bacterial inoculants named 'Alnit' from the mixture of useful bacteria and compost. This proved efficient for the growth of non-leguminous crops. These bacteria were identified to be common "anmonifiers". The discovery of *Clostridium* and *Azotobacter* opened a new field for the search of cheap bacterial fertilizer (Ashby 1907). Ghosh (2000) reported that the use of biofertilizer alone showed significant improvement in plant height and number of tillers per plant. The rhizosphere of plants contains various species of soil bacteria which may stimulate plant growth by various and different mechanisms. These bacteria are collectively known as plant growth-promoting rhizobacteria (PGPR). One of the mechanisms by which they function is through fixing of atmospheric N_2, which increases the availability of usable form of N_2 in the rhizosphere and which in turn helps in the better growth of plant roots. They are also known to increase the yield attributes and seed yield over control. They also promote beneficial plant and microbe symbiosis and therefore are more widespread and utilized as biofertilizers. However, not all the PGPR are utilized as biofertilizers.

In this chapter PGPR that include different types of biofertilizers, microbes involved in it and their impact on different crops have been described; in the last we have described the various mechanisms that are involved in the activity of PGPR (Table 15.1). The various biofertilizers which are described are *Azotobacter*, *Azospirillum*, *Rhizobium*, Blue green algae, phosphorus and potassium solubilizing micro-organisms (KSM), mycorrhizae and vermicompost.

2 Plant Growth-Promoting Rhizobacteria

PGPR are naturally occurring soil bacteria that remain in the vicinity of roots of plants for the safety and availability of nutrients to the plants. There is a symbiotic relationship between these bacteria and plants. The plethora of micro-organisms that benefit plants is termed as PGPR; they influence their growth in many ways. In any given situation plants cannot grow in isolation, they require micro-organism to support their life and vice versa (Doyle 1998). It has been seen that certain strains of PGPR help in the improvement of biomass either in root or shoot growth (Karlidag et al. 2007). They not only help plants in providing the necessary conditions for easy uptake of nutrients but also help them in defending themselves from attack of various pathogens. Moreover, they help plants in combating various biotic

Table 15.1 List of growth promoting rhizobacteria and their relation to host plants

Name of PGPR	Function	Crops	Relationship to host	References
Azotobacter sp.	Nitrogen fixation	Wheat, Oat, Barley Mustard, Seasum Rice, Linseeds, Sunflower Castor, Maize, Sorghum Cotton, Jute, Sugarbeats Tabacco, Tea, Coffee Rubber, Coconuts	Free-living	Burgmann et al., (2003)
Azospirillum sp.	Nitrogen fixation	Potato, Radish, Spinach Turnip, Carrot, Perwal Onion, Brinjal, Cauliflower, Cabbage Tomato, Chillies, Pearl millets, Fingermillets Kodomillet, Rice, Wheat, Oat, Barley	Free-living	Kanan et al. (2010), Dobereiner and Day, (1976) and Lakshmi-kumari et al. (1976)
Rhizobium sp.	Nitrogen fixation	Chickpea, Pea, Groundnut, Soyabean, Beans, Lentil, Lucern, Berseem, Green gram, Black gram, Cowpea, Pigeon pea	Symbiosis	Bajpai et al. (1974)
Bacillus sp.	Phosphorus/ Potash Solubilizer	Cotton, Jute, Banana, Potato	Free-living	Sheng and He (2006)
Aspergillus sp.	Phosphorus/ Potash Solubilizer	Black gram, Ground nut	Free-living	Kundu and Gaur (1980)
Penicillium sp.	Phosphorus Solubilizer	Green gram, Soyabean	Free-living	Kucey (1988)
Tolypothrix	Nitrogen fixation	Rice	Symbiosis	Kaushik (1998)
Scytonema	Nitrogen fixation	Rice	Symbiosis	Kaushik (1998)
Nostoc	Nitrogen fixation	Rice	Symbiosis	Kaushik (1998)
Anabaena	Nitrogen fixation	Rice	Symbiosis	Kaushik (1998)
Plectonema	Nitrogen fixation	Rice	Symbiosis	Kaushik (1998)
Ectomycorrhiza	Phosphorus uptake	Chickpea, mungbean, wheat Rice, Sorghum, Barley, Onions, Cowpea, rubber Coffee, Sugarcane	Symbiosis	Lamabam et al. (2011), Singh et al. (1991)
Endomycorrhiza	Phosphorus uptake	Chickpea, mungbean, wheat Rice, Sorghum, Barley, Onions, Cowpea, rubber Coffee, Sugarcane	Symbiosis	Lamabam et al. (2011)

and abiotic stresses (Saravanakumar et al. 2010). The various advantageous effects of PGPRs are efficient seed germination, plant height, increased chlorophyll content and nodulation in legumes (Holzinger et al. 2011; Tittabutr et al. 2008). They ensure the availability of certain important macronutrients that includes nitrogen, phosphorous, sulphur, iron and copper. They help in the induction of various growth regulators (Ahmad et al. 2008). They enhance the growth of other beneficial bacteria and fungi.

Some of the PGPR that have been identified in the last few years and that add to the natural flora of soil to change complex matter into simple and usable form are *Arthrobacter, Alcaligenes, Azospirillum, Azotobacter, Bacillus, Burkholderia, Enterobacter, Klebsiella, Pseudomonas, Serratia,* etc. We can further classify these PGPR into bio-protectants, bio-stimulants and biofertlizers. Some of the organisms that have been used at commercial level as bio-protectants are *Bacillus*, *Streptomyces*, *Pseudomonas*, *Burkholderia* and *Agrobacterium*. Bio-protectants induce SAR and production of siderophore. SAR is one of the methods that plant acquire to protect themselves. PGPR triggers defence-related genes with or without the involvement of SA and MeJA (Zhang et al. 2002; Sjoerd et al. 2009). The up-regulation of defence-related genes suppress the growth of deleterious micro-organisms that affects the growth of plants.

3 Types of Biofertlizers

4 Nitrogenous Biofertilizer

Nitrogen (N_2) is a key component of the constituents of life such as DNA, RNA, vitamins, hormones, proteins and enzymes which regulate the various metabolic processes. The application of nitrogen to rice in particular influences crop yield in different ways. For example, the addition of nitrogenous fertilizer increases the leaf area, tiller formation, photosynthetic productivity, etc., which in turn improves the biomass, grain yield and quality of production (Lampayan et al. 2010; Jiang et al. 2008). Deficiency of nitrogen is one of the current problems in agricultural soils (Abrol et al. 1999). Nitrogen is significant for rice in the initial days of its cultivation. Moreover, rice requires 1 kg of nitrogen to produce 15–20 kg of grain. In the tropics, lowland rice yields 2–3.5 Mg/ha using naturally available N_2 derived from biological nitrogen fixation (BNF) by free-living, plant-associated diazotrophs and from mineralization of soil N_2. Nitrogen-rich food is still the most favourite among consumers and its management is necessary for sustainable agriculture and food security (Spiertz 2010). Presently, higher yields are necessary to support the unprecedented growth in population. During the green revolution and since the 1960s, the application of nitrogenous fertilizers boosted rice yields by 200.5 million tonnes to match increasing demands (Lin et al. 2009). In the next 25 years, production will have to increase by nearly 70% more than the 460 million Mg harvested (IRRI 1993).

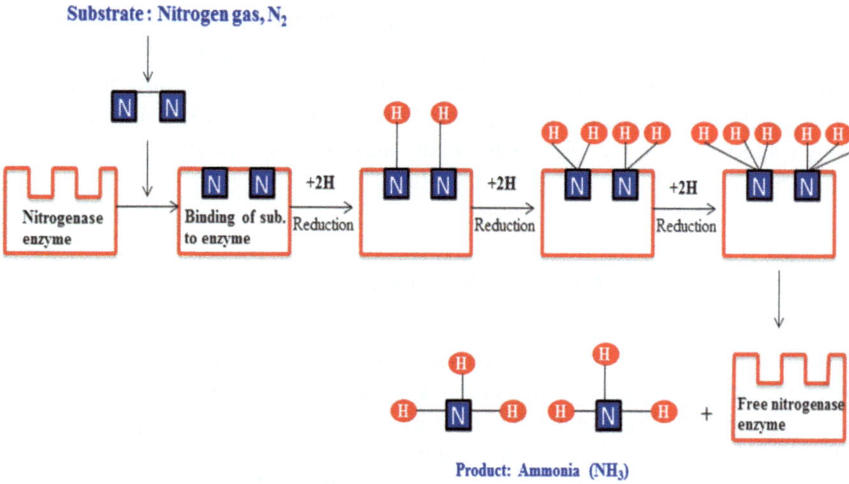

Fig. 15.1 Model to show the reduction of atmospheric nitrogen to ammonia by nitrogenase enzyme. N, Nitrogen; H, Hydrogen; and NH_3, Ammonia

Enhancing rice production from the present 8–12 Mg/ha in 2020 (Green Revolution 2) would require an increase in fertilizer application from 220 to 400 kg/ha. At current levels of N_2 use efficiency, it would require approximately double the volume of 10 million mg of nitrogenous fertilizer. It is in this context that biofertilizer-derived BNF gained importance.

4.1 Nitrogen Fixation Systems and Organisms

Nitrogen fixation is the conversion of atmospheric nitrogen (N_2) to ammonia (NH_3). Nitrogenase is the key enzyme for biological nitrogen fixation, which is possessed by diverse group of micro-organisms. Nitrogenase catalyses the reduction of nitrogen gas to ammonia in the absence of oxygen (Fig. 15.1). In the natural habitat the biological N_2-fixation is the most important source of nitrogen. Many bacteria and archea bacteria can fix biological nitrogen. It is estimated that free living nitrogen fixing prokaryotes (Table 15.1) of soil fix approximately 60 kg/ha nitrogen in a year (Burgmann et al. 2003). Biological N_2-fixation is gaining importance in the rice ecosystem because of environment degradation caused by the excessive usage of nitrogenous fertilizers to increase rice productivity. Thus, biological fixation of atmospheric N_2, especially non-symbiotic N_2-fixation in the soil, has drawn the attention of various agriculture scientists in recent decades especially for improving the plight of agriculture.

Azotobacter, *Azospirillum* and *Rhizobium* play a significant role in supply of atmospheric nitrogen to plants. In other words they help in the availability of atmospheric nitrogen available that is about 80,000 exact term over a hectare of land. Efficient strains of *Azotobacter* and *Azospirillum* and *Rhizobium* fix nitrogen significantly in comparison to inorganic fertilizers, and help in plant growth. In addition to N_2-fixation, they are supposed to promote the physiology of plants (Table 15.2) or improve the root morphology of the rice plant (Choudhury and Kennedy 2004).

4.2 *Azotobacter as Efficient Nitrogen Fixer*

Azotobacter is usually an aerobic, free-living, motile, oval or spherical gram negative (−ve) soil bacteria, which produce capsular slime (Tejera et al. 2005). An obligate diazotroph soil-dwelling organism is used by one of the important steps of nitrogen cycle with wide variety of metabolic capabilities, which include the capability to fix atmospheric nitrogen by converting it to ammonia (Gaur 2006). There are different strains of *Azotobacter* which can be distinguished on the basis of chemical and biological characters. However, some strains have higher nitrogen fixing capability than others (Burgmann et al. 2003). *Azotobacter chlorococcum* is a commonly occurring species of *Azotobacter* that can be found in most of the agricultural lands; however, *Azotobacter* is less versatile in the rhizosphere of crop plants and uncultivated land. *Azotobacter* is used as a biofertilizers for different economically important plants like wheat, oat, barley, mustard, sesame, rice, linseeds, sunflower, castor, maize, sorghum, cotton, jute, bajra, sugar beats, tobacco, sugarcane, tea, coffee, rubber and coconuts (Table 15.1).

Besides nitrogen fixation, *Azotobacter* also produces thiamine, riboflavin, indole acetic acid (IAA) and Gibberellins (GA). *Azotobacter* when applied to seeds can improve seed germination to a considerable extent; moreover, owing to its antifungal nature it also protects young seedlings from being attacked by fungal pathogens. It also releases vitamins and phytohormones that help plants in combating plant diseases; therefore it plays an important role in biotic stress tolerance (Kader et al. 2002). Azotobacter was also found to possess glucose dehydrogenase enzyme for the symbolizations of minerals. Some strains of *Azotobacter* also produce extracellular polysaccharides which protect the cell form desiccation and protozoan attack. These strains can be used as biofertilizers as well as for phytoremediation (Looijesteijn et al. 2001; Aguilera et al. 2008). The different species of the genus are *A. insignis, A. macrocytogenes, Azotobacter vinelandii, Azotobacter beijerinckii, Azotobacter nigricans, Azotobacter armeniacus* and *Azotobacter paspali.*

4.3 *Effects of Azotobacter Biofertilizer on Crop*

Field experiments conducted with *Azotobacter* inoculums showed that the crop yield can be increased in a few days (Shende 1987; Kızılkaya 2008). The inoculums of *Azotobacter* not only fix nitrogen but it also produces some growth-promoting

Table 15.2 Some common microbes used as biofertilizer and their effect against stress (biotic)

Microbes	Plant growth promoting substances	Antibiotics against pathogen (biotic stress)	Crop	References
Azotobacter	Vitamins Nicotinic acid Panthothenic acid Biotin Heteroauxins Gibberellins	Anisomycin (active against pathogens like Alternaria, Fusarium, Colletotrichum, Rhizoctonia, Microfomina, Diplodia, Botryadiplodia and Cephalosporium)	Maize Sorghum Wheat Cotton Rice Potato Onion Brinjal	Shende et al. (1975), Meshram et al. (2004)
Azospirillum	Indole acetic acid Indole lactic acid Gibberellins		Rice	Okon (1994); Umali-Garcia et al. (1980); Lin et al. (1983), Hubbel et al. (1979); Tien et al. 1979
Blue-green algae	Nicotinic acid Panthothenic acid Folic acids		Rice	Venkatraman and Neelakantan (1967); Kaushik (1998)
Phosphate solubilizing Micro-organisms	Thiamin Biotin Riboflavin Vitamin		Maize Sorghum Wheat Cotton Rice	Lockhead (1957), Barea et al. (1976)
Mycorrhizae	Gibberellin Cytokinin	Diaretyrenitrile, Polyacetylene (active against Phytophthora, Pythium, Rhizoctonia, Fusarium)	Maize Sorghum Wheat Cotton Rice	Sasek and Musilek (1968)

substances like gibberellins and vitamins which increase the growth of plants (Table 15.2).

Bukatsch and Heitzer in Germany at the Institute of Pasteur observed wide variation in the nitrogen fixing power of 11 strains of *Azotobacter* isolated from rhizosphere of wild and crop plants. Filtrates of *Azotobacter* cultures were observed to stimulate root growth at low concentration and cause retardation at high concentrations. Interestingly, they were found in the roots of *Zea mays* (Toledo et al. 1985). In sand cultures of peas, German workers obtained increased yields and nitrogen uptake due to inoculation with *Azotobacter*. Kurguzov (1954) observed that inoculation with *Azotobacter* at sowing time increased the available NO_3^-, P_2O_5 and K_2O in root zone. It was found that *Azotobacter* cannot grow in isolation and it requires other complements which include organic matter and composts; moreover, addition of fresh plants residues improves its growth (Table 15.2). Majority of fungi tested belonging to genera of *Alternaria*, *Fusarium*, *Colletotrichum*, *Rhizoctonia*, *Microfomina*, *Diplodia*, *Botryadiplodia* and *Cephalosporium* were found to be suppressed by *Azotobacter* (Table 15.2).

4.4 Azospirillum as Efficient Nitrogen Fixer

Azospirillum is also another free-living motile, gram variable bacterium (Table 15.1). It is microaerobic bacteria which perform well in flooded conditions. *Azospirillium* not only fixes nitrogen but it also releases plant growth-promoting substances (Okon and Labandera-Gonzalez 1994). They were isolated from the rhizosphere of many grasses and cereals all over the world, in tropical as well as in temperate climates (Dobereiner et al. 1976). *Azospirillum* was shown to exert beneficial effects on plant growth and crop yields both in greenhouses and fields (Boddey et al. 1986; Okon and Labandera-Gonzalez 1994). Different species of the genus are *Azospirillum lipoferum, Azospirillum brasilense, Azospirillum amazonense* (Magalhaes et al. 1983), *Azospirillum halopraeferens* (Reinhold et al. 1987) and *Azospirillum irakense* (Khammas et al. 1989).

Under certain environmental and soil conditions, *Azospirillum* can positively influence plant growth, crop yields and N-content of the plant. The plant stimulatory effect exerted by *Azospirillum* has been attributed to several mechanisms, including biological nitrogen fixation and auxin production (Table 15.2). Moreover, it was found that *Azospirillum* produces auxin-type phytohormones and does not release significant amounts of ammonium under diazotrophic growth; therefore it is considered more important due to its hormone-releasing activity than nitrogen fixation (Umali-Garcia et al. 1980; Lin et al. 1983; Dobbelaere et al. 2003). Interestingly, it was observed that *Azospirillium* colonies on the roots of one crop can easily colonize the roots of other crops too where it helps in increasing the rate of mineral uptake by plant roots (Lin et al. 1983) thus enhancing crop productivity (Lin et al. 1983). Upon *Azospirillum* inoculation an alteration in root morphology was observed, which has been ascribed to the bacterial production of plant growth regulating substances (Table 15.2) (Pacovsky et al. 1985). An increased number of lateral roots and root hairs increase the root surface available for nutrients (Sarig et al. 1992). *Azospirillium* is versatile in nature and can be found in temperate to desert environment though one species is also reported from saline soil (Berkum and Bohlool 1980; Rahman et al. 2007) and experimental evidences indicated that *Azospirillum* inoculation to seed, root and soil significantly increased the straw yield of rice over control and maximum straw yields of 9.34 t/ha and 9.22 t/ha were observed in kharif and rabi season respectively (Gopalswamy et al. 1989).

4.5 Effects of *Azospirillum* Biofertilizer on Crop

The effect of *Azospirillum* inoculation on a number of crop plants has been recently well documented by Dobereiner and Day (1976) and Lakshmi-kumari et al. (1976) (Table 15.1). By the use of *Azospirillum* as a seed inoculants, savings of 20–30 kg N/ha equivalents could be achieved in crops like barley, sorghum and millets (SubbaRao et al. 1980; Tilak and Murthy 1983). The affinity of *Azospirillum* to plant roots were carried out by Lakshmi et al. (1977) under aseptic conditions by growing plants on nitrogen-free seedling agar and inoculating them with 48-h-old culture of *Azospirillum*. The first sign of their adaptability to the root system was the proliferation and colonization of the bacteria around the root hairs, followed by their entry into the cortical layers often extending into the xylem. Dewan and SubbaRao (1979) showed that the application of *Azospirillum* cultures to rice and wheat increases the root biomass; plant growth responses observed after inoculation of *Azospirillum* have been explained by hormone production by this organism (Table 15.2) (Hubbel et al. 1979; Tien et al. 1979).

Kannan and Ponmurugan (2010) showed the percentage of seed germination was higher in *Azospirillum*-treated seeds than in control. Similarly, shoot and root lengths and fresh and dry weights of paddy varieties treated with *Azospirillum* inoculation showed better response than the untreated plants due to the secretion of plant growth hormones by *Azospirillum*. The biochemical parameters such as total chlorophyll, carotenoid, soluble protein and sugar and physiological parameters like photosynthetic rate were also increased to varying level in *Azospirillum*-treated plants. The overall studies indicated that the growth of *Azospirillum*-treated paddy seedlings excelled over the untreated ones due to biofertilizer effect upon nitrogen fixation.

Swedrzynska and Sawicka (2001) studied the changes in the numbers of *Azospirillum* bacteria growing in the soil along with winter wheat, oat and maize crops. The population of bacteria was estimated at different developmental stages of plants. Inoculation of cereals with *Azospirillum brasilense* bacteria contributed to the increase of their numbers in soil. No significant influence of fungicidal seed dressings on the numbers of *Azospirillum* bacteria has been noted till date. The application of mineral nitrogen to the crops was favourable for the multiplication of *Azospirillum* bacteria.

Gopalswamy et al. (1989) reported that *Azospirillum* inoculation to rice seed plus root plus soil significantly increased the plant height as against un-inoculated control; a maximum plant height of 88.9 and 77.6 cm was observed in kharif and rabi season respectively. More soil application was ineffective and it was on par with the un-inoculated control for plant height. Experimental evidences indicated that application of 50% of N as inorganic fertilizer plus 25% N thorough *Ipomoea carnea* plus *Azospirillum* as seed and soil application recorded maximum plant height of 96.4 cm (Balasubramanian and Veerabadran 1997).

Significant response to *A. brasilense* inoculation in the presence of limited supply of N was reported in the maize dry mass which was found to be increased by 64.0 g/plant as compared to un-inoculated condition where the dry mass was only 55.0 g/plant. Gopalswamy et al. (1989) observed that application of *Azospirillum* or combination of seed + root + soil treatment in rice crop significantly increased the panicle length as against the un-inoculated plots. Maximum panicle length of 19.1 and 22.4 cm was observed in kharif and rabi season respectively. Balasubramanian and Veerabadran (1997) found significant improvement in panicle length by application of N through inorganic fertilizer with green leaf manure and *Azospirillum* compared with inorganic N and control. Application of 50% N as inorganic fertilizer + 25% N in the form of prickly *Sesbania* plus *Azospirillum* as seed and soil treatment resulted in higher panicle length (24.2 cm). Experimental evidences indicated that application of *Azospirillum* along with sub-optional dose of either 75 or 50 kg N/ha showed an increasing trend in the number of filled grains as compared to application of chemical fertilizer alone. Balasubramanian and Veerabadran (1997) were of the opinion that application of 50% N as inorganic fertilizer plus 25% in the form of prickly *Sesbania* plus *Azospirillum* as seed and soil treatment in rice resulted maximum number of grains/panicle as compared with only inorganic N application and the control. Experimental results revealed that application of 50% N as inorganic fertilizer plus 25% N in the form of prickly *Sesbania* plus *Azospirillum* as seed and soil treatment in rice gave highest 1,000 grain weight of 23.1 g as compared with application of only inorganic fertilizer N and the control (Balasubramanian and Veerabadran 1997). Kumar et al. (1989) showed that seed/seedlings or soil application of *Azospirillum* plus 50% of N recommended dose to rice crop produced grain yield of 4.07 and 2.90 t/ha in low land and up land condition respectively, which was at par with 100% N dose.

Balasubramanian and Veerabadran (1997) stated that application of 50% of N as inorganic fertilizer plus 25% of N as prickly *Sesbania* plus *Azospirillum* as seed and soil application recorded higher grain yield of 588.7 q/ha in rice. Experimental evidence indicated that application of 75 kg N/ha plus *Azospirillum* inoculation to rice significantly increased the grain yield (3.5 t/ha) being 188% higher over control, which was at par with application of 100 kg N/ha alone. Therefore, saving of 25% of fertilizer N is possible due to inoculation with *Azospirillium* (Balasubramanian and Veerabadran 1997; Gopalswamy et al. 1989). It was observed that in rice, maize, sorghum and bajra use of *Azospirillum* without basal application of nitrogen was more desirable than applying 30 kg N/ha (Panwar 1991). Tien et al. (1979) found increased yield of pearl millet and attributed this increase due to IAA, GA and

Cytokinin-like substances. These growth promoting substances were produced by *Azospirillum.* It was shown that *Azospirillium* can promote root growth through nitric oxide (NO) mediated pathway (Molina-Favero et al. 2007).

4.6 Combined Effect of *Azotobacter* and *Azospirillum* on Crops

Seed inoculation in combination with *Azospirillum brasilense* and *Azotobacter chroococcum* produced synergistic effect on yield of maize, sorghum and barley (Tilak and Murthy 1983). Zambre et al. (1984) stated that tiller numbers of wheat increased by the application of nitrogenous fertilizer (up to 120 kg N/ha) along with *Azotobacter chroococcum* and *Azospirillum brasilense.* Similarly, wheat seeds inoculated with *Azotobacter*, *Azospirillum* and composted refuse stimulated tiller number and plant growth. Effect is more with *Azotobacter* than *Azospirillum* (Ishac et al. 1986). Zambre et al. (1984) are of the opinion that inoculation of wheat seeds with *Azotobacter chroococcum* increased the number of effective tiller per plant. Similarly they also stated that *Azospirillum brasilense* when inoculated with wheat seeds produced more number of effective tillers per plant.

Wani et al. (1988) found that continued inoculation of *Azotobacter* and *Azospirillum* in pearl millet plants for 2 or 3 years increased plant biomass yield. Dewan and SubbaRao (1979) are of the opinion that root biomass of rice seedling increased due to inoculation with *Azospirillum brasilense* and *Azotobacter chroococcum* alone or in combination. The increase in biomass of root was better in unsterilized soil than in sterilized soil with or without inorganic N applied as urea. Experimental evidences indicated that root length of wheat 35 days after seedling was largest when inoculated with *Azotobacter* plus *Azospirillum* but in maize root length increased only in sterilized soil when inoculated with *Azotobacter* alone and also in combination with *Azospirillum.* Wani et al. (1988) stated that application of *Azotobacter* plus *Azospirillum* plus Cyanobacteria along with one third of the recommended dose of chemical –N dose to rice variety giza-172 produced bigger size of grains.

Grain yields of rice and wheat were increased by inoculation with *Azotobacter* and *Azospirillum* along with the application of up to 120 kg N/ha N fertilizer. *Azospirillum* gave better yields than *Azotobacter* (Zambre et al. 1984; Wani et al. 1988; Gopalswamy et al. 1989). Increased grain yields of >10% (up to 33%) over the un-inoculated control were observed in pearl millet and maize plants when inoculation with *Azotobacter* and *Azospirillum* was carried out (Wani et al. 1988; Pandey et al. 1998).

Application of *Azotobacter* and *Azospirillum* to wheat crop gave the grain yield of 3.05–3.85 and 3.16–4.04 t/ha respectively over the yield of 2.90 to 3.22 t/ha without N. Application of 40 kg N plus *Azotobacter* was reported to be the most efficient fertilizer for wheat (Zambre et al. 1984). Experimental evidence indicated that combined inoculation of *Azotobacter* and *Azospirillum* produced higher grain yield of sorghum (3.32 t/ha) than inoculation with *Azotobacter* (2.53 t/ha) or *Azospirillum* alone (2.97 t/ha) or from control (2.27 t/ha). Experimental results revealed that when maize cv. Vijay composite seeds inoculated with 0.5 kg/ha *Azospirillum* plus 50% recommended dose of NPK and 0.5 kg/ha *Azotobacter* gave highest overall grain yields of 3.26 and 2.42 t/ha respectively which were 32.5 and 44% increase in yield over the respective controls. Inoculation of maize seeds with *Azotobacter chroococcum* plus *Azospirillum lipoferum* and or *Bradyrhizobium japonicum* along with 45 kg N (50% of recommended dose) produced higher straw yield than with 90 kg N and or no inoculation.

Balasubramanian and Veerabadran (1997) were of the opinion that combined application of *Azospirillum* along with 50% N as inorganic fertilizer and 25% N in the form of prickly *Sesbania* significantly increased the straw yield (588.99 t/ha) of rice over other treatments. Experimental results revealed that rice and wheat seeds inoculated with *Azotobacter* and *Azospirillum* along with increased rates of N fertilizer produced higher straw yield. *Azospirillum* gave better result than *Azotobacter* (Wani et al. 1988). Experimental evidences indicated that continued seed inoculation for 2 or 3 years with *Azotobacter* and *Azospirillum* to wheat, maize and pearl millet plants increased N uptake by the plants (Wani et al. 1988).

Wani et al. (1988) reviewed that combined application of three N fixers viz: *Azotobacter*, *Azospirillum* and Cyanobacteria along with third of the chemical N to the rice var. Giza-172 stimulated highest N content in the plants. It was observed that bacterial inoculation of *Azotobacter* and *Azospirillum* to wheat and maize seeds resulted in significantly higher values for nitrogen content of plant components viz: grain and straw. Zambre et al. (1984) and Pandey et al. (1998) found that bacterial inoculation of *Azotobacter* and *Azospirillum* to maize seeds resulted in significantly higher values of phosphorous contents of plant components.

Ishac et al. (1986) found that wheat seeds inoculated with *Azotobacter* and *Azospirillum* and composted refused amendment stimulated the nitrogenous activity in the soil. Mixture of *Azotobacter tropicalis* carrying high N_2 fixing ability, phosphate solubilizing bacteria (*Burkhoderia unamae*), potassium solubilizing bacteria (*Bacillus subtilis*) and produce auxin (KJB9/2 strain) increased the yield by seven times in corn and vegetables as compared to control (Leaungvutiviroj et al. 2010). Zambre et al. (1984) reported an increase in the N content of soil during harvest time after inoculation of *Azotobacter* and *Azospirillum* to wheat seeds. *Azospirillum* sp. was shown to withstand high salt or osmotic condition due to the accumulation of compatible solutes. *Rhodobacter capsulatus* reduced the need of CNF by 50% when it was added in combination with 50% CNF in the rice fields. Gamal-Eldin and Elbanna (2011) reported an increase in the N-content of soil during harvest time after inoculation of *Azotobacter* and *Azospirillum* to wheat seeds.

4.7 *Rhizobium* as Efficient Nitrogen Fixer and its Effect on Crops

Rhizobium has the ability to fix atmospheric nitrogen in symbiotic association with legumes and non-leguminous plant (Table 15.1). Generally, it enters the root hair, multiply there and resides in a special structure called root nodule. The amount of nitrogen fixed is dependent on the strain of *Rhizobium*, host and prevailing environmental conditions. Nonetheless, substantial increases in yield are often obtained from inoculating even in the fields which have grown the particular leguminous crop for several years. The response of seed inoculation with specific *Rhizobium* culture on grain yield of pigeon pea, green gram, black gram, cowpea, gram and lentil, conducted at different locations at farmer's fields (Table 15.1). The increase in grain yield due to *Rhizobium* inoculation over control ranged from 2 to 65%. In infertile soil, *Rhizobium* inoculation resulted in an increase in the total nitrogen content. *Rhizobium* inoculants in different locations and soil types were reported to significantly increase the grain yields of Bengal gram (Bajpai et al. 1974; Patil and Medhane 1974; Chundawat et al. 1976); lentil (Bagyaraj and Hedge 1978); pea (Rosendahl and Jakobsen 1987); berseem (Bajpai et al. 1974); and ground nut (Bajpai et al. 1974). There are several reports on the favourable effects of *Rhizobium* in soybean cultivation (Singh and Saxena 1973; Bajpai et al. 1974; Tripathi and Edward 1978).

4.8 Blue-Green Algae as Efficient Nitrogen Fixer and its Effect on Crops

Blue-green algae (BGA) or cyanobacteria are the most primitive organisms probably the first among those that started evolving oxygen. They exist in many forms which include single celled to branched or unbranched filaments. Many of them possess a peculiar structure called 'heterocyst' which is known to fix free nitrogen from the air. Recently, some BGA without heterocyst have been reported that can also fix atmospheric nitrogen. The algae that are commonly used in field application are genus *Aulosira*, *Tolypothrix*, *Scytonema*, *Nostoc*, *Anabaena* and *Plectonema*. BGA being from the class of algae can trap sunlight and converts it into usable form that can be used to fix nitrogen. These algae can thus be used as biological input in rice cultivation. Extensive field trials conducted on the use of BGA in the rice fields indicated that one third of the recommended nitrogen fertilizer could be observed without affecting crop productivity through its inoculation (Gaur 2006). Besides the contribution of nitrogen, growth-promoting substances are liberated by BGA. Production of auxin-like substances and vitamins by *Cylindrospermum musicola* increased the root growth and yield of rice crops (Table 15.2) (Venkataraman and Neelakantan 1967). A number of growth-promoting substances such as amino acids, sugars, polysaccharides, nicotinic acid, pantothenic acid, Folic acids and IAA are secreted by BGA (Table 15.2) (Gaur 2006; Kaushik 1998).

4.9 Azolla-Anabaena as Efficient Nitrogen Fixer

Azolla is a tiny freshwater, sessile fern, whose leaf has a prominent dorsal and ventral lobe. The dorsal lobe which is green in colour can harbour BGA (*Anabaena azollae*) as a symbiont within a central cavity. The heterocyst of the symbiont

Anabaena is the site of nitrogen fixation. Not only the wild varieties of *Anabaena* but the improved varieties of *Anabaena* sp. Strain PCC7120 also confer higher nitrogenase activity when it was transformed of hetR gene (Chaurasia and Apte 2011). *Azolla* application with basal dose of manure can meet general nitrogen requirement of rice crop (Gaur 2006).

5 Phosphorous Biofertilizer

Phosphorus (P) is one of the most important plant nutrients which is required in optimum amount for proper growth of plants and also as a co-factor for soil micro-organisms. Being a constituent of ATP, it is involved in various processes such as cell division, energy transduction through photosynthesis and biological oxidations and nutrient uptake. The average soil contains 0.05% phosphorus but only one tenth of this is available to plants due to its poor solubility and chemical fixation in soil (Barber 1984). The problem of P fertilization may become serious in coming years because of the fact that manufacturer of phosphatic fertilizers requires the use of non-renewable resources such as high grade rock phosphate and sulphur which are getting depleted progressively and becoming costlier. The situation is further aggravated by the fact that P is readily fixed in the soil and the average utilization efficiency of added P fertilizer by plants ranges from 15 to 25%. In this context, the role of efficient rock phosphate dissolving micro-organisms assumes greater importance for augmenting crop productivity.

Phosphate solubilizing micro-organisms (PSM) particularly bacteria and fungi have been reported to solubilize inorganic phosphatic compounds. The main advantage of these micro-organisms is that they assimilate phosphorus for their own requirement and release sufficient amount to the soil in soluble form. Plant can easily take the phosphate dissolved in soil water. *Burkholderia vietnamiensis* M6, which can survive in the stressed soil, showed rapid solubilization of insoluble phosphorous and proved itself as a potent biofertilizer. Interestingly, two fungi called *Aspergillus fumigatus* and *Aspergillus niger* isolated from decaying cassava peels could solubilize $Ca_3 (PO_4)_2$, $AlPO_4$ and $FePO_4$ in liquid Pikovskaya medium were reported to be potent candidates for biofertilizers (Ogbo 2010). The genera of bacteria such as *Pseudomonas*, *Bacillus*, Micrococcus, Flavobacterium, *Fusarium*, *Sclerotium*, *Aspergillus* and *Penicillium* have been reported to be active in the solubilization process (Table 15.3).

5.1 *Effect of Phosphate Solubilizing Micro-organisms on Crops (PSM)*

The available P content of soil inoculated with PSM has been found to increase in many studies. It was found that soluble P content increased due to inoculation

Table 15.3 Important genera of phosphate solubilizing micro-organisms (PSM)

Type of organism	Genera	Important species
Bacteria	*Bacillus*	*Bacillus megatarium*
		Bacillus circulans
		Bacillus subtilis
		Bacillus polymyxa
		Bacillus pulvifaciens
		Bacillus pumilus
	Pseudomonas	*Pseudomonas srtriata*
		Pseudomonas rathonis
		Pseudomonas putida
		Pseudomonas aeruginosa
		Pseudomonas liquifaciens
Fungi	*Aspergillus*	*Aspergillus awamori*
		Aspergillus carbonum
		Aspergillus fumigates
		Aspergillus flavus
		Aspergillus niger
	Penicillium	*Penicillium digitatum*
		Penicillium liiacinum
		Penicillium balaji
		Penicillium funicul
Yeast	*Schwaniomyces*	*Schwaniomyces occidentalis*

of soil with *Penicillium bilaji* (Kucey 1988). Similarly inoculation with *Bacillus lichenformis* in sandy soil significantly increased P in the soil both in the presence or absence of rock phosphate (Gupta et al. 1993). Gerretsen (1948) reported the increased phosphorus uptake and yield of oat plant inoculated with pure cultures of phosphate dissolving micro-organisms compared to control. Rao (1965) reported beneficial effect of PSM on berseem (Egyptian clover). Significant increase in yield of maize and wheat was obtained with Fosfo24 (Czechoslovakian culture). Sharma and Singh (1971) observed that phosphobacterin along with bone meal when applied to the soil can increase the nitrogen phosphorus content and in turn the grain yield of rice as compared to nitrogen in combination with bone meal treatments. Nair and SubbaRao (1977) recorded many phosphate solubilizing *Pseudomonas* and *Aspergillus* in the rhizosphere of coconut and cocoa and the phosphorus availability to plant was related with their occurrence.

In addition to several basic studies, the effect was tested on several crops such as wheat, Paddy, Bengal gram, soybean, potato and cotton by Gaur and co-workers (Table 15.1). The grain yield of soybean in sandy loam alluvial soil was increased by 2.4 q/ha due to rock phosphate plus *Pseudomonas striata* treatment whereas with 80 kg P_2O_5/ha as superphosphate the increase was hardly 1 q/ha. In medium black soil, grain yield of Bengal gram was increased by 33 % (4 q/ha additional) due to treatment of rock phosphate and culture of *Aspergillus awamori*. Potato tuber yields were increased by 60% due to the treatment of seeds with *Pseudomonas striata* and 52% due to treatment with *Bacillus polymyxa* without any application of phosphatic fertilizers in the hilly soil at Shimla, India. A significant increase in the grain yield of wheat crop was recorded when Mussoorie rock phosphate was applied in the soil along with the seeds, that were inoculated with *Pseudomonas striata*. Moreover, yield obtained in this treatment was comparable to 50 kg P2O5 /ha as superphosphate. Micrococcus strain NII-0909 showed many useful attributes like phosphate solubilizing properties, auxin production and siderophore production. These attributes increase the growth of cowpea (Dastager et al. 2010).

The PSM also showed response on the yield of cotton where *Pseudomonas striata* and *Aspergillus awamori* increased the yield of cotton by 71% over control (Kundu and Gaur 1980). There are some heat tolerant phosphate solubilizing microbes which include some particular strains of bacteria, actinomycetes and fungi that can act as a multi-functional biofertilizer due to their ability to solubilize calcium phosphate and Israel rock phosphate; moreover, they also possess amylase, CMCase, chitinase, pectinase, protease, lipase and nitrogenase activities (Chang et al. 2009). PSM inoculation alone or in combination with other bacteria was found effective and contributed significantly to the productivity of crop plants when compared with control over control (Meshram et al. 2004). PSM synthesize thiamine, biotin, riboflavin, Vitamin B (Table 15.2) (Lockhead 1957). Barea et al. (1976) reported that many PSM synthesize IAA, gibberellins and cytokinin (Table 15.2).

6 Potassium Biofertilizer

Potassium (K) plays an important role in the growth and development of plants. It activates enzymes, maintains turgor pressure of cell, enhances photosynthesis, reduces respiration, helps in transport of sugars and starches, helps in nitrogen uptake and is essential for protein synthesis. In addition to plant metabolism, potassium improves crop quality because it helps in grain filling and kernel weight, strengthens straw, increases disease resistance and helps the plant to better withstand stress. Potassium is applied externally to the soil in the form of potassic fertilizers. After USA, China, and Brazil India ranks fourth as far as the total consumption of potassium fertilizers in the world is concerned (FAI 2007). However, there is no reserve of K-bearing minerals in India for the production of commercial K-fertilizers and expensive K-fertilizers are imported in the form of muriate of potash (MOP) and sulphate of potash (K_2SO_4). This necessitates the search for an alternate indigenous source of K for plant growth and maintaining K status in the soil for sustaining crop production.

Waste materials like mica can effectively be used as a source of potassium, if modified or altered by some suitable chemical or biological means. One of the possible means of utilizing waste mica is by mobilizing their K through composting technology where unavailable K is converted into plant available form because of the acidic environment available during composting. KSM play a key role in the natural K cycle. Some species of rhizobacteria are capable of mobilizing potassium in accessible form in the soil. There are considerable population of K solubilizing bacteria in the soil and rhizosphere. Silicate bacteria were found to dissolve potassium, silicon and aluminium from insoluble minerals. Several micro-organisms like genus *Aspergillus*, *Bacillus* (Badr 2006) and *Clostridium* are found to be efficient in process of potassium solubilization in different crops (Table 15.1).

6.1 Effect of Potassium Solubilizing Micro-organisms on Crops

Phosphorus solubilizing bacteria and silicate bacteria play an important role in plant nutrition through the increase in P and K uptake by plant (Datta et al. 1982; Nianikoval et al. 2002). Zahro and Monib (1984) studied the effect of soil inoculation of the silicate bacteria. *Bacillus circulans* on the release of K and Si from different minerals and in different soil proved that bacteria could persist for a longer time where high population density could be detected after 14 months particularly in the soil containing higher levels of organic matter. An increased yield in rice crop was observed due to inoculation of silicate solubilizing bacteria. Xue et al. (2000) and Sheng (2005) reported silicate dissolving bacteria could improve soil P, K and Si reserves and promote plant growth. Lin et al. (2002) recorded increase in the biomass by 125%. K and P uptake were more than 150% in tomato plant due to inoculation of silicate dissolving bacteria (*B. mucilaginosus*) than the non-inoculation. Thus, there is a potential in applying RCBC13 for improving K and P nutrition.

The effects of plant growth PGPR including phosphate and potassium solubilizing bacteria (PSB and KSB) as biofertlizers are solutions to improve plant nutrient availibility and productivity (Vessey 2003). Park et al. (2003) reported that bacterial inoculation could improve phosphorus and potassium availability in the soils by producing organic acid and other chemicals by stimulating growth and mineral uptake of plants. Sheng (2005) studied the effect of inoculation of SSB (*Bacillus edaphicus*) on chilli and cotton which resulted in increased levels of available P and K contents in the plant biomass. In the study to assess the weathering of finely ground phogopite trictahedral mica by placing it in contact with heterotrophic bacteria *Bacillus cereus* and acidophilic (*Acidothiobacillus ferroxidans*) cultures enhanced the chemical dissolution of the mineral. The X-ray diffraction analysis of the phologopite sample before and after 24 weeks of contact with *Bacillus cereus* cultures revealed a decrease in the characteristic peak intensities of phologopite indicating destruction of individual structural planes of the mica; on the other hand, *Acidothiobacillus ferroxidans* cultures enhanced the chemical dissolution of the mineral and formed partial interlayer from where K was expelled. This was coupled with the precipitation of K and Jarosite (Styriakova et al. 2004).

Zhang et al. (2004) reported that the effect of potassic bacteria on sorghum resulted in the increased biomass and contents of P and K in plants than the control. The increased uptake of K coupled with increased yield while treating the plants with potassium mobilizer in conjunction with biofertilizers and chemical fertilizers has been reported in yam and tapioca (Clarson 2004). Chandra et al. (2005) reported increase in the yield of yam and tapioca by 15–20% due to the application of potash solubilizer in combination with other biofertilizer like *Rhizobium*, *Azospirillum*, *Azotobacter*, *Acetobacter* and PSM. Han and Lee (2005) found that the co-inoculation of PSB and KSB in combination with direct application of rock P and K materials into the soil resulted in the increased levels of N, P and K uptake. They also reported an enhancement in the level of photosynthesis and the yield of eggplant grown on P and K deficient soil. Ramarethinam and Chandra (2005) in a field experiment recorded significant increment in the yield, height and K uptake of brinjal plant when it was compared with control. This increment was due to inoculation of potash solubilizing bacteria (*Frateuria aurantia*). Mikhailouskaya and Tcherhysh (2005) reported that effect of inoculation of K immobilizing bacteria on severally eroded soil which is comparable with yields on moderately eroded soil without bacterial inoculation resulted in increased yield of up to 1.04 t/ha in wheat.

Potassium releasing bacterial strain of *Bacillus edaphicus* was found to increase the root and shoot length of cotton and rape seed due to increase in the uptake of soluble potassium (Sheng 2005). Increase in biomass and K uptake was reported in chilli due to inoculation of potash solubilization (Ramarethinam and Chandra 2005). Christophe et al. (2006) reported that *Burechulderia glathei* in association with pine roots significantly increased weathering of biotite and concluded that there was the effect of *B. glathei* PMB (7) and PML1 (12) on pine growth and its root morphology and which was attributed to the release of K from the mineral. Sheng and He (2006) recorded an increased root and shoot growth and also showed significantly higher N, P and K contents in wheat plants due to inoculation of *B. edaphicus* growth in a yellow brown soil that had low available K. And in the field experiment they recorded increased yield in tomato crop due to inoculation of silicate dissolving bacteria *B. cereus* as bio-inoculant along with feldspar and rice straw on K releasing capacity (Badr 2006). Han and Supanjani (2006) evaluated the potential of PSB and KSB inoculated in nutrient. Limited soil planted with pepper and cucumber showed that co-inoculation of PSB and KSB showed high rate of plant growth and P and K content when it was compared with control. Supanjani et al. (2006) reported that integration of P and K rocks with inoculation of phosphorus and potassium solubilizing bacteria increased P availability from 12 to 21% and K availability from 13 to 15% in the soil as compared with control. Subsequently, this combination improved the nutrient uptake of N, P and K uptake in *Capsicum annuum*. The integration also increased plant photosynthesis by 16% and leaf area by 35% as compared to control. On the other hand the biomass harvest and fruit yield of the treated plants were increased by 23–30% respectively. Overall results of this finding is that the treatment of P and K rocks with P and K solubilizing bacterial strain were sustainable alternative to the use of chemical fertilizer.

Badr et al. (2006) studied the effect of bacterial inoculation combined with K and P bearing minerals on sorghum plants. They later reported an increase in later reported increase in dry matter yield and P and K uptake in three different soils like clay, sandy and calcareous soils. They found 48, 65 and 58% of increase in dry matter, 71, 110 and 116% uptake of P and 41, 93 and 79% uptake of K and improved fertility through inoculation of SDB. The increased rice grain yield in a field experiment due to effect of silicate solubilizing bacteria recorded 5,218 kg/ha grain yield than control 4,419 kg/ha (Balasubramaniam and Subramanian 2006). The potential phosphate solubilizing bacteria (PSB) *B. megaterium* var., Phosphaticum and potassium solubilizing bacteria (KSB) *B. mucilaginosus* were evaluated using pepper and cucumber as test crops. The outcome of the experiment showed that rock phosphorus and potassium applied either singly or in combination do not significantly enhance availability of soil phosphorus and potassium indicating that their unsuitability for direct application co-incubation of PSB and KSB resulted in consistently higher P and K available than in the control (Vassilev et al. 2006).

7 Mycorrhizae as Biofertilizer

In 1885, A.B. Frank found that the roots of most plants are colonized by fungi transformed into fungus roots organ which he called mycorrhizae. Mycorrhizae are an excellent instance of symbiotic relation between fungus and the roots of higher plants. The successful establishment of this mutualistic association constitutes a strategy to fulfil the nutritional demands of both partners (Kogel et al. 2006). This requires a balance between the defence responses of the host plant and the nutrient demands of the endophyte, resulting in an altered defence-related gene expression. In mutualistic association between both the partners, fungal endophyte can enhance growth, increase reproduction and provide biotic and abiotic stress tolerance to its host plant (Lamabam et al. 2011). Roots of most plants live in mutual symbiosis with mycorrhizae which bio-tropically colonize the root cortex and extra-metrical mycelia help the plants to obtain plant nutrients from soil (Barea 1991). These fungi are ubiquitous in soil and are found in the roots of many Angiosperms, Gymnosperms, Pteridophytes and Thallophytes (Mosse et al. 1981). The mycorrhizal fungi perform the function similar to root hairs. The fungus derives carbohydrates from plants and in turn provides them with nutrients, hormones and also protects them from root pathogens. The mycorrhizal plants have greater tolerance to toxic heavy metals, high soil temperature, soil salinity, unfavourable soil pH and to transplantation shocks. They play an important role in increasing plant growth and nutrient uptake (Bagyaraj and Hedge 1978; VasanthaKrishna and Bagyaraj 1993). There are seven types of mycorrhiza, such as ectomycorrhiza, endomycorrhiza, ectendomycorrhiza, arbutoid mycorrhiza, monotropoid mycorrhiza, orchidaceous mycorrhiza and ericoid mycorrhiza.

8 Effect of Mycorrhizae on Crops

Mycorrhizae protect plant from pathogens and play an important role in biotic stress tolerance. It inhibits root pathogens such as *Rhizoctonia solani*, *Pythium* spp. and *Fusarium oxysporum* (Table 15.2) (Sasek and Musilek 1968). Pathogen like *Fomes annosus* was inhibited by antibiotics produced by several ecto-mycorrhizal fungi (Table 15.2) (Sasek and Musilek 1968). It enhances plant growth by improving mineral nutrition by providing soluble phosphorous (Barea and Azcon-Aguilar 1983; Hayman 1982; Smith and Gianinazzi-Pearson 1988). The increased uptake of phosphorus by plants occurs in two steps which involves absorption of phosphate by hyphae from outside to internal cortical mycelia and then transference of phosphate to root cortical cells (Barea 1991). Allen et al. (1982) showed that mycorrhiza directly affects the level of plant hormones such as cytokinins and gibberellin-like substances (Table 15.2).

The beneficial effects of mycorrhizae have also been reported under drought and saline conditions (Nelson and Safir 1982; Lamabam et al. 2011). Mycorrhizal roots have the ability to tap additional water sources unavailable to non-mycorrhizal plant roots under drought stress (Allen and Boosalis 1983). Such effect is apparent under phosphate limitation as additional phosphate to non-mycorrhizal plants boosted their performance under drought and salinity (Nelson and Safir 1982). Analytical and physiological studies have shown that mycorrhizal plants have increased rates of respiration and photosynthesis, higher levels of sugar, amino acids, RNA, etc. and larger or more number of chloroplasts, mitochondria, xylem vessels, motor cells, etc. (Hayman 1983). Fungi after colonization cause several changes in the rhizosphere and the type of microbes it carries due to alteration in root (Sattar and Gaur 1989).

The mycorrhizal inoculum increased the root colonization of garlic, horse bean, soybean, chickpea, melon, watermelon, cucumber, maize, cotton, pepper, eggplant and tomato plants compared with non-inoculated treatments (Ortas 2012).

9 Vermicomposting

Organic farming is one of the healthy approaches to green agriculture which includes bacterial biofertilizers, vermicomposting and use of manure. Tropical soils are mainly deficient in organic compounds but they can be replenished by applying organic wastes of the household. Nature has provided us with smart organisms (e.g., earthworm) which can convert complex organic refuse of domestic waste into simpler organic compound or humus. Recently, it was reported that earthworms are not only friends to farmer but they also help environmentalists by contributing to bioremediation. They sequester wastes like heavy metals and harmful pathogens from the soil and household wastes. Earthworms can minimize environmental degradation due to the rampant use of chemical fertilizers. They maintain the level of humus in the soil and can serve as a protectant of top soil. They not only prevent the soil degradation but also increase the productivity of crops.

Vermicompost contains higher level of nitrate or ammonium nitrogen, soluble phosphorous and potassium, calcium and magnesium derived from the wastes (Buchanan et al. 1988). Increase in the yield of *Morus* sp. was found when it was treated with full dose NPK fertilizers plus vermicompost and half dose of farmyard manure.

Along with the yield this combination also improved the growth of plants in terms of height of plants, number of leaves and leaf yield per plant (Murarkar et al. 1998). Application of vermicompost along with nitrogen fertilizer resulted in the more dry matter (16.2 g/plant) and grain yield (3.6 t/ha) of wheat (*Triticum aestivum*) and higher dry matter yield (0.66 g/plant); it also gave the similar result in case of *Coriandrum sativum* which was planted under sequential cropping system (Desai et al. 1999). Vermicompost increased the germination rate of seeds of *Vigna radiata* by 93% as compared to 84% in control; it also increased yield. Similar results have been observed in *Vigna unguiculata* when the soil was amended with vermicompost and biodigested slurry (Karmegam et al. 1999; Karmegam and Daniel 2000). Vermicompost was found efficient enough to enhance the biomass of plant right from root to shoot. The results were best in case of vegetables and ornamental plants (Grappelli et al.; Atiyeh et al. 1999; Cheema et al. 2001).

The efficacy of vermicompost was tested in gram *Cicer arietinum* and it was found that application of vermicompost can increase the seed yield up to 2 t/ha. This result was similar to that in case of *Vigna,* because the increase in yield was due to increased secondary branches per plant, pods per plant and seed index. Both the results were compared with control (Siag and Yadav 2004). They also mineralize nutrients like nitrogen, phosphorous and potassium and have converted sludge into useful vermicompost. In one study, an increase was found in the yield of two leguminous plants when the soil in which they were grown was pre-treated with vermicompost (Saha et al. 2010).

10 Biofertilizers: Environmental Stresses

Environmental stresses are limiting the productivity of crops worldwide. Biofertilizers are multifunctional agents that not only provide nutrients to the plants but also help in the alleviation of various stresses. Stresses such as salt, drought, metal and pathogenesis affect crops and also lead to the wastage of manpower of the farmers who work day and night for good production. Food security has become a major issue with the increase in the effects of global warming and other natural calamities. There is a great demand of strong and successful crop seeds that can bear harsh conditions like drought and salt stress. There must be an eco-friendly approach to deal with the situation where natural resources like biofertilizers can be used as stress releasing agents.

Biofertlizers have proved efficient in helping against various kinds of stresses including salt, drought, heavy metal and biotic stress. Several micro-organisms and mycorrhizae have been characterized for their extraordinary property of growing under beyond normal conditions. Interestingly, diazotrophic bacteria *Rhizobium* found near the coastal areas was shown to increase the productivity even in saline soil (Zahran 1999). *Phaseolus vulgaris* under salt stress in combination with *Azospirillum brasilense* co-inoculated with *Rhizobium* produces flavonoids and Nod factor (Dardanelli et al. 2008). AM produces natural glue-like substance called glomalin which is a glycoprotein that helps in the aggregation of soil particle and also for the retention of water. *Pseudomonas fluorescens* was found to be effective in the amelioration of drought stress by enhancing its growth and amount of ajmalicine production under water deficit condition (Jaleel et al. 2007).

Piriformospora indica, an endophytic arbuscular mycorrhiza which belongs to class basidiomycetes, lives in reciprocally beneficial relationships with plants, providing them with both biotic and abiotic stress tolerance including salinity. The interaction of barley with *P. indica* presents an ideal system to show resistance against systemic disease in cereals (Waller et al. 2005). *Arabidopsis* roots interact with *P. indica* which resulted in a considerable requisition of nitrogen from the environment by the plant (Peskan-Berghofer et al. 2004). In barley, *P. indica* induces resistance to *Fusarium culmorum*, one of the fungal species that causes head blight, as well as systemic resistance to barley (Waller et al. 2005).

Interestingly, it was shown that some bacteria belonging to PGPR produces an enzyme called ACC deaminase (ACCD) (Glick et al. 1998). This enzyme helps in the metabolizing precursor of ethylene, 1-aminocyclopropane-1-carboxylate (ACC), produced by plants under heavy metal stress. Generally, all bacterial strains that have ACCD activity can prevent the ill effects of ethylene on plants up to some extent (Penrose and Glick 2003; Mayak et al. 2004). Sahni et al. (2008) showed the role of vermicompost on the soil-borne pathogen *Sclerotium rolfsii* which causes collar rot of chickpea. Recently, AM fungus *Glomus intraradices* proved effective in rice production owing to its role in increasing photosynthetic efficiency and activating antioxidant machinery in drought stressed rice plant (Ruiz-Sanchez et al. 2010). Co-innoculation of bacteria *Pseudomonas mosselii* and vermicompost proved beneficial for potato crop which is prone to scab disease of potato (Singhai and Sarma 2011). Rhizobacterial strain S2BC-2 (*Bacillus atrophaeus*) and strain mixture, S2BC-2+TEPF-Sungal (*Burkholderiacepacia*), found inhibitory to the growth of *Fusarium oxysporum* f. sp. gladioli which causes vascular wilt and corm rot of gladiolus (Shanmugam et al. 2011). Root nodulating *Sinorhizobium fredii* KCC5 and *Pseudomonas fluorescens* LPK2, which were found as disease suppressive agents, were isolated from nodules of *Cajanus cajan* and tomato rhizosphere, respectively. Both strains showed anti-fungal properties against *Fusarium udum*. Recently, some defence-related enzymes were found to get induced by the application of class II d bacteriocins (thuricin 17 and bacthuricin F4) purified from Bacillus strains (Jung et al. 2011).

The study of PGPR-elicited responses should now target various other crops to test their range of activity against various abiotic and biotic stresses. Also the mechanism involved in the activation of defence signals and various proteins has to be understood to use this tool as a powerful measure against conventional and chemical methods to overcome environmental stresses.

11 Mechanism of Action of Various Biofertilizers

Any external element that promotes plant growth is also involved in the induction of various genes in the plants. These elements once recognized by a particular receptor can turn on the signal transduction pathway that shifts the internal metabolism of plant towards the growth and maintenance of plant despite adverse conditions of nutrient deficiency and various stresses. Symbiosis or association between two or more organisms is a necessary part of the ecosystem (Callaghan and Conrad 1992). An organism's interaction with other organisms and its surrounding environment is vital for its survival and the functioning of the ecosystem as a whole. Organisms support each other to adapt to a particular situation largely for their own benefit, and seek out better environments where they can flourish. Nevertheless, they contribute substantially by providing suitable conditions for the growth of plants. Beneficial bacteria and mycorrhiza not only thrive in close proximity to plants but also contribute to the functioning of their cellular system. *Rhizobium* lives in the root nodules of leguminous plants and its machinery related to nitrogen fixation can only be operated with the symbiosis with the host plant. In turn, they also bring about several changes in the cellular environment of plants (David et al. 1988). Mycorrhiza dwells in the cortical cells of roots of higher plants and influences the cellular machinery till the time its life cycle is completed. Substantial research has been carried out on the identification of these bacteria and mycorrhiza but the mechanism of their association is not very clear.

Phytohormones are known to be produced by plants but there are several bacteria those produce them in minute quantities (Dobbelaere et al. 2003). Many genes related to auxin biosynthesis and root morphogenesis showed up-regulation during mycorrhizal colonization (Dutra et al. 1998). Ethylene is responsible for the inhibition of growth of dicot plants and interestingly, it was found that PGPR could enhance the growth of plant by suppressing the expression of ethylene (Abeles et al. 1992; Glick et al. 1998; Holguin and Glick 2003).

It was found during the genome sequencing of EM fungi that genes encoding for transporter proteins are a special feature of the EM genome and their function as effectors and facilitators has been established in the light of this information (Bonfante and Requena 2011). A special compound strigolactones can cause branching of AM mycelium (Rani et al. 2008). Some chemical signals allow the fungus to develop contact with the plant root epidermal cell.

Nitrogen-fixation genes are popularly used by scientists today to create engineered plants that can fix atmospheric nitrogen. *Nif* genes are the chromosomal genes that act both as positive and negative regulators. They encode the proteins of nitrogenase enzyme complex and other proteinaceous machineries that are involved in nitrogen fixation. The *nif* genes are induced by low concentration of nitrogen and oxygen in the rhizosphere.

There are several other mechanisms which include solubilization of phosphates by phytase, reduction by phenazines and lumichromes that provide a basis for nutrient for the nutrient availability through PGPR or biofertilizers (Idriss et al. 2002; Greiner et al. 1997; Kerouvo et al. 1998).

12 Conclusion and Future Prospective

The positive role of biofertilizers in plant growth, productivity and protection against some stresses makes them a very important and powerful eco-friendly nutrient supplement to plants. They have the potential for wide-ranging impact. Usually, soils are inhabited by various types of microbial species. The co-existence of these species is determined by ecological factors prevailing in the soil. Many of these species have been used in biofertilizers and they have been shown to improve seed germination and plant growth, tolerance towards high salt conditions. They are beneficial for crops with regard to N_2-fixation and produce growth promoting substances and fungicidal substances. Overall, we can say that the biofertilizers are live formulates of micro-organisms which improve the quality of the soil and the plant species by increasing the nutrient availability for the soil, seeds and roots. Some of the species, like *Azotobacter*, thrive even in alkaline soils. *Azotobacter* also acts as a biological control agent against plant pathogens such as *Alternaria*, *Fusarium* and *Helminthosporium*. *Azotobacter* produces thiamine, Riboflavin, nicotin, indole-acetic acid and gibberellin, which also help to control plant diseases. Some biofertilizers are also known to destroy disease-causing components from the soil. In general, the nitrogen biofertilizers help to correct the nitrogen levels of the soil while the phosphate biofertilizers enhance the phosphorus levels of the soil. The biofertilizers are also very cost-effective in comparison to chemical fertilizers and are eco-friendly.

These days biofertilizers are considered a part of advanced biotechnology which is required for the development of clean, green and sustainable agriculture.

References

Abeles FB, Morgan PW, Saltveit ME (1992) Ethylene in plant biology, 2nd edn. Academic, New York

Abrol YP, Chatterjee SR, Kumar PA, Jain V (1999) Improvement in nitrogen use efficiency: Physiological and molecular approaches. Curr Sci 76:1357–1364

Aguilera M, Quesada MT, Aguila D, Morillo JA, Rivadeneyra MA, Ramos-Cormenzana A, Monteoliva-Sánchez M (2008) Characterisation of *Paenibacillus jamilae* strains that produce exopolysaccharide during growth on and detoxification of olive mill waste waters. Bioresour Technol 99:5640–5644

Allen MF, Boosalis MG (1983) Population dynamics of sugar beets, *Rhizoctonia solani*, and Laetisaria arvalis: responses of a host, plant pathogen, and hyperparasite to perturbation in the field. New Phytopathol 93:61–71

Allen MF, Moore JS, Christensen M (1982) Phytohormone changes in *Bouteloua gracilis* infected by vesicular-arbuscular mycorrhizae. II. Altered levels of gibberellin-like substances and abscisic acid in the host plant. Can J Bot 60:468–471

Ashby SF (1907) Some observations on assimilation of atmospheric nitrogen by a free living soil organism *Azotobacter chroococcum*. J Agric Sci 2:35–51

Atiyeh RM, Subler S, Edwards CA, Metzger J (1999) Growth of tomato plants in horticulture potting media amended with vermicompost. Pedobiologia 43:724–728

Badr MA (2006) Efficiency of K-feldspar combined with organic materials and silicate dissolving bacteria on tomato yield. J Appl Sci Res 2:1191–1198

Badr MA, Shafei AM, Sharaf SH (2006) The dissolution of K and phosphorus bearing minerals by silicate dissolving bacteria and their effect on sorghum growth. J Agric Biol Sci 2:5–11

Bagyaraj DJ, Hedge SV (1978) Interaction between a VA mycorrhiza and *Azotobacter* and their effects on rhizosphere microflora and plant growth. Curr Sci 47:548–549

Bajpai PD, Lehri LK, Pathak AN (1974) Effect of seed inoculation with *Rhizobium* strains on the yield of leguminous crops. Proc Ind Natl Sci Acad B Biol Sci 40:571–575

Balasubramaniam P, Subramanian S (2006) Assessment of soil test based potassium requirement for low land rice in udic haplustalf under the influence of silicon fertilization, Tamil Nadu Agricultural. University, Kumulur, Tiruchirapalli, pp. 621–712

Balasubramanian R, Veerabadran V (1997) Substitution of inorganic nitrogen through local green-leaf manure and *Azospirillum* in rice (*Oryza sativa* L.). Ind J Agron 42:285–287

Barber SA (1984) Soil nutrient bioavailability. A mechanistic approach. Wiley-Interscience, New York

Barea JM (1991) Vesicular-arbuscular mycorrhizae as modifiers of soil fertility. In: Stewart BA (ed) Advances in soil science. Springer, New York, pp 1–40

Barea JM, Azcon-Aguilar C (1983) Mycorrhizas and their significance in nodulating nitrogen fixing plants. Adv Agron 36:1–54

Barea JM, Navare E, Motoya E (1976) Production of plant growth regulators by rhizosphere phosphate-solubilizing bacteria. J Appl Microbiol 40:129–134

Berkum VP, Bohlool BB (1980) Evaluation of nitrogen fixation by bacteria in association with roots of tropical grasses. Microbiolol Rev 44:491–517

Boddey RM, Baldani VLD, Baldani JI, Dobereiner J (1986) Eject of inoculation of *Azospirillum* spp. on nitrogen accumulation by field-grown wheat. Plant Soil 95:109–121

Bonfante P, Requena N (2011) Dating in the dark: how roots respond to fungal signals to establish arbuscular mycorrhizal symbiosis. Curr Opin Plant Biol 14:451–457

Buchanan MA, Russelli E, Block SD (1988) Chemical characterization and nitrogen mineralization potentials of vermicomposts derived from differing organic wastes. In: Edwards CA, Neuhauser EF (eds) Earthworms in environmental and waste management. SPB Academic, The Hague, pp 231–239

Bukatsch F (1952) Contributions to the knowledge of the physiology of *Azotobacter*. Soils Fertilizer 16:79–96

Burgmann H, Widmer F, Sigler WV, Zeyer J (2003) Detection of *nifH* gene expression by *Azotobacter vinellandii*. Soil Appl Environ Microbiol 69:1928–1935

Callaghan JO, Conrad M (1992) Symbiotic interactions in the evolve III ecosystem model. Biosystem 26:199–209

Chandra K, Greep S, Ravindranath P, Sivathsa RSH (2005) Liquid biofertilizers. Regional center for organic farming, Hebbal, Bangalore, pp 22–35

Chaurasia AK, Apte SK (2011) Improved eco-friendly recombinant Anabaena sp. strain PCC7120 with enhanced nitrogen biofertilizer potential. Appl Environ Microbiol 77:395–399

Cheema MA, Malik MA, Hussain A, Saha SH, Basra AMA (2001) Effects of time and rate of nitrogen and phosphorus application on the growth and the seed and oil yields of Canola (*Brassica napus* L.). J Agron Crop Sci 186:103–110

Choudhury MA, Kennedy IR (2004) Prospects and potentials for system of biological nitrogen fixation in sustainable rice production. Biol Fertil Soils 39:219–227

Christophe C, Turpault MP, Freyklett P (2006) Root associated bacteria contribute to mineral weathering and to mineral nutrition in trees and budgeting analysis. Appl Environ Microbiol 72:1258–1266

Chundawat GS, Sharma R, Shekhawat GS (1976) Effect of nitrogen, phosphorus and bacterial fertilization on growth and yield of gram in Rajasthan. Indian J Agron 21:127

Clarson D (2004) Potash biofertilizer for eco-friendly agriculture. Agro-clinic and Research Centre, Poovanthuruthu, Kottayam (Kerala), India, pp 98–110

Dardanelli MS, Fernandez de Cordoba FJ, Espuny MR, Carvajal MAR, Diaz MES, Serrano AMG, Okon Y, Megeas M (2008) Effect of *Azospirillum brasilense* co-inoculated with *Rhizobium* on *Phaseolus vulgaris* flavonoids and Nod factor production under salt stress. Soil Biol Biochem 40:2713–2721

Dastager SG, Deepa CK, Pandey A (2010) Isolation and characterization of novel plant growth promoting *Micrococcus sp* NII-0909 and its interaction with cowpea. Plant Physiol Biochem 48:987–992

Datta M, Banish S, Dupta RK (1982) Studies on the efficacy of a phytohormone producing phosphate solubilizing *Bacillus firmus* in augmenting paddy yield in acid soils of Nagaland. Plant Soil 69:365–373

David M, Daveran ML, Batut J, Dedieu A, Domergue O, Ghai J, Hertig C, Boistard P, Kahn D (1988) Cascade regulation of *nif* gene expression in *Rhizobium meliloti*. Cell 54:671–683

Desai VR, Sabale RN, Raundal PV (1999) Integrated nitrogen management in wheat coriander cropping system. J Maharashtra Agric Univ 24:273–275

Dewan GI, SubbaRao NS (1979) Seed inoculation with *Azospirillum brasilense* and *Azotobacter chroococcum* and the root biomass of rice (*Oryza sativa* L.). Plant Soil 53:295–302

Dobbelaere S, Vanderleyden J, Okon Y (2003) Plant growth-promoting effects of diazotrophs in the rhizosphere. CRC Crit Rev Plant Sci 22:107–149

Dobereiner J, Day JM (1976) Associative symbioses in tropical grasses: characterization of microorganisms and dinitrogen-fixing sites. In: Newton WE, Nyman CJ (eds) Proceedings of the 1st international symposium on N_2 Fixation. Washington State University Press, Pullman, pp 518–538

Doyle J (1998) Phylogenetic perspectives on nodulation: evolving views of plants and symbiotic bacteria. Trends Plant Sci 12:473–478

Gamal-Eldin H, Elbanna K (2011) Field evidence for the potential of *Rhodobacter capsulatus* as Biofertilizer for flooded rice. Curr Microbiol 62:391–395

Gaur AC (2006) Blue- green algae as biofertilizer: biofertilizers in sustainable agriculture: Directorate of Information and Publications of Agriculture, Indian Council of Agricultural Research. Krishi Anusandhan Bhavan I, Pusa, New Delhi, pp 148–155

Gerretsen FC (1948) The influence of microorganisms on the phosphate intake by the plant. Plant Soil 1:51–81

Ghosh DC (2000) Growth, productivity of summers. Sesamum (*Sesamum indicum*) as influenced by biofertilizer and growth regulator. Indian J Agron 45:389–394

Glick BR, Penrose DM, Li J (1998) A model for the lowering of plant ethylene concentrations by plant growth promoting rhizobacteria. J Theor Biol 190:63–68

Gopalswamy G, Vidhyasekharan P, Chelliah S (1989) Effect of *Azospirillum lipoforum* incoulation and inorganic nitrogen on the wet land rice. Oryzae 26:378–380

Greiner R, Haller E, Konietzny U, Jany KD (1997) Purification and characterization of a phytase from *Klebsiella terrigena*. Arch Biochem Biophys 341:201–206

Gupta R, Shanker AB, Saxena RK, Kuhad RC (1993) Solubilization of low grade Indian rock phosphate and inorganic phosphates by *Bacillus licheniformis*. Folia Microbiol 38:274–276

Han HS, Lee KD (2005) Phosphate and potassium solubilising bacteria effect on mineral uptake, soil availability and growth of eggplant. Res J Agric Biol Sci 1:176–180

Han HS, Supanjani KD (2006) Effect of co-inoculation with phosphate and potassium solubilising bacteria on mineral uptake and growth of pepper and cucumber. Soil Environ 52:130–136

Hayman DS (1982) Influence of soils and fertility on activity and survival of vesicular-arbuscular mycorrhizal fungi. Phytopathology 72:1119–1125

Hayman DS (1983) The physiology of vesicular arbuscular endomycorrhizal symbiosis. Can J Bot 61:944–963

Holguin G, Glick BR (2003) Transformation of *Azospirillum brasilense* Cd with an ACC deaminase gene from Enterobacter cloacae UW4 fused to the Tetr promoter improves its fitness and plant growth promoting ability, Microb. Ecology 46:122–133

Holzinger A, Nagendra-Prasad D, Huys G (2011) Plant protection potential and ultrastructure of *Bacillus subtilis* strain 3A25. Crop Prot 30:739–744

Idriss EE, Makarewicz O, Farouk A, Rosner K, Greiner R, Bochow H, Richter T, Borris R (2002) Extracellular phytase activity of *Bacillus amyloliquefaciens* FZB45 contributes to its plant-growth-promoting effect. Microbiology 148:2097–2109

Ishac YZ, El-Haddad ME, Daft MJ, Ramadana EM, El-Demerdash ME (1986) Effect of seed inoculation, mycorrhizal infection and organic amendment on wheat growth. Plant Soil 90:373–382

Jaleel CA, Gopi R, AlaguLakshmanan GM, Panneerselvam R (2007) Triadimefon induced changes in the antioxidant metabolism and ajmalicine production in *Catharanthus roseus* (L.) G. Don. Plant Sci 171:271–276

Jiang H, Jiang L, Guo L, Gao Z, Zeng D, Zhu L, Liang G, Qian Q (2008) Conditional and unconditional mapping of quantitative trait loci underlying plant height and tiller number in rice (*Oryza sativa* L.) grown at two nitrogen levels. Progr Nat Sci 18:1539–1547

Jung WJ, Mabood F, Souleimanov A, Smith DL (2011) Induction of defense-related enzymes in soybean leaves by class IId bacteriocins (thuricin 17 and bacthuricin F4) purified from *Bacillus* strains. Microbiol Res 167:14–19

Kader MA, Mian MH, Hoque MS (2002) Effects of *Azotobacter* inoculant on the yield and nitrogen uptake by wheat. J Biol Sci 2:259–261

Kannan T, Ponmurugan P (2010) Response of paddy (*Oryza sativa L.*) varieties to *Azospirillum brasilense* inoculation. J Phytol 2:08–13

Karlidag H, Esitken A, Turan M, Sahin F (2007) Effects of root inoculation of plant growth promoting rhizobacteria (PGPR) on yield, growth and nutrient element contents of leaves of apple. Sci Hortic 114:16–20

Karmegam N, Daniel T (2000) Effect of biodigested slurry and vermicompost on the growth and yield of cowpea [*Vigna unguiculata* (L.)]. Environ Ecol 18:367–370

Karmegam N, Alagermalai K, Daniel T (1999) Effect of vermicompost on the growth and yield of greengram (*Phaseolus aureus* Rob.). Tropic Agric 76:143–146

Kaushik BD (1998) Soil plant Microbe interaction in relation to Nutrient management. Venus, New Delhi, pp 55–63

Kerouvo J, Lauraeus M, Nurminen P, Kalkkinen N, Apajalathi J (1998) Isolation, characterization, molecular gene cloning, and sequencing of a novel phytase from *Bacillus subtilis*. Appl Environ Microbiol 64:2079–2085

Khammas KM, Ageron E, Grimont PAD, Kaiser P (1989) *Azospirillum irakense* sp. nov., a nitrogen-fixing bacterium associated with rice roots and rhizosphere soil. Microbiology 140:679–693

Kızılkaya R (2008) Yield response and nitrogen concentrations of spring wheat (*Triticum aestivum*) inoculated with *Azotobacter chroococcum* strains. Ecol Eng 33:150–156

Kogel KH, Franken P, Huckelhovenl R (2006) Endophyte or parasite – what decides? Current Opinion. Plant Biol 9:358–363

Korth H, Benoni H (1990) Isolation of *Pseudomonas fluorescens* producing phenazine derivatives exclusively under strains conditions of iron deficiency. Zentralbl Bacteriol 274:433–435

Kucey RMN (1988) Plant growth-altering effects of *Azospirillum brasilense* and Bacillus species. on two wheat cultivars. J Appl Bacteriol 64:187–196

Kundu BS, Gaur AC (1980) Establishment of nitrogen fixing and phosphate solubilizing bacteria in rhizosphere and their effect on yield and nutrient uptake of wheat crop. Plant Soil 57:223–230

Kurguzov P (1954) The effect of boron on yield and quality of spring wheat. Soil Fertil 17:1187

Lakshmi V, Satyanarayan A, Vijayalakshmi K, Lakshmi Kumari M, Tilak KVBR, SubbaRao NS (1977) Establishment and Survival of *Spirillum lipoferum*. Proc Indian Acad Sci 86:397–404

Lakshmi-Kumari M, Kavimandan SK, SubbaRao NS (1976) Occurrence of nitrogen fixing *Spirillum* in roots of sorghum, maize and other plants. Indian J Exp Biol 14:638–649

Lamabam PS, Gill SS, Tuteja N (2011) Unraveling the role of fungal symbionts in plant abiotic stress tolerance. Plant Signal Behav 6:175–191

Lampayan RM, Bouman BAM, Dios JL, Soriano JB, Lactaoen AT, Faronilo JE, Thant KM (2010) Yield of aerobic rice in rainfed lowlands of the Philippines as affected by nitrogen management and row spacing. Field Crop Res 116:165–174

Lin W, Okon Y, Hardy RWF (1983) Enhanced mineral uptake by *Zea mays* and *Sorghum bicolor* roots inoculated with *Azospirillum brasilense*. Appl Environ Microbiol 45:1775–1779

Lin QM, Rao ZH, Sun YX, Yao J, Xing LJ (2002) Identification and practical application of silicate – dissolving bacteria. Agric Sci China 1:81–85

Lin XQ, Zhu DF, Chen HZ, Cheng SH, Uphoff N (2009) Effect of plant density and nitrogen fertilizer rates on grain yield and nitrogen uptake of hybrid rice (*Oryza sativa* L.). J Agric Biotechnol Sustain Dev 1:044–053

Lockhead AG (1957) Soil bacteria and growth-promoting substances. Bacteriol Rev 22:145–153

Looijesteijn PJ, Trapet L, de Vries E, Abee T, Hugenholtz J (2001) Physiological function of exo polysaccharides produced by *Lacto coccuslactis*. Int J Food Microbiol 64:71–80

Magalhaes FM, Baldani JI, Souto SM, Kuykendall JR, Dobereiner J (1983) A new acid-tolerant *Azospirillum* species. An Acad Bras Cienc 55:417–429

Mayak S, Tirosh T, Glick BR (2004) Plant growth-promoting bacteria confer resistance in tomato plants to salt stress. Plant Physiol Biochem 42:565–572

Meshram SU, Pande SS, Shanware AS, Kamdi RR (2004) Efficacy of biofertilizers integrated with chemical fertilizers in-vivo in soybean. Fertil Newslett 12:7–10

Mikhailouskaya N, Tcherhysh A (2005) K-mobilizing bacteria and their effect on wheat yield. Latvian J Agron 8:154–157

Molina-Favero C, MónicaCreus C, Luciana Lanteri M, Correa-Aragunde N, Lombardo MC, Barassi AC, Lamattina L (2007) Nitric Oxide and Plant Growth Promoting Rhizobacteria: Common Features Influencing Root Growth and Development. Adv Bot Res 46:1–33

Mosse B, Stribley DP, Tacon F (1981) Ecology of mycorrhiza and mycorrhizal fungi. Adv Microbiol Ecol 5:137–210

Murarkar SR, Tayade AS, Bodhade SN, Ulemale RB (1998) Effect of vermicompost on mulberry leaf yield. J Soils Crops 8:85–87

Nair NS, SubbaRao NS (1977) Distribution and activity of phosphate solubilising microorganisms in the rhizosphere of coconut and cacao under mixed cropping. J Plant Crops 5:67–70

Nelson CE, Safir GR (1982) Increased drought tolerance of mycorrhizal onion plants caused by improved phosphorus nutrition. Planta 154:407–413

Nianikoval GG, Kuprina EE, Pestova OV, Vodolazhskaia SV (2002) Immobilizing of *Bacillus muciloginosus* a producer of exopolysaccharides, on chitin. Prikladnaia Biokhimiia Mikrobiologiya 38:300–304

Ogbo FC (2010) Conversion of cassava wastes for biofertilizer production using phosphate solubilizing fungi. Bioresour Technol 101:4120–4124

Okon Y, Labandera-Gonzalez CA (1994) Agronomic applications of *Azospirillum*: an evaluation of 20 years worldwide field inoculation. Soil Biol Biochem 26:1551–1601

Ortas I (2012) The effect of mycorrhizal fungal inoculation on plant yield, nutrient uptake and inoculation effectiveness under long-term field conditions. Field Crop Res 125:35–48

Pacovsky RS, Fuller G, Paul EA (1985) Influence of soil on the interactions between endomycorrhizae and Azospirillum in sorghum. Soil Biol Biochem 17:523–531

Pandey A, Sharma E, Palni L (1998) Influence of bacterial inoculation on maize in upland farming systems of the Sikkim Himalaya. Soil Biol Biochem 3:379–384

Panwar JD (1991) Effect of VAM and *Azospirillum brasilense* on photosynthesis nitrogen metabolism and grain yield in wheat. Indian J Plant Physiol 34:357–361

Park M, Singvilay D, Seok Y, Chung J, Ahn K (2003) Effect of phosphate solubilizing fungi on 'P' uptake and growth of tobacco in rock phosphate. Korean J Soil Sci Fertil 36:233–238

Patil PL, Medhane NS (1974) Seed inoculation studies in gram (*Cicer arietinum*) with different strains of Rhizobium sp. Plant Soil 40:221–223

Pedraza RO (2008) Recent advances in nitrogen-fixing acetic acid bacteria. Int J Food Microbiol 125:25–35

Penrose DM, Glick BR (2003) Methods for isolating and characterizing ACC deaminase-containing plant growth-promoting rhizobacteria. Physiol Plant 118:10–15

Peskan-Berghofer TB, Shahollari PH, Giong S, Hehl C, Markert V, Blanke G, Kost A, Varma R (2004) Oelmuller, Association of *Piriformospora indica* with *Arabidopsis thaliana* roots represents a novel system to study beneficial plant-microbe interactions and involves early

plant protein modifications in the endoplasmic reticulum and at the plasma membrane. Plant Physiol 122:465–477
Rahman MM, Hoque S, Khan ZUM (2007) Nitrogen fixation and respiratory activity of *Azospirillum spp.* isolated from saline habitat of Bangladesh. Bangladesh J. Life Sci 19:55–60
Ramarethinam S, Chandra K (2005) Studies on the effect of potash solubilizing/mobilizing bacteria *Frateuria aurantia* on brinjal growth and yield. Pestology 11:35–39
Rani K, Zwanenburg B, Sugimoto Y, Yoneyama K, Bouwmeester HJ (2008) Biosynthetic considerations could assist the structure elucidation of host plant produced rhizosphere signalling compounds (strigolactones) for arbuscular mycorrhizal fungi and parasitic plants. Plant Physiol Biochem 46:617–626
Reinhold B, Hurek T, Fendrik I, Pot B, Gillis M, Kersters K, Thielemans S, De Ley J (1987) *Azospirillum halopraeferens* sp. nov., a nitrogen-fixing organism associated with roots of Kallar Grass (*Leptochloa fusca* L. Kunth). Int J Syst Bacteriol 37:43–51
Righi S, Lucialli P, Bruzzi L (2005) Health and environmental impacts of a fertilizer plant – Part I: Assessment of radioactive pollution. J Environ Radioact 82:167–182
Rosendahl L, Jakobsen I (1987) *Rhizobium* strain effects on pea: The relation between nitrogen accumulation, phosphoenolpyruvate carboxylase activity in nodules and asparagine in root bleeding sap. Physiol Planta 71:281–286
Saravanakumar D, Vijayakumar C, Kumar N, Samiyappan R (2010) PGPR-induced defense responses in the tea plant against blister blight disease. Crop Prot 26:556–565
Sarig S, Blum A, Okon Y (1992) Improvement of the water status and yield of field-grown grain sorghum (*Sorghum bicolor*) by inoculation with *Azospirillum brasilense*. J Agric Sci 110:271–277
Sasek V, Musilek M (1968) Two antibiotic compounds from mycorrhizal Basidiomycetes. Folia Microbiol 13:43–51
Sattar MA, Gaur AC (1989) Effect of VA-mycorrhiza and phosphate dissolving microorganism on the yield and phosphorus uptake of wheat (*Triticum vulgare*) in Bangladesh. Bangl J Agric Sci 14:233–239
Shanmugam V, Kanoujia N, Singh M, Singh S, Prasad R (2011) Biocontrol of vascular wilt and corm rot of gladiolus caused by *Fusarium oxysporum* f. sp. gladioli using plant growth promoting rhizobacterial mixture. Crop Protect 30:807–813
Sharma JP, Singh M (1971) Response of rice to phosphatic and nitrogenous fertilizers with and without phosphobacteria in culture. Indian J Agron 16:15–18
Shende ST (1987) Significance of *Azotobacter* inoculation for agricultural crops. In: Organic manures and biofertilizers. Division of Microbiology, IARI, New Delhi, pp. 38–45
Shende ST, Apte RG, Singh T (1975) Multiple action of *Azotobacter*. Indian J Genet Plant Breed 3:3–14
Sheng XF (2005) Growth promotion and increased potassium uptake of cotton and rape by a potassium releasing strain of *Bacillus edaphicus*. Soil Biol Biochem 37:1918–1922
Sheng XF, He LY (2006) Solubilization of potassium bearing minerals by a wild type strain of *Bacillus edaphicus* and its mutants and increased potassium uptake by wheat. Can J Microbiol 52:66–72
Siag RK, Yadav BS (2004) Effect of vermicompost and fertilizers on productivity of gram (*Cicer arietinum*) and soil fertility. Indian J Agric Sci 74:613–615
Singh NP, Saxena MC (1973) Phosphatic fertilization of soya bean. Indian J Agric Sci 43:62–66
Singhai PK, Sarma BK (2011) Srivastava JSBiological management of common scab of potato through Pseudomonas species and vermicompost. Biol Control 57:150–157
Smith SE, Gianinazzi-Pearson V (1988) Physiological interactions between symbionts in vesicular-arbuscular mycorrhizal plants. Annual Review Plant Molecular Biology 39:221–244
Spiertz J (2010) Nitrogen, sustainable agriculture and food security. A review Agron Sustain Dev 30:43–55
Sturz AV, Christie BR (2000) Bacterial endophytes: potential role in developing sustainable systems of crop production. Crit Rev Plant Sci 19:1–30
Styriakova I, Bhatti TM, Bigham JM, Styriak I, Vourniess A (2004) Tuoviness OH Weathering of phlogopite by *Bacillus cereus* and *Acidiothiobacillus ferroxidans*. Can J Microbiol 50:213–219

SubbaRao NS, Tilak KVBR, Lakshmi-Kumari M, Singh CS (1980) *Azospirillum* a new bacterial fertilizer. Indian Farming 30:3–5

Supanjani HHS, Jung SJ, Lee KD (2006) Rock phosphate potassium and rock solubilizing bacteria as alternative sustainable fertilizers. Agro Sustainable Development 26:233–240

Swedrzynska D, Sawicka A (2001) Effects of inoculation on population numbers of *Azospirillum* bacteria under winter wheat, oat and maize. Polish Journal of Environmental Studies 10:21–25

Tejera N, Luch C, Martinez-Toledo MV, Gonzalez-Lopez J (2005) Isolation and characterization of *Azotobacter* and *Azospirillum* strains from the sugarcane rhizosphere. Plant Soil 270:223–232

Tien TM, Gaskins MH, Hubbell DH (1979) Plant growth substances produced by *Azospirillum brasilense* and their effect on the growth of pearl millet (*Pennisetum americanum* L.). Appl Environ Microbiol 37:1016–1024

Tilak KVBR, Murthy BN (1983) Response of barley (*Hordeum vulgare*) to inoculation with *Azospirillum brasilense*. Curr Sci 52:257–258

Timonin MI (1948) *Azotobacter* preparation (Azotogen) as a fertilizer for cultivated plants. Soil Science Society of America 13:246–249

Tittabutr P, Awaya DJ, Li XQ, Borthakur D (2008) The cloned 1-aminocyclopropane-1-carboxylate (ACC) deaminase gene from *Sinorhizobium* sp. strain BL3 in Rhizobium sp. strain TAL1145 promotes nodulation and growth of *Leucaena leucocephala*. Syst Appl Microbiol 31:141–150

Tripathi SK, Edward JC (1978) Response of *Rhizobium* culture inoculation: zinc and molybdenum application to soybean Glycine max. Indian Journal of Plant Physiology 21:248–252

Umali-Garcia M, Hubbell DH, Gaskins MH, Dazzo FB (1980) Association of *Azospirillum* with grass roots. Appl Environ Microbiol 39:219–226

VasanthaKrishna M, Bagyaraj DJ (1993) Selection of efficient VA mycorrhizal fungi for inoculating *Casuarina equisetifolia*. Arid soil Research Rehabilitation 7:377–380

Vassilev N, Medina A, Azcon R, Vassilev M (2006) Microbial solubilization of rock phosphate as media containing agro industrial wastes and effect of the resulting products on plant growth and phosphorus uptake. Plant Soil 287:77–84

Venkataraman GS, Neelakantan S (1967) Effect of cellular constituents of the nitrogen fixing blue green algae. *Cylindrospermum nusciola* on the root growth of rice seedlings. J Gen Appl Microbiol 13:53–61

Vessey KJ (2003) Plant growth promoting rhizobacteria as biofertilizers. Plant Soil 25:557–586

Waller FB, Achatz H, Baltruschat J, Fodor K, Becker M, Fischer T, Heier R, Huckelhoven C, Neumann D, Wettstein P (2005) Franken Kogel KH, The endophytic fungus *Piriformospora indica* reprograms barley to salt-stress tolerance, disease resistance, and higher yield. Proc Natl Acad Sci 102:13386–13391

Wani SP, Chandrapalaih S, Zambre MA, Lee KK (1988) Association between N_2-fixing bacteria and pearl millet plants: Responses, mechanisms and persistence. Plant Soil 110:289–302

Xue QH, Sheng JW, Tang L (2000) Effect of K bacteria on nutrients activation in Lou Soil Acta. Agriculture Boreali-Occidentalis Sinica 9:67–71

Yuan BC, Li ZZ, Liu H, Gao M (2007) Zhang YY Microbial biomass and activity in salt affected soils under arid conditions. Applied Soil Ecology 35:319–328

Zahran HH (1999) Rhizobium-legume symbiosis and nitrogen fixation under severe conditions and in an arid climate. Microbiol Mol Biol Rev 63:968–989

Zahro MK, Monib M (1984) Abdel-Al Sl, Significance of soil inoculation with silicate bacteria. Zentralblatt fur Mikrobiologi 139:349–357

Zambre MA, Konde BK, Sonar KR (1984) Effect of *Azotobacter chrococcum* and *Azospirillum brasilense* inoculation under graded levels of nitrogen on growth and yield of wheat. Plant Soil 79:61–67

Zhang S, Moyne A, Reddy MS, Kloepper JW (2002) The role of salicylic acid induced systemic resistance elicited by plant growth-promoting rhizobacteria against blue mold of tobacco. Biological Control 25:288–296

Zhang CJ, Tu GQ, Cheng CJ (2004) Study on potassium dissolving ability of silicate bacteria. Shaguan College Journal 26:1209–1216

Chapter 16
Plant Pathogen Interactions: Crop Improvement Under Adverse Conditions

Kamal Kumar and Praveen Kumar Verma

1 Introduction

In the face of an expanding world population, we need more quantity of food, especially when the cultivated land resources are shrinking. It is estimated that to feed the world population by 2050 our food requirement will be 70% more, which means an increase in crop production at the rate of 44 million metric tons per year is required (Tester and Langridge 2010). The increasing food and energy demand calls for intensive crop production but it is also visualized that in intensive cropping systems the growth of plant pathogens is rapid and new virulence mechanisms appear in pathogen population, and minor pathogens become a major production constrain. Therefore, the incorporation of resistance is a major focus of many breeding programs. However, certain limitations like lack of resistance against many diseases in the primary gene pool, difficulty in transfer of resistance in desired host due to crossability barriers, rapid evolution of virulent pathogens, existence of high pathogenic variability amongst the pathogens, etc. override the advantages of traditional breeding. It is the consensus of plant breeders, geneticists, and other biologists that biotechnological approaches can play an important role in alleviating some of these problems.

Research on host–pathogen interaction in crop plants mainly has been focused on production of toxic substances. Recent advances in molecular biology, however, have offered efficient and precise tools for better understanding of plant–pathogen interactions. In the first half of this chapter, recent developments towards understanding of molecular aspect of plant immunity, mostly against the bacteria and fungi, have been described, although many of these pathways play an important role against other

K. Kumar • P.K. Verma (✉)
Plant Immunity Laboratory, National institute of Plant Genome Research,
New Delhi 110 067, India
e-mail: pkv@nipgr.ac.in

N. Tuteja and S. Singh Gill (eds.), *Plant Acclimation to Environmental Stress*,
DOI 10.1007/978-1-4614-5001-6_16,

pathogens also. This part is further divided into sections and subsections to provide clearly outlined apprehension of the topic. In the second half of the chapter various methods to achieve resistance against pathogens in crop plants have been discussed.

2 Plant–Pathogen Interactions

Plants are rich source of sugar, minerals, and water that attracts various organisms with heterotrophic lifestyle. The pathogenic organisms use host plant to serve basic aim of life, i.e., grow and reproduce. Plant diseases are comparatively less than the number and variety of potential pathogens in the surrounding environment of plants; this is due to the fact that they have developed a highly complex and multi-layered immune system while coevolving with pathogens. The outcome of a plant–pathogen interaction can be either an incompatible (disease resistance/tolerance) or a compatible (pathogen infection and disease) interaction, which is governed by the genetic makeup of both the plant and pathogen. The ability of a pathogen to infect plants depends upon the repertoire of its effectors to suppress or evade plant immune responses and modulate host cellular metabolism for its own benefit. The plant resistance against a potential pathogen depends upon the capacity of the plant to recognize this pathogen as nonself and induce immune response to restrict its growth.

In an ecosystem, pathogens pass their life on host plants in different modes. Many pathogens have evolved to infect only a single plant species (narrow host range) while a few of these pathogens can implicate more plant species for their survival (broad host range). Based on their lifestyle on host, pathogens are classified as biotrophs, hemibiotrophs, and necrotrophs. The biotrophs are entirely dependent upon living host and keep their host alive throughout their life cycle, the hemibiotrophs keep host alive for some period and then kill them, and the necrotrophs feed on host plants by killing them. The evolution of such lifestyles in filamentous pathogens was correlated with gain/loss of genes by comparative analysis (Dodds 2010).

Molecular plant pathologists have broadly classified plant disease resistance operating in natural habitats into two categories: the host resistance and the nonhost resistance (Heath 2000). The nonhost resistance dominates in nature as every plant withstands the injurious effect of most of the potential pathogens while host resistance against a particular pathogen species is shown by the some genotypes of an otherwise host species. To define, the nonhost resistance is the ability of an entire plant species to resist infection by all isolates of a pathogen species. Many reports suggest that the defense signaling against host and nonhost pathogens is similar and many components of these resistance mechanisms are common but the final outcome of their interaction with pathogen or pathogen effectors differs (Thordal-Christensen 2003). It is opined that the components of host resistance were isolated earlier through forward genetics in many known plant–pathogen pairs; hence data towards host resistance seem to be unfair. Therefore, with the advancements in biotechnology biologists are prompted to use the components of nonhost resistance due to its effectiveness and durability.

3 Multilayered Plant Immune System

A simple way to define plant immune system is to define the obstacles that a pathogen must overcome to invade host tissue, proliferate, and cause disease (Thordal-Christensen 2003). Bacterial pathogens get access to host tissue through stomata or wounds. The filamentous pathogens make their entry in host through stomata and may even directly penetrate the cuticle layer. Plants try to restrict pathogens by preformed and induced defenses. The induced defense responses are controlled by PTI (pathogen-associated molecular patterns-triggered immunity) and ETI (effector-triggered immunity). Only pathogens that can evade/suppress/manipulate these defensive layers can cause disease.

3.1 *Preformed Structural and Chemical Barriers*

The cuticle covers the epidermal cell wall and functions as the first barrier for pathogens. It is composed of polysaccharides, cutin, and waxes, whose composition changes within each species and according to environmental conditions (Shepherd and Wagner 2007). After landing of pathogen on plant surface, the cuticle plays an important part in the plant–pathogen communications. Generally cuticle is considered as a barrier for the entry of pathogens but now it is clear that many pathogens like *Uromyces appendiculatus*, *Fusarium solani* f. sp. *pisi*, *Ustilago maydis*, *Magnoporthe oryzae*, *Colletotrichum gloeosporioides*, *Puccinia graminis* f. sp. *tritici*, etc. require signals from the host plant surface to differentiate and penetrate the host (Mendoza et al. 2009; Reina-Pinto and Yephremov 2009; Liu et al. 2011). Thus, the cuticle's role is important towards resistance against nonadapted pathogens. The phytopathogenic fungi secrete cutinase to liberate cutin that serves as a signal for differentiation in *M. grisea* and *Erysiphe graminis* but not in *Botrytis cinerea* (Bessire et al. 2007). In case of necrotrophic fungi like *B. cinerea*, *Alternaria brassicicola*, and *Fusarium graminearum* secreted lipases are important for pathogenicity. The *Blumeria graminis* release lipolytic activity containing protein, Lip1, to release cues from the wheat plant surface for promoting pathogen development and infection (Feng et al. 2009). The *Arabidopsis* CUTE plants with cell wall targeted fungal cutinase, lipase, and mutants with altered cuticle showed higher resistance to *B. cinerea* but not to other necrotrophs like *Plectosphaerella cucumerina*, *A. brassicicola,* and *Sclerotinia sclerotiorum* (Bessire et al. 2007; Chassot et al. 2008). The increased resistance against *B. cinerea* in these plants was correlated with the induction of few genes and higher fungitoxic activity. Clearly a single mechanism cannot be charted out for the role of cuticle against diverse pathogens but the studies on

Table 16.1 Mutants and transgenic plants with altered cuticle composition affecting plant–pathogen interaction

Mutants/ Overexpressed gene	Plant	Features	Reference
sma4 mutant and *lacs2*	*Arabidopsis*	Enhanced susceptibility to *Pst* DC3000 strain with *avr* genes but resistance to *B. cinerea*	Tang et al. (2007)
att1 mutant	*Arabidopsis*	Susceptible to *Pst*	Xiao et al. (2004)
Yeast D-9 desaturase overexpression	*Solanum lycopersicum*	Higher resistance to *Erysiphe polygoni*	Wang et al. (1998); Wang et al. (2000)
bodyguard (*bdg*) mutant and *Fusarium solani* f. sp. *pisi* cutinase A and *B. cinerea LIP1* overexpression	*Arabidopsis*	Increased resistance to *B. cinerea*	Chassot et al. (2007)
permeable cuticle (*pec*) mutant	*Arabidopsis*	Increased resistance to *B. cinerea*	L'Haridon et al. (2011)
botrytis-resistant 1 (*bre1*)/*lacs2* mutant	*Arabidopsis*	Increased resistance to *B. cinerea*	Bessire et al. (2007)
gpat4/*gpat8* mutant	*Arabidopsis*	Sensitive to *A. brassicicola*	Li et al. (2007)
delayed fruit deterioration (*DFD*) mutant	*S. lycopersicum*	Resistance against *B. cinerea*	Saladie et al. (2007)
acyl carrier protein4 (*acp4*), *long-chain acyl-CoA synthetase2* (*lacs2*), and *lacs9*	*Arabidopsis*	Compromised systemic acquired resistance	Xia et al. (2009)
glabra1(*gl1*), *gl3*, *and ttg1*	*Arabidopsis*	Compromised systemic acquired resistance	Xia et al. (2010)
sitiens	*S. lycopersicum*	Resistance against *B. cinerea* due to ABA deficiency leading to cuticle permeability	Curvers et al. (2010)

various cuticle-related mutants have shown that its composition affects the final outcome of plant–pathogen interaction (Table 16.1). The adhesion level of cuticle with cell wall also modulates the defense responses of plants. The glandular trichomes also release antimicrobial substances that can inhibit pathogen growth.

After alteration of cuticle, the pathogens adopt a course of action to break the host plant cell wall by mechanical force and degrading enzymes such as polygalacturonases, xylanases, cellulases, and proteinases. Changes in the host cell wall components like less *O*-acetylation of cell wall polysaccharides in *Arabidopsis thaliana's Reduced Wall Acetylation* (*RWA2*) mutant plants displayed increased tolerance towards *Botrytis cinerea*, but mutation had no effect on infection by powdery mildew (*Golovinomyces cichoracearum*) suggesting differential mechanisms of fungal infection and plant resistance against these pathogens (Manabe et al. 2011). In another case, the *atmyb46* mutants have high level of cell wall-associated peroxidases that are involved in phenolic cross-linking at cell wall and ROS scavenging leading to enhanced resistance against *B. cinerea* (Ramirez et al. 2011). Many other cell wall-associated genes had been reported to influence resistance and susceptibility to pathogens (Hückelhoven 2007; Cantu et al. 2008). The molecules released by cell wall breakdown of the host (i.e., endogenous elicitors) and the pathogen can induce plant defense response, which has been discussed under induced defenses.

The apoplastic space is the site where many pathogen and plant-derived molecules counteract each other. Molecules having antimicrobial activity are secreted in the apoplastic space constitutively by plant or they can be induced after pathogen perception. Many enzyme inhibitors block the activity of pathogen-released enzymes and the plant-derived lipid transfer proteins (LTPs) have inhibitory effects on fungal growth (Molina and Garcia-Olmedo 1997; Patkar and Chattoo 2006). The *sad* mutants of *Avena strigosa* can be infected by the nonhost fungal pathogens *Gaeumannomyces graminis* var. *tritici* and *Fusarium culmorum* due to the lack of avenacins, a type of saponin with antifungal activity (Papadopoulou et al. 1999).

3.2 Pathogen-Associated Molecular Pattern-Triggered Immunity

When pathogens are able to breach the constitutive defensive layers then they are recognized as nonself by plant cell membrane receptors, which recognize conserved microbial components (flagellin and chitin in bacteria and fungi respectively) or motifs present in them and molecules released by pathogen. These molecules, termed as PAMPs/MAMPs (microbe-associated molecular patterns), are mostly conserved within a class of microbes and are essential for microbial survival and fitness (Bent and Mackey 2007). They are non-race-specific inducers of plant defense so are often mentioned as exogenous elicitors in contrast to the endogenous elicitors, called damage-associated molecular patterns (DAMPs), released from the host plant by virtue of pathogen attack (Lotze et al. 2007). Some of the known pathogen-associated molecular pattern (PAMPs)/DAMPs are listed in Table 16.2. The importance of PAMP-triggered immunity (PTI) in plant defense is manifested from the fact that the adapted pathogens have evolved effectors to suppress it or they have evolved mechanisms to mask the recognition of PAMPs/DAMPs but in nonhost resistance growth of a nonadapted pathogen is effectively restricted by PTI. The PTI in plants is very similar to that of animals.

The typical responses initiated in plant cell after PAMP/DAMP perception are generation of ion fluxes across plasma membrane, enhanced Ca^{2+} concentration in cytosol, protein phosphorylation, GTPases activation, rapid increase in reactive oxygen species (ROS), generation of nitric oxide (NO) and ethylene (ET), and many more associated changes (Garcia-Brugger et al. 2006; Boller and Felix 2009). These

Table 16.2 PAMPs/MAMPs perceived by plant cells

PAMPs	Active motif	Pathogen	Reference
Flagellin	flg22	Bacteria	Gomez-Gomez and Boller (2000)
Lipopolysaccharides (LPSs) and peptidoglycan	-	Bacteria	Erbs and Newman (2012)
Harpin	-	Gram-negative bacteria	Lee et al. (2001), Kim et al. (2004)
Cold shock protein	RPN-1 motif	Bacteria	Felix and Boller (2003)
N-glycosylated peptide	-	Yeast	Boller (1995)
Sulphated fucans	Fucan oligosaccharide	Brown Algae	Klarzynski et al. (2003)
Transglutaminase	Pep13 motif	*Phytophthora* spp.	Brunner et al. (2002)
Elicitins (sterol binding proteins)	-	*Phytophthora* spp., *Pythium* spp.	Osman et al. (2001)
Cellulose binding lectin	-	*Phytophthora* spp.	Gaulin et al. (2006)
Arachidonic acid	-	Oomycetes	Boller (1995)
β (Beta)-glucans	Oligomeric and multimeric-β (Beta)-glucosides	Filamentous pathogens	Yamaguchi et al. (2000), Fliegmann et al. (2004)
Ethylene inducing xylanase (EIX)	TKLGE pentapeptide	*Trichoderma* spp.	Ron and Avni (2004)
Chitin	-	Fungi	Wan et al. (2008)
Ergosterol	-	Fungi	Granado et al. (1995), Laquitaine et al. (2006), Lochman and Mikes (2006)
Cerebrosides A and C	-	*Magnaporthe* spp.	Koga et al. (1998)

changes lead to the activation of calcium-dependent protein kinases (CDPKs), calmodulins, and mitogen-activated protein kinases (MAPKs) through cascade of events that ultimately activates the transcription of numerous defense-related genes (Boudsocq et al. 2010). Scientists generally use alkalization of the growth medium, MAPK activation, hydrogen peroxide (H_2O_2) generation, callose deposition, and expression of early induced genes as markers for the flagellin, chitin, and other PAMP-activated responses (Asai et al. 2002; Denoux et al. 2008). In terms of the quality, responses elicited by various PAMPs from virus, bacteria, oomycetes, fungi, and other pathogens are same but quantitatively they may differ. The cumulative effect of these responses can often lead to hypersensitive response (HR) that is characterized by localized cell death at the site of attack to limit the pathogen spread (Heath 1998; Bolwell 1999).

Many PAMPs have been defined at molecular level based on the activation of PTI responses but their corresponding plant receptors working as sentinels at plasma membrane are not so well defined (Zipfel 2009). The first PAMP receptor cloned from plants was for flagellin (flg22). It is *FLAGELLIN-SENSING 2* (*FLS2*) that encodes for a leucine-rich repeat receptor-like kinase (LRR-RK) (Gomez-Gomez and Boller 2000). The orthologs of *FLS2* are present in other higher plants also suggesting that flagellin-mediated signaling is present in both monocot and dicot branches (Takai et al. 2008). Unlike flg22 responsiveness seen in many higher plants, the Brassicaceae family is only responsive to the N-terminus (elf18/26) of a highly conserved and abundant bacterial protein Elongation factor Tu (a GTPase). Its receptor in *Arabidopsis*, EFR, is also an LRR-RK (Kunze et al. 2004). Such is also the case with the recognition of Ax21 by some specific rice cultivars. It is thus apparent that each plant does not recognize every PAMP and not every pathogen displays all PAMPs (Zipfel and Robatzek 2010).

The nonhost interaction of *Arabidopsis thaliana* with *Blumeria graminis* f. sp. *hordei* (Bgh) has emerged as an excellent system to study the role of early induced genes as the infection is localized at the epidermal cells. Analysis of mutant plants for the various genes like *PENETRATION* (*PEN1*- a syntaxin, *PEN2*-a glycosyl hydolase, and *PEN3*-an ABC transporter) have suggested their role in plant immunity towards nonadapted pathogens (Ellis 2006).

3.3 *Effector-Triggered Immunity*

To suppress the PTI and to modulate host metabolism for their own benefit, pathogens secrete a variety of effector molecules inside the host cell (Hok et al. 2010). Bacteria mainly use type III secretion system while filamentous pathogens utilize host machinery to deliver effectors into the plant cell (Göhre and Robatzek 2008; Chibucos et al. 2009). These effectors can be proteases, toxins, transcriptional activators, etc. suggesting that diverse pathogens have evolved various strategies to subvert plant responses (de Jonge et al. 2011; Gheysen and Mitchum 2011; Hogenhout and Bos 2011; Stassen and Van den Ackerveken 2011). In a recent study, related to the interaction of pathogenic effectors with their target plant proteins, it was concluded that two diverse pathogens have evolved their effectors to target a selected set of plant proteins besides other individual targets. These common plant protein targets, in general, form large interaction networks in plants suggesting that pathogens target those proteins inside a host plant that are important for a signaling or interaction hub (Mukhtar et al. 2011). In response to effectors, plants have evolved an array of *R* (resistance) genes that recognizes these effectors directly or indirectly to rapidly induce a strong defense response. Many of the R proteins are associated with multi-protein immune complexes (Friedman and Baker 2007). Models have been proposed and experimentally verified to explain the evolution of *R* genes and the recognition of pathogen effectors by R proteins. Relevant among them are gene-for-gene, guard model, and decoy model (van der Hoom and Kamoun 2008).

Most of the known R proteins are multidomain NB-LRR (Nucleotide binding site and leucine-rich repeat) type but other types of R proteins are also known like protein kinase (Rpg1), LRR-receptor-like kinase (Xa21), LRR-TM (Cf's), etc. and

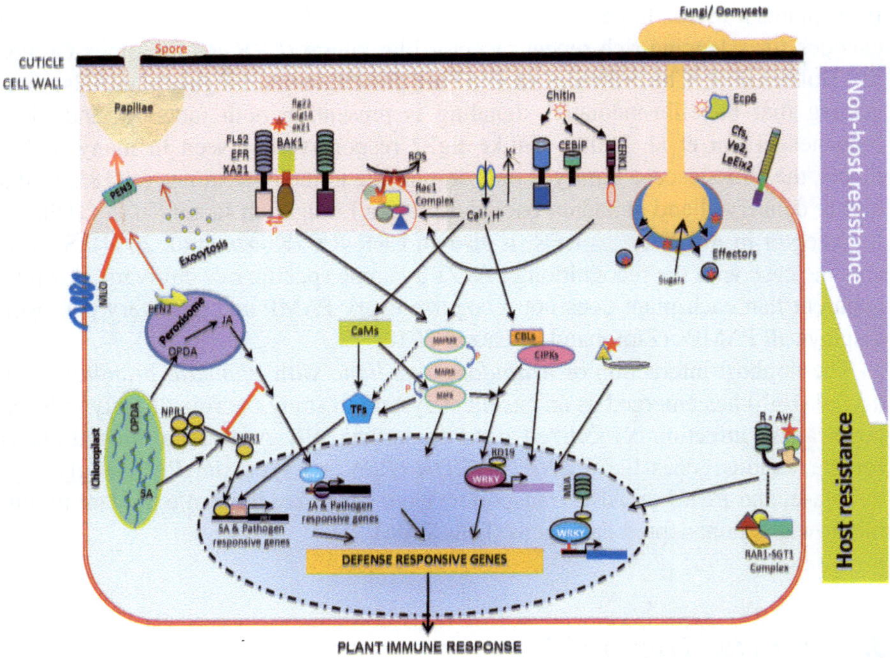

Fig. 16.1 The multilayered plant immune system

in some genes the promoter polymorphisms also genetically suggest it as *R* gene (Liu et al. 2007; Bogdanove et al. 2010; Chen et al. 2010). The NB-LRR proteins can be further subdivided based on N-terminal homology to TIR (Toll and Interleukin-1 Receptor; RPS4, SSI4, L6, etc.), CC (coiled-coil; RPM1, RCY1, Mi-1, etc.) or LZ (leucine-zipper; RPS5), and non-motif groups. The C-terminal LRR region binds to the decoy or the effector (direct Avr-R interaction) while the N-terminal is involved in transducing signals to the downstream components to initiate defense signaling. It is suggested that the intra-domain interaction or interaction with associated proteins keeps NB-LRR proteins under resting condition and with the perception of effectors or their activity the signaling is initiated (Caplan et al. 2008; Collier and Moffett 2009; Lukasik and Takken 2009). The signaling downstream to R-proteins is very complex, as some group of R-genes requires NDR1 (non-race-specific disease resistance 1) or EDS1 (enhanced disease susceptibility 1) or some are independent of these two parallel pathways. Further complexity appears in the requirement of RAR1 and SGT1 proteins (Thatcher et al. 2005; Shirasu 2008).

In model plant *Arabidopsis* and other crop plants various components of preformed and induced (PAMP and effector recognition-based) immunity have be isolated and from these analyses emerges a complex picture of plant immune responses (Fig. 16.1) (Thatcher et al. 2005; Chisholm et al. 2006; Knepper and Day, 2010; Nishimura and Dangl 2010; Zhang and Zhou 2010; Chen and Ronald 2011).

The signaling initiated by ETI and PTI shares many common points (Thomma et al. 2011) but the final outcome of defense response, i.e., plant immunity is brought by the cumulative effects of all these components some of which may also be involved in primary and secondary metabolism. A common feature associated with resistance against biotrophic pathogen is the development of hypersensitive response (HR) and systemic acquired resistance (SAR) along with some associated processes (Durrant and Dong 2004; Vlot et al. 2008). Plant hormones like salicylic acid (SA), jasmonic acid (JA), ethylene (ET), auxin, etc. also play an important role along with a myriad of small molecules and proteins in this complex plant response. Role of these components in plant immunity has been extensively reviewed (Lorenzo and Solano 2005; Roberts-Seilaniantz et al. 2007; Spoel and Dong 2008; Bari and Jones 2009; Pieterse et al. 2009; Ton et al. 2009).

4 Strategies to Develop Biotic Stress-Tolerant Crops

Since a number of crop species are cultivated under adverse stress conditions, Varshney et al. (2011) emphasized that the scientists should take up multiple approaches to develop biotic and abiotic stress-tolerant crops with adequate nutritional food value. This will be useful in meeting the food and biofuel security with the growing population and changing environment. As discussed earlier, the plant breeding has played a significant role in crop improvement; still we need to do more. In this context the impact of agrobiotechnology is both productive and benign. We can utilize the most cutting edge works associated with genetic mapping, molecular markers, and biotechnology to accelerate the crop development process. Methods through which crops with enhanced immunity can be generated are discussed in the following sections.

4.1 Molecular Plant Breeding

The plant breeding was the basis of the green revolution that led to increase in wheat and rice production in the twentieth century. The merger of biotechnology with conventional plant breeding techniques along with increase in our knowledge about basic plant biology has led to evolution of molecular plant breeding. Many reviews have discussed the molecular techniques and essential requirements for efficient use of molecular plant breeding in future crops (Jauhar 2006; Wenzel 2006; Moose and Mumm 2008; Hospital 2009; Torres 2010). A number of molecular markers based on simple sequence repeats (SSRs), single nucleotide polymorphism (SNPs), insert-deletions, and candidate gene markers are being developed in several crop species that will assist in genetic analysis and breeding programs (Feuillet et al. 2010). In recent years the next-generation sequencing (NGS) technologies have positively influenced the breeding programs (Varshney et al. 2005, 2010). A greater impact of NGS is noted on the comparative genomic studies which is expected to facilitate breeding programs.

The breeding for disease resistance is the greatest challenge because there is great variability both in plants and pathogens. Although our knowledge about disease resistance mechanisms has increased but still its application for developing resistant varieties is not an easy task because only the genes responsible for species level resistance (host resistance) can be transferred to elite varieties through breeding. Against many pathogens the plant resistance is a complex trait governed by QTLs having major or minor roles; with the advancement of molecular breeding technologies it will be possible to transfer many of the QTLs in elite varieties (Poland et al. 2009).

In breeding programs the field trials need to be well designed as various others environmental factors can also influence the final outcome of plant–pathogen interactions. It is visualized that next decade will be dominated by the high yielding and stress-tolerant varieties developed through traditional and molecular breeding due to the sociopolitical reasons associated with genetically modified (GM) crops.

4.2 Induction of Plant Immunity

Although breeding strategies are useful in enhancing species level resistance, they are time-consuming and have some drawbacks like linkage drag and nonavailability of effective resistant germplasms. The crop production can improve if we espouse environment friendly chemicals that enhance plant immunity, use nonpathogenic microbes as biocontrol agents that induce SAR, and raise transgenic plants with greater potential to recognize the pathogens and execute defense responses (Mourgues et al. 1998; Dita et al. 2006; Collinge et al. 2010; Gust et al. 2010; Shoresh et al. 2010; Wulff et al. 2011).

The initial transgenic crops were developed to overcome pathogen infestations and herbicide tolerance for industrial (ethanol, oil, textile, sugar) use of crops like corn, cotton, sugarcane, soybean, etc (Marshall 2010). When this trend shifted to crops for food consumption then various biosafety and ethical issues were raised, which were also raised for industrial crops but to a lower level. These issues were successfully overcome by the use of marker free transgenic, field trials, and well-designed experiments on animal models, so GM crops are making greater impact on the economy and accepted by people now (Carpenter 2010). Several genes are regularly being tried to get biotic stress-tolerant plants. Transgenic approaches to control herbivore pests are mainly expression of recombinant protease inhibitors and *Bacillus thuringiensis* endotoxins along with some alternate strategies (Bravo and Soberon 2008; Gatehouse 2008; Schlüter et al. 2010; Sanahuja et al. 2011). Some of the recent publications in this regard are mentioned in Table 16.3. The *cis*-engineering has provided promoters that precisely express the useful genes in an organ-specific and pathogen-inducible manner depending upon mode of pathogen infection (Venter 2006).

Table 16.3 Recent reports of enhanced tolerance against biotic stress by overexpression of gene(s)

Host transgenic plant	Gene(s) transformed	Tolerance against	Reference
Amorphophallus konjac	*Bacillus thuringiensis AiiA*	*Erwinia carotovora* subsp. *Carotovora* (Ecc) SCG1	Ban et al. (2009)
Arabidopsis thaliana	Pepper Mannose-binding lectin 1 (*CaMBL1*)	*Pseudomonas syringae* and *Alternaria brassicicola*	Hwang and Hwang (2011)
	B. vulgaris germin-like protein 1	*Verticillium longisporum* and *Rhizoctonia solani*	Knecht et al. (2010)
	Capsicum annuum ANTIMICROBIAL PROTEIN1	*P. syringae, H. parasitica, F.o.* f. sp. *matthiolae*, and *Alternaria brassicicola*	Lee et al. (2008)
	Solanum lycopersicumVe1	Race 1 of *Verticillium dahliae* and *V. albo-atrum*	Fradin et al. (2011)
	OsBSR1	*Colletotrichum higginsianum* and *Pst* DC3000	Dubouzet et al. (2011)
	Stellaria media SmAMP1 & 2	*Bipolaris sorokiniana*	Shukurov et al. (2012)
Arachis hypogaea	*Brassica juncea* defensin	*P. personata* and *Cercospora arachidicola*	Anuradha et al. (2008)
	OsChit-3	*Cercospora arachidicola*	Iqbal et al. (2012)
Brassica napus	*BnMPK4*	*Sclerotinia sclerotiorum* and *Botrytis cinerea*	Wang et al. (2009)
Brassica napus	*Triticum aestivum OXO*	*Sclerotinia sclerotiorum*	Dong et al. (2008)
Carica papaya	*Dahlia merckii DmAMP1*	*Phytophthora palmivora*	Zhu et al. (2007)
Colocynthis citrullus	*Wasabia japonica* defensin	*Alternaria solani* and *Fusarium oxysporum*	Ntui et al. (2010)
Daucus carota	*AtNPR1*	*Erysiphe heraclei*, *Xanthomonas hortorum*, *Botrytis cinerea*, *Alternaria radicina,* and *S. sclerotiorum*	Wally et al. (2009)
	HvChi-2 and *TaLTP*	*Alternaria radicicola* and *Botrytis cinerea*	Jayaraj and Punja (2007)
Gossypium hirsutum	*AtNPR1*	*V. dahliae* isolate TS2, *F.o.* f. sp. *vasinfectum*, *R. solani*, *Alternaria alternate*, and *Rotylenchulus reniformis*	Parkhi et al. (2010)
	NaPI and *StPin1A*	*Helicoverpa* spp.	Dunse et al. (2010)
	Xanthomonas oryzae pv. *oryzae Hpa1Xoo*	*Verticillium dahliae*	Miao et al. (2010)
	Momordica charantia McChit1	*Verticillium wilt*	Xiao et al. (2007)

(continued)

Table 16.3 (continued)

Host transgenic plant	Gene(s) transformed	Tolerance against	Reference
Hordeum vulgare	*D. melanogaster* metchnikowin	*F. graminearum*	Rahnamaeian et al. (2009)
L. esculentum	*CaMi*	*Meloidogyne incognita*	Chen et al. (2007)
	Bacillus thuringiensis Cry6A	*Meloidogyne incognita*	Li et al. (2007b)
Malus × domestica Borkh	*Nicotiana alata* proteinase inhibitor	*Epiphyas postvittiana*	Maheswaran et al. (2007)
Medicago sativa	*M. truncatula RCT1*	*Colletotrichum trifolii*	Yang et al. (2008)
Musa acuminata	*C. annuum* Hrap	*X. campestris* pv. *musacearum*	Tripathi et al. (2010)
N. benthamiana	*Momordica charantia McChit1*	*Phytophthora nicotianae*	Xiao et al. (2007)
	A. thaliana EFR	*Pss* B728a, *Pta* 11528, and *A. tumefaciens* A281	Lacombe et al. (2010)
N. tabacum	Protease inhibitor	*Spodoptera litura* and *Helicoverpa armigera*	Srinivasan et al. (2009)
	Brassica juncea defensin	*Phytophthora parasitica* pv. *Nicotianae* and *Fusarium moniliforme*	Anuradha et al. (2008)
	Gastrodia antifungal protein (GAFP-1)	*P. nicotianae*, *Meloidogyne incognita*, and *Rhizoctonia solani*	Cox et al. (2006)
	Hydronyche versuta Hvt1	*Heliothis armigera*	Shah et al. (2011)
	Stellaria media SmAMP1 & *2*	*Thielaviopsis basicola*	Shukurov et al. (2012)
	Nicotiana megalosiphon NmDef02	*Phytophthora parasitica* var. *nicotianae* and *Peronospora hyoscyami* f. sp. *tabacina*	Portieles et al. (2010)
	Metarhizium anisopliae Chit1	*Rhizoctonia solani*	Kern et al. (2010)
	Zephyranthes grandiflora, Agglutinin	*Myzus nicotianae*	Ye et al. (2009)
	MsrA2 and Temporin A	*Fusarium solani*, *F. oxysporum*, *Alternaria alternata*, *Botrytis cinerea*, *Sclerotinia sclerotiorum*, *Pythium aphanidermatum*, and *Pectobacterium carotovorum*	Yevtushenko and Misra (2009)

Oryza sativa	*Trichoderma virens* endochitinase	*Rhizoctonia solani*	Shah et al. (2009)
	Xanthomonas hrf1	*Magnaporthe grisea*	Shao et al. (2008)
	OsBSR1	*Xanthomonas oryzae*, *Magnaporthe grisea*,	Dubouzet et al. (2011)
	Pleurotus cornucopiae tamavidin 1	*Magnaporthe oryzae*	Takakura et al. (2012)
	Allium sativum and *Galanthus nivalis* lectin genes	Brown planthopper , Green leafhopper, and Whitebacked planthopper	Bharathi et al. (2011)
	Raphanus sativus AFP2	*Magnaporthe oryzae* and *Rhizoctonia solani*	Jha and Chattoo (2010)
	Chimeric Cry1Ab/Vip3H	*Chilo suppressalis* and *Sesamia inferens*	Chen et al. (2010)
	Podisus maculiventris thanatin	*Magnaporthe oryzae*	Imamura et al. (2010)
	Allium sativum leaf agglutinin	Green leafhopper and Brown planthopper	Yarasi et al. (2008); Sengupta et al. (2010)
	B. rapa BrD1	*Nilaparvata lugens*	Choi et al. (2009)
	Potato carboxypeptidase inhibitor	*Magnaporthe oryzae* and *Fusarium verticillioides*	Quilis et al. (2007)
Ricinus communis	*cry1EC*	*Spodoptera litura*, Fabr and *Achoea janata*	Sujatha et al. (2009)
Saccharum officinarum	*B. thuringiensis mCry1Ac*	*Proceras venosatus*	Weng et al. (2011)
S. lycopersicum	Pepper methionine sulfoxide reductase B2 (*CaMsrB2*)	*Phytophthora capsici* and *P. infestans*	Oh et al. (2010)
	A. thaliana EFR	*R. solanacearum* GMI1000 and *X. perforans* T4-4B	Lacombe et al. (2010)
Solanum tuberosum	*Nicotiana megalosiphon NmDef02*	*Alternaria solani* and *P. infestans*	Portieles et al. (2010)
	S. chacoense SN1	*Rhizoctonia solani* and *Erwinia carotovora*	Almasia et al. (2008)

(continued)

Table 16.3 (continued)

Host transgenic plant	Gene(s) transformed	Tolerance against	Reference
Triticum aestivum	*TaPIEP1*	*Bipolaris sorokiniana*	Dong et al. (2010)
	Actinidia chinensis pectin methyl esterase inhibitor	*Bipolaris sorokiniana* and *Fusarium graminearum*	Volpi et al. (2011)
	Thinopyrum intermedium ERF1	*Rhizoctonia cerealis*	Chen et al. (2008)
	Stpk-V	*Blumeria graminis* f. sp. *tritici*	Cao et al. (2011)
	Barley class II chitinase	*Fusarium graminearum*	Shin et al. (2008)
	Raphanus sativus AFP2	*Fusarium graminearum* and *Rhizoctonia cerealis*	Li et al. (2011)
	PvPGIP2	*Fusarium moniliforme* and *Bipolaris sorokiniana*	Janni et al. (2008)
Vigna radiata	*BjNPR1*	*Rhizoctonia solani*	Vijayan and Kirti (2012)
Vigna unguiculata	*Phaseolus vulgaris αAI-1*	*Callosobruchus maculatus* and *C. chinensis*	Solleti et al. (2008)
Zea mays	UMV4 virus modified KP4	*Ustilago maydis*	Allen et al. (2011)
	HvCPI-6	*Tetranychus urticae*	Carrillo et al. (2011)

4.3 Manipulation of Susceptibility Factors

It is now very clear that for pathogenesis, plant pathogens manipulate host metabolism and suppress plant defense. In some cases plant proteins behave as susceptibility factors, i.e., plant proteins help in pathogen growth and reproduction leading to disease establishment. The role of a gene in susceptibility can be either because of its own function as negative regulator of plant defense or plant effectors may target its protein product for their own growth, although the gene may have role in plant growth and development in normal conditions (Eckardt 2002; De Almeida et al. 2005; Pavan et al. 2010). The elimination or modification of such plant factors from crop plants can also be a method to achieve resistance against pathogens, although modifications of gene should not have obvious negative consequences on plant health and yield. Many recessive genes that act as negative regulators provide resistance by activating the cell death (*cpr*, *lsd*, *cim*, *acd*, and *mlo*) or by unknown mechanisms independent of salicylic acid, jasmonic acid, and ethylene signaling pathways (*pmr6*).

In one of the best examples of a susceptibility gene, barley's *Mlo* (*Mildew Resistance Locus o*) gene is required for successful colonization by the ascomycete *B. graminis* f. sp. *hordei* (Humphry et al. 2006). Nonfunctional mutant alleles of this gene provide durable resistance in many elite varieties of barley after their introgression into them. Its role in powdery mildew pathogenesis has also been found in *Arabidopsis*, tomato, and pea plants (Consonni et al. 2006; Bai et al. 2008; Humphry et al. 2011). The gene seems to function as a suppressor of nonhost defense response components/signaling as resistance in *mlo* mutant plants and nonhost resistance share analogous features (Humphry et al. 2006). The *pmr6* mutants showed enhanced recessive resistance to *Erysiphe orontii* and *E. cichoracearum* but these mutant plants were susceptible to *P. parasitica* (Vogel and Somerville 2000; Vogel et al. 2002). The *pmr6* gene encodes for a pectate lyase-like protein with extended C-terminal, the mutations in this gene show pleiotropic effects on plant growth, and the cell wall of these plants have high pectin content. The eukaryotic translation initiation factor subunits (mostly elF4E and elF4G) act as susceptibility factors for viral infections mainly potyviruses (Robaglia and Caranta 2006; Piron et al. 2010). In *Arabidopsis* a pathogen-inducible patatin-like lipid acyl hydrolase (*PLP2*) facilitates fungal and bacterial colonization (La Camera et al. 2005) and in rice loss of a proline-rich protein (Pi21) confers durable disease resistance (Fukuoka et al. 2009). The transcription-activator-like (TAL) effector proteins of bacteria target many susceptible factors and in resistant plants they are recognized by many *R*-genes (Lewis et al. 2009; Bogdanove et al. 2010). A group of 'SWEET' sugar efflux transporters are induced by several pathogens and it was shown that TAL effectors in case of *Xanthomonas* spp. regulate their induction for pathogen growth (Chen et al. 2010).

The availability of genome editing in plants and further technology improvements will help scientists to manipulate the pathogen-induced expression or the whole susceptibility gene from plant. Thus, this powerful method can also increase the hope for improved GM crops with durable disease resistance (Weinthal et al. 2010).

4.4 Host-Induced Gene Silencing in Pathogens

The sequencing projects of various pathogens especially filamentous pathogens have revealed that their effectors are rapidly evolving as compared to other genes and their genomes are rich in transposons (Dodds 2010). This suggests that in near future more virulent strains of a pathogen will emerge like the highly virulent strain of *Puccinia graminis* f. sp. *tritici* Ug99 and events of host jumps may also be seen. In the long run, breeding and induced defense-based approaches will work only against pathogens that will evolve slowly but approaches that target the basic cellular and pathogenicity mechanisms of pathogens would provide long-lasting resistance. The RNA interference (RNAi; RNA-guided regulation of RNA transcripts) based approach would make an ideal choice against rapidly evolving pathogens, as it is known to provide resistance against viral infection in natural environment (Baulcombe 2004). Transgenic plants with RNAi constructs targeting specific genes of pathogens have shown resistance against viruses, parasitic nematodes, herbivorous insects, and parasitic weeds in many plants (Huang et al. 2006; Frizzi and Huang 2010; Niu et al. 2010; Wani et al. 2010). In an unsuccessful attempt, the *Plasmodiophora brassicae* gene was also checked for downregulation on transgenic *Arabidopsis thaliana* plants as this phytomyxea pathogen remains in intimate contact with host cell (Bulman 2006).

Considering the situation that ~70% of all major crop diseases are caused by fungal pathogens (Agrios 2005), this RNAi technology against fungi would greatly help to increase crop yield. Two prerequisites for successful silencing of fungal genes on transgenic plants would be the transfer of silencing-RNAs from host plant cell to the fungi and a functional RNAi machinery of the pathogenic fungi. Many independent groups have reported the silencing of genes using RNAi constructs in fungi suggesting that the RNAi machinery works in many fungi. The uptake of dsRNA from outside the fungal cells and subsequent silencing of the targeted fungal gene transcripts were claimed in two US patents (Van De Craen et al. 2006; Roberts et al. 2008). Tinoco et al. (2010) reported silencing of the *gus* transcripts in transgenic *Fusarium verticillioides* when it was inoculated on transgenic tobacco plants expressing RNAi construct against *gus* gene. Nowara et al. (2010) also showed that dsRNA or siRNA molecules were exchanged between cereal hosts and the obligate biotrophic fungal pathogen *Blumeria graminis* and they called this technique of downregulating pathogen genes as host-induced gene silencing (HIGS). Using transient expression, virus-induced gene silencing (VIGS), and transgenic plants with RNAi constructs it was proved that HIGS could be an effective tool to study the role of fungal genes in pathogenesis and it has the potential of disease control against biotrophic fungal pathogens (Fig. 16.2). Using VIGS the genes that are expressed in haustorial cells were silenced efficiently in *Puccinia striiformis* f. sp. *tritici* rather than the genes that are constitutively expressed in whole pathogen, probably pointing towards the fact that tissue which remains in intimate contact with host will receive more silencing-RNAs (Yin et al. 2011). More experiments with other systems are needed to standardize this technology before engineering at mass level and

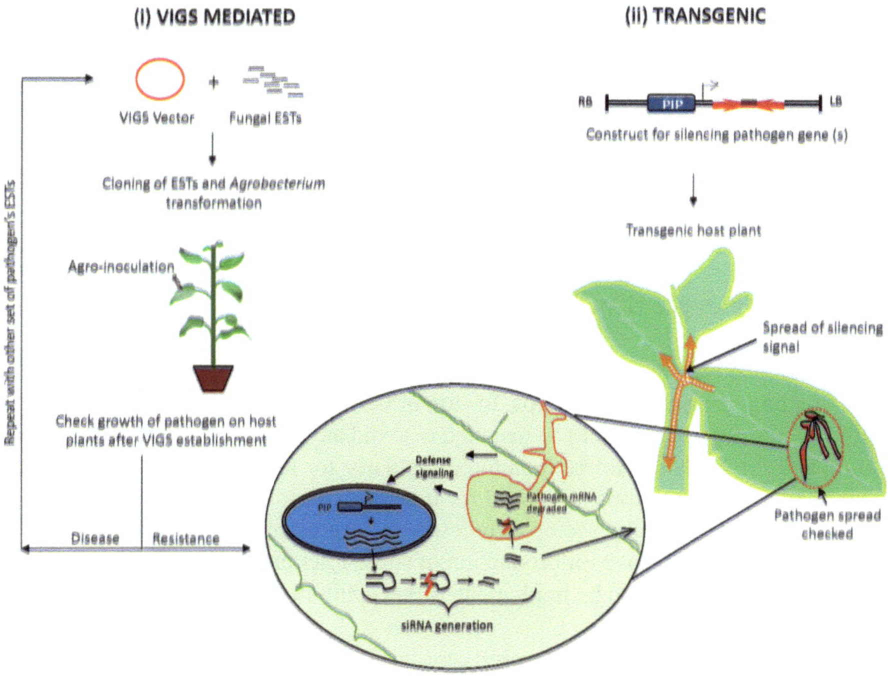

Fig. 16.2 Host-induced gene silencing (HIGS). Genes essential for pathogen growth on host plants can be downregulated by RNAi approach to limit the pathogen growth. (*i*) High-throughput approaches like virus-induced gene silencing (VIGS) can be used to identify the genes involved in pathogen growth and reproduction on host plants and (*ii*) the pathogen inducible promoter (PIP) can be used to generate transgenic plants having RNAi constructs against the gene/or genes of a pathogen

also the questions regarding the silencing of genes in hemibiotrophs and necrotrophs need to be answered. The usefulness of fungal inducible promoters to drive the RNAi constructs should help but the most important thing is to check for RNAi constructs off-targets and avoid it inside the plant cell. Overall the HIGS technology holds promise for generating fungal-tolerant crops leading to higher grain yield and it is believed that in future a common terminology of HIGS will be followed to make scientific literature retrieval easy regarding this type of silencing.

5 Conclusions and Future Prospects

We have come a long way in crop improvement from traditional elite variety selection to the development molecular breeding and transgenic crops. But our demand of food supply still needs rapid progress with growing population and nemesis of adverse environmental conditions. Also the increase in demand for biofuels will add more pressure on arable land. In this decade a great deal of information has been achieved about molecular aspects of plant–pathogen interactions. The technological advancements have certainly played a major role in this regard. Now, every aspect of plant–pathogen interaction is studied and sequencing of many crop plants and their pathogens will help in pyramiding various genes through marker-assisted selection especially against notorious pests and necrotrophic fungi where resistance is governed by many QTLs. Contrary to the biosafety-related opinions raised regarding GM crops, molecular plant biologists are optimistic about the need to incorporate GM crops in our crop improvement chain as it can be applied to all the crops outside the limits of species. Already more that 20% of arable land is under the GM crops in countries like USA, Brazil, and Argentina, which dictates the success story of GM crops.

We still need to study and effectively use the nonhost resistance components for high yielding disease-tolerant crops. In case of GM crops effective regulatory mechanisms and safeguards need to be installed to avoid any biosafety-related problem in future and the fields should be monitored regularly for the evolution of new pathogens against resistant crops. The need for translational of basic research to the field crops is more from public sector as investments are more in this sector. The areas where still we can improve for production of stress tolerance crops need to be evaluated and programs need to be implemented especially in developing countries.

Acknowledgements This work is supported partially by research grant provided by Department of Biotechnology, Government of India and National Institute of Plant Genome Research, New Delhi. We acknowledge Dr. K. D. Srivastava, Indian Agricultural Research Institute, New Delhi for valuable suggestions and critically editing the manuscript. K.K. acknowledges NIPGR for postdoctoral fellowship.

References

Agrios GN (2005) Plant pathology, 5th edn. Academic, San Diego

Asai T, Tena G, Plotnikova J, Willmann MR, Chiu WL (2002) MAP kinase signalling cascade in *Arabidopsis* innate immunity. Nature 415:977–983

Allen A, Islamovic E, Kaur J, Gold S, Shah D, Smith TJ (2011) Transgenic maize plants expressing the Totivirus antifungal protein, KP4, are highly resistant to corn smut. Plant Biotechnol J 9:857–864

Almasia NI, Bazzini AA, Hopp HE, Rovere CV (2008) Overexpression of snakin-1 gene enhances resistance to *Rhizoctonia solani* and *Erwinia carotovora* in transgenic potato plants. Mol Plant Pathol 9:329–338

Anuradha S, Divya K, Jami SK, Kirti PB (2008) Transgenic tobacco and peanut plants expressing a mustard defensin show resistance to fungal pathogens. Plant Cell Rep 27:1777–1786

Bai Y, Pavan S, Zheng Z, Zappel NF, Reinstädler A, Lotti C, De Giovanni C, Ricciardi L, Lindhout P, Visser R, Theres K, Panstruga R (2008) Naturally occurring broad-spectrum powdery mildew resistance in a Central American tomato accession is caused by loss of *Mlo* function. Mol Plant Microbe Interact 21:30–39

Ban H, Chai X, Lin Y, Zhou Y, Peng D, Zhou Y, Zou Y, Yu Z, Sun M (2009) Transgenic *Amorphophallus konjac* expressing synthesized acyl-homoserine lactonase (*aiiA*) gene exhibit enhanced resistance to soft rot disease. Plant Cell Rep 28:1847–1855

Bari R, Jones JDG (2009) Role of plant hormones in plant defense responses. Plant Mol Biol 69:473–488

Baulcombe DC (2004) RNA silencing in plants. Nature 431:356–363

Bent AF, Mackey D (2007) Elicitors, effectors, and *R* genes: the new paradigm and a lifetime supply of questions. Annu Rev Phytopathol 45:399–436

Bessire M, Chassot C, Jacquat AC, Humphry M, Borel S, MacDonald-Comber Petétot J, Métraux JP, Nawrath C (2007) A permeable cuticle in *Arabidopsis* leads to a strong resistance to *Botrytis cinerea*. EMBO J 26:2158–2168

Bharathi Y, Kumar SV, Pasalu IC, Balachandran SM, Reddy VD, Rao KV (2011) Pyramided rice lines harbouring *Allium sativum* (*asal*) and *Galanthus nivalis* (gna) lectin genes impart enhanced resistance against major sap-sucking pests. J Biotechnol 152:63-71

Bogdanove AJ, Schornack S, Lahaye T (2010) TAL effectors: finding plant genes for disease and defense. Curr Opin Plant Biol 13:394–401

Boller T (1995) Chemoperception of microbial signals in plant cells. Annu Rev Plant Physiol Plant Mol Biol 46:189–214

Boller T, Felix G (2009) A renaissance of elicitors: perception of microbe-associated molecular patterns and danger signals by pattern-recognition receptors. Annu Rev Plant Biol 60: 379–406

Bolwell GP (1999) Role of active oxygen species and NO in plant defense responses. Curr Opin Plant Biol 2:287–294

Boudsocq M, Willmann MR, McCormack M, Lee H, Shan L, He P, Bush J, Cheng SH, Sheen J (2010) Differential innate immunity signaling via Ca^{2+} sensor protein kinases. Nature 464:418–422

Bravo A, Soberon M (2008) How to cope with insect resistance to Bt toxins? Trends Biotechnol 26:573–579

Brunner F, Rosahl S, Lee J, Rudd JJ, Geiler C, Kauppinen S, Rasmussen G, Scheel D, Nurnberger T (2002) Pep1-13, a plant defense-inducing pathogen associated pattern from *Phytophthora* transglutaminases. EMBO J 21:6681–6688

Bulman SR (2006) Testing the effect of in planta RNA silencing on Plasmodiophora brassicae infection. Ph.D thesis. Lincoln University, Chester County

Cantu D, Vicente AR, Labavitch JM, Bennett AB, Powell ALT (2008) Strangers in the matrix: plant cell walls and pathogen susceptibility. Trends Plant Sci 13:610–617

Cao A, Xing L, Wang X, Yang X, Wang W, Sun Y, Qian C, Ni J, Chen Y, Liu D, Wang X, Chen P (2011) Serine/threonine kinase gene *Stpk-V*, a key member of powdery mildew resistance gene *Pm21*, confers powdery mildew resistance in wheat. Proc Natl Acad Sci USA 108:7727–7732

Caplan J, Padmanabhan M, Dinesh-Kumar SP (2008) Plant NB-LRR immune receptors: from recognition to transcriptional reprogramming. Cell Host Microbe 3:126–135

Carrillo L, Martinez M, Ramessar K, Cambra I, Castanera P, Ortego F, Diaz I (2011) Expression of a barley cystatin gene in maize enhances resistance against phytophagous mites by altering their cysteine-proteases. Plant Cell Rep 30:101–112

Carpenter JE (2010) Peer-reviewed surveys indicate positive impact of commercialized GM crops. Nat Biotechnol 28:319–321

Chassot C, Nawrath C, Metraux JP (2007) Cuticular defects lead to full immunity to a major plant pathogen. Plant J 49:972–980

Chassot C, Nawrath C, Metraux JP (2008) The cuticle: not only a barrier for plant defense. Plant Signal Behav 3:142–144

Chen L, Zhang ZY, Liang HX, Liu HX, Du LP, Xu H, Xin Z (2008) Overexpression of *TiERF1* enhances resistance to sharp eyespot in transgenic wheat. J Exp Bot 59:4195–4204

Chen R, Li H, Zhang L, Zhang J, Xiao J, Ye Z (2007) CaMi, a root-knot nematode resistance gene from hot pepper (*Capsium annuum* L.) confers nematode resistance in tomato. Plant Cell Rep 26:895–905

Chen X, Ronald PC (2011) Innate immunity in rice. Trends Plant Sci 16:451–459

Chen Y, Tian JC, Shen ZC, Peng YF, Hu C, Guo YY, Ye GY (2010b) Transgenic rice plants expressing a fused protein of Cry1Ab/Vip3H has resistance to rice stem borers under laboratory and field conditions. J Econ Entomol 103:1444–1453

Chen L-Q, Hou B-H, Lalonde S, Takanaga H, Hartung ML, Qu X-Q, Guo W-J, Kim J-G et al (2010) Sugar transporters for intercellular exchange and nutrition of pathogens. Nature 468:527–532

Chibucos MC, Tseng TT, Setubal JC (2009) Describing commonalities in microbial effector delivery using the Gene Ontology. Trends Microbiol 17:312–319

Chisholm ST, Goaker G, Day B, Staskawicz BJ (2006) Host-Microbe interactions: shaping the evolution of the plant immune system. Cell 124:803–814

Choi MS, Kim YH, Park HM, Seo BY, Jung JK, Kim ST, Kim MC, Shin DB, Yun HT, Choi IS, Kim CK, Lee JY (2009) Expression of *BrD1*, a plant defensin from *Brassica rapa*, confers resistance against brown plant hopper (*Nilaparvata lugens*) in transgenic rices. Mol Cells 28:131–137

Collier SM, Moffett P (2009) NB-LRRs work a "bait and switch" on pathogens. Trends Plant Sci 14:521–529

Collinge DB, Jorgensen HJL, Lund OS, Lyngkjaer MF (2010) Engineering pathogen resistance in crop plants: current trends and future prospects. Annu Rev Plant Pathol 48:269–291

Consonni C, Humphry ME, Hartmann HA, Livaja M, Durner J, Westphal L, Vogel J, Lipka V, Kemmerling B, Schulze-Lefert P, Somerville SC, Panstruga R (2006) Conserved requirement for a plant host cell protein in powdery mildew pathogenesis. Nat Genet 38:716–720

Cox KD, Layne DR, Scorza R, Schnabel G (2006) Gastrodia anti-fungal protein from the orchid *Gastrodia elata* confers disease resistance to root pathogens in transgenic tobacco. Planta 224:1373–1383

Curvers K, Seifi H, Mouille G, de Rycke R, Asselbergh B, Van Hecke A, Vanderschaeghe D, Höfte H, Callewaert N, Van Breusegem F, Höfte M (2010) Abscisic acid deficiency causes changes in cuticle permeability and composition that influence tomato resistance to *Botrytis cinerea*. Plant Physiol 154:847–860

De Almeida EJ, Favery B, Engler G, Abad P (2005) Loss of susceptibility as an alternative for nematode resistance. Curr Opin Biotechnol 16:112–117

De Jonge R, Bolton MD, Thomma BPHJ (2011) How filamentous pathogen co-opt plants: the ins and outs of fungal effector. Curr Opin Plant Biol 14:400–406

Denoux C, Galletti R, Mammarella N, Gopalan S, Werck D, De Lorenzo G, Ferrari S, Ausubel FM, Dewdney J (2008) Activation of defense response pathways by OGs and flg22 elicitors in *Arabidopsis* seedlings. Mol Plant 1:423–445

Dita MA, Rispail N, Prats E, Rubiales D, Singh KB (2006) Biotechnological approaches to overcome biotic and abiotic stress constraints in legumes. Euphytica 147:1–24

Dodds PN (2010) Genome evolution in plant pathogens. Science 330:1486–1487

Dong N, Liu X, Lu Y, Du L, Xu H, Liu H, Xin Z, Zhang Z (2010) Overexpression of *TaPIEP1*, a pathogen-induced ERF gene of wheat, confers host-enhanced resistance to fungal pathogen *Bipolaris sorokiniana*. Funct Integr Genomics 10:215–226

Dong X, Ji R, Guo X, Foster SJ, Chen H, Dong C, Liu Y, Hu Q, Liu S (2008) Expressing a gene encoding wheat oxalate oxidase enhances resistance to *Sclerotinia sclerotiorum* in oilseed rape (*Brassica napus*). Planta 228:331–340

Dubouzet JG, Maeda S, Sugano S, Ohtake M, Hayashi N, Ichikawa T, Kondou Y, Kuroda H, Horii Y, Matsui M, Oda K, Hirochika H, Takatsuji H, Mori M (2011) Screening for resistance against *Pseudomonas syringae* in rice-FOX Arabidopsis lines identified a putative receptor-like cytoplasmic kinase gene that confers resistance to major bacterial and fungal pathogens in Arabidopsis and rice. Plant Biotechnol J 9:466–485

Dunse KM, Stevens JA, Lay FT, Gaspar YM, Heath RL, Anderson MA (2010) Coexpression of potato type I and II proteinase inhibitors gives cotton plants protection against insect damage in the field. Proc Natl Acad Sci USA 107:15011–15015

Durrant WE, Dong X (2004) Systemic acquired resistance. Annu Rev Phytopathol 42:185–209

Eckardt NA (2002) Plant disease susceptibility genes? Plant Cell 14:1983–1986

Ellis J (2006) Insights into nonhost disease resistance: can they assist disease control in agariculture? Plant Cell 18:523–528

Erbs G, Newman MA (2012) The role of lipopolysaccharide and peptidoglycan, two glycosylated bacterial microbe-associated molecular patterns (MAMPs), in plant innate immunity. Mol Plant Pathol 13:95–104

Felix G, Boller T (2003) Molecular sensing of bacteria in plants. The highly conserved RNA-binding motif RNP-1 of bacterial cold shock proteins is recognized as an elicitor signal in tobacco. J Biol Chem 278:6201–6208

Feng J, Wang F, Liu G, Greenshields D, Shen W, Kaminskyj S, Hughes GR, Peng Y, Selvaraj G, Zou J, Wei Y (2009) Analysis of *Blumeria graminis*-secreted lipase reveals the importance of host epicuticular wax components for fungal adhesion and development. Mol Plant Microbe Interact 22:1601–1610

Feuillet C, Leach JE, Rogers J, Schnable PS, Eversole K (2010) Crop genome sequencing: lessons and rationales. Trends Plant Sci 16:77–88

Fliegmann J, Mithofer A, Wanner G, Ebel J (2004) An ancient enzyme domain hidden in the putative β-glucan elicitor receptor of soybean may play an active part in the perception of pathogen-associated molecular patterns during broad host resistance. J Biol Chem 279:1132–1140

Fradin EF, Abd-El-Haliem A, Masini L, van den Berg GCM, Joosten MHAG, Thomma BPHJ (2011) Interfamily transfer of tomato *Ve1* mediates *Verticillium* resistance in Arabidopsis. Plant Physiol 156:2255–2265

Friedman AR, Baker BJ (2007) The evolution of resistance genes in multi-protein plant resistance systems. Curr Opin Genet Dev 17:493–499

Frizzi A, Huang S (2010) Tapping the silencing pathways for plant biotechnology. Plant Biotechnol J 8:655–677

Fukuoka S, Saka N, Koga H, Ono K, Shimizu T, Ebana K, Hayashi N, Takahashi A, Hirochika H, Okuno K, Yano M (2009) Loss of function of a proline-containing protein confers durable disease resistance in rice. Science 325:998–1001

Garcia-Brugger A, Lamotte O, Vandelle E, Bourque S, Lecourieux D, Poinssot B, Wendehenne D, Pugin A (2006) Early signaling events induced by elicitors of plant defenses. Mol Plant Microbe Interact 19:711–724

Gatehouse JA (2008) Biotechnological prospects for engineering insect-resistant plants. Plant Physiol 146:881–887

Gaulin E, Drame N, Lafitte C, Torto-Alalibo T, Martinez Y, Torregrosa CA, Khatib M, Mazarguil H, Villalba-Mateos F, Kamoun S, Mazars C, Dumas B, Bottin A, Esquerre-Tugaye MT, Rickauer M (2006) Cellulose binding domains of a *Phytophthora* cell wall protein are novel pathogen-associated molecular patterns. Plant Cell 18:1766–1777

Gheysen G, Mitchum MG (2011) How nematodes manipulate plant development pathways for infection? Curr Opin Plant Biol 14:415–421

Göhre V, Robatzek S (2008) Breaking the barriers: microbial effector molecules subvert plant immunity. Annu Rev Phytopathol 46:189–215

Gomez-Gomez L, Boller T (2000) FLS2: An LRR receptor-like kinase involved in the perception of the bacterial elicitor flagellin in *Arabidopsis*. Mol Cell 5:1003–1011

Granado J, Felix G, Boller T (1995) Perception of fungal sterols in plant (subnanomolar concentrations of ergosterol elicit extracellular alkalization in tomato cells). Plant Physiol 107:485–490

Gust AA, Brunner F, Nürnberger T (2010) Biotechnological concepts for improving plant innate immunity. Curr Opin Biotechnol 21:204–210

Heath MC (1998) Apoptosis, programmed cell death and the hypersensitive response. Eur J Plant Pathol 104:117–124

Heath MC (2000) Nonhost resistance and nonspecific plant defenses. Curr Opin Plant Biol 3:315–319

Hogenhout SA, Bos JIB (2011) Effector proteins that modulate plant-insect interaction. Curr Opin Plant Biol 14:422–428

Hok S, Attard A, Keller H (2010) Getting the most from the host: how pathogens force plants to cooperate in disease. Mol Plant Microbe Interact 23:1253–1259

Hospital F (2009) Challenges for effective marker-assisted selection in plants. Genetica 136:303–310

Huang G, Allen R, Davis EL, Baum TJ, Hussey RS (2006) Engineering broad root-knot resistance in transgenic plants by RNAi silencing of a conserved and essential root-knot nematode parasitism gene. Proc Natl Acad Sci USA 103:14302–14306

Hückelhoven R (2007) Cell wall-associated mechanisms of disease resistance and susceptibility. Annu Rev Phytopathol 45:101–127

Humphry M, Consonni C, Panstruga R (2006) *mlo*-based powdery mildew immunity: silver bullet or simply non-host resistance? Mol Plant Pathol 7:605–610

Humphry M, Reinstädler A, Ivanov S, Bisseling T, Panstruga R (2011) Durable broad-spectrum powdery mildew resistance in pea *er1* plants is conferred by natural loss-of-function mutations in *PsMLO1*. Mol Plant Pathol 12:866–878

Hwang IS, Hwang BK (2011) The pepper mannose-binding lectin gene *CaMBL1* is required to regulate cell death and defense responses to microbial pathogens. Plant Physiol 155:447–463

Imamura T, Yasuda M, Kusano H, Nakashita H, Ohno Y, Kamakura T, Taguchi S, Shimada H (2010) Acquired resistance to the rice blast in transgenic rice accumulating the antimicrobial peptide thanatin. Transgenic Res 19:415–442

Iqbal MM, Nazir F, Ali S, Asif MA, Zafar Y, Iqbal J, Ali GM (2012) Over expression of rice chitinase gene in transgenic peanut (*Arachis hypogaea* L.) improves resistance against leaf spot. Mol Biotechnol 50:129–136

Janni M, Sella L, Favaron F, Blechl AE, Lorenzo GD, D'Ovidio R (2008) The expression of a bean PGIP in transgenic wheat confers increased resistance to the fungal pathogen *Bipolaris sorokiniana*. Mol Plant Microbe Interact 21:171–177

Jayaraj J, Punja ZK (2007) Combined expression of chitinase and lipid transfer protein genes in transgenic carrot plants enhances resistance to foliar fungal pathogens. Plant Cell Rep 26:1539–1546

Jha S, Chattoo BB (2010) Expression of a plant defensin in rice confers resistance to fungal phytopathogens. Transgenic Res 19:373-384

Jauhar P (2006) Modern biotechnology as an integral supplement to conventional plant breeding: the prospects and challenges. Crop Sci 46:1841–1859

Kim JG, Jeon E, Oh J, Moon JS, Hwang I (2004) Mutational analysis of *Xanthomonas* hairpin HpaG identifies a key functional region that elicits the hypersensitive response in nonhost plants. J Bacteriol 186:6239–6247

Klarzynski O, Descamps V, Plesse B, Yvin JC, Kloareq B, Fritiq B (2003) Sulfated fucan oligosaccharides elicit defense responses in tobacco and local systemic resistance against tobacco mosaic virus. Mol Plant Microbe Interact 16:115–122

Kern MF, Maraschin SF, Endt DV, Schrank A, Vainstein MH, Pasquali G (2010) Expression of a chitinase gene from *Metarhizium anisopliae* in tobacco plants confers resistance against *Rhizoctonia solani*. Appl Biochem Biotechnol 160:1933-1946

Knecht K, Seyffarth M, Desel C, Thurau T, Sherameti I, Lou B, Oelmuller R, Cai D (2010) Expression of BvGLP-1 encoding a germin-like protein from sugar beet in *Arabidopsis thaliana* leads to resistance against phytopathogenic fungi. Mol Plant Microbe Interact 23:446–457

Knepper C, Day B (2010) From perception to activation: the molecular-genetic and biochemical landscape of disease resistance signaling in plants. In: The Arabidopsis book. American Society of Plant Biologists, Rockville

Koga J, Yamauchi T, Shimura M, Ogawa N, Oshima K, Umemura K, Kikuchi M, Ogasawara N (1998) Cerebrosides A and C, sphingolipid elicitors of hypersensitive cell death and phytoalexin accumulation in rice plants. J Biol Chem 273:31985–31991

Kunze G, Zipfel C, Robatzek S, Niehaus K, Boller T, Felix G (2004) The N terminus of bacterial elongation factor Tu elicits innate immunity in *Arabidopsis* plants. Plant Cell 16:3496–3507

Lacombe S, Rougon-Cardoso A, Sherwood E, Peeters N, Dahlbeck D, van Esse HP, Smoker M, Rallapalli G, Thomma BP, Staskawicz B, Jones JD, Zipfel C (2010) Interfamily transfer of a plant pattern-recognition receptor confers broad-spectrum bacterial resistance. Nat Biotechnol 28:365–369

Lee SC, Hwang IS, Choi HW, Hwang BK (2008) Involvement of the pepper antimicrobial protein *CaAMP1* gene in broad spectrum disease resistance. Plant Physiol 148:1004–1020

Li XQ, Wei JZ, Tan A, Aroian RV (2007b) Resistance to root-knot nematode in tomato roots expressing a nematicidal *Bacillus thuringiensis* crystal protein. Plant Biotechnol J 5:455–464

Li Z, Zhou M, Zhang Z, Ren L, Du L, Zhang B, Xu H, Xin Z (2011) Expression of a radish defensin in transgenic wheat confers increased resistance to *Fusarium graminearum* and *Rhizoctonia cerealis*. Funct Integr Genomics 11:63–70

L'Haridon F, Besson-Bard A, Binda M, Serrano M, Abou-Mansour E, Balet F, Schoonbeek HJ, Hess S, Mir R, Leon J, Lamotte O, Metraux JP (2011) A permeable cuticle is associated with the release of reactive oxygen species and induction of innate immunity. PLoS Pathog 7:e1002148

La Camera S, Geoffroy P, Samaha H, Ndiaye A, Rahim G, Legrand M, Heitz T (2005) A pathogen-inducible patatin-like lipid acyl hydrolase facilitates fungal and bacterial host colonization in Arabidopsis. Plant J 44:810–825

Laquitaine L, Gomes E, Francois J, Marchive C, Pascal S, Hamdi S, Atanassova R, Delrot S, Coutos-Thevenot P (2006) Molecular basis of ergosterol-induced protection of grape against Botrytis cinerea: induction of type I LTP promoter activity, WRKY, and stilbene synthase gene expression. Mol Plant Microbe Interact 19:1103–1112

Lee J, Klessig DF, Nurnberger T (2001) A Harpin binding site in tobacco plasma membranes mediates activation of the pathogenesis-related gene *HIN1* independent of extracellular calcium but dependent on mitogen-activated protein kinase activity. Plant Cell 13:1079–1093

Lewis JD, Guttman DS, Desveaux D (2009) The targeting of plant cellular systems by injected type III effector proteins. Semin Cell Dev Biol 20:1055–1063

Li Y, Beisson F, Koo AJK, Molina I, Pollard M, Ohlrogge J (2007) Identification of acyltransferases required for cutin biosynthesis and production of cutin with suberin-like monomers. Proc Natl Acad Sci USA 104:18339–18344

Liu J, Liu X, Dai L, Wang G (2007) Recent progress in elucidating the structure, function and evolution of disease resistance genes in plants. J Genet Genomics 34:765–776

Liu W, Zhou X, Li G, Li L, Kong L, Wang C, Zhang H, Xu JR (2011) Multiple plant surface signals are sensed by different mechanisms in the rice blast fungus for appressorium formation. PLoS Pathog 7:e1001261

Lochman J, Mikes V (2006) Ergosterol treatment leads to the expression of a specific set of defense-related genes in tobacco. Plant Mol Biol 62:43–51

Lorenzo O, Solano R (2005) Molecular players regulating the jasmonate signaling network. Curr Opin Plant Biol 8:532–540

Lotze MT, Zeh HJ, Rubartelli A, Sparvero LJ, Amoscato AA, Washburn NR, DeVera ME, Liang X, Tör M, Billiar T (2007) The grateful dead: damage associated molecular pattern molecules and reduction/oxidation regulate immunity. Immunol Rev 220:60–81

Lukasik E, Takken FLW (2009) STANDing strong, resistance proteins instigators of plant defense. Curr Opin Plant Biol 12:427–436

Maheswaran G, Pridmore L, Franz P, Anderson MA (2007) A proteinase inhibitor from *Nicotiana alata* inhibits the normal development of light-brown apple moth, *Epiphyas postvittana* in transgenic apple plants. Plant Cell Rep 26:773–782

Manabe Y, Nafisi M, Verhertbruggen Y, Orfila C, Gille S, Rautengarten C, Cherk C, Marcus SE, Somerville S, Pauly M, Knox JP, Sakuragi Y, Scheller HV (2011) Loss-of-function mutation of *REDUCED WALL ACETYLATION2* in Arabidopsis leads to reduced cell wall acetylation and increased resistance to *Botrytis cinerea*. Plant Physiol 155:1068–1078

Marshall A (2010) 2nd-generation GM traits progress. Nat Biotech 28:306

Mendoza AM, Berndt P, Djamei A, Linne U, Marahiel M, Vranes M, Kämper J, Kashmann R (2009) Physical-chemical plant-derived signals induce differentiation in *Ustilago maydis*. Mol Microbiol 71:895–911

Miao W, Wang X, Li M, Song C, Wang Y, Hu D, Wang J (2010) Genetic transformation of cotton with a harpin-encoding gene *hpaXoo* confers an enhanced defense response against different pathogens through a priming mechanism. BMC Plant Biol 10:67

Molina A, Garcia-Olmedo F (1997) Enhanced tolerance to bacterial pathogens caused by the transgenic expression of barley lipid transfer protein LTP2. Plant J 12:669–675

Moose SP, Mumm RH (2008) Molecular plant breeding as the foundation of 21st century crop improvement. Plant Physiol 147:969–977

Mourgues F, Brisset M-N, Chevreau E (1998) Strategies to improve plant resistance to bacterial diseases through genetic engineering. Trends Biotechnol 6:203–210

Mukhtar MS, Carvunis A-R, Dreze M, Epple P, Steinbrenner J, Moore J, Tasan M et al (2011) Independently evolved virulence effectors converge onto hubs in a plant immune system network. Science 333:596–601

Nishimura MT, Dangl JL (2010) Arabidopsis and the plant immune system. Plant J 61:1053–1066

Niu JH, Jian H, Xu JM, Guo YD, Liu Q (2010) RNAi technology extends its reach: engineering plant resistance against harmful eukaryotes. Afr J Biotechnol 9:7573–7582

Nowara D, Gay A, Lacomme C, Shaw J, Ridout C, Douchkov D, Hensel G, Kumlehn J, Schweizer P (2010) HIGS: Host-induced gene silencing in the obligate biotrophic fungal pathogen *Blumeria graminis*. Plant Cell 22:3130–3141

Ntui VO, Thirukkumaran G, Azadi P, Khan RS, Nakamura I, Mii M (2010) Stable integration and expression of wasabi defensin gene in "Egusi" melon (*Colocynthis citrullus* L.) confers resistance to Fusarium wilt and Alternaria leaf spot. Plant Cell Rep 29:943–954

Oh SK, Baek KH, Seong ES, Joung YH, Choi GJ, Park JM, Cho HS, Kim EA, Lee S, Choi D (2010) *CaMsrB2*, Pepper methionine sulfoxide reductase B2, is a novel defense regulator against oxidative stress and pathogen attack. Plant physiol 154:245–261

Osman H, Vauthrin S, Mikes V, Milat ML, Panabieres F, Marais A, Brunie S, Maume B, Ponchet M, Blein JP (2001) Mediation of elicitin activity on tobacco is assumed by elicitin-sterol complexes. Mol Biol Cell 12:2825–2834

Papadopoulou K, Melton RE, Leggett M, Daniels MJ, Osbourn AE (1999) Compromised disease resistance in saponin-deficient plants. Proc Natl Acad Sci USA 96:12923–12928

Parkhi V, Kumar V, Campbell LM, Bell AA, Shah J, Rathore KS (2010) Resistance against various fungal pathogens and reniform nematode in transgenic cotton plants expressing Arabidopsis NPR1. Transgenic Res 19:959–975

Patkar RN, Chattoo BB (2006) Transgenic *indica* rice expressing ns-LTP-like protein shows enhanced resistance to both fungal and bacterial pathogens. Mol Breed 17:159–171

Pavan S, Jacobsen E, Visser RGF, Bai Y (2010) Loss of susceptibility as a novel breeding strategy for durable and broad-spectrum resistance. Mol Breed 25:1–12

Pieterse CMJ, Leon-Reyes A, Van der Ent S, Van Wees SCM (2009) Networking by small-molecule hormones in plant immunity. Nat Chem Biol 5:308–316

Piron F, Nicolai M, Minoïa S, Piednoir E, Moretti A, Salgues A, Zamir D, Caranta C, Bendahmane A (2010) An induced mutation in tomato eIF4E leads to immunity to two potyviruses. PLoS One 5:e11313

Poland JA, Balint-Kurti PJ, Wisser RJ, Pratt RC, Nelson RJ (2009) Shades of gray: the world of quantitative disease resistance. Trends Plant Sci 14(1):21–29

Portieles R, Ayra C, Gonzalez E, Gallo A, Rodriguez R, Chacon O, Lopez Y, Rodriguez M, Castillo J, Pujol M, Enriquez G, Borroto C, Trujillo L, Thomma BP, Hidalgo OB (2010) *NmDef02*, a novel antimicrobial gene isolated from *Nicotiana megalosiphon* confers high-level pathogen resistance under greenhouse and field conditions. Plant Biotechnol J 8:678–690

Quilis J, Meynard D, Vila L, Aviles FX, Guiderdoni E, Segundo BS (2007) A potato carboxypeptidase inhibitor gene provides pathogen resistance in transgenic rice. Plant Biotechnol J 5:537–553

Rahnamaeian M, Langen G, Imani J, Khalifa W, Altincicek B, von Wettstein D, Kogel KH, Vilcinskas A (2009) Insect peptide metchnikowin confers on barley a selective capacity for resistance to fungal ascomycetes pathogens. J Exp Bot 60:4105–4114

Ramirez V, Agorio A, Coego A, Andrade JG, Hernandez MJ, Balaguer B, Ouwerkerk PBF, Zarra I, Vera P (2011) MYB46 modulates disease susceptibility to *Botrytis cinerea* in Arabidopsis. Plant Physiol 155:1920–1935

Reina-Pinto JJ, Yephremov A (2009) Surface lipids and plant defenses. Plant Physiol Biochem 47:540–549

Robaglia C, Caranta C (2006) Translation initiation factors: a weak link in plant RNA virus infection. Trends Plant Sci 11:40–45

Roberts JK, Pitkin JW, Adams TH. 2008; USA patent publication no. 2008/0022423.
Roberts-Seilaniantz A, Navarro L, Bari R, Jones JDG (2007) Pathological hormone imbalances. Curr Opin Plant Biol 10:372–379
Ron M, Avni A (2004) The receptor for the fungal elicitor ethylene-inducing xylanase is a member of a resistance-like gene family in tomato. Plant Cell 16:1604–1615
Saladie M, Matas AJ, Isaacson T, Jenks MA, Goodwin SM, Niklas KJ, Xiaolin R, Labavitch JM, Shackel KA, Fernie AR, Lytovchenko A, O'Neill MA, Watkins CB, Rose JKC (2007) The reevaluation of the key factors that influence tomato fruit softening and integrity. Plant Physiol 144:1012–1028
Sanahuja G, Banakar R, Twyman RM, Capell T, Christou P (2011) *Bacillus thuringiensis*: a century of research, development and commercial applications. Plant Biotechnol 9:283–300
Schlüter U, Benchabane M, Munger A, Kiggundu A, Vorster J, Goulet M-C, Cloutier C, Michaud D (2010) Recombinant protease inhibitors for herbivore pest control: a multitrophic perspective. J Exp Bot 61:4169–4183
Sengupta S, Chakraborti D, Mondal HA, Das S (2010) Selectable antibiotic resistance marker gene-free transgenic rice harbouring the garlic leaf lectin gene exhibits resistance to sap-sucking plant hoppers. Plant Cell Rep 29:261–271
Shah AD, Ahmed M, Mukhtar Z, Khan SA, Habib I, Malik ZA, Mansoor S, Saeed NA (2011) Spider toxin (*Hvt*) gene cloned under phloem specific *RSs1* and *RolC* promoters provides resistance against American bollworm (*Heliothis armigera*). Biotechnol Lett 33:1457–1463
Shah JM, Raghupathy V, Veluthambi K (2009) Enhanced sheath blight resistance in transgenic rice expressing an endochitinase gene from *Trichoderma virens*. Biotechnol Lett 31:239–244
Shao M, Wang J, Dean RA, Lin Y, Gao X, Hu S (2008) Expression of a harpin-encoding gene in rice confers durable nonspecific resistance to *Magnaporthe grisea*. Plant Biotechnol J 6:73–81
Shepherd RW, Wagner GJ (2007) Phylloplane proteins: emerging defenses at the aerial frontline? Trends Plant Sci 12:51–56
Shirasu K (2008) The HSP90-SGT1 chaperone complex for NLR immune sensors. Annu Rev Plant Biol 60:139–164
Shin S, Mackintosh CA, Lewis J, Heinen SJ, Radmer L, Macky RD, Baldridge GD, Zeyen RJ, Muehlbauer GJ (2008) Transgenic wheat expressing a barley class II chitinase gene has enhanced resistance against *Fusarium graminearum*. J Exp Bot 59:2371–2378
Shoresh M, Harman GE, Mastouri F (2010) Induced systemic resistance and plant responses to fungal biocontrol agents. Annu Rev Phytopathol 48:21–43
Shukurov R, Voblikova V, Nikonorova AK, Komakhin RA, Komakhina V, Egorov T, Grishin E, Babakov A (2012) Transformation of tobacco and Arabidopsis plants with *Stellaria media* genes encoding novel hevein-like peptides increases their resistance to fungal pathogens. Transgenic Res 21:313–325
Solleti SK, Bakshi S, Purkayastha J, Panda SK, Sahoo L (2008) Transgenic cowpea (*Vigna unguiculata*) seeds expressing a bean alpha-amylase inhibitor 1 confer resistance to storage pests, bruchid beetles. Plant Cell Rep 27:1841–1850
Spoel SH, Dong X (2008) Making sense of hormone crosstalk during plant immune responses. Cell Host Microbe 3:348–351
Srinivasan T, Kumar KRR, Kirti PB (2009) Constitutive expression of a trypsin protease inhibitor confers multiple stress tolerance in transgenic tobacco. Plant Cell Physiol 50:541–553
Stassen JHM, Van den Ackerveken G (2011) How do oomycete effector interfere with plant life? Curr Opin Plant Biol 14:407–414
Sujatha M, Lakshminarayana M, Tarakeswari M, Singh PK, Tuli R (2009) Expression of the cry1EC gene in castor (*Ricinus communis L.*) confers field resistance to tobacco caterpillar (*Spodoptera litura* Fabr) and castor semilooper (*Achoea janata* L.). Plant Cell Rep 28:935–946
Takai R, Isogai A, Seiji S, Che FS (2008) Analysis of flagellin perception mediated by flg22 receptor OsFLS2 in rice. Mol Plant Microbe Interact 12:1635–1642
Takakura YY, Oka NN, Suzuki JJ, Tsukamoto HH, Ishida YY (2012) Intercellular production of tamavidin 1, a biotin-binding protein from Tamogitake mushroom, confers resistance to the blast fungus *Magnaporthe oryzae* in transgenic rice. Mol Biotechnol 51:9–17

Tang D, Simonich MT, Innes RW (2007) Mutations in *LACS2*, a long-chain acyl-coenzyme A synthetase, enhance susceptibility to avirulent *Pseudomonas syringae* but confer resistance to *Botrytis cinerea* in Arabidopsis. Plant Physiol 144:1093–1103

Tester M, Langridge P (2010) Breeding technologies to increase crop production in a changing world. Science 327:818–822

Thatcher LF, Anderson JP, Singh KB (2005) Plant defense responses: what have we learnt from *Arabidopsis*? Funct Plant Biol 32:1–19

Thomma BPHJ, Nürnberger T, Joosten MHAJ (2011) Of PAMPs and effectors: the blurred PTI-ETI dichotomy. Plant Cell 23:4–15

Thordal-Christensen H (2003) Fresh insights into processes of nonhost resistance. Curr Opin Plant Biol 6:351–357

Tinoco MLP, Dias BBA, Dall'Astta RC, Pamphile JA, Aragäo FJL (2010) *In vivo* trans-specific gene silencing in fungal cells by *in planta* expression of a double-stranded RNA. BMC Biol 8:27

Ton J, Flors V, Mauch-Mani B (2009) The multifaceted role of ABA in disease resistance. Trends Plant Sci 14:310–317

Torres AM (2010) Application of molecular markers for breeding disease resistant varieties in crop plants. In: Jain SM, Brar DS (eds.) Molecular techniques in crop improvement. Springer Science, Dordrecht, pp. 185–205

Tripathi L, Mwaka H, Tripathi JN, Tushemereirwe WK (2010) Expression of sweet pepper *Hrap* gene in banana enhances resistance to *Xanthomonas campestris* pv. *musacearum*. Mol Plant Pathol 11:721–731

van de Craen M, Goh PY, Logghe MG, Khu YL, Mortier K, Bogaert TAOE (2006) USA Patent Publication No. 2006/0247197A1

van der Hoom RAL, Kamoun S (2008) From guard to decoy: a new model for perception of plant pathogen effectors. Plant Cell 20:2009–2017

Varshney RK, Graner A, Sorrells ME (2005) Genomics-assisted breeding for crop improvement. Trends Plant Sci 10:621–630

Varshney RK, Nayak SN, May GD, Jackson SA (2010) Next-generation sequencing technologies and their implications for crop genetics and breeding. Trends Biotechnol 27:522–530

Varshney RK, Bansal KC, Aggarwal PK, Datta SK, Craufurd PQ (2011) Agricultural biotechnology for crop improvement in a variable climate: hope or hype? Trends Plant Sci 16:363–371

Venter M (2006) Synthetic promoters: genetic control through *cis* engineering. Trends Plant Sci 12:118–124

Vijayan S, Kirti PB (2012) Mungbean plants expressing BjNPR1 exhibit enhanced resistance against the seedling rot pathogen, *Rhizoctonia solani*. Transgenic Res 21:193–200

Vlot AC, Klessig DF, Park S-W (2008) Systemic acquired resistance: the elusive signal(s). Curr Opin Plant Biol 11:436–442

Vogel JP, Somerville SC (2000) Isolation and characterization of powdery mildew-resistant *Arabidopsis* mutants. Proc Natl Acad Sci USA 97:1897–1902

Vogel JP, Raab TK, Schiff C, Somerville SC (2002) *PMR6*, a pectate lyase-like gene required for powdery mildew susceptibility in Arabdiopsis. Plant Cell 14:1–13

Volpi C, Janni M, Lionetti V, Bellincampi D, Favaron F, D'Ovidio R (2011) The ectopic expression of a pectin methyl esterase inhibitor increases pectin methyl esterification and limits fungal diseases in wheat. Mol Plant Microbe Interact 24:1012–1019

Wally O, Jayaraj J, Punja ZK (2009) Broad-spectrum disease resistance to necrotrophic and biotrophic pathogens in transgenic carrots (*Daucus carota* L.) expressing an Arabidopsis *NPR1* gene. Planta 231:131–141

Wan J, Zhang XC, Neece D, Ramonell KM, Clough S, Kim SY, Stacey MG, Stacey G (2008) A LysM receptor-like kinase plays a critical role in chitin signaling and fungal resistance in *Arabidopsis*. Plant Cell 20:471–481

Wang C, Chin CK, Chen A (1998) Expression of the yeast Δ-9 desaturase gene in tomato enhances its resistance to powdery mildew. Physiol Mol Plant Pathol 52:371–383

Wang C, Chin CK, Gianfagna T (2000) Relationship between cutin monomers and tomato resistance to powdery mildew infection. Physiological Mol Plant Pathol 57:55–61

Wang Z, Mao H, Dong C, Ji R, Cai L, Fu H, Liu S (2009) Overexpression of *Brassica napus* MPK4 enhances resistance to *Sclerotinia sclerotiorum* in oilseed rape. Mol Plant Microbe Interact 22:235–244

Wani SH, Sanghera GS, Singh NB (2010) Biotechnology and plant disease control-role of RNA interference. Am J Plant Sci 1:55–68

Weinthal D, Tovkach A, Zeevi V, Tzfira T (2010) Genome editing in plant cells by zinc finger nucleases. Trends Plant Sci 15:308–321

Weng LX, Deng HH, Xu JL, Li Q, Zhang YQ, Jiang ZD, Li QW, Chen JW, Zhang LH (2011) Transgenic sugarcane plants expressing high levels of modified *cry1Ac* provide effective control against stem borers in field trials. Transgenic Res 20:759–772

Wenzel G (2006) Molecular plant breeding: achievements in green biotechnology and future perspectives. Appl Microbiol Biotechnol 70:642–650

Wulff BBH, Horvath DM, Ward ER (2011) Improving immunity in crops: new tactics in an old game. Curr Opin Plant Biol 14:468–476

Xia Y, Gao QM, Yu K, Lapchyk L, Navarre D, Hildebrand D, Kachroo A, Kachroo P (2009) An intact cuticle in distal tissues is essential for the induction of systemic acquired resistance in plants. Cell Host Microbe 5:151–165

Xia Y, Yu K, Navarre D, Seebold K, Kachroo A, Kachroo P (2010) The *glabra1* mutation affects cuticle formation and plant responses to microbes. Plant Physiol 154:833–846

Xiao F, Goodwin SM, Xiao Y, Sun Z, Baker D, Tang X, Jenks MA, Zhou JM (2004) Arabidopsis CYP86A2 represses Pseudomonas syringae type III genes and is required for cuticle development. EMBO J 23:2903–2913

Xiao YH, Li XB, Yang XY, Luo M, Hou L, Guo SH, Luo XY, Pei Y (2007) Cloning and characterization of a balsam pear class I chitinase gene (Mcchit1) and its ectopic expression enhances fungal resistance in transgenic plants. Biosci Biotechnol Biochem 71:1211–1219

Yamaguchi T, Yamada A, Hong N, Ogawa T, Ishii T, Shibuya N (2000) Differences in the recognition of glucan elicitors signals between rice and soybean: β-glucan fragments from the rice blast disease fungus *Pyricularia oryzae* that elicit phytoalexin biosynthesis in suspension cultured rice cells. Plant Cell 12:817–826

Yang S, Gao M, Xu C, Gao J, Deshpande S, Lin S, Roe BA, Zhu H (2008) Alfalfa benefits from *Medicago truncatula*: The *RCT1* gene from *M. truncatula* confers broad-spectrum resistance to anthracnose in alfalfa. Proc Natl Acad Sci USA 105:12164-12169

Yarasi B, Sadumpati V, Immanni CP, Vudem DR, Khareedu VR (2008) Transgenic rice expressing *Allium sativum* leaf agglutinin (ASAL) exhibits high-level resistance against major sap-sucking pests. BMC Plant Biol 8:102

Ye SH, Chen S, Zhang F, Wang W, Tian Q, Liu JZ, Chen F, Bao JK (2009) Transgenic tobacco expressing *Zephyranthes grandiflora* agglutinin confers enhanced resistance to aphids. Appl Biochem Biotechnol 158:615–630

Yevtushenko DP, Misra S (2007) Comparison of pathogen-induced expression and efficacy of two amphibian antimicrobial peptides, MsrA2 and temporin A, for engineering wide-spectrum disease resistance in tobacco. Plant Biotechnol J 5:720–734

Yin C, Jurgenson JE, Hulbert SH (2011) Development of a host-induced RNAi system in the wheat stripe rust fungus *Puccinia striiformis* f. sp. *tritici*. Mol Plant Microbe Interact 24:554–561

Zhang J, Zhou J-M (2010) Plant immunity triggered by microbial molecular signatures. Mol Plant 3:783–793

Zhu YJ, Agbayani R, Moore PH (2007) Ectopic expression of Dahlia merckii defensin *DmAMP1* improves papaya resistance to *Phytophthora palmivora* by reducing pathogen vigor. Planta 226:87–97

Zipfel C (2009) Early molecular events in PAMP-triggered immunity. Curr Opin Plant Biol 12:414–420

Zipfel C, Robatzek S (2010) Pathogen-associated molecular pattern-triggered immunity: Veni, vidi…? Plant Physiol 154:551–554

Chapter 17
Can G-Proteins be the Key Proteins for Overcoming Environmental Stresses and Increasing Crop Yield in Plants?

Deepak Bhardwaj, Suman Lakhanpaul, and Narendra Tuteja

1 Introduction

Plants are important natural resources source which can convert the solar energy into various usable forms including energy substance glucose and storage carbohydrate like starch. However, time to time, the potential of plants is challenged by many environmental fluctuations including various biotic and abiotic stresses. Excess and deficit of any physical or biological factor can cause stress to the plant. Recently, the world is witnessing climatic changes such as increase in temperature, irregular rains, drought, excessive salt, flooding, excessive chilling, cyclones, tornedos, and various other natural calamities.

Stress causes changes in plant cellular functioning; as it prepares the plant for the incoming dangers. All the inactive protector genes become active due to sudden shift in the biochemistry and molecular biology of plant cell. The key organ in plants that acts as a first line of defense is cell wall. Which is useful to protect plant from pest or pathogen invasion and chilling or mechanical stress. However, it is unable to secure plants from stresses like water deficit, salt stress etc.

D. Bhardwaj
International Centre for Genetic Engineering and Biotechnology (ICGEB), Aruna Asaf Ali Marg, New Delhi, 110 067, India

Department of Botany, University of Delhi, New Delhi, India

S. Lakhanpaul
Department of Botany, University of Delhi, New Delhi, India

N. Tuteja (✉)
Plant Molecular Biology, International Centre for Genetic Engineering and Biotechnology (ICGEB), Aruna Asaf Ali Marg, New Delhi, India
e-mail: narendra@icgeb.res.in

N. Tuteja and S. Singh Gill (eds.), *Plant Acclimation to Environmental Stress*,
DOI 10.1007/978-1-4614-5001-6_17,

Recently, G-proteins and related machinery has been noted for its role in regulating stress related pathways, including ROS production, stomatal regulation and process related to plant water relations. We hereby present the role of G-proteins in all aspects of plant life and its metabolism (AM Jones 1994).

2 What is Stress? Perception and Signal Transduction Pathway

Any object that is affected by stress whether living or non-living undergoes definite change in response to stress which is known as strain. In case of plants it is hard to measure the propensity of stress. However, during stressed conditions plants show many molecular and cellular changes.

Water is a primary requirement of plant, and its excess as well as deficit affect the growth of plants. Water is absorbed by roots, and root hairs serve as channels to increase the surface area of absorption. The whole physiology of stress can only be understood if one knows the parameters that decide the water status of plant. These parameters are water potential and relative water content.

Presence of cell wall in case of plants protects them from outer dangers including pathogens and adverse conditions. Plant cell membrane is a semi-permeable, quasi-fluid structure which can allow only selective substances to pass through it; they serve as a second wall having many gates in the form of receptors and ion channels. Receptors being sensitive in nature are vulnerable to several elicitors that direct plant cell to respond against various stimuli. The cellular responses are initiated primarily by interaction of the extracellular material with a plasma membrane protein. This extracellular molecule is called a ligand (or an elicitor) and the plasma membrane protein, which binds and interacts with this molecule, is called a receptor. Interestingly, various stress factors, both abiotic as well as biotic, act as elicitors for the plant cell. Plant growth and development are mediated by complex array of signaling pathways co-ordinated by exogenous factors, which regulate all phases of growth including cell division, differentiation, and cell death. During evolution all living organisms including plants have been subjected to continuous variation of external environmental stimuli such as heat, cold, light, darkness, and other rhythmic parameters. In response to these external stimuli cells start many reactions to regulate their physiological processes. The cells of multicellular organisms also need to communicate with each other in order to co-ordinate their growth and differentiation. All fundamental processes of biology including growth and development rely on the proper response to environmental signals (Bowler and Chua 1994). The molecular mechanism by which these signals are perceived by cells, transduced toward the proper targets, and integrated into a biological response is known as signal transduction. Basically, two functional principles are involved in the extracellular signal (primary stimulus) as the first messenger itself penetrates the cell (probably through specific receptors) and finds its way to the nucleus. Alternatively, the signal remains outside the cell and is converted at plasma membrane into intracellular signals called second messenger (Hucho and Buchner 1997). Subsequent, sites are in the cytoplasm, at organelle membranes such as nuclear envelope and inside the nucleus.

Cell is like a universe where proteins like humans interact with each other, and one protein can interact with several and so on. These connections and interactions create a network or web of many signaling cascades which finally lead to the differential regulation of genes. Receptors are like sensors that sense the external environment and send the information to the cell machinery through various effectors, carriers, modulators, transcription factors, secondary messengers, enzymes etc. No information can be transcribed into response until receptor sends the signals to nucleus which is harboring plethora of dormant genes. Many plant receptors have been reported till date.

Stress is first recognized by the receptors present on the plasma membranes. Many membrane receptors are characterized till date which includes serine threonine kinase (Afzal et al. 2008), receptor-like protein kinase (Walker 1994), calcineurin B-like sensors (Cheong et al. 2007), G-protein-coupled receptor (Jones and Assmann 2004), etc. After recognition, there is an induction of the relay of various proteins along with the generation of second messengers including calcium, calmodulin, calcium-induced protein kinases, calcium-dependent protein kinases, G-proteins, MAP kinase, (ROS), and inositol phosphates. Various proteins in a given cell may or may not work in isolation; they interact with several other proteins to bring about certain changes that could be the need of time. This initiates the activity of various kinases and phosphatases which ultimately targets various genes and trans-elements. The up-regulation of certain dormant genes is required for the survival of cell under stressed conditions. This gives plant the potential to adapt or acclimatize to the particular environment. There are many hormones related to stress that add to this effect and activate many stress-induced genes. Their level generally increases during stressed conditions and they further start different signaling pathways to resist stressed conditions.

There is a different group of molecules that initiates and catalyses the modification of certain signaling proteins. They help in the modification or assembly of signaling components. Myristoylation, glycosylation, methylation and ubiquitination are some common modifications that give stress-related proteins some special features so that they can perform functions under alternate situations. Stress response could be early or late, and, on the basis of this, various stress-responsive genes can be broadly categorized as early and late induced genes. Early genes are induced within minutes and these genes belong to transcription factors, whereas other genes like RD (responsive to dehydration) (Kariola et al. 2006) and KIN (cold induced)/COR (cold responsive) (Thomashow 1998), which encode and modulate the proteins needed for synthesis of LEA-like proteins (late embryogenesis abundant) (Hundertmark 2008), antioxidants, membrane-stabilizing proteins. However synthesis of osmolytes take hours to get induced.

There are many signature elements on the promoter of genes that also tell the importance of particular gene in the alleviation of various stresses such as. DRE (dehydration-responsive elements) (Sun et al. 2008) or CRT (C-repeats) and some of them which contain ABRE (ABA-responsive element) (Gómez-Porras et al. 2007). Stress-induced transcription factors bind to this element and their overexpression brings about the tolerance against various stresses. Same way, there are many nuclear genes whose overexpression can give genetic potential to the plant against any kind of environmental challenges.

Description of all the genes of all the genes, transcription factors, chaperones and related proteins are beyond the scope of this chapter and we will only focus on the G-proteins and their role in various plant processes including stress.

3 What are G-proteins and Their Role in Plant Metabolism, Growth, and Development?

3.1 GPCR and G-proteins

Plant membranes like animal membranes are studded by G-protein-coupled receptor (GPCR), RGS and heterotrimeric G-proteins (Gα, Gβ/Gγ subunits) that are often link to the GPCR, constituting one of the most important components of cell signaling cascade. Signals are perceived by GPCR or RGS and later on transduced to inner part of the cell through G-proteins. This constitutes GPCR/G-protein-induced signaling cascade. Alfred G. Gilman and Martin Rodbell were awarded Nobel prize in 1994 for their contribution in the field of cell signal transduction; they proved the role of G-proteins as signaling molecules. Martin Rodbell and his collaborators linked the transducer with hormones, whereas Alfred G. Gilman and his co-worker characterized and purified the G-protein. Hundreds of chemicals and physical signals that can be stress-causing elicitors constantly bombard the surface of all cells. Some of these do not enter the cell, but, instead, bind to receptors at the cell surface and initiate a flow of information that moves to the cell interior. The receptors for many hormones (such as catecholamine, gonadotropin, parathyroid hormone, glucagon and ABA), odorants and light are heptahelical structures. Stimulation of these receptors causes activation of special proteins that are linked to the G-proteins. G-protein performs various functions including regulating few types of enzymes and ion channel. The target enzymes or ion channels are called effectors and they are generally proteinaceous in nature; G-protein after activation becomes effector itself and it interacts with other proteins to bring about changes in their activity causing alterations in ionic composition or in second messenger level that ultimately leads to the cellular response (Neer 1995).

GPCR is an integral membrane protein receptor that contains seven-transmembrane α-helical regions and this portion of receptor binds to a wide range of ligands like hormones. Interestingly, binding of ligands can bring about conformational changes in the structure of GPCR activating G-proteins. Binding of ligands is a switching on of various steps that happen in tandem. This includes the exchange of GDP/GTP associated with the Gα subunit and disassociation of alpha subunit from beta gamma dimer. The separated subunits individually act as effectors and interact with other intracellular proteins. This whole process can be switched off by simply replacing GTP attached to alpha subunit by GDP. This creates the earlier situation when the heterotrimer of G-proteins was attached to GPCR as a dormant unit. Nevertheless, GPCR can also work independent of G-proteins (AM Jones 2002).

3.2 *Historical Background of G-proteins*

Historically, it is work on adenylyl cyclase since 1950 that finally resulted in the discovery of G-protein. Now, it is well known that G-proteins are involved in broad range of cellular regulatory activities. Nobel Prizes has been awarded in this field to Sutherland and Rall (1958), who discovered the cAMP and adenylyl cyclase as a second messenger; Fischer and Krebs (1992), whose extensive studies of reversible protein phosphorylation began with cAMP-dependent protein kinase; and Gilman and Rodbell (1994) for their work on heterotrimeric G-proteins (Vaugham 1998). Here, a brief history of G-proteins has been described.

1957–1958	The G-proteins field originated. Sutherland and Rail described the basic properties of the enzyme adenylyl cyclase (AC), which is involved in hormone (epinephrine)-regulated synthesis of second messenger (cAMP) (Rall 1957; Sutherland and Rall 1958)
1969	Fat cell adenylyl cyclase was reported to be activated by multiple hormones: the hormone receptors were distinct from the enzyme. ATP could reverse the binding action of glucagon to the rat liver cell membrane receptor and thus dissociate the glucagon from the cell (Birnbaumer and Rodbell 1969)
1971–1974	Role of GTP was first reported by Martin Rodbell. GTP required for glucagon stimulated adenylyl cyclase in liver. GTP enhances dissociation of glucagon from its receptor (Rodbell et al. 1971)
1975	Henery Bourne and co-workers isolated S49 mouse lymphoma cell line deficient in adenylyl cyclase activity, named cyc. These cells were later used as assay tool to isolate coupling factors (receptor, G-proteins, adenylyl cyclase) (Bourne et al. 1975)
	Pfeuffer and Helmreish (1975) separated a GTP-binding protein from the adenylyl cyclase complex
1976	Orly and Schramm (1976) directly demonstrated the independence of receptor and cyclase enzyme

1977	E.M. Robs and A.G. Gilman in 1977 demonstrated the resolution of some components of adenylyl cyclase necessary for catalytic activity. They reported the 4G kDa GTP-binding protein, now, known as $G_{\alpha s}$
	Rodbell discovered that G-proteins at the cell receptor could both inhibit and activate transduction, often at the same time
1978	Robert Lefkowitz in 1978 demonstrated the conformational change occurred to receptor after binding to ligands, and GTP regulated the association with macromolecule
	Cassel and Selingen in 1978 showed that activation of a GTPase (GDP was replaced by GTP) in membrane is important in hormonal activation of adenylyl cyclase
1979	Rodbell reported that G-proteins can carry out primary and secondary processes of signal transduction
1981	R.G.L. Shorr and co-workers in 1981 reported purification of a β-adrenergic receptor, the first G-protein-coupled receptor with seven-membrane-spanning α-helices
1983	A.F. Gilman resolved Gas from Gβγ. Gα binds GTP and activates adenylyl cyclase. Gβγ deactivates Gα-GTP
1984	Gilman identified Gαi protein (inhibitory G-protein), which inhibits the activity of adenylyl cyclase
1986	cDNAs encoding Gas, Gczt (transducin), Gαi, and Gαq were cloned
	Stryer and Bourne predicted the imminent rewards from X-ray crystallographic analysis of G-protein structure and site-specific mutagenesis for functional studies
	G-protein-coupled receptor (GPCR) was cloned from hamster and terkey (Dixon et al. 1986; Yarden et al. 1986)
1990	First plant Gα was isolated from *Arabidopsis thaliana* (Ma et al. 1990)
	Large G-proteins, Gha (74 kDa) and Gh7a (78 kDa), were reported in mammals (Im and Graham 1990)
	Potential G-protein-coupled receptor was shown in barely (Devoto and Turner 2003)
1994	First plant Gβ gene was isolated from *Arabidopsis thaliana* and maize (Weiss et al. 1993)
1995	AGS (activator of G-protein signaling), which activated heterotrimeric G-protein signaling in the absence of a cell surface receptor, was reported (Sato et al. 1995)
1999	Extra-large G-protein (AtXLG 1, 99 kDa) was isolated from plant (Lee and Assmann 1999)
	ATDRG1, *Arabidopsis thaliana* developmentally regulated G-proteins, was reported (Etheridge et al. 1999)

2000	The first high-resolution structure of a GPCR, Rhodopsin, the visual light receptor, was solved (Palczewski et al. 2000)
	First plant Gγ gene was isolated from *Arabidopsis thaliana* (Mason and Botella 2000)
2002	Plant Gα, β, and γ subunits interact with each other in vitro and in vivo, and also interact with PLC-δ protein
2003	A seven-transmembrane RGS protein function in plant was discovered
2004	Putative GPCR called GCR1 and its interaction with G-alpha with regard to ABA was discovered
2005	Role of G-protein in Arabidopsis seed germination
	Role of G-proteins in giving resistance against rice blast fungus
2006	Role of G-proteins in regulating cell divisions in roots of Arabidopsis
	Interaction between G-alpha and THF1 was shown in Arabidopsis
	Interaction between G-alpha and prephenatedehydratase 1 was shown in Arabidopsis
2007	GPCR was proved to be a receptor of ABA
	Gα and Gβ subunits of *Pisum sativum* and their roles in salt and heat stress, respectively
2008	Role of G-protein in regulating ion channels of stomata guard cells
2009	NDL is shown as interacting partners of beta subunit of G-protein of Arabidopsis
2010	G-protein alpha subunit involved in controlling TE transpiration efficiency
2011	G-protein intractome in Arabidopsis was elucidated

3.3 G-Proteins and Various Plant Processes

Life of all plants starts from seed germination which is supposed to be a complex mechanism involving many genes, enzymes and proteins. No seed can germinate properly without the acquisition and availability of nutrients stored in the seed. Seed reserve mobilization is largely dependent on cells which in turn are controlled by gibberellin and abscisic acid (ABA). Gibberellic acid (GA) being a phytohormone is involved in so many biological processes including stem elongation, flowering, senescence, and seed germination. GA helps in the induction of *de novo* synthesis of α amylase and other enzymes in the aleurone layer of barley. GA is supposed to be perceived by the membranes of aleurone cell, from where it induces the transcription of genes of α-amylase (Mirbahar and Laidman 1982). Role of

G-protein in gibberellin-induced α-amylase gene expression was shown by Huw et al. in 1998. G-proteins are connected with membrane-bound GPCR through Gα subunit which contains intrinsic GTPase activity (Misra et al. 2007). In the activated form Gα-GTP acts as a signaling molecule which can catalyze several downstream actions. Interestingly, Jones used mastoparan analogue Mas7 which stimulates GDP-to-GTP exchange in the same fashion as activated GPCR performs when elicited by ligands. It was found that when the protoplast of aleurone was incubated with Mas7 it can secrete *α*-amylase. Further, this was confirmed by using α-Amy2/54: GUS promoter: reporter construct and the hydrolysis-resistant guanine nucleotide analogues GTP-γ-S and GDP-β-S. GTP-γ-S and GDP-β-S bind to Gα subunits and keep them in either the activated (GTP-γ-S-bound) or inactivated (GDP-β-S-bound) form. Expectedly, when GDP-β-S was introduced into aleurone protoplasts along with reporter gene constructs, the GUS activity was found to be completely abolished. In the other case, GTP-γ-S caused the slight expression of reporter gene construct. Therefore, G-protein has been shown to be clearly linked with GA signaling and germination. Next to the germination is development of the seedling which requires sunlight and pigments like chlorophylls, phytochrome, and cryptochrome.

In early 1991, it was reported G-protein is associated with the plasma membranes of the apical bud of etiolated peas and the GTPase activity is induced by low influences of blue light (Warpeha et al. 1991). G-protein mediation of blue-green light perception by rhodopsin in flagellate green algae also has been proposed (Calenberg et al. 1998).

Deng's group (Okamoto et al. 2001) generated transgenic Arabidopsis expressing the Gα subunit under the control of a glucocorticoid-inducible promoter. With the conditional overexpression of either the wild type or a constitutively active version of Arabidopsis Gα, transgenic seedlings exhibited a hypersensitive response to light. This enhanced light sensitivity was more exaggerated at a relatively lower intensity of light and was observed in white as well as far-red, red, and blue light conditions.

Many research workers suggested the involvement of G-proteins in phytochrome-mediated light. In the early 1990s it was shown that dark grown soybean cells when treated with cholera or pertussis toxins could uncouple phytochrome-dependent expression of chlorophyll a/b-binding protein (cab) (Romero and Lam 1993). Interestingly, tomato photo-morphogenetic mutant can mock the effect of phytochrome-mediated responses like anthocyanin production and chloroplast development, when treated with elicitors of G-proteins (Neuhaus et al. 1993; Bowler et al. 1994a, b). Cross-talk between G-protein could be Ca^{2+} dependent or Ca^{2+} independent (Neuhaus et al. 1993; Bowler et al. 1994a, b). The link between G-protein in phyA signaling supported the fact that various other proteins are also involved to finally convey message from photoreceptors to these two proteins (Neuhaus et al. 1993; Bowler et al. 1994a, b). Chua and colleagues (Neuhaus et al. 1993; Bowler et al. 1994a, b) combined all biological, molecular, and genetic approaches to more thoroughly address the role of heterotrimeric G-protein in phytochrome responses.

They used the phytochrome A (Phy A)-deficient tomato mutant *aurea* to investigate whether G-protein activation could initiate known phytochrome responses like chloroplast development, anthocyanin biosynthesis, and CAB gene expression in a Phy-deficient genetic background. Microinjection of purified oat PhyA into individual hypocotyl cells was shown to partially restore this response in a cell-autonomous and red/far-red light-reversible manner. Co-injection of either GDPβS or PTX with PhyA eliminated the responses, whereas injection of GTPγS alone initiated them. Analogue manipulations implicated calcium/calmodulin as acting downstream of the G-protein.

G-proteins are also involved in brassinosteroids signaling. In 1997, Wang et al. characterized G-alpha subunit of a heterotrimer G-protein from rice. This subunit was found to be involved in disease resistance. Later on, the same subunit was discovered to be involved in brassinosteroids signaling (Wang et al. 2006; Oki et al. 2009a, b). Rice harbors one alpha (RGA1) and beta (RGB1) subunit and two gamma subunits (RGG1 and RGG2). The dwarf mutant of rice, often known as D1 mutant, is actually identified as mutant of RGA1 (Izawa et al. 2010).

Arabidopsis G-proteins were studied in great detail owing to the availability of its diverse natural and mutant genetic resources. It has one alpha subunit (GPA1), one beta (AGB1) and three gamma subunits (AGG1, AGG2, and AGG3). Interestingly, the results of rice have been confirmed in Arabidopsis though later is a dicotyledonous model system. The G-proteins of Arabidopsis remain intact till the GCR/RGS gets activated by a ligand. ABA was found to bring about the conformational changes in GCR thus causing the release of activated GPA1 and AGB1 and AGG1 or AGG2 or AGG3 dimer. ABA plays an important role in seed germination, seedling development, and guard cell regulation including various environmental stresses. In one of the study, the G-protein knock-out mutants were used to see the effect on ABA signaling on seed germination; interestingly, it was found that agb1 mutant is more sensitive to ABA than gpa1 (Pandey and Assmann 2004; Pandey et al. 2006). Null mutants of GPA1 were found to be insensitive to gibberellin and BRs and highly sensitive to glucose (Ullah et al. 2002). The same mutants behaved as wild type in response to ABA and ethylene.

Cytokinin, one of the plant hormones, is shown to induce stomata opening, cause cell division, and release plant from the effect of apical dominance. They are also involved in the photosensory responses. In fact, it was thought at one time that light-induced effectors activated by cytokinin have some interaction with G-proteins or related effectors. Though it is not yet proved that whether GCR1 is a receptor for cytokinin or not, some reports of late 90s proved the probable connection between GCR- and cytokinin-based signaling. GCR1 mutant of Arabidopsis was found insensitive to benzyl adenine, which is synthetic cytokinin (Plakidou-Dymock and Dymock 1998; Mark Estelle 1998). G-proteins are also influenced by Methyl jasmonate and related compounds (Trusov et al. 2006; Bhardwaj et al. 2011) as α subunit of Arabidopsis G-protein affects the jasmonate responses and its mutants are not responsive to JA (Okamoto et al. 2009).

3.4 Organ Development

G-protein regulates cell growth, differentiation, and transformation in animal cells (Gutkind 1998). In *Drosophila,* distinct mechanism orients asymmetric cell division along the apical basal axis in neuroblasts and along the anterior-posterior axis in sensory organ precursor (SOP) cells. The G-protein subunit Gαl localized apically in neuroblasts and anteriorly in SOP cells before and during mitosis. Interfering with G-protein function by Gαl overexpression or depletion of heterotrimeric G-protein complexes causes defects in spindle orientation and asymmetric localization of determinants (Schaffer et al. 2001). In *C. elegans,* a Gβγ subunit is required for correct orientation of mitotic spindles during early development (Gotta and Ahringer 2001; Zwaal et al. 1996), and two Gα subunits function redundantly in asymmetric spindle positioning and generation of daughter cells of different sizes (Gotta and Ahringer 2001). Signaling by heterotrimeric G-protein is also involved in the control of cell polarity in unicellular organisms such as yeast and *Dictyostelium* (Bähler and Peter 2000; Weiner et al. 2000). Taken together, it suggests that G-protein regulates cell polarity and asymmetric cell division. Additionally, not only Gα but Gβγ subunits were involved in Golgi organization (Jamora et al. 1997, 1999).

In plants, G-proteins play role in the development of roots, leaves, and reproductive organs (Lease et al. 2001; Chakravorty et al. 2011; Shengjun et al. 2012). G-protein OE lines and mutant lines are different from control plants in terms of morphology and physiological processes, yet these mutants are not lethal. This suggests the significance but not the necessity of G-proteins in plants. Mutants of G-beta are distinguishable from rest of the mutants of G-protein subunits and GCR mutants. AGB1 mutants have round spatulate leaves, the plant and floral bud are shorter and their siliques are different in shape from the normal wild-plant siliques, causing increased weight of fruit and seed (Lease et al. 2001). Interestingly, the sepals are longer in case of gpa1 mutant when they are compared with wild plants but sepals of agb1 mutants are longer than both wild-type and gpa1 mutant. G-beta mutants are abnormal in terms of flower development and inflorescence architecture as evident from the mutants in case of Arabidopsis, rice, and Tobacco. Interestingly, the effect was species specific. In Arabidopsis, it was the female gametophyte where the changes were conspicuous and visible, whereas in tobacco the male gametophyte was affected in terms of shape and viability of pollen grains. In tobacco, the beta null mutants were dysfunctional in anther development (Peskan-Berghofer et al. 2005). The plants raised by antisense technology were found to have aberrant anther shape with inviable pollen grains. Mutant of beta gene also affected vegetative organs including the panicle which was branched and stem that was short and branched unlike Arabidopsis in which beta mutant was not smaller and highly branched. These studies prove the role of beta subunit in plant morphology which is directly related to plant's growth and development. In rice, the well-studied RGA1 gene regulates shape and size of seed and also controls the length of internode through GA signaling (Ashikari et al. 1999; Oki et al. 2009a).

Mutants of all the three subunits have distinct phenotypes, especially gpa1 and agb1, in which gpa1 mutant lines have fewer lateral roots but normal primary roots, whereas agb1 mutant is hypersensitive to D-glucose, causing increased cell division and thus have longer primary roots with numerous lateral roots (Ullah et al. 2002). Auxin plays an important role in root growth and development and its signaling is linked with G-protein and glucose. Recently, role of glucose was discovered in relation to AtRGS and G-proteins. Glucose serves as a food for yeast and animals and plants are the direct source of this important organic compound. It is also a signaling molecule in Arabidopsis and yeast. NDL1 which interacts with AGB1 increases the basipetal movement of auxin, whereas AGB1 is considered to attenuate this effect by binding with NDL1 (Mudgil et al. 2009). In essence, understanding the role of G-proteins in the development of root can greatly benefit agriculture in many ways.

Recently, new gamma subunit was discovered (Chakravorty et al. 2011) which has DNA/protein sequence similarity with gamma subunit and is functionally analogous to AGB1. This was termed as AGG3. Phenotypes of mutants of both AGB1 and AGG3 are similar in terms of shape of leaves and reproductive organs, and their wider siliques are more or less similar and blunt from the tip unlike tapering in case of mutants of GPA1 and GCR1.

4 Interacting Partners of G-proteins

α, β, and γ subunits of G-proteins regulate the activities of a structurally diverse group of effector molecules. These include enzymes engaged in the synthesis and degradation of intracellular second messengers, as well as ion-selective channels.

The first effector that was discovered long ago was adenylyl cyclase (AC), an enzyme (AC) which catalyzes the formation of cAMP from the substrate Mg^{2+}-ATP (Tang and Gilman 1991). The G-protein-regulated AC isoform identified in higher eukaryotes share a common motif. These enzymes are large, single polypeptides (molecular mass of 120 kDa). Generally, all forms of AC-identified mammalian systems are stimulated by GTP-ligand Gαs.

In case of animals eye being a sensory organ was always favorite for the research on GPCR- and G-protein-regulated pathways. It harbors distinct cyclic phosphodiesterase (PDE) which is found in the vertebrate photoreceptor cells (rods and cones). Phosphodiesterase catalyzes the degradation of cyclic cGMP in response to light which in turn causes the closure of cyclic GMP-gated cation channels. This retinal PDE is stimulated by the active form of Gt (transducin), which, in turn, is activated by the photobleached (meta II) form of rhodopsin, a receptor whose ligand is covalently bound to the light-sensing chromopore of retina. Gt-sensitive PDE is a heterotetramer of a βγ catalytic dimer and two inhibitory γ subunits, to which activated Gt binds. In contrast to adenylyl cyclase, which is activated by the binding of Gαs, the PDE is stimulated by the dissociation of the γ subunit when it is sequestered by Gαt-GTP. The rhodopsin-Gt-phosphodiesterase pathway is also unusual in that its components are expressed at extraordinarily high levels in photoreceptors, where rhodopsin represents about 70 % of the total membrane protein and Gt is about half of the remainder (Arshavsky et al. 1992).

The other special class of enzyme that serves as an effector for G-proteins is PIP2-specific phospholipase C-β (PLC-β). Isozymes are peripheral members stimulated either by Gαq family membranes, by Gβγ, or by both, depending on the PLC-β isoform (Montell 2000). The PLC-βs contain a long, C-terminal extension that binds Gαq and mediates its stimulatory interaction, and Gβγ subunits bind at sites in the N-terminal half of the molecule. Both in vitro and in vivo, PLCβ-1 are primarily Gαq-sensitive; PLC-β-2 and β-3 are also markedly stimulated by Gβy. PLCδ, another isoform of phospholipase C, was reported as an effector of Gh, high-molecular-weight G-protein.

G-protein-activated pathways operating via Phospholipase C and Phospholipase D have been reported in animals and plants. Often the two are activated together, with PLC producing a short pulse (seconds) and PLD a more prolonged pulse (minutes) of lipid-derived second messengers (Munnik et al. 1995). Recently, both Gα and PLC genes were cloned and proteins were overexpressed in bacteria. When the G-protein was included in the PLD assay, a strong dosage-dependent inhibition of the PLD activity was observed (Lein and Saalbach 2001).

Direct regulation of protein kinases by G-protein was first demonstrated in the yeast S. *cerevisiae*, where Gβγ (Step4.Ste 18p) released from the Gα subunit (Gpa1p) in response to a pheromone receptor activates an ERK signaling cascade to control the mating response. An increasing number of protein kinase is now known to be regulated by G-protein; for example, Phosphoinositide 3-kinase (P1 3-kinase), tyrosine kinase, C-terminal Src kinase, and MAP kinase have been reported as effectors of G-protein (Neves et al. 2002).

G-protein-gated, inward rectifier K^+ channels (GIRKs), which are found in heart, neurons, atrial myocytes, and endocrine cells, cause a cellular hyperpolarization that is particularly important in the control of neurons, cardiac muscle, and smooth muscle (Medina et al. 2000). GIRKs are stimulated by Gβγ subunits that in most case are released by stimulation of Gi, so that GIRK activation is usually sensitive to pertussis toxin. A Gαi-regulated K^+ channel is either a homotetramer or heterotetramer of members of the GIRK promoter family. Free Gβγ binds directly to the GIRK channel, and addition of GDP-bound (inactive) Gα can inhibit GIRK stimulation by binding Gβγ and blocking its action (Leaney and Tinker 2000; Lei et al. 2000). It has been reported that activated Gαi can inhibit GIRK-mediated currents, although it is not clear how its inhibition relates to the ability of non-activated Gαi to chelate βγ and thus terminate its activating effect.

In plants, evidence of G-protein regulation of inward K^+ channel was reported very early (Fairley-Grenot and Assmann 1991; Armstrong and Blatt 1995; Kelly et al. 1995). Using Arabidopsis as model system, mutation of Gα subunit showed lacks both ABA inhibitions of guard cell inward K^+ channel and pH-independent ABA activation of anion channels. Stomatal opening in mutant plants is insensitive to inhibition by ABA, and the rate of water loss from mutants is greater than that from wild-type plants (Wang et al. 2006). It has been known for years that the rate of intrinsic GTPase activity of Gα *in vitro* is much lesser than the rate of termination

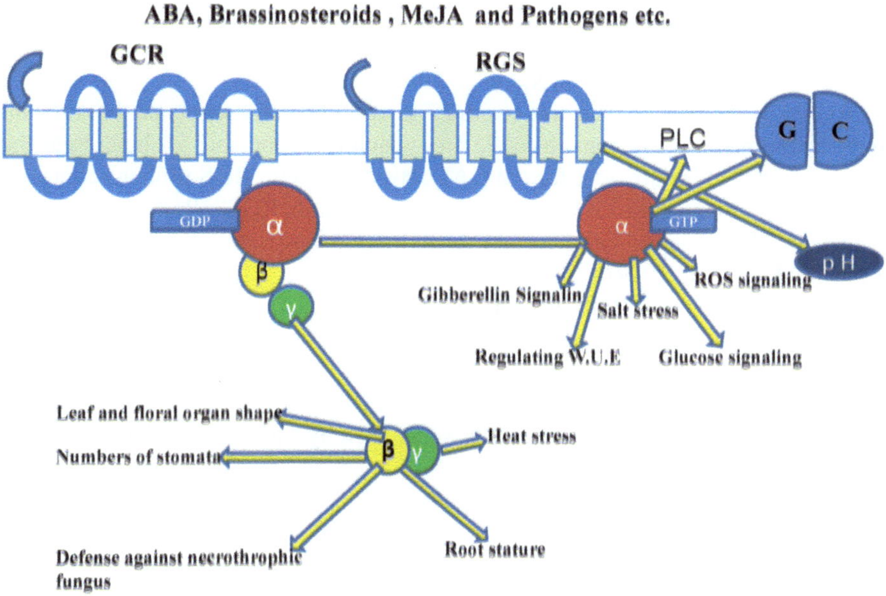

Fig. 17.1 A hypothetical model showing the various functions performed by GPCR, RGS, and G-proteins (Gα, Gβ, and Gγ). Elicitors like ABA, methyl jasmonate MeJA, brassinosteroids, and pathogens sensitize 7TM seven-transmembrane receptors like GCR (Pandey and Assmann 2004; Chen et al. 2004; Liu et al. 2007) and RGS (Plakidou-Dymock and Dymock 1998; Colucci et al. 2002; Chen et al. 2003, 2004), causing the G-protein heterotrimer to disintegrate forming GTP-Gα and Gβ-Gγ monomer and dimer, respectively. Each subunit plays significant role in various metabolic processes including environmental stress. Gα whose role has been shown in signaling cascades related to Gibberellins (Ashikari et al. 1999; Oki et al. 2009a, b), glucose (Grigston et al. 2008), salt (Misra et al. 2007), ROS (Zhang et al. 2011), water use efficiency (WUE) (Nilson and Assmann 2010), and stomatal regulation (Fan et al. 2008) (G.C) also serves as a interacting unit for enzyme like phospholipase (PLC). Gβ-Gγ dimer also plays significant role in maintaining shapes of petal and floral organs (Lease et al. 2001), root stature, number of stomata (Zhang et al. 2008a, b), defense against necrotrophic fungi (Trusov et al. 2006, 2009; Delgado-Cerezo et al. 2012), and heat stress (Misra et al. 2007)

of some physiological responses. Therefore, it has been proposed that additional factors accelerate GTPase activity *in vivo*. One class of G-protein GTPase-activating proteins (GAPs) is G-protein effectors such as cGMP phosphodiesterase and phospholipase C. Most of the G-protein effector molecules, however, do not possess GAP activity. In the recent years, a new class of GAPs, termed regulators of G-protein signaling (RGS), has emerged (Guan 1999; De Vries et al. 2000; Ross and Wilkie 2000; Cowan et al. 2000).

RGS protein family consists of at least 20 mammalian gene products that act as GTPase activating proteins on the α-subunits of heterotrimetric G-protein (Zhang et al. 1999). By accelerating the inactivation of GTP-bound Gα subunits, RGS serve as negative regulators of G-protein-mediated signaling pathways (Fig. 17.1). Additionally, two RGS proteins, p115-RhoGEF and PDZ (PSD-95, Disc-large, and

Z0-1)-RhoGEF, can also act as effectors (Cowan et al. 2000). On the other hand, RGS can also interact with targets other than G-proteins, for example the phosphoserine-binding protein, 14-3-3 (Niu et al. 2002).

RGS vary dramatically in size (from 23 to 160 kDa) and sequence, but they all have a common "RGS" domain (=125 aa), which is responsible for binding to the Gα subunits and is sufficient for the GAP activity. Several RGS proteins contain a G-protein γ-subunit-like (GGL) domain; work on RGS6, RGS7, RGS11, and RGS9 showed GGL domain to be important for RGS/Gβ5 association (Snow et al. 1998, 1999; Makino et al. 1999).

The GAP activity of RGS proteins has been demonstrated for all Gα subgroups except Gαs. RGS proteins appear to enhance the GTPase activity by binding to and stabilizing the transition state (Berman et al. 1996). The three-dimensional structure of an RGS and Gα complex demonstrates that RGS stabilizes the flexible switch region of G to resemble the transition state, thereby facilitating GTP hydrolysis (Tesmer et al. 1997).

The expression of more than 20 members of the mammalian family of RGS proteins suggests that they are likely to display marked selectivity of expression pattern and/or function, for example, RGS4 selectively enhances the GTPase activity of GO1 and Gi2 (Cavalli et al. 2000), and RGS11 selectively interacts with Gβ5 to act as GAPs on Gαo (Snow et al. 1998). Among the higher eukaryotes, certain RGS have been cloned, for example human (21 RGS genes), yeast (*S. cerevisiae;* 2 RGS genes), fruit fly (*Drosophila;* 5 RGS genes), and round worms (*C. elegans;* 12 RGS genes). Many studies on RGS are also available in plants including *Arabidopsis thaliana* and Glycine max. However, among the monocots only *Setaria italica* has been reported to possess RGS protein homolog (Chen & jones 2004; Urano et al. 2012; Choudhury et al. 2012).

Heterotrimers of G-proteins are like proteins that can be kept inactivated or activated depending upon the requirements of the cell at a given condition. Once activated, heterotrimer dissociates into single G-alpha subunit with GTPase activity and G-beta-gamma dimer. These uncoupled subunits then act as secondary messengers and they further catalyze the downstream processes. GPA1 physically interacts with plasidial protein called Thylakoid formation1 (THF1), which has been considered significant in the sugar signaling (Zhang et al. 2009). Arabidopsis plant grown on high level of glucose has expanded root growth which proves the role of sugar in cell division and cell expansion. Role of G-proteins has already been shown in root cell division. This supports the link between G-proteins and sugar signaling in plant root cells. Interestingly, the phenotypes of mutants of G-protein are highly distinguishable on the grounds of root and shoot stature.

GPA1 is also known to interact with PLDα1 at the DRY motif, and this interaction helps in increasing the efficacy of G-protein by influencing the GTPase activity of Gα and modulating its intracellular location (Zhao and Wang 2004). There are reports that support the involvement of Gα and Phospholipase A_2 in the biosynthesis of phytoelexin in *Eschscholzia californica* (Viehweger et al. 2006). G-proteins are supposed to be involved in retrograde signaling through the interaction of Gα and THF1, and any damage to the photosynthetic machinery specially D1 protein of

PS II can be renewed by the involvement of FtsH protease which in turn degrades the defective D1 protein. This FtsH protease is shown to be regulated by G-proteins with the involvement of other secondary messengers (Zhang et al. 2009).

5 Involvement of G-protein in Stomatal Functioning Suggests its Role in Environmental Stresses

Stomata are often considered as nostrils of plants. Present on the leaves, they perform the function of gaseous exchange and transpiration. Their role becomes evident during temperature and drought stress in order to prevent the excessive loss of water. Transpiration is a necessary evil; it is required to transport the nutrient from soil to the higher parts of the plant. Nutrients travel through the transpiration stream; indirectly it also maintains the turgidity and causes the cooling effect, saving plants from heat stress. Excessive transpiration can be dangerous as it leads to the wilting ultimately causing senescence of the plant. Plants have acquired smart mechanism to control the excessive loss of water by regulating the opening and closing of stomata. Role of ABA was reported in stomatal closure, as it induces the signal transduction pathway that finally causes the loss of certain osmolytes that causes the turgidity of guard cell thereby resulting in closing of stomata. The interplay between the membrane of guard cells and ion channels present on them is crucial in this regard. GCR, the seven-trans-membrane proteins, were proved to be the receptor for ABA (Chen et al. 2004). It transduces the signal by uncoupling G-proteins and activating several downstream pathways including activation of MAP kinase pathway and Ca^{2+} signaling pathways. Moreover, *Pisum sativum* Gβ subunit was shown to play a role in heat stress when overexpressed in Tobacco lines (Misra et al. 2007). In fact, there are many gaps in related to G-protein and other guard cell signaling proteins that need to be bridged to decipher a meaningful pathway out of it. In the wake of global warming, it is required to understand the exact mechanism of regulation of stomata. Guard cells are sensitive to high temperature and low CO_2 levels and they operate in connection with several molecular cues including K^+ channels. They act as balloons that can be swelled or diffused depending upon the concentration of K^+ ion inside them. Influx of K^+ ion through K^+ channels causes stomatal opening but there are several environmental factors like light, low CO_2 level and humidity that can cause the opening of stomata. Overall, stomatal closure is more or less dependent on factors like high temperature, drought, pathogen attack, and levels of phytohormones like ABA and MeJA. ROS are one of the byproducts of oxidative stress in plants; heat can also result in the release of these reactive oxygen free radicals. These radicals not only damage the cell membrane but they also can lead to cell death. G-proteins are also involved in the regulation of ROS signaling (Zhang et al. 2011).

Stomatal functioning is a complex phenomenon which includes several molecular proteins and secondary messengers. It was suggested that G-protein can be a link between receptor of ABA and ion channels on the guard cells (Fan et al. 2008); several researches in the past have put some light on the involvement of G-proteins in the stomatal functioning and related processes such as immunity against pathogens and relief against oxidative stress (Zhang et al. 2008a, b). Role of G-proteins was showed in the regulation of inward K^+ channel in fava bean (Fairley-Grenot and Assmann 1991).

In 1998, it was reported that beta subunit is involved in the voltage-dependent modulation of N-type Ca^{2+} channels (García et al. 1998). The direct proof of interaction of beta subunit with inward rectified K^+ channel Kir3 was given in 2002; this interaction was found to be significant for slow hyperpolarization of cardiac myocytes and neurons. On the same line, plant scientists also looked for the involvement of beta subunit in modulating K^+ channels and in this regard the first breakthrough was reported in the mutant lines of Arabidopsis. It was found that unlike in the case of animals, Gβγ has no direct control over the K^+ channels in plants and it may require the involvement of other proteins to influence the channel proteins (Fan et al. 2008). Indirect involvement of G-proteins in ion channel control was further supported by one of the findings in which the null mutant of Gα failed to show flg22-regulated stomatal responses. Flg22 is a bacterial protein which causes the inhibition of light-induced stomatal opening (Zhang et al. 2008a, b). Interestingly, agb1 mutants have more number of stomata per unit area, whereas gpa1 mutants have low number as compared to control plant (Zhang et al. 2008a, b). This property could be the reason of gpa1 mutant being more efficient with regard to transpiration especially during drought stress (Nilson and Assmann 2010). *Pisum sativum* beta subunit PsGβ showed interaction with PsMPK3, further strengthening the possibility of involvement of G-proteins with stomatal functioning though more experimentation is required to prove this fact (Bhardwaj et al. 2011). Recently, it was shown that one of the interacting proteins AGB1 in Arabidopsis named AGG3 is involved in the regulation of guard cell K^+ channels. AGG3 is predicted to be the third gamma subunit in Arabidopsis, with AGG1 and AGG2 being the other two and mostly involved in the developmental processes (Chakravorty et al. 2011; Shengjun et al. 2012).

All these reports indicate the involvement of G-protein in stomatal regulations which seem to be operating through the interactions with other proteins.

6 G-proteins and their Role in Pathogenesis and Biotic Stress

Evidence is accumulating for the importance of plant heterotrimeric G-protein in response to both bacterial and fungal pathogens. Beffa et al. (1995) showed that stable transformation of tobacco plants with the A1 subunit of CTX resulted in reduced susceptibility to *Pseudomonas tabaci,* accumulation of salicylic acid, and constitutive expression of pathogenesis-related genes. The role of plant G-proteins in response to fungal pathogens was suggested by the work of Legendre and Heinstein (1992) who showed that mastoparan elicits an oxidative burst in soybean cell suspensions, whereas CTX enhances the burst initiated by the elicitor *Verticillium dahliae.* They demonstrated subsequently that both the elicitor polygalacturonic acid and the G-protein activator mastoparan stimulate PLC activity in this preparation, leading to an increase in inositol triphosphate (Legendre et al. 1993).

Many studies using inhibitors and mutants have pinpointed the role of G-proteins in biotic stress and defense-related signaling pathway. Gα mutant of rice which is commonly known as D1 mutant was found ineffective in fighting against blast disease due to delayed response of pathogenesis-related genes (Suharsono et al. 2002).

G-protein, especially Gβγ dimer, provides resistance against necrotrophic fungi through jasmonate-mediated pathway in Arabidopsis, while their mutants are susceptible to pathogenesis (Trusov et al. 2009; Trusov et al. 2006). The reduced resistance of mutant plants could be because of difference in cell wall composition, especially xylose content (Delgado-Cerezo et al. 2012). Recently, up-regulation of gene of Gβ subunit of *Pisum sativum* was also reported after treating the plant with methyl jasmonate (Bhardwaj et al. 2011).

7 Conclusions

Climate change is a major concern of every nation but no country is in the position of cutting down their greenhouse emissions due to the challenges of accelerated development. Climate change has adverse effects on the production and yield of various crops, in turn creating alarming situations for everyone including agricultural scientists. Plant breeders and genetic engineers have to look for useful genes and genetic systems whose fundamentally correct manipulations can help ameliorate the status quo. Responding to the situation, G-proteins have been identified as a genetic tool whose manipulations could be useful for protecting plants from various abiotic and biotic stresses under the heat of climate change. G-proteins are smart molecular machines whose structure and functions are already well known in animals but their functions are still to be fully understood in plants. Recently, their overexpression in Arabidopsis and tobacco has given these plants genetic potential to overcome many abiotic and biotic stresses, and the same technology is now applied on several economically important plants such as rice, pea, tomato, and soybean. The cell machinery is endowed in its ability to allow various pathways to work simultaneously under various stimuli of nature, some act independently and some open the gates for cross-talks. Interestingly, G-proteins linked to the GPCRs the receptors find their way in after getting proper signals. They interact with several other proteins to bring about desirable changes in the plant cell environment. G-proteins also interact with several other proteins in the cellular environment. Based on the above-mentioned information we can conclude that GCR, RGS, and G-proteins play significant role in the biological processes with GCR and RGS acting as a sensor protein and G-protein as the molecular machines that execute the downstream cascade.

Therefore, studies leading to their effective manipulations can result in plants that are genetically better equipped to perceive and combat biotic as well as abiotic stresses thereby leading to increased crop productivity.

Acknowledgements The authors acknowledge the University Grant Commission and Department of Science and Technology, New Delhi, India, for funding their research.

References

Afzal AJ, Wood AJ, Lightfoot DA (2008) Plant receptor-like serine threonine kinases: roles in signaling and plant defense. Mol Plant Microbe Interact 21:507–517

Armstrong F, Blatt MR (1995) Evidence for K+ channel control in *Vicia* guard cells coupled by G-proteins to a 7TMS receptor mimetic The. Plant J 8:187–198

Arshavsky VY, Dumke CL, Bownds MD (1992) Noncatalyticc GMP binding sites of amphibian rod cGMP phosphodiesterase control interaction with its inhibitory γ-subunits. A putative regulatory mechanism of the rod photoresponse. J Biol Chem 267:24501–24507

Ashikari M, Wu J, Yano M, Sasaki T, Yoshimura A (1999) Rice gibberellin insensitive dwarf mutant gene Dwarf 1 encodes the alpha-subunit of GTP-binding protein. Proc Natl Acad Sci USA 96:10284–10289

Bähler J, Peter M (2000) Cell polarity in yeast. In: Drubin DG (ed) Frontiers in molecular biology: cell polarity. Oxford University Press, Oxford

Beffa R, Szell M, Meuwly P, Pay A, Vogeli-Lange R, Metraux JP, Neuhaus G, Meins F Jr, Nagy F (1995) Cholera toxin elevates pathogen resistance and induces pathogenesis-related gene expression in tobacco. EMBO J 14:5753–5761

Berman DM, Kozasa T, Gilman AG (1996) The GTPase activating protein RGS4 stabilizes the transition state for nucleotide hydrolysis. J Biol Chem 271:27209–27212

Bhardwaj D, Sheikh AH, Sinha AK, Tuteja N (2011) Stress induced β subunit of heterotrimeric G-proteins from Pisum sativum interacts with mitogen activated protein kinase. Plant Signal Behav 6:287–292

Birnbaumer L, Rodbell M (1969) Adenyl cyclase in fat cells. II. Hormone receptors. J Biol Chem 224:3477–3482

Bourne HR, Coffino P, Tompkins GM (1975) Selection of a variant lymphoma cell deficient in adenylate cyclase. Science 187:750–752

Bowler C, Chua NH (1994) Emerging themes of plant signal transduction. Plant Cell 6:1529–1541

Bowler C, Neuhaus G, Yamagata H, Chua NH (1994a) Cyclic GMP and calcium mediate phytochromephototransduction. Cell 77:73–81

Bowler C, Yamagata H, Neuhaus G, Chua NH (1994b) Phytochrome signal transduction pathways are regulated by reciprocal control mechanisms. Genes Dev 8:2188–2202

Calenberg M, Brohsonn U, Zedlacher M, Kreimer G (1998) Light and Ca^{2+}–modulated GTPase in the eyespot apparatus of a flagellate green alga. Plant Cell 10:91–103

Cassel D, Selinger Z (1978) Mechanism of adenylate cyclase activation through the β-adrenergic receptor: catecholamine-induced displacement of bound GDP by GTP. Proc Natl Acad Sci USA 75:4155–4159

Cavalli A, Druey KM, Milligan G (2000) The regulator of G protein signalling RGS4 selectively enhances alpha 2A-adreoreceptor stimulation of the GTPase activity of Go1alpha and Gi2alpha. J Biol Chem 275:23693–23699

Chakravorty D, Trusov Y, Zhang W, Acharya BR, Sheahan MB, McCurdy DW, Assmann SM, Botella JR (2011) An atypical heterotrimeric G-protein γ-subunit is involved in guard cell K-channel regulation and morphological development in Arabidopsis thaliana. http://www.ncbi.nlm.nih.gov/pubmed/21575088 Plant J 67:840–51

Chen JG, Jones AM (2004) AtRGS1 function in Arabidopsis thaliana. Methods Enzymol 389:338–350

Chen JG, Willard FS, Huang J, Liang J, Chasse SA, Jones AM, Siderovski DP (2003) A seven-Tran membrane RGS protein that modulates plant cell proliferation. Science 301:1728–1731

Chen JG, Pandey S, Huang J, Alonso JM, Ecker JR, Assmann SM, Jones AM (2004) GCR1 can act independently of heterotrimeric G-protein in response to brassinosteroids and gibberellins in Arabidopsis seed germination. Plant Physiol 13:907–915

Cheong YH, Pandey GK, Grant JJ, Batistic O, Li L, Kim BG, Lee SC, Kudla J, Luan S (2007) Two calcineurin B-like calcium sensors, interacting with protein kinase CIPK23, regulate leaf transpiration and root potassium uptake in Arabidopsis. Plant J 52:223–239

Choudhury SR, Westfall CS, Pandey S (2012) The RGS proteins add to the diversity of soybean heterotrimeric G-protein signaling. Plant Signal Behav. 1; http://www.ncbi.nlm.nih.gov/pubmed/22899066 [Epub ahead of print]

Colucci G, Apone F, Alyeshmerni N, Chalmers D, Chrispeels MJ (2002) GCR1, the putative Arabidopsis G protein-coupled receptor gene is cell cycle regulated, and its overexpression abolishes seed dormancy and shortens time to flowering. Proc Natl Acad Sci USA 99:4736–4741

Cowan CW, Wensel TG, Arshavsky VY (2000) Enzymology of GTPase acceleration in phototransduction. Methods Enzymol 315:524–538

De Vries L, Zheng B, Fischer T, Elenko E, Farquhar MG (2000) The regulator of G protein signaling family. Annu Rev Pharmacol Toxicol 40:235–271

Delgado-Cerezo M, Sánchez-Rodríguez C, Escudero V, Miedes E, Fernández PV, Jordá L, Hernández-Blanco C, Sánchez-Vallet A, Bednarek P, Schulze-Lefert P, Somerville S, Estevez JM, Persson S, Molina A (2012) Arabidopsis heterotrimeric G-protein regulates cell wall defense and resistance to necrotrophic fungi. Mol Plant 5:98–114

Devoto A, Turner JG (2003) Regulation of jasmonate-mediated plant responses in Arabidopsis. Ann Bot 92:329–337

Dixon RA, Kobilka BK, Strader DJ, Benovic JL, Dohlman HG, Frielie T, Bolanowski MA, Bennett CD, Rands E, Diehl RE (1986) Cloning of the gene and c DNA for mammalian β-adrenergic receptor and homology with rhodopsin. Nature 321:75–79

Estelle M (1998) Cytokinin action: two receptors better than one? Curr Biol 8:539–541

Etheridge N, Trusov Y, Verbelen JP, Botella JR (1999) Characterization of ATDRG1, a member of a new class of GTP-binding proteins in plants. Plant Mol Biol 39:1113–1126

Fairley-Grenot KA, Assmann SM (1991) Evidence for G-protein regulation of inward K+ channel current in guard cells of fava bean. Plant Cell 3:1037–1044

García DE, Li B, García-Ferreiro RE, Hernández-Ochoa EO, Yan K, Gautam N, Catterall WA, Mackie K, Hille B (1998) G-protein beta-subunit specificity in the fast membrane-delimited inhibition of Ca^{2+} channels. J Neurosci 18:9163–9170

Gómez-Porras JL, Riaño-Pachón DM, Dreyer I, Mayer JE, Mueller-Roeber B (2007) Genome-wide analysis of ABA-responsive elements ABRE and CE3 reveals divergent patterns in Arabidopsis and rice. BMC Genomics 8:260

Gotta M, Ahringer J (2001) Distinct roles for Galpha and G-beta-gamma in regulating spindle position and orientation in *Caenorhabditis elegans* embryos. Nat Cell Biol 3:297–300

Grigston JC, Osuna D, Scheible WR, Liu C, Stitt M, Jones AM (2008) D-Glucose sensing by a plasma membrane regulator of G signaling protein, AtRGS1. FEBS Lett 582:3577–3584

Guan KL, Han M (1999) G-protein signaling network mediated by an RGS protein. Genes Dev 13:1763–1767

Gutkind JS (1998) The pathways connecting G protein-coupled receptors to the nucleus through divergent mitogen-activated protein kinase cascades. J Biol Chem 273:1839–1842

Hucho F and Buchner K (1997) Signal transduction and protein kinases: the long way from the plasma membrane into the nucleus. Naturwissenschaften. 84, 281–90

Hundertmark M, Dirk KH (2008) LEA (late embryogenesis abundant) proteins and their encoding genes in *Arabidopsis thaliana*. BMC Genomics 9:118

Huw D, Jones HD, Smith SJ, Desikan R, Plakidou-Dymock S, Alison Lovegrove A, Hooley R (1998) Heterotrimeric G proteins are implicated in gibberellin induction of α-amylase gene expression in wild oat aleurone. Plant Cell 10:245–253

Im MJ, Graham RM (1990) A novel guanine nucleotide-binding protein coupled to the a1-adrenergic receptor. I. Identification by photolabeling of membrane and ternary complex preparations. J Biol Chem 265:18952–18960

Jamora C, Takizawa PA, Zaarour RF, Denesvre C, Faulkner DJ, Malhotra V (1997) Regulation of Golgi structure through heterotrimeric G proteins. Cell 91:617–626

Jones AM (2002) G-protein-coupled signaling in Arabidopsis. Curr Opin Plant Biol 5(5):402–407

Jones AM, Assmann SM (2004) Plants: the latest model system for G-protein research. EMBO Rep 5:572–578

Kariola T, Brader G, Helenius E, Li J, Heino P, Palva ET (2006) *EARLY RESPONSIVE TO DEHYDRATION 15*, a Negative Regulator of Abscisic Acid Responses in Arabidopsis. Plant Physiol 142:1559–1573

Kelly WB, Esser JE, Schroeder JI (1995) Effects of cytosolic calcium and limited, possible dual, effects of G protein modulators on guard cell inward potassium channels. Plant J 8:479–489

Leaney JL, Tinker A (2000) The role of members of the pertussis toxin sensitive family of G proteins in coupling receptors to the activation of the G protein-gated inwardly rectifying potassium channel. Proc Natl Acad Sci USA 97:5651–5656

Lease KA, Wen J, Li J, Doke JT, Liscum E, Walker JC (2001) A mutant Arabidopsis heterotrimeric G-protein beta subunit affects leaf, flower, and fruit development. Plant Cell 13:2631–2641

Lee YRJ, Assmann SM (1999) *Arabidopsis thaliana* 'extra-large GTP-binding protein (AtXLG1): A new class of G-protein. Plant Mol Biol 40:55–64

Legendre L, Heinstein PF, Low PS (1992) Evidence for participation of GTP-binding proteins in elicitation of the rapid oxidative burst in cultured soybean cells. J Biol Chem 267: 20140–22014

Legendre L, Heinstein PF, Crain RC, Low PS (1993) Phospholipase C activation during elicitation of the oxidative burst in cultured plant cells. J Biol Chem 268:24559–24563

Lei Q, Jones MB, Talley EM, Schrier AD, McIntire WE, Garrison JC, Bayliss DA (2000) Activation and inhibition of G protein-coupled inwardly rectifying potassium (Kir3) channels by G protein beta gamma subunits. Proc Natl Acad Sci USA 97:9771–9776

Lein W, Saalbach G (2001) Cloning and direct G-protein regulation of phospholipase D from tobacco. Biochim BiophyS Acta 1530:172–183

Liu X, Yue Y, Li B, Nie Y, Li W, Wu WH, Ma L (2007) A G protein-coupled receptor is a plasma membrane receptor for the plant hormone abscisic acid. Science 315:1712–1716

Ma H, Yanofsky M, Meyerowitz EM (1990) Molecular cloning and characterization of *GPA1*, a G protein α subunit gene from *Arabidopsis thaliana*. Proc Natl Acad Sci USA 87:3821–3825

Makino ER, Handy JW, Li T, Arshavsky VY (1999) The GTPase activating factor for transducin in rod photoreceptors is the complex between RGS9 and type 5G protein subunit. Proc Natl Acad Sci USA 96:1947–1952

Mason MG, Botella JR (2000) Completing the heterotrimer: isolation and characterization of an Arabidopsis thaliana G protein gamma-subunit cDNA. Proc Natl Acad Sci USA 97: 14784–14788

Medina JF, Garcia KS, Nores WL, Taylor NM, Mauk MD (2000) Timing mechanisms in the cerebellum: testing predictions of a large-scale computer simulation. J Neurosci 20:5516–5525

Mirbahar RB, Laidman DL (1982) Gibberellic acid-stimulated alpha-amylase secretion and phospholipid metabolism in wheat aleurone tissue. Biochem J 208:93–100

Misra S, Wu Y, Venkataraman G, Sopory S, Tuteja N (2007) Heterotrimeric G-protein complex and G-protein-coupled receptor from a legume (Pisum sativum): role in salinity and heat stress and cross-talk with Phospholipase C. Plant J 52:656–669

Montell C (2000) PLC fills a GAP in G-protein-coupled signalling. Nat Cell Biol 2:82–83

Munnik T, Arisz SA, de Vrije T, Musgrave A (1995) G protein activation stimulates phospholipase D signaling in plants. Plant Cell 7:2197–2210

Neer EJ (1995) Heterotrimeric G-proteins: organizers of transmembrane signals. Cell 80:249–257

Neuhaus G, Bowler C, Kern R, Chua NH (1993) Calcium/calmodulin-dependent and -independent phytochrome signal transduction pathways. Cell 73:937–952

Neves SR, Ram PT, Iyengar R (2002) G protein pathways. Science (Wash DC) 296:1636–1639

Nilson SE, Assmann SM (2010) The alpha-subunit of the Arabidopsis heterotrimeric G protein, GPA1, is a regulator of transpiration efficiency. http://www.ncbi.nlm.nih.gov/pubmed/20200073 Plant Physiol.152:2067–77

Niu J, Scheschonka A, Druey KM, Davis A, Reed E, Kolenko V, Bodnar R, Voyno-Yasenetskaya T, Du X, Kehrl J, Dulin NO (2002) RGS3 interacts with 14-3-3 via the N-terminal region distinct from the RGS (regulator of G-protein signaling) domain. Biochem J 365:677–684

Okamoto H, Matsui M, Deng XW (2001) Overexpression of the heterotrimeric G-protein α-subunit enhances phytochrome-mediated inhibition of hypocotyl elongation. Plant Cell 13:1639–1652

Okamoto H, Gobel C, Capper RG, Saunders N, Feussner I, Knight MR (2009) The α-subunit of the heterotrimeric G-protein affects jasmonate responses in Arabidopsis thaliana. J Exp Bot 60:1991–2003

Oki K, Inaba N, Kitagawa K, Fujioka S, Kitano H, Fujisawa Y, Kato H, Iwasaki Y (2009a) Function of the alpha subunit of rice heterotrimeric G-protein in brassinosteroid signaling. Plant Cell Physiol 50:161–172

Orly J, Schramm M (1976) Coupling of catecholamine receptor from one cell with adenylate cyclase from another cell by cell fusion. Proc Natl Acad Sci USA 73:4410–4414

Palczewski K, Polans AS, Baehr W, Ames JB (2000) Ca (2+)-binding proteins in the retina: structure, function, and the etiology of human visual diseases. Bioessays 22:337–350

Pandey S, Assmann SM (2004) The Arabidopsis putative G protein-coupled receptor GCR1 interacts with the G protein α subunit GPA1 and regulates abscisic acid signaling. Plant Cell 16:1616–1632

Pandey S, Chen JG, Jones AM, Assmann SM (2006) G-protein complex mutants are hypersensitive to abscisic acid regulation of germination and post germination development. Plant Physiol 141:243–256

Peskan-Berghofer T, Neuwirth J, Kusnetsov V, Oelmeller R (2005) Suppression of heterotrimeric G-protein β-subunit affects anther shape, pollen development and inflorescence architecture in tobacco. Planta 220:737–746

Pfeuffer T, Helmreish EJM (1975) Activation of pigeon erythrocytes membrane adenylate cyclase by guanyl nucleotide analogues and separation of nucleotide-binding protein. J Biol Chem 250:867–876

Plakidou-Dymock S, Dymock D, Hooley R (1998) A higher plant seven-transmembrane receptor that influences sensitivity to cytokinins. Curr Biol 8:315–324

Rall K, Sutherland EW, Berthet J (1957) The relationship of epinephrine and glucagon to liver phosphorylase. IV. Effect of epinephrine and glucagons on the reactivation of phosphorylase in liver homogenates. J Biol Chem 224:463–3475

Rodbell M, Birnbaumer L, Pohl SL, Krans HMJ (1971) The glucagons sensitive adenylyl cyclase in plasma membrane of rat liver. V. An obligatory role of guanyl nucleotides in glucagons action. J Biol Chem 246:1877–1882

Romero LC, Lam E (1993) Guanine nucleotide binding protein involvement in early steps of phytochrome regulated gene expression. Proc Natl Acad Sci USA 90:1465–1469

Ross EM, Wilkie TM (2000) GTPase-activating proteins for heterotrimeric G proteins: regulators of G protein signaling (RGS) and RGS-like proteins. Annu Rev Biochem 69:795–827

Sato M, Kataoka R, Dingus J, Wilcox M, Hildebrandt J, Lanier SM (1995). Factors determining specificity of signal transduction by G-protein-coupled receptors. J. Biol. Chem. 270, 15269–15276

Schaffer R, Landgraf J, Accerbi M, Simon V, Larson M, Wisman E (2001) Microarray analysis of diurnal and circadian-regulated genes in Arabidopsis. Plant Cell 13:113–123

Shorr RGL, Lefkowitz RJ, Caron MG (1981) Purification of the β-adrenergic receptor. Identification of the hormone-binding subunit. J Biol Chem 256:5820–5826

Snow BE, Krumins AM, Brothers GM, Lee S, Wall MA, Chung S, Mangion J, Arya S, Gilman AG, Siderovske DP (1998) A G protein g subunit-like domain shared between RGS11 and other RGS proteins specifies binding to Gb5 subunits. Proc Natl Acad Sci USA 95:13307–13312

Snow BE, Betts L, Mangion J, Sondek J, Siderovski DP (1999) Fidelity of G protein b-subunit association by the G protein g-subunit-like domains of RGS6, RGS7, and RGS11. Proc Natl Acad Sci USA 96:6489–6494

Suharsono U, Fujisawa Y, Kawasaki T, Iwasaki Y, Satoh H, Shimamoto K (2002) The heterotrimeric G protein alpha subunit acts upstream of the small GTPase Rac in disease resistance of rice. Proc Natl Acad Sci USA 99:13307–13312

Sun S, Yu JP, Chen F, Zhao TJ, Fang XH, Li YQ, Sui SF (2008) TINY, a dehydration-responsive element (DRE)-binding protein-like transcription factor connecting the DRE- and ethylene-responsive element-mediated signaling pathways in Arabidopsis. J Biol Chem 283: 6261–6271

Sutherland EW, Rall TW (1958) Fractionation and characterization of a cyclic adenine ribonucleotide formed by tissue particle. J Biol Chem 232:1077–1091

Tang WJ, Gilman AG (1991) Type-specific regulation of adenylyl cyclase by G protein βγ subunits. Science (Wash DC) 254:1500–1503

Tesmer JJ, Sunahara RK, Gilman AG, Sprang SR (1997) Crystal structure of the catalytic domains of adenylyl cyclase in a complex with Gsa-GTPgS. Science 278:1907–1916

Thomashow MF (1998) Role of cold-responsive genes in plant freezing tolerance. Plant Physiol 118:1–8

Trusov Y, Rookes JE, Chakravorty D, Armour D, Schenk PM, Botella JR (2006) Heterotrimeric G proteins facilitate Arabidopsis resistance to necrotrophic pathogens and are involved in jasmonate signaling. Plant Physiol 140:210–220

Trusov Y, Sewelam N, Rookes JE, Kunkel M, Nowak E, Schenk PM, Botella JR (2009) Heterotrimeric G proteins-mediated resistance to necrotrophic pathogens includes mechanisms independent of salicylic acid-, jasmonic acid/ethylene- and abscisic acid-mediated defense signaling. Plant J 58:69–81

Ullah H, Chen JG, Wang S, Jones AM (2002) Role of a heterotrimeric G protein in regulation of Arabidopsis seed germination. Plant 129:897–907

Urano D, Jones JC, Wang H, Matthews M, Bradford W, Bennetzen JL, Jones AM (2012) G protein activation without a GEF in the plant kingdom. http://www.ncbi.nlm.nih.gov/pubmed/22761582 PLoS Genet. 8:e1002756.

Vaugham M (1998) Signaling by heterotrimeric G proteins minireview series. J Biol Chem 273:667–668

Walker JC (1994) Structure and function of the receptor-like protein kinases of higher plants. Plant Mol Biol 26:1599–1609

Wang L, Xu YY, Ma QB, Li D, Xu ZH, Chong K (2006) Heterotrimeric G protein a subunit is involved in rice brassinosteroid response. Cell Res 16:916–922

Warpeha KMF, Hamm HE, Rasenick MM, Kaufman LS (1991) A blue-light activated GTP binding protein in the plasma membrane of etiolated pea. Proc Natl Acad Sci USA 88:8925–8929

Weiner OD, Neilsen PO, Prestwich GD, Kirschner MW, Cantley LC, Bourne HR (2002) A PtdInsP(3)- and Rho-GTPase-mediated positive feedback loop regulates neutrophil polarity. Nat Cell Biol 4:509–513

Weiss CA, Huang H, Ma H (1993) Immunolocalization of the G-protein alpha subunit encoded by the GPA1 gene in Arabidopsis. Plant Cell 5:1513–1528

Yarden Y, Rodriguez H, Wong SK, Brandt DR, May DC, Burnier JR, Harkinns RN, Chen EY, Ramachandran J, Ullrich A (1986) The avian beta-adrenergic receptor: primary structure and membrane topology. Proc Natl Acad Sci 83:6795–6799

Zhang Y, Neo SY, Han J, Yaw LP, Lin SC (1999) RGS16 attenuates G-αq-dependent p38 mitogen-activated protein kinase activation by platelet-activating factor. J Biol Chem 274:2851–2857

Zhang L, Hu G, Cheng Y, Huang J (2008a) Heterotrimeric G protein alpha and beta subunits antagonistically modulate stomatal density in Arabidopsis thaliana. Dev Biol 324:68–75

Zhang W, He SY, Assmann SM (2008b) The plant innate immunity response in stomatal guard cells invokes G-protein-dependent ion channel regulation. Plant J 56:984–996

Zhang L, Wei Q, Wu W, Cheng Y, Hu G, Hu F, Sun Y, Zhu Y, Sakamoto W, Huang J (2009) Activation of the heterotrimeric G-protein α-subunit gpa1 suppresses the ftsh-mediated inhibition of chloroplast development in Arabidopsis. Plant J 58:1041–1053

Zhang W, Jeon BW, Assmann SM (2011) Heterotrimeric G-protein regulation of ROS signaling and calcium currents in *Arabidopsis* guard cells. J Exp Bot 62:2371–2379

Zhao J, Wang X (2004) Arabidopsis phospholipase Dα1 interacts with the heterotrimeric g-protein α-subunit through a motif analogous to the dry motif in G-protein-coupled receptors. J Biol Chem 279:1794–1800

Zwaal RR, Ahringer J, van Luenen HGAM, Rushfort A, Anderson P, Plasterk RHA (1996) G proteins are required for spatial orientation of early cell cleavage in *C. elegans* embryos. Cell 86:1–20

Index

N. Tuteja and S. Singh Gill (eds.), *Plant Acclimation to Environmental Stress*,
DOI 10.1007/978-1-4614-5001-6,

MIX
Papier aus verantwortungsvollen Quellen
Paper from responsible sources
FSC® C105338
FSC
www.fsc.org

If you have any concerns about our products,
you can contact us on
ProductSafety@springernature.com

In case Publisher is established outside the EU,
the EU authorized representative is:
Springer Nature Customer Service Center GmbH
Europaplatz 3, 69115 Heidelberg, Germany

Printed by Libri Plureos GmbH
in Hamburg, Germany